Robust Cluster Analysis and Variable Selection

MONOGRAPHS ON STATISTICS AND APPLIED PROBABILITY

General Editors

F. Bunea, V. Isham, N. Keiding, T. Louis, R. L. Smith, and H. Tong

Monographs on Statistics and Applied Probability 137

Robust Cluster Analysis and Variable Selection

Gunter Ritter

Universität Passau

Germany

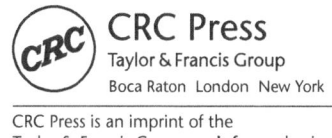

CRC Press
Taylor & Francis Group
Boca Raton London New York

CRC Press is an imprint of the
Taylor & Francis Group, an **informa** business

A CHAPMAN & HALL BOOK

First published in paperback 2024

Published 2015
by **Chapman Hall\CRC Press**
2385 NW Executive Center Drive, Suite 320, Boca Raton FL 33431
4 Park Square, Milton Park, Abingdon, Oxon, OX14 4RN

CRC Press is an imprint of Taylor & Francis Group, LLC

Library of Congress Cataloging-in-Publication Data

Ritter, Gunter.
   Robust cluster analysis and variable selection / Gunter Ritter.
       pages cm. -- (Chapman & Hall/CRC monographs on statistics &
   applied probability ; 137) "A CRC title."
   Includes bibliographical references and index.
   ISBN 978-1-4398-5796-0 (hardcover : alk. paper) 1. Cluster analysis. I. Title.

QA278.R58 2015
519.5'3--dc23   2014023649

ISBN: 978-1-03-292066-5 (pbk)
ISBN: 978-1-4398-5796-0 (hbk)
ISBN: 978-0-429-06343-5 (ebk)

DOI: 10.1201/b17353

Visit the Taylor & Francis Web site at
http://www.taylorandfrancis.com

and the CRC Press Web site at
http://www.crcpress.com

Contents

Preface

These days there is no lack of clustering methods. The novice in search of an appropriate method for his or her classification problem is confronted with a confusing multitude. Many methods are heuristic, and clustering data is often considered an algorithmic activity of geometric and exploratory nature. In Iven Van Mechelen's address at the biannual 2013 conference of the International Federation of Classification Societies, in Tilburg, the Netherlands, he was even speaking of a "classification jungle." While there is no single method that will solve *all* kinds of clustering problems, this is no excuse not to look for methods that cover broad ranges of applications.

Data may often be perceived as realizations of random variables. In these cases, it is best to start from a probabilistic model of the data, deriving from it estimators and criteria by application of well-founded inferential methods. Probabilistic cluster analysis has a long history dating back to Newcomb [387] 1886, Karl Pearson [404] 1894 and Charlier [86] 1906. These authors were far ahead of their time. It was only in the second half of the twentieth century that cluster analysis picked up speed. Yet the design of automatic methods is still considered a true challenge, in particular, in view of real applications. Still, the field doesn't have a good reputation among practitioners. Dougherty and Brun [131] write: "*Although used for many years, data clustering has remained highly problematic – and at a very deep level.*" Earlier, Milligan [376], p. 358, had complained about the poor performance of clustering methods based on multivariate, normal mixture models. He conjectured that the crux might be model over-parameterization. However, this is not the point. Identifiability shows that all parameters in a normal mixture are actually needed. One reason is the sheer multitude of local solutions. Another is the possible existence of multiple reasonable solutions. A third is sensitivity to contamination. In the meantime, methods have been developed that remedy these shortcomings.

The present text presents a probabilistic approach to robust mixture and cluster analysis with likelihood-based methods. It thus views mixture and cluster analysis as an instance of statistical inference. The relevant material is spread across various journals, over a broad time frame, and deserves to be unified in one volume. This book has a special focus on robustness. Of course, it isn't possible to write about robust cluster analysis without writing about cluster analysis itself. Despite the book's general title, it is by no means intended as an overview of the literature on cluster analysis. Other authors have served this purpose. Instead, the focus is on the methods that the author found to be most useful on simulated data and real applications. Different applications have different needs. The analysis of gene expression data, for instance, requires mainly accuracy. Real-time applications such as the analysis of magnetic resonance images also require speed. These are antagonistic assets. Although some remarks on speed will be made in Sections 3.1.5 and 3.1.6, the main emphasis will be on accuracy.

The theoretical parts of this book are written in the "definition – theorem – proof" style, which has been customary in mathematics for a long time now. This is intended to facilitate reading. It makes it easier to follow the main theory without having to plunge deeply into details, such as proofs, that are not essential for a basic understanding. The alternative "novel style" makes it difficult to highlight the important issues, which are the definitions and theorems that make up the basis of the theory, and the resulting methods.

Proofs begin with the word Proof and end with a square $\square$. Although parts of the text are highly technical, the researcher in applied fields who is not interested in theory will still find something of interest. The practitioner who needs some guidance in useful methods, and who intends to implement one or another algorithm, will find procedures and algorithms applicable without even having to understand their probabilistic fundamentals. I hope that the book will provide a basis for further research in this important, yet difficult field.

In conclusion, I would like to thank everyone who has contributed to this book. First, the authors whose substantial contributions I am able to report here. Without their contributions this book would not exist. With great gratitude I thank my wife, Freia, for her patience during the last three years. With immense pleasure I thank my longtime co-author, Dr. María Teresa Gallegos, for her interest in this book, her critical reading of large parts of the manuscript, as well as for her contributions, programs, and comments. I also appreciate the valuable conversations offered by Dr. Hans-Hermann Bock, Dr. Siegfried Graf, Dr. Christian Hennig, Dr. Claude Portenier, and Dr. Thomas Weber concerning various aspects of this book. Additionally, thanks go to Elsevier for permitting their authors to reuse portions of their communications in other works, and to Springer Science+Business Media, Heidelberg, Germany, for their kind permission to reuse excerpts from [183]. Last, but not least, I thank Rob Calver, and his team at Chapman & Hall, for their cooperation.

List of Figures

List of Tables

Glossary of notation

To facilitate the reading of formulæ, notation will adhere to some general rules. A boldface letter will indicate an array of variables of the same letter, for instance, $\mathbf{x} = (x_1, \ldots, x_n)$, $\mathbf{m} = (m_1, \ldots, m_g)$, $\mathbf{V} = (V_1, \ldots, V_g)$, $\mathbf{\Lambda} = (\Lambda_1, \ldots, \Lambda_g)$. An r-element subset R of the data set $\mathbf{x}$, written $R \subseteq \mathbf{x}$, is of the form $R = (x_{i_1}, \ldots, x_{i_r})$ with $1 \leq i_1 < \cdots < i_r$.

A parameter identifier with one or more arguments attached to it will denote the maximum likelihood estimate (MLE) of the parameter w.r.t. to the data specified by the arguments. Thus, in the framework of a normal model, $m(T)$ $(V(T))$ will stand for the sample mean vector $\overline{x}_T$ (scatter matrix S_T) of $T \subseteq \mathbf{x}$ and $m_j(\boldsymbol{\ell})$ for the sample mean vector of cluster j w.r.t. the labeling $\boldsymbol{\ell}$. The letter Λ will stand for the inverse matrix V^{-1} and will be used as a text substitution, e.g., $\Lambda_j(\boldsymbol{\ell}) = V_j^{-1}(\boldsymbol{\ell})$. Upright letters indicate operators: D, E, V, etc. Local notation will be explained where it appears. *A number is positive if it is* ≥ 0. Such numbers are often called nonnegative, but this could create confusion since the sine function is certainly not negative. Numbers > 0 will be called *strictly* positive.

The view about the notation of partial derivatives will be pragmatic. Functions will bear with them generic names of their variables. So, if the generic variables of a function f are x and ϑ (think of a model function), then $\mathrm{D}_\vartheta f(y; \theta)$ will mean the partial derivative of f w.r.t. the second variable evaluated at the pair (y, θ). Likewise, $\mathrm{D}_x f(y; \theta)$ would be the partial derivative of f w.r.t. the first variable evaluated at this pair.

The following is a list of some notation appearing ubiquitously in the text.

Notation	Meaning	Example
$\ll$	absolutely continuous (measures)	$\mu \ll \varrho$
$\lVert \cdot \rVert$, $\lVert\!\lvert \cdot \rvert\!\rVert$	norms	$\lVert x \rVert$, $\lVert\!\lvert A \rvert\!\rVert$
$\lVert \cdot \rVert_1$	sum norm	$\lVert x \rVert_1 = \sum_{k=1}^d \lvert x_k \rvert$
$\lVert \cdot \rVert_2$	Euclidean norm	$\lVert x \rVert_2 = \sqrt{\sum_{k=1}^d x_k^2}$
$\lVert \cdot \rVert_\infty$	uniform norm	$\lVert x \rVert_\infty = \max_{1 \leq k \leq d} \lvert x_k \rvert$
$a_{k,+}$	sum of the kth row in the matrix $(a_{k,\ell})_{k,\ell}$	
$\lvert A \rvert$	size (or cardinality) of the set A	
B^A	collection of all maps $A \to B$	
$a \vee b$	maximum of the numbers a and b	
$a \wedge b$	minimum of the numbers a and b	
$\succeq, \preceq, \succ, \prec$	Löwner order on the symmetric matrices	see Definition A.8
$a .. b$	interval of all integers between a and b	$= \{i \in \mathbb{N} \mid a \leq i \leq b\}$
$\complement$	complement of a set	
$\backslash k$	entry or index k missing	$x_{\backslash k}$
$\binom{T}{q}$	collection of all subsets of T of size q	
$B_r(a)$	open ball about a with radius r	$= \{x \mid d(x, a) < r\}$
$\overline{B}_r(a)$	closed ball about a with radius r	$= \{x \mid d(x, a) \leq r\}$
const	a constant (it may vary from line to line)	
$\mathcal{C}$	a partition or clustering	
$C_j(\boldsymbol{\ell})$	cluster j w.r.t. assignment $\boldsymbol{\ell}$	

Notation	Meaning	Example
D	differential operator	$D_\vartheta f(x; \eta)$
D_{KL}	Kullback–Leibler divergence	
Diag	diagonal matrix defined by some vector	$\mathrm{Diag}(a_k)_k$
Diag A	square matrix A reduced to its diagonal	
$\Delta_a, \overset{\circ}{\Delta}_a$	closed and open unit simplices of dimension a	see Section C.1
d, D	dimension of Euclidean sample spaces	$\mathbb{R}^d$
d	some metric or distance	$d(x, y)$
d_V	Mahalanobis distance w.r.t. V	$d_V(x, y) = \sqrt{(x - y)^\top \Lambda (x - y)}$
E	expectation of a random vector	EX
E	a general sample space, often a topological space with special properties	
F	a subset of indices of variables	$k \in F$
f	model, likelihood, or density function	$f_\gamma(x) = f(x; \gamma) = $ probability density function of μ_γ at x
g	number of components or clusters (groups)	
Γ	parameter space of the basic population	
γ	parameter of the basic population	γ_j (of component j)
γ_T	MLE of parameter γ for the sample T of observations	m_T, V_T
$\gamma_j(\ell)$	MLE of parameter γ for cluster j w.r.t. ℓ	$m_j(\ell), V_j(\ell)$
$\boldsymbol{\gamma}$	array $(\gamma_1, \ldots, \gamma_g)$ of population parameters	$= (\gamma_1, \ldots, \gamma_g)$
H	entropy	$= -\sum_j \pi_j \log \pi_j$
$\mathcal{I}$	expected Fisher information	
I	subset of indices i	$I \subseteq 1..n$
I_d	d-dimensional unit matrix	
i	generic index of an observation and imaginary unit of the complex plane	x_i $i^2 = -1$
j	generic index of a component or cluster	C_j
k	index of a coordinate, variable, or feature	$x_{i,k}$
L	random cluster index or label (in the Bayesian context)	$L : \Omega \to 1..g$
$\mathcal{L}^p(\varrho)$	vector space of all ϱ-integrable functions	
$\mathbb{L}^p(\varrho)$	Banach space of all equivalence classes of ϱ-integrable functions	
Λ	inverse V^{-1} of V	$\Lambda_j = V_j^{-1}$
$\boldsymbol{\ell}$	assignment of all objects to clusters	$= (\ell_1, \ldots, \ell_n)$
ℓ	generic index of a component or cluster	
ℓ_i	cluster of object i if $\neq 0$ or discarded if $= 0$	
M	EM operator	$M\vartheta$
M	a modified data set (in robustness)	
$\mathbb{M}(\varrho)$	linear space of ϱ-equivalence classes of real-valued measurable functions	
m_j	location parameter of component j (normal or elliptical mixture or classification model)	
$m_j(\ell)$	MLE of the parameter m_j w.r.t. ℓ (classification model)	
μ	probability distributions on sample space	

Notation	Meaning	Example
$\mathbb{N}$	set of all natural numbers	$= \{0, 1, 2, \dots\}$
n	number of observations	
n_j	size (cardinality) of cluster j	$n_j(\boldsymbol{\ell})$ (w.r.t. $\boldsymbol{\ell}$)
$N_{m,V}^{(d)}$	d-dimensional normal distribution and	$N_{0,1}^{(1)}(x) = \frac{1}{\sqrt{2\pi}} e^{-x^2/2}$
	its Lebesgue density	
ν	probability distribution on sample space	
Ω	universal measurable space	
P	a probability on Ω	
Ψ	a constrained parameter space	
	a map	
$\mathrm{PD}(d)$	convex cone of positive definite $d \times d$ matrices	
$\boldsymbol{\pi}$	mixing rates $(\pi_1, \dots, \pi_g)$	
π_j	mixing rate of component j	$\sum_j \pi_j = 1$
φ	density generator (elliptical distribution)	
ϕ	radial function (elliptical distribution)	
R	subset of "retained" elements (trimming)	
$\mathbb{R}$	real line	
$\mathbb{R}_>$	open ray of real numbers > 0	
$\mathbb{R}_+$	closed ray of real numbers ≥ 0	
$\mathbb{R}^d, \mathbb{R}^D$	Euclidean spaces	
r	size (cardinality) of R	
ϱ	reference measure for density functions	
	of distributions on sample space	
$S_r(a)$	sphere about a with radius r	$= \{x \mid \|x - a\| = r\}$
S_T	scatter (sample covariance) matrix of $T \subseteq \mathbf{x}$	
$S(\boldsymbol{\ell})$	pooled scatter (sample covariance) matrix w.r.t.	
	assignment $\boldsymbol{\ell}$	
$S_j(\boldsymbol{\ell})$	scatter (sample covariance) matrix	
	of cluster j w.r.t. $\boldsymbol{\ell}$	
$S_T(w)$	weighted scatter matrix w.r.t. $T \subseteq \mathbf{x}$	
	and weight vector w	
$S(\mathbf{w})$	pooled weighted scatter matrix w.r.t. the	
	weight matrix $\mathbf{w}$	
$\mathrm{SYM}(d)$	the vector space of symmetric $d \times d$ matrices	
s	score functional	$s_\gamma(x) = s(x; \gamma)$
s_T	scatter value (sample variance) of $T \subseteq \mathbf{x} \subseteq \mathbb{R}$	
$s_j(\boldsymbol{\ell})$	scatter value (sample variance) of cluster j w.r.t. $\boldsymbol{\ell}$	
$\top$	transposition of a vector or matrix	$A^\top$
Θ	parameter space of a general statistical model,	
	often topological	
ϑ	parameter of a general statistical model	$\vartheta = (\boldsymbol{\pi}, \boldsymbol{\gamma})$
$\mathrm{V}X$	covariance matrix of the random vector X	
V_j	scale parameter of component j (normal or	
	elliptical mixture or classification model)	
$V_j(\boldsymbol{\ell})$	MLE of the parameter V_j w.r.t. $\boldsymbol{\ell}$	
	(classification model)	
$\mathcal{V}_c$	array of positive definite matrices under HDBT	see p. 32
	constraints	
$W_{\mathbf{x}}, W_T$	total sum of squares and products (SSP) matrix	see p. 33
	of $\mathbf{x}$, $T \subseteq \mathbf{x}$	

Notation	Explanation	Example
$W_j(\boldsymbol{\ell})$	total SSP matrix of cluster j w.r.t. $\boldsymbol{\ell}$	see p. 55
$W(\boldsymbol{\ell})$	pooled within-groups SSP matrix w.r.t. $\boldsymbol{\ell}$	$W(\boldsymbol{\ell}) = \sum_j W_j(\boldsymbol{\ell})$
X	random variable or vector defined on Ω	$X_i : (\Omega, P) \to \mathbb{R}^d$
X_1^n	n-tuple $(X_1, \ldots, X_n)$ of random variables	
Ξ	parameter space of a spurious outlier	Ξ_i
x	observation or data point	x_i
x_F	observation x restricted to the variables in F	$x_{i,F}$
$\mathbf{x}$	data set (data matrix)	$= (x_1, \ldots, x_n)$
$\mathbf{x}_F$ ($\mathbf{x}_I$)	data set $\mathbf{x}$ restricted to the variables in F (indices in I)	$x_{I,F}$
$\overline{x}_T$	sample mean vector of $T \subseteq \mathbf{x}$	
$\overline{x}_j(\boldsymbol{\ell})$	sample mean vector of cluster j w.r.t. $\boldsymbol{\ell}$	
$\overline{x}_j(\mathbf{w})$	weighted sample mean vector of cluster j w.r.t. $\mathbf{w}$	
Y	random variable or vector defined on Ω	$Y_i : (\Omega, P) \to \mathbb{R}^d$

Introduction

Probabilistic data analysis assumes that data is the result of repeated, often independent, sampling from some population. The information contained in the data is used to infer some of its properties or even a description. This text deals with mixture and cluster analysis, also called partitioning or unsupervised classification. In these disciplines the data is assumed to emanate from mixtures consisting of unimodal subpopulations which correspond to different *causes*, *sources*, or *classes* and which all contribute to the data. Of course, there are data sets that emerge neither from a unimodal population nor from such a mixture. For instance, self-similar data sets are different. Consider the data set $1, 2, 4, 8, 16, \ldots, 2^k$ for some natural number $k \geq 10$. This set is completely deterministic and does not show any subpopulation structure except, perhaps, one of $k+1$ singletons. Other data sets are chaotic, composed solely of outliers. Again, others are geometric, such as a double helix. They belong to the realm of image processing rather than to data analysis. Therefore, it is appropriate to first say some words about the populations and models relevant to cluster analysis.

Fortunately, the numerical data sets appearing in a mixture or cluster analysis are often of a special kind. The components, or subpopulations, can often be described by a typical member, a set point or center, of which the other members are modifications randomly scattered about in all directions. Sometimes a simple transformation may be needed to achieve this, for instance, logging when the variables are inherently positive. In Euclidean space, populations of this type may frequently be described as unimodal, in particular, Gaussian distributions. They are adapted to the symmetries of Euclidean space and used as ideal models of real-world populations. They keep statements and algorithms simple and allow us to use their efficient and convenient apparatus.

In 1947, Richard Geary [192] stated *"Normality is a myth; there never was, and never will be, a normal distribution."* Of course, there is a normal distribution, namely, the right-hand side of Laplace's Central Limit Theorem. What Geary was actually saying is that exact mathematical normality is not part of the real world. For this reason, Karl Pearson had earlier introduced his more complex elliptically symmetric, type-VII distributions. Some real-world populations are even at least slightly asymmetric. Therefore, asymmetric distributions, such as skew normal and t-distributions, have also been proposed in the past. However, no model is identical with reality, and real-world populations are not even distributions in the mathematical sense. There is always something beyond our control. Consequently, the adjectives "correct" and "wrong" are not applicable to models. Models can only be "good" or "poor" in a given situation. What matters is that the data has not come from populations so different from the assumed ones as to invalidate their use. Although some estimators and procedures will be formulated for mixtures of general components, they are most efficient when specialized for normal or elliptically symmetric ones. Generalization to richer models is often possible and sometimes straightforward. Submodels, such as ones with spherical, diagonal, or equal scale parameters, will also be considered. In particular situations, the complexity of the model to be chosen also depends on data set size.

The boss in the present text is a mixture of $g \geq 1$ unimodal components that stand for one cause each. Given a sample of the mixture, the Bayesian discriminant rule, based on the mixture parameters, defines clusters and a partition manifesting the components of the mixture. The clusters witness the components. They appear as "cohesive" subsets of the data, and "isolated" when they are separated by location. (Components in real data are rarely separated by scale, for example, in the form of an X.) The main goal is to detect and describe the original subpopulations, or mixture components. They represent the different causes responsible for the data at hand. This task is as important as it is difficult. It would be easier if we possessed the whole mixture; however, it is actually only a sample that we observe. When there is heavy overlap, detecting causes and their number may become impossible. Under this circumstance, the number of detectable clusters will be smaller than the number of causes underlying the data set. On the other hand, clustering algorithms tend to partition data sets even if their parents are not composed of several unimodal components. For instance, this may be the case when the data comes from a uniform distribution on a box.

There is no single clustering algorithm that will produce a reasonable result on *any* data set. All methods depend on the data model on which they are based. Estimates depend on the assumptions one is willing to accept. This is the price to be paid for trying to extract the characteristics of a population from a *finite* sample. Here, as in all probabilistic approaches to inference, model specification plays a crucial role. It's an important, as well as a notorious, problem in estimation practice which generally requires methods from systems analysis. One of the aims of cluster analysis is the data model. However, it needs some prior information on the components. This problem has been pointed out by many authors. In contrast, when a cluster is very well isolated, it can be recovered with almost any method, even if cluster shapes are odd and do not conform to the model. The better the separation is, the less compliance with assumptions is needed.

However, in real-world applications, separation is generally rather poor. In these cases, the assumptions underlying the method should be general enough to avoid conflict with the parent population. Otherwise, the partitions created may be biased toward the assumptions. The richer the model is, and the looser the assumptions are, the better the chance that the parent is close to the model, and the estimate is close to the parent. This supports using a rich model in cases where the data set is large enough to allow estimation of its parameters. On the other hand, when reliable a priori information on the parent is available, it should be used. A narrow, yet suitable model can still estimate overlapping components with a fairly small data set. Therefore, submodels will also be addressed. Their use requires special care, as an inadequate, too narrow model may grossly distort the solution in its own direction. Refer also to the discussion in Gordon [208], Chapter 6. In the extreme case, a submodel is a priori completely specified. This is actually discriminant, rather than mixture or cluster analysis.

Data emanating from multiple sources occurs in important applications of statistics. Take, for example, administration, archeology, astronomy, biology, biomedicine, economics, image processing, medicine, pattern and speech recognition, psychology, and sociology. In marketing research, for instance, it is used for dividing potential customers into "natural" groups. More examples and related literature are offered in the Introduction of Redner and Walker [435]. Mixture models are useful for describing such distributions. Their decomposition in components plays a major role in the above fields. The analysis has various goals, besides detection of causes, reduction of the data to the essentials (prototypes) and establishing of classification rules.

Mixture and cluster analysis should not be confused with vector quantization or segmentation; see the special issue [263]. Optimal vector quantization is the subdivision or segmentation of an entity in several metrically similar parts. It is quantitative and geometric, depending on an a priori given metric. Nearby objects are grouped together; distant ones are separated. The result may change after selecting other units of measurement. The main tool of optimal quantization is the k-means algorithm. In this context it's commonly referred to as Lloyd's [333] algorithm and its ramifications. In quantization, even the normal distribution is decomposed in $g \geq 2$ parts. The result of cluster analysis would be a "single source." Quantization is used, for instance, for organizational purposes. A related reference appears in the beginning of Gaius Julius Caesar's [75] famous first report to the Roman Senate, "Gallia est omnis divisa in partes tres." By contrast, cluster analysis is qualitative and concerned with detecting causes in data. It should not depend on the units of measurement used.

Mixtures are also applied in approximation theory. It is well known that normal mixtures can approximate *any* distribution in the weak sense with an arbitrarily small error. Once again, here, this is not the objective of mixture decomposition.

Probabilistic clustering methods start from a probabilistic model of the data. These days, the ML and MAP paradigms are the two preferred approaches to estimating its parameters. In mixture and cluster analysis, various problems, both expected and unexpected, are encountered. First, likelihood functions and criteria do not possess maxima in general, so special precaution has to be taken. Nevertheless they remain prime tools. Second, likelihood functions and criteria usually possess many so-called "local maxima" (mixture model) and "steady" partitions (classification model). The selection of the desired ones is not straightforward. Most of them are unreasonable. Third, the clustering method obtained may not be robust in the sense that deviations from the model may grossly falsify the result. Fortunately, solutions to most of these problems are currently available.

In order to design effective methods for mixture decomposition and cluster analysis, it is necessary to follow the "instructions," important facts and acknowledged principles, as opposed to opinions or heuristics. The facts are the theorems. Principles are invariance and equivariance under natural transformations of the sample space. For example, consider a data set from physics with two variables: time and distance. There is no natural relation between these two quantities. The choice of their units of measurement is largely conventional. It makes quite a difference whether we measure time in seconds and length in inches, or time in minutes and length in centimeters. If components look spherical in the first case, they'll look like needles in the second. A method that lacks equivariance under variable scaling runs the risk of producing different solutions in the two set-ups, which it should not. It makes no sense to obtain different partitions of the same phenomenon in London and in Paris. This may happen with the k-means criterion. It is biased toward spherical clusters of about equal size.

Sometimes, assumptions have to be made to assert the very existence of likelihood maxima, consistency, robustness, and stability. One should then be aware of their invariance and equivariance properties. They should comply with prior information. Equivariance w.r.t. translation is rarely disputed. If we are willing to assume a spherical model, then we get rotational equivariance. However, we must not expect invariance of our clustering result w.r.t. variable scaling. By contrast, when we use a diagonal model because it is known that variables are independent in each class, then we buy equivariance w.r.t. variable scaling, but not w.r.t. rotations. Often no such prior information is available. It is then safest to "let the data speak for themselves," assuming full affine equivariance of estimators and constraints. This is equivalent to having equivariance w.r.t. both rotations and variable scaling. It is the richest class.

The importance of equivariance has also been pointed out in other areas. Refer to Thode [501] in normality testing, Lehmann and Casella [315] and Barndorff-Nielsen [24], p. 261, in point estimation, and Tyler et al. [514]. Everitt et al. [147] discuss some scaling methods proposed in the literature based on prior metrics. Obviously discouraged by their properties, the authors state on p. 68: "*In the end, the best way of dealing with the problem of the appropriate unit of measurement might be to employ a cluster method which is invariant under scaling ....*"

Note that mixture components bear metrics of their own. They are known only *after* a successful analysis and not before. Methods based on prior metrics lack equivariance and cannot adapt to different scales. It's true that the design of scale invariant estimators, criteria, and algorithms as well as their analysis are more demanding. According to Lopuhaä and Rousseeuw [336], *"the design of equivariant estimators with a high breakdown point is not trivial."* Fortunately, one of the advantages of likelihood methods in equivariant models is their equivariance. Previous data sphering renders all methods affine equivariant. However, equivariance is not everything. Otherwise we could always use the normal model with full scale matrices. Using sphering for this purpose has often been criticized. It may not produce the desired result; see, for instance, Gnanadesikan et al. [201].

The presence of outliers in a data set degrades all statistical methods. The results of mixture and cluster analysis may be completely upset. Here, it is imperative to have protection against observations that deviate from the probabilistic assumptions. Mild outliers that do not conform exactly to the posited model can be dealt with by using heavy-tailed models, such as elliptical symmetries. They are more indulgent than members of the light-tailed normal family, which are not protective at all. Trimming is probably the earliest safeguard against outliers. Newcomb [387] used it already toward the end of the nineteenth century. The method is data driven and effective in the presence of both mild and gross outliers. Of course, the latter are much more harmful. The underlying idea is to decompose the data set in high-density regions, plus a rest that consists mainly of outliers; see Hartigan [228].

The following is a brief description of the six chapters. Chapters 1 and 2 constitute the basis for the methods reported in the book. In Chapter 1, the parametric and nonparametric finite mixture models and the classification model are described. A major portion is dedicated to the consistency theorems of likelihood estimators for elliptical mixture models in various settings. Section 1.1 is a prologue containing three generally applicable theorems, one for local likelihood maxima in the parametric setting based on Cramér's [105] theory, Theorem 1.2, and two for constrained maxima in the nonparametric and parametric settings, Theorems 1.5 and 1.7, respectively. They are based on Kiefer and Wolfowitz' [291] seminal paper. The former rests on differentiability, the latter hinge on topology, that is, continuity and compactness, and on integrability. Besides consistency, the theorems also guarantee the existence of solutions.

All theorems are conveniently applied in Section 1.2 to finite elliptical mixtures. See Theorems 1.19, 1.22, and 1.27. Elliptically symmetric distributions with density generators of different decay, from light to heavy tails, offer an entire scale of robustness to certain mild outliers. Special emphasis will be put on the estimation of arbitrary locations and scales. That is, full and distinct scale matrices will be employed. In contrast to the existence of local likelihood maxima, that of global maxima needs the application of constraints to the scale parameters. Because of the required affine equivariance, the Hathaway–Dennis–Beale–Thompson (HDBT) constraints (1.27) will be used here. Three more subjects are identifiability, asymptotic normality, and the phenomenon of the so-called spurious solutions. The third part of Chapter 1 is devoted to the related classification models and their ML and MAP criteria. Here, steady partitions replace (local) likelihood maxima. Special cases are the classical sum-of-squares and determinant criteria. Theorem 1.40 on consistency of the HDBT constrained MAP criterion of the classification model does not need specified parents. The consistency theorems are not just of theoretical interest. They rather offer a basis for the estimation methods described in Section 4.1.

Chapter 2 is focused on outlier robustness. After a discussion regarding the outlier concept, there are sensitivity studies with examples. They show that some well-known robustness schemes, effective in the case of unimodal populations, fail in mixture and cluster analysis. However, it turns out that trimming safeguards against all kinds of outliers, theoretically as well as practically. Besides sensitivity studies, the theoretical properties of outlier protection methods can be fathomed out with breakdown points. Two estimators are analyzed, the trimmed MLE for the normal mixture model and the Trimmed Determinant Criterion, TDC, both under the HDBT scale constraints. The restriction to normality has no adverse effects – after trimming tails are light. The algorithms need the number of discards as a parameter. In practice, analyses for various numbers have to be performed and validated.

Theorems 2.5 and 2.22 show that the universal asymptotic breakdown points of scale matrices are high without any assumptions. The same cannot be expected for mean vectors and mixing rates, Theorems 2.9 and 2.25. However, they, too, share a positive breakdown point when components are well separated and the proper number of components is used for the analysis, Theorems 2.16 and 2.32. Again, the HDBT constraints are a crucial assumption for these results. Thus, the HDBT constraints are responsible for the existence of solutions, their consistency and robustness, and they are a key to feasible solutions.

Statistics is becoming more and more algorithmic, using portions of computer science. In particular, this observation applies to clustering algorithms. Chapter 3 presents some algorithms for computing the solutions promised in the first two chapters, mainly by iterative methods. A standard tool for computing local likelihood maxima and for mixture decomposition is the EM algorithm. Extended to trimming, it becomes the EMT algorithm. It iteratively computes weights (E) and parameters (M), and performs a trimming step (T). Its properties will be analyzed. Section 3.1.5 presents an analysis of the order of (local) convergence of EM. Some methods for acceleration are discussed in Section 3.1.6.

Methods for computing steady solutions of cluster criteria are the k-parameters algorithms. They resemble the EM, but substitute a partitioning step for the E-step. The weights determined in the E-step actually define a fuzzy partition. The special case for Ward's sum-of-squares criterion reduces to the classical k-means algorithm. It is spherically equivariant but does not comply with the requirement of scale equivariance. When it is a priori known that the parent is homoscedastic, isobaric, and spherical, there is nothing better.

The k-parameters algorithm can create deficient clusters when no protection is provided. This corresponds to nonconvergence of the EM algorithm. The result is a program failure, since the parameters cannot be estimated in the subsequent step. Protection against deficient clusters leads to the realm of combinatorial optimization, Procedure 3.29. All procedures are presented in algorithmic form and readily implementable.

So far, the methods discussed hinge on three computational parameters – the number of clusters, the number of discards, and the HDBT scale constraint. All are in general ambiguous. However, Proposition 4.2 and Lemma 4.3 of Chapter 4 show that the scale constraint can be estimated along with a solution given the other two parameters. In unimodal data analysis, the likelihood principle suffices. In cluster analysis, we are led to a second principle, that of *scale balance*. The estimation problem is reduced to computing Pareto points w.r.t. likelihood (fit) and HDBT ratio (scale balance) and leads to biobjective optimization. The problem doesn't have a unique solution in general. However, some Pareto solutions may be deleted since they are either spurious or there is a cluster of poor quality. In complex situations, there remain a few solutions of comparable quality. Each of them can represent the parent population. The algorithms derived from the theory contain no parameters other than the number of clusters and outliers.

These two numbers remain to be determined. In normal and important elliptical mixtures, the number of components is well defined by identifiability, Section 1.2.3. If only a *sample* from the mixture is available, then the number of components of an estimated mixture is taken as the estimated number of components of the parent mixture. Some affine equivariant methods for uncontaminated data are compiled in Section 4.2, a modification of the likelihood-ratio test, tests based on cluster criteria, model selection criteria such as the Bayesian information criterion (BIC), and the ridge line manifold. Some simple and crude methods for determining the number of clusters and outliers, simultaneously, are compiled in Section 4.3. They work well on sufficiently clear data sets. All methods require that solutions for all relevant numbers of clusters and outliers be produced.

For several reasons it is not sufficient in mixture and cluster analysis to just fit a model. On the one hand, the model assumptions may be too far from reality. On the other hand, algorithms have a tendency to decompose everything, even when there is nothing to do. All this may lead to wrong conclusions. It is, therefore, imperative that every result of a mixture or cluster analysis be validated. This will be the subject of Section 4.4. Unfortunately, cluster validation is still in its infancy. It should be applicable to clustered data sets that the employed cluster algorithm could not cope with. In contrast to cluster analysis, we are now given a partition. We just wish to know whether it is reasonable, that is, whether the subsets found originate from different mixture components. Although disputable, this is often done by verifying isolation and cohesion. Four validation methods are separation indices such as the Kullback–Leibler divergence, the Hellinger distance, and indices of linear separation, Section 4.4.1, moreover significance tests, visualization, Section 4.4.3, and stability, Section 4.4.5. Yet the solution to a clustering task is often ambiguous.

There is no reason why *all* offered variables should be equally useful in view of the desired partition. Chapter 5, therefore, describes methods for enhancing mixture and cluster analysis by variable (or subset) selection. Roughly speaking, a subset of variables is *irrelevant* if it is devoid of information or if the information it contains on the cluster structure is already contained in its complement, Section 5.1. It is then possible, and even desirable, to remove this subset. This concept comprises in particular variables that are just noise degrading the analysis. The main Theorem 5.5 states that the variables uniquely disintegrate in a subset that contains the information on the partition and the irrelevant rest. Filter and wrapper methods are at our disposal for detecting irrelevant variables. Typical filters are tools from projection pursuit, often univariate or low-dimensional and mainly applied to data sets with many variables. Some more recent ones explore the whole space at once, Section 5.2. Filters do not use mixture or cluster analysis. In contrast, wrapper methods are controlled by some clustering method. Like trimming, variable selection is data driven. Four wrappers are presented. One uses a likelihood ratio test, Section 5.3.1, another one Bayes factors, Section 5.3.2, and the two others use the mixture and cluster analysis of the previous chapters, Section 5.3.3.

The theoretical part of the book concludes with guidelines for tackling a mixture or cluster analysis. Only in simple cases will it be possible to just press a button and get a reasonable solution in a few seconds.

The methods of the book are applicable to data sets from diverse fields. Some examples are analyzed in Chapter 6. They are the classical IRIS data set from botanics, the SWISS BILLS from criminology, the STONE FLAKES from archaeology, and the LEUKEMIA data set from biomedicine. With the exception of the archaeological data, all are now classical. None of them is trivial. The examples are chosen so that the true solutions are essentially known. They will, however, be used only for final comparisons. The methods themselves are unsupervised. The aim here is not to analyze the data, but to assess the efficacy of the methods which could not be convincingly presented with data of unknown structure.

Modern statistics draws on many disciplines. Material indispensable for the understanding of the main text, but not directly related to mixture or cluster analysis, is deferred to the appendix. Of its seven sections, one each is devoted to geometry, topology, analysis, measure and probability theory, statistics, and optimization theory. They testify to the diversity of the disciplines necessary to analyze and implement all methods. In the appendix, proofs available in the literature are omitted. Information on the origin of theorems and methods and some additional references are found in the Notes following most major sections.

Chapter 1

Mixture and classification models and their likelihood estimators

This chapter focuses on mixture and classification models and on asymptotic properties of their likelihood estimators. Section 1.1 covers convergence theorems general enough to be applicable to various nonparametric and parametric statistical models, not only to mixtures. The nonparametric framework for consistency has the advantage of needing no parameter space and thus no identifiability. Many models possess a natural parameter space; often it is Euclidean. We are then also interested in convergence w.r.t. to the topology of this space. This leads to the parametric theory. Asymptotic normality needs even differentiable models.

In Section 1.2, the general consistency theorems of Section 1.1 are applied to mixture models. The identifiability of mixture models will be carefully discussed, see Section 1.2.3. Section 1.3 deals with classification models. Their convergence properties need a different approach; see Section 1.3.7. In both cases, a focus is on Euclidean sample spaces since they are versatile enough to allow a broad range of applications. Robustness against outliers is a leitmotiv of this book. Likelihood estimators can be safeguarded against outliers by employing models with heavy tails; see Jeffreys [274, 275], Maronna [349], Lange et al. [307]. In this first chapter, systematic emphasis is therefore placed on clustering models over elliptically symmetric basic models. They are adapted to the symmetries of Euclidean space and, endowed with heavy tails, they offer the flexibility needed for *small outlier* robustness where the ubiquitous normal model lacks sufficient fit. Therefore, the multivariate *t*-distribution, a special case of Pearson's Type VII, plays a special role. In mixture analysis, in contrast to parameter estimation in unimodal families, these models do not yet enjoy *gross* outlier robustness. However, gross outliers appear almost always when data is automatically extracted from complex objects by a machine. Robustness to gross outliers will be the subject matter of Chapter 2.

1.1 General consistency and asymptotic normality

Theorems on consistency and asymptotic normality of likelihood estimates are an important part of statistical inference. They have a long history, dividing in two branches. The first goes back to Cramér [105]. It is applicable to *differentiable* statistical models and deals with *local* maxima of the likelihood function. The second needs only a *topological* parameter space. Milestones here are the theorems of Wald [520] and of Kiefer and Wolfowitz [291]. Under certain assumptions they yield consistency of a *global* maximum of the likelihood function – if one exists. The present section deals with theorems of both kinds. Section 1.1.1 is about the consistency theory of *local* maxima of the likelihood function and the asymptotic normality of the consistent local maximum. Section 1.1.2 is dedicated to consistency of *global* maxima. In Section 1.2, both theories will be applied to the likelihood functions of mixture models.

1.1.1   Local likelihood estimates

Let (E, ϱ) be a σ-finite measure space. Let $X_1, X_2, \ldots$ be an i.i.d. sequence of E-valued random observations distributed according to some unknown member of a ϱ-dominated $\mathcal{C}^2$ model $(\mu_\vartheta)_{\vartheta \in \Theta}$ with ϱ-densities f_ϑ; see Section F.1.2. The parameter ϑ is subject to estimation on the basis of the realizations $X_1, \ldots, X_n$ up to some n. They are drawn from the n-fold product model $(E^n, (\mu_\vartheta^{\otimes n})_{\vartheta \in \Theta})$. Its joint likelihood function is denoted by $f(\mathbf{x}; \vartheta) = f_n(\mathbf{x}; \vartheta) = \prod_i f(x_i, \vartheta)$, $\mathbf{x} = (x_1, \ldots, x_n) \in E^n$. Its score function, (expected) Fisher information, and Hessian matrix (see Section F.1.2) can be represented in a simple way as sums of elementary functions. They are $s_n(\mathbf{x}, \vartheta) = \sum_i s(x_i, \vartheta)$, $\mathcal{I}_n(\vartheta) = n\mathcal{I}(\vartheta)$, and $\mathrm{D}_\vartheta^2 \log f(\mathbf{x}; \vartheta) = \sum_i \mathrm{D}_\vartheta^2 \log f(x_i; \vartheta)$, respectively. Asymptotic statistics is concerned with a sequence $t_n : E^n \to \Theta$, $n \geq 1$, of parameter estimators, where t_n depends on the first n observations X_1^n of the sequence $X_1, X_2, \ldots$. Let $T_n = t_n(X_1, \ldots, X_n)$ be the nth estimate for $(X_1, X_2, \ldots)$. The sequence (T_n) is called **strongly consistent** at ϑ_0 in the classical sense if (T_n) converges P-a.s. to ϑ_0 as $n \to \infty$ whenever $X_i \sim f_{\vartheta_0}$.

Cramér [105] showed that, under some regularity conditions, the likelihood function possesses a *local* maximum which is consistent in the classical sense. The version below does not need a third derivative. It requires that the likelihood function satisfy the following second-order regularity conditions. They are motivated by Lemma D.4, which links the analysis of the models to their measure theory.

1.1 Regularity Conditions

(i)   *The likelihood function is strictly positive and twice continuously differentiable on Θ.*

(ii)   *There exists an open neighborhood $U \subseteq \Theta$ of ϑ_0, a ϱ-integrable function h, and a μ_{ϑ_0}-integrable function h_0 such that*

$$\|\mathrm{D}_\vartheta f(x; \vartheta)\| + \|\mathrm{D}_\vartheta^2 f(x; \vartheta)\| \leq h(x),$$
$$\|\mathrm{D}_\vartheta s(x; \vartheta)\| \leq h_0(x),$$

for all $x \in E$ and all $\vartheta \in U$.

(iii)   *The (expected) Fisher information exists and is positive definite at ϑ_0.*

Actually, since the parameter ϑ_0 is unknown, one will have to check Condition (iii) at every parameter in Θ. The regularity conditions depend solely on the statistical model. Regularity Condition (i) says in particular that the model is $\mathcal{C}^2$. Since we are here concerned with local properties of the parameter space Θ, it is sufficient to assume that it is an open subset of q-dimensional Euclidean space. The following theorem is applicable to *local* maxima of a likelihood function and will later be applied to mixtures.

1.2 Theorem (Existence and consistency of a local maximum) *Let Θ be an open subset of Euclidean space. If the statistical model meets the Regularity Conditions 1.1, then, P-a.s.,*

(a)   *there exists an open neighborhood U of ϑ_0 such that, if n is large enough, the joint likelihood function $f_n(X_1^n; \cdot)$ possesses P-a.s. exactly one local maximum T_n in U;*

(b)   *the (random) sequence (T_n) converges to ϑ_0, that is, (T_n) is strongly consistent at ϑ_0 in the classical sense.*

PROOF.   We apply Proposition C.11 with some open ball A about ϑ_0 such that $\bar{A} \subseteq \Theta$ and with the score function of the nth product model divided by n,

$$G_n(\vartheta) = \tfrac{1}{n} s_n(X; \vartheta) = \frac{1}{n} \sum_{i=1}^{n} s(X_i, \vartheta),$$

the arithmetic mean of the elementary score functions. All functions G_n are continuously differentiable by the Regularity Condition (i).

Regularity Condition (ii) asserts in particular that the statistic $s(\cdot, \vartheta_0)$ is μ_{ϑ_0}-integrable. Therefore, the Strong Law of Large Numbers (SLLN), Theorem E.31(b), applies to show

$$\lim_n G_n(\vartheta_0) = \lim_n \frac{1}{n}\sum_{i=1}^{n} s(X_i, \vartheta_0) = \mathrm{E}\, s(X_1, \vartheta_0) = \int_E s(x; \vartheta_0)\mu_{\vartheta_0}(\mathrm{d}x) = 0, \qquad (1.1)$$

P-a.s., by Lemma F.4(a). This is assumption (i) of Proposition C.11. By Regularity Conditions (i) and (ii), $G_n(\vartheta; x)$ is continuously differentiable w.r.t. ϑ and $\|\mathrm{D}_\vartheta s(x; \vartheta)\|$ is bounded by a μ_{ϑ_0}-integrable function independent of ϑ. Now, the Uniform SLLN, Theorem E.33, applies to $\mathrm{D}_\vartheta s$ to show that $\mathrm{E}\,\mathrm{D}_\vartheta s(X_1, \vartheta)$ is continuous and

$$\lim_n \mathrm{D}_\vartheta G_n(\vartheta) = \lim_n \frac{1}{n}\sum_{i=1}^{n} \mathrm{D}_\vartheta s(X_i, \vartheta) = \mathrm{E}\,\mathrm{D}_\vartheta s(X_1, \vartheta) = \int_E \mathrm{D}_\vartheta^2 \log f(x; \vartheta)\mu_{\vartheta_0}(\mathrm{d}x), \quad (1.2)$$

uniformly for $\vartheta \in \bar{A}$, P-a.s. This is assumption (ii) of Proposition C.11. Together with Regularity Condition (iii) and Lemma (F.4)(c) this also implies that $\lim_n \mathrm{D}_\vartheta G_n(\vartheta_0)$ is negative definite, P-a.s., that is, assumption (iii) of Proposition C.11.

Proposition C.11 now asserts that, P-a.s., there exists a neighborhood $W \subseteq A$ of ϑ_0 such that the restrictions of eventually all functions G_n to W are diffeomorphisms, in particular one-to-one. Moreover, for eventually all n, the open image $G_n(W)$ contains some open ball with center $G(\vartheta_0) = \lim G_n(\vartheta_0) = 0$ by (1.1). Both facts combine to show that G_n has exactly one critical point $T_n \in W$ for eventually all n.

Finally, this critical point must be a local maximum of G_n since $\mathrm{D}_\vartheta G_n(T_n) = \mathrm{D}_\vartheta G_n(G_n^{-1}(0)) = \mathrm{D}_\vartheta G_n(G_n^{-1}(G(\vartheta_0)))$ is negative definite by Proposition C.11(f). $\qquad\square$

Another important fact is that, after suitable scaling, the deviation of the estimate from the true value is asymptotically normal and the estimate is asymptotically efficient; see Section F.2.6. This is the content of the following lemma. It is applicable to the local maxima found in Theorem 1.2 and also to the parametric maximum likelihood (ML) estimators of Section 1.1.2.

1.3 Proposition (Asymptotic normality and efficiency) *Let Θ satisfy the assumptions of Theorem 1.2 and let T_n be a consistent zero of the nth score function s_n. Then*

(a) $\sqrt{n}(T_n - \vartheta_0)$ converges in distribution to $N_q\big(0, \mathcal{I}(\vartheta_0)^{-1}\big)$;

(b) the sequence $(T_n)_n$ is asymptotically efficient.

PROOF. Let G_n be defined as in the proof of the theorem. Applying the mean value theorem to each of its q components $G_{n,k}$, $1 \le k \le q$, we obtain a q-vector $(\theta_{n,1}, \ldots, \theta_{n,q}) = \boldsymbol{\theta}_n$ of convex combinations $\theta_{n,k}$ of T_n and ϑ_0 such that

$$-G_n(\vartheta_0) = G_n(T_n) - G_n(\vartheta_0) = \begin{pmatrix} \mathrm{D}_\vartheta G_{n,1}(\theta_{n,1}) \\ \vdots \\ \mathrm{D}_\vartheta G_{n,q}(\theta_{n,q}) \end{pmatrix}(T_n - \vartheta_0).$$

Abbreviate the $q \times q$ matrix by $\mathrm{D}_\vartheta G_n(\boldsymbol{\theta}_n)$. By the assumption $T_n \to \vartheta_0$, the random parameters $\theta_{n,k}$ converge to ϑ_0 and, by Eq. (1.2), the random matrices $\mathrm{D}_\vartheta G_n(\vartheta)$ converge to $\mathrm{E}\,\mathrm{D}_\vartheta s(X_1, \vartheta)$ uniformly for ϑ close to ϑ_0, P-a.s. Therefore, $\mathrm{D}_\vartheta G_n(\boldsymbol{\theta}_n) \longrightarrow_n \mathrm{E}\,\mathrm{D}_\vartheta s(X_1, \vartheta_0) = -\mathcal{I}(\vartheta_0)$, P-a.s. by Lemma F.4(c). In particular, $\mathrm{D}_\vartheta G_n(\boldsymbol{\theta}_n)$ is regular for eventually all n and the equality above can be written

$$\sqrt{n}(T_n - \vartheta_0) = -\big(\mathrm{D}_\vartheta G_n(\boldsymbol{\theta}_n)\big)^{-1}\sqrt{n}G_n(\vartheta_0) = -\big(\mathrm{D}_\vartheta G_n(\boldsymbol{\theta}_n)\big)^{-1}\frac{1}{\sqrt{n}}\sum_{i=1}^{n} s(X_i, \vartheta_0). \qquad (1.3)$$

Since $s(X_1, \theta)$ is centered and square integrable with covariance matrix $\mathcal{I}(\vartheta_0)$ by Regularity Condition (iii), the central limit theorem (CLT) shows that $\frac{1}{\sqrt{n}}\sum_{i=1}^{n} s(X_i, \vartheta_0)$ converges in distribution to a random vector $Y \sim N_q(0, \mathcal{I}(\vartheta_0))$. Finally, Slutsky's lemma E.41 applies to Eq. (1.3), showing that $\sqrt{n}(T_n - \vartheta_0)$ converges in distribution to $-(\mathcal{I}(\vartheta_0))^{-1}Y$. This proves part (a), and part (b) follows in the sense of the remark after Theorem F.22.   □

Theorem 1.2 and its corollary leave open a number of questions that are important in practice when nothing is known about ϑ_0:

(i) Given a data set $(x_1, \ldots, x_n)$, is n large enough to allow a local maximum close to ϑ_0?

(ii) If the answer is yes, which one is it?

(iii) If the product likelihood function possesses a *global* maximum, is it the consistent local maximum promised in Theorem 1.2?

Question (i) seems to be notorious in statistics: Is my data set large enough to reveal the desired information? Also an answer to question (ii) is not easy. Something will be said in Chapter 4. The answer to the third question is yes, as we will now see.

1.1.2   Maximum likelihood estimates

The preceding matter is applicable to likelihood functions irrespective of the existence of global maxima. This is its advantage. If one or more maxima exist, then more can be said. Whereas Theorem 1.2 was based on differentiability, the theorems of this section will rest on topology and compactness. The ideas go back to Wald [520] and Kiefer and Wolfowitz [291]. In order to take advantage of the properties of compact topological spaces, Wald [520] restricted matters to a closed *subset* of some Euclidean parameter space. This excludes equivariant scale models. Kiefer and Wolfowitz overcame this limitation with a seminal idea. Instead of resorting to a closed *subset* of the parameter space, they compactify it.

The first aim of this section is a completely nonparametric consistency theorem for absolute maxima of the likelihood function. The nonparametric treatment removes the influence of parameters, concentrates on what is really important, the populations and their densities, and avoids the somewhat inconvenient requirement of identifiability that also plays an important part in Kiefer and Wolfowitz [291]. We will follow these authors compactifying the parameter space but the construction will be different. We will use the real vector space $\mathbb{M}(\varrho)$ of ϱ-equivalence classes of real-valued, measurable functions on E. As shown in Proposition D.26, it bears a complete metric structure that induces the ϱ-stochastic convergence. Moreover, it is large enough to contain all probability density functions w.r.t. ϱ. Therefore, its structure makes it an ideal basis for compactifying the parameter space and thus for consistency.

Let $\mathcal{D} \subseteq \mathbb{M}(\varrho)$ be some set of probability densities w.r.t. ϱ. Assume that the observations $X_1, X_2, \ldots$ are sampled from some population with ϱ-density $f \in \mathcal{D}$. Instead of parameter-valued statistics, we consider now the $\mathcal{D}$-valued statistics $t_n : E^n \to \mathcal{D}$, $n \geq 1$. Everything else is as before. In particular, $T_n = t_n(X_1, \ldots, X_n)$.

1.4 Definition (Nonparametric consistency) The sequence (T_n) is **strongly consistent** at $f_0 \in \mathcal{D}$ if (T_n) converges P-a.s. to f_0 as $n \to \infty$ w.r.t. the topology of ϱ-stochastic convergence in $\mathbb{M}(\varrho)$.

The announced nonparametric consistency theorem reads as follows. Under an additional assumption it asserts at the same time the *existence* of the ML estimator. The notation $\log^+ a$ ($\log^- a$) stands for the positive (negative) part of $\log a$, $a > 0$.

1.5 Theorem (Existence and consistency of the nonparametric ML estimator) *Let E be a Hausdorff space and let ϱ be a σ-finite Borel measure on E. Let $\mathcal{D}$ be some set of ϱ-probability densities on E and let $f_0 \in \mathcal{D}$ be strictly positive, $\mu_0 = f_0\varrho$. Assume that there is an open subset $E_0 \subseteq E$, $\varrho(E \backslash E_0) = 0$, such that*

(i) all functions in $\mathcal{D}$ are continuous on E_0;

(ii) each sequence in $\mathcal{D}$ contains a subsequence that converges pointwise everywhere on E_0 to a continuous function on E_0;

(iii) there is a μ_0-integrable function h_0 on E_0 such that, for all $f \in \mathcal{D}$,

$$\log^+ \frac{f}{f_0} \le h_0$$

(for this to be true it suffices that $\log^+ f + \log^- f_0 \le h_0$, $f \in \mathcal{D}$).

Then:

(a) Any sequence (T_n) of ML estimators is strongly consistent at f_0; see Definition 1.4.

(b) If, in addition, $\mathcal{D}$ is locally compact w.r.t. the topology of ϱ-stochastic convergence on $\mathbb{M}(\varrho)$, then, P-a.s., the nth product likelihood function has a maximum T_n in $\mathcal{D}$ for eventually all n.

PROOF. Let $\overline{\mathcal{D}}$ be the closure of $\mathcal{D}$ in $\mathbb{M}(\varrho)$ w.r.t. the topology of ϱ-stochastic convergence. By assumption (ii), each sequence in $\mathcal{D}$, and hence in $\overline{\mathcal{D}}$, has a cluster point in $\overline{\mathcal{D}}$. Theorem B.10 therefore asserts that $\overline{\mathcal{D}}$ is a compact metric space and hence separable. The same assumption implies that each function $g \in \overline{\mathcal{D}}$ is the pointwise limit of a *sequence* of functions in $\mathcal{D}$. Note that the functions $g \in \overline{\mathcal{D}} \backslash \mathcal{D}$ need not be probability densities w.r.t. ϱ – there may be a loss of mass.

By the entropy inequality, Lemma D.17, the Kullback–Leibler divergence satisfies for all $f \in \mathcal{D} \backslash \{f_0\}$,

$$\int_E \log \frac{f}{f_0} \, \mathrm{d}\mu_0 < 0. \tag{1.4}$$

This inequality is valid even for $g \in \overline{\mathcal{D}} \backslash \{f_0\}$. Indeed, let $(f_n) \subseteq \mathcal{D}$ such that $f_n \to g$ pointwise. Using Jensen's inequality and Fatou's lemma, we estimate

$$\int_E \log \frac{g}{f_0} \, \mathrm{d}\mu_0 \le \log \int_E \frac{g}{f_0} \, \mathrm{d}\mu_0 \le \log \liminf_n \int_E \frac{f_n}{f_0} \, \mathrm{d}\mu_0 \le 0.$$

In the first inequality, there is equality only if $g = c \cdot f_0$ with some constant $c \le 1$. But $c = 1$ is ruled out since $g \ne f_0$. This is the claim.

Let B_r stand for r-balls in $\overline{\mathcal{D}}$ w.r.t. some metric that induces the topology on $\mathbb{M}(\varrho)$, for instance, d_ν; see Proposition D.26. Since all $f \in \overline{\mathcal{D}}$ are continuous on E_0, the least upper bound of B_r is lower semi-continuous and hence Borel measurable. By monotone convergence and assumption (iii), we conclude for each $g \in \overline{\mathcal{D}} \backslash \{f_0\}$

$$\mathrm{E} \sup_{f \in B_r(g)} \log \frac{f(X_1)}{f_0(X_1)} \xrightarrow[r \to 0]{} \mathrm{E} \log \frac{g(X_1)}{f_0(X_1)} = -\gamma_g < 0$$

as $r \to 0$. For each such g, choose a ball $B(g) \subseteq \overline{\mathcal{D}}$ with center g such that $\mathrm{E} \sup_{f \in B(g)} \log \frac{f(X_1)}{f_0(X_1)} < -\gamma_g/2$. The SLLN, Theorem E.31(a), along with assumption (iii), then asserts

$$\limsup_n \sup_{f \in B(g)} \frac{1}{n} \sum_{i=1}^n \log \frac{f(X_i)}{f_0(X_i)} \le \limsup_n \frac{1}{n} \sum_{i=1}^n \sup_{f \in B(g)} \log \frac{f(X_i)}{f_0(X_i)}$$

$$\le \mathrm{E} \sup_{f \in B(g)} \log \frac{f(X_1)}{f_0(X_1)} < -\gamma_g/2,$$

P-a.s. Hence, for each $g \in \overline{\mathcal{D}} \backslash \{f_0\}$, there is a natural number N_g such that, P-a.s.,

$$\sup_{f \in B(g)} \frac{1}{n} \sum_{i=1}^n \log \frac{f(X_i)}{f_0(X_i)} \le -\gamma_g/2$$

whenever $n \geq N_g$.

Now, let U be any open neighborhood of f_0 in $\overline{\mathcal{D}}$. A finite number of balls $B(g_1), \ldots, B(g_\ell)$ cover the compact set $\overline{\mathcal{D}} \backslash U$. Letting $N_U = \max N_{g_j}$ and $\gamma = \min \gamma_{g_j}$, we infer for $n \geq N_U$

$$\sup_{f \notin U} \frac{1}{n} \sum_{i=1}^n \log \frac{f(X_i)}{f_0(X_i)} \leq \sup_{j \in 1..\ell} \sup_{f \in B(g_j)} \frac{1}{n} \sum_{i=1}^n \log \frac{f(X_i)}{f_0(X_i)} \leq -\gamma/2.$$

Rewritten slightly, we have P-a.s.

$$\sup_{f \notin U} \prod_{i=1}^n \frac{f(X_i)}{f_0(X_i)} \leq e^{-n\gamma/2}$$

for all $n \geq N_U$ with some $\gamma > 0$.

Since the product attains the value 1 at $f = f_0$, the estimate implies that $\overline{\mathcal{D}} \backslash U$ contains no maximum. In part (a) we assume that the likelihood function $f \mapsto \prod_{i=1}^n f(X_i)$ has a maximum in $\mathcal{D}$. It follows that it must be in $U \cap \mathcal{D}$. Since U was an arbitrary open neighborhood of f_0 w.r.t. $\overline{\mathcal{D}}$, $U \cap \mathcal{D}$ is an arbitrary open neighborhood of f_0 w.r.t. $\mathcal{D}$ and we have proved part (a).

Under the additional assumption of (b) $\mathcal{D}$ is open in $\overline{\mathcal{D}}$ (see Proposition B.20) and we insert $\mathcal{D}$ for U above. There is a sequence $(f^{(k)})_k \subseteq \mathcal{D}$, depending on X, such that

$$\prod_{i=1}^n f^{(k)}(X_i) \xrightarrow[k \to \infty]{} \sup_{f \in \mathcal{D}} \prod_{i=1}^n f(X_i).$$

By assumption (ii), we may assume that $(f^{(k)})$ converges pointwise on E_0 to a continuous function $g \in \overline{\mathcal{D}}$. It follows that

$$\prod_{i=1}^n g(X_i) = \sup_{f \in \mathcal{D}} \prod_{i=1}^n f(X_i),$$

that is, g is maximal. We have seen above that $\overline{\mathcal{D}} \backslash \mathcal{D}$ contains no maximum and so g must be in $\mathcal{D}$. This is claim (b). □

1.6 Remarks (a) The proof of Theorem 1.5 requires a metrizable space $\mathcal{D}$. This was used twice, namely, in the application of Fatou's lemma and of monotone convergence, and seems to be unavoidable. The space $\mathbb{M}(\varrho)$ has this property. On the other hand, the space must be compact since we need to select a finite number of the neighborhoods constructed to cover $\overline{\mathcal{D}}$. This requires the assumption (ii). The functions in (i) and the limits in (ii) do not have to be continuously extensible to all of E. Assumption (iii) is Wald's [520] assumption 2 on integrability and sharpens the existence of the Kullback–Leibler divergence. It is localized there. In applications one will usually get it globally right away when it is true.

(b) In applications of Theorem 1.5, the subset $E_0 \subseteq E$ is defined by properties intrinsic to the sample space E, often to its geometry. Assumption (ii) is often no problem. All that is needed is a set $\mathcal{D}$ of continuous density functions that is small enough to be relatively compact in $\mathbb{M}(\varrho)$. It turns out that many models are benign in this respect. More problematic is the bound (iii). As Example 1.9(b) below shows, it is not even satisfied in the univariate normal case. However, there is a simple remedy, namely, using a small product of the model. For example, in the case of the normal model on Euclidean d-space the theorem can be applied with its $d+1$-fold product $E = \mathbb{R}^{(d+1)d}$ (see Examples 1.9(b) and (c)) and E_0 consists of all affine independent $d+1$-tuples of vectors in $\mathbb{R}^d$. The points in the complement of E_0 are singularities – they do not allow parameter estimation since their scatter matrix is singular. The reader may notice that the singularities have not been completely suppressed by restricting the sample space to E_0. They strike back in assumption (iii), which is mainly concerned with integrability close to the singularities. Its verification plays a major role in applications of the theorem; see, for instance, the proofs of Examples 1.9 and of Theorem 1.22.

Many statistical models can actually be considered as parametric and the question arises whether convergence in Theorem 1.5 is related to convergence in parameter space Θ. Of course, convergence of the parameters cannot be expected when there is more than one parameter that represents the sampling density f_{ϑ_0}. It is here that identifiability is needed. There is the following theorem.

1.7 Theorem (Existence and consistency of the parametric ML estimator)
Let E be a Hausdorff space, let Θ be locally compact and metrizable, let $\mathcal{D}$ be some set of ϱ-probability densities on E, let $\Theta \to \mathcal{D}$ be some parametric model on E, and let $\vartheta_0 \in \Theta$ such that the sampling density f_{ϑ_0} is strictly positive. Assume that

(i) the model $\Theta \to \mathcal{D}$ is identifiable.

Moreover, assume that there exists an open subset $E_0 \subseteq E$, $\varrho(E \backslash E_0) = 0$, such that

(ii) if $(\vartheta_n)_n \subseteq \Theta$ converges to ∞ (the Alexandrov point of Θ) while $(f_{\vartheta_n})_n$ converges pointwise on E_0, then the limit is not in $\mathcal{D}$ ("identifiable at the boundary");

(iii) the associated nonparametric model $\mathcal{D}$ satisfies the assumptions (i)–(iii) of Theorem 1.5 with E_0 and $f_0 = f_{\vartheta_0}$.

Then:

(a) The map $\Theta \to \mathcal{D}$ is a homeomorphism with the topology of ϱ-stochastic convergence on $\mathcal{D}$.

(b) Any sequence (T_n) of ML estimators in Θ is strongly consistent at ϑ_0.

(c) P-a.s., the nth product likelihood function has a maximum T_n in Θ for eventually all n.

PROOF. Part (a) claims that the map $\psi : \Theta \to \mathcal{D}$ defined by $\vartheta \mapsto f_\vartheta$ is a homeomorphism, where $\mathcal{D}$ is equipped with the topology of ϱ-stochastic convergence. Since the map ψ is continuous, onto and, by the assumption (i), one-to-one, we have only to show that it is closed. Now, let $A \subseteq \Theta$ be closed and let $(\vartheta_n)_n \subseteq A$ be such that $(f_{\vartheta_n})_n$ converges in $\mathcal{D}$. We have to show that the limit is an element of $\psi(A)$. By assumption (ii) of Theorem 1.5, the sequence $(f_{\vartheta_n})_n$ has a subsequence $(f_{\vartheta_{n_k}})_k$ that converges pointwise on E_0. Assumption (ii) implies that (ϑ_{n_k}) does not converge to the Alexandrov point of Θ. It has thus a cluster point $\vartheta \in \Theta$ and a subsequence, say $(\vartheta_{n_k})_k$, converges to ϑ. By closedness of A, $\vartheta \in A$ and we have shown that $\psi(\vartheta_{n_k}) = f_{\vartheta_{n_k}} \to f_\vartheta = \psi(\vartheta) \in \psi(A)$. This implies the claim.

As a consequence, the associated nonparametric statistical model $\mathcal{D}$ satisfies all assumptions of Theorem 1.5, including local compactness of $\mathcal{D}$. Hence, the densities f_{T_n} are maximal w.r.t. the associated nonparametric likelihood function. By Theorem 1.5(a), we have $\psi(T_n) = f_{T_n} \to f_{\vartheta_0} = \psi(\vartheta_0)$ and homeomorphy implies $T_n \to \vartheta_0$. This is (b). Part (c) follows directly from Theorem 1.5(b) applied to the associated nonparametric statistical model. □

1.8 Remarks (a) Besides identifiability (i), Theorem 1.7 also requires that the model should be identifiable "at the boundary" (ii). The latter means that the mapping $\vartheta \to f_\vartheta$ is natural in the sense of assigning densities far away from f_{ϑ_0} to parameters far away from ϑ_0. It avoids that f_ϑ is close to f_{ϑ_0} for ϑ near the ideal boundary of Θ. Such a "reentrance" of f_{ϑ_0} through the "back door" would be harmful because the estimator would run the risk of coming up with a parameter near the boundary instead of one close to ϑ_0.

To make this point clear, an example of an unnatural parameterization is presented in Fig. 1.1. The parametric model lacks consistency because it is not identifiable "at the boundary." The parameter space is the open interval $]-2, 2.5[$ and the associated probability measures are bivariate, spherically normal of unit variance with means on the solid curve shown in the figure. More specifically, $\mu_\vartheta = N_{m_\vartheta, I_2}$ with

$$m_\vartheta = \begin{cases} (-\vartheta, 1), & -2 < \vartheta \le 0, \\ (-\sin \pi \vartheta, \cos \pi \vartheta), & 0 \le \vartheta \le 1.5, \\ (1, \vartheta - 1.5), & 1.5 \le \vartheta < 2.5. \end{cases}$$

By Steiner's formula A.11, the nth log-likelihood function is

$$-n \log 2\pi - \tfrac{1}{2} \sum_{i=1}^{n} \|X_i - m_\vartheta\|^2$$

$$= -n \log 2\pi - \tfrac{1}{2} \left(\sum_{i=1}^{n} \|X_i - \overline{X}_n\|^2 - n\|\overline{X}_n - m_\vartheta\|^2 \right).$$

Thus, the ML estimator T_n is the projection of the mean $1/n \sum_{i=1}^{n} X_i$ to the closest point on the curve $\vartheta \mapsto m_\vartheta$ shown in Fig. 1.1. Now, assume that the data is sampled from the population with center $(1, 1)$, that is, $\vartheta_0 = -1$. If the nth sample mean lies in the open cone with tip $(1, 1)$ indicated in the figure, then $T_n \in]1.5, 2.5[$. By recurrence of the random process $\frac{1}{\sqrt{n}} \sum_{i=1}^{n} (X_i - (1, 1))$, this happens infinitely often. Thus, the ML estimator is not consistent; it does not even converge. The reason is that the model is not identifiable at the boundary of the parameter space, more precisely, at its Alexandrov point. The densities f_ϑ converge to f_{-1} as ϑ approaches the boundary point 2.5. Note that this problem completely disappears when the nonparametric consistency theorem is applied.

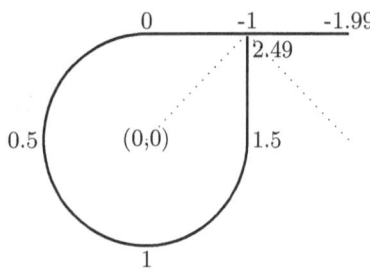

Figure 1.1 *An unnatural parameterization of the statistical model.*

(b) Assumption 1.7(ii) does not require that the limit of (f_{ϑ_n}) should exist when $(\vartheta_n)_n$ converges to ∞. If it does not, then the assumption is satisfied. If the limit exists, one has to show that it is not a member of $\mathcal{D}$. The assumption 1.7(ii) and the assumption 1.5(ii) appearing in 1.7(iii) can be combined in the following way: Each sequence $(f_{\vartheta_n})_n$ has a subsequence that converges pointwise to a continuous function on E_0 and, if the parameters of the subsequence are unbounded, then its limit is not in $\mathcal{D}$.

(c) The consistency theorems have two weak points that should be addressed. First, they are asymptotic and do not tell us from which n on we may rely on their assertions. This question is answered by investigations on the rate of convergence as $n \to \infty$; see Bahadur [19] and Wong and Severini [533]. Second, and more importantly, the consistency theorems need a specified parent, that is, they make a statement about a data sequence sampled from some member of the assumed model. If this assumption is violated, they are not applicable. ML estimators do not necessarily exist and even if they do, they do not have to converge. Even if they converge, the limit, of course, cannot be the parent distribution. However, Foutz and Srivastava [166] and White [526] deal with quasi maximum likelihood estimation for unspecified parents and Nishi [394] studies their rates of convergence.

As an example, consider the bivariate normal model $\mathcal{D} = \{N_{m, I_2} \mid m \in \mathbb{R}^2\}$ with variable mean and fixed variance I_2. The unique ML estimator for all n is the sample mean, irrespective of the data, be it sampled from some member of $\mathcal{D}$, some other distribution, or from no statistical model at all. The situation changes dramatically as we punch out the *closed* unit disk $\overline{B}_1(0)$ from the parameter space (not the sample space) but draw our data from the standard bivariate normal N_{0, I_2}. This violates the assumption of sampling from a member of the model. A minimum of the negative log-likelihood function

$$-\log f_n(X_1^n) = \tfrac{n}{2}(2 \log 2\pi + s_n(X) + \|\overline{X}_n - m\|^2)$$

does not exist when the nth mean $\overline{X}_n$ is contained in the removed disk. By the SLLN this is the case for eventually all n. In a second example, punch out the *open* unit disk $B_1(0)$ from the parameter space $\mathbb{R}^2$, leaving a closed subset of the plane. Sample again from the standard bivariate normal. This time, the ML estimator $T_n = \text{argmin}_{\|m\| \geq 1} \|\overline{X}_n - m\|$ exists for all n. Being the projection of $\overline{X}_n$ to the unit circle when $\|\overline{X}_n\| < 1$, it is P-a.s. unique. The sequence $(T_n)_n$ is P-a.s. dense there by the CLT and does not converge. Note that, in both examples, the other assumptions of Theorem 1.5 are met. In order to mitigate this robustness problem, it is useful to employ rich or heavy tailed models as alternatives to the normal. Elliptical symmetries such as Pearson's Type-VII family are useful in this respect; see Examples E.22 and 1.9(c) and Theorems 1.22 and 1.27.

The hypothesis of a specified parent seems particularly disquieting in view of real applications. In general, it is satisfied only more or less. However, if model densities are > 0 on the sample space, then each of them has a strictly positive chance of being the sampling density for any *finite* section of the data. Fortunately, the existence of ML estimates in $\mathcal{D}$ can often be guaranteed by other means; see, for instance, Propositions F.11 and 1.24. Even if they cannot be the "true" parent distribution, there is a relationship. The ML estimate, if it exists, answers the question: If my data had come from some member of the model, which one would it most likely be? Hjort [248] states that the ML estimate consistently estimates the density f closest to the "true" density f_0 in the sense of the Kullback–Leibler divergence (D.8), $D_{KL}(f, f_0)$.

1.9 Examples If there is an explicit representation of the maximum likelihood estimator, then consistency can be often verified by applying the Strong Law to the representation. Appealing to Theorem 1.5 or 1.7 is not needed in such cases. The strength of the theorems becomes apparent when such a representation is lacking or unknown. Before applying Theorem 1.7 to mixture models, it is illustrated with three examples. The second one is classical. They show how the theorems work.

(a) (Triangular densities) Let $E =]0, 1[$ be endowed with Lebesgue measure ϱ, let the graph of the ϱ-density f_ϑ, $\vartheta \in \Theta =]0, 1[$, be specified by the triangle with vertices $(0,0)$, $(1,0)$, and $(\vartheta, 2)$, that is,

$$f_\vartheta(x) = \begin{cases} \frac{2}{\vartheta} x, & x \leq \vartheta, \\ \frac{2}{1-\vartheta}(1-x), & x \geq \vartheta, \end{cases}$$

and let $\vartheta_0 \in \Theta$. The identifiabilities (i) and (ii) of Theorem 1.7 and the continuity (i) of Theorem 1.5 are plainly satisfied with $E_0 = E$. We verify the assumptions (ii) and (iii) of the latter theorem. If (ϑ_n) is a sequence in $]0, 1[$, then three cases can occur: $\inf_n \vartheta_n = 0$, $\sup_n \vartheta_n = 1$, or the sequence is relatively compact in $]0, 1[$. In the first case, there is a subsequence $(\vartheta_{n(k)})_k$ that converges to 0 and so $f_{\vartheta_{n(k)}}$ converges to $2(1-x)$ pointwise for $x \in E$ as $k \to \infty$. The second case is similar. In the third case, the theorem of Bolzano-Weierstraß shows that the sequence (ϑ_n) has a cluster point $\gamma \in]0, 1[$ and so there is a subsequence of (f_{ϑ_n}) that converges even uniformly to f_γ. This is assumption (ii) of Theorem 1.5. Note that $\vartheta \mapsto f_\vartheta$ has a continuous extension to a one-to-one map from $[0, 1]$ onto $\overline{\mathcal{D}}$. It is, therefore, a homeomorphism between the compact interval $[0, 1]$ and $\overline{\mathcal{D}}$ and the closure of $\mathcal{D}$ in $\mathbb{M}(\varrho)$ is homeomorphic with the compact interval $[0, 1]$ here. In view of (iii), we have

$$2(x-1) \leq f_\vartheta(x) \leq 2, \text{ if } \vartheta \leq x \quad \text{and} \quad 2x \leq f_\vartheta(x) \leq 2, \text{ if } \vartheta \geq x.$$

It follows $2(1-x)x \leq 2\min\{(1-x), x\} \leq f_\vartheta(x) \leq 2$. Since $2(1-x)x \leq 1/2$, we have

$$|\log f_\vartheta(x)| \leq -\log(1-x) - \log x - \log 2.$$

This function is even ϱ-integrable, that is, assumption (iii) of Theorem 1.5, too, is satisfied.

(b) (The multivariate normal model) This example may be viewed as a warm-up exercise for the theorems on mixtures in Section 1.2. It is well known that the ML estimator of the d-variate normal model does not exist for less than $d+1$ observations. The failure of assumption (iii) of Theorem 1.5 reflects this: $N_{m,V}(x)$ is unbounded for all x. We have recourse to the $d+1$-fold tensor product $f_{d+1}(\,\cdot\,;m,V) = N_{m,V}^{d+1}$. So let $E = \mathbb{R}^{(d+1)d}$ with Lebesgue measure ϱ and let $\Theta = \mathbb{R}^d \times \mathrm{PD}(d)$. We verify the assumptions of Theorem 1.7. Identifiability (i) is satisfied since the parameters m and V of the normal density function $N_{m,V}$ are determined by the location of its maximum and by the Hessian there.

Now let $(m,V) \in \Theta$, let $\mathbf{x} = (x_0,\cdots,x_d) \in \mathbb{R}^{(d+1)d}$, let $S_{\mathbf{x}} = W_{\mathbf{x}}/(d+1)$ be the scatter matrix of $\mathbf{x}$, and let ζ be its smallest eigenvalue. Writing $\Lambda = V^{-1}$, we use Steiner's Formula A.11 to estimate

$$2\log f_{d+1}(\mathbf{x};m,V) = 2\log N_{m,V}^{(d+1)}(\mathbf{x})$$

$$= \mathrm{const} + (d+1)\log\det\Lambda - \sum_{k\leq d}(x_k - m)^\top\Lambda(x_k - m)$$

$$= \mathrm{const} + (d+1)\big(\log\det\Lambda - \mathrm{tr}\,S_{\mathbf{x}}\Lambda - (\overline{x} - m)^\top\Lambda(\overline{x} - m)\big) \qquad (1.5)$$

$$\leq \mathrm{const} + (d+1)(\log\det\Lambda - \mathrm{tr}\,S_{\mathbf{x}}\Lambda)$$

$$\leq \mathrm{const} + (d+1)(\log\det\Lambda - \zeta\,\mathrm{tr}\,\Lambda). \qquad (1.6)$$

If the $d+1$-tuple $(x_0,\cdots,x_d)$ is affine dependent in $\mathbb{R}^d$, then $S_{\mathbf{x}}$ is singular (see F.2.4) and by (1.5) the functional $\Lambda \mapsto f_{d+1}(\mathbf{x};\overline{x},V)$ is unbounded since the determinant can tend to ∞ while the rest remains constant. Therefore, we collect all affine independent $d+1$-tuples $\mathbf{x}$ in the open subset $E_0 \subseteq \mathbb{R}^{(d+1)d}$. It is the complement of the differentiable manifold defined by

$$\det\begin{pmatrix} 1 & 1 & \cdots & 1 \\ x_0 & x_2 & \cdots & x_d \end{pmatrix} = 0$$

(see Corollary F.15). The manifold is closed and Lebesgue null. For $\mathbf{x} \in E_0$ we have $\zeta > 0$.

We now verify identifiability at the boundary along with assumption (ii) of Theorem 1.5, showing first that $f_{d+1}(\mathbf{x};m,V)$ vanishes at all points $\mathbf{x} \in E_0$ as $(m,V) \to \infty$, the Alexandrov point of the locally compact space Θ; see Section B.5. A point $(m,V) \in \Theta$ is near the Alexandrov point if an eigenvalue of V is large or close to zero or if $\|m\|$ is large. The representation (1.5) and the estimate (1.6) exhibit the behavior of $f_{d+1}(\mathbf{x};m,V)$ w.r.t. V and m. Let $\lambda_1 \leq \cdots \leq \lambda_d$ be the ordered eigenvalues of Λ. If $\lambda_d \to \infty$, then the estimate (1.6) tends to $-\infty$. The same happens as λ_d remains bounded and $\lambda_1 \to 0$. If all eigenvalues remain bounded and bounded away from zero, then $\|m\| \to \infty$ and $f_{d+1}(\mathbf{x};m,V) \to 0$ by (1.5). This is the behavior as (m,V) approaches the Alexandrov point. If $\big(m^{(k)},\Lambda^{(k)}\big)$ is any bounded sequence of parameters, then the theorem of Bolzano and Weierstraß yields a cluster point $\gamma \in \mathbb{R}^d \times \mathrm{PD}(d)$ and so some subsequence of $f_{d+1}(\mathbf{x};m^{(k)},V^{(k)})$ converges to $f_{d+1}(\mathbf{x};\gamma)$. We have thus verified assumption (ii) of Theorem 1.5 and identifiability at the boundary. Moreover, we have shown that the map Φ defined by $\Phi(m,\Lambda) = f_{m,V}$ and $\Phi(\infty) = 0$ is continuous and one-to-one from the Alexandrov compactification of Θ onto $\overline{\mathcal{D}}$. It is therefore a homeomorphism, $\mathcal{D}$ is homeomorphic with $\mathbb{R}^d \times \mathrm{PD}(d)$, and $\overline{\mathcal{D}}$ is its Alexandrov compactification.

Finally, (1.5) is quadratic in $\mathbf{x}$ and hence integrable w.r.t. $N_{m_0,V_0}^{(d+1)}$ for all m_0 and V_0. Moreover, the maximizer of (1.5), $\mathbf{x} \in E_0$, has the classical representation $m = \overline{x}$ and $\Lambda = S_{\mathbf{x}}^{-1}$; the maximum is $\log f_{\overline{x},S_{\mathbf{x}}}(\mathbf{x}) = \mathrm{const} - \frac{d+1}{2}\log\det S_{\mathbf{x}}$. Hence, assumption (iii) of Theorem 1.5 is satisfied if $\mathrm{E}\log^-\det W_{\mathbf{x}} < \infty$. Now, if $(X_0,\ldots,X_d) \sim N_{m_0,V_0}^{(d+1)}$, we know from Property E.12(iii) that $W_{(X_0,\ldots,X_d)}$ has a Wishart distribution and the claim follows from Proposition E.13. This establishes assumption (iii) of Theorem 1.5 and terminates the proof of example (b).

(c) (Multivariate elliptical symmetry with variable density generator) This example includes the previous one. Let $\beta > (d+1)d$ and $C > 0$ and let Φ be a pointwise closed, equicontinuous set of decreasing density generators $\varphi : \mathbb{R}_+ \to \mathbb{R}_+$ such that $\varphi(t) \leq C(1+t)^{-\beta/2}$. We consider the parametric, elliptically symmetric, statistical model with parameter space $\Theta = \Phi \times \mathbb{R}^d \times \text{PD}(d)$ and the generic density

$$f_{\varphi,m,V}(x) = \sqrt{\det \Lambda} \, \varphi\big((x-m)^\top \Lambda(x-m)\big), \quad \Lambda = V^{-1}.$$

Its parameter space consists of three factors, the shape parameters $\varphi \in \Phi$, the location parameters $m \in \mathbb{R}^d$, and the scale parameters $V \in \text{PD}(d)$; see Appendix E.3.2. Let us use Theorem 1.7 to show that ML estimators exist and that they are consistent at all parameters (φ_0, m_0, V_0) such that $\varphi_0 > 0$ everywhere.

Identifiability of the model has been shown in Proposition E.27. For the same reason as in the normal case (b) we next resort to the $d+1$-fold product model and let E_0 be the set of all affine independent $d+1$-tuples $(x_0, \ldots, x_d)$ of elements of $\mathbb{R}^d$. Assumption (i) of Theorem 1.5 is clear. Let us turn to identifiability at the boundary along with assumption (ii) of Theorem 1.5. According to the theorem of Ascoli-Arzela, B.26, the set Φ is compact in $\mathcal{C}(\mathbb{R}_+)$ w.r.t. compact convergence. Moreover, it is metrizable; see Proposition B.28. If $\big(\varphi^{(n)}, m^{(n)}, V^{(n)}\big)_n$ is any parameter sequence, then, by compactness of Φ, there is a subsequence $(n_k)_k$ such that $(\varphi^{(n_k)})_k$ converges compactly on $\mathbb{R}_+$ to a function in Φ. In order to analyze the behavior w.r.t. (m, V), we need some preparation. For $\mathbf{x} \in E_0$ we have

$$f_{d+1}(\mathbf{x}; \varphi, m, V) = \prod_{i=0}^d \sqrt{\det \Lambda} \, \varphi\big((x_i - m)^\top \Lambda(x_i - m)\big)$$

$$\leq \text{const} \cdot (\det \Lambda)^{(d+1)/2} + \prod_i \varphi\big((x_i - m)^\top \Lambda(x_i - m)\big)$$

$$\leq \text{const} \cdot (\det \Lambda)^{(d+1)/2} \prod_i \big(1 + (x_i - m)^\top \Lambda(x_i - m)\big)^{-\beta/2}$$

$$\leq \text{const} \cdot (\det \Lambda)^{(d+1)/2} \Big(1 + \sum_i (x_i - m)^\top \Lambda(x_i - m)\Big)^{-\beta/2}. \tag{1.7}$$

By Steiner's Formula A.11,

$$\sum_{i=0}^d (x_i - m)^\top \Lambda(x_i - m) \geq \sum_{i=0}^d (x_i - \overline{x})^\top \Lambda(x_i - \overline{x}) = (d+1)\text{tr}\, S_\mathbf{x}\Lambda \geq \kappa(d+1)\text{tr}\,\Lambda,$$

where κ is the smallest eigenvalue of $S_\mathbf{x}$. Using the geometric-arithmetic inequality A.5, we continue

$$(1.7) \leq \text{const} \cdot (\det \Lambda)^{(d+1)/2} (\text{tr}\,\Lambda)^{-\beta/2} \leq \text{const} \cdot (\text{tr}\,\Lambda)^{((d+1)d-\beta)/2}. \tag{1.8}$$

Let us next analyze the behavior of $f_{d+1}(\mathbf{x}; \varphi, m, V)$ as (m, V) tends to the Alexandrov point of $\mathbb{R}^d \times \text{PD}(d)$. As in the previous example, consider three mutually exclusive cases: The largest eigenvalue of Λ tends to ∞, the eigenvalues are bounded and the smallest one converges to zero, and all eigenvalues of Λ are bounded and bounded away from zero and $\|m\| \to \infty$. If the largest eigenvalue λ tends to ∞, then the last expression in (1.8) converges to 0 by $\beta > (d+1)d$. If the smallest eigenvalue tends to zero and if all eigenvalues remain bounded, then the first expression in (1.8) converges to 0. In the third case, (1.7) converges to 0. In all cases, the $d+1$-fold joint likelihood function converges to zero. Finally, if $(m^{(k)}, V^{(k)})$ is bounded in $\mathbb{R}^d \times \text{PD}(d)$, then the theorem of Bolzano-Weierstraß yields a convergent subsequence and the related subsequence of density functions converges, even uniformly. We have verified identifiability at the boundary and assumption (ii) of Theorem 1.5.

It remains to verify assumption (iii) of Theorem 1.5. In order to see that

$$\mathrm{E} \log^- f_{d+1}(X_0^d; \varphi_0, m_0, V_0) \leq \sum_{i=0}^d \mathrm{E} \log^- f(X_i; \varphi_0, m_0, V_0)$$

$$= (d+1)\mathrm{E} \log^- f(X_1; \varphi_0, m_0, V_0) = (d+1) \int_{\mathbb{R}^d} \log^- f(x; \varphi_0, m_0, V_0) f(x; \varphi_0, m_0, V_0) \mathrm{d}x$$

is finite, it is sufficient to refer to Lemma E.26 and to the a priori bound on the density generators. This is the $\log^-$ part of assumption (iii) of Theorem 1.5.

For the $\log^+$ part we seek an upper bound of (1.8) that does not depend on Λ and is integrable w.r.t. the $d+1$-fold product of f_{φ_0,m_0,V_0}. The shape and location parameters have already been eliminated in (1.8). The geometric-arithmetic inequality A.5 implies

$$\operatorname{tr} S_{\mathbf{x}}\Lambda = \operatorname{tr}\left(\sqrt{\Lambda}S_{\mathbf{x}}\sqrt{\Lambda}\right) \geq d\left(\det \sqrt{\Lambda}S_{\mathbf{x}}\sqrt{\Lambda}\right)^{1/d} = d(\det S_{\mathbf{x}}\Lambda)^{1/d}.$$

Therefore

$$
\begin{aligned}
(1.8) &\leq \text{const} + \tfrac{d+1}{2}\log\det\Lambda - \tfrac{(d+1)d}{2}\log\left(d(\det S_{\mathbf{x}}\Lambda)^{1/d}\right)\\
&\leq \text{const} + \tfrac{d+1}{2}\log\det\Lambda - \tfrac{d+1}{2}\log\det S_{\mathbf{x}}\Lambda\\
&= \text{const} - \tfrac{d+1}{2}\log\det S_{\mathbf{x}}.
\end{aligned}
$$

This is the desired upper bound. Up to constants, its positive part is $\log^-\det S_{\mathbf{x}}$. In order to see that it is integrable as requested, let $X_i = \sqrt{V_0}Y_i + m_0$ with i.i.d. spherical random vectors $Y_i \sim f_{\varphi_0,0,I_d}$, $Y = (Y_0,\ldots,Y_d)$, and write

$$\operatorname{E}\log^-\det S_X = \operatorname{E}\log^-\det V_0\det S_Y \leq \log^-\det V_0 + \operatorname{E}\log^-\det S_Y.$$

The claim finally follows from Corollary E.14. Thus, Theorem 1.7 asserts the existence of ML estimates and their consistency.

As a special case, consider Pearson's Type-VII family of density generators with indices $\geq \beta$, where again $\beta > (d+1)d$; see Appendix E.22. It is not pointwise closed but becomes so if the normal density generator is added. Therefore the conclusions just stated apply to Pearson's Type-VII elliptical model combined with the normal model.

The normal distribution is defined on Euclidean space. Therefore, in contrast to consistency in Theorems 1.5 and 1.7, asymptotic normality needs the tangent space of a *differentiable* parametric model (without boundary). It serves to equip $\mathcal{D}$ with a Euclidean structure. (Of course there are other spaces that allow the definition of a normal distribution, for instance, Hilbert space, but these are not considered here.) The following consequence of Proposition 1.3 uses a Euclidean parameterization of the family of density functions.

1.10 Corollary (Asymptotic normality of the ML estimator) *Let* Θ *be an open subset of some Euclidean space and assume that the parametric model meets the Regularity Conditions 1.1. If* $(T_n)_n$ *is a sequence of ML estimators consistent at* $\vartheta_0 \in \Theta$, *then* $\sqrt{n}(T_n - \vartheta_0)$ *converges to* $N_q(0, \mathcal{I}(\vartheta_0)^{-1})$ *in distribution. In particular,* $(T_n)_n$ *is asymptotically efficient.*

1.1.3 Notes

Cramér [105] treated consistency of local likelihood estimates in the sense of Fisher [154] in the univariate case. Huzurbazar [261] showed uniqueness of the consistent estimates. Chanda [84] extended the results to more general sample spaces. His proof contains an incorrect application of Rolle's theorem. This was pointed out by Tarone and Gruenhage [496], who also corrected the proof. Theorem 1.2 is a version of these theorems but its statement is new since it does not need a third derivative.

The study of the consistency of ML estimates was initiated by Doob [130]. It was continued by Wald [520] and Kiefer and Wolfowitz [291]. Wald's theorem is parametric. It makes an assumption that restricts its application essentially to compact subsets of the parameter space, preventing equivariance. Kiefer and Wolfowitz [291] present a theorem based on two components: a parametric, Euclidean statistical model and a completely general latent distribution on the parameter space. It forgoes a compactness assumption. Instead, the authors introduced ad hoc a totally bounded metric on the parameter space. Its completion is compact; see Section B.5. With their approach, a prior assumption on location and scale of the estimate is unnecessary. However, their use of the arctan as a shrinking transformation restricts their theorem to the real line and related parameter spaces. Under certain assumptions their theorem yields consistency of the ML estimate of the latent distribution – if a maximum exists. This statement reduces to consistency of an ML estimate if the admissible latent distributions are restricted to point masses.

Kiefer and Wolfowitz [291] make five assumptions. The first is a Euclidean model. This is needed neither in Theorem 1.5 nor in 1.7. Their second and third assumptions require that the likelihood function be continuously extensible to the compactification of the parameter space and that certain (denumerable) suprema of the density functions be measurable. Both are for free in the present set-up since it assumes continuity of the functions in $\overline{\mathcal{D}}$. This is no restriction in view of later applications. Kiefer and Wolfowitz' [291] assumption 4, "identifiability," means uniqueness of the parameters on the *compactification* of the parameter space. It is not needed in the nonparametric Theorem 1.5 since there are no parameters. Their assumption 4 is not to be confused with mere uniqueness of the parameters in Θ. Some identifiability "at the boundary," too, is needed. In Theorem 1.7 it is divided in the two assumptions (i) and (ii).

The assumption (iii) of Theorem 1.5 is the fifth assumption in [291] and essentially the combination of Wald's [520] second and sixth assumptions. Wald and Kiefer and Wolfowitz localize it, but since compactness ensures that it holds globally, it is here stated as a global assumption. Note that the present theorems are formulated completely in terms of the original model. They do not refer to the compactification of $\mathcal{D}$, which appears only in the proof. The way $\mathcal{D}$ is compactified is actually the main difference between the present theorems and Kiefer and Wolfowitz [291]. Finally, besides showing consistency, the present theorems also provide existence of an ML estimator.

Asymptotic normality in the univariate case goes back to Doob [130]. Proposition 1.3 is fairly standard. A very general version of asymptotic normality is due to LeCam [311]. The asymptotic distribution in more general situations with dependent observations is a consequence of the martingale property of the score function; see Barndorff-Nielsen and Cox [25], p. 85. Fundamental properties of martingales are the decomposition and the inequalities due to Doob [129]; see also Bauer [31].

1.2 Mixture models and their likelihood estimators

After this digression to more general models we now turn to one of the main subjects of this book: mixture models of the observed data. They are the basis for most of the estimators and algorithms presented in this text, in particular mixtures of normal and more general elliptically contoured distributions. As a link between Section 1.1.2 and data clustering and to widen the scope of Theorem 1.5, let us first consider ML estimation of latent distributions on the parameter space.

1.2.1 Latent distributions

Let (E, ϱ) be a σ-finite measure space and let $(\mu_\gamma)_{\gamma \in \Gamma}$ be a ϱ-dominated, parametric model on E with a Hausdorff parameter space Γ. Denote the ϱ-density of μ_γ by $f_\gamma = f(\cdot\,; \gamma)$. Sometimes, one wishes to estimate a (Borel) *probability* on Γ, called **latent** or **mixing**, instead of a single, deterministic parameter $\gamma \in \Gamma$. This means to extend the parameter space from Γ to the convex set $\widehat{\Gamma} = \mathcal{P}(\Gamma)$ of all Borel probability measures on Γ (or to a subset thereof). Each original parameter γ is represented in $\widehat{\Gamma}$ as the "point mass" δ_γ. If all original likelihood functions $f(x; \cdot)$ are bounded, then Fubini's theorem D.5 shows that

$$f_\pi(x) = f(x; \pi) = \int_\Gamma f_\gamma(x) \pi(\,\mathrm{d}\gamma), \quad x \in E, \ \pi \in \widehat{\Gamma},$$

defines a ϱ-probability density f_π on E for *all* π. Thus, $(f_\pi)_{\pi \in \widehat{\Gamma}}$ is a *nonparametric* model that extends the original one. The density f_π is called a general mixture of the f_γ's. Since π is a probability on the parameter space Γ, estimation of π is also called **empirical Bayes** estimation; see Maritz [347], Robbins [441], and Maritz and Lwin [348]. The problem was stated by Robbins [438, 439, 440] and solved in various special cases by Simar [475] (compound Poisson), Laird [305] (analytical likelihoods), and Jewell [277] (exponential distributions). Lindsay [323, 324] treated the problem in high generality on a *compact* parameter space. He gives some examples where this restriction is relaxed and recommends to compactify the parameter space on a case-by-case basis. His approach was novel in the sense that it utilized the close relationship between ML estimation of latent distributions and convex analysis; see Section C.1.

Given observations $x_1, \ldots, x_n \in E$, the maximum likelihood paradigm for $(f_\pi)_{\pi \in \widehat{\Gamma}}$ requires to determine a Borel probability π on Γ such that the joint log-likelihood function of the extended model,

$$\prod_{i=1}^n f_\pi(x_i) = \prod_{i=1}^n \int_\Gamma f_\gamma(x_i)\pi(\,\mathrm{d}\gamma),$$

be maximal. It follows from finiteness of the data set that there is an optimal mixing distribution supported by only finitely many points $\gamma \in \Gamma$. The following theorem on the ML estimate due to Lindsay [323] makes this precise.

1.11 Theorem *Assume that*

 (i) Γ is compact;

 (ii) all likelihood functions $\gamma \mapsto f_\gamma(x_i)$, $1 \leq i \leq n$, are continuous;

 (iii) there is $\gamma \in \Gamma$ such that $f(x_i; \gamma) > 0$ for all i.

Then there exists an ML estimate of π supported by at most n parameters. In other words, the estimated density has the representation

$$\sum_{j=1}^n \pi_j f_{\gamma_j}, \quad \gamma_j \in \Gamma, \ \pi_j \geq 0, \ \sum_{j=1}^n \pi_j = 1.$$

PROOF. Since Γ is compact, so is $\widehat{\Gamma}$ with the weak topology induced by the continuous real-valued functions on Γ; see Choquet [93], p. 217. By linearity and continuity of $\pi \mapsto f_\pi(x)$, $x \in E$, the image K of $\widehat{\Gamma}$ w.r.t. the map Ψ defined by

$$\Psi(\pi) = (f(x_i; \pi))_i \in \mathbb{R}^n$$

is a compact and convex subset of $\mathbb{R}^n$. Note that K is finite dimensional. Maximization of $\prod_{i=1}^n f_\pi(x_i)$ over $\widehat{\Gamma}$ is tantamount to maximizing $\prod_{i=1}^n t_i$ over all $\mathbf{t} = (t_1, \ldots, t_n) \in K$. The equation $\prod_{i=1}^n t_i = \alpha$ describes a "pseudohyperboloid" (it is a hyperbola if $n = 2$) and the space it encloses is strictly convex. By assumption (iii), the pseudohyperboloid touches the compact set K at a point $\mathbf{t}^* \in K$ for some $\alpha > 0$. (Since K is convex, this point is even unique but this is not important here.) Since $\mathbf{t}^*$ lies on the boundary of K, the smallest face $F \subseteq K$ that contains it is of dimension $< n$. According to a theorem of Carathéodory's [78] (see Theorem C.4(a)) $\mathbf{t}^*$ can be represented as the convex combination $\sum_j \pi_j \mathbf{s}_j^*$ of at most n extreme points $\mathbf{s}_j^* \in F$. But each of them is also extreme in K, Property C.2(c). By Lemma C.6, $\mathbf{s}_j^*$ is the Ψ-image of an extreme point of $\widehat{\Gamma}$. By Choquet [94], p. 108, this is a point mass at some $\gamma_j^* \in \Gamma$. Thus,

$$\Psi\left(\sum_j \pi_j \delta_{\gamma_j^*}\right) = \sum_j \pi_j \Psi(\delta_{\gamma_j^*}) = \sum_j \pi_j \mathbf{s}_j^* = \mathbf{t}^*$$

and $\sum_j \pi_j \delta_{\gamma_j^*}$ is hence maximal. This proves the theorem. $\square$

The space Γ used in common ML estimation is here replaced with its "convex hull" $\widehat{\Gamma}$; see Fig. 1.2. The figure shows the likelihood curve $\Psi(\Gamma)$ for the Cauchy location model

$\frac{1}{\pi(1+(x-m)^2)}$ and two observations $x_1 = 1$, $x_2 = -1$, its convex hull $K = \Psi(\widehat{\Gamma})$ (shaded), and the touching hyperbola; see the proof of Theorem 1.11. The touching point $\mathbf{t}^*$ determines the discrete mixing distribution. The compact, convex set K is the convex hull of $\Psi(\Gamma)$.

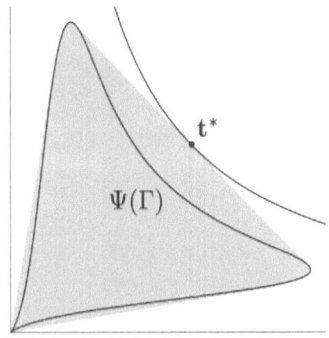

The face generated by the point where the hyperbola touches K is here a line segment. Its tips lie on the likelihood curve making up the support of the mixing distribution.

Algorithmic consequences of the theorem are discussed in Lindsay [324], Chapter 6. The problem is essentially one of maximizing the increasing, differentiable function $\mathbf{t} \mapsto \prod_{i=1}^n t_i$ on the convex, compact subset $K \subseteq \mathbb{R}^n$, the feasible region. It is, therefore, a constrained numerical optimization problem. The feasible region is the convex hull of the manifold $\Psi(\Gamma)$. A representation in implicit form, that is, by equalities and inequalities standard in constrained optimization, does not seem to be known. Since the target function increases in each coordinate, gradient methods lead to the boundary of K. Of course, a necessary and sufficient condition for $\mathbf{t}^*$ to be maximal is that the derivative of the target function at $\mathbf{t}^*$ in each direction interior to K be ≤ 0. Once

Figure 1.2 *The likelihood curve $\Psi(\Gamma)$ for the Cauchy location model at the points $x_1 = 1$ and $x_2 = -1$.*

a point close to the maximum is found, the equations $f(x_i; \pi) = t_i^*$, $1 \leq i \leq n$, have to be solved for $\pi \in \widehat{\Gamma}$.

Leroux [316] notices that Lindsay's theorem can be extended to a *locally* compact parameter space Γ if the likelihood functions vanish at infinity. If Γ is not compact, then the compactification appearing in the proof of Theorem 1.5 may sometimes be used to prove an extension of Lindsay's theorem. The theorem is a second reason why it is sufficient to deal with *finite* mixtures. The first is that the cluster analyst wishes to determine causes in data, of which there are, in general, only finitely many.

1.2.2 Finite mixture models

Let E be a measurable space. Any (finite) convex combination

$$\sum_{j=1}^g \pi_j \mu_j, \quad \pi_j \geq 0, \quad \sum_{j=1}^g \pi_j = 1, \tag{1.9}$$

of g probabilities μ_j on E is again a probability. It is called a **mixture** of the g **components** μ_j. The numbers π_j are the **mixing rates** or **mixing proportions**. If the measures μ_j possess density functions f_j w.r.t. the same σ-finite reference measure ϱ on E, then the mixture (1.9) has the ϱ-density $\sum_{j=1}^g \pi_j f_j$. The densities f_j are generally not arbitrary but members of some "basic" nonparametric or parametric model; see Definition F.1. Mixtures are often used as models for observations that do not obey one of the standard unimodal models. Diagnostics to detect the adequacy of a mixture assumption are presented in Lindsay and Roeder [325]. A mixture can be given the following important stochastic interpretation.

1.12 Lemma *Let $Y^{(j)} : \Omega \to E$, $1 \leq j \leq g$, be random variables, $Y^{(j)} \sim \mu_j$, and let $L : \Omega \to 1..g$ be independent of $(Y^{(j)})_j$ and distributed according to $\boldsymbol{\pi}$. The distribution of the random variable $Y^{(L)} : \Omega \to E$ is the mixture $\sum_j \pi_j \mu_j$.*

PROOF. The formula of total probability shows, for any measurable subset $B \subseteq E$,

$$P[Y^{(L)} \in B] = \sum_j P[Y^{(j)} \in B \mid L = j] P[L = j]$$

$$= \sum_j P[Y^{(j)} \in B] P[L = j] = \sum_j \pi_j \mu_j(B). \qquad \square$$

The moments of mixtures possess simple representations. Let X be distributed according to the mixture (1.9) and let $Y^{(j)}$ be as in Lemma 1.12. If the expectations of the components μ_j exist, then the expectation of X is

$$\mathrm{E}X = \int x \sum_j \pi_j \mu_j(\mathrm{d}x) = \sum_j \pi_j \int x \mu_j(\mathrm{d}x) = \sum_j \pi_j \mathrm{E}Y^{(j)}.$$

This is the convex combination of the g expectations. Analogous results hold for all moments, but not for the covariance matrix. In fact, if $\mathrm{E}\|Y^{(j)}\|^2 < \infty$, we have with $m = \mathrm{E}X$ and $m_j = \mathrm{E}Y^{(j)}$

$$\mathrm{V}X = \mathrm{E}(X-m)(X-m)^\top = \sum_j \pi_j \int (x-m)(x-m)^\top \mu_j(\mathrm{d}x)$$

$$= \sum_j \pi_j \int \big((x-m_j)+(m_j-m)\big)\big((x-m_j)+(m_j-m)\big)^\top \mu_j(\mathrm{d}x)$$

$$= \sum_j \pi_j \big(\mathrm{V}Y^{(j)}+(m_j-m)(m_j-m)^\top\big)$$

$$= \sum_j \pi_j \mathrm{V}Y^{(j)} + \sum_j \pi_j (m_j-m)(m_j-m)^\top, \qquad (1.10)$$

since the integral $\int (x-m_j)\mu_j(\mathrm{d}x)$ vanishes. Thus, the covariance matrix of a mixture is the mixture of the component covariance matrices *plus* the covariance matrix of the discrete probability $\sum_j \pi_j \delta_{m_j}$. This is Steiner's formula for integrals; cf. the discrete case, Lemma A.11.

Mixtures are the members of mixture models. We distinguish between two types: mixture models with a **variable** and with a **fixed number of components**. Let $\mathcal{Q} \subseteq \mathbb{M}(\varrho)$ be the set of probabilities of some nonparametric or parametric model on E. A **nonparametric mixture model** over the **basic model** $\mathcal{Q}$ with **variable number of components** $\leq g$ is the set

$$\mathcal{D}_{\leq g} = \left\{ \sum_{1 \leq j \leq g} \pi_j \mu_j \mid \boldsymbol{\pi} \in \Delta_{g-1},\ \mu_j \in \mathcal{Q},\ 1 \leq j \leq g \right\} \subseteq \mathbb{M}(\varrho). \qquad (1.11)$$

Since some mixing rates π_j can vanish and since components can be collected if f_j's are identical, each member has a "genuine," minimal, number of different components between one and g. Hence the adjective "variable." As prior information in statistical inference, the model with a variable number of components needs only the *maximum* possible number, g.

Besides nonparametric mixture models, there are parametric ones. However, their parameter space is more complex and we confine ourselves to considering **parametric mixture models** with a *fixed* number of components, g, over an *identifiable*, parametric basic model $(\mu_\gamma)_{\gamma \in \Gamma}$ with a Hausdorff parameter space Γ. Their parameter space is the open subset

$$\Theta_g = \overset{\circ}{\Delta}_{g-1} \times \{\boldsymbol{\gamma} \in \Gamma^g \mid \gamma_j \in \Gamma \text{ pairwise distinct}\} \qquad (1.12)$$

of the product $\overset{\circ}{\Delta}_{g-1} \times \Gamma^g$; see also Li and Sedransk [319], Formula (2.1). Of course, the model itself is the mapping $(\boldsymbol{\pi}, \boldsymbol{\gamma}) \mapsto \sum_{j=1}^g \pi_j \mu_{\gamma_j}$. Since the basic model $(\mu_\gamma)_{\gamma \in \Gamma}$ is assumed to be identifiable, since no mixture rate vanishes, and since no two parameters are equal, each member of the mixture model is in fact composed of g different components. Yet the mixture could be represented by fewer than g. An example will be presented in Section 1.2.3. If the basic model is Euclidean or differentiable, then so is the parametric mixture model. Its dimension is then $g - 1 + g \dim \Gamma$.

The two definitions above are general enough to cover a myriad of different instances. Basic models of various types and parsimonious submodels defined by constraints within or between components can be considered. Only a few will be described here. All mixture models used in this book are derived from parametric basic models. Of particular interest are, for instance, elliptical mixtures derived from elliptically symmetric basic models on Euclidean space; see Appendix E.3.2. They are adapted to the symmetries of the sample space and allow the statistician to fit data with heavy and light tails. In this sense they are often used when robustness is an issue. Let Φ be some set of density generators and consider the basic parametric elliptical model

$$\Theta = \Phi \times \mathbb{R}^d \times \mathrm{PD}(d) \to \mathbb{M}(\mathbb{R}^d),$$

$$(\varphi, m, V) \mapsto E_{\varphi,m,V}(x) = \sqrt{\det \Lambda}\, \varphi\big((x - m)^\top \Lambda (x - m)\big).$$

The density generator $\varphi \in \Phi$ serves as **shape parameter**, and m and V are the location and scale parameters, respectively. Special properties of Φ will be specified as they are needed. The basic model gives rise to a *nonparametric* elliptical mixture model with a variable number of components,

$$\mathcal{D}_{\leq g}$$

$$= \big\{ f_{\boldsymbol{\pi},\boldsymbol{\varphi},\mathbf{m},\mathbf{V}} \mid \boldsymbol{\pi} \in \Delta_{g-1}, \boldsymbol{\varphi} \in \Phi^g, \mathbf{m} \in \mathbb{R}^{gd}, \mathbf{V} \in \mathrm{PD}(d)^g \big\} \subseteq \mathbb{M}(\lambda^d), \qquad (1.13)$$

$$f_{\boldsymbol{\pi},\boldsymbol{\varphi},\mathbf{m},\mathbf{V}}(x) = f(x; \boldsymbol{\pi}, \boldsymbol{\varphi}, \mathbf{m}, \mathbf{V}) = \sum_j \pi_j E_{\varphi_j, m_j, V_j}(x). \qquad (1.14)$$

Besides the mixing rates $\boldsymbol{\pi}$, the location parameters $\mathbf{m} = (m_1, \ldots, m_g)$, and the scale parameters $\mathbf{V} = (V_1, \ldots, V_g)$, it contains the density generators $\boldsymbol{\varphi} = (\varphi_1, \ldots, \varphi_g) \in \Phi^g$ as shape parameters. The related *parametric* elliptical mixture model with a fixed number of components is slightly smaller. It has the parameter space

$$\Theta_g = \overset{\circ}{\Delta}_{g-1} \times \big\{ (\boldsymbol{\varphi}, \mathbf{m}, \mathbf{V}) \in \Phi^g \times \mathbb{R}^{gd} \times \mathrm{PD}(d)^g \mid (\varphi_j, m_j, V_j) \text{ p.w. distinct} \big\}. \qquad (1.15)$$

It is an open subset of the product $\overset{\circ}{\Delta}_{g-1} \times \Phi^g \times \mathbb{R}^{gd} \times \mathrm{PD}(d)^g$. The associated map is $(\boldsymbol{\pi}, \boldsymbol{\varphi}, \mathbf{m}, \mathbf{V}) \mapsto f_{\boldsymbol{\pi},\boldsymbol{\varphi},\mathbf{m},\mathbf{V}}$. Its identifiability is discussed in Section 1.2.3. This parametric model is Euclidean or differentiable if Φ is.

Often Φ is a singleton. The parameter space reduces then to the open set

$$\Theta_g = \overset{\circ}{\Delta}_{g-1} \times \big\{ (\mathbf{m}, \mathbf{V}) \in \mathbb{R}^{gd} \times \mathrm{PD}(d)^g \mid (m_j, V_j) \text{ p.w. distinct} \big\} \qquad (1.16)$$

of Euclidean space and the map is

$$(\boldsymbol{\pi}, \mathbf{m}, \mathbf{V}) \mapsto \sum_{1 \leq j \leq g} \pi_j E_{\varphi, m_j, V_j}(x), \qquad (1.17)$$

with a fixed density generator $\varphi \in \Phi$. This location and scale model is differentiable if φ is. Dimension is $(g-1) + gd + g\binom{d+1}{2} = (g-1) + g\frac{d(d+3)}{2}$, that is, quadratic in d.

Examples flow from the family of d-variate Pearson Type-VII (or t-) distributions; see Section E.3.2. The basic density is

$$f_{\boldsymbol{\pi},\boldsymbol{\eta},\mathbf{m},\mathbf{V}}(x) = \sum_j \pi_j \sqrt{\det \Lambda_j}\, \varphi_{\eta_j, d}\big((x - m_j)^\top \Lambda_j (x - m_j)\big), \qquad (1.18)$$

$\boldsymbol{\eta} = (\eta_1, \ldots, \eta_g)$, $\eta_j > d$. The heavy tails of these mixtures can be used for modeling small (local) outliers. The smaller the index is, the heavier the tail. A graphic is presented in Fig. E.3. The parameter set of the parametric Pearson Type-VII mixture model with a variable index is an open subset of $\overset{\circ}{\Delta}_{g-1} \times \mathbb{R}_+^g \times \mathbb{R}^{gd} \times \mathrm{PD}(d)^g$. This model is differentiable of dimension $(g-1) + g + gd + g\binom{d+1}{2} = (2g-1) + g\frac{d(d+3)}{2}$.

If we let $\eta_j \to \infty$, then the Pearson Type-VII density generator tends to the normal $\varphi(t) = (2\pi)^{-d/2} e^{-t/2}$. It generates the **normal mixture model** with the normal model $N_{m,V}^{(d)}$, $m \in \mathbb{R}^d$, $V \in \mathrm{PD}(d)$, for a basis. Written explicitly, the density of its generic member w.r.t. d-dimensional Lebesgue measure has the form

$$\sum_{j=1}^{g} \pi_j N_{m_j,V_j}(x) = \sum_{j=1}^{g} \frac{\pi_j}{\sqrt{\det 2\pi V_j}} e^{-\frac{1}{2}(x-m_j)^\top V_j^{-1}(x-m_j)}. \tag{1.19}$$

It is worthwhile pointing out two normal submodels, diagonal and spherical. If variables are independent (or equivalently, uncorrelated) given the component, then $V_j = \mathrm{Diag}(v_{j,1}, \ldots, v_{j,d})$ is diagonal and the normal mixture assumes the form

$$f_{\boldsymbol{\pi},\mathbf{m},\mathbf{V}}(x) = \sum_{j=1}^{g} \frac{\pi_j}{\sqrt{\prod_k 2\pi v_{j,k}}} e^{-\sum_k \frac{(x_k - m_{j,k})^2}{2 v_{j,k}}}. \tag{1.20}$$

Its dimension reduces to $g(2d+1)-1$, a number linear in d. If variables are, in addition, identically distributed given the component, then $V_j = v_j I_d$, $v_j \in \mathbb{R}$, is even spherical and the normal mixture becomes

$$\sum_{j=1}^{g} \frac{\pi_j}{(2\pi v_j)^{d/2}} e^{-\frac{1}{2v_j} \|x - m_j\|^2}. \tag{1.21}$$

The dimension of this model is $g(d+2)-1$, that is, still linear in d. Diagonal and spherical submodels also exist for elliptical families but here they do not mean independence of the coordinates. This is a particularity of the normal distribution; see Fang et al. [148]. All models that allow their scale parameters to vary in all of $\mathrm{PD}(d)$ will be given the adjective "full." Otherwise, they are "diagonal" or "spherical."

The elliptical and normal mixture models above are **heteroscedastic**, that is, each component has a shape and scale parameter of its own. If they are equal across components, then the model is called **homoscedastic**. The shape and scale parameters, that is, the density generators and scale matrices in Eqs. (1.13), (1.15), and (1.17), the covariance matrices in Eq. (1.19), and the variances in Eqs. (1.20) and (1.21) lose their indices j.

Submodels are coarser than full models. They are nevertheless useful in at least two situations. First, when it is a priori known that their defining constraints are met by the application. For instance, a homoscedastic model should be applied in situations where each component arises from "noisy" versions of one of g prototypes and the noise affects each prototype in the same way. Examples are the transmission of signals through a noisy channel and certain object recognition problems. In the former case, the prototypes are the ideal, unperturbed, signals and in the latter they are the clean images of the different objects. Analogously, a diagonal normal model is appropriate when the entries of the observations may be assumed to be independent. Submodels are also useful when there are too few cases to allow estimation of all parameters of the full model. Examples occur in medicine when the cases are patients suffering from a rare disease.

The piece of information in our hands is a **data set** $\mathbf{x} = (x_1, \ldots, x_n)$ sampled from n i.i.d. random variables $X_i : \Omega \to E$, each distributed according to the mixture $f_{\boldsymbol{\gamma}} = \sum_{j=1}^{g} \pi_j f_{\gamma_j}$ for unknown g-tuples $\boldsymbol{\pi} = (\pi_1, \ldots, \pi_g) \in \Delta_{g-1}$ and $\boldsymbol{\gamma} = (\gamma_1, \ldots, \gamma_g) \in \Gamma^g$. By independence, the density of the joint random variable $X = (X_1, \ldots, X_n)$ w.r.t. the n-fold product measure $\varrho^{\otimes n}$ is the (tensor) product

$$\mathbf{x} \mapsto \prod_{i=1}^{n} \sum_{j=1}^{g} \pi_j f_{\gamma_j}(x_i).$$

In the normal case $\gamma_j = (m_j, V_j)$, for instance, the generic density reads

$$\mathbf{x} \mapsto \prod_{i=1}^{n} \sum_{j=1}^{g} \frac{\pi_j}{\sqrt{\det 2\pi V_j}} e^{-\frac{1}{2}(x_i - m_j)^\top V_j^{-1}(x_i - m_j)}. \tag{1.22}$$

The next proposition analyzes the equivariance groups of the six elliptical mixture models; see Section F.1.3.

1.13 Proposition (Equivariance of elliptical mixture models)
(a) *All elliptical mixture models are equivariant w.r.t. the translation group on $\mathbb{R}^d$.*
(b) *There are the following scale equivariances:*

  (i) *The fully elliptical mixture model is equivariant w.r.t. the general linear group.*

 (ii) *The diagonally elliptical mixture model is equivariant w.r.t. the group of all scale matrices equal to the product of a nonsingular diagonal and a permutation matrix.*

(iii) *The spherically elliptical mixture model is equivariant w.r.t. the group of all nonzero scalar multiples of orthogonal matrices.*

PROOF. The claim about translations follows from the identity

$$\sum_j \pi_j E_{\varphi, m_j + a, V_j}(x + a) = \sum_j \pi_j E_{\varphi, m_j, V_j}(x).$$

Now let A be a scale transformation as in part (b). It is clear that the full model allows any nonsingular matrix A. In order to determine the subgroup of scale transformations in the parameter space of the diagonal model, we have to determine all A such that $ADA^\top$ is diagonal with diagonal entries > 0 for all such D. The condition means

$$ADA^\top(i, k) = \sum_j A(i, j) D(j, j) A(k, j) \begin{cases} = 0, & i \neq k, \\ > 0, & i = k. \end{cases}$$

It holds true if and only if, for each i, there is j_i such that $A(i, j_i) \neq 0$ and $A(k, j_i) = 0$ for all $k \neq i$. In other words, A must contain exactly one nonzero entry in each row and each column. Such an A is the product of a nonsingular diagonal and a permutation matrix. Finally, a scale transformation A in the spherical parameter space satisfies $AI_d A^\top = cI_d$ for some constant $c > 0$, that is, $(\frac{1}{\sqrt{c}} A)(\frac{1}{\sqrt{c}} A)^\top = I_d$. It follows that $\frac{1}{\sqrt{c}} A$ is orthogonal.

After having determined the transformation groups, we show that equivariance holds w.r.t. all these scale transformations. Put $m_j^{(A)} = Am_j$ and $V_j^{(A)} = AV_j A^\top$. The claim follows from

$$\sum_j \pi_j E_{\varphi, m_j^{(A)}, V_j^{(A)}}(Ax) = \sum_j \frac{\pi_j}{\sqrt{\det(AV_j A^\top)}} \varphi\big((Ax - Am_j)^\top (AV_j A^\top)^{-1}(Ax - Am_j)\big)$$

$$= \frac{1}{\det A} \sum_j \frac{\pi_j}{\sqrt{\det V_j}} \varphi\big((x - m_j)^\top V_j^{-1}(x - m_j)\big) = \frac{1}{\det A} \sum_j \pi_j E_{\varphi, m_j, V_j}(x). \qquad \square$$

Two issues of mixture models deserve special attention: the nonexistence of the MLE and, to a lesser degree, the general lack of identifiability. The former generally occurs when a model has the power to fit parts of the data too closely. Even if the likelihood function of the basic statistical model possesses a maximum the same can, unfortunately, not be said about the related mixture models ($g \geq 2$). The standard counterexample is due to Kiefer and Wolfowitz [291], Section 6; see also Day [113], Section 7. The trick is to center one component at a single data point and to let its variance tend to zero. This makes the likelihood blow up. Nevertheless, the situation is not as bad as it may look at first sight. There are two remedies: recurring to *local* optima of the likelihood function and to Theorem 1.2 (see Section 1.2.4) and scale constraints; see Sections 1.2.5 and 1.2.6.

Although Theorem 1.2 and Proposition 1.3 are parametric, they do not need identifiable parameters. They are applicable to any parameter ϑ_0 parental to the data. If there are two (or more), then each of them possesses a neighborhood that contains exactly one consistent *local* maximum of the likelihood function if n is large enough. Their application to special models, however, usually requires that parameters be identifiable; see Theorem 1.19. Identifiability is also a main prerequisite of Theorem 1.7. We study next this issue.

1.2.3 Identifiable mixture models

In the context of mixture models, identifiability (see Definition F.1(e)(i)) could mean uniqueness of the parameters for a *fixed* number of components. It could also mean uniqueness independent of the number of components. Of course, the latter is more restrictive. The following is an example over an identifiable basic model where the former holds but the latter does not. Consider $E = \mathbb{R}$ and $\Gamma = \{(a, b, c) \in \mathbb{R}^3 \mid a < b < c\}$. For each parameter $(a, b, c) \in \Gamma$ there is exactly one "triangular" probability measure $\mu_{(a,b,c)}$ whose Lebesgue density vanishes off the interval $[a, c]$ and is linear on the intervals $[a, b]$ and $[b, c]$. Now, for any two points $r \in]0, 1[$ and $t \in]1, 2[$, there are mixing proportions π_1, π_2, π_3 such that the probability $\mu_{(0,1,2)}$ is also represented by the three-component mixture $\pi_1 \mu_{(0,r,1)} + \pi_2 \mu_{(r,1,t)} + \pi_3 \mu_{(1,t,2)}$ specified by r and t, as a simple geometric argument shows. Thus, this model admits a continuum of three-component mixtures representing the same single-component probability. It is common to stick to the more restrictive notion.

Mixture models with fixed g over identifiable basic models, too, are not identifiable in general; see Teicher [499]. A classical example is provided by multivariate Bernoulli mixtures. Consider the bivariate case with sample space $E = (0 . . 1)^2$ and two-component mixtures. For $\mathbf{x} \in E$, let $\mathrm{ones}(\mathbf{x}) = \sum_k x_k$ and $\mathrm{zeros}(\mathbf{x}) = 2 - \sum_k x_k$. Given four parameters $\gamma_1, \gamma_2, \eta_1, \eta_2 \in [0, 1]$ and mixing rates $u, v \in [0, 1]$, the equation

$$(1 - u)(1 - \gamma_1)^{\mathrm{zeros}(\mathbf{x})} \gamma_1^{\mathrm{ones}(\mathbf{x})} + u(1 - \gamma_2)^{\mathrm{zeros}(\mathbf{x})} \gamma_2^{\mathrm{ones}(\mathbf{x})}$$
$$= (1 - v)(1 - \eta_1)^{\mathrm{zeros}(\mathbf{x})} \eta_1^{\mathrm{ones}(\mathbf{x})} + v(1 - \eta_2)^{\mathrm{zeros}(\mathbf{x})} \eta_2^{\mathrm{ones}(\mathbf{x})}$$

reduces to the pair of equations

$$(1 - u)\gamma_1 + u\gamma_2 = (1 - v)\eta_1 + v\eta_2,$$
$$(1 - u)\gamma_1^2 + u\gamma_2^2 = (1 - v)\eta_1^2 + v\eta_2^2.$$

This system has, for instance, the nontrivial solution

$$\gamma_1 = 0.1, \ \gamma_2 = 0.2, \ \eta_1 = 0.15, \ \eta_2 = 0.05, \ u = v = 1/4.$$

In the parametric models (1.12), (1.15), and (1.16), the parameters are tuples of mixing rates and population parameters. Strictly speaking, no mixture model is actually identifiable w.r.t. these parameters. For example, the parameters $((\pi_1, \pi_2), (\gamma_1, \gamma_2))$ and $((\pi_2, \pi_1), (\gamma_2, \gamma_1))$ both represent the same mixture $\pi_1 \mu_{\gamma_1} + \pi_2 \mu_{\gamma_2}$ but are different when $\gamma_1 \neq \gamma_2$. This "label switching" causes each mixture over Θ_g to appear $g!$ times. Often, it is the only source of nonidentifiability. An example is the normal mixture model; see Proposition 1.16. We therefore define identifiability of mixture models "up to label switching." The simplest way of expressing this is by declaring the mixing distributions $\sum_j \pi_j \delta_{\gamma_j}$ the parameters.

1.14 Definition A finite parametric mixture model over $(\mu_\gamma)_{\gamma \in \Gamma}$ is **identifiable (up to label switching)** if the canonical map that takes each mixing distribution $\sum_j \pi_j \delta_{\gamma_j}$ to the associated mixture $\sum_j \pi_j \mu_{\gamma_j}$ is one-to-one.

The definition means that, for any $g, h \in \mathbb{N}$, for any two tuples $(\gamma_j)_{1 \leq j \leq g} \in \Gamma^g$, $(\eta_\ell)_{1 \leq \ell \leq h} \in \Gamma^h$, and for any $\pi \in \Delta_{g-1}$, $\tau \in \Delta_{h-1}$ such that $\sum_{j=1}^g \pi_j \mu_{\gamma_j} = \sum_{\ell=1}^h \tau_\ell \mu_{\eta_\ell}$, the mixing distributions $\sum_{j=1}^g \pi_j \delta_{\gamma_j}$ and $\sum_{\ell=1}^h \tau_\ell \delta_{\eta_\ell}$ are equal. If each tuple $(\gamma_j)_{1 \leq j \leq g} \in \Gamma^g$ and $(\eta_\ell)_{1 \leq \ell \leq h} \in \Gamma^h$ has pairwise distinct entries, then it means that $g = h$ and that there is a permutation $\sigma \in \mathcal{S}(g)$ such that $(\pi_j, \gamma_j) = (\tau_{\sigma(j)}, \eta_{\sigma(j)})$ for all j. This is label switching. Note that identifiability of a mixture model implies identifiability of the basic model.

The difference $\mu - \nu$ of two finite measures on a measurable space E is called a **signed measure**. For example, if ϱ is a (reference) measure on E and $f \in \mathcal{L}^1(\varrho)$, then $\mu = f\varrho$ $(= f^+ \varrho - f^- \varrho)$ defines a signed measure on E. The set of signed measures on E is a real vector space. Identifiability of a family $(\mu_\gamma)_\gamma$ of probabilities just means its affine independence there; see Section A.1. The affine span of all probabilities does not contain the zero measure. Indeed, if $\mu_1, \ldots, \mu_h$ are probabilities and $\lambda_1, \ldots, \lambda_h$ are real numbers such that $\sum_j \lambda_j = 1$, then $\sum_j \lambda_j \mu_j(E) = 1$. Lemma A.3(b),(c) therefore says that affine and linear independence of $(\mu_\gamma)_\gamma$ are the same. We have obtained the following well-known proposition by Yakowitz and Spragins [535].

1.15 Proposition (Identifiability of finite mixtures) *A parametric mixture model (1.12) over a basic model $(\mu_\gamma)_{\gamma \in \Gamma}$ is identifiable if and only if the family $(\mu_\gamma)_{\gamma \in \Gamma}$ is linearly independent in the vector space of all signed measures on E.*

Identifiability of a mixture model thus puts an exacting requirement on the basic model. Its members must be linearly independent. If $\mu_\gamma = f_\gamma \varrho$, then verification of identifiability resembles the first-year exercise that 1, x, and x^2 are linearly independent functions on $\mathbb{R}$. As a counterexample, consider the triangular densities of Section 1.2.2. Although this (basic) model is identifiable, the identifiability of the associated mixture model is violated, not only by label switching. The reason is that the triangles are linearly dependent. It is noted in passing that there is again a relationship with simplexes in *Choquet's theory* on compact convex sets in locally convex topological vector spaces (see Choquet [94] and Phelps [413]). It pertains in particular to mixtures $\int_{\mathcal{D}} f_\gamma \pi(\mathrm{d}\gamma)$ w.r.t. arbitrary mixing distributions π instead of discrete ones with finite support; see also Section 1.2.1.

Linear independence of densities can be verified by using their local properties in the finite domain or their decay at infinity. Moreover, transforms such as the Fourier or the Laplace transform can be used in the same way; see the proofs of Propositions 1.16–1.18 below. Yakowitz and Spragins [535] applied the foregoing proposition to normal mixtures. It is a typical example.

1.16 Proposition *Finite normal mixture models are identifiable.*

PROOF. Yakowitz and Spragins [535] credit the anonymous referee with the short and elegant geometric proof which replaced their own. Since the diagonal and the spherical cases are submodels of the full model, it is sufficient to consider the latter. We verify the condition of Proposition 1.15. So let $g \in \mathbb{N}$, let $(m_1, V_1), \ldots, (m_g, V_g) \in \mathbb{R}^d \times \mathrm{PD}(d)$ be pairwise distinct, and let $\lambda_1, \ldots, \lambda_g \in \mathbb{R}$ be such that $\sum_{j=1}^g \lambda_j N_{m_j, V_j} = 0$. We have to show that all coefficients λ_j vanish. The moment generating function, Example E.6, shows for all $y \in \mathbb{R}^d$

$$\sum_{j=1}^g \lambda_j e^{m_j^\top y + y^\top V_j y/2} = 0. \tag{1.23}$$

Since the pairs (m_j, V_j) are pairwise distinct, for any two different indices j and ℓ the set of points $y \in \mathbb{R}^d$ for which $m_j^\top y + y^\top V_j y/2 = m_\ell^\top y + y^\top V_\ell y/2$ is either a hyperplane or a (possibly degenerate) conic in $\mathbb{R}^d$. Thus, there is a point y such that all pairs $(y^\top V_j y, m_j^\top y/2)$, $1 \le j \le g$, are distinct. (In fact most y's share this property.) Assume without loss of generality that the tuple $(y^\top V_1 y/2, m_1^\top y), \ldots, (y^\top V_g y/2, m_g^\top y)$ increases lexically. Now, inserting ry for y in Eq. (1.23) and letting $r \to \infty$, the behavior of the exponential function for large arguments shows that the last summand cannot be balanced by the others unless $\lambda_g = 0$. Hence, λ_g vanishes and the proof proceeds in the same way until $j = 1$. Finally, $\lambda_1 = 0$ since the exponential is not zero. $\square$

The proof of Proposition 1.16 uses the moment generating function. This tool does not work for all distributions and the more versatile Fourier transform is often preferable. Proposition 1.16 can also be proved with the following proposition.

1.17 Proposition (Identifiability of mixtures of location families) *Let $(\mu_\gamma)_{\gamma \in \Gamma}$ be some family of probability measures on $\mathbb{R}^d$. Assume that, for each pair $\gamma_1 \neq \gamma_2$ and for Lebesgue* *a.a.* $y \in \mathbb{R}^d$, *the quotient* $\widehat{\mu}_{\gamma_1}(ry)/\widehat{\mu}_{\gamma_2}(ry)$ *tends to* ∞ *or to zero as* $r \to \infty$. *Then finite mixtures over the location model generated by* $(\mu_\gamma)_\gamma$ *are identifiable.*

PROOF. Let $(\mu_{m,\gamma})_{m \in \mathbb{R}^d, \gamma \in \Gamma}$ be this location model. We use Proposition 1.15. So let (m_j, γ_j), $1 \le j \le g$, be pairwise distinct such that $\sum_j \lambda_j f_{m_j, \gamma_j} = 0$. Choose a vector y that satisfies the hypothesis of the proposition for all pairs $\gamma_j \ne \gamma_\ell$ and such that the products $m_j^\top y$ are pairwise distinct. Then

$$\gamma_j \sqsubset \gamma_\ell \quad \Leftrightarrow \quad \widehat{\mu}_{\gamma_\ell}(ry)/\widehat{\mu}_{\gamma_j}(ry) \to 0$$

defines a total order on $\{\gamma_1, \dots, \gamma_g\}$ (it depends on y). Without loss of generality assume $\gamma_1 = \dots = \gamma_\ell \sqsubset \gamma_{\ell+1} \sqsubseteq \dots \sqsubseteq \gamma_g$. Taking the Fourier transform at ry, $r \in \mathbb{R}$, we find with (E.10) $\sum_j \lambda_j e^{-ir(m_j^\top y)} \widehat{\mu}_{\gamma_j}(ry) = 0$, and dividing by $\widehat{\mu}_{\gamma_1}$ gives

$$\sum_{j \le \ell} \lambda_j e^{-ir(m_j^\top y)} + \sum_{j > \ell} \lambda_j e^{-ir(m_j^\top y)} \frac{\widehat{\mu}_{\gamma_j}(ry)}{\widehat{\mu}_{\gamma_1}(ry)} = 0.$$

By assumption the trigonometric polynomial $r \to \sum_{j \le \ell} \lambda_j e^{-ir(m_j^\top y)}$ vanishes at ∞. Note that it is the Fourier transform of the discrete measure $\kappa = \sum_{j \le \ell} \lambda_j \delta_{m_j^\top y}$ on the real line and that its frequencies $m_j^\top y$ are pairwise distinct. The proof will be finished by induction as soon as we have shown that $\lambda_1 = \dots = \lambda_\ell = 0$. This can be accomplished in two ways. First remember that any trigonometric polynomial is almost periodic; see Katznelson [287]. As such it can vanish at ∞ only if it is altogether zero and the claim follows from the identity theorem for Fourier transforms. Second, there is Wiener's identity $\sum_{j \le \ell} \lambda_j^2 = \lim_{s \to \infty} \frac{1}{2s} \int_{-s}^s |\widehat{\kappa}(r)|^2 dr$; see again [287], Chapter VI, 2.9. But the limit is zero. $\square$

The total, lexical, order appearing in the proofs of Propositions 1.16 and 1.17 expresses dominance of a parameter over another and of a value over another. It appears for the first time in Teicher [500]. The normal density generator is the limit of Pearson's Type-VII family of density generators as the index tends to ∞; see Example E.22(b). The following proposition makes a statement on their combination. The Type-VII part is due to Holzmann et al. [250].

1.18 Proposition *The family of all finite mixtures over the extended Pearson Type-VII family E.22(b) is identifiable.*

PROOF. In a vanishing linear combination of Type-VII and normal densities, both parts must vanish separately since the exponential normal tails cannot balance the polynomial Type-VII tails. By Proposition 1.16, it is sufficient to treat the Type-VII case. Let $I =]d, \infty[$ be the index set of the Pearson Type-VII family of density generators. We apply Proposition 1.17 to the centered family $(f_{\eta,0,V})_{(\eta,V) \in \Gamma}$ with $\Gamma = I \times \mathrm{PD}(d)$. With the abbreviation $\alpha = y^\top V y$ the asymptotic formula (E.8) shows

$$\widehat{f}_{\eta,0,V}(ry) = \tau_{\eta,d}(\alpha r^2) = \sqrt{\pi} \frac{(\sqrt{\eta\alpha} r/2)^{\frac{\eta-d-1}{2}} e^{-\sqrt{\eta\alpha} r}}{\Gamma(\frac{\eta-d}{2})} \{1 + \mathcal{O}(1/r)\}.$$

Thus, if we also put $\alpha' = y^\top V' y$,

$$\frac{\widehat{f}_{\eta,0,V}(ry)}{\widehat{f}_{\eta',0,V'}(ry)} = \frac{(\eta\alpha)^{(\eta-d-1)/4}}{(\eta'\alpha')^{(\eta'-d-1)/4}} \frac{\Gamma(\frac{\eta'-d}{2})}{\Gamma(\frac{\eta-d}{2})} \left(\frac{r}{2}\right)^{(\eta-\eta')/2} e^{(\sqrt{\eta\alpha} - \sqrt{\eta'\alpha'})r} \{1 + \mathcal{O}(1/r)\}.$$

If $(\eta, V) \ne (\eta', V')$ and $y \ne 0$, then three cases can occur: (i) $\eta\alpha \ne \eta'\alpha'$, (ii) $\eta\alpha = \eta'\alpha'$ and $\eta \ne \eta'$, (iii) $\eta\alpha = \eta'\alpha'$, $\alpha = \alpha'$, and $V \ne V'$. In the first two cases, the quotient above tends to ∞ or to zero as $r \to \infty$ and the last one occurs for y in the (possibly degenerate) conic $y^\top(V - V')y = 0$ only. We have verified the hypothesis of Proposition 1.17 and the claim follows. $\square$

1.2.4   Asymptotic properties of local likelihood maxima

A standard method of parameter estimation in mixture models is the likelihood paradigm. Day [113] already noted that, under mild conditions, the likelihood function possesses *local* maxima. Unless mixture components are well separated, local maxima are almost always not unique. However, the theory of consistency, asymptotic normality, and efficiency presented in Section 1.1.1 is applicable.

This section deals with local optima in the unconstrained heteroscedastic case where global maxima of the likelihood functions are not available. Kiefer [292] and Peters and Walker [412] applied Cramér's [105] consistency theorem to mixtures of normals in the univariate and multivariate cases, respectively. Such mixtures satisfy the assumptions of Theorem 1.2 and of its corollary and so do, more generally, elliptically symmetric distributions under mild conditions. This is made precise in the following theorem.

1.19 Theorem (Consistency and asymptotic efficiency of unconstrained elliptical mixtures) *Consider the g-component mixture model* (1.17) *based on the elliptical location and scale family with (fixed) density generator* φ,

$$f(x; \boldsymbol{\pi}, \mathbf{m}, \mathbf{V}) = \sum_{1 \leq j \leq g} \pi_j E_{\varphi, m_j, V_j}(x),$$

$\boldsymbol{\pi} \in \overset{\circ}{\Delta}_{g-1}$, $m_j \in \mathbb{R}^d$, $V_j \in \mathrm{PD}(d)$, (m_j, V_j) *p.w. distinct. Assume that*

 (i) φ *is strictly positive and twice continuously differentiable on* $\mathbb{R}_+$;

 (ii) $\varphi(t) \leq C(1+t)^{-\beta/2}$ *for some* $\beta > d + 4$;

 (iii) *the first and second derivatives of* $\phi = -\log\varphi$ *are bounded;*

 (iv) *the mixture model is identifiable.*

Let $(\boldsymbol{\pi}_0, \mathbf{m}_0, \mathbf{V}_0)$ *be the parameters of the sampling distribution. Then*

 (a) *there exists an open neighborhood* U *of* $(\boldsymbol{\pi}_0, \mathbf{m}_0, \mathbf{V}_0)$ *such that the n-fold likelihood function of the mixture model possesses P-a.s. exactly one local maximum* T_n *in* U *for eventually all* n.

 (b) *The sequence* $(T_n)_n$ *is strongly consistent, asymptotically normal, and asymptotically efficient.*

 (c) *The same assertions hold true for the homoscedastic model* $V_j = V$.

PROOF.   We apply Theorem 1.2 and verify first that the mixture model (1.14) satisfies the Regularity Conditions. In order to avoid messy notation, let us restrict the proof to the univariate case. Thus, consider $f(x; \boldsymbol{\pi}, \mathbf{m}, \mathbf{v}) = \sum_{1 \leq j \leq g} \pi_j E_{\varphi, m_j, v_j}$ with $\mathbf{m} = (m_1, \ldots, m_d) \in \mathbb{R}^d$ and $\mathbf{v} = (v_1, \ldots, v_d) \in \mathbb{R}^d_>$, (m_j, V_j) p.w. distinct. It is clear that Regularity Condition (i) is satisfied. The fact that the mixing rates π_j are actually constrained to the manifold $\sum_j \pi_j = 1$ may be ignored in proving Regularity Condition (ii). In view of the first line, observe that mixed derivatives of $f(x; \boldsymbol{\pi}, \mathbf{m}, \mathbf{v})$ w.r.t. parameters of different components vanish. Moreover, one shows by induction on the total order $r + s + t$ that, for $r + s + t > 0$, the "pure" partial derivative

$$\frac{\partial^{r+s+t}}{\partial \pi_j^r \partial m_j^s \partial v_j^t} f(x; \boldsymbol{\pi}, \mathbf{m}, \mathbf{v})$$

is of the form

$$Q^{(r,s,t)}(x; \pi_j, m_j, v_j, \phi', \ldots, \phi^{(r+s+t)}) E_{\varphi, m_j, v_j}(x),$$

where the factor $Q^{(r,s,t)}(x; p, m, v, \phi', \ldots, \phi^{(r+s+t)})$ is a polynomial in $x, p, m, 1/v$, and the derivatives $\phi', \ldots, \phi^{(r+s+t)}$. For instance,

$$Q^{(0,0,1)}(x; p, m, v, \phi') = p\left(-\frac{1}{2v} + \frac{(x-m)^2}{v^2}\phi'\right)$$

and
$$Q^{(0,0,2)}(x;p,m,v,\phi',\phi'') = p\Big(\frac{3}{4v^2} - 3\frac{(x-m)^2}{v^3}\phi' + \frac{(x-m)^4}{v^4}((\phi')^2 - \phi'')\Big).$$

Of course $Q^{(r,s,t)} = 0$ if $r \geq 2$. The maximum degree of x among the factors $Q^{(r,s,t)}(x;p,m,v,\phi',\dots,\phi^{(r+s+t)})$ with constant sum $r+s+t = a$ is attained at $Q^{(0,0,a)}$; it is $2a$. For $\vartheta = (\pmb{\pi},\mathbf{m},\mathbf{v})$ in the closed box defined by $\pi_j \geq \varepsilon$, $m_j \in [-r,r]$, and $v_j \in [\varepsilon, 1/\varepsilon]$ for some $0 < \varepsilon < 1/2$ and $r > 0$, assumption (iii) implies

$$\|\mathrm{D}_\vartheta f(x;\vartheta)\| \leq \sum_j \Big(\sum_{k=0}^2 A_k(\pi_j,m_j,v_j)|x|^k\Big) E_{\varphi,m_j,v_j}(x)$$
$$\leq \mathrm{const} \cdot (1+x^2)\sum_j E_{\varphi,m_j,v_j}(x) \tag{1.24}$$

and

$$\|\mathrm{D}_\vartheta^2 f(x;\vartheta)\| \leq \sum_j \Big(\sum_{k=0}^4 B_k(\pi_j,m_j,v_j)|x|^k\Big) E_{\varphi,m_j,v_j}(x)$$
$$\leq \mathrm{const} \cdot (1+x^4)\sum_j E_{\varphi,m_j,v_j}(x), \tag{1.25}$$

with functions A_k and B_k bounded and continuous on the box. Now, for $|x| \geq 2r$, we have $|x-m| \geq |x| - |m| \geq |x| - r \geq |x|/2$ and hence by $v_j \leq 1/\varepsilon$ and assumption (ii),

$$\varphi((x-m)^2/v) \leq (1 + (x-m)^2/v)^{-\beta/2} \leq (1 + \varepsilon x^2/4)^{-\beta/2}.$$

Since $\beta > d+4 = 5$, a multiple of the Lebesgue integrable function $(1+x^4)/(1+\varepsilon x^2/4)^{\beta/2}$ dominates both derivatives uniformly for π_j, m_j, and v_j in the closed set. It serves as the majorant requested in the first line of Regularity Condition (ii).

In view of the second line of Regularity Condition (ii), we use again Eqs. (1.24) and (1.25) to estimate for $\vartheta = (\pmb{\pi},\mathbf{m},\mathbf{v})$ in the box above

$$\|\mathrm{D}_\vartheta^2 \log f(x;\vartheta)\| \leq \frac{\|\mathrm{D}_\vartheta^2 f(x;\vartheta)\|}{f(x;\vartheta)} + \frac{\|(\mathrm{D}_\vartheta f(x;\vartheta))^\top \mathrm{D}_\vartheta f(x;\vartheta)\|}{f^2(\vartheta;x)}$$
$$\leq \mathrm{const} \cdot (1+x^4)\Big(\frac{\sum_j E_{\varphi,m_j,v_j}(x)}{f(x;\vartheta)} + \Big(\frac{\sum_j E_{\varphi,m_j,v_j}(x)}{f(x;\vartheta)}\Big)^2\Big).$$

The claim now follows from the estimate $\sum_j E_{\varphi,m_j,v_j}(x)/f(x;\vartheta) \leq 1/\min \pi_j \leq 1/\varepsilon$ and from assumption (ii).

In order to have positive definiteness of the Fisher information, we show that the $3g-1$ derivatives $\frac{\partial \log f(x;\pmb{\pi},\mathbf{m},\mathbf{v})}{\partial \pi_j}$, $2 \leq j \leq g$, $\frac{\partial \log f(x;\pmb{\pi},\mathbf{m},\mathbf{v})}{\partial m_j}$, $1 \leq j \leq g$, and $\frac{\partial \log f(x;\pmb{\pi},\mathbf{m},\mathbf{v})}{\partial v_j}$, $1 \leq j \leq g$, are lineraly independent functions of x. The derivatives are

$$\begin{cases} \frac{\partial \log f(x;\pmb{\pi},\mathbf{m},\mathbf{v})}{\partial \pi_j} = \frac{1}{f(x;\pmb{\pi},\mathbf{m},\mathbf{v})}(E_{\varphi,m_j,v_j}(x) - E_{\varphi,m_1,v_1}(x)), & 2 \leq j \leq g, \\ \frac{\partial \log f(x;\pmb{\pi},\mathbf{m},\mathbf{v})}{\partial m_j} = \frac{2}{f(x;\pmb{\pi},\mathbf{m},\mathbf{v})}\frac{\pi_j}{v_j}(x-m_j)\phi' E_{\varphi,m_j,v_j}(x), & 1 \leq j \leq g, \\ \frac{\partial \log f(x;\pmb{\pi},\mathbf{m},\mathbf{v})}{\partial v_j} = \frac{1}{f(x;\pmb{\pi},\mathbf{m},\mathbf{v})}\frac{\pi_j}{v_j^2}\big((x-m_j)^2\phi' - \frac{v_j}{2}\big)E_{\varphi,m_j,v_j}(x), & 1 \leq j \leq g. \end{cases}$$

So let α_j, $2 \leq j \leq g$, and β_j,γ_j, $1 \leq j \leq g$, be real coefficients such that $\sum_{j=2}^g \alpha_j \frac{\partial \log f(x;\pmb{\pi},\mathbf{m},\mathbf{v})}{\partial \pi_j} + \sum_{j=1}^g \beta_j \frac{\partial \log f(x;\pmb{\pi},\mathbf{m},\mathbf{v})}{\partial m_j} + \sum_{j=1}^g \gamma_j \frac{\partial \log f(x;\pmb{\pi},\mathbf{m},\mathbf{v})}{\partial v_j} = 0$ for all x, that is,

$$\sum_{j=2}^g \alpha_j(E_{\varphi,m_j,v_j}(x) - E_{\varphi,m_1,v_1}(x)) + 2\sum_{j=1}^g \beta_j \frac{\pi_j}{v_j}(x-m_j)\phi' E_{\varphi,m_j,v_j}(x) \tag{1.26}$$
$$+ \sum_{j=1}^g \gamma_j \frac{\pi_j}{v_j^2}\big((x-m_j)^2\phi' - \frac{v_j}{2}\big)E_{\varphi,m_j,v_j}(x) = 0.$$

Since the parameters (v_j, m_j) are pairwise distinct, it follows from identifiability (iv) along with Proposition 1.15

$$
\begin{cases}
\sum_{j=2}^{g} \alpha_j + 2\beta_1 \frac{\pi_1}{v_1}(x - m_1)\phi' + \gamma_1 \frac{\pi_1}{v_1^2}\left((x - m_1)^2\phi' - \frac{v_1}{2}\right) = 0, \\
\alpha_j + 2\beta_j \frac{\pi_j}{v_j}(x - m_j)\phi' + \gamma_j \frac{\pi_j}{v_j^2}\left((x - m_j)^2\phi' - \frac{v_j}{2}\right) = 0, & j \geq 2.
\end{cases}
$$

Substituting t for $(x - m_j)/\sqrt{v_j}$, we infer from the second line for all $t \in \mathbb{R}$

$$
\alpha_j + 2\beta_j \frac{\pi_j}{\sqrt{v_j}} t\phi'(t^2) + \gamma_j \frac{\pi_j}{v_j}\left(t^2\phi'(t^2) - \tfrac{1}{2}\right) = 0.
$$

Evaluating at $t = 0$ yields $\alpha_j = \frac{\gamma_j}{2}\frac{\pi_j}{v_j}$ and hence $2\beta_j \frac{\pi_j}{\sqrt{v_j}} t\phi'(t^2) = -\gamma_j \frac{\pi_j}{v_j}t^2\phi'(t^2)$ or $2\beta_j\sqrt{v_j} = -\gamma_j t$ at all points t such that $t\phi'(t^2) \neq 0$. It follows $\beta_j = \gamma_j = 0$ for $j \geq 2$. Therefore also $\alpha_j = 0$ for these j. Finally, the first line above also implies $\beta_1 = \gamma_1 = 0$ and we have independence.

The claims on asymptotic normality and efficiency follow from Proposition 1.3.   □

Mixture models based on Pearson's robust Type-VII and normal families are special cases of Theorem 1.19. It also includes multiples of $e^{-\beta t^\alpha}$, $\alpha, \beta > 0$, as density generators.

1.20 Corollary *The claims of Theorem 1.19 are true for Type-VII mixture models* (1.18) *with index* $\eta > d + 4$,

$$
\sum_{1 \leq j \leq g} \frac{\pi_j}{\sqrt{\det V_j}} \varphi_\eta\big((x - m_j)^\top V_j^{-1}(x - m_j)\big),
$$

$\pi \in \overset{\circ}{\Delta}_{g-1}$, $m_j \in \mathbb{R}^d$, $V_j \in \mathrm{PD}(d)$, $(m_j, V_j))$ *p.w. distinct, as well as for the normal mixture model* (1.19).

A serious drawback of free heteroscedastic mixtures has been pointed out by many authors beginning with Day [113]. It is the occurrence of many so-called "spurious" solutions. They contain components with scales (or spreads) too different to be credible. The framed point triple in Fig. 1.3 generates a spurious component which, in turn, gives rise to the solution with the largest local maximum of the likelihood. There are many more point constellations like the one displayed. Which are they? Random data sets are rather unstable concerning the presence or absence of such spurious components. It is sufficient to just slightly modify a single data point to create a spurious component or to have it disappear. A remedy is to use scale constraints as in the following section.

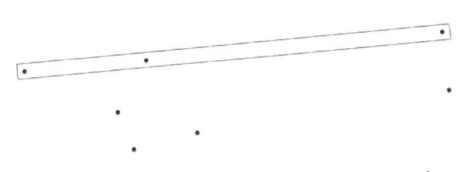

Figure 1.3 *A data set of 20 points randomly sampled from the normal mixture* $\frac{1}{2}N_{-2e_1, I_2} + \frac{1}{2}N_{2e_1, I_2}$.

1.2.5   Asymptotic properties of the MLE: constrained nonparametric mixture models

We have seen at the end of Section 1.2.2 that global maxima of the likelihood functions of location and scale mixtures do not exist straightforwardly. Therefore, the focus in Section 1.2.4 has been on *local* maxima in parameter space. The reason for the nonexistence was components with arbitrarily small variances. However, maxima may be forced by various types of scale constraint. ML estimation, too, may then be applied to mixtures. A popular constraint is compactness of the parameter space. Indeed, its combination with continuity of the likelihood function guarantees the existence of a maximum. However, when the natural parameter space of a model is not compact, this imposes on the analyst to first prove that the sequence of ML estimates remains in a compact subset. This fact is often overlooked. A related problem is that a model with compact scale space cannot be equivariant w.r.t. variable scaling, let alone w.r.t. affine transformations in sample space. In other words, artificially restricting the parameter space to a compact subset has serious drawbacks. The same is true if lower bounds are put on covariance matrices, another popular constraint.

Homoscedastic models are affine equivariant. Yet, they are not rich enough and are unable to adapt to many real data sets. While affine equivariance is a necessary condition, it is often not sufficient. Real data often require the estimation of *full* location and scale structures. Fortunately, heteroscedastic mixtures with full scales allow constraints less restrictive than compactness that circumvent the nonexistence of ML estimates, remove many of the undesirable, spurious local solutions, and keep the model even affine equivariant. These **HDBT constraints** were first investigated in depth by Hathaway [232] in the case of heteroscedastic, univariate normal mixtures.[1] They will be a key theoretical assumption on heteroscedastic models in the present text.

Denote the positive semi-definite (or Löwner) order on the space of symmetric matrices by $\succeq$; see also Section A.3. A g-tuple or a set of matrices $V_j \in \mathrm{PD}(d)$ satisfies the HDBT (cross-component scale) constraints with **HDBT constant** c if

$$V_j \succeq cV_\ell \quad \text{for all} \quad j, \ell \in 1..g. \tag{1.27}$$

The constant c is necessarily ≤ 1 and the constraints are affine invariant. They restrict the domain of definition of the likelihood function. While restricting the relationships between sizes and shapes of the g scale parameters V_j, they impose no restriction on their absolute values. In other words, V_1 may be any positive definite matrix. Figure 1.4 shows three pairs of ellipses representing three pairs of covariance matrices. The pairs on the left and in the center satisfy the HDBT constraints with $c = 1/100$, their scales being sufficiently similar. The pair on the right does not. The HDBT constraints are applied to the scale matrices of mixture models. Two choices of the constant c should be noted: $c = 1$ specifies homoscedasticity and the constrained heteroscedastic case with $c \ll 1$ allows much freedom between scale parameters. We denote the collection of all tuples (or subsets) over $\mathrm{PD}(d)$ that satisfy the HDBT constraints with constant $c > 0$ by

$$\mathcal{V}_c = \{\mathbf{V} = (V_1, \ldots, V_g) \mid V_j \succ 0, \ V_j \succeq cV_\ell \ \text{for all} \ 1 \leq j, \ell \leq g\}.$$

If the scale parameters are restricted in this way, the nonparametric elliptical model (1.13), for instance, adopts the form

$$\mathcal{D}_{\leq g, c} = \{f_{\boldsymbol{\pi}, \boldsymbol{\varphi}, \mathbf{m}, \mathbf{V}} \mid \boldsymbol{\pi} \in \Delta_{g-1}, \boldsymbol{\varphi} \in \Phi^g, \mathbf{m} \in \mathbb{R}^{gd}, \mathbf{V} \in \mathcal{V}_c\} \subseteq \mathbb{M}(\lambda^d). \tag{1.28}$$

Besides the full scale $\mathrm{PD}(d)$, the definition may also be applied to diagonal and spherical models; see, for instance, (1.20) and (1.21). Theorems 1.22 and 1.27 below show that the likelihood function of the HDBT constrained elliptical model possesses a maximum and its consistency. Moreover, Proposition 1.24 will give simpler conditions for the MLE to exist.

---

[1] He refers to Dennis [116], who, in turn, gives credit to Beale and Thompson (oral communications). On account of this origin and since the constraints play a key rôle in the theory of Euclidean data clustering, Gallegos and Ritter [182, 183] called them the Hathaway–Dennis–Beale–Thompson (HDBT) constraints.

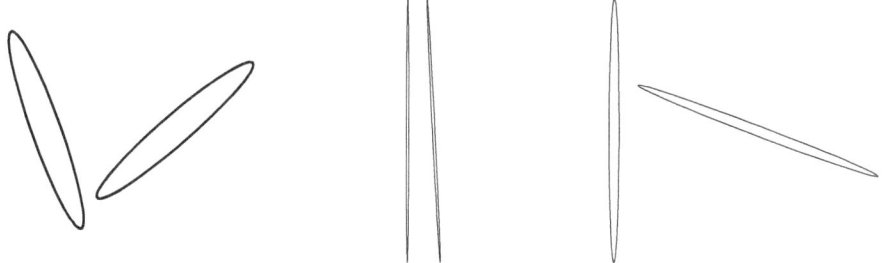

Figure 1.4: *Illustration of the HDBT constraints.*

To this end we apply Theorem 1.5 to HDBT constrained elliptical mixtures. In order to discuss the assumption (iii) of the theorem we need some notation. According to Lemma 1.12, each observation X_i, $1 \leq i \leq n$, can be represented as $Y_i^{(L_i)}$, where the sequence $(Y_i^{(j)})_i$ is i.i.d. $\sim \mu_j$ for all j, and where $(L_i)_i$ is i.i.d. $\sim \boldsymbol{\pi}$ and independent of $Y_i^{(j)}$. Define the weight $w_j(x)$ as the posterior probability of an observation x to come from component j w.r.t. the parameters $\boldsymbol{\pi}$ and γ. By Bayes' formula,

$$w_j(x) = P\big[L = j \mid Y^{(L)} = x\big] = \frac{\pi_j f_{\gamma_j}(x)}{\sum_\ell \pi_\ell f_{\gamma_\ell}(x)}. \tag{1.29}$$

The weights sum up to 1 w.r.t. j, that is, $\mathbf{w} = (w_j(x_i))_{i,j}$ is a stochastic matrix. We also abbreviate $w_j(T) = \sum_{i \in T} w_j(x_i)$, $T \subseteq 1..n$. The weighted mean vector and sum of squares and products (SSP) matrix of T w.r.t. the weights $w = (w(x))_{x \in T}$ are

$$\overline{x}_T(\mathbf{w}) = \sum_{i \in T} w(x_i)(x_i) \quad \text{and}$$

$$W_T(\mathbf{w}) = \sum_{i \in T} w(x_i)(x_i - \overline{x}_T(w))(x_i - \overline{x}_T(w))^\top, \quad \text{respectively.}$$

The proof of the existence of an ML estimate and of its consistency for HDBT constrained, heteroscedastic, elliptical mixture models needs a preparatory lemma. It is crucial for our theoretical analyses and makes up for the missing representation of an ML estimate – it is in fact needed for proving its existence and for showing that the HDBT constraints establish the integrability assumption (iii) of Theorem 1.5. Recall $\Lambda_j = V_j^{-1}$.

1.21 Lemma *Consider an HDBT constrained, elliptical mixture model* (1.28) *with density generators* $\varphi_j \leq \psi$, $1 \leq j \leq g$, *for some bounded, decreasing function* $\psi : \mathbb{R}_+ \to \mathbb{R}_+$. *For any data set* $\mathbf{x}$ *in* $\mathbb{R}^d$ *of length* $n \geq gd + 1$ *there exist* $k = \lceil \frac{n-gd}{d+1} \rceil$ *disjoint subsets* $I_1, \dots, I_k \subseteq 1..n$ *of size* $d+1$ *such that, for all* j,

$$f_n(\mathbf{x}; \boldsymbol{\pi}, \boldsymbol{\varphi}, \mathbf{m}, \mathbf{V}) \leq \text{const} \cdot (\det \Lambda_j)^{n/2} \prod_{p=1}^{k} \psi(c \cdot \operatorname{tr} S_{I_p} \Lambda_j). \tag{1.30}$$

PROOF. Let $(R_1, \dots, R_g)$ be the ML partition of $1..n$ w.r.t. $(\boldsymbol{\varphi}, \mathbf{m}, \mathbf{V})$, that is, $i \in R_\ell$ if and only if $\ell \in \operatorname{argmax}_j E_{\varphi_j, m_j, V_j}(x_i)$ (ties are broken arbitrarily). Then

$$\log f_n(\mathbf{x}; \boldsymbol{\pi}, \boldsymbol{\varphi}, \mathbf{m}, \mathbf{V}) = \sum_i \log \sum_j \pi_j E_{\varphi_j, m_j, V_j}(x_i)$$

$$= \sum_\ell \sum_{i \in R_\ell} \log \sum_j \pi_j E_{\varphi_j, m_j, V_j}(x_i) \leq \sum_\ell \sum_{i \in R_\ell} \log E_{\varphi_\ell, m_\ell, V_\ell}(x_i)$$

$$= \tfrac{1}{2} \sum_\ell \sum_{i \in R_\ell} \log \det \Lambda_\ell + \sum_\ell \sum_{i \in R_\ell} \log \psi\big((x_i - m_\ell)^\top \Lambda_\ell (x_i - m_\ell)\big).$$

An elementary combinatorial reasoning shows that the clusters R_ℓ contain at least k disjoint batches $I_1, \ldots, I_k$ of size $d+1$. The second term on the right has the upper bound

$$(n - k(d+1)) \log \psi(0) + \sum_{p=1}^{k} \sum_{i \in I_p} \log \psi((x_i - m_{\ell_p})^\top \Lambda_{\ell_p} (x_i - m_{\ell_p})),$$

where $1 \leq \ell_p \leq g$ for all p. Now Steiner's formula A.11 asserts

$$\sum_{i \in I_p} (x_i - m_{\ell_p})^\top \Lambda_{\ell_p} (x_i - m_{\ell_p}) \geq \sum_{i \in I_p} (x_i - \bar{x}_{\ell_p})^\top \Lambda_{\ell_p} (x_i - \bar{x}_{\ell_p}) = (d+1) \operatorname{tr} S_{I_p} \Lambda_{\ell_p}.$$

Hence, each I_p contains an index i such that $(x_i - m_{\ell_p})^\top \Lambda_{\ell_p} (x_i - m_{\ell_p}) \geq \operatorname{tr} S_{I_p} \Lambda_{\ell_p}$. It follows

$$\log f_n(\mathbf{x}; \boldsymbol{\pi}, \boldsymbol{\varphi}, \mathbf{m}, \mathbf{V}) \leq \tfrac{1}{2} \sum_{\ell} \sum_{i \in R_\ell} \log \det \Lambda_\ell + \sum_{p=1}^{k} \log \psi(\operatorname{tr} S_{I_p} \Lambda_{\ell_p}) + \text{const}$$

$$\leq \tfrac{n}{2} \log \det \Lambda_j + \sum_{p=1}^{k} \log \psi(c \operatorname{tr} S_{I_p} \Lambda_j) + \text{const}$$

by the HDBT constraints. This concludes the proof. $\qquad \square$

The lemma develops its full power when the data is in general position since the scatter matrices in (1.30) are then nonsingular. When diagonal and spherical submodels are considered, "general position" is replaced with "pairwise difference of the entries in each coordinate" and "pairwise difference of all data points," respectively. In these cases, the requirement $n \geq gd + 1$ can be relaxed to $n \geq g + 1$. If the diagonal and spherical scale matrices are again HDBT constrained, there exist $k = \left\lceil \frac{n-g}{2} \right\rceil$ disjoint two-element subsets $I_1, \ldots, I_k \subseteq 1..n$ satisfying (1.30), where the scatter matrices S_{I_p} may be replaced with their diagonal and spherical versions, respectively.

The following theorem uses the lemma to establish conditions for the ML estimate of a constrained nonparametric elliptical mixture model (1.28) to be consistent.

1.22 Theorem (Existence and consistency of the nonparametric MLE for constrained elliptical mixtures) *Let $\beta > (gd+1)d$ and $C > 0$. Let Φ be some equicontinuous set of density generators such that $\varphi(t) \leq C(1+t)^{-\beta/2}$, $t \geq 0$, $\varphi \in \Phi$. Let*

$$\mathcal{D}_{\leq g, c} = \left\{ f_{\boldsymbol{\pi}, \boldsymbol{\varphi}, \mathbf{m}, \mathbf{V}} \mid \boldsymbol{\pi} \in \Delta_{g-1}, \boldsymbol{\varphi} \in \Phi^g, \mathbf{m} \in \mathbb{R}^{gd}, \mathbf{V} \in \mathcal{V}_c \right\} \subseteq \mathrm{M}(\lambda^d)$$

be the associated nonparametric, HDBT constrained, elliptical model with variable number of components (see Eq. (1.28)) and let $f_0 \in \mathcal{D}_{\leq g, c}$ be strictly positive. Then:

(a) Any sequence $(T_n)_n$ of ML estimates is strongly consistent at f_0.

(b) If the map $\Delta_{g-1} \times \Phi^g \times \mathbb{R}^{gd} \times \mathcal{V}_c \to \mathcal{D}_{\leq g, c}$, $(\boldsymbol{\pi}, \boldsymbol{\varphi}, \mathbf{m}, \mathbf{V}) \mapsto f_{\boldsymbol{\pi}, \boldsymbol{\varphi}, \mathbf{m}, \mathbf{V}}$, is open, then P-a.s., an ML estimate T_n exists in $\mathcal{D}_{\leq g, c}$ for eventually all n.

PROOF. By Lebesgue's dominated convergence with dominating function $\psi(t) = C(1 + t)^{-\beta/2}$, the pointwise closure $\bar{\Phi}$ of Φ consists of density generators and shares all properties of Φ. By Ascoli-Arzela's Theorem B.26, $\bar{\Phi}$ is compact w.r.t. compact convergence on $\mathbb{R}_+$. Therefore, we assume without loss of generality that Φ has this property.

Let $n = gd + 1$. We first define the set E_0 needed in Theorem 1.5. For any subset $I \subseteq 0..gd$ of size $d+1$, let E_I be the set of all $\mathbf{x} = (x_0, \ldots, x_{gd}) \in \mathbb{R}^{nd}$ for which the vectors x_i, $i \in I$, are affine independent and let

$$E_0 = \bigcap_{I \in \binom{0..gd}{d+1}} E_I.$$

Any $d+1$ elements of $\mathbf{x} \in E_0$ are affine independent, that is, the set E_0 consists of all data sets of length n in general position. The complement of each E_I is a differentiable manifold and therefore Lebesgue null and closed in $\mathbb{R}^{nd}$ and so is the complement of E_0.

We next verify assumption (ii) utilizing the parameters $\boldsymbol{\pi}$, $\boldsymbol{\varphi}$, $\mathbf{m}$, and $\Lambda_j = V_j^{-1}$ of the elliptical mixture model as indices. Let $(\boldsymbol{\pi}^{(k)}, \boldsymbol{\varphi}^{(k)}, \mathbf{m}^{(k)}, \boldsymbol{\Lambda}^{(k)})$ be any sequence. By applying Bolzano-Weierstraß and successively selecting subsequences, we may and do assume that $(\boldsymbol{\pi}^{(k)}, \boldsymbol{\varphi}^{(k)})_k$ converges in the compact product $\Delta_{g-1} \times \Phi^g$, that each $m_j^{(k)}$ either converges in $\mathbb{R}^d$ or $\|m_j^{(k)}\|$ converges to ∞, and that $\Lambda_1^{(k)}$ either converges in $\mathrm{PD}(d)$ or converges to the Alexandrov point of $\mathrm{PD}(d)$. The HDBT constraints imply that the other scale parameters Λ_j behave correspondingly. We begin with analyzing the parameter Λ_1. For the same reason as in Examples 1.9(b) and (c), we have to study a product model of fixed length, here the joint density of n observations

$$\prod_{i=0}^{gd} \sum_{j=1}^{g} \pi_j \sqrt{\det \Lambda_j}\, \varphi_j((x_i - m_j)^\top \Lambda_j (x_i - m_j)). \tag{1.31}$$

Now use Lemma 1.21 (with $n = gd+1$) to show that $\log f_n(\mathbf{x}; \boldsymbol{\pi}, \boldsymbol{\varphi}, \mathbf{m}, \mathbf{V})$ converges to $-\infty$ as the largest eigenvalue λ of Λ_1 converges to ∞. Since $\left\lceil \frac{n-gd}{d+1} \right\rceil = 1$, we have for $\mathbf{x} \in E_0$

$$\log f_n(\mathbf{x}; \boldsymbol{\pi}, \boldsymbol{\varphi}, \mathbf{m}, \mathbf{V}) \le \tfrac{n}{2} \log \det \Lambda_j + \log \psi(c \operatorname{tr} S_{\mathbf{x}_I} \Lambda_j) \tag{1.32}$$

$$\le \tfrac{n}{2} \log \det \Lambda_j - \tfrac{\beta}{2} \log(\operatorname{tr} S_{\mathbf{x}_I} \Lambda_j) + \mathrm{const} \le \tfrac{nd}{2} \log \lambda - \tfrac{\beta}{2} \log \lambda + \mathrm{const}$$

for some $d+1$-element subset $I \subseteq 0..gd$ since the smallest eigenvalue of $S_{\mathbf{x}_I}$ is strictly positive. The claim follows from the definition of β. If Λ is bounded and the smallest eigenvalue converges to zero, then the estimate (1.32) shows again that $\log f_n(\mathbf{x}; \boldsymbol{\pi}, \boldsymbol{\varphi}, \mathbf{m}, \mathbf{V})$ converges to $-\infty$ on E_0. This implies that the n-fold likelihood function converges to $-\infty$ as Λ approaches the Alexandrov point of $\mathcal{V}_c$.

In order to analyze the behavior w.r.t. m_j, assume now that $\boldsymbol{\Lambda}^{(k)}$ converges to $\boldsymbol{\Lambda}^* \in \mathrm{PD}(d)^g$ and return to Eq. (1.31). Collect all components j for which $m_j^{(k)}$ converges in $\mathbb{R}^d$ in the (possibly empty) set J, let m_j^*, $j \in J$, be the limits, and let $\boldsymbol{\pi}^*$ and $\boldsymbol{\varphi}^*$ be the limits of $(\boldsymbol{\pi}^{(k)})_k$ and $(\varphi_j^{(k)})_k$, respectively. Since φ_j vanishes at ∞, (1.31) converges pointwise (even locally uniformly) everywhere to the continuous function

$$\prod_{i=0}^{gd} \sum_{j \in J} \pi_j^* \sqrt{\det \Lambda_j^*}\, \varphi_j^*\big((x_i - m_j^*)^\top \Lambda_j^* (x_i - m_j^*)\big)$$

and we have shown assumption (ii) of Theorem 1.5.

It remains to verify the two parts of assumption (iii). We have

$$\mathrm{E}\log^- f_n(X_1^n; \boldsymbol{\pi}_0, \boldsymbol{\varphi}_0, \mathbf{m}_0, \mathbf{V}_0) \le \sum_{i=0}^{gd} \mathrm{E}\log^- f(X_i; \boldsymbol{\pi}_0, \boldsymbol{\varphi}_0, \mathbf{m}_0, \mathbf{V}_0).$$

Each summand on the right equals

$$\sum_j \pi_{0,j} \int \log^- f(x; \boldsymbol{\pi}_0, \boldsymbol{\varphi}_0, \mathbf{m}_0, \mathbf{V}_0) f(x; \varphi_{0,j}, m_{0,j}, V_{0,j}) \mathrm{d}x$$

$$\le \sum_j \pi_{0,j} \int \log^-(\pi_{0,j} f(x; \varphi_{0,j}, m_{0,j}, V_{0,j})) f(x; \varphi_{0,j}, m_{0,j}, V_{0,j}) \mathrm{d}x$$

and the first half of assumption (iii) follows from Lemma E.26 and the bound on φ.

By Eq. (1.32), there exists a subset $I \subseteq 0..gd$ of size $d+1$ such that

$$\log f_n(\mathbf{x}; \boldsymbol{\pi}, \mathbf{m}, \mathbf{V}) \le \tfrac{n}{2} \log \det \Lambda_j - \tfrac{\beta}{2} \log(1 + \operatorname{tr} S_{\mathbf{x}_I} \Lambda_j) + \mathrm{const}$$

$$\le \tfrac{n}{2} \log \det \Lambda_j - \tfrac{nd}{2} \log(1 + \operatorname{tr} S_{\mathbf{x}_I} \Lambda_j) + \mathrm{const},$$

where $\mathbf{x}_I = (x_i)_{i \in I}$ and I depends on $\mathbf{x}$. The geometric-arithmetic inequality A.5 implies $\operatorname{tr} S_{\mathbf{x}_I} \Lambda_j \geq d(\det S_{\mathbf{x}_I} \Lambda_j)^{1/d}$ and we continue

$$\cdots \leq \tfrac{n}{2} \log \det \Lambda_j - \tfrac{n}{2} \log(\det S_{\mathbf{x}_I} \Lambda_j) + \text{const}$$

$$= -\tfrac{n}{2} \log \det S_{\mathbf{x}_I} + \text{const} = -\tfrac{n}{2} \log \det W_{\mathbf{x}_I} + \text{const}$$

$$\leq \tfrac{n}{2} \sum_{T \in \binom{0 \cdots gd}{d+1}} \log^- \det W_{\mathbf{x}_T} + \text{const} = h_0(\mathbf{x}).$$

We show $\operatorname{E} \log^- \det W_{X_T} < \infty$ for all $T \subseteq 0 \mathinner{\ldotp\ldotp} gd$ of size $d+1$. (The value does not depend on T since the X_i are i.i.d.) For all $x \in \mathbb{R}^d$, we have $\|x\|^2 \leq (\|x - m_j\| + \|m_j\|)^2 \leq 2\|x - m_j\|^2 + 2\|m_j\|^2$. Denoting the largest eigenvalue of all $V_{0,j}$ by v it follows

$$(x - m_{0,j})^\top V_{0,j}^{-1}(x - m_{0,j}) \geq \tfrac{1}{v}\left(\|x - m_{0,j}\|^2\right) \geq \tfrac{1}{v}\left|\tfrac{1}{2}\|x\|^2 - \|m_{0,j}\|^2\right|$$

and along with the estimate $\psi(|r - a|)\psi(a) \leq \psi(0)\psi(r)$, valid for all $a, r \geq 0$, this shows

$$f(x; \varphi_{0,j}, m_{0,j}, V_{0,j}) = \varphi_{0,j}\left((x - m_{0,j})^\top V_{0,j}^{-1}(x - m_{0,j})\right)$$

$$\leq \psi\left((x - m_{0,j})^\top V_{0,j}^{-1}(x - m_{0,j})\right) \leq \psi\left(\big|\|x\|^2/(2v) - \|m_{0,j}\|^2/v\big|\right)$$

$$\leq \text{const} \cdot \psi\left(\|x\|^2/(2v)\right).$$

Hence,

$$\operatorname{E} \log^- \det W_{(X_0, \ldots, X_d)}$$

$$= \int_{\mathbb{R}^{(d+1)d}} (\log^- \det W_{\mathbf{x}}) \prod_{i \in T} \sum_j f(x_i; \varphi_{0,j}, m_{0,j}, V_{0,j})$$

$$\leq \text{const} \int_{\mathbb{R}^{(d+1)d}} (\log^- \det W_{\mathbf{x}}) \prod_{i \in T} \psi\left(\|x_i\|^2/(2v)\right)$$

$$= \text{const} \cdot \operatorname{E} \log^- \det W_{(Y_0, \ldots, Y_d)},$$

where Y_i is an i.i.d. sequence of multivariate t-distributed random vectors with index η. The right-hand side is finite by Corollary E.14. This is the second half of assumption (iii). The proof of part (a) is finished. Since the map $(\boldsymbol{\pi}, \boldsymbol{\varphi}, \mathbf{m}, \mathbf{V}) \mapsto f_{\boldsymbol{\pi}, \boldsymbol{\varphi}, \mathbf{m}, \mathbf{V}}$ is continuous, part (b) follows from part (a) of Theorem 1.5 and Proposition B.5. $\qquad \square$

1.23 Remarks (a) The constrained consistency Theorem 1.22 must not be misunderstood. It does not invite us to design constrained optimization algorithms. In fact, Theorem 1.19 rather tells us that there is an *unconstrained,* consistent local maximum of the likelihood function. This solution is also the constrained solution according to Theorem 1.22 (and 1.27 below) if c is chosen small enough (and n is large enough). Although the mixture likelihood contains some information about the parent mixture, it is not sufficient to provide an estimate. It needs a complementary piece of information. A solution to this problem is postponed to Section 4.1. It contains also an algorithm devoid of constraints.

(b) The consistency theorems of this section assume specified parents, that is, the data is assumed to come from a member of the statistical model. This is never exactly fulfilled in real applications (not even in simulations). This limits their usefulness. However, the present theorem applies to rich models. An example is the extended Pearson Type-VII model presented in (E.7). It contains members with heavy tails, the Type-VII distributions, and a member with a light tail, the normal. It provides, therefore, the flexibility that is needed for *robustness.* Its ML estimator can actually be considered a robust M estimator. The fact that robustness can be combined with consistency has already been pointed out by Huber [253, 254].

(c) Theorem 1.22 says that an ML estimator in the HDBT constrained model with variable number of components, $\mathcal{D}_{\leq g,c}$, is consistent. Since it contains models with between one and g components, it thereby shows that the ML estimator consistently estimates the *number* of components, too. This is generally considered a major problem in data analysis.

(d) Besides consistency, Theorem 1.5 provides the *asymptotic existence* of the ML estimator for a variable number of components. This holds under the condition that the subspace $\mathcal{D}_{\leq g,c} \subseteq \mathbb{M}(\varrho)$ is locally compact. The assumption of part (b) of Theorem 1.22 uses the map $(\boldsymbol{\pi}, \boldsymbol{\varphi}, \mathbf{m}, \mathbf{V}) \mapsto f_{\boldsymbol{\pi},\boldsymbol{\varphi},\mathbf{m},\mathbf{V}}$ from $\Delta_{g-1} \times \Phi^g \times \mathbb{R}^{gd} \times \mathcal{V}_c$ onto the subset $\mathcal{D}_{\leq g,c}$ in order to assure this condition. However, more can be said. A certain minimum size of the data set is sufficient for an ML estimator to exist in $\mathcal{D}_{\leq g,c}$. The following proposition states that the model $\mathcal{D}_{\leq g,c}$ is universal in the sense that it admits an ML estimator for all data sets irrespective of their origin. It also indicates its relationship with the sampling distribution. Such a result would not be possible in a model with a *fixed* number of components such as Θ_g; see (1.12) above. For instance, a model with two components can, with positive probability, produce a data set of any size that looks sampled from a single component. The estimator for $\mathcal{D}_g$ would tend to estimate just one component, but this is not admissible in Θ_g.

1.24 Proposition *Let Φ be some equicontinuous, pointwise closed set of density generators φ such that $\varphi(t) \leq C(1+t)^{-\beta/2}$, $t \geq 0$.*

(a) If $\beta > \frac{nd}{k}$ with $k = \lceil \frac{n-gd}{d+1} \rceil$, then the nonparametric, HDBT constrained, elliptical mixture model (1.28) possesses an ML estimate for all data sets of length $n \geq gd+1$ in general position.

(b) If $\beta > \frac{nd}{k}$ with $k = \lceil \frac{n-g}{2} \rceil$, then the nonparametric, HDBT constrained, diagonal submodel possesses an ML estimate for all data sets of length $n \geq g+1$ if the data points are pairwise different in each coordinate.

(c) The assertion (b) is valid for spherical submodels even if the data points are only pairwise distinct.

PROOF. Let us first show that it is sufficient to consider mean values m_j in the (compact) convex hull of $\mathbf{x} = (x_1, \ldots, x_n)$, conv $\mathbf{x}$. Let $V_j \in \mathrm{PD}(d)$, let $m_j \in \mathbb{R}^d \setminus \mathrm{conv}\,\mathbf{x}$, and let m'_j be the Euclidean projection of $V_j^{-1/2} m_j$ to the compact, convex set $V_j^{-1/2} \mathrm{conv}\,\mathbf{x}$. Then $V_j^{1/2} m'_j \in \mathrm{conv}\,\mathbf{x}$ and we have

$$\left(x_i - V_j^{1/2} m'_j\right)^\top V_j^{-1}\left(x_i - V_j^{1/2} m'_j\right) = \|V_j^{-1/2} x_i - m'_j\|^2$$
$$< \|V_j^{-1/2}(x_i - m_j)\|^2 = (x_i - m_j)^\top V_j^{-1}(x_i - m_j).$$

The first claim follows from (1.28) and the strict decrease of φ_j.

Now let c be the HDBT constant, let κ be $c \times$ the maximum of all eigenvalues of the scatter matrices of all $d+1$-element subsets of $\mathbf{x}$, and let $\psi(t) = C(1+t)^{-\beta/2}$. Lemma 1.21 implies

$$f_n(\mathbf{x}; \boldsymbol{\pi}, \boldsymbol{\varphi}, \mathbf{m}, \mathbf{V}) \leq \mathrm{const} \cdot (\det \Lambda_1)^{n/2} \psi^k(\kappa \operatorname{tr} \Lambda_1).$$

We use this estimate in order to analyze the behavior of $f_n(\mathbf{x}; \boldsymbol{\pi}, \boldsymbol{\varphi}, \mathbf{m}, \mathbf{V})$ as $\mathbf{V}$ approaches the Alexandrov point of $\mathcal{V}_c$. By the HDBT constraints this means that $\Lambda_1 = V_1^{-1}$ (or any other Λ_j) approaches the boundary of $\mathrm{PD}(d)$. We have to consider two cases: either the largest eigenvalue of Λ_1 approaches ∞ *or* all eigenvalues remain bounded and the smallest one approaches zero. In the first case let λ be the largest eigenvalue of Λ_1. We have $\kappa \operatorname{tr} \Lambda_1 \geq \kappa\lambda$ and hence

$$f_n(\mathbf{x}; \boldsymbol{\pi}, \boldsymbol{\varphi}, \mathbf{m}, \mathbf{V}) \leq \mathrm{const} \cdot \lambda^{nd/2}(\kappa\lambda)^{-k\beta/2}.$$

It follows from the assumption on β that the quantity on the right converges to 0 in this case. In the second case, the eigenvalues of Λ_1 are bounded and the smallest one converges to zero. This time, $\det \Lambda_1$ converges to zero while ψ remains bounded. We have thus shown that the likelihood function vanishes as $\mathbf{V}$ approaches the Alexandrov point. Part (a) now follows from continuity of the function $(\boldsymbol{\pi}, \boldsymbol{\varphi}, \mathbf{m}, \mathbf{V}) \to f_n(\mathbf{x}; \boldsymbol{\pi}, \boldsymbol{\varphi}, \mathbf{m}, \mathbf{V})$ and the fact that the sets Δ_{g-1}, Φ, and conv $\mathbf{x}$ are all compact.

Parts (b) and (c) are obtained in the same way using the paragraph following Lemma 1.21 instead of the lemma itself. $\square$

The proposition applies to any individual density generator with sufficient decay, in particular to the normal mixture model.

1.25 Corollary *If the data is in general position containing at least $gd+1$ data points, then the HDBT constrained, heteroscedastic, fully normal mixture model with variable number of components possesses an ML estimate w.r.t. $\boldsymbol{\pi} \in \Delta_{g-1}$, $\mathbf{m}$, and $\mathbf{V} \in \mathcal{V}_c$.*

ML estimation in homo- and constrained heteroscedastic *diagonal* models needs at least $g+1$ different entries in each coordinate. Hence, the minimum number of pairwise different data points is $g+1$. In homo- and constrained heteroscedastic *spherical* models $g+1$ pairwise different data points are sufficient.

1.2.6   Asymptotic properties of the MLE: constrained parametric mixture models

In order to handle asymptotic normality of the ML estimator of (constrained) elliptical mixtures, we return to the parameter space (1.12) of the mixture model with g components over

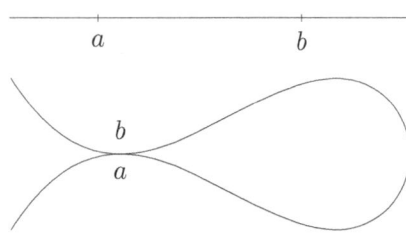

$(f_\gamma)_{\gamma \in \Gamma}$, transferring it to a form that renders the model identifiable (without the attribute "up to label switching") while it remains $\mathcal{C}^2$ parametric. Taking into account the unordered structure of the mixture components makes it necessary to pass to a quotient space. Quotient spaces appear quite commonly in statistics; see Redner [434] and McCullagh [356]. In the present context it is, however, not sufficient to just "weld" together all parameters that represent the sampling mixture μ_0. The parameter space would no longer be locally Euclidean at the welding

Figure 1.5 *A Euclidean 1-manifold before (top) and after welding two points.*

point; see Fig. 1.5. Assume that a mixture model with the parameter space above is identifiable (up to label switching). Then each mixture is represented by exactly $g!$ parameters in Θ_g. Consider the collection of sets

$$\widetilde{\Theta}_g = \{\{(\pi_1, \gamma_1), \ldots, (\pi_g, \gamma_g)\} \mid \pi_j > 0, \sum \pi_j = 1, \gamma_j \in \Gamma \text{ pairwise distinct}\}.$$

Its members are g-element subsets of $]0, 1[\times\Gamma$. The set $\widetilde{\Theta}_g$ is the quotient space of Θ_g sought and $\psi : \Theta_g \to \widetilde{\Theta}_g$ defined by $\psi(\boldsymbol{\pi}, \boldsymbol{\gamma}) = \{(\pi_1, \gamma_1), \ldots, (\pi_g, \gamma_g)\}$ is the quotient map; see Section B.13. The following lemma analyzes the new parameter space $\widetilde{\Theta}_g$.

1.26 Lemma *Let Γ be metric and endow $\widetilde{\Theta}_g$ with the Hausdorff metric; see Section B.6.3.*

(a) The Hausdorff metric on $\widetilde{\Theta}_g$ induces the quotient topology of ψ.

(b) $\widetilde{\Theta}_g$ and Θ_g are locally homeomorphic via the quotient map ψ.

(c) If Γ is even an open $\mathcal{C}^k$-manifold, then $\widetilde{\Theta}_g$ is again $\mathcal{C}^k$.

PROOF.   (a) By Proposition B.16 we have to show that the quotient map $\psi : \Theta_g \to \widetilde{\Theta}_g$ is closed w.r.t. the Hausdorff metric. Let $\mathcal{A} \subseteq \Theta_g$ be a closed subset and let $(\boldsymbol{\pi}^{(n)}, \boldsymbol{\gamma}^{(n)})$ be a sequence in $\mathcal{A}$ such that $\psi(\boldsymbol{\pi}^{(n)}, \boldsymbol{\gamma}^{(n)})$ converges to $A \in \widetilde{\Theta}_g$. By definition of Θ_g, the set A contains g different parameters. Sort the g elements of A in an arbitrary way to obtain a g-tuple $a \in \Theta_g$. For eventually all n, there is a natural one-to-one assignment of the elements of $\psi\big(\boldsymbol{\pi}^{(n)},\boldsymbol{\gamma}^{(n)}\big)$ to those of A. It defines a reordering $a^{(n)}$ of $\big(\boldsymbol{\pi}^{(n)},\boldsymbol{\gamma}^{(n)}\big)$. The g-tuples $a^{(n)}$ are not necessarily members of $\mathcal{A}$ but in its closure $\mathcal{A}'$ w.r.t. all permutations. Moreover, by $\psi\big(\boldsymbol{\pi}^{(n)},\boldsymbol{\gamma}^{(n)}\big) \to A$, we have $a^{(n)} \to a$ as $n \to \infty$. Since $\mathcal{A}'$, too, is closed in Θ_g, we have $a \in \mathcal{A}'$. Hence $A = \psi(a) \in \psi(\mathcal{A}') = \psi(\mathcal{A})$.

In order to study claim (b), let $(\boldsymbol{\pi},\boldsymbol{\gamma}) \in \Theta_g$. By definition of Θ_g, the inverse image $\psi^{-1}(\psi(\boldsymbol{\pi},\boldsymbol{\gamma}))$ consists of $g!$ distinct points, one of which is $(\boldsymbol{\pi},\boldsymbol{\gamma})$. They all differ in the permutation of $\boldsymbol{\gamma}$. There is $\delta > 0$ such that the δ-balls about the $g!$ points do not intersect. Then $\psi(B_\delta(\boldsymbol{\pi},\boldsymbol{\gamma}))$ is the homeomorphic image of $B_\delta(\boldsymbol{\pi},\boldsymbol{\gamma})$. $\qquad\square$

There is another representation of the quotient space $\widetilde{\Theta}_g$ endowed with the Hausdorff

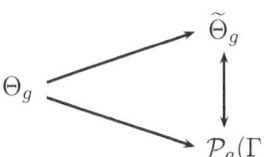

Figure 1.6 *Arrow diagram.*

metric. Let $\mathcal{P}_g(\Gamma)$ be the set of all probability measures on Γ supported by exactly g points. Each element in $\mathcal{P}_g(\Gamma)$ has a representation $\sum_{1\le j\le g}\pi_j\delta_{\gamma_j}$ with $\boldsymbol{\pi} = (\pi_1,\dots,\pi_g) \in \mathring{\Delta}_{g-1}$ and $\gamma_j \in \Gamma$ pairwise distinct. This set of discrete probabilities is endowed with the weak topology. The assignment $(\boldsymbol{\pi},\boldsymbol{\gamma}) \mapsto \sum_{1\le j\le g}\pi_j\delta_{\gamma_j}$

defines a natural one-to-one map from $\widetilde{\Theta}_g$ onto $\mathcal{P}_g(\Gamma)$ giving rise to the commutative diagram in Fig. 1.6. For each bounded, continuous, real-valued function h on Γ, the map

$$(\boldsymbol{\pi},\boldsymbol{\gamma}) \mapsto \sum_j \pi_j h(\gamma_j) = \int_\Gamma h(\gamma)\sum_j \pi_j \delta_{\gamma_j}(\mathrm{d}\gamma)$$

is continuous, that is, the function $\widetilde{\Theta}_g \to \mathcal{P}_g(\Gamma)$ is continuous w.r.t. the Hausdorff metric. The inverse map $\sum_{1\le j\le g}\pi_j\delta_{\gamma_j} \mapsto \sum_{1\le j\le g}\pi_j f_{\gamma_j}$, too, is continuous, that is, the map $\widetilde{\Theta}_g \to \mathcal{P}_g(\Gamma)$ is even a homeomorphism.

We are now prepared to prove the main consistency theorem for ML estimators of parametric elliptical mixtures (1.15) under HDBT constraints (1.27),

$$\Theta_{g,c} = \mathring{\Delta}_{g-1} \times \big\{(\boldsymbol{\varphi},\mathbf{m},\mathbf{V}) \in \Phi^g \times \mathbb{R}^{gd} \times \mathcal{V}_c \,|\, (\varphi_j, m_j, V_j) \text{ pairwise distinct}\big\},$$

$$(\boldsymbol{\pi},\boldsymbol{\varphi},\mathbf{m},\mathbf{V}) \mapsto \sum_j \frac{\pi_j}{\sqrt{\det V_j}}\,\varphi_j\big((x-m_j)^\top V_j^{-1}(x-m_j)\big). \qquad (1.33)$$

1.27 Theorem (Consistency of the parametric MLE for constrained elliptical mixtures) *Let $\Theta_{g,c} \to \mathcal{D}_{g,c}$ be some parametric, HDBT constrained, elliptical, shape, location, and scale mixture model (1.33) with g components (and full, diagonal, or spherical scale parameters) over the set of density generators Φ. Assume that*

(i) Φ is equicontinuous on $\mathbb{R}_+$;

(ii) there are constants $\beta > (gd+1)d$ and $C > 0$ such that $0 < \varphi(t) \le C(1+t)^{-\beta/2}$ for all $t \ge 0$, $\varphi \in \Phi$;

(iii) the model is identifiable (up to label switching).

Let $(\boldsymbol{\pi}_0, \boldsymbol{\varphi}_0, \mathbf{m}_0, \mathbf{V}_0) \in \Theta_{g,c}$, $\varphi_{0,j} > 0$, be the parameter of the sampling distribution.

(a) If $(T_n)_n$ is a sequence of ML estimates in $\Theta_{g,c}$, then a representative $\widetilde{T}_n \in \Theta_{g,c}$ of each T_n w.r.t. label switching can be chosen in such a way that the sequence $(\widetilde{T}_n)$ is strongly consistent at $(\boldsymbol{\pi}_0, \boldsymbol{\varphi}_0, \mathbf{m}_0, \mathbf{V}_0) \in \Theta_{g,c}$ w.r.t. the Euclidean metric on $\Theta_{g,c}$.

(b) P-a.s., an ML estimate T_n exists in $\Theta_{g,c}$ for eventually all n.

PROOF. By Lemma 1.26(a), the mixture model $\Theta_{g,c} \to \mathcal{D}_{g,c}$ can be lifted via the quotient map $\psi : \Theta_{g,c} \to \widetilde{\Theta}_{g,c}$ to an identifiably parametric statistical model $\widetilde{\Theta}_{g,c} \to \mathcal{D}_{g,c}$. Here, $\widetilde{\Theta}_{g,c}$ is endowed with the Hausdorff metric. We apply Theorem 1.7 with $\Theta = \widetilde{\Theta}_{g,c}$ and $\mathcal{D} = \mathcal{D}_{g,c}$ to this model. Lemma 1.26(b) asserts that the Hausdorff metric makes $\widetilde{\Theta}_{g,c}$ a locally compact space along with $\Theta_{g,c}$. Plainly, all density functions are continuous.

Let us turn to the assumptions (ii) of Theorems 1.5 and 1.7. As in the proof of Theorem 1.22, consider the $gd+1$-fold product model and let E_0 be the set of all affine independent $gd+1$-tuples of vectors in $\mathbb{R}^d$. A sequence $(\vartheta^{(k)})_k$ in $\Theta_{g,c}$ converges to the Alexandrov point of $\Theta_{g,c}$ if and only if the sequence $(\psi(\vartheta^{(k)}))_k$ converges to the Alexandrov point of $\widetilde{\Theta}_{g,c}$. We have to show that any sequence $\vartheta^{(k)} = \left(\boldsymbol{\pi}^{(k)}, \boldsymbol{\varphi}^{(k)}, \mathbf{m}^{(k)}, \mathbf{V}^{(k)}\right) \in \Theta_{g,c}$ contains a subsequence, again denoted by $\vartheta^{(k)}$, such that $f_{gd+1}\left(\mathbf{x}; \boldsymbol{\pi}^{(k)}, \boldsymbol{\varphi}^{(k)}, \mathbf{m}^{(k)}, \mathbf{V}^{(k)}\right)$ converges pointwise and if $(\vartheta^{(k)})_k$ converges to the Alexandrov point of $\Theta_{g,c}$, then the limit is not in $\mathcal{D}_{g,c}$. From the definition (1.33) of $\Theta_{g,c}$ it follows that, in addition to the cases already treated in Theorem 1.22, the sequence $(\vartheta^{(k)})$ also converges to the Alexandrov point when $\pi_j^{(k)}$ converges to zero for some j or when all triples $(\boldsymbol{\varphi}^{(k)}, \mathbf{m}^{(k)}, \mathbf{V}^{(k)})$ remain in a compact subset of $\Phi^g \times \mathbb{R}^{gd} \times \mathcal{V}_c$ and the triples $(\varphi_j^{(k)}, m_j^{(k)}, V_j^{(k)})$ for two different j's approach each other as $k \to \infty$. But in these cases the joint densities converge to a density not contained in $\mathcal{D}_{g,c}$ by (the strong form of) identifiability up to label switching. This shows the two assumptions (ii). Finally, assumption (iii) of Theorem 1.5, too, has been verified in the proof of Theorem 1.22.

We have confirmed all conditions of Theorem 1.7. It follows that the map $\widetilde{\Theta}_{g,c} \to \mathcal{D}_{g,c}$ is a homeomorphism, where $\mathcal{D}_{g,c}$ is endowed with the topology of stochastic convergence w.r.t. Lebesgue measure. By Lemma 1.26(b), $\widetilde{\Theta}_{g,c}$ is locally homeomorphic with $\Theta_{g,c}$ and the stochastic, the pointwise, and the compact convergences coincide on $\mathcal{D}_{g,c}$. Parts (a) and (b) follow from Theorem 1.7. □

Part (a) of the foregoing theorem says that everything is as we would expect: After appropriate label switching we have convergence w.r.t. the usual topology on $\overset{\circ}{\Delta}_{g-1} \times \mathbb{R}^{gd} \times \mathrm{PD}(d)^g$ and pointwise (or compact) convergence on Φ.

A fine point has to be addressed. Theorems 1.22 and 1.27 say that if the parental scale parameters satisfy the HDBT constraints with a constant $\geq c_0$, then, asymptotically, a consistent maximum of the likelihood function exists in Θ_{g,c_0}. However, there may be local maxima that violate the constraints. These belong often to "spurious" solutions arising from data points that happen to be almost coplanar, that is, lie almost on a hyperplane of the sample space. In the univariate case, $d = 1$, "coplanar" means equal, on the plane collinear, and so on. Creating large likelihood values, such constellations mask the consistent maximum. They exist with positive probability for any n.

The following discussion needs the **HDBT ratio** $r_{\mathrm{HDBT}}(\mathbf{V})$ of a tuple $\mathbf{V} = (V_1, \ldots, V_g)$ (or of a collection $\{V_1, \ldots, V_g\}$) of positive definite matrices; see Gallegos and Ritter [182]. It is the largest number c such that $(V_1, \ldots, V_g)$ lies in $\mathcal{V}_c$,

$$r_{\mathrm{HDBT}}(\mathbf{V}) = \max_{\mathbf{V} \in \mathcal{V}_c} c. \qquad (1.34)$$

The concept applies mainly to the scale matrices of mixture models and solutions. It is a measure of their balance and will be used extensively later in theory and applications. By Lemma A.10 it is computed as

$$r_{\mathrm{HDBT}}(\mathbf{V}) = \min_{j,\ell,k} \lambda_k \left(V_\ell^{-1/2} V_j V_\ell^{-1/2}\right),$$

where $\lambda_1(A), \ldots, \lambda_d(A)$ denote the d eigenvalues of a symmetric $d \times d$ matrix. The larger the HDBT ratio of a mixture is the more similar the component scales. Now, if the data is sampled from Θ_{g,c_0}, then it is also sampled from $\Theta_{g,c}$ for all $c < c_0$ since the parameter spaces $\Theta_{g,c}$ increase with decreasing c. Again, the theorem says that, asymptotically, a consistent maximum of the likelihood function exists in $\Theta_{g,c}$. It must, therefore, be the very maximum found in Θ_{g,c_0} if n is large enough. In other words, in the long run there is no maximum with HDBT ratio in the interval $]c_0, c[$. This statement is true for all $c > c_0$. Of course, "in the long run" depends on c. For a c much larger than c_0 it may take a long time

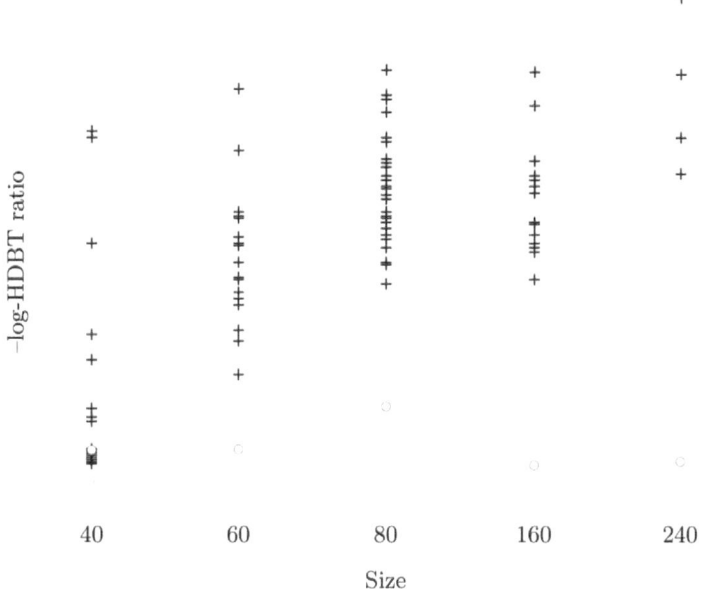

Figure 1.7: *The phenomenon of spurious solutions and the growing gaps.*

until maxima with HDBT ratio in the gap $]c_0, c[$ are no longer observed. This phenomenon is illustrated in Fig. 1.7 with a plane data set $X_1, X_2, \ldots$ sampled from $\frac{3}{4} N_{2e_1, I_2} + \frac{1}{4} N_{-2e_1, I_2}$. In the vertical direction the negative log-HDBT ratios of local maxima of the likelihood function for sections $X_1, \ldots, X_n$ of the data set are plotted. The lengths n are marked on the abscissa. Plus signs stand for local maxima and the circle for the (consistent) maximum in Θ_{g,c_0}. It is clearly seen that the gap increases with the number of observations. In this context, Hathaway [232] asks two questions. (1) Do all spurious local maxima disappear as the sample size increases to ∞ while c is kept fixed? Figure 1.7 seems to support this conjecture. (2) Is it possible to decrease c to zero while maintaining consistency?

Theorem 1.27 is rather general. It is worthwhile deriving some corollaries for elliptical and normal mixture models and submodels; see Section 1.2.2. Density generators of Pearson's Type-VII are popular for their robustness properties. The resulting models are then C^∞ and we can also talk about asymptotic normality. There is the following corollary.

1.28 Corollary *Let I be the open interval $](gd+1)d, \infty[\subseteq \mathbb{R}_+$ and let $\varphi_{\eta,d}$ be the Pearson Type-VII density generator with index $\eta \in I$. Let*

$$\Theta_{g,c} = \overset{\circ}{\Delta}_{g-1} \times \big\{ (\boldsymbol{\eta}, \mathbf{m}, \mathbf{V}) \in I^g \times \mathbb{R}^{gd} \times \mathcal{V}_c \,|\, (\eta_j, m_j, V_j) \text{ p.w. distinct} \big\}$$

and $f_{\boldsymbol{\pi},\boldsymbol{\eta},\mathbf{m},\mathbf{V}} = f_{\boldsymbol{\pi},\boldsymbol{\varphi},\mathbf{m},\mathbf{V}}$, $\varphi_j = \varphi_{\eta_j,d}$, $1 \leq j \leq g$, and let $\Theta_{g,c} \to \mathcal{D}_{g,c}$ be the associated constrained elliptical C^∞ mixture model (1.33) with g components (and full, diagonal, or spherical scale parameters).

(a) P-a.s., an ML estimate T_n exists in $\Theta_{g,c}$ for eventually all n.

(b) A representative $\widetilde{T}_n \in \Theta_{g,c}$ of T_n w.r.t. label switching can be chosen in such a way that the sequence $(\widetilde{T}_n)$ is strongly consistent at any $(\boldsymbol{\pi}_0, \boldsymbol{\eta}_0, \mathbf{m}_0, \mathbf{V}_0) \in \Theta_{g,c}$ w.r.t. the Euclidean metric on $\Theta_{g,c}$.

(c) If $c < 1$ and if the scale matrices of the sampling distribution satisfy the HDBT constraints even with strict inequality, then the sequence $(\widetilde{T}_n)$ is asymptotically normal.

That is, we have in distribution

$$\sqrt{n}(\widetilde{T}_n - (\boldsymbol{\pi}_0, \boldsymbol{\eta}_0, \mathbf{m}_0, \mathbf{V}_0)) \xrightarrow[n\to\infty]{} N_q(0, \mathcal{I}(\boldsymbol{\pi}_0, \boldsymbol{\eta}_0, \mathbf{m}_0, \mathbf{V}_0)^{-1}).$$

Moreover, the sequence $(\widetilde{T}_n)$ is asymptotically efficient.

PROOF. Parts (a) and (b) follow from Theorem 1.27. Part (c) follows from Corollary 1.10 since the conditions on the scale matrices describe an open subset of Euclidean $g\binom{d+1}{2}$-space. □

Corollary 1.28 for a single density generator reads as follows.

1.29 Corollary *Statements (a)–(c) of Corollary 1.28 hold true for the normal and Pearson Type-VII mixture models and a single density generator. The parameter space of this C^∞ model is*

$$\overset{\circ}{\Delta}_{g-1} \times \{(\mathbf{m}, \mathbf{V}) \in \mathbb{R}^{gd} \times \mathcal{V}_c \,|\, (m_j, V_j) \text{ p.w. distinct}\}.$$

1.30 Remarks (a) Once the parameters π_j and γ_j of the mixture have been estimated, it is easy to create a partition of the given data set $\mathbf{x}$. It is reasonable to employ the Bayesian discriminant rule with the estimated parameters,

$$i \in C_\ell \quad \Leftrightarrow \quad \ell = \operatorname*{argmax}_j \pi_j f_{\gamma_j}(x_i).$$

Ties are broken, for instance, by favoring the smallest index.

(b) Asymptotic normality as stated in Corollaries 1.28 and 1.29 is traditionally used for assessing the accuracy of parameter estimates. Boldea and Magnus [54] nicely discuss this matter for three consistent estimates of the Fisher information matrix, the empirical versions of the expected information matrix and of the expected Hessian,

$$\mathcal{I}_1 = \sum_i s(x_i; T_n)s(x_i; T_n)^\top \quad \text{and} \quad \mathcal{I}_2 = -\sum_i \mathrm{D}_\vartheta^2 \log f(x_i; T_n),$$

respectively, and the so-called "sandwich" version, $\mathcal{I}_3 = \mathcal{I}_2^{-1}\mathcal{I}_1\mathcal{I}_2^{-1}$. Their simulations also show that $\mathcal{I}_2$ is most reliable, outperforming several bootstrap methods. Huber [253] noted that $\mathcal{I}_3$ is consistent even when the model is not correctly specified. Boldea and Magnus [54], Theorem 2, apply White's [526] information matrix test for misspecification of a parametric model to mixtures. Basford et al. [29] computed the score functional of a mixture model with a fixed number of components needed for $\mathcal{I}_1$. The Hessian needed for $\mathcal{I}_2$ was computed by Boldea and Magnus [54], Theorem 1.

(c) Some results on the rate of convergence of ML and other estimators for normal mixtures are available. Chen [88] shows that the optimal rate, $\sqrt{n}$, for a finite mixture can be obtained only if the number of components is correctly specified. Otherwise, it is at most $\sqrt[4]{n}$. This is attained for a certain minimum distance estimator; see [88], Theorem 2, and the remark on the bottom of p. 227. Ghosal and van der Vaart [197], Theorem 4.1, consider univariate normal mixtures over a compact subset of the parameter space. They show that the MLE of the mixture density attains a rate of a.s. convergence in Hellinger distance of $\log^{3/2} n/\sqrt{n}$. See also the beginning of Remark 1.8(c).

(d) Consider the parametric model (1.12) with g components. If Γ is furnished with a total order (in any way), then a complete system of representatives of the quotient set $\widetilde{\Theta} = \Theta_g/\sim$ w.r.t. label switching $\sim$ can easily be given. It is sufficient to use the lexical order on Γ^g and to define a representative $\widetilde{\gamma}$ of each γ by ordering the γ_j's, that is, $\widetilde{\gamma} = \gamma_\uparrow$. This idea would avoid speaking of equivalence classes; however, it would upset the given topological structure on Γ^g. By way of illustration assume $\Gamma = \mathbb{R}^2$ and $g = 2$. Then the parameter $\gamma = \left(\binom{1-\varepsilon}{10}, \binom{1}{1}\right)$ is close to $\zeta = \left(\binom{1}{10}, \binom{1-\varepsilon}{1}\right)$ w.r.t. the Euclidean metric on $\mathbb{R}^4$ if ε is small. However, their representatives $\gamma_\uparrow = \left(\binom{1-\varepsilon}{10}, \binom{1}{1}\right)$ and $\zeta_\uparrow = \left(\binom{1-\varepsilon}{1}, \binom{1}{10}\right)$ are no longer close. The parameters of normal families and Pearson Type-VII shape, location, and scale models can be ordered lexically in many ways but the example shows that it would be useless, at least as far as consistency in Corollaries 1.28 and 1.29 is concerned.

(e) One of the advantages of the ML estimate in the nonparametric mixture model is its consistency for the number of components; see Remark 1.23. The space $\mathcal{D}_{\leq g,c}$ of mixture densities is appropriate for this purpose. If we wished to treat consistency for the number of components in a *parametric* model, then we would have to identifiably parameterize $\mathcal{D}_{\leq g,c}$ by a manifold. In rare cases this can indeed be done. The presentation needs a little algebra. Two forms of a complex polynomial are common, the factorized and the expanded form,

$$(z - z_1) \cdot \ldots \cdot (z - z_g) = z^g - a_1 z^{g-1} + - \ldots + (-1)^{g-1} a_{g-1} z + (-1)^g a_g, \qquad (1.35)$$

with the g **roots** $z_j \in \mathbb{C}$ and the g **coefficients** $a_j \in \mathbb{C}$, both possibly multiple. Transition from the roots to the coefficients, is carried out by expanding the polynomial. It yields the **elementary symmetric functions**

$$a_j(\mathbf{z}) = \sum_{1 \leq i_1 < \cdots < i_j \leq n} z_{i_1} \cdot \ldots \cdot z_{i_j}.$$

Examples are $a_1(\mathbf{z}) = \sum_{j=1}^g z_j$ and $a_g(\mathbf{z}) = \prod_{j=1}^g z_j$. **Factorization**, that is, the transition $\mathbf{a} \to (z_1(\mathbf{a}), \ldots, z_g(\mathbf{a}))$ from the coefficients to the roots is always possible by the **Fundamental Theorem of Algebra**. Closed expressions exist up to $g = 4$, the Cardano formulæ. For larger g there are numerical methods; see Press et al. [423]. The coefficients are unique and the roots are unique up to permutation, that is, they uniquely determine an element in the quotient space $\mathbb{C}^{(g)} = \mathbb{C}^g / \sim$ with the equivalence relation $\sim$ defined by permuting the entries. The space $\mathbb{C}^{(g)}$ is a manifold called a **permutation product** of $\mathbb{C}$ and the map $\mathbb{C}^g \to \mathbb{C}^{(g)}$, $\mathbf{a} \to \mathbf{z}$, is a homeomorphism.

These facts can be used to parameterize mixture models if each component has just *two* real parameters. An example is an isobaric mixture of g univariate normals, $\frac{1}{g} N_{m_1, v_1} + \cdots + \frac{1}{g} N_{m_g, v_g}$. Like the roots of a polynomial, the components are uniquely determined up to permuting the components. Let $m_j(\mathbf{a})$ and $v_j(\mathbf{a})$ be the real and imaginary parts, respectively, of the jth root in $\mathbb{C} = \mathbb{R}^2$ of the polynomial with coefficients $\mathbf{a}$. Then the parametric model

$$\mathbb{C}^g \to \mathcal{D}_{\leq g,c}, \quad \mathbf{a} \mapsto \frac{1}{g} \sum_j N_{m_j(\mathbf{a}), v_j(\mathbf{a})},$$

is well defined. Being identifiable by Proposition 1.16 and by what was said above, it allows the application of parametric consistency theorems. Unfortunately, the method works only for mixtures with two real parameters per component. In fact, Wagner [518] showed that $\left(\mathbb{R}^d\right)^{(g)}$ is a manifold only for $d = 2$, that is, in the complex case above.

(f) The homoscedastic model can be regarded as unconstrained parametric with the parameter space $\overset{\circ}{\Delta}_{g-1} \times \mathbb{R}^{gd} \times \mathrm{PD}(d)$. It can be treated with the method of Theorem 1.19. The MLE exists and is the consistent solution.

1.2.7 Notes

The application of mixture analysis to classification problems can be traced back at least to Simon Newcomb [387] in 1886 and Karl Pearson [404] in 1894.[2] Pearson was then ahead of his time, using his method of moments for decomposing Weldon's Naples crabs. Mixture analysis was taken up again only half a century later by William Feller [150], who proved that arbitrary mixtures of Poisson distributions are identifiable. Estimation by local likelihood

---

[2]Pearson's paper is contained in the historical archive of the Royal Society, royalsociety.org. It has been made permanently free to access online.

maxima was first proposed by Hasselblad [230] and Wolfe [530]. The problem of multiple solutions was known to Day [113].

A systematic investigation into the identifiability of mixtures was initiated by Henry Teicher [498, 499, 500]. The communications [498, 499] are devoted to arbitrary mixtures. The identity $\int N_{m,1}(x)N_{0,1}(m)\,dm = \int N_{0,1}(x-m)N_{0,1}(m)\,dm = N_{0,1} * N_{0,1}(x) = N_{0,\sqrt{2}}$ shows that *general* normal mixtures are not identifiable. If either the mean or the variance is kept fixed, then they are. Certain univariate, one-parameter semi-groups and location and scale mixtures are identifiable. The binomial model with a fixed number of trials and variable success rate is not. *Finite* mixtures are the subject matter of Teicher [500]. He uses the behavior of the Fourier transforms at ∞ required in Proposition 1.17 and the lexical order appearing in its proof, concluding that finite mixtures of *univariate* normals are identifiable. Proposition 1.15 and identifiability of multivariate normal mixtures, Proposition 1.16, were shown by Yakowitz and Spragins [535]. That of Pearson Type-VII mixtures, Proposition 1.18, was settled much later by Holzmann et al. [250].

The utility of elliptical mixtures in the context of robust clustering has been pointed out by McLachlan and Peel [365]. They deal mainly with algorithmic aspects. An early consistency theorem for disjoint elliptical distributions appears in Davies [110]. Theorem 1.19 is new and an extension of Kiefer [292] and Peters and Walker [412] to multivariate elliptical mixtures. Hathaway [232] initiated the study of ML estimators of HDBT constrained heteroscedastic, univariate, normal mixtures. The HDBT constraints (1.27) assume here the simple form $v_j \geq cv_\ell$. Lemma 1.21 is an extension of a crucial lemma appearing in Gallegos and Ritter [182]. Theorem 1.22 is new. Theorem 1.27 and its Corollaries 1.28 and 1.29 are extensions of Hathaway's consistency theorem [232] from univariate normal to multivariate elliptical mixtures.

Henna [239] gives a condition for finiteness of an arbitrary univariate mixture, presenting in this case a strongly consistent estimator of the number of components. Leroux [316] studies weak consistency of ML estimators of general and finite mixing distributions and of their number of components. He needs a strong compactness condition, his assumption 2, that excludes scale families. Instead of HDBT constraints on the scale parameters, Ciuperka et al. [96], Chen and Tan [89], and Tanaka [495] impose penalty terms on the likelihood function so that it becomes bounded. The penalty functions depend on scales and data set size. They obtain again consistency of the penalized ML estimates.

Marin et al. [346] review Bayesian inference on mixture models. These models allow us to implement prior information and expert opinion. Among others, the authors cover Markov chain Monte Carlo (MCMC) methods to determine the posterior density as the equilibrium distribution of a Markov chain. Examples are Gibbs sampling (see Geman and Geman [195] and Diebolt and Robert [125]) and the Metropolis-Hastings [375, 231] algorithm. See also Gelman et al. [193], and Kaipio and Somersalo [284].

The present text deals mainly with Euclidean data. Vermunt and Magidson [516] propose a mixture model that simultaneously includes Euclidean and discrete data. Each class-conditional random variable is of the form $(X, Z_1, \ldots, Z_q)$, where X is normal, the Z_k's are ordinal or nominal, and X and all Z_k's are independent. Thus, the mixture density is of the form

$$(x, z_1, \ldots, z_q) \mapsto \sum_{j=1}^{g} \pi_j N_{m_j, V_j}(x) \prod_{k=1}^{q} h_{j,k}(z_k);$$

here $\pi_j > 0$, $\sum \pi_j = 1$, $h_{j,k} \geq 0$, $\sum_z h_{j,k}(z_k) = 1$ for all j and k.

1.3   Classification models and their criteria

In parallel with the mixture model, literature also knows the so-called *classification model*, Bock [47, 48], John [279], Scott and Symons [461], and Symons [492]. When all components are well separated, the solutions obtained from both models are similar. Fundamental to classification models are "partitions," "assignments," "clusters," and "cluster criteria." Let $\mathcal{D}$ be a subset of some set $\mathcal{S}$. A function $\mathcal{D} \to \mathbb{R}$ is called a **partial function** on $\mathcal{S}$ with **domain of definition** $\mathcal{D}$.

1.31 Definition Consider the data $\mathbf{x} = (x_i)_{i \in I}$ with index set I and let $g \geq 1$ be a natural number.

(a) A g-**clustering** $\mathcal{C}$ of I is a collection of g pairwise disjoint subsets $C \subseteq I$, called **clusters**.

(b) A g-**partition** $\mathcal{C}$ of I is a g-clustering such that $\bigcup_{C \in \mathcal{C}} C = I$.

(c) Any map $\ell : I \to 0..g$ is called an **assignment** or a **labeling**.

(d) A **cluster criterion** is a partial, real-valued map on the Cartesian product $E^I \times \{\text{all } g\text{-partitions of } I\}$.

(e) A clustering $\mathcal{C}$ is called **admissible** for $\mathbf{x}$ w.r.t. a given criterion if $(\mathbf{x}, \mathcal{C})$ lies in its domain of definition.

The definition is deliberately loose. It allows, for instance, a clustering of two elements in three clusters (at least one of them empty). Various situations will need further specifications. The definition refers purposely to the index set of the data and not to the data themselves. It would, otherwise, not be possible to speak about *invariance* of a partition, for instance, w.r.t. translations of the data. (The data changes but the indices remain.) However, it is common to also speak about a partition of the data $(x_i)_{i \in I}$, meaning a partition of the index set I. This is fair when there are no multiple points. The label $0 \in 0..g$ accounts for potential outliers.

Criteria rate all admissible partitions in view of a desired partition. Being a partial map, each cluster criterion has a domain of definition that has to be carefully determined. It depends on the special form of the criterion. In rare cases it is the *whole* product; often it is smaller. An advantage of using cluster criteria over a direct approach via the mixture model is the finiteness of the set of partitions. They serve as the possible solutions of a clustering problem and correspond to the continuum of possible parameter sets in the mixture model.

Sometimes, assignments offer a more convenient way of describing partitions. Any assignment defines a unique clustering $\mathcal{C} = \{C_j \mid j \in \ell(I), j \geq 1\}$ of I in clusters C_j via

$$i \in C_j \quad \Leftrightarrow \quad \ell_i = j \text{ for } j \in \ell(I), \ j \neq 0.$$

Conversely, let $\mathcal{C}$ be a partition of I in $g' \leq g$ nonempty clusters. Then $\mathcal{C}$ gives rise to $\binom{g}{g'} g'!$ assignments $I \to 1..g$ that create it. Two assignments are *equivalent* if they create the same partition. When a criterion 1.31(c) uses assignments instead of partitions, care must be taken so that equivalent assignments lead to equal criteria. In most cases, the index set I of the data will be $1..n$. If the data points are pairwise distinct, then the points themselves may be taken as I. There are g^n assignments $1..n \to 1..g$. The number of partitions of $1..n$ in g nonempty clusters equals that of all maps of $1..n$ onto $1..g$ divided by $g!$. By Sylvester-Poincaré's sieve formula, this number is the alternating sum $\sum_{j=0}^{g-1} (-1)^j \binom{g}{j} (g-j)^n$.

After establishing the classification model for general parametric basic populations, we will derive its cluster criterion. It will subsequently be specialized to elliptical and normal models and submodels with different parameter and size constraints. In special cases, we retrieve criteria that have been proposed earlier on intuitive, geometric grounds, Ward's [522] pooled trace or sum-of-squares criterion and Friedman and Rubin's [172] pooled determinant criterion. In the context of the mixture model, local maxima of the likelihood function are crucial. Analogs here are steady partitions, Section 1.3.3. Their properties are similar to those of local maxima. Some invariance properties and a consistency analysis will conclude this section.

1.3.1 Probabilistic criteria for general populations

This section introduces classification models and their probabilistic criteria. Let E be a measurable sample space endowed with some σ-finite reference measure ϱ. As before, the observations X_i will be sampled from a finite mixture $\sum_j \pi_j \mu_{\gamma_j}$ over some basic parametric model $(\mu_\gamma)_{\gamma \in \Gamma}$ on E, dominated by ϱ. We will again need the stochastic representation $X_i = Y_i^{(L_i)}$, $1 \leq i \leq n$, of the data with random variables $Y_i^{(j)} : \Omega \to E$, $\sim \mu_{\gamma_j}$, $1 \leq j \leq g$, and a random label $L : \Omega \to 1..g$, $\sim \pi$, independent of all $Y_i^{(j)}$; see Lemma 1.12. The variables $Y_i^{(j)}$ are independent across i but not necessarily across j.

The labels L_i were implicit in Section 1.2. We make them now explicit. They assume here the rôle of the mixing rates. In this context, the question arises whether they are variables or parameters. They share with *variables* the increase of their number with the size of the data. Their trait in common with *parameters* is that they are not observed and hence unknown and we wish to estimate them. They are thus hybrid and this is why they are called *hidden* or *latent* variables (see also Everitt [146]) and also *incidental* parameters.[3] This term emerged from a paper of Neyman and Scott's [392]. More recent presentations of the idea appear in Kumon and Amari [303], Amari and Kumon [11], and Moreira [383].

It is not in general possible to "estimate" hidden variables by ML. Therefore, the derivation of the estimator of the classification model is less straightforward than that of the mixture model, where the ML paradigm can be directly applied. There are two approaches to the estimator: the *maximum a posteriori* (MAP) approach and the *approximation* approach. The latter has the advantage of establishing a connection to the ML estimator of the mixture model, the former does not need an approximation.

The *MAP approach* resorts to the interpretation of the labels as incidental parameters. We want to decompose the data set in at most g clusters, described by an assignment $1..n \to 1..g$. The unknown mixing probability π appears as a latent distribution on the common "parameter space" $1..g$ of the labels L_i. The MAP approach uses an empirical Bayes procedure declaring the mixing probability π the realization of a random probability $U : \Omega \to \Delta_{g-1}$ on $1..g$. It thus turns to the prior probability of the random labels L_i.

Now, consider $\gamma = (\gamma_1, \ldots, \gamma_g) \in \Gamma^g$ the realization of a random parameter $G : \Omega \to \Gamma^g$. The parameter space of the classification model is $(1..g)^{1..n} \times \Delta_{g-1} \times \Gamma^g$ and the conditional density

$$f_{L,X}[\boldsymbol{\ell}, \mathbf{x} \,|\, U = \boldsymbol{\pi}, G = \boldsymbol{\gamma}],$$

$\boldsymbol{\ell} \in (1..g)^{(1..n)}$, $\boldsymbol{\pi} \in \Delta_{g-1}$, and $\boldsymbol{\gamma} \in \Gamma^g$ is central to the derivation. It is the mixture of a likelihood function w.r.t. $\boldsymbol{\pi}$ and $\boldsymbol{\gamma}$ and a posterior density w.r.t. $\boldsymbol{\ell}$. Assuming conditional independence of (X, G) and U given L and independence of (U, L) and G, Bayesian calculus shows

$$f_{L,X}[\boldsymbol{\ell}, \mathbf{x} \,|\, U = \boldsymbol{\pi}, G = \boldsymbol{\gamma}] = f_X[\mathbf{x} \,|\, L = \boldsymbol{\ell}, U = \boldsymbol{\pi}, G = \boldsymbol{\gamma}] f_L[\boldsymbol{\ell} \,|\, U = \boldsymbol{\pi}, G = \boldsymbol{\gamma}]$$
$$= f_X[\mathbf{x} \,|\, L = \boldsymbol{\ell}, G = \boldsymbol{\gamma}] P[L = \boldsymbol{\ell} \,|\, U = \boldsymbol{\pi}]. \qquad (1.36)$$

This is a weighted likelihood function, the weight $P[L = \boldsymbol{\ell} \,|\, U = \boldsymbol{\pi}]$ appearing as the prior probability of $\boldsymbol{\ell}$. Because of the independence of the labels L_i given U, the weight is

$$P[L = \boldsymbol{\ell} \,|\, U = \boldsymbol{\pi}] = \prod_{i=1}^n P[L_i = \ell_i \,|\, U = \boldsymbol{\pi}] = \prod_{i=1}^n \pi_{\ell_i} = \prod_{j=1}^g \pi_j^{n_j}.$$

Here $n_j = n_j(\boldsymbol{\ell}) = |\boldsymbol{\ell}^{-1}(j)|$ is the number of occurrences of j in $\boldsymbol{\ell}$. If this number is strictly positive, then $C_j = \boldsymbol{\ell}^{-1}(j)$ is a nonempty cluster. The first term in (1.36) is

$$f_{\boldsymbol{\ell},\boldsymbol{\gamma}}(\mathbf{x}) = f_X[\mathbf{x} \,|\, L = \boldsymbol{\ell}, G = \boldsymbol{\gamma}] = \prod_{j=1}^g \prod_{i \in C_j} f_{\gamma_j}(x_i).$$

---

[3] Different from a *nuisance* parameter, that is, one of only secondary interest.

Passing to the logarithm in (1.36), we find the conditional log-density

$$\log f_{L,X}[\boldsymbol{\ell}, \mathbf{x} \,|\, U = \boldsymbol{\pi}, G = \boldsymbol{\gamma}] = \sum_{j=1}^{g} \sum_{i \in C_j} \log f_{\gamma_j}(x_i) + \sum_{j=1}^{g} \log \left(\pi_j^{n_j} \right)$$

$$= \sum_{j=1}^{g} \sum_{i \in C_j} \log(\pi_j f_{\gamma_j}(x_i)) \qquad (1.37)$$

$$= \sum_{j=1}^{g} \sum_{i \in C_j} \log f_{\gamma_j}(x_i) + n \sum_{j=1}^{g} \frac{n_j(\boldsymbol{\ell})}{n} \log \pi_j.$$

The first term on the right does not depend on $\boldsymbol{\pi}$, the second not on $\boldsymbol{\gamma}$; both depend on $\boldsymbol{\ell}$. According to the entropy inequality D.17, the sum on the right assumes its partial maximum w.r.t. $\boldsymbol{\pi}$ at $\pi_j = \frac{n_j(\boldsymbol{\ell})}{n}$. The maximum is the negative entropy –H of the relative frequencies. We have derived the **MAP classification model**

$$\sum_{j=1}^{g} \sum_{i \in C_j} \log f_{\gamma_j}(x_i) - n \cdot \mathrm{H}\left(\frac{n_1(\boldsymbol{\ell})}{n}, \ldots, \frac{n_g(\boldsymbol{\ell})}{n} \right). \qquad (1.38)$$

It depends on the data, the assignments $\boldsymbol{\ell}$, and on the population parameters $\boldsymbol{\gamma} \in \Gamma^g$. We put $\sum_{i \in \emptyset} = 0$ in order to account for empty clusters. For fixed data $\mathbf{x}$, it is called the **MAP criterion**. The criterion is a mixture of an MAP and an ML estimator, MAP w.r.t. $\boldsymbol{\ell}$ and ML w.r.t. $\boldsymbol{\gamma}$. It does not depend on the specific assignment $\boldsymbol{\ell}$ but only on the induced partition. Like the mixture likelihood, the criterion contains information on the correct clustering – if there is one. However, this information does not suffice to extract it. This will need a second constituent and is postponed to Section 4.1; see also Remark 1.23(a). Sometimes the criterion is divided by $-n$; see, for instance, Section 1.3.7. It is then minimized. This concludes the first derivation of the criterion.

The *approximation approach* to the MAP criterion starts from the ML estimates $\boldsymbol{\pi}^* = (\pi_1^*, \ldots, \pi_g^*)$ and $\boldsymbol{\gamma}^* = (\gamma_1^*, \ldots, \gamma_g^*)$ of the mixing rates and population parameters in the associated constrained mixture model. If separation is good, then the posterior probabilities

$$P[L_i = \ell \,|\, U = \boldsymbol{\pi}^*] = \frac{\pi_\ell^* f_{\gamma_\ell^*}(x_i)}{\sum_j \pi_j^* f_{\gamma_j^*}(x_i)}$$

are almost point masses. This means that, for each i, one of the terms $\pi_j^* f_{\gamma_j^*}(x_i)$ in the denominator is dominant. Call its index ℓ_i^*. We estimate

$$\prod_i \sum_j \pi_j^* f_{\gamma_j^*}(x_i) - \varepsilon \leq \prod_i \pi_{\ell_i^*}^* f_{\gamma_{\ell_i^*}^*}(x_i) \leq \max_{\boldsymbol{\ell}, \boldsymbol{\pi}, \boldsymbol{\gamma}} \prod_i \pi_{\ell_i} f_{\gamma_{\ell_i}}(x_i)$$

$$\leq \max_{\boldsymbol{\pi}, \boldsymbol{\gamma}} \prod_i \sum_j \pi_j f_{\gamma_j}(x_i) = \prod_i \sum_j \pi_j^* f_{\gamma_j^*}(x_i).$$

A comparison with Eq. (1.37) shows that the third term in this chain is just the MAP criterion maximized under the constraints. For well separated clusters, the constrained MAP criterion thus almost reproduces the constrained maximum mixture likelihood.

The criteria could now be maximized w.r.t. $\boldsymbol{\ell}$ for each parameter $\boldsymbol{\gamma}$. This would lead to the so-called *best-location problem*. It will be central to the consistency proof of the classification model in Section 1.3.7 below. We take here another approach, dealing with the population parameters first. However, as in the mixture model, it is generally neither possible nor useful to try the optimization w.r.t. the population parameters for all $\boldsymbol{\ell}$. The reasons are quite similar. When a cluster is deficient, the likelihood maximum w.r.t. to its population parameters does not exist. Even if all clusters allow ML estimation of their parameters, the solution may be spurious and useless. In other words, admitting *all* parameters and *all* assignments at the same time is generally not possible. There are four remedies:

(i) Imposing appropriate constraints on the parameters, specified by a submodel $\Psi \subseteq \Gamma^g$. For instance, in the normal and elliptical cases, HDBT scale constraints will turn out to be a *theoretical* tool as effective here as they have proved useful in the mixture context. In Sections 1.3.4 and 1.3.5 below, the HDBT constraints will guarantee a maximum of the criterion on the whole partition space to exist. Moreover, they keep the partitions invariant w.r.t. certain sample space transformations (see Proposition 1.36 and Corollary 1.37 below) and, if suitably chosen, they avoid spurious solutions.

The reader should, however, not receive the impression that the obvious theoretical importance of the HDBT constraints implies their practical importance in an estimation process. This is not the case. The HDBT constraints are just a powerful theoretical tool. They guarantee the existence of the MLE and its consistency, and they are responsible for its stability. (Spurious solutions are extremely unstable.) Moreover, the constraints would pose a major problem in an estimation process since they are unknown. It would not even make sense to compute the *precise* optimum of the criterion under unknown constraints. It is the steady solutions introduced in Section 1.3.3 below that will assume the main rôle in the estimation process in Section 4.1.

(ii) Using submodels such as homoscedasticity or spherical populations. They lead to reasonable results when they conform to the parent or when its components are (very) well separated.

(iii) Restricting the set of partitions by imposing lower bounds on cluster sizes so that the MLE's of the population parameters exist for all clusters. This will be the point of view of Section 1.3.2.

(iv) Forgoing global maxima. In the mixture model, *local* maxima of the likelihood function are most important. In the classification model, the aforementioned steady partitions assume their role. They are of most practical importance.

1.32 Remarks (a) A naïve algorithm based on the cluster criterion would require us to run across all Ψ-admissible partitions, to compute their parameters, and to select a large one. In view of the large number of partitions, this is a formidable task, calling for approximation algorithms. They will be a subject matter of Chapter 3.

(b) When it is a priori known that the mixture is **isobaric**, that is, mixing rates π_j are equal, then the information $\pi = (\frac{1}{g}, \ldots, \frac{1}{g})$ should be introduced in the deduction above. It implies $P[L = \ell \mid U = \pi] = g^{-n}$ for all ℓ, the entropy term turns to a constant and can be dropped, and the ML criterion **ML classification model**

$$\sum_{j=1}^{g} \sum_{i \in C_j} \log f_{\gamma_j}(x_i) \tag{1.39}$$

remains. For fixed $\mathbf{x}$, it is called the **ML criterion**. It tends to equalize cluster sizes and should be applied when cluster sizes are a priori known to be about equal. The entropy term in the MAP criterion (1.38) penalizes uniform cluster sizes, thus counteracting the equalizing tendency of the ML.

Table 1.1 warns us what happens when the ML criterion is applied in a situation where it should not. It shows the MAP and ML cluster criteria (1.38) and (1.39) w.r.t. the heteroscedastic univariate normal model with two clusters, applied to the data set consisting of two clusters of unequal size (and scale) presented in Fig. 1.8. It is known that the optimal partition must be defined by a cut between the two clusters; see Bock [49]. In Table 1.1, cluster means, variances, criteria, and cluster sizes are shown for a large number of cuts. Note that the minimum of the ML criterion is attained for the cut at the value 8.5, whereas that for the MAP criterion is at 1.5. The latter value is reasonable, whereas the former is grossly misleading, as a comparison with the histogram Fig. 1.8 shows.

Table 1.1 *Parameters and negative ML and MAP criteria, Eq. (1.39) and Eq. (1.38), respectively, for two-cluster partitions of the univariate data with the histogram of Fig. 1.8. Each "cut" separates the data set in two clusters.*

cut	mean1	var1	mean2	var2	ML	MAP	n1	n2
0.0	-0.420	0.090	9.656	12.285	2479.13	2927.57	57	1043
0.5	-0.207	0.160	9.916	10.103	2191.95	2790.48	85	1015
1.0	-0.072	0.241	10.055	8.959	2050.80	2721.00	100	1000
1.5	-0.058	0.257	10.063	8.891	2045.77	2720.57	101	999
2.0	-0.058	0.257	10.063	8.891	2045.77	2720.57	101	999
2.5	-0.013	0.357	10.079	8.787	2060.82	2744.74	103	997
3.0	0.089	0.616	10.109	8.604	2085.37	2787.28	107	993
3.5	0.259	1.099	10.150	8.372	2107.99	2836.26	113	987
4.0	0.590	2.030	10.229	7.958	2110.94	2889.85	125	975
4.5	0.842	2.773	10.284	7.705	2109.23	2924.40	134	966
5.0	1.208	3.796	10.366	7.355	2097.07	2965.93	148	952
5.5	1.667	5.007	10.471	6.959	2079.10	3015.97	167	933
6.0	2.294	6.439	10.626	6.436	2048.27	3082.29	197	903
6.5	2.718	7.216	10.747	6.060	2020.58	3124.23	221	879
7.0	3.214	8.088	10.893	5.677	1999.34	3183.33	252	848
7.5	3.840	8.986	11.101	5.203	1977.09	3262.24	298	802
8.0	4.387	9.598	11.313	4.787	1963.23	3333.15	346	754
8.5	4.901	10.056	11.543	4.390	1958.05	3398.99	399	701
9.0	5.440	10.434	11.829	3.957	1963.02	3460.93	464	636
9.5	5.831	10.705	12.063	3.658	1981.88	3502.84	517	583
10.0	6.278	11.021	12.366	3.329	2022.17	3542.89	584	516
10.5	6.634	11.300	12.639	3.087	2073.10	3567.10	642	458
11.0	7.153	11.732	13.124	2.710	2173.71	3571.72	735	365
11.5	7.458	12.025	13.482	2.435	2247.16	3547.86	794	306
12.0	7.731	12.354	13.855	2.164	2326.44	3510.43	848	252
12.5	7.980	12.751	14.234	1.941	2418.11	3470.18	897	203
13.0	8.179	13.132	14.586	1.754	2502.45	3428.94	936	164
13.5	8.390	13.597	15.046	1.482	2598.27	3368.93	977	123
14.0	8.527	13.951	15.413	1.238	2664.17	3320.46	1003	97
14.5	8.661	14.375	15.793	1.053	2741.31	3278.40	1027	73
15.0	8.760	14.733	16.113	0.929	2804.40	3247.00	1044	56
15.5	8.882	15.239	16.580	0.821	2891.12	3208.14	1064	36
16.0	8.965	15.616	17.054	0.654	2950.21	3173.63	1077	23
16.5	9.011	15.847	17.447	0.428	2981.55	3148.69	1084	16
17.0	9.054	16.093	17.829	0.287	3015.97	3129.89	1090	10
17.5	9.077	16.235	18.051	0.238	3036.40	3121.15	1093	7
18.0	9.108	16.446	18.534	0.132	3065.64	3107.06	1097	3
18.5	9.116	16.503	18.780	0.018	3070.37	3099.61	1098	2

(c) Bock [49, 51] discusses Bayesian models for various prior measures on the set $\mathcal{C}$ of all partitions. For one of his models he requires that all partitions in g clusters be equally likely. He obtains again an entropy term as in the MAP criterion but with the opposite sign! This criterion favors equally sized clusters even more than the ML criterion does. The criteria depend heavily on prior measures on the partitions $\mathcal{C}$.

1.3.2 *Admissibility and size constraints*

The MAP criterion (1.38) does not yet make sense per se. It has to be filled with life by special conditions so that assignments and parameters fit each other. A combination of 1.3.1(i)–(iv) will now be formulated in order to keep the flexibility needed. So, let the

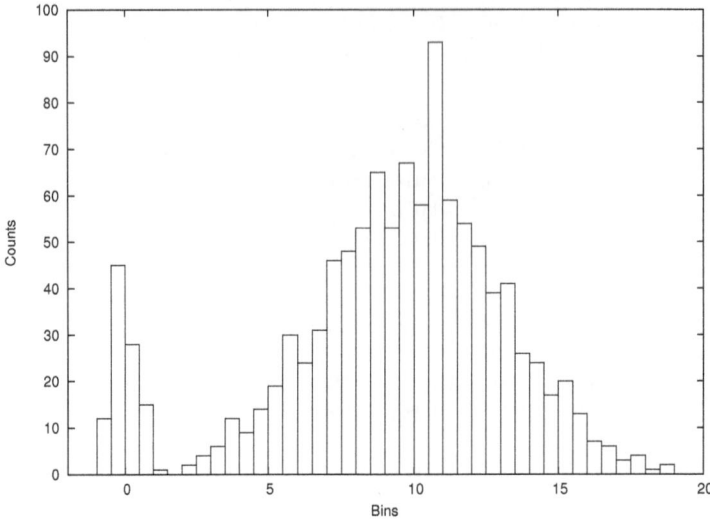

Figure 1.8 *Histogram of 1100 simulated data points, 100 from $N_{0,1/4}$ and 1000 from $N_{10,9}$; see also Table 1.1.*

subset $\Psi \subseteq \Gamma^g$ define a submodel. The main example is provided by the HDBT constraints applicable to elliptical families. We define the following.

1.33 Definition Let a data set $\mathbf{x}$ and a submodel $\Psi \subseteq \Gamma^g$ be given. An assignment $\boldsymbol{\ell}$ that admits the joint ML estimate $\boldsymbol{\gamma}(\boldsymbol{\ell}) \in \Psi$ is called Ψ-**admissible**.

In other words, if $\boldsymbol{\ell}$ is Ψ-admissible, then the maximum of the MAP criterion (1.38) and its ML counterpart (1.39) exist in Ψ. There are two extreme cases. At one end, we admit *all assignments* in $1 .. g^{(1 \cdots n)}$. Besides mild conditions on the data and on n, this is possible under appropriate cross-cluster constraints. For instance, if data points are pairwise distinct, then $n \geq d+1$ data points suffice in a homoscedastic, spherical normal model. The maximizer is then the *joint*, Ψ-constrained ML estimate $\boldsymbol{\gamma}(\boldsymbol{\ell}) = (\gamma_1(\boldsymbol{\ell}), \ldots, \gamma_g(\boldsymbol{\ell}))$ of $\boldsymbol{\gamma} \in \Gamma^g$ given $\boldsymbol{\ell}$.

At the other end, a method that allows unconstrained parameter estimation, $\Psi = \Gamma^g$, with the cluster criterion (1.38) is also desirable. It restricts the set of admissible assignments $\boldsymbol{\ell}$. Let us take a closer look at this case. Many basic models allow estimation of their parameters when there is at least a minimum number b of data points. The normal model, for instance, takes $b = d + 1$ if data is in general position. This is the minimum number needed for scatter matrices to be regular. Thus, a way of forcing ML estimates $\boldsymbol{\gamma}(\boldsymbol{\ell})$ even with $\Psi = \Gamma^g$ is the application of *size constraints* to all clusters. However, size constraints destroy the conditional independence of the labels L_i given U. The derivation of the entropy term in the MAP criterion in Section 1.3.1 is no longer exact. This poses, however, no big problem in practice.

By dynamic optimization, an assignment $\boldsymbol{\ell}$ is admissible w.r.t. the whole product $\Psi = \Gamma^g$ if and only if the products $\prod_{i \in C_j} f_\gamma(x_i)$ have maxima for all j, that is, if the individual ML estimates of γ exist w.r.t. all C_j. It follows then

$$\max_{(\gamma_1, \ldots, \gamma_g)} \prod_{j=1}^{g} \prod_{i \in C_j} f_{\gamma_j}(x_i) = \prod_{j=1}^{g} \max_{\gamma \in \Gamma} \prod_{i \in C_j} f_\gamma(x_i) = \prod_{j=1}^{g} \prod_{i \in C_j} f_{\gamma_{C_j}}(x_i).$$

Thus, each parameter γ_j in the MAP criterion (1.38) may be estimated in its own factor Γ separately. The ML estimate of $\gamma \in \Gamma$ w.r.t. cluster C_j is $\gamma_j(\boldsymbol{\ell})$ and the MAP criterion becomes

$$\sum_{j=1}^{g}\sum_{i\in C_j} \log f_{\gamma_j(\boldsymbol{\ell})}(x_i) - n\cdot \mathrm{H}\Big(\tfrac{n_1(\boldsymbol{\ell})}{n},\ldots,\tfrac{n_g(\boldsymbol{\ell})}{n}\Big).$$

Admissibility depends on the model and on the data. There are even data sets for which no admissible assignment exists. On the left of Fig. 1.9 there exists a Γ^2-admissible assignment

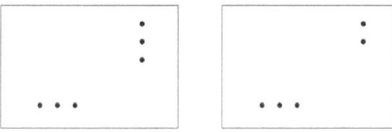

w.r.t. the fully normal basic model $\Gamma = \mathbb{R}^2 \times \mathrm{PD}(2)$, although the obvious partition is not admissible. The two linear clusters do not admit ML estimation of their covariance matrices. For the data set on the right an assignment admissible w.r.t. this model does not even exist, since one of the clusters would have ≤ 2 elements. The diagonally normal model does not help

Figure 1.9: *Nonadmissibility.*

either. It is only the spherically normal model that is successful in bot cases and supplies the expected partitions. It is interesting to characterize a collection of data sets

(i) that has probability one for the parent distribution and

(ii) for which admissibility does not depend on the special data but rather on the numbers d, n, and n_j, only.

Answers for elliptical and normal families will be given in Sections 1.3.4 and 1.3.5. Algorithms for size-constrained clustering will be a subject matter of Chapter 3.

1.3.3 Steady partitions

We have seen in Section 1.2.2 that local maxima of the likelihood function play a central rôle in mixture analysis. The classification model knows an analogous concept: *steady partitions,* called "éléments non biaisés" or "points fixes" in Diday [123] and "éléments stables" in Schroeder [459]. Some of them are promising candidates for the desired solution. Like local likelihood maxima, they are not unique, and the question of how to choose between them will be dealt with in Section 4.1. See also Remark 1.23(a). The following defines an analog of local maxima in mixture analysis.

1.34 Definition

(a) An index $i \in 1..n$ is called a **misfit** in a Ψ-admissible assignment $\boldsymbol{\ell}$ if there exists $j \neq \ell_i$, $C_j \neq \emptyset$, such that

$$\begin{aligned} |C_j|\, f_{\gamma_j(\boldsymbol{\ell})}(x_i) &> |C_{\ell_i}|\, f_{\gamma_{\ell_i}(\boldsymbol{\ell})}(x_i), \quad \text{(MAP)} \\ f_{\gamma_j(\boldsymbol{\ell})}(x_i) &> f_{\gamma_{\ell_i}(\boldsymbol{\ell})}(x_i). \qquad\quad \text{(ML)} \end{aligned} \tag{1.40}$$

That is, the Bayesian or ML discriminant rule does not assign i to ℓ_i. (Here, $(\gamma_j(\boldsymbol{\ell}))_j$ is the Ψ-constrained ML estimate w.r.t. $\boldsymbol{\ell}$.)

(b) A Ψ-admissible assignment and its associated partition are called **steady** for the MAP (ML) criterion (w.r.t. the data set $\mathbf{x}$) if there is no misfit.

Note that the parameters remain unchanged when an index is checked for misfit. The condition for steadiness is $|C_j|\, f_{\gamma_j(\boldsymbol{\ell})}(x_i) \leq |C_{\ell_i}|\, f_{\gamma_{\ell_i}(\boldsymbol{\ell})}(x_i)$ (MAP) and $f_{\gamma_j(\boldsymbol{\ell})}(x_i) \leq f_{\gamma_{\ell_i}(\boldsymbol{\ell})}(x_i)$ (ML) for all i and all j.

There is a simple and natural undirected graph structure on the set $(1..g)^{(1..n)}$ of all assignments. It arises from linking two assignments if they differ at exactly one site (index) $i \in 1..n$. This is the usual **spin flip** known from statistical mechanics. Steady assignments and partitions must *not* be confused with locally optimal ones w.r.t. this graph structure. In order to test an assignment for local optimality, the parameters n_j and γ_j would have to be recomputed after each flip, whereas they remain unchanged when tested for steadiness. This is different.

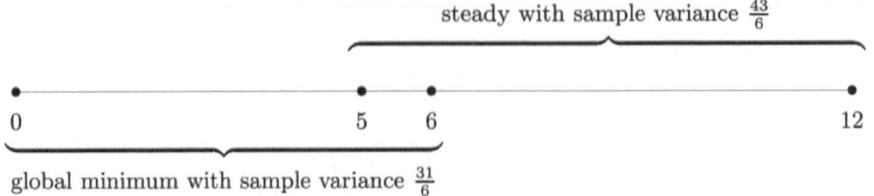

Figure 1.10 *A nonoptimal steady partition for the ML criterion* (1.39) *with the homoscedastic normal model and two clusters.*

If the basic model is spherically normal with common covariance matrix vI_d, the conditions (1.40) for steadiness can be much simplified. Consider the ML case. We have $\gamma_j = (m_j, vI_d)$ and $2\log f_{m_j,vI_d}(x) = -d\log 2\pi v - \|x - m_j\|^2/v$. Since v does not depend on j, the left-hand side of the ML part of condition (1.40) reduces to $-\|x_i - m_j\|^2$. Thus, i is a misfit in ℓ if *the cluster center closest to x_i is not that of its own cluster.* In other words, ℓ is steady if each data point is closest to the center of its own cluster. This justifies the name "steady." In this special case a partition is steady if it is the Voronoi decomposition of its centers; see the literature on algorithmic geometry, for instance, Boissonnat and Yvinec [53].

To appreciate the concept of steadiness, we take another look at optimal solutions. See also Mardia et al. [345], p. 362.

1.35 Proposition *Assume that the joint likelihood function $f(\mathbf{x}; \ell, \boldsymbol{\gamma})$ does not vanish for all Ψ-admissible ℓ and all $\boldsymbol{\gamma} \in \Psi$. Let ℓ^* be Ψ-admissible and maximal w.r.t. the MAP (or ML) criterion. If all assignments obtained from ℓ^* by spin flips are again Ψ-admissible, then ℓ^* is steady.*

PROOF. Rewrite the MAP criterion (1.38) as $n^{-n} \prod_i n_{\ell_i}(\ell) f_{\gamma_{\ell_i}(\ell)}(x_i)$. By assumption, it is > 0 for some ℓ. An optimal ℓ^* is optimal w.r.t. any spin flip and $\boldsymbol{\gamma} \in \Psi$. If ℓ^* contained a misfit i, then i could be relabeled to achieve a pair $(\ell', \boldsymbol{\gamma}(\ell^*))$ (different assignment but same parameters) with a Ψ-*admissible* assignment ℓ' and a larger value of the criterion. (Note that parameters $\gamma_j(\ell^*)$ are also defined for empty clusters j because of Ψ-admissibility. Thus, the criterion is defined even if the misfit was moved to an empty cluster.) This would contradict optimality. The proof for the ML criterion is similar.   □

The converse is not true in general, that is, a steady assignment need not be maximal. A counterexample in the ML case is shown in Fig. 1.10. The univariate data set $\{0, 5, 6, 12\}$ is to be decomposed in two clusters w.r.t. the normal model with unknown locations m_1, m_2 and unknown, common variance v. This case was discussed in the paragraph preceding Proposition 1.35. The negative log-likelihood function of an assignment ℓ with partition $\{C_1, C_2\}$ is $2\log 2\pi v + \frac{1}{2v}\left(\sum_{i \in C_1}(x_i - m_1)^2 + \sum_{i \in C_2}(x_i - m_2)^2\right)$; see (1.39). It is well known and easily seen that the ML estimates of the parameters are the sample means $m_j(\ell) = \overline{x}_j$ and the pooled sample variance $v(\ell) = s(\ell)$. The related value of the negative log-likelihood function is $2(1 + \log 2\pi + \log s(\ell))$. Its minimum is attained for $C_1^* = \{0, 5, 6\}$ and $C_2^* = \{12\}$ and we have $s(\ell^*) = 31/6$. The assignment ℓ with partition $C_1 = \{0\}$, $C_2 = \{5, 6, 12\}$ is suboptimal and steady. Indeed, $v(\ell) = s(\ell) = 43/6$. Further, the simplification in the aforementioned paragraph shows that ℓ is indeed steady.

In the following, we will focus on constrained or unconstrained steady partitions rather than on maxima. The standard iterative relocation algorithm for computing steady partitions will be a subject matter of Chapter 3.

1.3.4   Elliptical models

In applications to basic elliptical shape, location, and scale families, submodels Ψ needed in the MAP classification model (1.38) can be specified by constraints *between* scale parameters that leave it sufficiently rich, in particular affine equivariant. As in the mixture model, the HDBT constraints (1.27) can and will be used as a theoretical tool here, too. Simple additional bounds on the size of the data and on the decay of the density generator ensure admissibility. We have the following proposition for HDBT constrained elliptical shape, location, and scale families.

1.36 Proposition (constrained elliptical MAP criterion) *Let $n \geq gd + 1$ and let the data be in general position. Assume an elliptical shape, location, and HDBT constrained scale family with an equicontinuous, pointwise closed set Φ of density generators φ such that $\varphi(t) \leq C(1 + t)^{-\beta/2}$, $t \geq 0$, for some $\beta > \frac{n}{n - gd}(d + 1)d$.*

(a) The ML estimates $\varphi_j(\ell)$, $m_j(\ell)$, and $\mathbf{V}(\ell)$ of the population parameters $\varphi_j \in \Phi$, $m_j \in \mathbb{R}^d$, $1 \leq j \leq g$, and $\mathbf{V} \in \mathcal{V}_c$, respectively, exist for all $\ell \in (1..g)^{(1 \cdots n)}$. That is, all assignments are admissible w.r.t. the HDBT constrained elliptical model.

(b) The MAP criterion (1.38) turns to the **elliptical MAP criterion**

$$\sum_{j=1}^{g} \left\{ -\tfrac{n_j(\ell)}{2} \log \det V_j(\ell) + \sum_{i \in C_j} \log \varphi_j(\ell)\big((x_i - m_j(\ell))^\top V_j(\ell)^{-1}(x_i - m_j(\ell))\big) \right\}$$

$$- n \cdot \mathrm{H}\big(\tfrac{n_1(\ell)}{n}, \ldots, \tfrac{n_g(\ell)}{n}\big).$$

Here, $n_j(\ell) = |C_j|$.

PROOF. (a) Let $\ell \in (1..g)^{(1 \cdots n)}$. As in the first paragraph of the proof of Proposition 1.24, one shows that optimal means m_j are contained in the convex hull $\mathrm{conv}\,\mathbf{x}$. Moreover, for all φ, $\mathbf{m}$, and $\mathbf{V}$, the classification likelihood (1.39) assumes the form

$$f_{\ell,\varphi,\mathbf{m},\mathbf{V}}(\mathbf{x}) = \prod_i \sqrt{\det \Lambda_{\ell_i}}\, \varphi_{\ell_i}\big((x_i - m_{\ell_i})^\top \Lambda_{\ell_i}(x_i - m_{\ell_i})\big)$$

$$\leq g^n \prod_i \tfrac{1}{g} \sum_{1 \leq j \leq g} \sqrt{\det \Lambda_j}\, \varphi_j\big((x_i - m_j)^\top \Lambda_j(x_i - m_j)\big) = g^n f_n(\mathbf{x}; \boldsymbol{\pi}, \boldsymbol{\varphi}, \mathbf{m}, \mathbf{V})$$

with the uniform $\boldsymbol{\pi}$. By Lemma 1.21, there exist $k = \lceil \frac{n - gd}{d + 1} \rceil$ disjoint subsets $I_1 \ldots, I_k \subseteq 1..n$ of size $d + 1$ such that we have for all j the estimate

$$f_n(\mathbf{x}; \boldsymbol{\pi}, \boldsymbol{\varphi}, \mathbf{m}, \mathbf{V}) \leq \mathrm{const} \cdot (\det \Lambda_j)^{n/2} \prod_{p=1}^{k} \big(1 + c\,\mathrm{tr}\,(S_{I_p}\Lambda_j)\big)^{-\beta/2}.$$

Now, let κ be the smallest eigenvalue of all scatter matrices of $d+1$-element subsets of $\mathbf{x}$. By general position we have $\kappa > 0$. The geometric-arithmetic inequality A.5 finally yields the two estimates

$$f_{\ell,\varphi,\mathbf{m},\mathbf{V}}(\mathbf{x}) \leq \mathrm{const} \cdot (\det \Lambda_j)^{n/2}\big(\mathrm{tr}\,\Lambda_j\big)^{-k\beta/2} \tag{1.41}$$

$$\leq \mathrm{const} \cdot (\mathrm{tr}\,\Lambda_j)^{(nd - k\beta)/2}. \tag{1.42}$$

From

$$\beta > \frac{nd}{(n - gd)/(d + 1)} \geq \frac{nd}{k}$$

and (1.42), it follows that $f_{\ell,\varphi,\mathbf{m},\mathbf{V}}(\mathbf{x})$ vanishes as some eigenvalue of Λ_j converges to ∞. From (1.41), it follows that $f_{\ell,\varphi,\mathbf{m},\mathbf{V}}(\mathbf{x})$ vanishes as some eigenvalue of Λ_j converges to zero while all eigenvalues remain bounded. We have thus shown that $f_{\ell,\varphi,\mathbf{m},\mathbf{V}}(\mathbf{x})$ converges to zero as $\mathbf{V}$ approaches the Alexandrov point of $\mathcal{V}_c$. Since we consider m_j in the compact set $\mathrm{conv}\,\mathbf{x}$, only, and since Φ is compact, this means that the continuous function $(\varphi, \mathbf{m}, \mathbf{V}) \mapsto f_{\ell,\varphi,\mathbf{m},\mathbf{V}}(\mathbf{x})$ vanishes at infinity and has thus a maximum. This is part (a) of the proposition. Part (b) follows from the MAP criterion (1.38) with $\Gamma = \Phi \times \mathbb{R}^d \times \mathrm{PD}(d)$, $\Psi = \Phi^g \times \mathbb{R}^{gd} \times \mathcal{V}_c$, and $\gamma = (\varphi, \mathbf{m}, \mathbf{V})$.   □

A standard robust instance for Proposition 1.36 is the HDBT constrained Pearson Type-VII family; see Appendix E.22. There is the following corollary.

1.37 Corollary (Constrained Pearson Type-VII criterion) *Let $n \geq gd+1$ and let the data $\mathbf{x}$ be in general position. Assume an HDBT constrained extended Pearson Type-VII model with index $\geq \beta$ for some $\beta > \frac{n}{n-gd}(d+1)d$.*

(a) The ML estimates $\eta_j(\boldsymbol{\ell})$, $m_j(\boldsymbol{\ell})$, and $\mathbf{V}(\boldsymbol{\ell})$ of the population parameters $\eta_j \in [\beta, \infty]$, $m_j \in \mathbb{R}^d$, $1 \leq j \leq g$, and $\mathbf{V} \in \mathcal{V}_c$, respectively, exist for all $\boldsymbol{\ell} \in (1 \mathbin{..} g)^{(1 \mathbin{..} n)}$.

(b) The MAP criterion for an assignment $\boldsymbol{\ell}$ assumes the form

$$-\sum_{j=1}^{g}\left\{\tfrac{n_j(\boldsymbol{\ell})}{2}\log\det V_j(\boldsymbol{\ell}) + h(\mathbf{x}; \eta_j(\boldsymbol{\ell}), m_j(\boldsymbol{\ell}), V_j(\boldsymbol{\ell}))\right\} - n\cdot \mathrm{H}\left(\tfrac{n_1(\boldsymbol{\ell})}{n}, \ldots, \tfrac{n_g(\boldsymbol{\ell})}{n}\right).$$

Here $n_j(\boldsymbol{\ell}) = |C_j|$ and

$$h(\mathbf{x}; \eta, m, V) = \begin{cases} \frac{\eta}{2}\sum_{i \in C_j}\log\left(1 + \frac{1}{\eta}(x_i - m)^\top V^{-1}(x_i - m)\right), & \eta < \infty, \\ \frac{1}{2}\sum_{i \in C_j}(x_i - m)^\top V^{-1}(x_i - m), & \eta = \infty. \end{cases}$$

Since the constraints in Proposition 1.36 and its corollary affect neither the scale nor the location parameters, both $\varphi_j(\boldsymbol{\ell})$ and $m_j(\boldsymbol{\ell})$ are *unconstrained* maximizers. However, they depend on the *constrained* maximizer of $\mathbf{V}$. It is not in general possible to apply proposition and corollary in practice since an HDBT constant is unknown. Even if it were known, the constrained optimization in their parts (a) would not be straightforward. Instead of forcing parameter constraints it is preferable to use admissible partitions realized by constraints on cluster sizes and unconstrained steady partitions; see Sections 1.3.2, 1.3.3, and 4.1.

1.38 Corollary (Unconstrained elliptical MAP criterion) *Let the data be in general position. Assume an elliptical shape, location, and* unconstrained *scale family with an equicontinuous, pointwise closed set Φ of density generators φ such that $\varphi(t) \leq C(1+t)^{-\beta/2}$, $t \geq 0$, for some index $\beta > \frac{n}{n-d}(d+1)d$. Then an assignment $\boldsymbol{\ell}$ is admissible if and only if $n_j(\boldsymbol{\ell}) \geq d+1$ for all j. The criterion equals formally that shown in Proposition 1.36(b).*

PROOF. An application of Proposition 1.24 or Proposition 1.36(a) applied with $g = 1$ shows that parameters can be estimated separately for any cluster j with $n_j(\boldsymbol{\ell}) \geq d+1$. Note that the HDBT constraints are void if $g = 1$. □

1.3.5 Normal models

In Section 1.3.7 below it will be shown that the HDBT constrained optimum of elliptical criteria is consistent. The optimization of the location parameters is unconstrained since they are not affected by the constraints. In the context of *normal* models, optimization w.r.t. the location parameters can even be performed analytically. This far-reaching simplification is due to the algebraic structure of the exponential function and to Steiner's formula A.11. Let us therefore rewrite the MAP and ML criteria (1.38) and (1.39) for the six hetero- and homoscedastic normal models with full, diagonal, and spherical covariance matrices after partial maximization w.r.t. to the location parameters. It is appropriate to first discuss admissibility here. The following lemma states conditions for a given partition to be admissible. Its proof uses Proposition F.11.

1.39 Lemma *A partition is admissible if its clusters satisfy the following conditions depending on the* normal *submodels.*

(a) Normal models without cross-cluster constraints.

   (i) Full covariance matrices: It is necessary, and sufficient, that each cluster contain (at least) $d+1$ points in general position.

    (ii) Diagonal covariance matrices*: It is necessary, and sufficient, that each cluster contain at least two distinct entries in each coordinate.*

    (iii) Spherical covariance matrices*: It is necessary, and sufficient, that each cluster contains at least two distinct points.*

(b) Normal submodels subject to HDBT constraints.

    (i) Full covariance matrices*: It is sufficient that one cluster contain (at least) $d+1$ points in general position. If data is in general position, then $n \geq gd + 1$ suffices.*

    (ii) Diagonal covariance matrices*: It is necessary, and sufficient, that, for each coordinate, there is a cluster with two distinct entries there.*

    (iii) Spherical covariance matrices*: It is necessary, and sufficient, that one cluster contain at least two distinct points. If all data points are pairwise distinct, then $n \geq g + 1$ is sufficient.*

Besides general position, the conditions for full and spherical scales in the lemma depend only on cluster sizes. In the diagonal cases they also depend on the data.

It is interesting to ask for conditions that guarantee an admissible partition of a data set in g clusters to *exist*. They can be derived from the lemma. If there are no parameter constraints between clusters, the size $g(d+1)$ is necessary and sufficient in the full case if data is in general position. The number $2g$ is necessary and sufficient in the spherical case if data points are pairwise distinct.

If the model is HDBT constrained, in particular homoscedastic, there are simple sufficient conditions on the number of data points for *all* assignments to be admissible. If covariance matrices are full or diagonal, it reads $n \geq gd + 1$; if it is spherical, the condition is $n \geq g + 1$. This is a consequence of the pigeon hole principle of combinatorics.[4]

The derivation of the optimal mixing rates $\boldsymbol{\pi}$ in the MAP criterion (1.38) was possible analytically. It led to the entropy term. In the normal cases further far-reaching simplifications are possible since optimization w.r.t. some or all population parameters, too, can be performed analytically. In the following theorems, a bar indicates sample means and the letters W and S denote within-groups SSP and scatter matrices, respectively. Thus, $\overline{x}_j(\boldsymbol{\ell}) = \sum_{i \in C_j} x_i$ is the sample mean of cluster j, $W_j(\boldsymbol{\ell}) = \sum_{i \in C_j}(x_i - \overline{x}_j(\boldsymbol{\ell}))(x_i - \overline{x}_j(\boldsymbol{\ell}))^\top$ $(S_j(\boldsymbol{\ell}) = W_j(\boldsymbol{\ell})/n_j)$ is its SSP (scatter) matrix, and $W(\boldsymbol{\ell}) = \sum_j W_j$ $(S(\boldsymbol{\ell}) = W(\boldsymbol{\ell})/n)$ is the pooled within-groups SSP (scatter) matrix, all w.r.t. the assignment $\boldsymbol{\ell}$. Note that the pooled matrices differ from the total ones, that is, those of the whole data set.

The first three criteria concern variable scale matrices, that is, the heteroscedastic cases. Since steady solutions are of major practical interest, we restrict attention to assignments admissible without parameter constraints between clusters; see Lemma 1.39.

1.40 Theorem (Unconstrained heteroscedastic, normal MAP criteria)
The following heteroscedastic MAP criteria arise from the MAP criterion (1.38) after partial maximization w.r.t. the location parameters.

(a) Fully normal case: The **heteroscedastic MAP determinant criterion**

$$-\tfrac{1}{2}\sum_{j=1}^{g} n_j(\boldsymbol{\ell}) \log \det S_j(\boldsymbol{\ell}) - n \cdot \mathrm{H}\big(\tfrac{n_1(\boldsymbol{\ell})}{n}, \ldots, \tfrac{n_g(\boldsymbol{\ell})}{n}\big).$$

The ML estimates of the scale parameters V_j, $1 \leq j \leq g$, for a steady solution $\boldsymbol{\ell}^*$ are $V_j^* = S_j(\boldsymbol{\ell}^*)$.

(b) Diagonally normal case:

$$-\tfrac{1}{2}\sum_{j} n_j(\boldsymbol{\ell}) \log \prod_{k=1}^{d} S_j(\boldsymbol{\ell})(k,k) - n \cdot \mathrm{H}\big(\tfrac{n_1(\boldsymbol{\ell})}{n}, \ldots, \tfrac{n_g(\boldsymbol{\ell})}{n}\big).$$

---

[4]It states that there is no one-to-one map from a $(g+1)$- to a g-element set.

The ML estimates of the diagonal variances $v_{j,k}$, $1 \leq j \leq g$, $1 \leq k \leq d$, for a steady solution $\boldsymbol{\ell}^*$ are $v_{j,k}^* = S_j(\boldsymbol{\ell}^*)(k,k)$.

(c) Spherically normal case: The **heteroscedastic MAP trace criterion**

$$-\frac{d}{2} \sum_j n_j(\boldsymbol{\ell}) \log \frac{1}{d} \text{tr}\, S_j(\boldsymbol{\ell}) - n \cdot \text{H}\Big(\frac{n_1(\boldsymbol{\ell})}{n}, \ldots, \frac{n_g(\boldsymbol{\ell})}{n}\Big).$$

The ML estimates of the variances v_j, $1 \leq j \leq g$, for a steady solution $\boldsymbol{\ell}^*$ are $v_j^* = \frac{1}{d}\text{tr}\, S_j(\boldsymbol{\ell}^*) = \frac{1}{n_j(\boldsymbol{\ell}^*)d} \sum_{i \in C_j^*} \|x_i - \overline{x}_j(\boldsymbol{\ell}^*)\|^2$.

(d) In all cases, the ML estimates of the means m_j and the mixing rates π_j, $1 \leq j \leq g$, are $m_j^* = \overline{x}_j(\boldsymbol{\ell}^*)$ and $\pi_j^* = \frac{n_j(\boldsymbol{\ell}^*)}{n}$, respectively.

PROOF. Apply the MAP criterion (1.38) to the three models. According to Proposition F.11(a), the ML estimates in case (a) exist if and only if all scatter matrices are regular. By Corollary F.15, this is the case if and only if each cluster spans a $d+1$-dimensional affine space. The estimates are $m_j = \overline{x}_j(\boldsymbol{\ell})$ and $V_j = S_j(\boldsymbol{\ell})$. The resulting value of the log-likelihood function is $-\frac{n_j(\boldsymbol{\ell})}{n}\{d(\log 2\pi + 1) - \log \det S_j(\boldsymbol{\ell})\}$. Parts (a) and (d) follow. Part (b) follows in the same way from Proposition F.11(b) and part (c) from Proposition F.11(c). □

The previous cluster criteria discriminate between clusters by location, scale, or both. The following ones assume a common covariance matrix. That is, the covariance matrices are HDBT constrained with constant $c = 1$. The MAP criterion (1.38) applies again. Some of the related ML criteria are classical and had been put forward on an intuitive basis by early authors; see Remark 1.42(a).

1.41 Theorem (Homoscedastic normal MAP criteria) *The following homoscedastic MAP criteria arise from the MAP criterion (1.38) after partial maximization w.r.t. the location parameters.*

(a) Fully normal case: The **MAP pooled determinant criterion**

$$-\frac{n}{2} \log \det S(\boldsymbol{\ell}) - n \cdot \text{H}\Big(\frac{n_1(\boldsymbol{\ell})}{n}, \ldots, \frac{n_g(\boldsymbol{\ell})}{n}\Big).$$

The ML estimate of the common scale parameter V for a steady solution $\boldsymbol{\ell}^*$ is $V^* = S(\boldsymbol{\ell}^*)$.

(b) Diagonally normal case:

$$-\frac{n}{2} \log \prod_{k=1}^{d} S(\boldsymbol{\ell})(k,k) - n \cdot \text{H}\Big(\frac{n_1(\boldsymbol{\ell})}{n}, \ldots, \frac{n_g(\boldsymbol{\ell})}{n}\Big).$$

The ML estimates of the common variances $v_1, \ldots, v_d$ for a steady solution $\boldsymbol{\ell}^*$ are $v_k^* = S(\boldsymbol{\ell}^*)(k,k)$.

(c) Spherically normal case: The **MAP pooled trace criterion**

$$-\frac{nd}{2} \log \frac{1}{d} \text{tr}\, S(\boldsymbol{\ell}) - n \cdot \text{H}\Big(\frac{n_1(\boldsymbol{\ell})}{n}, \ldots, \frac{n_g(\boldsymbol{\ell})}{n}\Big).$$

The ML estimate of the common variance v for a steady solution $\boldsymbol{\ell}^*$ is $v^* = \frac{1}{d}\text{tr}\, S(\boldsymbol{\ell}^*) = \frac{1}{nd} \sum_{j=1}^{g} \sum_{i \in C_j} \|x_i - \overline{x}_j(\boldsymbol{\ell}^*)\|^2$.

(d) In all cases, the ML estimates of the means m_j and of the mixing rates π_j, $1 \leq j \leq g$, are $m_j^* = \overline{x}_j(\boldsymbol{\ell}^*)$ and $\pi_j^* = \frac{n_j(\boldsymbol{\ell}^*)}{n}$.

PROOF. Apply the MAP criterion (1.38) with HDBT constant $c = 1$ to the three models. We have to maximize the classification likelihood (1.39)

$$f_{\boldsymbol{\ell},\mathbf{m},V}(\mathbf{x}) = \prod_j \prod_{i \in C_j} N_{m_j,V}(x_i)$$

w.r.t. $\mathbf{m}$ and V. After an application of Steiner's formula A.11, its negative logarithm becomes

$$\frac{n}{2}\log\det 2\pi V + \frac{1}{2}\sum_j\sum_{i\in C_j}(x_i - m_j)^\top V^{-1}(x_i - m_j)$$

$$= \frac{n}{2}\log\det 2\pi V + \frac{1}{2}\sum_j\left\{\operatorname{tr}(\Lambda W_j(\boldsymbol{\ell})) + n_j(\overline{x}_j(\boldsymbol{\ell}) - m_j)^\top\Lambda(\overline{x}_j(\boldsymbol{\ell}) - m_j)\right\}.$$

Up to the additive constant $\frac{nd}{2}\log 2\pi$ this is

$$\frac{n}{2}\left\{\log\det V + \operatorname{tr}\left(V^{-1}S(\boldsymbol{\ell})\right)\right\} + \sum_j\frac{n_j}{2}(\overline{x}_j(\boldsymbol{\ell}) - m_j)^\top V^{-1}(\overline{x}_j(\boldsymbol{\ell}) - m_j).$$

The claims will follow from dynamic optimization G.9. The minimum w.r.t. m_j is attained at $\overline{x}_j(\boldsymbol{\ell})$. It remains to minimize the expression in curly brackets w.r.t. V. First note that the conditions along with Corollary F.15 imply that the SSP matrix of one cluster, say $W_j(\boldsymbol{\ell})$, is regular in all three cases. Therefore, so is $W(\boldsymbol{\ell}) \succeq W_j(\boldsymbol{\ell})$. Now distinguish between the three cases. In case (a), according to Lemma F.10, the minimum exists since the pooled scatter matrix $S(\boldsymbol{\ell})$ is regular and the minimizer is $S(\boldsymbol{\ell})$. In case (b), $V = \operatorname{Diag}(v_1, \ldots, v_d)$ and V^{-1} are diagonal and hence

$$\log\det V + \operatorname{tr}\left(V^{-1}S(\boldsymbol{\ell})\right)$$

$$= \log\det V + \operatorname{tr}\left(V^{-1}\operatorname{Diag}(S(\boldsymbol{\ell})(1,1), \ldots, S(\boldsymbol{\ell})(d,d))\right) = \sum_k\left(\log v_k + \frac{S(\boldsymbol{\ell})(k,k)}{v_k}\right).$$

The minimum can be determined in each summand separately. It is attained at $v_k = S(\boldsymbol{\ell})(k,k)$. In case (c), $V = vI_d$ is spherical and $\log\det V + \operatorname{tr}\left(V^{-1}S(\boldsymbol{\ell})\right) = d\log v + \operatorname{tr}S(\boldsymbol{\ell})/v$. The minimizer is $v = \operatorname{tr}S(\boldsymbol{\ell})/d$. Finally, it is sufficient to insert these values in the likelihood function to obtain the three criteria. $\qquad\square$

1.42 Remarks (a) The criteria contain the constants d and n as they appear in the likelihood function. This enables us to compare the values across the various models. Some of them could be dropped when this is not requested. For instance, the ML version of the pooled trace criterion reduces to $\operatorname{tr}S(\boldsymbol{\ell})$ (to be minimized). This is the **sum-of-squares criterion** intuitively obtained by Ward [522]. A similar criterion results when the common cluster scale is known up to a factor. In the same way, the ML version of the pooled determinant criterion reduces to $\det S(\boldsymbol{\ell})$ (again to be minimized). Friedman and Rubin [172] have been the first authors to detect this criterion, again on heuristic grounds.

(b) Friedman and Rubin's criterion should be used only when the parent mixture is expected to be not too far away from homoscedastic and mixing rates about equal (or when components are very well separated). Ward's criterion assumes in addition (about) spherical components. The probabilistic derivation of his criterion starts from "spherical" clusters of equal extent and about equal size. Being based on a very narrow model, Ward's criterion is insensitive. It is stable, leading often to a single solution which may, however, be unrealistic. The criterion tries to impose its own structure upon the data set even when it is not natural. An example is shown in Fig. 1.11. The data is sampled from plane normal populations with means $(0, 3)$ and $(0, -3)$ and with common covariance matrix $\operatorname{Diag}(25, 1)$. The clusters of the correct solution are thus elongated in the horizontal direction and well separated by the abscissa. The estimated means and scales are indicated by the two dotted circles. Ward's criterion forces sphericity, resulting in a vertical separator. Although there is a minute probability that this

Figure 1.11 *A partition resulting from improper use of Ward's criterion.*

is the correct partition, the result obtained represents a worst case of a clustering result. The error rate is that of a random assignment of the points. The separator is vertical instead of horizontal! Not only does the wrong scale assumption force wrong scale estimates, the location estimates too are unreasonable.[5] Some heuristic clustering methods proposed more recently share this weakness; see Handl et al. [227]. See also the counterexample in Table 1.1.

(c) Optimization w.r.t. γ_j, m_j and V_j, still required in the general criterion (1.38) and in the elliptical MAP (and ML) cluster criteria in Proposition 1.36, disappears in the normal criteria, Theorems 1.40 and 1.41. This is a rare case where the partial optimum w.r.t. the parameters can be computed analytically. The dependence on the data is thereby wrapped up in the scatter matrix. Minimization of the normal criteria is thus reduced to discrete optimization w.r.t. ℓ and to numerical computation of traces and determinants.

(d) Theorem 1.40 shows that the determinant criterion turns to the diagonal criterion after replacing the scatter matrices $S_j(\ell)$ with their diagonals. In the same way, the determinant and diagonal criteria turn to the trace criterion after inserting the spherical matrices $\frac{1}{d}\mathrm{tr}\, S_j(\ell)I_d$ for $S_j(\ell)$. The homoscedastic case is analogous, Theorem 1.41.

(e) The optimal clustering under HDBT constraints may contain empty clusters, even without trimming. This would indicate that the number of clusters g has been chosen too large. For instance, if a data set is a clear sample from a single univariate normal population, then the optimal partition in two clusters will leave one cluster empty. A simple example is $n = r = 4$, $\mathbf{x} = \{0, 3, 4, 7\}$, and $c = 1$. Some values of the MAP pooled trace criterion are

$$\begin{cases} 3.66516, & \text{for the partition } \{\mathbf{x}, \emptyset\}, \\ 3.79572, & \text{for the partition } \{\{0, 3, 4\}, \{7\}\}, \\ 4.39445, & \text{for the partition } \{\{0, 3\}, \{4, 7\}\}. \end{cases}$$

The remaining partitions need not be considered, either by symmetry or since they cannot be optimal. Hence, the method returns a single nonempty cluster. Empty clusters become less likely as c decreases.

1.3.6 Geometric considerations

Equivariance w.r.t. inherent classes of sample space transformations (see Section F.1) is an important asset of a cluster criterion. The determinant criteria, for instance, are affine equivariant, the diagonal criteria are equivariant w.r.t. variable scaling, and the spherical criteria w.r.t. orthogonal transformations. Given a transformation t on E, we abbreviate $t(\mathbf{x}) = (t(x_1), \ldots, t(x_n))$.

1.43 Definition A cluster criterion K is called **equivariant** w.r.t. the transformation t on E if

  (i) an assignment is admissible w.r.t. the transformed data $t(\mathbf{x}) \in E^n$ if and only if it is admissible w.r.t. $\mathbf{x}$;

  (ii) there exists a strictly increasing function $\Phi_{\mathbf{x},t} \colon \mathbb{R} \to \mathbb{R}$ such that $K(t(\mathbf{x}), \mathcal{C}) = \Phi_{\mathbf{x},t}(K(\mathbf{x}, \mathcal{C}))$ for all clusterings $\mathcal{C}$.

It follows that the steady clusterings (of the index set) derived from $\mathbf{x}$ and $t(\mathbf{x})$ are equal. If a criterion is equivariant w.r.t. t_1 and t_2, then it is also equivariant w.r.t. the composition $t_2 \circ t_1$ and w.r.t. the inverse t_1^{-1}. The set of all transformations which render the criterion equivariant thus forms a composition group. The criteria derived from the normal

---

[5]This is unlike the one-component case, where the MLE of the location (the sample mean) does not depend on a scale assumption.

models cannot be expected to be equivariant w.r.t. *nonlinear* transformations. For *linear* transformations, we have the following proposition.

1.44 Proposition (Equivariance of normal criteria)

(a) *All criteria in Theorems 1.40 and 1.41 are equivariant w.r.t. the translation group on $\mathbb{R}^d$.*

(b) *There are the following scale equivariances:*

    (i) *The determinant criteria are equivariant w.r.t. the general linear group.*

    (ii) *The criteria for the diagonally normal models are equivariant w.r.t. the group generated by all nonsingular diagonal matrices and all variable permutations.*

    (iii) *The trace criteria are equivariant w.r.t. the orthogonal group.*

PROOF. Let $(x_1, \ldots, x_n)$ be a Euclidean data set. We show that the difference between the criterion applied to $(x_1, \ldots, x_n)$ and to the transformed data is constant. It is sufficient to consider the terms $\log \det S_j(\ell)$, $1 \le j \le g$.

The claim is clear in case (a) since a translation leaves the scatter matrices unchanged. For part (b), let $A \in \mathbb{R}^{d \times d}$ be a regular matrix. The jth mean of the transformed data set $(Ax_1, \ldots, Ax_n)$ is $\overline{x}_j^A(\ell) = \frac{1}{n_j} \sum_{i \in C_j} Ax_i = A\overline{x}_j(\ell)$ and hence its jth scatter matrix w.r.t. ℓ is

$$S_j^A(\ell) = \frac{1}{n_j} \sum_{i \in C_j} (Ax_i - \overline{x}_j^A(\ell))(Ax_i - \overline{x}_j^A(\ell))^\top$$

$$= \frac{1}{n_j} \sum_{i \in C_j} (Ax_i - A\overline{x}_j(\ell))(Ax_i - A\overline{x}_j(\ell))^\top = AS_j(\ell)A^\top.$$

It follows

$$\log \det S_j^A(\ell) - \log \det S_j(\ell) = \log \det AS_j(\ell)A^\top - \log \det S_j(\ell) = \log \det AA^\top,$$

a number independent of ℓ. This proves part (b)(i). The proofs of the remaining cases are similar. □

1.45 Illustrations (a) Consider the data set

$$\mathbf{x} = (x_1, \ldots, x_6) = \left(\begin{pmatrix} 0 \\ 0 \end{pmatrix}, \begin{pmatrix} 1 \\ \varepsilon \end{pmatrix}, \begin{pmatrix} 10 \\ 0 \end{pmatrix}, \begin{pmatrix} 0 \\ 1 \end{pmatrix}, \begin{pmatrix} 9 \\ 1 - \varepsilon \end{pmatrix}, \begin{pmatrix} 10 \\ 1 \end{pmatrix} \right) \tag{1.43}$$

shown in Fig. 1.12. The six points are to be decomposed in two clusters. The intuitive decomposition separates two clusters $\{x_1, x_2, x_4\}$ and $\{x_3, x_5, x_6\}$ by a vertical straight line between them. In fact, as we use the hetero- or homoscedastic trace criterion, we get just this if $\varepsilon > 0$ is small, as shown in the graphic. This criterion is equivariant w.r.t. Euclidean motions and corresponds to intuition. If we measured time in minutes instead of seconds, the data set would look quite different and the intuitive decomposition would strongly favor

Figure 1.12 *Plot of the data set (1.43). What changes as time is measured in minutes instead of seconds?*

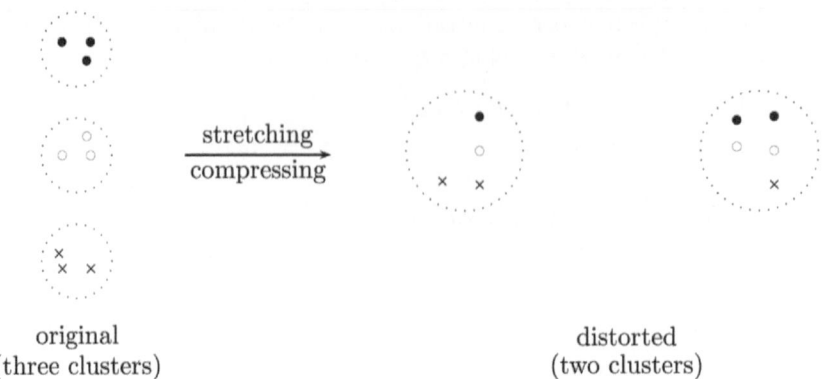

original   distorted
(three clusters)   (two clusters)

Figure 1.13 *Seemingly unclear number of clusters. Stretching the abscissa and compressing the ordinate reduces the three original spherical clusters to two.*

$\{x_1, x_2, x_3\}$ and $\{x_4, x_5, x_6\}$. Now, there is no natural scale comparison between length and time. Therefore, a criterion equivariant w.r.t. diagonal or even general linear transformations should be preferred here. Indeed, the hetero- or homoscedastic determinant criterion yields the latter solution, and so do the "diagonal" criteria (b) in Theorems 1.40 and 1.41. Note that the determinants of the scatter matrices of the two clusters $\{x_1, x_2, x_3\}$ and $\{x_4, x_5, x_6\}$ converge to zero as $\varepsilon \to 0$, whereas those of the two intuitive clusters do not. Of course, if there is prior information that the clusters should be spherical at the given scale, then the "intuitive" solution is the preferred one.

A further complication arises as ε approaches $1/10$ (but is still different from $1/10$). Then four points are almost collinear and the heteroscedastic determinant criterion provides the two equivalent admissible solutions $(\{x_1, x_2, x_5\}, \{x_3, x_4, x_6\})$ and $(\{x_2, x_5, x_6\}, \{x_1, x_3, x_4\})$. If we assume that the data comes from a Lebesgue absolutely continuous distribution, then exact collinearity is almost surely impossible.

(b) Scale transformations may alter the number of clusters w.r.t. the spherical and diagonal models. An example is shown in Fig. 1.13. After stretching the abscissa and compressing the ordinate, the three original clusters of three points, each, turn into two clusters of four and five points if spherical clusters are requested. The spherical model is not equivariant w.r.t. variable scaling!

The ML and Bayesian discriminant rules applied with the parameters of a mixture decompose the sample space in cells. If the mixture is elliptical, then the cells are geometric objects in Euclidean space. The following propositions describe them in some cases. They apply in particular to mixtures estimated from given data sets (for instance, by the ML or classification paradigm) and show that the clusters created by them according to Remark 1.30(a) are separated by geometric manifolds. In normal cases these are conics.

1.46 Proposition *Consider a normal mixture*

$$f(x; \mathbf{m}, \mathbf{V}) = \sum_{j=1}^{g} \pi_j N_{m_j, V_j}(x)$$

with pairwise distinct mean pairs (m_j, V_j). Let $\{C_1, \ldots, C_g\}$ be the related decomposition of the sample space $\mathbb{R}^d$ in cells according to the Bayesian discriminant rule.

(a) If $V_j \neq V_\ell$, then C_j and C_ℓ are separated by a conic.

(b) Otherwise, they are separated by a hyperplane perpendicular to the vector $V_j^{-1}(m_j - m_\ell)$.

PROOF. The points in $x \in C_\ell$ are characterized by the relation $\pi_j N_{m_j, V_j}(x) < \pi_\ell N_{m_\ell, V_\ell}(x)$

for all $j \neq \ell$. This is equivalent with

$$(x - m_\ell)^\top V_\ell^{-1}(x - m_\ell) - (x - m_j)^\top V_j^{-1}(x - m_j) < 2 \log \frac{\pi_\ell}{\pi_j} + \log \det V_j V_\ell^{-1}$$

and describes the interior of a conic if $V_j \neq V_\ell$. If $V_j = V_\ell = V$, then $m_j \neq m_\ell$ and the left-hand side reduces to $(m_j - m_\ell)^\top V^{-1}(2x - (m_j + m_\ell))$. Thus, the separating manifold is a hyperplane.   □

The homoscedastic elliptical ML and the pooled trace criteria deserve special mention. The proposition asserts in these cases that the partition has the property of **convex admissibility** introduced by Fisher and Van Ness [153]. It means that the convex hulls of clusters do not intersect. There is the following proposition.

1.47 Proposition *Consider a homoscedastic, isobaric, elliptical mixture*

$$f(x; \varphi, \mathbf{m}, V) = \sum_{j=1}^{g} \tfrac{1}{g} E_{\varphi, m_j, V}(x)$$

with pairwise distinct mean vectors m_j and a strictly decreasing density generator φ. Let $\{C_1, \ldots, C_g\}$ be the related decomposition of the sample space $\mathbb{R}^d$ in cells according to the ML discriminant rule.

(a) Two different cells C_j, C_ℓ are separated by the hyperplane

$$h_{j,\ell}(x) = (m_\ell - m_j)^\top V^{-1}\big[x - \tfrac{1}{2}(m_\ell + m_j)\big] = 0.$$

(b) In the spherical case $V = vI_d$, the sets C_j are the Voronoi cells associated with the points m_j.

PROOF. (a) Recall that $x \in C_\ell$ if and only if

$$\sqrt{\det \Lambda}\, \varphi\big((x - m_\ell)^\top \Lambda(x - m_\ell)\big) = E_{\varphi, m_\ell, V}(x) > E_{\varphi, m_j, V}(x)$$
$$= \sqrt{\det \Lambda}\, \varphi\big((x - m_j)^\top \Lambda(x - m_j)\big)$$

for all $j \neq \ell$. Since φ is strictly decreasing, this is tantamount to $(x - m_\ell)^\top \Lambda(x - m_\ell) < (x - m_j)^\top \Lambda(x - m_j)$. Expanding this we obtain

$$2(m_j - m_\ell)^\top \Lambda x < m_j^\top m_j - m_\ell^\top m_\ell = (m_j - m_\ell)^\top (m_j + m_\ell)$$

and claim (a) follows.

(b) The reasoning above shows that, in the spherical case, $x \in C_\ell$ if and only if $\|x - m_\ell\| < \|x - m_j\|$ for all $j \neq \ell$. This relation describes the Voronoi cells.   □

1.3.7 Consistency of the MAP criterion

An estimator would not make sense if it did not behave stably as the size of the data set increases. For the mixture model, positive answers are found in Sections 1.2.4, 1.2.5, and 1.2.6. There are analogs for the classification model. Pollard [416, 417] proved convergence of the means for the ML criteria of *homoscedastic*, isobaric, spherical classification models and for a wide class of sampling distributions. He also identified their limits as the solutions to the population criterion related to the sampling criterion (the canonical functional). This means that the *global* maximum is the favorite solution if the data set is large. The following theory due to Gallegos and Ritter [185] extends Pollard's result to the HDBT constrained *heteroscedastic*, elliptical classification model (1.38) (with a fixed density generator).

Let $X = (X_i)_{i \geq 1}$ be an i.i.d. sequence of $\mathbb{R}^d$-valued random variables with common parent distribution μ. In the present context, the most interesting μ's are mixtures of unimodal distributions. We start from a basic model with density function f_γ, $\gamma \in \Gamma$. It will later be elliptical. The distribution $\mu = P_{X_i}$ does not have to be related to this model; it may even be unspecified and rather general. The related classification model in the form (1.37) is

$$\sum_{j=1}^{g} \sum_{i \in C_j} \log(\pi_j f_{\gamma_j}(x_i)).$$

Up to here, we have used the Principle of Dynamic Optimization, Lemma G.9, in order to optimize the MAP criterion (1.38) first w.r.t. the parameter γ and then w.r.t. the assignment ℓ. This was possible for the admissible assignments. Reversing the order of optimization yields the so-called **best-location problem** described in the following lemma.

1.48 Lemma *If the submodel $\Psi \subseteq \Gamma^g$ is chosen so that all assignments $1..n \to 1..g$ become Ψ-admissible, then the maximum of*

$$\frac{1}{n} \sum_{i=1}^{n} \max_{1 \le j \le g} \log\left(\pi_j f_{\gamma_j}(X_i)\right) \tag{1.44}$$

w.r.t. $\pi \in \Delta_{g-1}$ and $\gamma \in \Psi$ exists and equals the maximum of the MAP criterion (1.38) w.r.t. γ and ℓ divided by n.

PROOF. We know from Section 1.3.1 that the maximum of $\prod_i \pi_{\ell_i} f_{\gamma_{\ell_i}}(X_i)$ w.r.t. $\ell \in (1..g)^{(1..n)}$, $\pi \in \Delta_{g-1}$, and $\gamma \in \Psi$ exists. Its logarithm equals the maximum of the MAP criterion. By finiteness, each maximum $\max_{1 \le j \le g} \pi_j f_{\gamma_j}(X_i)$, too, exists. Hence, the Principle of Dynamic Optimization, Lemma G.9, applies again, this time by first maximizing w.r.t. ℓ and then w.r.t. π and γ. We obtain

$$\max_{\ell,\pi,\gamma} \prod_i \pi_{\ell_i} f_{\gamma_{\ell_i}}(X_i) = \max_{\pi,\gamma} \max_{\ell} \prod_i \pi_{\ell_i} f_{\gamma_{\ell_i}}(X_i) = \max_{\pi,\gamma} \prod_i \max_{1 \le j \le g} \pi_j f_{\gamma_j}(X_i)$$

and the claim follows after taking the logarithm and dividing by n. $\square$

The lemma offers a completely different view of the criterion. The question arises whether the expression (1.44) is stable w.r.t. increasing n, that is, whether there is convergence of the optimal π and γ and of the whole expression as $n \to \infty$. The natural limit in the consistency theorems for the mixture model, Section 1.2, is the parent distribution $\mu = P_{X_i}$ of the observations X_i. This cannot be expected here; see Remark 1.57(c) below. But if we tentatively assume that the limit w.r.t. n and the maximum w.r.t. π and γ in (1.44) commute, then the Strong Law suggests that the limit might be

$$\max_{\pi,\gamma} \mathrm{E} \max_{1 \le j \le g} \log\left(\pi_j f_{\gamma_j}(X_1)\right).$$

This is what Pollard [416] confirmed in the case of the homoscedastic, isobaric, spherical classification model. In order to do this he replaced g-tuples $(a_1, \ldots, a_g) \in (\mathbb{R}^d)^g$ of parameters (means in his case) with finite subsets $A \subseteq \mathbb{R}^d$ of size $\le g$. The collection of subsets is endowed with the Hausdorff metric; see Appendix B.6.3. It turns out that Pollard's idea is fruitful in the present context of more general basic models and mixing rates, too. Thus, let Γ be metric and let Θ be the collection of all nonempty, compact (or finite) subsets of the Cartesian product $]0,1] \times \Gamma$ endowed with the Hausdorff metric. This structure is similar to the parameter set $\tilde{\Theta}_g$ used in Section 1.2.6 in the context of the parametric mixture model with a fixed number of components. There, the subsets take account of the unordered structure of mixture components. Here, they are used in a form that also enables easy transition between different numbers of clusters. For $a = (\pi, \gamma) \in]0,1] \times \Gamma$, we write $\pi_a = \pi$ and $\gamma_a = \gamma$. For an integer $g \ge 1$, define the subset

$$\Theta_{\le g} = \left\{ A \in \Theta \,\middle|\, 1 \le |A| \le g, \ \sum_{a \in A} \pi_a \le 1 \right\} \subseteq \Theta.$$

The relation "≤ 1" (instead of "$= 1$") makes it an extension of the solution space. The reader may wonder why the definition of $\Theta_{\leq g}$ requires the sum to be ≤ 1 and not $= 1$. The reason is that the modification of the space $\Theta_{\leq g}$ defined with $\sum_{a \in A} \pi_a = 1$ lacks the important compactness property that will be established in Lemma 1.49(b) below. To see this, consider for simplicity a pure location model $\Gamma = \mathbb{R}$. Let $a_\gamma = (1/2, \gamma)$, $\gamma \in \Gamma$. For $\gamma \neq 0$, the set $A_\gamma = \{a_0, a_\gamma\} \in \Theta_{\leq 2}$ has the property $\pi_{a_0} + \pi_{a_\gamma} = 1$ and converges w.r.t. the Hausdorff metric to the singleton $A_0 = \{a_0\}$ as $\gamma \to 0$. But $\pi_{a_0} = 1/2$ and so the limit is not a member of the modified space. To counteract this defect we complete the parameter space for deficient mixing rates.

Now, define the functional

$$t_a(x) = -\log \pi_a - \log f_{\gamma_a}(x), \ a = (\pi_a, \gamma_a) \in]0,1] \times \Gamma, \tag{1.45}$$

and denote the *sampling criterion* by

$$\Phi_n(A) = \frac{1}{n} \sum_{i=1}^{n} \min_{a \in A} t_a(X_i), \quad A \in \Theta_{\leq g}. \tag{1.46}$$

Up to the opposite sign this is the criterion introduced in (1.44). Using the empirical measure $\mu_n = \frac{1}{n} \sum_{i=1}^{n} \delta_{X_i}$ it can be written $\Phi_n(A) = \int \min_{a \in A} t_a(x) \mu_n(\mathrm{d}x)$. It equals the MAP criterion (1.38) when all assignments are Ψ-admissible. We will also need the *population criterion*

$$\Phi(A) = \mathrm{E} \min_{a \in A} t_a(X_1) = \int \min_{a \in A} t_a(x) \mu(\mathrm{d}x), \quad A \in \Theta_{\leq g}. \tag{1.47}$$

It is well defined if $\log f_\gamma$ is μ-integrable for all $\gamma \in \Gamma$. Although the definition of $\Theta_{\leq g}$ allows deficient mixing rates, it is reassuring to observe that optimal solutions A^* for Φ_n and Φ do enjoy the property $\sum_{a \in A^*} \pi_a = 1$ and correspond to mixtures. Moreover, each parameter γ appears only once in A^*. The reason is the strict increase of the logarithm: As a typical example, let $A = \{(\pi_{11}, \gamma_1), (\pi_{12}, \gamma_1), (\pi_2, \gamma_2)\}$ with $\pi_{11} \neq \pi_{12}$ and $p = \pi_{11} + \pi_{12} + \pi_2 \leq 1$. The modification $A' = \{(\pi_1', \gamma_1), (\pi_2', \gamma_2)\}$ with $\pi_1' = (\pi_{11} + \pi_{12})/p$ and $\pi_2' = \pi_2/p$ satisfies $\pi_1' + \pi_2' = 1$. Moreover, $\log \pi_1' > \log \pi_{11} \vee \log \pi_{12}$ and $\log \pi_2' \geq \log \pi_2$ and hence $\Phi_n(A') < \Phi_n(A)$ and $\Phi(A') < \Phi(A)$:

$$\min_{a' \in A'} t_{a'} = \min_{a' \in A'}(-\log \pi_{a'} - \log f_{\gamma_{a'}}) = \left(-\log \pi_1' - \log f_{\gamma_1}\right) \wedge \left(-\log \pi_2' - \log f_{\gamma_2}\right)$$
$$< \left(-\log \pi_{11} - \log f_{\gamma_1}\right) \wedge \left(-\log \pi_{12} - \log f_{\gamma_1}\right) \wedge \left(-\log \pi_2 - \log f_{\gamma_2}\right) = \min_{a \in A} t_a.$$

The above set-up applies in particular to the (HDBT constrained) elliptical basic model $f_{m,V} = E_{\varphi,m,V}$, $m \in \mathbb{R}^d$, $V \in \mathrm{PD}(d)$ used in Proposition 1.36. We fix here a continuous, strictly positive density generator φ and the associated radial function $\phi = -\log \varphi$. The functional t_a becomes

$$t_a(x) = -\log \pi_a - \tfrac{1}{2} \log \det \Lambda_a + \phi\big((x - m_a)^\top \Lambda_a (x - m_a)\big);$$

here, $a = (\pi_a, m_a, V_a) \in]0,1] \times \mathbb{R}^d \times \mathrm{PD}(d)$ and, as before, $\Lambda_a = V_a^{-1}$. The related population criterion Φ is well defined when $\phi(\beta \|X_1\|^2)$ is P-integrable for all $\beta > 0$. If the scale matrices satisfy the HDBT constraints, if ϕ is large enough, if the data is in general position, and if $n \geq gd + 1$, then Proposition 1.36 is applicable, stating that all assignments $1..n \to 1..g$ are admissible w.r.t. this elliptical model. In this case, Lemma 1.48 says that Φ_n is its cluster criterion and that it has a minimum. Therefore, we restrict the set $\Theta_{\leq g}$ to its subset

$$\Theta_{\leq g, c} = \left\{ A \in \Theta \mid |A| \leq g, \ \sum_{a \in A} \pi_a \leq 1, \ (V_a)_{a \in A} \in \mathcal{V}_c \right\} \subseteq \Theta_{\leq g}$$

with the HDBT constant c. The parameter spaces $\Theta_{\leq g}$ and $\Theta_{\leq g, c}$ should not be mixed up with those introduced in Section 1.2 for the mixture model; see (1.12) and (1.33). The remainder of this chapter will be devoted to the question, under what conditions the minimizers π and $\gamma = (\mathbf{m}, \mathbf{V})$ of the sampling criterion Φ_n converge as $n \to \infty$ in the case described. We need a series of lemmas. The first one collects some basic properties of the quantities just defined.

1.49 Lemma

(a) The set $\Theta_{\leq g,c}$ is closed in Θ (and in $\Theta_{\leq g}$) w.r.t. the Hausdorff metric.

(b) Let $0 < \pi_0 \leq 1$ and $R, \varepsilon > 0$. The sub-collection of $\Theta_{\leq g,c}$ consisting of all sets of elements a such that $\pi_a \geq \pi_0$, $\|m_a\| \leq R$, and $\varepsilon I_d \preceq V_a \preceq \frac{1}{\varepsilon} I_d$ is compact.

(c) If ϕ is continuous, then the function $A \mapsto \min_{a \in A} t_a(x)$ is continuous on $\Theta_{\leq g,c}$ for all $x \in \mathbb{R}^d$.

PROOF. (a) By Lemma B.24(c), the limit of any sequence $(A_n)_n$ in $\Theta_{\leq g,c}$ convergent in Θ is a set in $\Theta_{\leq g}$. Moreover, the limit inherits the constraints $V_a \succeq cV_b$, $a, b \in A_n$.

   Part (b) follows from Lemma B.24(e). Continuity (c) follows from continuity of the minimum operation and of the mapping $a \mapsto t_a(x)$ which, in turn, follows from continuity of ϕ. □

1.50 Lemma Assume that

(i) ϕ is increasing and continuous;

(ii) $\phi(\beta\|X_1\|^2)$ is P-integrable for all $\beta \geq 1$.

Then, P-a.s., the sampling criterion (1.46), Φ_n, converges to the population criterion (1.47), Φ, locally uniformly on $\Theta_{\leq g,c}$. In particular, Φ is continuous.

PROOF. We verify the assumptions of the Uniform SLLN E.33 with $g(x, A) = \min_{a \in A} t_a(x)$. Continuity E.33(i) is just Lemma 1.49(c). In order to obtain the integrable upper bound E.33(iii) it is sufficient to consider elements a in the compact set $K_{\varepsilon,R} = [\varepsilon, 1] \times \overline{B}_R(0) \times \{V \in \mathrm{PD}(d) \mid \varepsilon I_d \preceq V \preceq \frac{1}{\varepsilon} I_d\}$ for $0 < \varepsilon \leq 1$ and $R > 0$. Since ϕ increases, we may estimate for $\|x\| \geq R$

$$t_a(x) = -\log \pi_a - \tfrac{1}{2} \log \det \Lambda_a + \phi\big((x - m_a)^\top \Lambda_a (x - m_a)\big)$$
$$\leq -\log \varepsilon - \tfrac{d}{2} \log \varepsilon + \phi\big(\|x - m_a\|^2/\varepsilon\big)$$
$$\leq \mathrm{const} + \phi\Big(\tfrac{2}{\varepsilon}\big(\|x\|^2 + \|m_a\|^2\big)\Big) \leq \mathrm{const} + \phi\Big(\tfrac{4}{\varepsilon}\|x\|^2\Big).$$

By assumption (ii), the right-hand side is the μ-integrable upper bound requested. □

   The assumption 1.50(ii) requires that the density generator φ should have a heavy tail if μ does. It is satisfied if the radial function ϕ grows at most polynomially and $\phi(\|X_1\|^2)$ is P-integrable. The main idea of the consistency proof is to show that all optimal sets remain in a compact subset of $\Theta_{\leq g,c}$ independent of n. The following lemma is a first step toward this goal and prepares Lemma 1.52, which shows that the optimal scale parameters are bounded.

1.51 Lemma Let $\alpha > 0$ and assume that $\|X_1\|^{2\alpha}$ is P-integrable. Then the minimum of

$$\mathrm{E} \min_{a \in A} \big|v^\top (X_1 - m_a)\big|^{2\alpha}$$

w.r.t. v, $\|v\| = 1$, and $A \in \Theta_{\leq g,c}$ exists. It is zero if and only if μ is supported by g parallel hyperplanes.

PROOF. Let $\Theta'_{\leq g}$ stand for the collection of all nonempty subsets of $\mathbb{R}$ with at most g elements and let μ_v be the projection of μ to the real line via $x \mapsto v^\top x$. We prove that the minimum of the finite functional

$$J(v, B) = \int_{\mathbb{R}} \min_{b \in B} |y - b|^{2\alpha} \mu_v(\mathrm{d}y)$$

w.r.t. v, $\|v\| = 1$, and $B \in \Theta'_{\leq g}$ exists. This is equivalent to the claim. Indeed, if B is related to A by $b = v^\top m_a$ or $m_a = bv$, then

$$\int_{\mathbb{R}} \min_{b \in B} |y - b|^{2\alpha} \mu_v(\mathrm{d}y) = \mathrm{E} \min_{a \in A} |v^\top (X_1 - m_a)|^{2\alpha}.$$

The proof proceeds in three steps. In step (α), it will be shown that sets B which contain only remote points cannot have small values of the criterion. In the main step (β), the same is shown for a mixture of nearby and remote points. Thus, sets of nearby points remain to be considered. The lemma follows in a third step. Let $r > 0$ be such that $\mu(B_r(0)) > 0$.

(α) If $R > r$ satisfies

$$|R - r|^{2\alpha} \mu(B_r(0)) \geq \int_{\mathbb{R}^d} \|x\|^{2\alpha} \mu(\mathrm{d}x), \tag{1.48}$$

then no set $B \in \Theta'_{\leq g}$ such that $B \cap [-R, R] = \emptyset$ can be optimal for J:
Note $\mu_v(]-r, r[) \geq \mu(B_r(0))$ and use the Cauchy-Schwarz inequality to estimate

$$J(v, B) \geq \int_{]-r, r[} \min_{b \in B} |y - b|^{2\alpha} \mu_v(\mathrm{d}y) > |R - r|^{2\alpha} \mu_v(]-r, r[)$$

$$\geq |R - r|^{2\alpha} \mu(B_r(0)) \geq \int_{\mathbb{R}^d} \|x\|^{2\alpha} \mu(\mathrm{d}x) \geq \int_{\mathbb{R}^d} |w^\top x|^{2\alpha} \mu(\mathrm{d}x) = J(w, \{0\}), \quad \|w\| = 1.$$

The rest of the proof proceeds by induction on g. The claim for $g = 1$ follows immediately from (α). Let $g \geq 2$ and assume that the claim is true for $g-1$. We distinguish between two cases. If $\inf_{v, B \in \Theta'_{\leq g}} J(v, B) = \min_{v, B \in \Theta'_{\leq g-1}} J(v, B)$, then nothing has to be proved since the largest lower bounds decrease with g. In the opposite case, let

$$\varepsilon = \min_{v, B \in \Theta'_{\leq g-1}} J(v, B) - \inf_{v, B \in \Theta'_{\leq g}} J(v, B) \quad (> 0)$$

and let $R > 0$ satisfy

$$\int_{\|x\| \geq 2R} \|2x\|^{2\alpha} \mu(\mathrm{d}x) < \varepsilon/2. \tag{1.49}$$

(β) If B contains elements in both $[-R, R]$ and the complement of $[-5R, 5R]$, then its J-value cannot be arbitrarily close to the largest lower bound:
We argue by contradiction. Assume that there is $(v, B) \in S_1(0) \times \Theta'_{\leq g}$ such that

  (i) B contains a point b_-, $|b_-| \leq R$, and a point b_+, $|b_+| > 5R$, and

  (ii) $J(B, v) - \inf_{v, B \in \Theta'_{\leq g}} J(v, B) \leq \varepsilon/2$.

We compare the J-value of B with that of its subset $\widetilde{B} = B \cap [-5R, 5R] \in \Theta'_{\leq g-1}$. As a consequence of removing points from B, some points have to be reassigned. Collect them in the set C. If $|y| < 2R$, then $|y - b_-| < 3R < |y - b_+|$. Thus, the deletion of points in the complement of $[-5R, 5R]$ from B does not affect the assignment of points $|y| < 2R$, that is, $C \subseteq \{y \in \mathbb{R} \mid |y| \geq 2R\}$. If $|y| \geq 2R$, then $|y - b_-| \leq |2y|$. It follows

$$J(v, \widetilde{B}) - J(v, B) = \int_C \min_{b \in \widetilde{B}} |y - b|^{2\alpha} \mu_v(\mathrm{d}y) - \int_C \min_{b \in B} |y - b|^{2\alpha} \mu_v(\mathrm{d}y)$$

$$\leq \int_C \min_{b \in \widetilde{B}} |y - b|^{2\alpha} \mu_v(\mathrm{d}y) \leq \int_{|y| \geq 2R} \min_{b \in \widetilde{B}} |y - b|^{2\alpha} \mu_v(\mathrm{d}y)$$

$$\leq \int_{|y| \geq 2R} |y - b_-|^{2\alpha} \mu_v(\mathrm{d}y) \leq \int_{|y| \geq 2R} |2y|^{2\alpha} \mu_v(\mathrm{d}y) \leq \int_{\|x\| \geq 2R} \|2x\|^{2\alpha} \mu(\mathrm{d}x) < \varepsilon/2.$$

Recalling $\widetilde{B} \in \Theta'_{\leq g-1}$, we have found a contradiction to the definition of ε and (ii) above. This proves (β).

Let R satisfy (1.48) and (1.49). Assertions (α) and (β) show that small values of J are attained on $\Theta_{\leq g}$ only if $B \subseteq [-5R, 5R]$. The collection $\mathcal{K}$ of these sets B is compact w.r.t. the Hausdorff metric. The first claim finally follows from continuity of J on the compact space $S_1(0) \times \mathcal{K}$.

The expectation is zero if and only if the μ-integral of $\min_{a \in A} v^{\top}(\cdot - m_a)$ vanishes for some pair (v, A). This means that the topological support of μ is contained in some set of the form $\bigcup_{a \in A}\{x \mid v^{\top}(x - m_a) = 0\}$. $\qquad\square$

1.52 Lemma *Assume that*

(i) μ *is not supported by* g *parallel hyperplanes;*

(ii) $\phi(t) \geq b_0 + b_1 t^{\alpha}$ *for some numbers* $b_0 \in \mathbb{R}$ *and* $b_1, \alpha > 0$;

(iii) $\phi(\|X_1\|^2)$ *is* P-*integrable.*

Then there are numbers $0 < c_1 \leq c_2$ *such that,* P-*a.s., all scale parameters* V^* *of sets optimal for* Φ_n *on* $\Theta_{\leq g,c}$ *(see (1.46)) satisfy* $I_d/c_2 \preceq V^* \preceq I_d/c_1$ *for eventually all* n.

PROOF. First note that (iii) along with (ii) implies integrability of $\|X_1\|^{2\alpha}$. Let us show that $\Phi_n(A)$ is large for eventually all n if an eigenvalue of the scale matrices of A is small or large. Let $A \in \Theta_{\leq g,c}$ and let $a_0 \in A$ such that $\det \Lambda_{a_0}$ is maximal. Using the SLLN, we estimate

$$\Phi_n(A) = \frac{1}{n} \sum_{i=1}^{n} \min_{a \in A} t_a(X_i)$$

$$= \frac{1}{n} \sum_{i=1}^{n} \min_{a \in A} \left\{ -\log \pi_a - \frac{1}{2}\log \det \Lambda_a + \phi\big((X_i - m_a)^{\top}\Lambda_a(X_i - m_a)\big) \right\}$$

$$\geq b_0 - \frac{1}{2}\log \det \Lambda_{a_0} + \frac{b_1}{n} \sum_{i=1}^{n} \min_{a \in A} \big(c(X_i - m_a)^{\top}\Lambda_{a_0}(X_i - m_a)\big)^{\alpha}$$

$$\xrightarrow[n \to \infty]{} b_0 - \frac{1}{2}\log \det \Lambda_{a_0} + b_1 c^{\alpha} \mathrm{E} \min_{a \in A} \big((X_1 - m_a)^{\top}\Lambda_{a_0}(X_1 - m_a)\big)^{\alpha},$$

P-a.s. Let $\Lambda_{a_0} = \sum_k \lambda_k v_k v_k^{\top}$ be the spectral decomposition (A.3). Since $t \mapsto t^{\alpha}$ is increasing and multiplicative,

$$\big((x - m_a)^{\top}\Lambda_{a_0}(x - m_a)\big)^{\alpha} = \left(\sum_k \lambda_k\big(v_k^{\top}(x - m_a)\big)^2\right)^{\alpha}$$

$$\geq \frac{1}{d}\sum_k \left(\lambda_k\big(v_k^{\top}(x - m_a)\big)^2\right)^{\alpha} = \frac{1}{d}\sum_k \lambda_k^{\alpha}\big|v_k^{\top}(x - m_a)\big|^{2\alpha}.$$

Inserting, we find

$$\lim_n \Phi_n(A) \geq b_0 + \sum_k \left\{ -\frac{1}{2}\log \lambda_k + \frac{b_1 c^{\alpha}}{d}\lambda_k^{\alpha} \mathrm{E} \min_{a \in A}\big|v_k^{\top}(X_1 - m_a)\big|^{2\alpha} \right\}$$

$$\geq b_0 + \sum_k \left\{ -\frac{1}{2}\log \lambda_k + \kappa\frac{b_1 c^{\alpha}}{d}\lambda_k^{\alpha} \right\} \qquad (1.50)$$

with the constant $\kappa = \min\limits_{v \in S_1(0), A} \mathrm{E} \min_{a \in A}\big|v^{\top}(X_1 - m_a)\big|^{2\alpha} > 0$; see Lemma 1.51. Note that the kth summand in (1.50) converges to ∞ as $\lambda_k \to 0$ and, since λ_k^{α} beats $\log \lambda_k$, also as $\lambda_k \to \infty$. It has thus a minimum for $\lambda_k > 0$. Therefore, if one summand converges to ∞, so does the whole sum. But one summand tends to ∞ under either of the two cases: (1) the largest of all eigenvalues λ_k exceeds a certain bound c_2 and (2) $\lambda_k < c_2$ for all k and the smallest eigenvalue drops below another bound c_1. We conclude that

$$\Phi_n(A) \geq \Phi(\{(1,0,I_d)\}) + 1 \tag{1.51}$$

for eventually all n if (1) or (2) is satisfied. Finally, a set A_n^* minimizing Φ_n satisfies

$$\Phi_n(A_n^*) \leq \Phi_n(\{(1,0,I_d)\}) \underset{n\to\infty}{\longrightarrow} \Phi(\{(1,0,I_d)\})$$

by (iii) and the SLLN. Combined with (1.51), this is the claim. $\qquad\square$

For $A \in \Theta_{\leq g,c}$ and $a \in A$, define the set $C_a(A) = \{x \in \mathbb{R}^d \mid t_a(x) \leq t_b(x) \text{ for all } b \in A\}$ by the Bayes discriminant rule. In order to obtain disjoint, measurable sets, ties are broken by ordering A and favoring the smallest a. We get a decomposition $\{C_a(A) \mid a \in A\}$ of $\mathbb{R}^d$ in at most g measurable "cells." The population criterion $\Phi(A)$ has the representation

$$\Phi(A) = \int_{\mathbb{R}^d} \min_{a\in A} t_a \, \mathrm{d}\mu = \sum_{a\in A} \int_{C_a(A)} t_a \, \mathrm{d}\mu.$$

The entropy inequality, Theorem D.17, shows

$$\begin{aligned}
\Phi(A) = \sum_{a\in A} \Big\{ &- \mu(C_a(A)) \log \pi_a - \frac{\mu(C_a(A))}{2} \log \det \Lambda_a \\
&+ \int_{C_a(A)} \phi\big((x-m_a)^\top \Lambda_a(x-m_a)\big)\mu(\mathrm{d}x) \Big\} \\
\geq \ \mathrm{H}&(\mu(C_a(A)) \mid a \in A) + \sum_{a\in A} \Big\{ -\frac{\mu(C_a(A))}{2} \log \det \Lambda_a \\
&+ \int_{C_a(A)} \phi\big((x-m_a)^\top \Lambda_a(x-m_a)\big)\mu(\mathrm{d}x) \Big\}.
\end{aligned} \tag{1.52}$$

In particular, an optimal set A^* satisfies $\pi_a = \mu(C_a(A^*))$ for all $a \in A^*$. Moreover, equality obtains in (1.52) if and only if $\pi_a = \mu(C_a(A^*))$ for all $a \in A^*$. The following lemma states conditions ensuring that optimal means and mixing rates will be bounded. In view of its part (b), note that, under the assumptions of Lemma 1.52, $\liminf_n \min_{A\in\Theta_{\leq g,c}} \Phi_n(A) > -\infty$ and $\limsup_n \min_{A\in\Theta_{\leq g,c}} \Phi_n(A) < \infty$ for eventually all n. Indeed, $\min_{A\in\Theta_{\leq g,c}} \Phi_n(A) \geq -\frac{d}{2}\log c_2 + b_0$ and $\min_{A\in\Theta_{\leq g,c}} \Phi_n(A) \leq \Phi_n(\{(1,0,I_d)\}) \to \Phi(\{(1,0,I_d)\})$ by the SLLN.

1.53 Lemma *Assume that*

(i) the parent distribution μ is not supported by g parallel hyperplanes;

(ii) ϕ is increasing and $\phi(t) \geq b_0 + b_1 t^\alpha$ for some numbers $b_0 \in \mathbb{R}$ and $b_1, \alpha > 0$;

(iii) $\phi(\|X_1\|^2)$ is P-integrable.

Let $g \geq 1$ and let A_n^*, $n \geq gd+1$, be optimal for Φ_n on $\Theta_{\leq g,c}$. Then we have, P-a.s.:

(a) If $a_n^* \in A_n^*$ is such that $\pi_{a_n^*} = \mu_n\big(C_{a_n^*}(A_n^*)\big) \geq \varepsilon > 0$, then the sequence of means $(m_{a_n^*})_n$ is bounded.

(b) If $g \geq 2$ and

$$\limsup_n \min_{A\in\Theta_{\leq g,c}} \Phi_n(A) < \limsup_n \min_{A\in\Theta_{\leq g-1,c}} \Phi_n(A),$$

then the mixing rates $\pi_{a_n^*}$ of all $a_n^* \in A_n^*$, $n \geq gd+1$, are bounded away from zero (and all means $m_{a_n^*}$ are bounded).

PROOF. Let c_1 and c_2 be as in Lemma 1.52.

(a) This claim states that means of sets A_n^* optimal for Φ_n with cells of lower bounded probabilities are small. Let $r > 0$ be so large that $\mu(B_r(0)) > 1 - \varepsilon/2$. Using the estimates $c_1 I_d \preceq \Lambda_a \preceq c_2 I_d$, we infer

$$\Phi_n(\{(1,0,I_d)\}) \geq \Phi_n(A_n^*) = \int \min_{a \in A_n^*} t_a \, \mathrm{d}\mu_n = \sum_{a \in A_n^*} \int_{C_a(A_n^*)} t_a \, \mathrm{d}\mu_n$$

$$\geq \sum_{a \in A_n^*} \int_{C_a(A_n^*)} \left\{ -\tfrac{1}{2} \log \det \Lambda_a + \phi\big((x - m_a)^\top \Lambda_a (x - m_a)\big) \right\} \mu_n(\mathrm{d}x)$$

$$\geq -\tfrac{d}{2} \log c_2 + b_0 + b_1 \sum_{a \in A_n^*} \int_{C_a(A_n^*)} \big(c_1 \|x - m_a\|^2\big)^\alpha \mu_n(\mathrm{d}x)$$

$$\geq -\tfrac{d}{2} \log c_2 + b_0 + b_1 c_1^\alpha \int_{C_{a_n^*}(A_n^*) \cap B_r(0)} \big(\|x - m_{a_n^*}\|^{2\alpha}\big) \mu_n(\mathrm{d}x)$$

$$\geq -\tfrac{d}{2} \log c_2 + b_0 + b_1 c_1^\alpha \big| \|m_{a_n^*}\| - r \big|^{2\alpha} \mu_n\big(C_{a_n^*}(A_n^*) \cap B_r(0)\big)$$

if $\|m_{a_n^*}\| \geq r$. By the SLLN, we have $\mu_n(B_r(0)) > 1 - \varepsilon/2$ for eventually all n and hence $\mu_n\big(C_{a_n^*}(A_n^*) \cap B_r(0)\big) \geq \varepsilon/2$ for these n. Therefore, an application of (iii) and the SLLN imply

$$\Phi(\{(1,0,I_d)\}) = \lim_n \Phi_n(\{(1,0,I_d)\}) \geq -\tfrac{d}{2} \log c_2 + b_0 + \frac{b_1 c_1^\alpha \varepsilon}{2} \limsup_n \big| \|m_{a_n^*}\| - r \big|^{2\alpha}.$$

This defines a bound for all $m_{a_n^*}$.

(b) Since $\sum_{a \in A_n} \pi_a = 1$, there is $a_n \in A_n^*$ such that $\pi_{a_n} \geq 1/g$, $n \geq gd + 1$. Let $R > 0$, $R' \geq 2R$, and $u' < 1/g$ be three constants to be specified later. According to part (a) we may and do assume $R \geq \|m_{a_n}\|$, $n \geq gd + 1$. Also assume that there is an element $a' \in A_n^*$ with the property $\|m_{a'}\| > R'$ or $\pi_{a'} < u'$ and delete all such elements from A_n^* to obtain a set $\widetilde{A}_n \in \Theta_{\leq g-1,c}$. Of course, $a_n \in \widetilde{A}_n$. Note that any x assigned to $a \in \widetilde{A}_n$ w.r.t. A_n^* is also assigned to a w.r.t. $\widetilde{A}_n$. Therefore, the sample space splits in two parts: The set $\bigcup_{a \in \widetilde{A}_n} C_a(A_n^*)$ of points assigned to elements in $\widetilde{A}_n$ w.r.t. both A_n^* and $\widetilde{A}_n$ and the set $C = \bigcup_{a \in A_n \setminus \widetilde{A}_n} C_a(A_n^*)$ of points reassigned w.r.t. $\widetilde{A}_n$ because they were originally assigned to points deleted from A_n^*.

We first show that the centered ball with radius $2R$ is contained in the complement of C. So let $\|x\| < 2R$ and let $a' \in A_n^* \setminus \widetilde{A}_n$. We have

$$t_{a_n}(x) = -\log \pi_{a_n} - \tfrac{1}{2} \log \det \Lambda_{a_n} + \phi\big((x - m_{a_n})^\top \Lambda_{a_n}(x - m_{a_n})\big)$$

$$\leq \log g - \tfrac{d}{2} \log c - \tfrac{1}{2} \log \det \Lambda_{a'} + \phi\big(c_2(\|x\| + R)^2\big)$$

$$\leq \log g - \tfrac{d}{2} \log c - \tfrac{1}{2} \log \det \Lambda_{a'} + \phi\big(9 c_2 R^2\big). \tag{1.53}$$

Now fix u' and R' in such a way that

$$\log g - \tfrac{d}{2} \log c + \phi\big(9 c_2 R^2\big) < (b_0 - \log u') \wedge \phi\big(c_1(R' - 2R)^2\big).$$

The element a' has either one of two properties. If $\pi_{a'} < u'$, then

$$(1.53) < b_0 - \log u' - \tfrac{1}{2} \log \det \Lambda_{a'} \leq b_0 - \log \pi_{a'} - \tfrac{1}{2} \log \det \Lambda_{a'}.$$

If $\|m_{a'}\| > R'$, then $R' - 2R \leq \|m_{a'}\| - \|x\| \leq \|x - m_{a'}\|$ and

$$(1.53) < -\tfrac{1}{2} \log \det \Lambda_{a'} + \phi\big(c_1(R' - 2R)^2\big) \leq -\tfrac{1}{2} \log \det \Lambda_{a'} + \phi\big(c_1 \|x - m_{a'}\|^2\big).$$

Hence, in both cases, $t_{a_n}(x) < t_{a'}(x)$, that is, x is not assigned to a' and $B_{2R}(0) \subseteq \complement C$ as claimed.

Observing the properties of the set C explained above, we find

$$\Phi_n(\widetilde{A}_n) - \Phi_n(A_n^*) = \int_C \Big(\min_{a \in \widetilde{A}_n} t_a - \min_{a \in A_n^*} t_a \Big) \mathrm{d}\mu_n$$

$$= \int_C \Big(\min_{a \in \widetilde{A}_n} t_a - \min_{a \in A_n^*} t_a \Big) \mathrm{d}\mu_n \leq \int_C t_{a_n} \, \mathrm{d}\mu_n - \int_C \min_{a \in A_n^* \setminus \widetilde{A}_n} t_a \, \mathrm{d}\mu_n.$$

Now we have

$$t_{a_n}(x) = -\log \pi_{a_n} - \tfrac{1}{2} \log \det \Lambda_{a_n} + \phi\big((x - m_{a_n})^{\top} \Lambda_{a_n}(x - m_{a_n})\big)$$
$$\leq \log g - \tfrac{d}{2} \log c_1 + \phi\big(c_2 \|x - m_{a_n}\|^2\big)$$

and $t_a(x) \geq -\tfrac{d}{2} \log c_2 + b_0$ for all a. Inserting and observing $C \subseteq \complement B_{2R}(0)$, we infer for all n

$$\min_{A \in \Theta_{\leq g-1,c}} \Phi_n(A) - \min_{A \in \Theta_{\leq g,c}} \Phi_n(A) \leq \Phi_n(\widetilde{A}_n) - \Phi_n(A_n^*) \tag{1.54}$$

$$\leq \int_C \Big\{ \log g - \tfrac{d}{2} \log c_1 + \phi\big(c_2 \|x - m_{a_n}\|^2\big) \Big\} \mu_n(\mathrm{d}x) + \int_C \Big\{ \tfrac{d}{2} \log c_2 - b_0 \Big\} \mu_n(\mathrm{d}x)$$

$$= \Big\{ \log g + \tfrac{d}{2} \log \tfrac{c_2}{c_1} - b_0 \Big\} \mu_n(\complement B_{2R}(0)) + \int_{\complement B_{2R}(0)} \phi\big(4c_2 \|x\|^2\big) \mu_n(\mathrm{d}x).$$

From the estimate $t_a(x) \geq -\tfrac{d}{2} \log c_2 + b_0$ and the assumption of part (b), we also have

$$\liminf_n \min_{A \in \Theta_{\leq g,c}} \Phi_n(A) \in \mathbb{R}.$$

Passing to the $\limsup_n$ in (1.54) with the aid of the SLLN, we therefore obtain

$$\limsup_n \min_{A \in \Theta_{\leq g-1,c}} \Phi_n(A) - \limsup_n \min_{A \in \Theta_{\leq g,c}} \Phi_n(A)$$

$$\leq \limsup_n \Big(\min_{A \in \Theta_{\leq g-1,c}} \Phi_n(A) - \min_{A \in \Theta_{\leq g,c}} \Phi_n(A) \Big)$$

$$\leq \Big\{ \log g + \tfrac{d}{2} \log \tfrac{c_2}{c_1} - b_0 \Big\} \mu(\complement B_{2R}(0)) + \int_{\|x\| \geq 2R} \phi\big(4c_2 \|x\|^2\big) \mu(\mathrm{d}x).$$

The assumption of part (b) says that the left-hand side is P-a.s. strictly positive. Since the right-hand side vanishes as $R \to \infty$, the assumption on the existence of a' made at the beginning cannot hold if R is large. This proves part (b). $\qquad\square$

1.54 Lemma *Let the assumptions (i),(ii) of Lemma 1.50 be satisfied. If $\mathcal{K} \subseteq \Theta_{\leq g,c}$ is compact and contains some minimizer of the sampling criterion Φ_n for all $n \geq gd+1$, then*

(a) $\mathcal{K}$ contains a minimizer of the population criterion Φ;

(b) P-a.s., $\min \Phi_n \underset{n \to \infty}{\longrightarrow} \min \Phi$ on $\Theta_{\leq g,c}$.

PROOF. Since Φ is continuous, the restriction $\Phi_{|\mathcal{K}}$ has a minimizer A^*. We have to show that A^* minimizes Φ on all of $\Theta_{\leq g,c}$. Now, let $A_n^* \in \mathcal{K}$ be some minimizer of Φ_n. The uniform convergence $\Phi_n \to \Phi$ on $\mathcal{K}$, Lemma 1.50, implies

$$\Phi(A^*) \leq \Phi(A) \leq \Phi_n(A) + \varepsilon, \ A \in \mathcal{K},$$

for eventually all n. Conversely, by optimality of A_n^*,

$$\Phi_n(A_n^*) \leq \Phi_n(A^*) \leq \Phi(A^*) + \varepsilon$$

if n is large. Hence, $\Phi_n(A_n^*) \to \Phi(A^*)$ as $n \to \infty$. Finally, the inequality $\Phi_n(A_n^*) \leq \Phi_n(A)$ for all $A \in \Theta_{\leq g,c}$ shows that A^* is a minimizer of Φ on all of $\Theta_{\leq g,c}$. $\qquad\square$

The main Theorem 1.56 of this section depends on the following notion introduced in Gallegos and Ritter [185].

1.55 Definition An integer $g \geq 2$ is a **drop point** of the population criterion Φ (under the HDBT constant c; see (1.27) and (1.47)), if

$$\inf_{A \in \Theta_{\leq g,c}} \Phi(A) < \inf_{A \in \Theta_{\leq g-1,c}} \Phi(A). \tag{1.55}$$

Also $g = 1$ is defined as a drop point.

A drop point depends on μ and on Φ restricted to $\Theta_{g,\leq c}$, that is, to the set of elliptical distributions with at most g components. It means that μ is in closer harmony with some g-component distribution than with any one with $g-1$ components. The number of drop points may be finite or infinite. We are now prepared to prove consistency of the HDBT constrained elliptical MAP classification model with a fixed radial function ϕ. An essential part of the theorem says that all minimizers of the sampling criteria remain in a compact set if g is a drop point. This is the basis for consistency. The theorem shows that the minimizers converge and identifies their limit.

1.56 Theorem Consistency of the elliptical MAP criterion *Let* $0 < c \leq 1$. *Let* $(X_i)_i$ *be i.i.d. with common distribution* μ. *Assume that*

(i) *hyperplanes in* $\mathbb{R}^d$ *are* μ-*null sets;*

(ii) ϕ *is continuous and increasing;*

(iii) $\phi(t) \geq b_0 + b_1 t^\alpha$ *for some numbers* $b_0 \in \mathbb{R}$ *and* $b_1, \alpha > 0$;

(iv) $\phi^p(\beta \|X_1\|^2)$ *is* P-*integrable for some* $p > 1$ *and all* $\beta \geq 1$.

Then the following claims hold true for all $g \geq 1$:

(a) P-*a.s., each sampling criterion* Φ_n, $n \geq gd+1$, *has a minimizer* $A_n^* \in \Theta_{\leq g,c}$;

(b) *the population criterion* Φ *has a minimizer* $A^* \in \Theta_{\leq g,c}$;

(c) *we have* $\sum_{a \in A_n^*} \pi_a = 1$, $n \geq gd+1$, *and* $\sum_{a \in A^*} \pi_a = 1$.

(d) P-*a.s.,* $\min \Phi_n \xrightarrow[n\to\infty]{} \min \Phi$ *on* $\Theta_{\leq g,c}$.

Moreover, if g *is a drop point, then*

(e) *there is a compact subset of* $\Theta_{\leq g,c}$ *that contains all minimizers* A_n^* *of* Φ_n *for all* $n \geq gd+1$;

(f) P-*a.s., any sequence of minimizers* A_n^* *of* Φ_n *on* $\Theta_{\leq g,c}$ *converges to the set of minimizers of* Φ *on* $\Theta_{\leq g,c}$.

(g) *In particular: If the minimizer* A^* *of* Φ *on* $\Theta_{\leq g,c}$ *is unique, then* (A_n^*) *converges* P-*a.s. to* A^* *for any choice of minimizers* A_n^*.

PROOF. By assumption (i), the data is P-a.s. in general position. Therefore, Lemma 1.36 shows that each assignment is admissible and so claim (a) is just Lemma 1.48. Claim (c) was discussed after the definition of Φ in (1.47). For claims (e), (b), and (d) we use induction on $g \geq 1$. Let $g = 1$. By Lemmas 1.52, 1.53(a), and 1.49(b), there exists a compact subset $\mathcal{K}_1 \subseteq \Theta_{\leq 1,c}$ that contains all minimizers of all sampling criteria Φ_n, that is, claim (e) for $g = 1$. Claims (b) and (d) for $g = 1$ follow from Lemma 1.54.

Now let $g \geq 2$. For the following arguments we will need the fact that $\inf_{A \in \Theta_{\leq g,c}} \Phi(A)$ is finite. The proof is similar to that of (1.50). Indeed, let $a_0 \in A$ such that $\det \Lambda_{a_0}$ is maximal and let $\Lambda_{a_0} = \sum_k \lambda_k v_k v_k^\top$ be the spectral decomposition (A.3). We have

$$\Phi(A) \geq -\tfrac{1}{2} \log \det \Lambda_{a_0} + \int \min_{a \in A} \phi\big((x - m_a)^\top \Lambda_a (x - m_a)\big) \mu(dx)$$

$$\geq \tfrac{d}{2} \log c + b_0 + \sum_k \Big(-\tfrac{1}{2} \log(c\lambda_k) + \kappa \tfrac{b_1}{d}(c\lambda_k)^\alpha \Big)$$

with the strictly positive constant $\kappa = \min_{\|v\|=1, A} \mathrm{E} \min_{a \in A} |v^\top (X_1 - m_a)|^{2\alpha}$. The claim follows from the fact that each summand on the right-hand side is bounded below as a function of λ_k.

In view of the induction step $g-1 \to g$, let A_n^* be minimal for Φ_n on $\Theta_{\leq g,c}$, $n \geq gd+1$. We distinguish between two cases. First, assume $\pi_a \geq \varepsilon > 0$ for all $a \in A_n^*$ and all such n. By Lemmas 1.52, 1.53(a), and 1.49(b), there exists a compact subset $\mathcal{K}_g \subseteq \Theta_{\leq g,c}$ which contains all minima A_n^*. This is one half of claim (e) and claims (b) and (d) follow again from Lemma 1.54.

In the second case we may and do assume that there are elements $a_n \in A_n^*$ such that $\pi_{a_n} = \mu_n(C_{a_n}(A_n^*)) \to 0$ as $n \to \infty$. Of course there is at least one element $a_n' \in A_n^*$ such that $\mu_n(C_{a_n'}(A_n^*)) \geq 1/g$ and so, Lemma 1.53(a) implies $\|m_{a_n'}\| \leq R$ for some R. By assumption (iv) and by Hölder's inequality with $\frac{1}{p} + \frac{1}{q} = 1$,

$$\int_{C_{a_n}(A_n^*)} \phi\big((x - m_{a_n'})^\top \Lambda_{a_n'}(x - m_{a_n'})\big) \mu_n(\mathrm{d}x)$$

$$\leq \int_{C_{a_n}(A_n^*)} \phi\big(c_2 \|x - m_{a_n'}\|^2\big) \mu_n(\mathrm{d}x) \leq \int_{C_{a_n}(A_n^*)} \phi\big(2c_2(\|x\|^2 + R^2)\big) \mu_n(\mathrm{d}x)$$

$$\leq \mu_n\big(C_{a_n}(A_n^*)\big)^{1/q} \Big(\int |\phi|^p \big(2c_2(\|x\|^2 + R^2)\big) \mu_n(\mathrm{d}x) \Big)^{1/p}$$

$$\leq \mu_n\big(C_{a_n}(A_n^*)\big)^{1/q} \Big(|\phi|^p \big(4c_2 R^2\big) + \int_{\|x\| > R} |\phi|^p \big(4c_2 \|x\|^2\big) \mu_n(\mathrm{d}x) \Big)^{1/p} \xrightarrow[n \to \infty]{} 0.$$

Since $\mu_n\big(C_{a_n}(A_n^*)\big)\big\{ \log \mu_n\big(C_{a_n'}(A_n^*)\big) + \frac{1}{2} \log \det \Lambda_{a_n'} \big\} \to 0$, we have $\int_{C_{a_n}(A_n^*)} t_{a_n'} \, \mathrm{d}\mu_n \to 0$ as $n \to \infty$. Now write

$$\Phi_n(A_n^*) = \sum_{a \in A_n^*} \int_{C_a(A_n^*)} t_a \, \mathrm{d}\mu_n = \sum_{a \neq a_n} \int_{C_a(A_n^*)} t_a \, \mathrm{d}\mu_n$$

$$+ \int_{C_{a_n}(A_n^*)} t_{a_n} \, \mathrm{d}\mu_n + \int_{C_{a_n}(A_n^*)} t_{a_n'} \, \mathrm{d}\mu_n - \int_{C_{a_n}(A_n^*)} t_{a_n'} \, \mathrm{d}\mu_n$$

and put $A_n' = \{a \in A_n^* \mid a \neq a_n\} \in \Theta_{\leq g-1, c}$. The sum of the first and the third terms on the right is the μ_n-integral of a function pieced together from the functions t_a, $a \in A_n'$. It is thus $\geq \int \min_{a \in A_n'} t_a \, \mathrm{d}\mu_n = \Phi_n(A_n')$. Since t_{a_n} is lower bounded, the $\liminf_n$ of the second term is ≥ 0. Moreover, we have already seen that the $\limsup_n$ of the last term vanishes. Therefore,

$$\liminf_n \min_{A \in \Theta_{\leq g, c}} \Phi_n(A) = \liminf_n \Phi_n(A_n^*) \geq \liminf_n \Phi_n(A_n')$$

$$\geq \liminf_n \min_{A \in \Theta_{\leq g-1, c}} \Phi_n(A) = \min_{A \in \Theta_{\leq g-1, c}} \Phi(A)$$

by the inductive hypotheses (b) and (d). Furthermore,

$$\limsup_n \min_{A \in \Theta_{\leq g, c}} \Phi_n(A) \leq \limsup_n \Phi_n(A_0) = \Phi(A_0)$$

for all $A_0 \in \Theta_{\leq g, c}$ by the SLLN and hence $\limsup_n \min_{A \in \Theta_{\leq g, c}} \Phi_n(A) \leq \inf_{A \in \Theta_{\leq g, c}} \Phi(A)$. We conclude

$$\limsup_n \min_{A \in \Theta_{\leq g, c}} \Phi_n(A) \leq \inf_{A \in \Theta_{\leq g, c}} \Phi(A) \leq \min_{A \in \Theta_{\leq g-1, c}} \Phi(A).$$

Both estimates combine to show $\inf_{A \in \Theta_{\leq g, c}} \Phi(A) = \min_{A \in \Theta_{\leq g-1, c}} \Phi(A)$, that is, g is no drop point in this case, the other half of claim (e). Moreover, claims (b) and (d) follow for g.

By Lemma 1.54, the compact set $\mathcal{K}_g$ appearing in (e) contains at least one minimum of Φ. Denote the set of these minima by K ($\subseteq \mathcal{K}_g$). For $\varepsilon > 0$ let $U_\varepsilon = \{A \in \mathcal{K}_g \mid \Phi(A) \leq \min \Phi + \varepsilon\}$. If U is any open neighborhood in $\mathcal{K}_g$ of the compact set K, then $\bigcap_{\varepsilon > 0} U_\varepsilon \backslash U = K \backslash U = \emptyset$. By compactness, $U_\varepsilon \backslash U = \emptyset$ for some $\varepsilon > 0$, that is, $U_\varepsilon \subseteq U$. Hence, U_ε forms a neighborhood base of K. Because of Lemma 1.50, all minima of Φ_n lie in U_ε for eventually all n. This is the consistency (f) and the consistency (g) is a direct consequence. □

1.57 Remarks (a) The parameter space consists of *sets* A (and not of tuples) and so, there is no label switching that can cause the usual nonuniqueness of the parameters. Yet, nonuniqueness arises whenever different sets A generate the same minimum $\min_{a \in A} t_a$. If sets are unique, then nonuniqueness of solutions occurs when the parent μ bears symmetries. For instance, bivariate standard normals centered at the four vertices of a square allow two equivalent minima of the population criterion (1.47) on $\Theta_{\leq 2, c}$. It is for these reasons that Theorem 1.56(f) states "converges to the set of minimizers of Φ."

(b) If there is an uninterrupted chain of drop points $1..g_{\max}$, then the claims of Theorem 1.56 hold true for $g \in 1..g_{\max}$ with assumption (iv) relaxed to the assumption (ii) of Lemma 1.50,

$\phi(\beta \|X_1\|^2)$ is P-integrable for all $\beta \geq 1$.

Indeed, in this case the proof of part (e) of Theorem 1.56, namely, the fact that the optimal sampling parameters remain in a compact subset of $\Theta_{\leq g, c}$, can be simplified. The proof proceeds again by induction on $g \in 1..g_{\max}$: The case $g = 1$ is as in Theorem 1.56. For $g \geq 2$, we verify the assumption of Lemma 1.53(b). Let $A_0 \in \Theta_{\leq g, c}$ such that $\Phi(A_0) \leq \inf_{A \in \Theta_{\leq g, c}} \Phi(A) + \varepsilon$. By convergence $\Phi_n \to \Phi$, Lemma 1.50,

$$\min_{A \in \Theta_{\leq g, c}} \Phi_n(A) \leq \Phi_n(A_0) \leq \Phi(A_0) + \varepsilon \leq \inf_{A \in \Theta_{\leq g, c}} \Phi(A) + 2\varepsilon$$

if n is large. Since g is a drop point, it follows

$$\limsup_n \min_{A \in \Theta_{\leq g, c}} \Phi_n(A) \leq \inf_{A \in \Theta_{\leq g, c}} \Phi(A) < \inf_{A \in \Theta_{\leq g-1, c}} \Phi(A) = \lim_n \min_{A \in \Theta_{\leq g-1, c}} \Phi_n(A).$$

The last equality follows from the inductive hypothesis and from Lemma 1.54(b). This is the hypothesis of Lemma 1.53(b) for g. The rest of the proof proceeds like the end of the proof of Theorem 1.56.

(c) In the consistency theorems of Section 1.1.2, the parent is specified. Theorem 1.56 is different. As in Pollard's [416, 417] theorem, the parent distribution μ does not have to belong to $\Theta_{\leq g, c}$. On the other hand, even if μ is an elliptical mixture, it cannot be the mixture associated with the limit. This is in contrast to the mixture model; see Theorems 1.19, 1.22, and 1.27. No matter how well the components are separated, think, for instance, of two, their portions in the tails on the opposite side of the separating hypersurface are assigned to the wrong cluster. Thus, the variances are underestimated and the distance between the mean values is overestimated in the limit as $n \to \infty$; see also Marriott [350] and Bryant and Williamson [73]. However, these negative and positive biases disappear as cluster separation grows, since the overlap decreases.

(d) Theorem 1.56 concerns the MAP criterion. It is, for instance, applicable to the normal model and needs in this case $\int \|x\|^p \mu(dx) < \infty$ for some $p > 4$. By restricting the parameter set $\Theta_{\leq g, c}$, it also applies to diagonal, spherical, and homoscedastic submodels. The related ML version maximizes the classification likelihood (1.39), $\prod_i f_{m_{\ell_i}, V_{\ell_i}}(X_i)$, w.r.t. ℓ, $\mathbf{m}$, and $\mathbf{V} \in \mathcal{V}_c$. Redefine the parameter set $\Theta_{\leq g, c}$ and the functionals t_a, omitting the mixing rates, and assume that Φ has a minimum on $\Theta_{\leq g, c}$ for all $g \in 1..g_{\max}$. (This is, for instance, true under the assumptions of the theorem.) If the topological support of μ (see Theorem D.2(a)) contains at least $g_{\max}$ points, then every point $g \leq g_{\max}$ is a drop point of the ML criterion. To see this, we proceed by induction on $g \geq 2$. Let $A^* \in \Theta_{\leq g-1, c}$ be optimal. The induction hypothesis implies $|A^*| = g - 1$. By assumption, there is $a^* \in A^*$ whose cell, $C_{a^*}(A^*)$, contains two elements of $\operatorname{supp} \mu$. Choose $m_{a'} \in C_{a^*}(A^*) \cap \operatorname{supp} \mu$, $m_{a'} \neq m_{a^*}$, and put $V_{a'} = V_{a^*}$. This defines an element $a' \notin A^*$ and a set $A' = A^* \cup \{a'\} \in \Theta_{\leq g, c}$, $|A'| = g$. Strict increase of ϕ and the choice of $m_{a'}$ and $V_{a'}$ entail $t_{a'} < t_{a^*} \leq t_a$ at $m_{a'}$ for all $a \in A^*$. Because of $A' \supseteq A^*$, we have $\min_{a \in A'} t_a \leq \min_{a \in A^*} t_a$ and conclude

$$\Phi(A') = \int \min_{a \in A'} t_a \, d\mu < \int \min_{a \in A^*} t_a \, d\mu = \Phi(A^*).$$

This shows that the optimal set in $\Theta_{\leq g, c}$ w.r.t. the ML criterion has in fact g elements and

Table 1.2 *Optimal values of the Ward and MAP criteria for the homoscedastic, isobaric, standard bivariate normal model;* $Q = \mathrm{E}\min_{a \in A^*} \|X_1 - m_a\|^2$.

g	Q	Ward $\log 2\pi + Q/2$	Population criterion $\log g + \log 2\pi + Q/2$
1	2	$2.8378\ldots$	$2.8378\ldots$
2	$2 - \frac{2}{\pi}$	$2.5195\ldots$	$3.2127\ldots$
3	$2 - \frac{27}{8\pi}$	$2.3007\ldots$	$3.3993\ldots$

that its value has improved.

The decrease of the criterion is actually not always desirable in cluster analysis. If μ is a mixture of g well-separated components, then we would prefer that any solution with more than g clusters be rejected; see Fig. 1.14. This would give us a hint to the number of components. As shown above, the ML criterion does not comply with this wish. In contrast to the ML criterion, the MAP criterion may possess non-drop points. Indeed, the examples in Table 1.2 say that the MAP criteria of the ML solutions

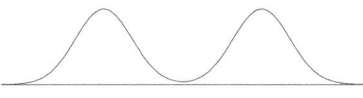

Figure 1.14 *A normal mixture with two well-separated components.*

for $g = 2, 3$ do not beat the optimum MAP criterion for $g = 1$. The reason is the unimodal distribution. More generally, if μ is a well-separated mixture of g unimodal distributions (see Fig. 1.14), then we expect g to be a drop point, but not $g + 1$. Examples are presented after the next proposition.

Ward's [522] criterion (see Remark 1.42(a)) is the standard tool in *optimal vector quantization* in order to subdivide a distribution into parts of about equal size. There are two reasons for its popularity: First, its compliance with the task of a uniform subdivision and, second, the stability of Ward's criterion and of the related k-means algorithm with only a few steady solutions. The only parameters are the means, here called **principal points**. Those of the standard multivariate normal distribution have been extensively studied. In two dimensions, the optimal solutions A^* of Ward's criterion $Q = \mathrm{E}\min_{a \in A^*}\|X_1 - m_a\|^2$ are, respectively for $g = 1$, 2 and 3, $\{(0,0)\}$, the antipodes $\{\pm(a,0)\}$, $a = \sqrt{2/\pi} = 0.7978\ldots$, and the vertices of a centered, equilateral triangle with circumradius $3\sqrt{3}/(2\sqrt{2\pi}) = 1.0364\ldots$; see Graf and Luschgy [212]. (The last solution is strongly conjectured by numerical evidence, only; see Flury [164].) Of course, rotations of the solutions about the origin are optimal, too. The optimal values of the ML criterion and the values of the solutions w.r.t. the isobaric MAP criterion are compiled in Table 1.2. While the ML values improve with increasing g, the MAP values go in the opposite direction, showing that these solutions are not optimal w.r.t. MAP; see also Example 1.59(a). Theorem 1.56(b) states that, under certain assumptions, the population criterion Φ has a minimum. The following proposition provides an alternative to compute it. We will use it to verify the subsequent examples but it is interesting in its own right since it is the population version of the negative elliptical MAP criterion, Proposition 1.36.

1.58 Proposition *Let μ and ϕ satisfy the assumptions 1.50(i),(ii) and 1.56(i),(iii). Denote partitions of the sample space $\mathbb{R}^d$ in at most g events C such that $\mu(C) > 0$ by the letter $\mathbf{P}$. Let $\mathbf{m} = (m_C)_{C \in \mathbf{P}}$ and $\mathbf{V} = (V_C)_{C \in \mathbf{P}}$.*

(a) For each such partition $\mathbf{P}$ the minimum of

$$\sum_{C \in \mathbf{P}} \left\{ \tfrac{\mu(C)}{2} \log \det V_C + \int_C \phi\big((x - m_C)^\top V_C^{-1}(x - m_C)\big)\mu(\mathrm{d}x)\right\}$$

w.r.t. **m** *and* $\mathbf{V} \in \mathcal{V}_c$ *exists.*

(b) The population criterion Φ *has a minimum on* $\Theta_{\leq g,c}$ *if and only if*

$$\mathrm{H}(\mu(C) \mid C \in \mathbf{P})$$
$$+ \min_{\mathbf{m}, \mathbf{V} \in \mathcal{V}_c} \sum_{C \in \mathbf{P}} \left\{ \tfrac{\mu(C)}{2} \log \det V_C + \int_C \phi((x - m_C)^\top V_C^{-1}(x - m_C)) \mu(\mathrm{d}x) \right\}$$

has a minimum w.r.t. all **P**. *In this case, the minima coincide.*

(c) In the homoscedastic, *normal case* $\phi(t) = \tfrac{t}{2} + \tfrac{d}{2} \log 2\pi$, *let* $V(\mathbf{P}) = \sum_{C \in \mathbf{P}} V[X_1; X_1 \in C]$ *denote the pooled covariance matrix of* μ *w.r.t.* **P**. *The minimum of* Φ *on* $\Theta_{\leq g,1}$ *exists if and only if the minimum of*

$$\mathrm{H}(\mu(C) \mid C \in \mathbf{P}) + \tfrac{d}{2}(1 + \log 2\pi) + \tfrac{1}{2} \log \det V(\mathbf{P})$$

over all **P** *exists and, in this case, the minima are equal.*

PROOF. (a) By the assumptions on μ and ϕ, the sum is continuous as a function of $\mathbf{m} = (m_C)_C$ and $\mathbf{V} = (V_C)_C$. Let $\Lambda_C = \sum_k \lambda_k v_k v_k^\top$ be the spectral decomposition (A.3). We have $(x - m_C)^\top \Lambda_C (x - m_C) = \sum_k \lambda_k (v_k^\top (x - m_C))^2$ and, by 1.56(iii) and the increase of $t \mapsto t^\alpha$,

$$\int_C \phi((x - m_C)^\top \Lambda_C (x - m_C)) \mu(\mathrm{d}x)$$

$$\geq b_0 \mu(C) + b_1 \int_C \Big(\sum_k \lambda_k (v_k^\top (x - m_C))^2 \Big)^\alpha \mu(\mathrm{d}x)$$

$$\geq b_0 \mu(C) + \tfrac{b_1}{d} \int_C \sum_k \big(\lambda_k (v_k^\top (x - m_C))^2 \big)^\alpha \mu(\mathrm{d}x)$$

$$= b_0 \mu(C) + \tfrac{b_1}{d} \sum_k \lambda_k^\alpha \int_C |v_k^\top (x - m_C)|^{2\alpha} \mu(\mathrm{d}x)$$

$$\geq b_0 \mu(C) + \tfrac{b_1}{d} \sum_k \lambda_k^\alpha \min_{\|v\|=1, m} \int_C |v^\top (x - m)|^{2\alpha} \mu(\mathrm{d}x).$$

The minimum exists since the integral is continuous as a function of m and v and converges to ∞ as $\|m\| \to \infty$. Moreover, by assumption (i) of Theorem 1.56 and by $\mu(C) > 0$ it is a strictly positive constant κ_C. It follows

$$\tfrac{\mu(C)}{2} \log \det V_C + \int_C \phi((x - m_C)^\top \Lambda_C (x - m_C)) \mu(\mathrm{d}x)$$

$$\geq b_0 \mu(C) + \sum_k \left\{ -\tfrac{\mu(C)}{2} \log \lambda_k + \kappa_C \tfrac{b_1}{d} \lambda_k^\alpha \right\}.$$

This expression converges to ∞ as $\lambda_k \to 0$ or $\lambda_k \to \infty$. It is thus sufficient to consider matrices V_C such that $c_1 I_d \preceq V_C \preceq c_2 I_d$ for two numbers $0 < c_1 \leq c_2$. Since $\phi(t) \to \infty$ as $t \to \infty$ by assumption (iii) of Theorem 1.56(iii) and since $\mu(C) > 0$, it follows that each integral tends to ∞ as $\|m_C\| \to \infty$. It is therefore sufficient to restrict the range of each m_C to a compact subset of $\mathbb{R}^d$ and the claim follows from continuity.

(b) We have to show that, to each $A \in \Theta_{\leq g,c}$, there corresponds some partition **P** for which the expression has a value $\leq \Phi(A)$ and vice versa. The first claim follows from (1.52). For the converse, let **P** be given and let $m(C)$ and $V(C)$ be the minimizers w.r.t. $m_C \in \mathbb{R}^d$ and $(V_C)_C \in \mathcal{V}_c$ in (a). The elements $a_C = (\mu(C), m(C), V(C))$, $C \in \mathbf{P}$, satisfy

$\{a_C \mid C \in \mathbf{P}\} \in \Theta_{\leq g,c}$ and we have

$$H(\mu(C) \mid C \in \mathbf{P}) + \min_{\mathbf{m}, \mathbf{V} \in \mathcal{V}_c} \sum_{C \in \mathbf{P}} \left\{ \tfrac{\mu(C)}{2} \log \det V_C + \int_C \phi((x - m_C)^\top \Lambda_C (x - m_C)) \mu(\mathrm{d}x) \right\}$$

$$= \sum_{C \in \mathbf{P}} \int_C \left\{ -\log \mu(C) + \tfrac{1}{2} \log \det V(C) + \phi((x - m(C))^\top \Lambda_C (x - m(C))) \right\} \mu(\mathrm{d}x)$$

$$= \sum_{C \in \mathbf{P}} \int_C t_{a_C} \, \mathrm{d}\mu \geq \int_{\mathbb{R}^d} \min_C t_{a_C} \, \mathrm{d}\mu.$$

This is the desired inequality.

(c) The proof in the homoscedastic, normal case is similar to that of the pooled determinant criterion, Theorem 1.41(a):

$$\min_{\mathbf{m}, V \in \mathrm{PD}(d)} \sum_{C \in \mathbf{P}} \left\{ \tfrac{\mu(C)}{2} \log \det V + \int_C \phi((x - m_C)^\top \Lambda (x - m_C)) \mu(\mathrm{d}x) \right\}$$

$$= \tfrac{1}{2} \min_{\mathbf{m}, V \in \mathrm{PD}(d)} \left\{ \log \det V + d \log 2\pi + \sum_{C \in \mathbf{P}} \int_C (x - m_C)^\top \Lambda (x - m_C) \mu(\mathrm{d}x) \right\}$$

$$= \tfrac{1}{2} \min_{V \in \mathrm{PD}(d)} \left\{ \log \det V + d \log 2\pi + \mathrm{tr} \, (\Lambda V(\mathbf{P})) \right\}$$

$$= \tfrac{1}{2} \left\{ d(1 + \log 2\pi) + \log \det V(\mathbf{P}) \right\}$$

by Lemma F.10. This is claim (c). □

In parts (b) and (c), it is sufficient to take the minimum w.r.t. a collection of partitions $\mathbf{P}$ that is known to contain the optimal one.

1.59 Examples (a) Let $\mu = N_{0,I_d}$ and let the approximating model be the normal location and scale family, that is, $\phi(t) = (d/2) \log 2\pi + t/2$. We determine the minimum of the population criterion. For $g = 1$, the entropy inequality shows that the optimal solution in $\Theta_{\leq 1,c}$ is $\{(1, 0, I_d)\}$. Let now $g \geq 2$, $A = \{(\pi_1, m_1, V_1), \ldots, (\pi_g, m_g, V_g)\} \in \Theta_{\leq g,c}$, (m_j, V_j) pairwise distinct, $\sum \pi_j = 1$, $\Lambda_j = V_j^{-1}$, $f_j(x) = (2\pi)^{-d/2} \sqrt{\det \Lambda_j} e^{-(x - m_j)^\top \Lambda_j (x - m_j)/2}$, and abbreviate $t_j = t_{(\pi_j, m_j, V_j)}$. We have $\sum_j \pi_j f_j > \max_j \pi_j f_j$ and hence $-\log \sum_j \pi_j f_j < \min_j t_j$. Again, the entropy inequality shows

$$\Phi(A) = \int \min_j t_j \, \mathrm{d}\mu > -\int \log \sum_j \pi_j f_j \, \mathrm{d}\mu \geq -\int \log N_{0,I_d} \, \mathrm{d}\mu = \Phi(\{(1, 0, I_d)\}).$$

Thus, the only optimal solution in $\Theta_{\leq g,c}$ is the singleton $\{(1, 0, I_d)\}$. No genuine mixture of normals is superior. This is true for any HDBT constant $c \leq 1$.

(b) Example (a) raises the question whether the criterion can decrease after it has been constant for at least two (consecutive) values of g. The answer is yes. Consider the homoscedastic normal classification model, $c = 1$, and the distribution μ on the real line with Lebesgue density

$$f_0(x) = \begin{cases} 1/(8\alpha), & |x \pm 1| < \alpha, \\ 1/(4\alpha), & |x| < \alpha, \\ 0, & \text{otherwise}, \end{cases}$$

for $0 < \alpha < 1/3$. The optimal solution for $g = 1$ w.r.t. the population criterion Φ is $\{(1, 0, v)\}$ with $v = \tfrac{1}{2} + \tfrac{\alpha^2}{3}$ and $\Phi(\{(1, 0, v)\}) = \tfrac{1}{2}(\log 2\pi + 1 + \log v)$.

In order to see that there is no better solution in $\Theta_{\leq 2,1}$ note that any solution $A^* = \{a_1, a_2\}$, $a_1 \neq a_2$, is specified by some cut $s^* \in]-1 - \alpha, 1 + \alpha[$ that separates $C_{a_1}(A^*)$ from $C_{a_2}(A^*)$. Let F be the c.d.f of f_0 and let $R = 1 - F$ be its tail distribution. According to Proposition 1.58(c), it is sufficient to run across all cuts s and to compute entropy and pooled variance

$$v(s) = \int_{-\infty}^{s} (x - m_1(s))^2 f_0(x) \mathrm{d}x + \int_{s}^{\infty} (x - m_2(s))^2 f_0(x) \mathrm{d}x,$$

with the conditional expectations

$$m_1(s) = \mathrm{E}[X_1 \mid X_1 < s] = \tfrac{1}{F(s)} \int_{-\infty}^{s} x f_0(x) \mathrm{d}x,$$

$$m_2(s) = \mathrm{E}[X_1 \mid X_1 > s] = \tfrac{1}{R(s)} \int_{s}^{\infty} x f_0(x) \mathrm{d}x.$$

Omitting the addend $\frac{1}{2} \log(2\pi)$ in Φ, the integral version of Lemma A.13, a formula of Steiner's type, asserts that

$$\Phi(A_s) = \mathrm{H}(F(s), R(s)) + \tfrac{1}{2}\big(1 + \log v(s)\big)$$
$$= \mathrm{H}(F(s), R(s)) + \tfrac{1}{2} + \tfrac{1}{2} \log \big(v - F(s)R(s)(m_2(s) - m_1(s))^2\big),$$

where $A_s = \{a_1(s), a_2(s)\}$, $a_1(s) = (F(s), m_1(s), v(s))$, $a_2(s) = (R(s), m_2(s), v(s))$, v is the total variance above, and H stands for the entropy. The difference between this value and the optimum for $g = 1$ is

$$\Phi(A_s) - \Phi(\{(1, 0, v)\})$$
$$= \mathrm{H}(F(s), R(s)) + \tfrac{1}{2} \log \Big(1 - F(s)R(s)\frac{(m_2(s) - m_1(s))^2}{v}\Big). \qquad (1.56)$$

We have to show that this number is strictly positive for all $s \in\,]-1-\alpha, 1+\alpha[$ and begin with $s \in\,]-1-\alpha, -1+\alpha[$. The conditional expectations are $m_1(s) = \frac{1}{2}(s - 1 - \alpha)$ and $m_2(s) = \frac{(1+\alpha)^2 - s^2}{14\alpha - 2(s+1)}$. Hence,

$$m_2(s) - m_1(s) = 4\alpha \frac{1 + \alpha - s}{7\alpha - (s+1)} \leq \tfrac{4}{3}$$

since $\alpha < 1/3$. Inserting in (1.56) and observing $v > 1/2$ yields

$$\Phi(A_s) - \Phi(\{(1, 0, v)\}) \geq \mathrm{H}(F(s), R(s)) + \tfrac{1}{2} \log \big(1 - \tfrac{32}{9} F(s)R(s)\big).$$

The derivatives w.r.t. F of the functions $-F \log F$ and $-R \log R$, where $R = 1 - F$, are strictly decreasing on $]0, 1/4[$. The same is true for $\log \big(1 - \tfrac{32}{9} FR\big)$ since

$$\frac{\mathrm{d}}{\mathrm{d}F} \log \big(1 - \tfrac{32}{9}(1 - F)F\big) = -\tfrac{32}{9} \frac{1 - 2F}{1 - \tfrac{32}{9}(1 - F)F}$$

and since $\tfrac{32}{9}(1 - F) \geq 2$ for $F \leq \tfrac{1}{4}$. Hence, the function $\mathrm{H}(F, R) + \tfrac{1}{2} \log \big(1 - \tfrac{32}{9} FR\big)$ is strictly concave. Since it vanishes at $F = 0$ and has at $F = 1/4$ the value $-\big(\tfrac{3}{4} \log \tfrac{3}{4} + \tfrac{1}{4} \log \tfrac{1}{4}\big) + \tfrac{1}{2} \log \tfrac{1}{3} = \log 4 - \tfrac{5}{4} \log 3 = 0.01302\ldots$, it is strictly positive for $0 < F \leq 1/4$. That is, $\Phi(A_s) > \Phi(\{(1, 0, v)\})$ for $-1 - \alpha < s \leq -1 + \alpha$, the first claim. The value obtained for $s = -1 + \alpha$ persists on the interval $[-1 + \alpha, -\alpha]$ since $F(s) = 1/4$, $m_1(s) = -1$ and $m_2(s) = 1/3$ do not depend on s.

For reasons of symmetry we are done after we have verified the claim for $s \in\,]-\alpha, 0]$. In this case, $F(s) = \tfrac{1}{2} + \tfrac{s}{4\alpha}$, $R(s) = \tfrac{1}{2} - \tfrac{s}{4\alpha}$, $F(s)m_1(s) = \tfrac{s^2 - \alpha^2}{8\alpha} - \tfrac{1}{4} = -R(s)m_2(s)$, and $m_2(s) - m_1(s) = \frac{R(s)m_2(s)}{F(s)R(s)}$. Hence,

$$R(s)F(s)(m_2(s) - m_1(s))^2 = \frac{(R(s)m_2(s))^2}{R(s)F(s)} = \frac{(s^2 - \alpha^2 - 2\alpha)^2}{4(4\alpha^2 - s^2)}.$$

For $0 < \alpha \leq 0.3$ the right-hand side is $\leq 1/3$. Indeed, $3(\alpha^2 - s^2) + 12\alpha < 4$ and hence

$$3(2\alpha + \alpha^2 - s^2)^2 = 12\alpha^2 + \big(3(\alpha^2 - s^2) + 12\alpha\big)(\alpha^2 - s^2)$$
$$< 12\alpha^2 + 4(\alpha^2 - s^2) = 4(4\alpha^2 - s^2).$$

Inserting in (1.56) yields

$$\Phi(A_s) - \Phi(\{(1,0,v)\})$$
$$\geq H(F(s), R(s)) + \tfrac{1}{2} \log\left(1 - 2F(s)R(s)(m_2(s) - m_1(s))^2\right)$$
$$\geq H\left(\tfrac{1}{4}, \tfrac{3}{4}\right) + \tfrac{1}{2} \log \tfrac{1}{3}.$$

This is again the strictly positive number computed above. We have shown that the optimum number of components up to two is one and that Φ is bounded below on $\Theta_{\leq 2,c}$ by a number independent of α. That is, $g = 2$ is no drop point. The minimum of Φ on $\Theta_{\leq 3,c}$ is smaller than that on $\Theta_{\leq 2,c}$, at least for small α. It is in fact unbounded below as $\alpha \to 0$.

Finally, note that the situation changes completely as we consider the uniform sampling distribution on the set $[-1-\alpha, -1+\alpha] \cup [-\alpha, \alpha] \cup [1-\alpha, 1+\alpha]$. Here the optimal solution for $g = 1$ is no longer optimal for $g \leq 2$. By weak continuity it is sufficient to study the weak limit as $\alpha \to 0$, the discrete probability $\mu = \tfrac{1}{3}(\delta_{-1} + \delta_0 + \delta_1)$. The optimal solution for $g = 1$ is $\left\{\left(1, 0, \tfrac{2}{3}\right)\right\}$, its population criterion being (up to the constant $(\log 2\pi)/2$) $\Phi\left(\left\{\left(1, 0, \tfrac{2}{3}\right)\right\}\right) = \tfrac{1}{2}\left(1 + \log \tfrac{2}{3}\right) = 0.2972\ldots$. A solution for $g = 2$ is $A_2 = \{a_1, a_2\}$ with $a_1 = \left(\tfrac{1}{3}, -1, \tfrac{1}{6}\right)$ and $a_2 = \left(\tfrac{2}{3}, \tfrac{1}{2}, \tfrac{1}{6}\right)$. Its criterion is $\Phi(A_2) = H\left(\tfrac{1}{3}, \tfrac{2}{3}\right) + \tfrac{1}{2}\left(1 + \log \tfrac{1}{6}\right) = 0.2406\ldots$.

(c) This example shows that h is a drop point of a homoscedastic mixture of $h \geq 2$ univariate, normal distributions if there is sufficient separation. It can be extended to more general mixtures but, for simplicity, let us use Proposition 1.58(c). For $v > 0$, $m_1 < m_2 < \cdots < m_h$ and $\pi_j > 0$ such that $\sum \pi_j = 1$, consider the homoscedastic, normal mixture $\mu_v = \sum_{j=1}^{h} \pi_j N_{m_j, v}$. We denote the population criterion (1.47) w.r.t. μ_v by $\Phi_v(A) = \int \min_{a \in A} t_a \, d\mu_v$ and show that its minimum over $A \in \Theta_{\leq h-1, 1}$ remains bounded below while that over $A \in \Theta_{\leq h, 1}$ becomes arbitrarily small as $v \to 0$. By Proposition 1.58(c), $\min_{A \in \Theta_{\leq g, 1}} \Phi_v(A)$ is the minimum of

$$H(\mu_v(C) \mid C \in \mathbf{P}) + \tfrac{1}{2}(1 + \log 2\pi) + \tfrac{1}{2} \log \det V_v(\mathbf{P}), \tag{1.57}$$

taken over all partitions $\mathbf{P}$ of $\mathbb{R}^d$ in at most g measurable subsets. Here, $V_v(\mathbf{P})$ is the pooled variance of $\mathbf{P}$ w.r.t. μ_v.

Any partition $\mathbf{P}$ of $\mathbb{R}$ in $g \leq h - 1$ subsets contains at least one subset C' where two different components of μ_v contain probability $\geq 1/(h-1)$. Indeed, the stochastic matrix $(N_{m_j, v}(C) \mid 1 \leq j \leq h, C \in \mathbf{P})$ with indices j and C has h rows and at most $h - 1$ columns. Each row contains an entry $\geq 1/(h-1)$ and, since there are more rows than columns, the pigeon hole principle shows that one column must contain two such entries. In other words, there are two different components that load some subset $C' \in \mathbf{P}$ with probability at least $1/(h-1)$ each. Since all elements m_j are different, a moment of reflection shows that the pooled variance $V_v(\mathbf{P})$ and hence (1.57) remain bounded below as $v \to 0$. By Proposition 1.58(c), the same is true for $\Phi_v(A)$ uniformly for $A \in \Theta_{\leq h-1, 1}$. This is the first half of the claim.

Now let $g = h$. We construct a partition with a small value of (1.57). There is $r > 0$ such that the open intervals $]m_j - r, m_j + r[$, $1 \leq j \leq g$, are pairwise disjoint. For $1 \leq j < g$, put $C_j =]m_j - r, m_j + r[$, let C_g be the complement of $\bigcup_{j<g} C_j$, and let $\mathbf{P}$ be the partition $\{C_1, \ldots, C_g\}$. Each point m_j is contained in the interior of C_j. As $v \to 0$, $N_{m_j, v}$ concentrates toward C_j and close to m_j; therefore, the pooled variance $V_v(\mathbf{P})$ converges to zero. Hence, (1.57) diverges to $-\infty$ as $v \to 0$ and, again by Proposition 1.58(c), so does the minimum of Φ_v on $\Theta_{\leq h, 1}$, the second half of the claim. We have seen that h is a drop point if v is small enough.

The asymptotic theory presented in this chapter will serve as an aid for recovering the true partition in Chapter 4.

1.3.8 Notes

Likelihood-based clustering consists of probabilistic *models*, *criteria*, and *algorithms*. Ironically, these components have been historically introduced in reverse order. The algorithms will be treated in Chapter 3. The classical sum-of-squares criterion 1.41 was first proposed on geometric grounds by Ward [522]. Bock [47, 48] derived it later from a Bayesian probabilistic model. Again later, Scott and Symons [461], obviously unaware of Bock's German dissertation, used the likelihood paradigm to propose a non-Bayesian, probabilistic approach to Ward's criterion. They extended it to the more general homo- and heteroscedastic ML determinant criteria, Theorems 1.40 and 1.41, also including Friedman and Rubin's [172] pooled determinant criterion. The heteroscedastic result was anticipated by John [279]. Introducing a Bayesian approach, Symons [492], criterion (11), derived the heteroscedastic MAP determinant criterion 1.40 with the entropy term.

Pollard [416, 417] was the first author to prove consistency and asymptotic normality for a classification model. In the terminology used here, he treated the homoscedastic, spherical ML criterion. His theorems are of particular practical importance since they deal with an unspecified parent. Earlier theorems about unspecified parents in different contexts are due to Foutz and Srivastava [166] and White [526].

Chapter 2

Robustification by trimming

Whereas Chapter 1 was concerned with mild outliers, this chapter will focus on gross outliers. Their major characteristic is that they do not appear to be sampled from a population. Otherwise, they could be modeled by an additional mixture component. A modification to the previous estimators will now be introduced that turns out to be effective not only against noise but also against gross contamination: trimming. Trimming is probably the oldest method of robustification, having already been known to Newcomb [387]. It provides an omnibus outlier protection. The data identifies its own outliers. Trimming was introduced to cluster analysis by Cuesta-Albertos et al. [106].

We will first take a look at outliers and at robustness measures for estimators, Section 2.1. In Sections 2.2 and 2.3, trimming methods for ML estimators of mixture and classification models will be introduced and their robustness analyzed. The main Theorems 2.5 and 2.16 state that HDBT constrained trimming methods provide substantial protection against gross outliers while preserving their equivariance properties. It will turn out that the robustness of covariance estimates differs from that of mean vectors and mixing rates. While the former are robust without any structural assumptions, the latter are more fragile. But they, too, are robust in the presence of data sets composed of substantial, well-separated clusters. This will be made precise by the Separation Properties 2.15 and 2.29.

2.1 Outliers and measures of robustness

2.1.1 Outliers

The concept of outlier was known at least to Benjamin Peirce [408]. While a cluster is defined as an element of a partition, this concept escapes a unanimously accepted, rigorous mathematical definition. Do we consider the observation 3.3 in a data set of 100 observations sampled from a standard normal an outlier? Roughly speaking, outliers are observations discordant with the posited population. They are known to severely hamper the performance of statistical methods; see Barnett and Lewis [26], Becker and Gather [34]. Since outliers indicate a disharmony between model and data, the two ways of dealing with them are modifying the model or modifying the data, that is, hiding the outliers. Outliers may roughly be divided into two types:

(i) The first type is sampled from some population different or even far from the assumed model. Such outliers document mainly the difficulty of the specification problem. They will be called **mild**. In their presence the statistician is recommended to choose a model flexible enough to accommodate all data points, including the outliers. Examples in the literature are elliptically symmetric distributions or mixture models with outliers; see Huber [254]. They have been a point of view of Chapter 1. The first probabilistic outlier models applied to mixture analysis are heavy-tailed elliptical symmetry, McLachlan and Peel [365], and the uniform model, Fraley and Raftery [170] and [171], Section 5.3. The outliers of the former authors are connected to the regular populations. The latter authors reserve a separate, uniform, component for the outliers, thereby reducing the influence of uniform "noise." Both approaches refer to special distributions and work well when they actually apply. Although they reduce the influence of certain outliers, both are not robust in the breakdown sense; see Hennig [241] and Section 2.1.3.

(ii) Probability offers models for handling uncertainties of the real world. But not every uncertainty can be described by probabilistic means. The second type of outlier is, therefore, an observation that cannot be modeled by a distribution. Such an observation is usually called a **gross outlier**. It is unpredictable and incalculable. When an experiment producing such outliers is repeated, then anything can happen – there may be fewer or more outliers and they may look very different. Such outliers typically appear when observations are automatically sampled by a machine. There are many examples. Consider an optical character recognition system (OCR) that has been designed to read the ten cyphers and 26 letters in handwritten or typed texts. First the text is segmented into separate visual objects, in general characters. This image processing phase is not perfect. It will now and then happen that two characters cling together so as to be recognized as a single object by the machine. Stains, too, may be recognized as objects. The subsequent image processing extracts features from all objects. It counts their number of pixels, takes linear measurements, and determines more sophisticated features. The data set thus created will contain the spurious objects as gross outliers. Their features will usually be far away from those of a character. Ironically, it may happen that the false measurements do have the characteristics of a letter. In this case we still have to speak of a gross outlier but it happens to appear as a regular observation. Whether a data set contains outliers at all and, if so, whether they are gross is not a priori known. In any case, we do not lose much if we safeguard our methods against the worst. A method of choice for suppressing gross outliers is trimming. Of course, for this method to be effective the estimator must recognize the outliers *automatically*. This will be the point of view of Sections 2.2 and 2.3.

Most methods presented in this text are based on the likelihood function. Although gross outliers are no probabilistic objects, we therefore need to interpret them in probabilistic terms. A proposal is found in Gallegos and Ritter [180, 183, 184] under the name of *spurious* outlier. The attribute "spurious" refers to the conception of an outlier created by a spurious effect. The observation itself does not have to be unreasonable, but it will usually be. In the communications above it is argued that the best way of handling such outliers in a statistical framework is by assuming that each of them, say i, comes from its own Bayesian model with random parameter ξ_i and prior measure τ_i. This leads to the following loose "spurious-outliers" model.

(SV$_{\mathrm{o}}$) A **spurious outlier** $X_i : \Omega \to E$, $i \in 1..n$, obeys a parametric model f_{ξ_i} with parameter $\xi_i \in \Xi_i$ such that the likelihood integrated w.r.t. some prior measure τ_i on Ξ_i satisfies

$$\int_{\Xi_i} f_{\xi_i}(x)\tau_i(\,\mathrm{d}\xi_i) = 1, \quad x \in E, \tag{2.1}$$

that is, does not depend on x. In some sense, the equality says that the density of X_i is flat. The model is probabilistic but not distributional. Its purpose is to get rid of the outliers. Since each spurious outlier is observed only once, we cannot, and do not wish to, estimate the parameters ξ_i and will treat them as nuisances. There are two important and sufficiently general situations where (SV$_{\mathrm{o}}$) holds.

(A) The sample space is Euclidean, $E = \mathbb{R}^d$, $\Xi_i = E$, the outliers obey some *location model*

$$X_i = U_i + \xi_i$$

with (unknown) random noise $U_i : (\Omega, P) \to E$, and τ_i is Lebesgue measure on Ξ_i. Here, the conditional Lebesgue density is $f_{\xi_i}(x) = f_{U_i}(x - \xi_i)$ and hence $\int_{\Xi_i} f_{\xi_i}(x)\,\mathrm{d}\xi_i = 1$.

(B) Each parameter set Ξ_i is a singleton and the distribution of X_i is the reference measure on E, so $f_{X_i} = 1$. This case includes the idea of irregular objects "uniformly distributed" on some domain.

Spurious outliers are not sampled from a probability and can be scattered anywhere in sample space. It is a most flexible outlier concept and will be used throughout. We will see that it leads to trimming. It turns out that trimming is robust, also in the context of cluster analysis. It also works in practice on a wide variety of examples. The spurious-outliers model has another advantage: After deletion of the outliers, light-tailed populations remain and so we may assume that they are *normal*. This allows application of the efficient apparatus developed for normal distributions.

2.1.2 The sensitivities

Let Θ be a locally compact parameter space. In order to assess the robustness of an estimator $t_n : E^n \to \Theta$, the addition and the replacement sensitivities are useful. Given a data set $\mathbf{x} = (x_1, \ldots, x_n)$ of points in E and an additional, arbitrary observation $x \in E$ (outlier), the **addition sensitivity** is the statistic defined by

$$x \to t_{n+1}(\mathbf{x}, x).$$

It registers to what extent the estimate changes under the influence of the outlier. The estimator is considered robust if the image $t_{n+1}(\mathbf{x}, E)$ is a relatively compact subset of Θ. This means that there are no choices of x such that the estimate $t_{n+1}(\mathbf{x}, x)$ for the augmented data set approaches the Alexandrov point of the locally compact Θ. In other words, one arbitrarily bad outlier cannot create an *arbitrarily* bad estimate. Of course, if Θ is compact, then any t_n is robust in this sense.

If Θ is even a Euclidean space (or at least a locally compact abelian group), then the addition sensitivity is intimately related to Tukey's [512] **sensitivity function**

$$SC_n(\mathbf{x}; x) = (n+1)(t_{n+1}(\mathbf{x}, x) - t_n(\mathbf{x}))$$

of which it is just an affine transformation (note that n and $\mathbf{x}$ are treated as constants here). Relative compactness in Euclidean space means boundedness. The sensitivity function is bounded as a function of x if and only if the addition sensitivity is. Boundedness of the sensitivity function therefore indicates robustness, and both concepts are equivalent if Θ is Euclidean. The sensitivity function can be derived from Hampel's [224, 225] **influence function** (influence curve); see also Huber [254]. However, the restriction to Euclidean parameter spaces is a severe loss of generality. Whereas the two concepts are applicable to location parameters, they fail when applied to scale parameters. Here the approach of an estimate to zero, too, must be considered an indication of nonrobustness.

In a similar way, the n **replacement sensitivities**, $1 \le i \le n$, are the statistics defined by

$$x \to t_n(x_1, \ldots, x_{i-1}, x, x_{i+1}, \ldots, x_n).$$

They answer the question of what the estimate would be if x_i had been replaced with an outlier x.

Let us take a look at two illustrative, classical examples. For the mean $t_n(\mathbf{x})$ of vectors $x_1, \ldots, x_n$, the ith replacement sensitivity is

$$t_n(x_1, \ldots, x_{i-1}, x, x_{i+1}, \ldots, x_n) = \frac{1}{n}\left(\sum_k x_k - x_i + x \right) = \overline{\mathbf{x}} + \frac{1}{n}(x - x_i).$$

Since this expression is not bounded in x, the sample mean is not robust. Now, let $t_n(\mathbf{x})$ be the median of the numbers $x_1, \ldots, x_n$, $n \ge 3$. As we replace any x_i with an arbitrary outlier x, the median stays within the convex hull of the original data set $\mathbf{x}$, that is,

$$t_n(x_1, \ldots, x_{i-1}, x, x_{i+1}, \ldots, x_n) \in \operatorname{conv} \mathbf{x}.$$

The convex hull does not depend on the outlier x. Thus, the median turns out robust.

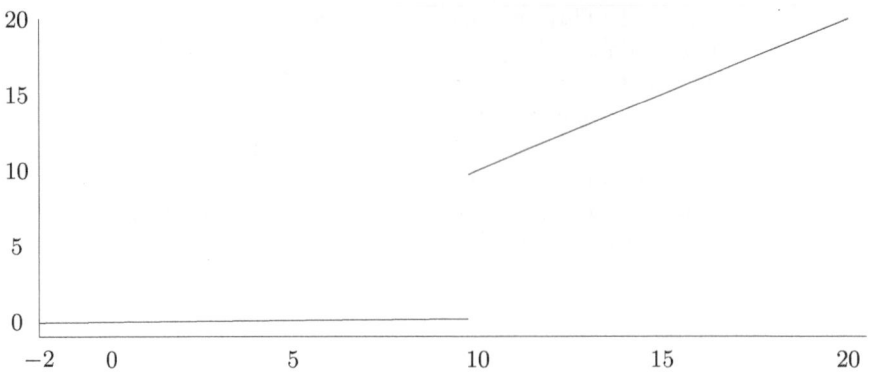

Figure 2.1: *Replacement sensitivity of the normal mixture model.*

2.1.3 Sensitivity of ML estimates of mixture models

Hennig [241] computed the sensitivity functions of the ML estimates of the means for two elliptical mixtures and for a mixture model with outliers. He found that all were unbounded. By means of illustration, let us take a look at the location estimates for a univariate, homoscedastic, normal and a Pearson Type-VII mixture. Graphical representations are customary and informative.

2.1 Examples We follow Hennig in using as a standard data set quantiles from the normal mixture $\frac{1}{2}N_{0,1} + \frac{1}{2}N_{-5,1}$. That is, both components are discretized by their 2%, 4%, ..., 98% quantiles and the data set is their union. It contains 98 points.

(a) We first study the ML parameter estimates of the homoscedastic normal mixture model with two components. Figure 2.1 shows the replacement sensitivity of the larger of the two estimated means. The largest data point is replaced with the outlier shown on the abscissa. Up to the value 9.741, the sensitivity curve is almost constant, increasing from -0.085 to 0.147. With the outlier in this range, the ML estimate separates the two clusters accommodating the outlier in the component with the larger mean, which is close to zero. At the point 9.742 a phase transition occurs. From here on, accommodation of the outlier becomes too expensive and the outlier establishes a component of its own. The other component comprises the two clusters. At this point, the estimated common variance jumps from 1.92 to 6.91 and the larger mean starts increasing to ∞.

(b) Let us next analyze the replacement sensitivity for the homoscedastic Pearson Type-VII mixture model with index 4,

$$f(x) = c_4\left(\frac{1 - \pi_2}{(1 + (x - m_1)^2/4v)^2} + \frac{\pi_2}{(1 + (x - m_2)^2/4v)^2}\right).$$

Again, the largest point in the data set is replaced with an outlier. The replacement sensitivity of the larger of the two estimated means decreases from zero to -0.025 as the outlier x moves from 2 to 30. It stays there up to about $x = 474,800$. At this point it jumps to the value x, indicating that the outlier has again established a component of its own. The estimated scale parameter jumps from 0.475 to 4.595. As in the normal model (a), the sensitivity is unbounded.

(c) In a third experiment we use again the elliptical model of example (b), but this time we shift the five (ten) largest data points to the right by the amount s. This provides information about how sensitive the estimator is w.r.t. a set of outliers. Here, phase transition occurs already at $s = 40.516$ ($s = 8.638$). At this point, the estimated scale parameter jumps from 0.555 to 4.07 (from 0.78 to 3.58).

These examples show that heavy-tailed elliptical symmetry adds some robustness to the ML estimator compared with the normal model but does not avoid unboundedness of the sensitivity function. The resistance against outliers weakens as their number increases. Thus, in the mixture context, elliptical symmetry does not create full robustness. This is in contrast to simple parameter estimation where elliptical symmetry is known to lead to robust estimates, Tyler [513]. It will turn out that trimming renders mixture models much more robust provided that a lower bound on the number of outliers is known. This phenomenon will be theoretically investigated in more generality in Sections 2.2 and 2.3. We will return to the foregoing examples in Remark 2.18 below.

2.1.4 Breakdown points

In Section 2.1.2, robustness was defined in terms of "boundedness" of the sensitivity function with respect to *one* outlier. An estimator would be even more useful if it remained stable in the presence of several outliers. This leads to the concept of the finite-sample **breakdown point** (or breakdown value) of an estimator, Hodges [249] and Donoho and Huber [128]. It measures the minimum fraction of gross outliers that can *completely* spoil the estimate. Again, two types of breakdown points are customary, the *addition* and *replacement* breakdown points. Given a natural number $r \leq n$, the former refers to the addition of $n - r$ arbitrary elements to a data set of r regular observations and the latter to $n - r$ replacements in a data set of n regular observations. The former needs a sequence of estimators since the addition increases the number of objects. The latter needs only the estimator for the given number of objects, n. For this reason, we will focus on replacements.

Let $\delta \colon E^n \to \Theta$ be a statistic partially defined on E^n with its natural domain of definition $\mathcal{A} \subseteq E^n$. The domain consists of all n-tuples for which δ is defined. For instance, for the normal MLE on $(\mathbb{R}^d)^n$ to exist we need $n \geq d + 1$ and $\mathcal{A}$ is defined by general position. Given a natural number $m \leq n$, we say that $M \in \mathcal{A}$ is an m-**modification** of $\mathbf{x} \in \mathcal{A}$ if it arises from $\mathbf{x}$ by modifying m entries in an (admissible but otherwise) arbitrary way. A statistic δ "breaks down with the (n-element) data set $\mathbf{x}$ under m replacements" if the set

$$\{\delta(M) \mid M \text{ is an } m\text{-modification of } \mathbf{x}\} \subseteq \Theta$$

is not relatively compact in Θ.[1] The **individual breakdown point** of δ at $\mathbf{x}$ is the number

$$\beta(\delta, \mathbf{x}) = \inf_{1 \leq m \leq n} \left\{ \frac{m}{n} \mid \delta \text{ breaks down with } \mathbf{x} \text{ under } m \text{ replacements} \right\}.$$

If there is such an m, then it is the minimal fraction of replacements in $\mathbf{x}$ that may cause δ to break down, otherwise it is ∞. The statistic δ does not approach the boundary of its target space even though $n\beta(\delta, \mathbf{x}) - 1$ replacements approach the boundary of the sample space. The individual breakdown point tells the statistician how many gross outliers the data set M under his or her study may contain without causing excessive damage if the imaginary "clean" data set that should have been observed had been $\mathbf{x}$. It is not an interesting concept per se since it depends on a single data set.

Donoho and Huber's breakdown point is the **universal breakdown point**

$$\beta(\delta) = \min_{\mathbf{x} \in \mathcal{A}} \beta(\delta, \mathbf{x}).$$

This concept depends solely on the statistic. A statistic with a large universal breakdown point acts robustly on any data set of size n. However, we cannot always expect the universal breakdown point to be the right instrument in order to assess robustness. Assume that we have a data set of two clear clusters but, for some reason, we insist on $g = 3$. As we replace part of one cluster with a remote cluster, then the data set consists of three clear clusters

---

[1] Of course, no breakdown is possible if Θ is compact.

and any reasonable estimator is, of course, expected to return them, also a robust one. (It cannot tell that the remote cluster is meant to consist of outliers.) But then the mean of the remote cluster breaks down. A concrete example is provided in Theorem 2.9. Thus, when dealing with location parameters in clustering, we need a more relaxed notion of a breakdown point; see Section 2.2.4. Therefore, Gallegos and Ritter [180] introduced the **restricted breakdown point** of δ w.r.t. some subclass $\mathcal{K} \subseteq \mathcal{A}$ of admissible data sets. It is defined as

$$\beta(\delta, \mathcal{K}) = \min_{\mathbf{x} \in \mathcal{K}} \beta(\delta, \mathbf{x}).$$

The restricted breakdown point depends on δ and on the subclass $\mathcal{K}$. It provides information about the robustness of δ if the hypothetic "clean" data set $\mathbf{x}$ that should have been observed instead of the contaminated data set M had been a member of $\mathcal{K}$. The restricted breakdown point is a relaxed version of the universal since we have the estimates

$$\beta(\delta) \leq \beta(\delta, \mathcal{K}) \leq \beta(\delta, \mathbf{x}), \quad \mathbf{x} \in \mathcal{K}.$$

Plainly, $\beta(\delta) = \beta(\delta, \mathcal{A})$. The asymptotic breakdown point of a sequence $(\delta_n)_n$ of statistics $\delta_n : \mathcal{A}_n \to \Theta$, $\mathcal{A}_n \subseteq E^n$, is $\liminf_{n\to\infty} \beta(\delta_n)$. If this value is strictly positive, then the sequence is called **asymptotically breakdown robust**.

As shown in Section 2.1.2, breakdown of the mean t_n occurs with one bad outlier, and this is independent of the data set. That is, all individual breakdown points are equal and equal to the universal breakdown point, $\beta(t_n) = \beta(t_n, \mathbf{x}) = \frac{1}{n}$ for all $\mathbf{x}$. The asymptotic breakdown point of the mean is zero. Section 2.1.2 also shows that the median does not break down with a single replacement. It is actually much more robust. Indeed, if n is even, then the median stays in the convex hull of the retained original observations as $\frac{n}{2} - 1$ observations are arbitrarily replaced. On the other hand, a remote cluster of $\frac{n}{2}$ replacements causes the median to break down. If n is odd, then there is no breakdown as $\frac{n-1}{2}$ observations are arbitrarily replaced and a remote cluster of $\frac{n+1}{2}$ makes the median break down. That is, $\beta(t_n) = \beta(t_n, \mathbf{x}) = \lfloor \frac{n+1}{2} \rfloor$ for all $\mathbf{x}$ of size n. The asymptotic breakdown point of the median is $1/2$.

According to Davies [109], the maximal asymptotic breakdown point of any affine equivariant covariance estimator is $1/2$. Lopuhaä and Rousseeuw [336] showed that the same value is true for any translation equivariant location estimator. Thus, the median enjoys optimal robustness. A unified upper bound for equivariant location, scale, and regression problems appears in Davies and Gather [112].

We have seen in the Examples 2.1 that the ML estimates for the mixture models treated in Chapter 1 break down with one bad outlier. García-Escudero and Gordaliza [187] showed that the influence function of the ML pooled trace criterion 1.41, too, is unbounded. Thus, the classical cluster criteria, too, are not robust. It is, however, pleasing to know that all these estimators can be modified by trimming methods so as to be even asymptotically breakdown robust. In the remainder of this chapter, the constrained, trimmed ML estimators will be introduced and breakdown points of the estimated means, mixing rates, and covariance matrices will be computed. In order to compute breakdown points in specific situations, one needs the system of compact subsets of the target space Θ of the statistic. The relatively compact subsets of the parameter space $\mathbb{R}^d$ of the means are the bounded ones. A subset of PD(d) is relatively compact if it is bounded above and below by positive definite matrices in the positive definite (or Löwner) order $\preceq$ on SYM(d). This is equivalent to saying that the eigenvalues of all its members are uniformly bounded and bounded away from zero.

2.2 Trimming the mixture model

We will first consider a trimmed version of the constrained mixture likelihood treated in Section 1.2. Since trimming removes remote data points, it is sufficient here to restrict attention to a normal basic model. Second, we will compute the breakdown points of mixing rates, means, and covariance matrices for the resulting MLEs. Conditions insuring that the estimators are asymptotically breakdown robust will be given. The presentation follows Gallegos and Ritter [182].

2.2.1 Trimmed likelihood function of the mixture model

Given a natural number $r \leq n$, we will now establish an ML estimator for normal mixtures which acts reasonably when the data contains at least r regular observations. The remaining $\leq n - r$ observations may (but do not have to) be gross, unpredictable outliers not sampled from a probability. We model them as "spurious;" see Section 2.1.1.

Let Γ be again the parameter space of an identifiably parametric basic model and let $\Psi \subseteq \Gamma^g$ be a subset of the joint parameter space representing parameter constraints. As in Chapter 1, they will again act as a theoretical panacea against any kind of problems such as missing likelihood maxima, missing consistency, and spurious solutions. It will turn out that they are also needed for robustness. The model of a *regular* observation X_i is a mixture $f(\,\cdot\,; \boldsymbol{\pi}, \boldsymbol{\gamma}) = \sum_{i=1}^{g} \pi_j f_{\gamma_j}$ of g densities $f_{\gamma_j} > 0$, $(\gamma_1, \ldots, \gamma_g) \in \Psi$, w.r.t. some fixed reference measure ϱ on E. The parametric mixture model is assumed to be identifiable (up to label switching). Outliers will be assumed to be spurious; see Eq. (2.1). The unknown quantities are the set $R \in \binom{\mathbf{x}}{r}$ of regular elements, the mixing rates, $\boldsymbol{\pi} = (\pi_1, \ldots, \pi_g)$, the population parameters $\boldsymbol{\gamma} = (\gamma_1, \ldots, \gamma_g)$, and the parameters of the spurious outliers $\boldsymbol{\xi} = (\xi_1, \ldots, \xi_n)$. Therefore, we take

$$\binom{\mathbf{x}}{r} \times \overset{\circ}{\Delta}_{g-1} \times \Psi \times \prod_{i=1}^{n} \Xi_i \tag{2.2}$$

as the parameter space of the complete mixture model with spurious outliers and trimming. The set $\binom{\mathbf{x}}{r}$ of all r-element subsets of $\mathbf{x}$ stands for the possible $\binom{n}{r}$ subsets R of regular observations. Depending on whether the ith observation x_i is regular or an outlier, the density w.r.t. ϱ is

$$f(x; R, \boldsymbol{\pi}, \boldsymbol{\gamma}, \boldsymbol{\xi}) = \begin{cases} f(x; \boldsymbol{\pi}, \boldsymbol{\gamma}), & x \in R, \\ f(x; \xi_i), & x \notin R. \end{cases}$$

We assume that the sequence of observations $X = (X_1, \ldots, X_n)$ is statistically independent but not necessarily i.i.d. unless there are no outliers, $r = n$. Therefore, the product formula implies that the joint likelihood for the data set $\mathbf{x} = (x_1, \ldots, x_n)$ is of the form

$$f_X[\mathbf{x} \mid R, \boldsymbol{\pi}, \boldsymbol{\gamma}, \boldsymbol{\xi}] = \prod_{x \in R} f(x; \boldsymbol{\pi}, \boldsymbol{\gamma}) \prod_{x \notin R} f(x; \xi_i).$$

Regarding the parameters ξ_i of the outliers as nuisances to be integrated out w.r.t. to the prior measures τ_i, we obtain along with Eq. (2.1) the **trimmed mixture likelihood**

$$f(\mathbf{x}_R; \boldsymbol{\pi}, \boldsymbol{\gamma}) = \prod_{x \in R} f(x; \boldsymbol{\pi}, \boldsymbol{\gamma}) = \prod_{x \in R} \Big(\sum_j \pi_j f(x; \gamma_j) \Big), \tag{2.3}$$

to be maximized w.r.t. the parameters $R \in \binom{\mathbf{x}}{r}$, $\boldsymbol{\pi} \in \overset{\circ}{\Delta}_{g-1}$, and $\boldsymbol{\gamma} \in \Psi$. This justifies the idea of trimming the likelihood (instead of the data), which goes back to Neykov and Neytchev [390]. In the context of the trimmed likelihood function, R will be called the set of **retained** data points and the points in its complement will be called **trimmed** or **discarded**. If, for each $R \in \binom{\mathbf{x}}{r}$, the MLE's $\boldsymbol{\pi}^*$ and $\boldsymbol{\gamma}^*$ of $\boldsymbol{\pi}$ and $\boldsymbol{\gamma}$ w.r.t. R exist, then, by the principle of dynamic optimization, the MLE of the parameters is defined by

$$\underset{R}{\operatorname{argmax}} \max_{\boldsymbol{\pi}, \boldsymbol{\gamma}} \log f(\mathbf{x}_R; \boldsymbol{\pi}, \boldsymbol{\gamma}) = \underset{R}{\operatorname{argmax}} \log f(\mathbf{x}_R; \boldsymbol{\pi}^*, \boldsymbol{\gamma}^*).$$

Its existence also depends on Ψ. The number of components, g, and the number of retained elements, r, are parameters of the model and of the estimators and the ensuing algorithms in Chapter 3. They may be used to estimate the number of components and outliers in the data, as will be shown in Chapter 4. Both are unknown in most applications.

The following result is an extension of a formula found in Hathaway [233]. Recall the definition of the weights w_j in (1.29) and abbreviate $w_j(R) = \sum_{x \in R} w_j(x)$.

2.2 Lemma *Let $(R, \boldsymbol{\pi}, \boldsymbol{\gamma})$ and $(R, \widetilde{\boldsymbol{\pi}}, \widetilde{\boldsymbol{\gamma}})$ be two parameter triples such that $\boldsymbol{\pi}, \widetilde{\boldsymbol{\pi}} \in \overset{\circ}{\Delta}_{g-1}$. Let $\mathbf{w}$ be the weight matrix (1.29) defined from $(\boldsymbol{\pi}, \boldsymbol{\gamma})$ and let $\widetilde{\mathbf{w}}$ be the corresponding one from $(\widetilde{\boldsymbol{\pi}}, \widetilde{\boldsymbol{\gamma}})$. There is the representation*

$$\log f(\mathbf{x}_R; \boldsymbol{\pi}, \boldsymbol{\gamma}) \tag{2.4}$$
$$= \sum_\ell \widetilde{w}_\ell(R) \log \pi_\ell - \sum_\ell \sum_{x \in R} \widetilde{w}_\ell(x) \log w_\ell(x) + \sum_\ell \sum_{x \in R} \widetilde{w}_\ell(x) \log f_{\gamma_\ell}(x).$$

Moreover, $\log f(\mathbf{x}_R; \boldsymbol{\pi}, \boldsymbol{\gamma}) \le \sum_\ell \sum_{x \in R} w_\ell(x) \log f_{\gamma_\ell}(x)$.

PROOF. Note that $\sum_j \pi_j f_{\gamma_j}(x) = \frac{\pi_\ell f_{\gamma_\ell}(x)}{w_\ell(x)}$ for all ℓ. Since $\sum_\ell \widetilde{w}_\ell(x) = 1$, Eq. (2.3) implies

$$\log f(\mathbf{x}_R; \boldsymbol{\pi}, \boldsymbol{\gamma}) = \sum_{x \in R} \log \sum_{j=1}^g \pi_j f_{\gamma_j}(x) = \sum_{x \in R} \sum_\ell \widetilde{w}_\ell(x) \log \sum_{j=1}^g \pi_j f_{\gamma_j}(x)$$
$$= \sum_{x \in R} \sum_\ell \widetilde{w}_\ell(x) \log \frac{\pi_\ell f_{\gamma_\ell}(x)}{w_\ell(x)}.$$

This is the first claim. For the second, let $(\boldsymbol{\pi}, \boldsymbol{\gamma}) = (\widetilde{\boldsymbol{\pi}}, \widetilde{\boldsymbol{\gamma}})$ in the first claim and apply the entropy inequality to the probabilities $(w_\ell(x))_\ell$ and $(\pi_\ell)_\ell$.   □

If the weights $w_j^*(x)$ are defined by an ML estimate $(R^*, \boldsymbol{\pi}^*, \boldsymbol{\gamma}^*)$, $\boldsymbol{\pi}^* \in \overset{\circ}{\Delta}_{g-1}$, then there is the representation

$$\pi_j^* = \tfrac{1}{r} w_j^*(R^*). \tag{2.5}$$

Indeed, $\boldsymbol{\pi}^*$ is a maximum of the function

$$\boldsymbol{\pi} \mapsto \sum_{x \in R} \log \left(\left(1 - \sum_{j=2}^g \pi_j \right) f_{\gamma_1}(x) + \sum_{j=2}^g \pi_j f_{\gamma_j}(x) \right).$$

The partial derivatives w.r.t. π_j, $j \ge 2$, yield

$$\sum_{x \in R} f_{\gamma_j^*}(x) \Big/ \sum_j \pi_j^* f_{\gamma_j^*}(x) = \sum_{x \in R} f_{\gamma_1^*}(x) \Big/ \sum_j \pi_j^* f_{\gamma_j^*}(x)$$

and hence $w_j^*(R)/\pi_j^* = w_1^*(R)/\pi_1^*$, from which the claim follows.

2.2.2 Normal components

We now specialize Section 2.2.1 to the HDBT constrained fully, diagonally, and spherically normal mixture models. Thus, E is here Euclidean d-space and the parameter space is $\Psi = \mathbb{R}^{gd} \times \mathcal{V}_c$ with $\mathcal{V}_c \subseteq \mathrm{PD}(d)^g$ as in Sections 1.2.2 and 1.2.5. The trimmed likelihood (2.3) becomes the **normal trimmed mixture likelihood**

$$f(\mathbf{x}_R; \boldsymbol{\pi}, \mathbf{m}, \mathbf{V}) = \prod_{x \in R} \sum_{j=1}^g \pi_j N_{m_j, V_j}(x). \tag{2.6}$$

In the homoscedastic case, we have $V_1 = \cdots = V_g = V$, and the HDBT constrained heteroscedastic case is characterized by $\mathbf{V} \in \mathcal{V}_c$ with $c < 1$.

Besides the *full* model, two submodels specified by special population scales are customary in each of the two cases, *diagonal* covariance matrices $V_j = \mathrm{Diag}(v_{j,1}, \ldots, v_{j,d})$ and *spherical* covariance matrices $V_j = v_j I_d$. We will make the following *standard assumptions*.

(i) The data set $\mathbf{x}$ is in general position.

(ii) The number r of retained elements satisfies $r \geq gd + 1$ in the full case, and $r > g + 1$ in the diagonal and spherical cases.

General position could have different meanings in the three normal cases. For the sake of simplicity we will use the most stringent one – any $d + 1$ points in $\mathbf{x}$ are affine independent.

Let us agree on the following notation. Given a subset $T \subseteq \mathbf{x}$, the symbols $\bar{x}_T$, W_T, and S_T designate the *sample mean vector*, the *SSP matrix*, and the *scatter matrix* of T, respectively. We will also need *weighted* analogs of these statistics w.r.t. a weight vector $w = (w(x))_{x \in T}$ of real numbers $w(x) \geq 0$ (in most cases $w = w_j$, the jth column in a stochastic weight matrix $\mathbf{w} = (w_j(x))_{x \in T, j \in 1 \ldots g}$, $\sum_j w_j(x) = 1$ for all $x \in T$). Writing $w(T) = \sum_{x \in T} w(x)$, we define them, respectively, as

$$\bar{x}_T(w) = \tfrac{1}{w(T)} \sum_{x \in T} w(x)x \quad (= 0, \text{ if } w(T) = 0),$$

$$W_T(w) = \sum_{x \in T} w(x)(x - \bar{x}_T(w))(x - \bar{x}_T(w))^\top, \text{ and}$$

$$S_T(w) = \tfrac{1}{w(T)} W_T(w) \quad (= I_d, \text{ if } w(T) = 0).$$

The *pooled weighted SSP matrix* and the *pooled weighted scatter matrix* w.r.t. a stochastic weight matrix $\mathbf{w}$ are, respectively,

$$W_T(\mathbf{w}) = \sum_j W_T(w_j) \text{ and } S_T(\mathbf{w}) = \tfrac{1}{|T|} W_T(\mathbf{w}).$$

If weights are binary, then $\mathbf{w}$ defines in a natural way a partition of T and $W_T(\mathbf{w})$ and $S_T(\mathbf{w})$ reduce to the ordinary pooled quantities.

We will repeatedly use the MAP partition $\{T_1, \ldots, T_g\}$ of some subset $T \subseteq \mathbf{x}$ w.r.t. some stochastic weight matrix $\mathbf{w} = (w_j(x))_{x \in T, j \in 1 \ldots g}$: $x \in T_\ell \Leftrightarrow \ell = \operatorname{argmax}_j w_j(x)$. The obvious estimate

$$w_j(x) \geq 1/g \quad \text{for all } x \in T_j \tag{2.7}$$

and the weighted Steiner formula A.12 imply for all j

$$W_T(w_j) \succeq \sum_{x \in T_j} w_j(x)(x - \bar{x}_T(w_j))(x - \bar{x}_T(w_j))^\top \succeq \tfrac{1}{g} W_{T_j}. \tag{2.8}$$

The optimal parameters in the HDBT *constrained heteroscedastic* case can be computed from their weights. Thus, knowing the optimal weights is tantamount to knowing the optimal parameters.

2.3 Proposition *Let* $(R^*, \boldsymbol{\pi}^*, \mathbf{m}^*, \mathbf{V}^*)$ *be a constrained ML estimate and let*

$$w_\ell^*(x) = \frac{\pi_\ell^* N_{m_\ell^*, V^*}(x)}{\sum_{j=1}^g \pi_j^* N_{m_j^*, V^*}(x)}$$

be the posterior distribution for $x \in \mathbb{R}^d$ *to come from component* ℓ; *see* (1.29). *Then*

$$\pi_j^* = \tfrac{1}{r} w_j^*(R^*);$$

$$m_j^* = \bar{x}_{R^*}(w_j^*) = \frac{1}{w_j^*(R^*)} \sum_{x \in R} w_j^*(x)x; \tag{2.9}$$

$$\mathbf{V}^* = \underset{\mathbf{V} \in \mathcal{V}_c}{\operatorname{argmin}} \sum_j w_j^*(R^*) \big(\log \det V_j + \operatorname{tr} S_{R^*}(w_j^*) V_j^{-1} \big). \tag{2.10}$$

PROOF. The first claim was already shown in more generality in Section 2.2.1. For the second claim we use the fact that $\mathbf{m}^*$ is a critical point of the function $\mathbf{m} \mapsto$

$\sum_{x \in R^*} \log \sum_j \pi_j^* N_{m_j, V_j^*}(x)$. We obtain that $\sum_{x \in R} w_j^*(x)(x - m_j^*)^\top (V_j^*)^{-1}$ vanishes for all j and the claim follows.

In order to see (2.10) denote the sum there by $h(\mathbf{V})$, $\mathbf{V} \in \mathcal{V}_c$. Lemma 2.2 applied with $(R, \boldsymbol{\pi}, \mathbf{m}, \mathbf{V}) = (R^*, \boldsymbol{\pi}^*, \mathbf{m}^*, \mathbf{V})$ and $(R, \widetilde{\boldsymbol{\pi}}, \widetilde{\mathbf{m}}, \widetilde{\mathbf{V}}) = (R^*, \boldsymbol{\pi}^*, \mathbf{m}^*, \mathbf{V}^*)$ shows

$$\log f(\mathbf{x}_{R^*}; \boldsymbol{\pi}^*, \mathbf{m}^*, \mathbf{V})$$

$$= \sum_\ell w_\ell^*(R^*) \log \pi_\ell^* - \sum_\ell \sum_{x \in R^*} w_\ell^*(x) \log w_\ell(x) + \sum_\ell \sum_{x \in R^*} w_\ell^*(x) \log N_{m_\ell^*, V_\ell}(x)$$

$$= \sum_\ell w_\ell^*(R^*) \log \pi_\ell^* - \sum_\ell \sum_{x \in R^*} w_\ell^*(x) \log w_\ell(x) + \tfrac{rd}{2} \log 2\pi - \tfrac{1}{2} h(\mathbf{V}),$$

where the weights $w_j(x)$ and $w_j^*(x)$ are defined with $(\boldsymbol{\pi}^*, \mathbf{m}^*, \mathbf{V})$ and $(\boldsymbol{\pi}^*, \mathbf{m}^*, \mathbf{V}^*)$, respectively. Now, put $\mathbf{V} = \mathbf{V}^*$ to also obtain

$$\log f(\mathbf{x}_{R^*}; \boldsymbol{\pi}^*, \mathbf{m}^*, \mathbf{V}^*) \tag{2.11}$$

$$= \sum_\ell w_\ell^*(R^*) \log \pi_\ell^* - \sum_\ell \sum_{x \in R^*} w_\ell^*(x) \log w_\ell^*(x) + \tfrac{rd}{2} \log 2\pi - \tfrac{1}{2} h(\mathbf{V}^*).$$

Subtracting these two equalities, we find for an arbitrary $\mathbf{V} \in \mathcal{V}_c$

$$\tfrac{1}{2}(h(\mathbf{V}^*) - h(\mathbf{V}))$$

$$= \log f(\mathbf{x}_{R^*}; \boldsymbol{\pi}^*, \mathbf{m}^*, \mathbf{V}) - \log f(\mathbf{x}_{R^*}; \boldsymbol{\pi}^*, \mathbf{m}^*, \mathbf{V}^*) + \sum_{x \in R^*} \sum_\ell w_\ell^*(x) \log \frac{w_\ell(x)}{w_\ell^*(x)} \le 0$$

by maximality and by the entropy inequality. This implies (2.10). $\square$

For $c < 1$, there does not seem to be a representation of $\mathbf{V}^*$ in closed form. In general, it depends on the (unknown) constant c. In the *homoscedastic*, normal case, $c = 1$, the optimal estimate of the common scale matrix V, too, can be represented in closed form. Here, the sum in (2.10) simplifies to

$$\sum_j w_j^*(R^*) \big(\log \det V + \operatorname{tr} S_{R^*}(w_j^*) V^{-1}$$

$$= r \log \det V + \operatorname{tr} \Big(\sum_j w_j^*(R^*) S_{R^*}(w_j^*) V^{-1} \Big) = r \log \det V + \operatorname{tr} W_{R^*}(\mathbf{w}^*) V^{-1}.$$

Normal estimation theory shows that the minimizer of this expression given R^* and $\mathbf{w}^*$ is

$$\begin{cases} V^* = S_{R^*}(\mathbf{w}^*) & \text{(full)}, \\ v_k^* = V^*(k, k) & \text{(diagonal)}, \\ v^* = \tfrac{1}{d} \sum_k v_k^* & \text{(spherical)}, \end{cases} \tag{2.12}$$

$k \in 1..d$. Furthermore, the trimmed likelihood (2.6) of $(\boldsymbol{\pi}^*, \mathbf{m}^*, V^*)$ w.r.t. R^* assumes the form

$$\log f(\mathbf{x}_{R^*}; \boldsymbol{\pi}^*, \mathbf{m}^*, V^*)$$

$$= c_{d,r} + r \sum_{j=1}^g \pi_j^* \log \pi_j^* - \sum_{j=1}^g \sum_{x \in R^*} w_j^*(x) \log w_j^*(x) - \tfrac{r}{2} \log \det V^*, \tag{2.13}$$

with $c_{d,r} = -\frac{dr}{2}(1 + \log 2\pi)$ and $w_\ell^*(x) = \frac{\pi_\ell^* N_{m_\ell^*, V^*}(x)}{\sum_{j=1}^g \pi_j^* N_{m_j^*, V^*}(x)}$. Just insert π_j^* and $\mathbf{m}^*$ from Proposition 2.3 and V^* from (2.12) into the trimmed likelihood (2.11) and apply standard matrix analysis.

2.2.3   Universal breakdown points of covariance matrices, mixing rates, and means

We next investigate the universal breakdown points of all parameters of a mixture. The following lemma will be crucial for the theoretical analyses. It makes up for the missing representation of the minimizer of (2.10). Part (a) is a special case of Lemma 1.21.

2.4 Lemma *Let R be some data set in $\mathbb{R}^d$ of cardinality r and let $\mathbf{V} \in \mathcal{V}_c$.*
   (a) With $w_\ell(x) = \frac{\pi_\ell N_{m_\ell, V_\ell}(x)}{\sum_{j=1}^g \pi_j N_{m_j, V_j}(x)}$, $x \in R$, $1 \le \ell \le g$, we have for all j

$$2 \log f(\mathbf{x}_R; \boldsymbol{\pi}, \mathbf{m}, \mathbf{V}) \le -r \log \det 2\pi c V_j - c \operatorname{tr}\left(W_R(\mathbf{w}) V_j^{-1}\right)$$

and there exists j such that

$$2 \log f(\mathbf{x}_R; \boldsymbol{\pi}, \mathbf{m}, \mathbf{V}) \le -r \log \det 2\pi V_j - c \operatorname{tr}\left(W_R(\mathbf{w}) V_j^{-1}\right).$$

(b) If $|R \cap \mathbf{x}| \ge gd + 1$ and if $\mathbf{x}$ is in general position, there is a constant $K > 0$ that depends only on $\mathbf{x}$ such that, for all j,

$$2 \log f(\mathbf{x}_R; \boldsymbol{\pi}, \mathbf{m}, \mathbf{V}) \le -r \log \det 2\pi c V_j - cK \operatorname{tr} V_j^{-1}. \tag{2.14}$$

PROOF. (a) By Lemma 2.2, the HDBT constraints, and by Steiner's formula A.12, we have

$$2 \log f(\mathbf{x}_R; \boldsymbol{\pi}, \mathbf{m}, \mathbf{V}) \le 2 \sum_\ell \sum_{x \in R} w_\ell(x) \log N_{m_\ell, V_\ell}(x)$$

$$= - \sum_\ell \sum_{x \in R} w_\ell(x)\left(\log \det 2\pi V_\ell + (x - m_\ell)^\top V_\ell^{-1}(x - m_\ell)\right)$$

$$\le - \sum_\ell \sum_{x \in R} w_\ell(x)\left(\log \det 2\pi c V_j + c(x - m_\ell)^\top V_j^{-1}(x - m_\ell)\right)$$

$$= - r \log \det 2\pi c V_j - c \operatorname{tr} \sum_\ell \sum_{x \in R} w_\ell(x)(x - m_\ell)(x - m_\ell)^\top V_j^{-1}$$

$$\le - r \log \det 2\pi c V_j - c \operatorname{tr} W_R(\mathbf{w}) V_j^{-1}.$$

This is the first claim in (a). Now choose in the third line the index j that minimizes $\det V_j$ and drop the constant c under the determinant. The second claim follows.

(b) Let $\{R_1, \ldots, R_g\}$ be the MAP partition of R w.r.t. $\mathbf{w}$ in (a). By assumption on $|R \cap \mathbf{x}|$ there is a subset R_ℓ that contains at least $d + 1$ elements of $\mathbf{x}$. By general position, W_{R_ℓ} is regular and, by Eq. (2.8), $W_R(\mathbf{w}) \succeq W_{R,\ell}(\mathbf{w}) \succeq \frac{1}{g} W_{R_\ell} \succeq K I_d$ with some constant $K > 0$ that depends only on $\mathbf{x}$. The claim therefore follows from (a). $\square$

We infer from Proposition 1.24 that an MLE exists for the trimmed, normal model under scale constraints as long as $r \ge gd + 1$,

$$\underset{R}{\operatorname{argmax}} \underset{\boldsymbol{\pi}, \mathbf{m}, \mathbf{V} \in \mathcal{V}_c}{\max} \log f(\mathbf{x}_R; \boldsymbol{\pi}, \mathbf{m}, \mathbf{V}) = \underset{R}{\operatorname{argmax}} \log f(\mathbf{x}_R; \boldsymbol{\pi}^*, \mathbf{m}^*, \mathbf{V}^*). \tag{2.15}$$

We show next that the covariance matrices of this MLE are robust and compute their individual breakdown point.

2.5 Theorem (Individual breakdown point of the covariance matrices)

(a) Assume $2r \ge n + g(d + 1)$. The constrained, trimmed normal MLE's of all covariance matrices remain in a compact subset of $\mathrm{PD}(d)$ that depends only on $\mathbf{x}$ as at most $n - r + g - 1$ data points are replaced in an arbitrary but admissible way.

(b) They break down as $n - r + g$ data points are suitably replaced.

(c) Under the assumption of (a) the individual breakdown value of the estimates of the covariance matrices is for all data sets $\mathbf{x}$

$$\beta_{\mathrm{Cov}}(n, g, r, \mathbf{x}) = \tfrac{1}{n}(n - r + g).$$

PROOF. (a) We first show that, no matter what the modified data set M is, the constrained maximum of the trimmed mixture likelihood remains bounded below by a strictly positive constant. The constant is determined by a simple mixture that is sufficient for our purpose. We choose as R the remaining $n - (n - r + g - 1) = r - g + 1$ original observations and $g - 1$ of the replacements. Without loss of generality, let the original data be $x_1, \ldots, x_{r-g+1}$ and the replacements $y_1, \ldots, y_{g-1}$. Let $\pi_1 = \cdots = \pi_g = \frac{1}{g}$, $m_1 = 0$, $m_j = y_{j-1}$, $j \in 2..g$, and $V_j = I_d$. The trimmed mixture likelihood is

$$\prod_{i=1}^{r-g+1} \frac{1}{g} \left\{ (2\pi)^{-d/2} e^{-\|x_i\|^2/2} + \sum_{j=1}^{g-1} (2\pi)^{-d/2} e^{-\|x_i - y_j\|^2/2} \right\}$$

$$\times \prod_{i=1}^{g-1} \frac{1}{g} \left\{ (2\pi)^{-d/2} e^{-\|y_i\|^2/2} + \sum_{j=1}^{g-1} (2\pi)^{-d/2} e^{-\|y_i - y_j\|^2/2} \right\}$$

$$\geq \prod_{i=1}^{r-g+1} \frac{1}{g} \left\{ (2\pi)^{-d/2} e^{-\|x_i\|^2/2} \right\} \prod_{i=1}^{g-1} \frac{1}{g} (2\pi)^{-d/2} \geq (2\pi)^{-dr/2} g^{-r} e^{-\|\mathbf{x}\|^2/2}$$

$$= C_{\mathbf{x}},$$

a strictly positive constant.

Now, by assumption, any r-element subset R of the modified data set contains at least $r - (n - r + g - 1) = 2r - n - g + 1 \geq gd + 1$ original points. Therefore, Lemma 2.4(b) may be applied and we have thus shown

$$2 \log C_{\mathbf{x}} \leq 2 \log f(\mathbf{x}_R; \boldsymbol{\pi}^*, \mathbf{m}^*, \mathbf{V}^*) \leq -r \log \det 2\pi c V_j^* - K_{\mathbf{x}} \operatorname{tr}\left(V_j^*\right)^{-1}, \quad 1 \leq j \leq g,$$

with some constant $K_{\mathbf{x}}$ that depends only on $\mathbf{x}$. It is well known that the set of matrices $V_j \in \mathrm{PD}(d)$ for which the right side is bounded below is compact.

(b) Let M be the data set $\mathbf{x}$ modified by $n - r + g$ replacements to be specified later. Let $(R^*, \boldsymbol{\pi}^*, \mathbf{m}^*, \mathbf{V}^*)$ be the constrained maximum of the trimmed mixture likelihood for M with associated weight matrix $\mathbf{w}^*$ and let $(R_1, \ldots, R_g)$ be the MAP partition of the r-element subset $R^* \subseteq M$ obtained from the stochastic matrix $\mathbf{w}^*$. By assumption, R^* contains at least g replacements. Hence, *either* one cluster contains at least two replacements *or* each cluster contains at least one replacement, in particular some cluster with $\geq d + 1$ elements. In any case, the partition has a cluster R_ℓ containing a replacement y and some other element x. Equation (2.8) implies

$$W_{R^*}(w_\ell^*) \succeq \tfrac{1}{g} W_{R_\ell}$$

$$\succeq \tfrac{1}{g}\left((y - \tfrac{1}{2}(y + x))(y - \tfrac{1}{2}(y + x))^\top + (x - \tfrac{1}{2}(y + x))(x - \tfrac{1}{2}(y + x))^\top\right)$$

$$= \tfrac{1}{2g}(y - x)(y - x)^\top.$$

Now, since also $2\mathbf{V}^* \in \mathcal{V}_c$, we infer from minimality (2.10) of $(\boldsymbol{\pi}^*, \mathbf{m}^*, \mathbf{V}^*)$

$$0 \leq \sum_j w_j^*(R^*)\left\{ \log \det 2V_j^* + \operatorname{tr} S_{R^*}(w_j^*)(2V_j^*)^{-1} - \log \det V_j^* - \operatorname{tr} S_{R^*}(w_j^*)(V_j^*)^{-1} \right\}$$

$$= rd \log 2 - \tfrac{1}{2} \sum_j \operatorname{tr} W_{R^*}(w_j^*)(V_j^*)^{-1} \leq rd \log 2 - \tfrac{1}{2} \operatorname{tr} W_{R^*}(w_\ell^*)(V_\ell^*)^{-1}$$

$$\leq rd \log 2 - \tfrac{1}{4g}(y - x)^\top (V_\ell^*)^{-1}(y - x).$$

The estimate $(y-x)^\top (V_\ell^*)^{-1}(y-x) \leq 4grd \log 2$ obtained proves that the smallest eigenvalue of $(V_\ell^*)^{-1}$ approaches zero arbitrarily closely if the replacements are chosen so as to be far away from all original data and from each other.

Part (c) follows from (a) and (b).   □

The assumptions $2r \geq n + g(d+1)$ and $2r \geq n + 2g$ made in the theorem state that the number $n - r$ of discards must be somewhat smaller than $n/2$. It is interesting to remark that the estimates in Theorem 2.5 withstand $g - 1$ more outliers than they discard, $n - r$. This fact has a simple reason. The constraints effect that outliers that are spread out may each create a component of their own and outliers located close together may create a common component. In each case the covariance matrices of the optimal mixture do not completely break down.

2.6 Corollary (a) *The maximal number of outliers that the estimates of the (full) covariance matrices in Theorem 2.5 can resist is* $\lfloor \frac{n - g(d-1)}{2} \rfloor - 1$. *The parameter r has to be set to* $\lceil \frac{n + g(d+1)}{2} \rceil$.

(b) *The asymptotic breakdown point is $1/2$ in each case.*

PROOF. (a) We are asking for the largest integer $n - r + g - 1$ under the constraint $2r \geq n + g(d+1)$. This proves part (a) and (b) is immediate. □

Part (b) of the corollary says that the ML estimates of the covariance matrices attain the asymptotic breakdown point, $1/2$. Davies [109] has shown that this is the maximum any affine equivariant covariance estimator can attain. Despite trimming and the constraints, the *universal* breakdown point of the sample mean is small. To show this we first need two lemmas.

2.7 Lemma *Let $1 \leq q \leq r$, let $R = (x_1, \ldots, x_{r-q}, y_1, \ldots, y_q)$ consist of $r - q$ original data points x_i and q replacements y_i, and let $(\boldsymbol{\pi}^*, \mathbf{m}^*, \mathbf{V}^*)$ be parameters optimal for R. Then* $\max_j \|m_j^*\| \longrightarrow \infty$ *as* $\|y_1\| \to \infty$ *such that $y_i - y_1$, $2 \leq i \leq q$, remain bounded.*

PROOF. Let w_j^* be the weights induced by the optimal parameters. Equation (2.9) implies

$$\sum_j m_j^* \left(\sum_{i=1}^{r-q} w_j^*(x_i) + \sum_{i=1}^{q} w_j^*(y_i) \right) = \sum_{i=1}^{r-q} x_i + \sum_{i=1}^{q} y_i = \sum_{i=1}^{r-q} x_i + q y_1 + \sum_{i=1}^{q} (y_i - y_1)$$

and the claim follows since the quantities in parentheses on the left side remain bounded. □

2.8 Lemma *Let $g \geq 2$, $p \geq 2$, $q \geq g - 2$, and $r = p + g$ be natural numbers and let*

$$M = \{x_1, \ldots, x_p\} \cup \{y_1, y_2\} \cup \{z_1, \ldots, z_q\},$$

with pairwise disjoint elements x_i, y_h, and z_l. Any partition of a subset of M of size r in g clusters is either of the form

$$\mathcal{R}^\star = \{\{x_1, \ldots, x_p\}, \ \{y_1, y_2\}, \ g\text{–}2 \text{ singletons } \{z_l\}\}$$

or has a cluster C that contains some pair $\{x_i, y_h\}$ or some pair $\{z_l, u\}$, $u \neq z_l$.

PROOF. Let n_x, n_y, and n_z be the numbers of x_i's, y_h's, and z_l's, respectively, that make up the configuration $\mathcal{R}$. By assumption, $n_x + n_y + n_z = r$ and hence

$$n_z = r - n_x - n_y \geq r - p - 2 = g - 2.$$

The claim being trivial if $n_z > g$, we consider the three remaining cases $n_z = g$, $n_z = g - 1$, and $n_z = g - 2$ separately. Now, $n_z = g$ implies $n_x + n_y = r - g = p \geq 2$. Therefore, if no cluster contains two z_l's, then one cluster must contain some z_l together with an x_i or a y_h. If $n_z = g - 1$, then $n_x + n_y = r - g + 1 = p + 1$; since $p \geq 2$, at least one x_i and one y_h must belong to the configuration. A simple counting argument shows the claim in this case. Finally, if $n_z = g - 2$, then $n_x + n_y = r - g + 2 = p + 2$, that is, all x_i's and all y_h's belong to the configuration. If all elements z_l form one-point clusters, then the x_i's and y_h's must share the remaining two clusters. If they are separated, then $\mathcal{R} = \mathcal{R}^\star$. In the opposite case, some cluster contains both an x_i and a y_h. □

We next examine the breakdown point of the ML estimates of the mean vectors for the HDBT constrained, trimmed normal model.

2.9 Theorem (Universal breakdown point of the means) Let $g \geq 2$.

(a) Assume $r < n$ and $r \geq gd + 2$. The constrained, trimmed, normal MLE's of all mean vectors remain bounded by a constant that depends only on the data set $\mathbf{x}$ as *one* observation is arbitrarily replaced.

(b) Assume $r \geq g + 2$ and $r \geq gd + 1$. There is a data set in general position such that the MLE of one mean vector breaks down as *two* particular observations are suitably replaced.

(c) Under the assumptions of (a), we have $\beta_{\mathrm{mean}}(n, g, r) = \frac{2}{n}$.

PROOF. (a) Let $M = \{x_1, \ldots, x_{n-1}, y\}$ be a modification of $\mathbf{x}$ by one admissible replacement y. Let the solution $(\widetilde{R}, \widetilde{\boldsymbol{\pi}}, \widetilde{\mathbf{m}}, \widetilde{\mathbf{V}})$ be optimal for M with $r < n$ under the condition that y is not discarded. We will show that $(\widetilde{R}, \widetilde{\boldsymbol{\pi}}, \widetilde{\mathbf{m}}, \widetilde{\mathbf{V}})$ is inferior to some solution which discards y if y is sufficiently distant. Let $\widetilde{d}_j$ denote the Mahalanobis distance induced by $\widetilde{V}_j$, that is, $\widetilde{d}_j(u, v) = \sqrt{(u-v)^{\mathsf{T}} \widetilde{V}_j^{-1}(u-v)}$, let $\widetilde{d}_j(u, \mathbf{x}) = \min_{v \in \mathbf{x}} \widetilde{d}_j(u, v)$, and $\widetilde{d}_j(\mathbf{x}) = \max_{u, v \in \mathbf{x}} \widetilde{d}_j(u, v)$ denote the distance between u and $\mathbf{x}$ and the diameter of $\mathbf{x}$ w.r.t. $\widetilde{d}_j$, respectively.

Without loss of generality, $\widetilde{R} = (x_1, \ldots, x_{r-1}, y)$. Let $R = (x_1, \ldots, x_r)$ and let

$$
m_j = \begin{cases} x_r, & \text{if } \widetilde{d}_j(\widetilde{m}_j, \mathbf{x}) > \widetilde{d}_j(\mathbf{x}), \\ \widetilde{m}_j, & \text{otherwise.} \end{cases}
$$

We show that the solution $(\widetilde{R}, \widetilde{\boldsymbol{\pi}}, \widetilde{\mathbf{m}}, \widetilde{\mathbf{V}})$ is inferior to $(R, \widetilde{\boldsymbol{\pi}}, \mathbf{m}, \widetilde{\mathbf{V}})$ if y is such that $\widetilde{d}_j(y, \mathbf{x}) > 3\widetilde{d}_j(\mathbf{x})$ for all j. Comparing the trimmed likelihood of the former

$$
\left(\prod_{i=1}^{r-1} \sum_{j=1}^{g} \widetilde{u}_j N_{\widetilde{m}_j, \widetilde{V}_j}(x_i) \right) \sum_{j=1}^{g} \widetilde{u}_j N_{\widetilde{m}_j, \widetilde{V}_j}(y)
$$

termwise with that of the latter

$$
\left(\prod_{i=1}^{r-1} \sum_{j=1}^{g} \widetilde{u}_j N_{m_j, \widetilde{V}_j}(x_i) \right) \sum_{j=1}^{g} \widetilde{u}_j N_{m_j, \widetilde{V}_j}(x_r),
$$

we see that it is sufficient to show $\widetilde{d}_j(x_i, x_r) < \widetilde{d}_j(x_i, \widetilde{m}_j)$, $i < r$, if j is such that $\widetilde{d}_j(\widetilde{m}_j, \mathbf{x}) > \widetilde{d}_j(\mathbf{x})$ and $\widetilde{d}_j(x_r, \widetilde{m}_j) < \widetilde{d}_j(y, \widetilde{m}_j)$ in the opposite case.

Now, if $\widetilde{d}_j(\widetilde{m}_j, \mathbf{x}) > \widetilde{d}_j(\mathbf{x})$, then $\widetilde{d}_j(x_i, x_r) \leq \widetilde{d}_j(\mathbf{x}) < \widetilde{d}(\widetilde{m}_j, \mathbf{x}) \leq \widetilde{d}_j(x_i, \widetilde{m}_j)$; if $\widetilde{d}_j(\widetilde{m}_j, \mathbf{x}) \leq \widetilde{d}_j(\mathbf{x})$, then

$$
\widetilde{d}_j(y, \widetilde{m}_j) \geq \widetilde{d}_j(y, \mathbf{x}) - \widetilde{d}_j(\widetilde{m}_j, \mathbf{x}) > 3\widetilde{d}_j(\mathbf{x}) - \widetilde{d}_j(\widetilde{m}_j, \mathbf{x}) \geq \widetilde{d}_j(\mathbf{x}) + \widetilde{d}_j(\widetilde{m}_j, \mathbf{x}) \geq \widetilde{d}_j(x_r, \widetilde{m}_j).
$$

In order to prove that the means remain bounded, we still have to prove that the locations of the replacement y where it is not necessarily discarded are bounded by a constant that depends only on $\mathbf{x}$. (Note that $\widetilde{V}_j$ and hence the distance $\widetilde{d}_j$ depends on y!) In other words, we have to show that the sets $\{y \mid \widetilde{d}_j(y, \mathbf{x}) \leq 3\widetilde{d}_j(\mathbf{x})\}$ are bounded by constants that depend only on $\mathbf{x}$. To this end we next show that $\widetilde{V}_j$ is bounded below and above by positive definite matrices L_j and U_j that depend only on $\mathbf{x}$.

Indeed, the optimal parameters $(\widetilde{R}, \widetilde{\boldsymbol{\pi}}, \widetilde{\mathbf{m}}, \widetilde{\mathbf{V}})$ are superior to the parameters

$\left(\widetilde{R}, \left(\frac{1}{g}, \ldots, \frac{1}{g}\right), (0, \ldots, 0, y), I_d\right)$, that is,

$$f(\mathbf{x}_{\widetilde{R}}; \widetilde{\boldsymbol{\pi}}, \widetilde{\mathbf{m}}, \widetilde{\mathbf{V}}) \geq f(\mathbf{x}_{\widetilde{R}}; (1/g, \ldots, 1/g), (0, \ldots, 0, y), I_d)$$

$$= g^{-r} \prod_{i<r} \left(\sum_{j<g} N_{0,I_d}(x_i) + N_{y,I_d}(x_i)\right) \left(\sum_{j<g} N_{0,I_d}(y) + N_{y,I_d}(y)\right)$$

$$\geq g^{-r} \left(\prod_{i<r} \sum_{j<g} N_{0,I_d}(x_i)\right) N_{0,I_d}(0) =: c_{\mathbf{x}}.$$

The constant $c_{\mathbf{x}}$ does not depend on y. Since $r \geq gd+2$, $\widetilde{R}$ contains at least $gd+1$ original elements and Lemma 2.4(b) shows

$$2\log c_{\mathbf{x}} \leq 2\log f(\mathbf{x}_{\widetilde{R}}; \widetilde{\boldsymbol{\pi}}, \widetilde{\mathbf{m}}, \widetilde{\mathbf{V}}) \leq -r\log\det 2\pi c\widetilde{V}_j - K_{\mathbf{x}}\operatorname{tr}\widetilde{V}_j^{-1},$$

that is, the right side is bounded below by a constant that depends only on $\mathbf{x}$. Its behavior as a function of $\widetilde{V}_j$ provides two matrices L_j and U_j as required. Denoting the Mahalanobis distances w.r.t. L_j and U_j by d_{L_j} and d_{U_j}, respectively, the claim finally follows from

$$d_{U_j}(y, x_1) \leq \widetilde{d}_j(y, x_1) \leq \widetilde{d}_j(y, \mathbf{x}) + \widetilde{d}_j(\mathbf{x}) \leq 4\widetilde{d}_j(\mathbf{x}) \leq 4d_{L_j}(\mathbf{x}).$$

(b) We proceed in several steps.

(α) Construction of the data set $\mathbf{x}$ and its modification M:

Let $F = \{x_1, \ldots, x_{r-g}\}$ be any set of data points in general position. We complete F to a data set $\mathbf{x}$ by points which we control by a constant $K_1 > 0$ and we control the two replacements by another constant $K_2 > 0$. Both constants will be specified later. The idea will be to place points z_ℓ and the replacements y_k in directions of space so that general position is conserved as homotheties are applied. Using Lemma C.8, it is possible to inductively add points $z_1, \ldots, z_{n-r+g-2}$ to F such that

(i) $\|z_\ell - z_k\| \geq K_1$ for all $\ell \neq k$;

(ii) $\|x_i - z_k\| \geq K_1$ for all $i \in 1..(r-g)$ and all $k \in 1..(n-r+g-2)$;

(iii) $W_H \succeq c_F I_d$ for all $H \in \binom{F \cup \{z_1, \ldots, z_{n-r+g-2}\}}{d+1}$,

with some constant c_F that depends only on F. The set of x's and z's is of size $n-2 \geq d+1$. (For $d=1$ this estimate follows from $r \geq g+2$ and for $d \geq 2$ it follows from $r \geq gd+1$.) Thus, (iii) implies general position of the points so far constructed. The data set $\mathbf{x}$ is completed by two arbitrary points q_1, q_2 in general position. In order to obtain the modified data set

$$M = F \cup \{z_1, \ldots, z_{n-r+g-2}\} \cup \{y_1, y_2\}$$

we use again Lemma C.8, replacing the two points q_1 and q_2 with a twin pair $y_1 \neq y_2$ such that

(iv) $\|y_1 - y_2\| = 1$;

(v) $\|u - y_k\| \geq K_2$ for all $u \in F \cup \{z_1, \ldots, z_{n-r+g-2}\}$ and for $k = 1, 2$;

(vi) $W_T \succeq c_F I_d$ for all $T \in \binom{M}{d+1}$ that contain at least one y_k.

Conditions (iii) and (vi) taken together imply $W_T \succeq c_F I_d$ for all $d+1$-element subsets $T \subseteq M$. In view of Lemma 2.7, we will show that the optimal solution does not discard the outliers y_1 and y_2 if K_1 and K_2 are chosen large enough.

(β) The maximum of the trimmed likelihood for the modified data set M is bounded below by a constant that depends only on F, g, and r:

It is sufficient to construct a subset $R \subseteq M$, $|R| = r$, and parameters such that the likelihood is bounded below by a function of F, g, and r. Let $R = F \cup \{z_1, \ldots, z_{g-2}\} \cup \{y_1, y_2\}$, $\pi_j = 1/g$, $m_j = z_j$, $1 \leq j \leq g-2$, $m_{g-1} = 0$, $m_g = y_1$, and $V = I_d$. Using $g \geq 2$ and (iv), we have

$$f(\mathbf{x}_R; \boldsymbol{\pi}, \mathbf{m}, V) = \prod_{i=1}^{r-g} \frac{1}{g} \sum_{j=1}^{g} N_{m_j, I_d}(x_i) \prod_{i=1}^{g-2} \frac{1}{g} \sum_{j=1}^{g} N_{m_j, I_d}(z_i) \prod_{i=1}^{2} \frac{1}{g} \sum_{j=1}^{g} N_{m_j, I_d}(y_i)$$

$$\geq g^{-r} \prod_{i=1}^{r-g} N_{0, I_d}(x_i) \prod_{i=1}^{g-2} N_{z_i, I_d}(z_i) \prod_{i=1}^{2} N_{y_1, I_d}(y_i) \geq g^{-r} (2\pi)^{-\frac{gd}{2}} e^{-\frac{1}{2}} \prod_{i=1}^{r-g} N_{0, I_d}(x_i)$$

as required.

The set M satisfies the assumptions of the combinatorial Lemma 2.8. Hence, any g-partition $\mathcal{R}$ of any subset of M of size r is of one of two kinds: Either

$$\mathcal{R} = \{\{x_1, \dots, x_{r-g}\}, \ \{y_1, y_2\}, \ g\text{–2 singletons } \{z_k\}\} \quad or$$

$\mathcal{R}$ has a clustert R_ℓ, $|R_\ell| \geq 2$, containing some pair $\{x_i, y_h\}$ or some z_k.

(γ) The MAP partition $\mathcal{R}$ associated with the weight matrix $\mathbf{w}$ of an optimal solution $(R, \boldsymbol{\pi}^*, \mathbf{m}^*, \mathbf{V}^*)$ is of the first kind if K_1 and K_2 are sufficiently large:

Assume on the contrary that $\mathcal{R}$ is of the second kind. Choose R_ℓ containing a pair x_i, y_h or z_k, u with some $u \neq z_k$. By (i), (ii), and (v), R_ℓ contains two distant elements, so Eq. (2.8) implies

$$g \operatorname{tr} W_R(\mathbf{w}) \geq g \operatorname{tr} W_R(w_\ell) \geq \operatorname{tr} W_{R_\ell} \xrightarrow[K_1, K_2 \to \infty]{} \infty. \tag{2.16}$$

Moreover, by $r \geq gd + 1$ there exists j such that $|R_j| \geq d + 1$. We infer from Eq. (2.8), (iii), and (vi)

$$g W_R(\mathbf{w}) \succeq g W_R(w_j) \succeq W_{R_j} \succeq c_F I_d. \tag{2.17}$$

Now, Lemma 2.4(a), (β), and Eq. (2.17) show that the quantities

$$2 \log f(\mathbf{x}_R; \boldsymbol{\pi}^*, \mathbf{m}^*, \mathbf{V}^*) \leq - r \log \det 2\pi c V_j^* - c \operatorname{tr} W_{R^*}(\mathbf{w})(V_j^*)^{-1}$$

$$\leq - r \log \det 2\pi c V_j^* - \operatorname{const} \operatorname{tr} (V_j^*)^{-1}$$

all remain bounded below by a constant that depends only on F. The second estimate shows that V_j^* lies in a compact subset of $\operatorname{PD}(d)$ that is independent of the choice of the points z_k and of the replacements. Therefore, the first estimate shows that $\operatorname{tr} W_R(\mathbf{w})$ is bounded above by a constant that depends only on F. This contradiction to Eq. (2.16) proves (γ).

Finally, choose K_1 and K_2 so large that the MAP partition of any optimal solution is of the first kind. In particular, the solution does not discard the replacements. According to Lemma 2.7, at least one mean breaks down as $K_2 \to \infty$.

Claim (c) follows from (a) and (b). □

2.2.4 Restricted breakdown point of mixing rates and means

Theorem 2.9 states that the asymptotic breakdown point of the constrained, trimmed, normal MLE's of the means is zero. This is an at first sight disappointing result for a trimming algorithm. It is, however, not the estimator that is to be blamed for this weakness but the stringent *universal* breakdown point. Besides allowing any kind of contamination, it makes a statement about any data set, even if it does not come from a g-component model. Robustness of the means also depends on the structure of the data set. It must be endowed with a distinct cluster structure, as noted by Gallegos and Ritter [182] for the heteroscedastic normal mixture model under the HDBT constraints. In this section we will study the restricted breakdown point of the means w.r.t. a subclass of data sets with a clear cluster structure; see Section 2.15.

We will need more notation. Let $\mathcal{P} = \{P_1, \dots, P_g\}$ be a partition of $\mathbf{x}$ and let $\emptyset \neq T \subseteq \mathbf{x}$.

The partition $\mathcal{P} \cap T = \{P_1 \cap T, \ldots, P_g \cap T\}$ of T is the *trace* of $\mathcal{P}$ in T. Let $g' \geq 1$ be a natural number and let $\mathcal{T} = (T_1, \ldots, T_{g'})$ be a partition of T. The *common refinement* of $\mathcal{P}$ and $\mathcal{T}$ is denoted by $\mathcal{P} \sqcap \mathcal{T} = \{P_j \cap T_k \mid j \leq g, k \leq g'\}$, a partition of T; some clusters may be empty. The *pooled SSP matrix* $\sum_k W_{T_k}$ of $\mathcal{T}$ is denoted by $W_{\mathcal{T}}$ and the *pooled scatter matrix* of $\mathcal{T}$ is $S_{\mathcal{T}} = \frac{1}{|T|} W_{\mathcal{T}}$. We denote the set of all stochastic matrices over the index set $T \times (1..g')$ by $\mathcal{M}(T, g')$. For T and $\alpha \in \mathcal{M}(T, g')$, $W_T(\alpha) = \sum_{k \leq g'} W_T(\alpha_k)$ is the *pooled weighted SSP matrix*, cf. Section 2.2.2. Given a partition $\mathcal{Q} = \{Q_1, \ldots, Q_g\}$ of some subset $Q \subseteq \mathbf{x}$, we also define

$$W_{\mathcal{Q}}(\alpha) = \sum_{j \leq g} W_{Q_j}(\alpha) = \sum_{j \leq g} \sum_{k \leq g'} W_{Q_j}(\alpha_k), \quad S_{\mathcal{Q}}(\alpha) = \frac{1}{|Q|} W_Q(\alpha).$$

If α is binary, then $W_{\mathcal{Q}}(\alpha)$ is the pooled SSP matrix of the common refinement of $\mathcal{Q}$ and the partition defined by α. The proof of the theorem of this section depends on a series of lemmas. The first one concerns a basic condition which implies a solution in the parametric model (2.2) to exist and robustness of the mixture rates and means. In Theorem 2.16, we will see that it actually means a separation property of the data set.

2.10 Lemma *Let $g \geq 2$ and $gd + 1 < r < n$ and let $q \in \max\{2r - n, gd + 1\}..r$. Assume that the data set $\mathbf{x}$ possesses a partition $\mathcal{P}$ in g clusters such that for all $T \subseteq \mathbf{x}$, $|T| = q$, and all $\alpha \in \mathcal{M}(T, g - 1)$*

$$\det W_T(\alpha) \geq g^2 \max_{R \in \binom{\mathbf{x}}{r}, R \supseteq T} \det \tfrac{1}{c} W_{\mathcal{P} \cap R}. \tag{2.18}$$

Then:

(a) For every m-modification M of $\mathbf{x}$, $m \leq r - q$, there exists a constrained, trimmed, normal MLE in the parameter space

$$\binom{M}{r} \times \overset{\circ}{\Delta}_{g-1} \times \{(\mathbf{m}, \mathbf{V}) \in \mathbb{R}^{gd} \times \mathcal{V}_c \mid (m_j, V_j) \text{ pairwise distinct}\}.$$

(b) The individual breakdown point of the estimates of the mixture rates satisfies

$$\beta_{rate}(n, g, r, \mathbf{x}) \geq \tfrac{1}{n}(r - q + 1).$$

(c) The same lower bound holds for the individual breakdown point $\beta_{mean}(n, g, r, \mathbf{x})$ of the estimates of the means.

PROOF. First note that (2.18) is also true for all $T \subseteq \mathbf{x}$ such that $|T| \geq q$ since the left side increases with $|T|$ by Steiner's formula and the right side decreases.

Let M be any modification obtained from $\mathbf{x}$ by replacing $m \leq r - q$ elements. It follows from Proposition 1.24 that an optimal solution $(R^*, (\pi_j^*)_{j=1}^g, (m_j^*)_{j=1}^g, (V_j^*)_{j=1}^g)$ exists in the space $\binom{M}{r} \times \Delta_{g-1} \times \{(\mathbf{m}, \mathbf{V}) \in \mathbb{R}^{gd} \times \mathcal{V}_c \mid (m_j, V_j) \text{ pairwise distinct}\}$. Let $\mathbf{w}^* = (w_j^*(x))_{x,j}$ be the deduced posterior probabilities. The proof of the claims proceeds in several steps. We write $\overline{\mathbf{x}}_{\mathcal{P} \cap R} = (\overline{x}_{P_1 \cap R}, \ldots, \overline{x}_{P_g \cap R})$.

(α) $f(\mathbf{x}_{R^*}; \pi^*, \mathbf{m}^*, \mathbf{V}^*) \geq \max_{R \in \binom{M \cap \mathbf{x}}{r}} f(\mathbf{x}_R; g^{-1}, \overline{\mathbf{x}}_{\mathcal{P} \cap R}, S_{\mathcal{P} \cap R}) \geq \min_{R \in \binom{\mathbf{x}}{r}} f(\mathbf{x}_R; g^{-1}, \overline{\mathbf{x}}_{\mathcal{P} \cap R}, S_{\mathcal{P} \cap R})$:

Since $q \geq 2r - n$ we replaced at most $r - q \leq n - r$ points and so $|M \cap \mathbf{x}| \geq r$. Hence, there exists a subset $R \subseteq M \cap \mathbf{x}$ of size r. All sets R on the right side compete for the maximum and the claim follows from optimality of $(R^*, \pi^*, \mathbf{m}^*, \mathbf{V}^*)$.

(β) $\log f(\mathbf{x}_{R^*}; \pi^*, \mathbf{m}^*, \mathbf{V}^*) \leq c_{d,r} - \frac{dr}{2} \log c - \frac{r}{2} \log \det S_{R^*}(\mathbf{w}^*)$:

By 2.4(a), there exists j such that

$$2 \log f(\mathbf{x}_{R^*}; \pi^*, \mathbf{m}^*, \mathbf{V}^*) \leq -dr \log 2\pi - r \left[\log \det V_j^* + \operatorname{tr}\left(c S_{R^*}(\mathbf{w}^*)(V_j^*)^{-1}\right) \right].$$

Now, given $B \in \mathrm{PD}(d)$, standard normal estimation theory shows that the function

$$A \mapsto \log \det A + \operatorname{tr}\left(B A^{-1}\right), \quad A \succ 0,$$

attains its minimum at B with value $\log \det B + d$. This is the claim.

(γ) $\min_{R \in \binom{M \cap \mathbf{x}}{r}} \log \det S_{\mathcal{P} \cap R} \leq \max_{R \in \binom{\mathbf{x}}{r}, R \supseteq R^* \cap \mathbf{x}} \log \det S_{\mathcal{P} \cap R}$:

Since $|M \cap \mathbf{x}| \geq r$, the set $R^* \cap \mathbf{x}$ can be completed with points from $M \cap \mathbf{x}$ to an r-element subset $R \subseteq M \cap \mathbf{x}$ with the property $R \supseteq R^* \cap \mathbf{x}$. The estimate follows.

(δ) For all $j \in 1 .. g$, there exists $x \in R^* \cap \mathbf{x}$ such that $w_j^*(x) \geq \varepsilon$ for some constant $\varepsilon > 0$ that depends only on $\mathbf{x}$:

The proof is by contradiction. Thus, assume that, for all $\varepsilon > 0$, there exists some m-modification of $\mathbf{x}$ such that $w_g^*(x) < \varepsilon$ for all $x \in R^* \cap \mathbf{x}$. Putting $\alpha_j^*(x) = \frac{w_j^*(x)}{1 - w_g^*(x)}$, $x \in R^* \cap \mathbf{x}$, $j < g$, we have $\boldsymbol{\alpha}^* \in \mathcal{M}(R^* \cap \mathbf{x}, g-1)$ and

$$W_{R^*}(\mathbf{w}^*) \succeq \sum_{j=1}^{g-1} \sum_{x \in R^* \cap \mathbf{x}} w_j^*(x)(x - m_j^*)(x - m_j^*)^\top$$

$$\succeq \min_{x \in R^* \cap \mathbf{x}} (1 - w_g^*(x)) \sum_{j=1}^{g-1} \sum_{x \in R^* \cap \mathbf{x}} \alpha_j^*(x)(x - m_j^*)(x - m_j^*)^\top$$

$$\succeq (1 - \varepsilon) \sum_{j=1}^{g-1} W_{R^* \cap \mathbf{x}}(\alpha_j^*) = (1 - \varepsilon) W_{R^* \cap \mathbf{x}}(\boldsymbol{\alpha}^*)$$

by Steiner's formula A.12. We replaced at most $r - q$ elements. Hence, $|R^* \cap \mathbf{x}| \geq q$ and we may apply hypothesis (2.18) and the initial remark to continue

$$\det W_{R^*}(\mathbf{w}^*) \geq (1 - \varepsilon)^d \det W_{R^* \cap \mathbf{x}}(\boldsymbol{\alpha}^*) \geq (1 - \varepsilon)^d g^2 \max_{R \in \binom{\mathbf{x}}{r}, R \supseteq R^* \cap \mathbf{x}} \det \tfrac{1}{c} W_{\mathcal{P} \cap R}.$$

Both R^* and $\mathcal{P} \cap R$ consist of r elements, so this is equivalent to

$$d \log c + \log \det S_{R^*}(\mathbf{w}^*)$$
$$\geq d \log(1 - \varepsilon) + 2 \log g + \max_{R \in \binom{\mathbf{x}}{r}, R \supseteq R^* \cap \mathbf{x}} \log \det S_{\mathcal{P} \cap R}. \qquad (2.19)$$

Now, for all $R \subseteq \mathbf{x}$ such that $|R| = r$,

$$\log f(\mathbf{x}_R; g^{-1}, \overline{\mathbf{x}}_{\mathcal{P} \cap R}, S_{\mathcal{P} \cap R})$$

$$= \sum_{\ell=1}^{g} \sum_{x \in P_\ell \cap R} \log \tfrac{1}{g} \sum_{j=1}^{g} (\det 2\pi S_{\mathcal{P} \cap R})^{-1/2} e^{-1/2(x - \overline{x}_{P_j \cap R})^\top S_{\mathcal{P} \cap R}^{-1}(x - \overline{x}_{P_j \cap R})}$$

$$> -r \log g - \tfrac{r}{2} \log \det 2\pi S_{\mathcal{P} \cap R} - \tfrac{1}{2} \sum_{\ell=1}^{g} \sum_{x \in P_\ell \cap R} (x - \overline{x}_{P_\ell \cap R})^\top S_{\mathcal{P} \cap R}^{-1}(x - \overline{x}_{P_\ell \cap R})$$

$$= c_{d,r} - r \log g - \tfrac{r}{2} \log \det S_{\mathcal{P} \cap R}.$$

Due to finiteness of $\mathbf{x}$, there exists a constant $\delta > 0$ such that

$$\log f(\mathbf{x}_R; g^{-1}, \overline{\mathbf{x}}_{\mathcal{P} \cap R}, S_{\mathcal{P} \cap R}) \geq \delta + c_{d,r} - r \log g - \tfrac{r}{2} \log \det S_{\mathcal{P} \cap R} \qquad (2.20)$$

for all $R \subseteq \mathbf{x}$ such that $|R| = r$.

The estimates (2.20), (γ), (2.19), along with part (β) show

$$\max_{R \in \binom{M \cap \mathbf{x}}{r}} \log f(\mathbf{x}_R; g^{-1}, \overline{\mathbf{x}}_{\mathcal{P} \cap R}, S_{\mathcal{P} \cap R}) \geq \delta + c_{d,r} - r \log g - \tfrac{r}{2} \min_{R \in \binom{M \cap \mathbf{x}}{r}} \log \det S_{\mathcal{P} \cap R}$$

$$\geq \delta + c_{d,r} - r \log g - \tfrac{r}{2} \max_{R \in \binom{\mathbf{x}}{r}, R \supseteq R^* \cap \mathbf{x}} \log \det S_{\mathcal{P} \cap R}$$

$$\geq \delta + c_{d,r} + \tfrac{dr}{2} \log(1 - \varepsilon) - \tfrac{dr}{2} \log c - \tfrac{r}{2} \log \det S_{R^*}(\mathbf{w}^*)$$

$$\geq \delta + \tfrac{dr}{2} \log(1 - \varepsilon) + \log f(\mathbf{x}_{R^*}; \boldsymbol{\pi}^*, \mathbf{m}^*, \mathbf{V}^*),$$

a contradiction to (α) since $\varepsilon > 0$ is arbitrary. This proves claim (δ).

Now let $\mathcal{R}^* = (R_1^*, \ldots, R_g^*)$ be an MAP partition of R^* w.r.t. the optimal solution and let

$$\lambda_{\min} = \min\{\lambda \mid \lambda \text{ eigenvalue of } W_C, \ C \subseteq \mathbf{x}, \ |C| = d+1\},$$

a constant > 0 that depends only on the data set $\mathbf{x}$.

(ϵ) The matrices V_j^*, $j \in 1..g$, are bounded above and below by positive definite matrices that depend only on $\mathbf{x}$, not on the replacements:
Since $|R^*| = r$, $R^* = \bigcup_{j=1}^g R_j^*$ has at least $q \geq gd + 1$ original observations. By the pigeon hole principle, there exists $j \in 1..g$ such that $|R_j^* \cap \mathbf{x}| \geq d+1$. By Eq. (2.8), this implies the lower estimate

$$W_{R^*}(\mathbf{w}^*) \succeq W_{R^*}(\mathbf{w}_j^*) \succeq \tfrac{1}{g} W_{R_j^*} \succeq \tfrac{\lambda_{\min}}{g} I_d.$$

Applying Lemma 2.4(a) to the optimal solution, we infer from (α) that the expression $-\tfrac{r}{2} \log \det 2\pi c V_j^* - \tfrac{c}{2} \lambda_{\min} \operatorname{tr}(V_j^*)^{-1}$ remains bounded below by a constant which depends solely on $\mathbf{x}$, n, g, and r. The well-known behavior of this function of V_j^* implies that the matrices V_j^*, $j \in 1..g$, remain Löwner bounded above and below.

(ζ) If R_j^* contains some original observation, then m_j^* is bounded by a number that depends only on $\mathbf{x}$:
Let $x \in R_j^* \cap \mathbf{x}$. Since $m_j^* = \overline{x}_{R^*}(w_j^*)$ we have

$$W_{R^*}(\mathbf{w}^*) \succeq W_{R^*}(w_j^*) \succeq w_j^*(x)(x - \overline{x}_{R^*}(w_j^*))(x - \overline{x}_{R^*}(w_j^*))^\top \succeq \frac{1}{g}(x - m_j^*)(x - m_j^*)^\top$$

and hence $\|x - m_j^*\|^2 \leq g \operatorname{tr} W_{R^*}(\mathbf{w}^*)$. Part ($\zeta$) will be proved if we show that $\operatorname{tr} W_{R^*}(\mathbf{w}^*)$ has an upper bound that does not depend on the replacements. By Lemma 2.4(a),

$$c \operatorname{tr} W_{R^*}(\mathbf{w}^*)(V_j^*)^{-1} \leq -r \log \det 2\pi c V_j^* - 2 \log f(\mathbf{x}_{R^*}; \boldsymbol{\pi}^*, \mathbf{m}^*, \mathbf{V}^*),$$

and the claim follows from (α) and (ϵ).

(η) If R_ℓ^* contains some replacement, then $\|m_\ell^*\| \to \infty$ as the replacement tends to ∞:
This is proved like (ζ) with the replacement substituted for x.

It follows from (ζ) and (η) that, in the long run as all replacements tend to ∞, each set R_ℓ^* consists solely of original observations or solely of replacements, that is, we have the dichotomy either $R_\ell^* \subseteq \mathbf{x}$ or $R_\ell^* \cap \mathbf{x} = \emptyset$.

(θ) If R_ℓ^* contains some replacement, then $w_\ell^*(x) \to 0$ for all $x \in R^* \cap \mathbf{x}$ as the replacement tends to ∞:
Let $x \in R_j^* \cap \mathbf{x}$. By Eq. (2.7), we have $\pi_j^* = \tfrac{1}{r} \sum_{z \in R^*} w_j^*(z) \geq \tfrac{1}{r} w_j^*(x) \geq 1/gr$ and hence

$$w_\ell^*(x) = \frac{\pi_\ell^*(\det V_\ell^*)^{-1/2} e^{-\frac{1}{2}(x - m_\ell^*)^\top (V_\ell^*)^{-1}(x - m_\ell^*)}}{\sum_k \pi_k^*(\det V_k^*)^{-1/2} e^{-\frac{1}{2}(x - m_k^*)^\top (V_k^*)^{-1}(x - m_k^*)}}$$

$$\leq gr \sqrt{\frac{\det V_j^*}{\det V_\ell^*}} \frac{e^{-\frac{1}{2}(x - m_\ell^*)^\top (V_\ell^*)^{-1}(x - m_\ell^*)}}{e^{-\frac{1}{2}(x - m_j^*)^\top (V_j^*)^{-1}(x - m_j^*)}}.$$

The claim follows from (ϵ), (ζ), and (η).

We finally prove parts (a)–(c). By (δ), $w_j^*(R^*)$ remains bounded away from zero for all m-modifications and the same is true for $\pi_j^* = w_j^*(R^*)/r$. This proves parts (a) and (b). By (δ) and (θ) there is $K > 0$ such that R^* contains no replacement y, $\|y\| > K$. Therefore, the means m_j^* are convex combinations of elements from $\mathbf{x}$ and replacements in the centered ball of radius K. This proves part (c). $\qquad\square$

The dependence on $\boldsymbol{\alpha}$ in Lemma 2.10 is somewhat awkward. In order to remove it, we need some preparations.

2.11 Lemma *Let $g \geq 2$. For any stochastic matrix $(a_{h,j})_{h,j} \in \mathbb{R}^{(g-1)\times g}$ we have*

$$\sum_j \min_h \sum_{\ell \neq j} a_{h,\ell} \geq 1.$$

PROOF. For each column j, consider the row h with minimal sum $\sum_{\ell \neq j} a_{h,\ell}$. Since there are g columns but only g-1 rows, the pigeon hole principle implies that some row appears for two j's and the two related sums cover all elements in this row. $\qquad \square$

2.12 Lemma *Let $g \geq 2$.*

(a) The maximum of the function $\mathbf{a} \mapsto \sum_{1 \leq h < k \leq g} a_h a_k$ w.r.t. all vectors $\mathbf{a} = (a_1, \ldots, a_g) \in [0,1]^g$ is attained at the point $(1/g, \ldots, 1/g) \sum_k a_k$ and has the value $\frac{g-1}{2g}\left(\sum_k a_k\right)^2$.

(b) Let $\varrho \leq 1/g$ be a nonnegative real number. The minimum of the expression

$$\sum_{1 \leq k < g} \beta_k \sum_{1 \leq j < \ell \leq g} a_{k,j} a_{k,\ell}$$

w.r.t. all probability vectors $\beta \in \mathbb{R}^{g-1}$ and all stochastic matrices $A = (a_{k,j})_{k,j} \in \mathbb{R}^{(g-1)\times g}$ such that $\beta^{\top} A \geq \varrho$ (pointwise) is assumed at $\beta^* = (1)$ and $A^* = (1 - \varrho, \varrho)$ if $g = 2$ and at $\beta^* = \left(\frac{1-2\varrho}{g-2}, \ldots, \frac{1-2\varrho}{g-2}, 2\varrho\right)^{\top}$ and

$$A^* = \begin{pmatrix} 1 & & & & 0 & 0 \\ & 1 & & 0 & 0 & 0 \\ & & \ddots & & 0 & 0 \\ & 0 & & 1 & 0 & 0 \\ 0 & 0 & \cdots & 0 & 1/2 & 1/2 \end{pmatrix},$$

if $g \geq 3$. The minimum is the number κ_ϱ defined before Sect. 2.15.

PROOF. (a) It is sufficient to determine the maximum of the sum $\sum_{1 \leq h < k \leq g} a_h a_k$ subject to the constraint $\sum_t a_t = 1$. Now, $2 \sum_{1 \leq h < k \leq g} a_h a_k = \left(\sum_k a_k\right)^2 - \|\mathbf{a}\|^2 = 1 - \|\mathbf{a}\|^2$ and the point on the plane $\sum_t a_t = 1$ with minimal distance to the origin is $(1/g, \ldots, 1/g)$. The claim follows.

(b) The case $g = 2$ being simple, we let $g \geq 3$. Using Lemma 2.11, we estimate

$$2 \sum_{1 \leq k < g} \beta_k \sum_{1 \leq j < \ell \leq g} a_{k,j} a_{k,\ell} = \sum_{1 \leq k < g} \beta_k \sum_{j \neq \ell} a_{k,j} a_{k,\ell} = \sum_j \sum_k \beta_k a_{k,j} \sum_{\ell \neq j} a_{k,\ell}$$

$$\geq \sum_j \sum_k \beta_k a_{k,j} \min_h \sum_{\ell \neq j} a_{h,\ell} \geq \varrho \sum_j \min_h \sum_{\ell \neq j} a_{h,\ell} \geq \varrho = 2\kappa_\varrho.$$

That is, the least upper bound is $\geq \kappa_\varrho$. The remaining claims are plain. $\qquad \square$

The following lemma frees the estimates from the stochastic matrix $\boldsymbol{\alpha}$, replacing it with SSP matrices. Part (b) could be proved by the extremal property of the partitions contained in $\mathcal{M}(T, g')$. The present proof uses a disintegration technique that provides more insight.

2.13 Lemma *Let $\mathcal{P}$ be a partition of $\mathbf{x}$, let T be a nonempty subset of $\mathbf{x}$, let $g' \geq 1$, and let $\boldsymbol{\alpha} \in \mathcal{M}(T, g')$.*

(a) There exists $t \leq g'|T|$, a t-tuple $(r_m)_{m=1}^t$ of strictly positive numbers r_m such that $\sum r_m = 1$, and a finite sequence $\left(\mathcal{T}^{(m)}\right)_{m=1}^t$ of partitions $\mathcal{T}^{(m)} = \{T_1^{(m)}, \ldots, T_{g'}^{(m)}\}$ of T in g' subsets (some may be empty) such that $W_{\mathcal{P} \sqcap T}(\boldsymbol{\alpha}) \succeq \sum_{m=1}^t r_m W_{\mathcal{P} \sqcap T^{(m)}}$.

(b) *Assume that, for each m, there exist indices k and j such that $|P_j \cap T_k^{(m)}| > d$. Then*

(i) *The matrix $W_{\mathcal{P} \cap T}(\boldsymbol{\alpha})$ is regular and $W_{\mathcal{P} \cap T}(\boldsymbol{\alpha})^{-1} \preceq \sum_{m=1}^{t} r_m W_{\mathcal{P} \sqcap \mathcal{T}^{(m)}}^{-1}$.*

(ii) *Given $y \in \mathbb{R}^d$, there exists a partition $\mathcal{T}$ of T in g' clusters such that, for all $\boldsymbol{\alpha} \in \mathcal{M}(T, g')$, $y^\top W_{\mathcal{P} \cap T}(\boldsymbol{\alpha})^{-1} y \leq y^\top W_{\mathcal{P} \sqcap \mathcal{T}}^{-1} y$.*

PROOF. (a) Let r_1 be the smallest nonzero entry in the matrix $(\alpha_k(x))$. In each row x, subtract the number r_1 from any of its smallest nonzero entries to obtain a new matrix $\boldsymbol{\alpha}^{(1)}$ with entries ≥ 0. All its row sums are equal and it contains at least one additional zero. Let $T_k^{(1)} = \{x \in T \mid \alpha_k^{(1)}(x) \neq \alpha_k(x)\}$, $k \leq g'$. Now continue this procedure with $\boldsymbol{\alpha}^{(1)}$ instead of $\boldsymbol{\alpha}$ and so on. It stops after at most $g'|T|$ steps with the zero matrix and we have constructed a representation

$$\alpha_k(x) = \sum_{m=1}^{t} r_m \mathbf{1}_{T_k^{(m)}}(x)$$

of $\boldsymbol{\alpha}$. Moreover, $\sum r_m = 1$. Define $\mathcal{T}^{(m)} = (T_1^{(m)}, \ldots, T_{g'}^{(m)})$. Summing up over $x \in T$, we have

$$W_{P_j \cap T}(\alpha_k) = \sum_{x \in P_j \cap T} \alpha_k(x)(x - \bar{\mathbf{x}}_{P_j \cap T}(\alpha_k))(x - \bar{\mathbf{x}}_{P_j \cap T}(\alpha_k))^\top$$

$$= \sum_{m=1}^{t} r_m \sum_{x \in P_j \cap T} \mathbf{1}_{T_k^{(m)}}(x)(x - \bar{\mathbf{x}}_{P_j \cap T}(\alpha_k))(x - \bar{\mathbf{x}}_{P_j \cap T}(\alpha_k))^\top$$

$$= \sum_{m=1}^{t} r_m \sum_{x \in P_j \cap T_k^{(m)}} (x - \bar{\mathbf{x}}_{P_j \cap T}(\alpha_k))(x - \bar{\mathbf{x}}_{P_j \cap T}(\alpha_k))^\top$$

$$\succeq \sum_{m=1}^{t} r_m \sum_{x \in P_j \cap T_k^{(m)}} \left(x - \bar{\mathbf{x}}_{P_j \cap T_k^{(m)}}\right)\left(x - \bar{\mathbf{x}}_{P_j \cap T_k^{(m)}}\right)^\top$$

by Steiner's classical formula. Now sum up over j and k to obtain (a).

(b) By assumption, each SSP matrix $W_{\mathcal{P} \sqcap \mathcal{T}^{(m)}}$ is regular. Estimate (i) therefore follows from (a) and monotone decrease and convexity of matrix inversion on PD(d); cf. [351], E.7.c.

(ii) Among the finitely many partitions of T there is some $\mathcal{T}$ such that $y^\top W_{\mathcal{P} \sqcap \mathcal{T}}^{-1} y$ is maximal. Hence, by (i),

$$y^\top W_{\mathcal{P} \cap T}(\boldsymbol{\alpha}) y \leq \sum_{m=1}^{t} r_m y^\top W_{\mathcal{P} \sqcap \mathcal{T}^{(m)}}^{-1} y \leq y^\top W_{\mathcal{P} \sqcap \mathcal{T}}^{-1} y. \qquad \square$$

Assume $g \geq 2$, let $0 < \varrho < 1$, let $u \leq n/g$ be an integer, and define the real number

$$q_{u,\varrho} = \max\left\{2r - n, (g-1)gd + 1, \tfrac{n-u}{1-\varrho}\right\}.$$

Plainly, $q_{u,\varrho}/((g-1)g) > d$. Before stating the separation property, we prove a combinatorial lemma. One of its assumptions is $q_{u,\varrho} \leq r$. This is equivalent to $n \geq r \geq (g-1)gd+1$ and $n - (1-\varrho)r \leq u$. Combined with $u \leq n/g$, the last estimate also implies $\varrho \leq 1/g$.

2.14 Lemma *Assume $q_{u,\varrho} \leq r$. Let $\mathcal{P} = \{P_1, \ldots, P_g\}$ be a partition of $\mathbf{x}$ in clusters of size $\geq u$, let $T \subseteq \mathbf{x}$ such that $|T| \geq q_{u,\varrho}$ (the assumption on $q_{u,\varrho}$ implies the existence of such a subset T), and let $\mathcal{T} = \{T_1, \ldots, T_{g-1}\}$ be a partition of T; some T_k's may be empty. Then:*

(a) *For all j, we have $|P_j \cap T| \geq \varrho|T|$.*

(b) *There are clusters T_k and P_j such that $|T_k \cap P_j| \geq \frac{q_{u,\varrho}}{(g-1)g}$.*

PROOF. (a) Assume on the contrary that $|P_\ell \cap T| < \varrho|T|$. From $\mathbf{x} \supseteq T \cup P_\ell$ we infer
$$n \geq |T| + |P_\ell| - |P_\ell \cap T| > |T| + u - \varrho|T| = u + (1 - \varrho)|T| \geq u + (1 - \varrho)q_{u,\varrho} \geq u + n - u$$
by definition of $q_{u,\varrho}$. This is a contradiction.

(b) The observations in T are spread over the $(g-1)g$ disjoint subsets of the form $T_k \cap P_j$. If (b) did not hold, we would have $|T| = \sum_{k,j} |T_k \cap P_j| < q_{u,\varrho}$, a contradiction. $\qquad \square$

Define
$$\kappa_\varrho = \begin{cases} (1 - \varrho)\varrho, & g = 2, \\ \varrho/2, & g \geq 3. \end{cases}$$

Given two subsets $S, T \subseteq \mathbb{R}^d$, $d(S, T)$ denotes their Euclidean distance and $d(S)$ the Euclidean diameter of S.

2.15 Definition (Separation property for the mixture model) Assume $q_{u,\varrho} \leq r$ and let c be the HDBT constant. We denote by $\mathcal{L}_{u,\varrho,c}$ the system of all d-dimensional admissible data sets $\mathbf{x}$ of size n with the following **separation property**:

The data $\mathbf{x}$ possesses a partition $\mathcal{P}$ in g subsets of size at least u such that, for all subsets $T \subseteq \mathbf{x}$, $|T| = q_{u,\varrho}$,

$$1 + \kappa_\varrho \min_{\substack{s_h \in \text{conv } P_h \cap T \\ \ell \neq j}} (s_\ell - s_j)^\top S_{\mathcal{P} \cap T}^{-1} (s_\ell - s_j) \tag{2.21}$$

$$\geq g^2 \frac{\max_{R \in \binom{\mathbf{x}}{r}, R \supseteq T} \det \frac{1}{c^2} W_{\mathcal{P} \cap R}}{\det W_{\mathcal{P} \cap T}} \left\{ 1 + \frac{g-2}{2d(g-1)} \max_{\substack{s, s' \in P_j \\ 1 \leq j \leq g \\ \mathcal{T}}} (s - s')^\top S_{\mathcal{P} \cap T}^{-1} (s - s') \right\}^d,$$

where $\mathcal{T}$ runs over all partitions of T in $g-1$ clusters.

Lemma 2.14(b) shows that the inverse matrices appearing in Eq. (2.21) do exist. The factor on the right hand side of (2.21) may be replaced with the expression

$$\exp\left\{ \frac{g-2}{2(g-1)} \max_{\substack{s, s' \in P_j \\ 1 \leq j \leq g \\ \mathcal{T}}} (s - s')^\top S_{\mathcal{P} \cap T}^{-1} (s - s') \right\},$$

a number independent of dimension d. Note that condition (2.21) is affine invariant. The set $\mathcal{L}_{u,\varrho,c}$ increases with decreasing u and increasing $\varrho \leq 1/2$ and c.

The partition $\mathcal{P}$ appearing in the separation property plays the role of a partition of the data set in well-separated clusters. Therefore, roughly speaking, a data set $\mathbf{x}$ has the separation property if it is composed of well-separated clusters. Moreover, the larger c is, that is, the more balanced population scales are, the easier it is for the data to meet the separation property. In addition, it is of benefit if cluster sizes, too, are balanced, that is, if u is large and so κ_ϱ and ϱ may be chosen large. Note, however, that κ_ϱ is bounded since $\varrho \leq 1/g$.

If a data set has the separation property, then the constrained, trimmed, normal MLE's of the means are much more robust than predicted by the universal breakdown point in Theorem 2.9.

2.16 Theorem (Restricted breakdown point of rates and means)
Let $g \geq 2$ and $r < n$.
(a) Assume $r \geq (g-1)gd + 1$ and let $u \in \mathbb{N}$ s.th. $n - (1 - \varrho)r \leq u \leq n/g$ (hence $\varrho \leq 1/g$ by $n \geq r$). Then the restricted breakdown point of the constrained, trimmed, normal MLE's of the mixing rates w.r.t. $\mathcal{L}_{u,\varrho,c}$ satisfies
$$\beta_{\text{rate}}(n, g, r, \mathcal{L}_{u,\varrho,c}) \geq \frac{1}{n}(r + 1 - q_{u,\varrho}).$$
The same lower estimate holds for $\beta_{\text{mean}}(n, g, r, \mathcal{L}_{u,\varrho,c})$.

(b) The individual breakdown point for any data set $\mathbf{x}$ satisfies

$$\beta_{\text{mean}}(n, g, r, \mathbf{x}) \leq \frac{1}{n}(n - r + 1).$$

(c) Let $2r - n \geq (g - 1)gd + 1$ and assume $2(n - r) \leq n/g - 1$. Let $u \in \mathbb{N}$ be s.th. $2(n - r) < u \leq n/g$ and put $\varrho = \frac{u - 2(n-r)}{2r - n}$ (> 0). Then

$$\beta_{\text{mean}}(n, g, r, \mathcal{L}_{u,\varrho,c}) = \tfrac{1}{n}(n - r + 1).$$

(d) If the hypotheses of (a) are satisfied and if the data set is of the class $\mathcal{L}_{u,\varrho,c}$, then the constrained, trimmed, normal MLE discards all replacements that are large enough.

PROOF. (a) The assumptions imply $q_{u,\varrho} \leq r$. Let $T \subseteq \mathbf{x}$ such that $|T| = q_{u,\varrho}$ and let $\boldsymbol{\alpha} \in \mathcal{M}(T, g - 1)$. In order to shorten notation we use the abbreviation

$$d(h, j, k, \ell) = \overline{x}_{P_j \cap T}(\alpha_h) - \overline{x}_{P_\ell \cap T}(\alpha_k).$$

For $j \leq g$ such that $|P_j \cap T| > 0$, let

$$A_{T,j}(\boldsymbol{\alpha}) = \frac{1}{|P_j \cap T|} \sum_{1 \leq h < k < g} \alpha_h(P_j \cap T)\alpha_k(P_j \cap T)d(h, j, k, j)d(h, j, k, j)^\top.$$

Applying Lemma A.15 with $g - 1$ instead of g, we obtain $W_{P_j \cap T} = W_{P_j \cap T}(\boldsymbol{\alpha}) + A_{T,j}(\boldsymbol{\alpha})$ and

$$W_{\mathcal{P} \cap T} = W_{\mathcal{P} \cap T}(\boldsymbol{\alpha}) + \sum_j A_{T,j}(\boldsymbol{\alpha}) = W_{\mathcal{P} \cap T}(\boldsymbol{\alpha}) + A_T(\boldsymbol{\alpha}). \tag{2.22}$$

For $k < g$ such that $\alpha_k(T) > 0$, let

$$B_T(\alpha_k) = \frac{1}{\alpha_k(T)} \sum_{1 \leq j < \ell \leq g} \alpha_k(P_j \cap T)\alpha_k(P_\ell \cap T)d(k, j, k, \ell)d(k, j, k, \ell)^\top.$$

Applying Lemma A.13 with $w(x) = \alpha_k(x)$ and $T_j = P_j \cap T$, we have for $\alpha_k(T) > 0$ the identity $W_T(\alpha_k) = W_{\mathcal{P} \cap T}(\alpha_k) + B_T(\alpha_k)$ and hence

$$W_T(\boldsymbol{\alpha}) = W_{\mathcal{P} \cap T}(\boldsymbol{\alpha}) + \sum_{k:\alpha_k(T) > 0} B_T(\alpha_k). \tag{2.23}$$

According to Lemma A.6, $\det(A + \sum_h y_h y_h^\top) \geq (1 + \sum_h y_h^\top A^{-1} y_h) \det A$ for all $A \in \mathrm{PD}(d)$. Applying this estimate, and using Eqs. (2.23) and (2.22), we infer

$$\det W_T(\boldsymbol{\alpha})$$

$$\geq \det W_{\mathcal{P} \cap T}(\boldsymbol{\alpha})\Big(1 + \sum_{k:\alpha_k(T) > 0} \frac{1}{\alpha_k(T)} \sum_{1 \leq j < \ell \leq g} \alpha_k(P_j \cap T)\alpha_k(P_\ell \cap T) \times$$

$$d(k, j, k, \ell)W_{\mathcal{P} \cap T}^{-1}d(k, j, k, \ell)^\top\Big)$$

$$= \det W_{\mathcal{P} \cap T}(\boldsymbol{\alpha})(1 + r_T(\boldsymbol{\alpha})). \tag{2.24}$$

We next estimate the two factors in (2.24), rendering them devoid of $\boldsymbol{\alpha}$. Since

$$\sum_k \frac{\alpha_k(T)}{|T|} \frac{\alpha_k(P_j \cap T)}{\alpha_k(T)} = \sum_k \frac{\alpha_k(P_j \cap T)}{|T|} = \frac{|P_j \cap T|}{|T|} \geq \varrho$$

according to Lemma 2.14(a), Lemma 2.12(b) may be applied to the expression

$$\sum_{k:\alpha_k(T) > 0} \frac{1}{\alpha_k(T)} \sum_{1 \leq j < \ell \leq g} \alpha_k(P_j \cap T)\alpha_k(P_\ell \cap T)$$

$$= |T| \sum_{k:\alpha_k(T) > 0} \frac{\alpha_k(T)}{|T|} \sum_{1 \leq j < \ell \leq g} \frac{\alpha_k(P_j \cap T)}{\alpha_k(T)} \frac{\alpha_k(P_\ell \cap T)}{\alpha_k(T)}$$

implying

$$r_T(\boldsymbol{\alpha}) \geq \kappa_\varrho |T| \min_{\substack{s_h \in \text{conv } P_h \cap T \\ \ell \neq j}} (s_\ell - s_j)^\top W_{\mathcal{P}\cap T}^{-1}(s_\ell - s_j). \tag{2.25}$$

Next use Lemmas 2.12(a) and 2.13 to estimate

$$\text{tr } W_{\mathcal{P}\cap T}^{-1}(\boldsymbol{\alpha}) A_T(\boldsymbol{\alpha})$$

$$= \sum_{j \leq g} \frac{1}{|P_j \cap T|} \sum_{1 \leq h < k < g} \alpha_h(P_j \cap T)\alpha_k(P_j \cap T)\text{tr } W_{\mathcal{P}\cap T}^{-1}(\boldsymbol{\alpha})d(h,j,k,j)d(h,j,k,j)^\top$$

$$= \sum_{j \leq g} \frac{1}{|P_j \cap T|} \sum_{1 \leq h < k < g} \alpha_h(P_j \cap T)\alpha_k(P_j \cap T)d(h,j,k,j)^\top W_{\mathcal{P}\cap T}^{-1}(\boldsymbol{\alpha})d(h,j,k,j)$$

$$\leq \frac{|T|}{2} \frac{g-2}{g-1} \max_{\substack{s,s' \in P_j \\ 1 \leq j \leq g}} (s - s')^\top W_{\mathcal{P}\cap T}^{-1}(\boldsymbol{\alpha})(s - s')$$

$$\leq \frac{|T|}{2} \frac{g-2}{g-1} \max_{T} \max_{\substack{s,s' \in P_j \\ 1 \leq j \leq g}} (s - s')^\top W_{\mathcal{P}\cap T}^{-1}(s - s'). \tag{2.26}$$

The identity $\det(A + B) = \det A \cdot \det(I_d + A^{-1/2}BA^{-1/2})$ and the arithmetic-geometric inequality $\det C \leq \left(\frac{1}{d}\text{tr } C\right)^d$, valid for $A \succ 0$, $B \in \mathbb{R}^{d\times d}$, and $C \succeq 0$, yield the inequality $\det(A + B) \leq \det A\left(1 + \frac{1}{d}\text{tr } A^{-1}B\right)^d$. Its application to Eq. (2.22) shows, together with Eq. (2.26),

$$\det W_{\mathcal{P}\cap T} \leq \det W_{\mathcal{P}\cap T}(\boldsymbol{\alpha})\left\{1 + \frac{1}{d}\text{tr } W_{\mathcal{P}\cap T}^{-1}(\boldsymbol{\alpha})A_T(\boldsymbol{\alpha})\right\}^d \tag{2.27}$$

$$\leq \det W_{\mathcal{P}\cap T}(\boldsymbol{\alpha})\left\{1 + \frac{1}{2d}\frac{g-2}{g-1} \max_{T} \max_{\substack{s,s' \in P_j \\ 1 \leq j \leq g}} (s - s')^\top S_{\mathcal{P}\cap T}^{-1}(s - s')\right\}^d.$$

Equations (2.24) and (2.25), the separation property, and Eq. (2.27) finally combine to show Eq. (2.18) and claim (a) follows from Lemma 2.10.

(b) Let M be a set obtained from $\mathbf{x}$ by replacing $n - r + 1$ of its elements with a narrow and distant cluster. M contains only $r - 1$ original observations, so each r-element subset of M contains one replacement, in particular, each optimal set R^*. Let $\{R_1^*, \ldots, R_g^*\}$ be an associated MAP partition. Then some R_j^* contains at least one replacement and Lemma 2.7 shows that the norm of m_j^* tends to infinity together with the compact cluster of replacements.

(c) Since $2r - n \leq r$, the first condition in (a) is fulfilled and the given ϱ satisfies $1 - \varrho = (n - u)/(2r - n) \geq (n - u)/r$. It follows that the second condition in (a), too, is satisfied and that ϱ is the largest number s.th. $q_{u,\varrho} = 2r - n$. The claim on β_{mean} now follows from (a).

Claim (d) follows from the proof of Lemma 2.10(c).   □

The following corollary is a consequence of Theorem 2.16. Combined with Corollary 2.6, it says that the constrained, trimmed, normal MLE is asymptotically robust on data sets with a well-separated, balanced cluster structure if the natural parameter g is used.

2.17 Corollary *Let* $g \geq 2$, *let* $0 < \eta < \delta < \frac{1}{g}$, *let* $r = \lceil n\left(1 - \frac{1}{2g} + \frac{\delta}{2}\right)\rceil$, *let* $u = \lceil n\left(\frac{1}{g} - \eta\right)\rceil$, *and let* $\varrho = \frac{\delta - \eta}{1 - 1/g + \delta}$. *Then, asymptotically,*

$$\beta_{mean}(n, g, r, \mathcal{L}_{u,\varrho,c}) \xrightarrow[n\to\infty]{} \frac{1}{2}\left(\frac{1}{g} - \delta\right).$$

2.18 Remarks (a) The inequality $n - (1 - \varrho)r \leq u$ required in Theorem 2.16(a) implies $u \geq n - r + 1$. That is, the sizes of the natural clusters must exceed the number of discards

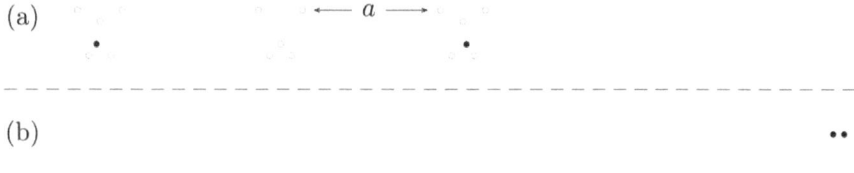

Figure 2.2 *(a) The data set* **x** *and (b) its modification shown at the critical distance* $a = 12.42$ *where transition between robustness and breakdown occurs.*

in Theorem 2.16(a). Moreover, the assumptions of part (c) imply that these sizes must even exceed twice the number of discards.

(b) Although the constrained trimmed mixture likelihood and its MLE were formulated for general statistical models, the robustness results were stated and proved for various normal models, only. Extensions to other models are possible but not straightforward, in general. In Gallegos and Ritter [182], Remark 2.14(b), it is indicated how the crucial Lemma 2.4 and much of the robustness theory can be extended to a whole family of light-tailed elliptical distributions including the normal case. Because of trimming, heavy-tailed models are only of little interest here.

(c) Theorem 2.16 requires well-separated components in order to guarantee robustness. We take first a look at an example which shows that such an assumption is necessary. Consider the two-dimensional data set **x** consisting of the $n = 17$ regular points

$$\mathbf{x} = \left\{ \begin{pmatrix} -a-6 \\ 4 \end{pmatrix}, \begin{pmatrix} -a-4 \\ 3 \end{pmatrix}, \begin{pmatrix} -a-2 \\ 4 \end{pmatrix}, \begin{pmatrix} -a-5 \\ 0 \end{pmatrix}, \begin{pmatrix} -a-4.3 \\ 1 \end{pmatrix}, \begin{pmatrix} -a-3 \\ 0 \end{pmatrix} \right\}$$
$$\cup \left\{ \begin{pmatrix} -2 \\ 4 \end{pmatrix}, \begin{pmatrix} 2 \\ 4 \end{pmatrix}, \begin{pmatrix} -1 \\ 0 \end{pmatrix}, \begin{pmatrix} 0 \\ 1 \end{pmatrix}, \begin{pmatrix} 1 \\ 0 \end{pmatrix} \right\}$$
$$\cup \left\{ \begin{pmatrix} a+2 \\ 4 \end{pmatrix}, \begin{pmatrix} a+4 \\ 3 \end{pmatrix}, \begin{pmatrix} a+6 \\ 4 \end{pmatrix}, \begin{pmatrix} a+3 \\ 0 \end{pmatrix}, \begin{pmatrix} a+4.3 \\ 1 \end{pmatrix}, \begin{pmatrix} a+5 \\ 0 \end{pmatrix} \right\}$$

displayed in Fig. 2.2(a). As expected, the classical MLE for the full *homoscedastic* model and $g = 3$ recovers the left, the middle, and right components. Now assume that the two solid data points in Fig. 2.2(a) have been grossly mismeasured, as shown in Fig. 2.2(b). Their distance is now 1. Then, for $a \geq 12.43$, the homoscedastic trimmed MLE with $g = 3$ and $r = 15$ produces a reasonable estimate, discarding both outliers. However, for $a \leq 12.42$, it discards the two (original) observations interior to the convex hulls of the right and middle clusters, thus producing an estimate with two slim horizontal components complemented by a component determined by the two outliers. Its negative log-likelihood is 73.3324, whereas that of the natural solution is 73.3355.

Upper and lower bounds on the Mahalanobis distance square appearing on the left side of Eq. (2.21) provide a lower bound on the point a for which the separation property is satisfied. The lower bound for the full model lies in the interval [1763,3841]. For the spherical model this interval is [60,61].

This special and contrived example represents the worst case. First, the two replaced elements are purposefully chosen. Second, they are replaced with two close outliers aligned horizontally. Third, a further removal of two points makes the remaining original points two slim horizontal clusters which determine the solution when there is not enough separation between the three original clusters.

The constrained, trimmed, normal MLE may also be robust w.r.t. the means even when applied to substantially overlapping components. To underpin this contention, we generate in a second experiment 54 four-dimensional data sets of 150+100+50 regular data points each. They are randomly drawn from N_{0,I_4}, N_{3e_1,V_2}, and N_{6e_2,V_3}, respectively, where

$$V_2 = \begin{pmatrix} 1 & & & \\ -1.5 & 4 & & \\ -1 & -2 & 9 & \\ 0 & -1 & 3 & 16 \end{pmatrix}, \ V_3 = \mathrm{Diag}(1,16,9,4).$$

To each data set we add 30 noise points randomly drawn from $N_{0,10^4 I_4}$. It turns out that the HDBT constrained estimates of full covariance matrices with $c = 200$ and with the parameters $g = 3$ and $r = 300$, 295, 290, 285, 280 correctly identify all outliers in all 270 cases. These findings support again the robustness of the trimmed, constrained, normal MLE. The assumption of too many outliers is not harmful. Of course, we must not assume less than 30 outliers.

(d) Let us finally complete the study in Example 2.1(a) with the trimmed, constrained, normal MLE. We shift the largest six observations to the right and first discard this number of data points. At the beginning (no shift) the MLE removes the largest two, the smallest two, and the middle two elements of the data set as "outliers." The estimated mean (variance) of the right component is –0.0383 (0.6746). For shifts less than 0.54, it removes more and more shifted elements, finally all six. At the value 0.53, the estimated mean (variance) is –0.1262 (0.7413). At 0.54, the two estimates jump to –0.2103 and 0.5976, respectively. From this value on, nothing changes, all shifted elements are recognized as outliers and, as a consequence, the estimated parameters remain constant. As we discard more than six elements, the behavior resembles that with six elements, since all outliers are discarded for large shifts. However, a disaster results as we discard fewer than six elements. The solution has to accommodate outliers. This effects that at least one mean breaks down.

2.2.5 Notes

A discussion of various outlier types appears in Ritter and Gallegos [437]. The idea of a "spurious" outlier emerged from Mathar [354], Section 5.2. It was later used by Pesch [411], Gallegos [179], Gallegos and Ritter [180, 183, 182, 184], and García-Escudero et al. [188]. It serves to embed gross outliers in a probabilistic framework. The trimmed likelihood goes back to Neykov and Neytchev [390]. It was placed into a broader context by Hadi and Luceño [222].

The main results of this section appear in Gallegos and Ritter [182]. They are Theorem 2.5 on the individual breakdown point of the constrained, trimmed, normal MLE's of the covariance matrices, Theorem 2.9 on the universal breakdown point of the means, and Theorem 2.16 on the restricted breakdown point of the means.

A clear decay of the maximum likelihood from g to $g-1$ components indicates superiority of g components over $g-1$. Hennig [241], Theorem 4.11, uses this property to analyze robustness of a mixture model that contains a component with an improper density. He uses a scale constraint more restrictive than HDBT.

2.3   Trimming the classification model – the TDC

Cuesta-Albertos et al. [106] proposed to use trimming to robustify Ward's sum-of-squares criterion, Remark 1.42(a). García-Escudero and Gordaliza [187] noticed that it breaks down under the influence of a single gross outlier. The simulations of the latter suggest that it is hard to break down clear cluster structures with a *trimmed* version of the criterion if the natural number of clusters is chosen as the parameter g. This section offers a theoretical underpinning of these findings. The program carried out in Section 2.2 for the mixture model will now be extended to the classification model of Section 1.3. The spurious-outliers model can again be employed in order to extend the MAP classification model (1.38) to trimming, Section 2.3.1. The trimmed MAP classification model uses the number of clusters and the number of discards as computational parameters. Their estimation is postponed to Chapter 4.

In the normal case, the criterion turns to a trimmed, heteroscedastic, fully normal, and affine invariant cluster criterion, the *Trimmed Determinant Criterion* (TDC), Gallegos and Ritter [180, 183]; see Section 2.3.2. In practice, certain steady solutions already introduced in Section 1.3.3 play the most prominent rôle. The theoretical properties of the TDC become most clearly visible under the HDBT constraints (1.27) on the covariance matrices. Its maximum exists under HDBT constraints; see Proposition 2.19. In practice, constraints on cluster sizes are again useful.

It the remaining Sections 2.3.3–2.3.5 it will be shown that the robustness properties of the TDC under HDBT constrained are much like those of the constrained, trimmed mixture model in Section 2.2. These results will flow again from a mathematical analysis of breakdown points. Whereas the covariance matrices turn out to be very robust under the constraints without any further structural assumptions (Section 2.3.3), the mixing rates and mean vectors are more fragile; see Section 2.2.4. Ensuring their robustness needs again a separation property like that used for the mixture model; compare Definitions 2.15 and 2.29. Without the HDBT constraints, the model would lack robustness. The larger the HDBT constant c is the more robust the method turns out to be. That is, robustness grows with scale balance.

2.3.1   Trimmed MAP classification model

We next establish the trimmed MAP cluster criterion proceeding along the lines of criterion (1.38), but now taking into account spurious outliers; see Section 2.1.1. In the trimming context, a labeling of the n objects is an array $\boldsymbol{\ell} = (\ell_1, \ldots, \ell_n)$, $\ell_i \in 0 \, .. \, g$, where we take $\ell_i = j \in 1 \, .. \, g$ to mean that object i is **retained** and assigned to cluster j and the label $\ell_i = 0$ **discards** it. Besides discriminating between retained and discarded elements, the labeling also creates a partition of the r retained elements R. Let $(f_\gamma)_{\gamma \in \Gamma}$ be the underlying, basic model. The joint population parameters of the g components will be contained in the Cartesian product Γ^g. The mixing rates $\boldsymbol{\pi} = (\pi_1, \ldots, \pi_g) \in \Delta_{g-1}$ appear again as prior probabilities of the labels. Moreover, we introduce the parameter spaces Ξ_i of the spurious outliers i. We arrive at the parameter space

$$(0 \, .. \, g)^{1 \, .. \, n} \times \Delta_{g-1} \times \Gamma^g \times \prod_{i=1}^n \Xi_i$$

and the conditional density $f_{L,X}[\boldsymbol{\ell}, \mathbf{x} \mid \boldsymbol{\pi}, \boldsymbol{\gamma}, \boldsymbol{\xi}]$. It is a posterior probability w.r.t. $\boldsymbol{\ell}$ and a likelihood function w.r.t. $\boldsymbol{\pi}, \boldsymbol{\gamma}, \boldsymbol{\xi}$. The natural independence assumptions used in Section 1.3.1 imply the conditional density

$$f_{L,X}[\boldsymbol{\ell}, \mathbf{x} \mid \boldsymbol{\pi}, \boldsymbol{\gamma}, \boldsymbol{\xi}] = f_X[\mathbf{x} \mid \boldsymbol{\ell}, \boldsymbol{\gamma}, \boldsymbol{\xi}] P[\boldsymbol{\ell}; \boldsymbol{\pi}] = f(\mathbf{x}_R; \boldsymbol{\ell}, \boldsymbol{\gamma}) \prod_{\ell_i = 0} f_{X_i}[x_i; \xi_i] P[\boldsymbol{\ell} \mid \boldsymbol{\pi}].$$

Using (SV$_o$) (see p. 80), the integral w.r.t. the nuisance parameters $\xi_i \in \Xi_i$ of the spurious outliers $\ell_i = 0$ becomes

$$f(\mathbf{x}_R; \boldsymbol{\ell}, \boldsymbol{\gamma}) P[\boldsymbol{\ell} \mid \boldsymbol{\pi}] = \prod_{j=1}^{g} \prod_{\ell_i=j} f_{\gamma_j}(x_i) \prod_{j=1}^{g} \pi_j^{n_j}.$$

Here, $n_j = n_j(\boldsymbol{\ell})$ is the size of cluster $C_j = \boldsymbol{\ell}^{-1}(j)$. Of course, $\sum_{j=1}^{g} n_j = r$. We now proceed as in Section 1.3.1 with R instead of $1..n$. After taking logarithms and partial maximization w.r.t. $\boldsymbol{\pi}$, we obtain the **trimmed MAP classification model**

$$\sum_{j=1}^{g} \sum_{i \in C_j} \log f_{\gamma_j}(x_i) - r \cdot \mathrm{H}\left(\tfrac{n_1(\boldsymbol{\ell})}{r}, \ldots, \tfrac{n_g(\boldsymbol{\ell})}{r}\right). \tag{2.28}$$

It depends on the data, the assignment, and on the population parameters. If the data is kept fixed, then it is called the **trimmed MAP criterion**. The first term here is called the **trimmed ML classification model (criterion)**. It is again neither desirable nor possible to maximize it w.r.t. all parameters $\boldsymbol{\gamma} \in \Gamma^g$ and all assignments $\boldsymbol{\ell} \in (1..g)^{1 \cdots n}$. Instead, we recur to parameter constraints Ψ, cluster size constraints, submodels, steady partitions, or combinations thereof. The discussion on p. 48 is relevant here, too.

If size constraints $|C_j| \geq b$ are imposed on clusters so that the MLE of the parameters exists for each cluster C_j separately, then parameter estimation may be extended to the whole product Γ^g. Dynamic optimization G.9 then applies and partial maximization w.r.t. the parameters yields $\max_{\boldsymbol{\gamma}} \sum_{j=1}^{g} \sum_{i:\ell_i=j} \log f_{\gamma_j}(x_i) = \sum_{j=1}^{g} \max_{\gamma_j} \sum_{i:\ell_i=j} \log f_{\gamma_j}(x_i) = \sum_{j=1}^{g} \sum_{i:\ell_i=j} \log f_{\gamma_{C_j}}(x_i)$ with the g individual MLE's γ_{C_j}. The trimmed MAP criterion turns into the criterion

$$\sum_{j=1}^{g} \sum_{i:\ell_i=j} \log f_{\gamma_{C_j}}(x_i) - r\mathrm{H}\left(\tfrac{n_1(\boldsymbol{\ell})}{r}, \ldots, \tfrac{n_g(\boldsymbol{\ell})}{r}\right)$$

for the assignment $\boldsymbol{\ell}$. Both, parameter and size constraints are unknown. Therefore, the most important method forgoes likelihood maxima using again steady solutions as introduced in Section 1.3.3. These issues will now be studied in more detail in the normal case.

2.3.2 Normal case – the TDC

We next specialize the trimmed MAP classification log-likelihood (2.28) to the normal case with parameters $\gamma_j = (m_j, V_j)$, $1 \leq j \leq g$. The fact that (unconstrained) maximization w.r.t. the parameters in (2.28) is neither possible nor desirable does not pertain to the location parameters; see also the beginning of Section 1.3.5. The problem arises only with the scale matrices. So let us maximize (2.28) partially w.r.t. the means m_j. For any assignment $\boldsymbol{\ell}$ and all scale matrices, the maximum exists. By Steiner's formula, Lemma A.11, it depends only on the sample means of the clusters $C_j(\boldsymbol{\ell})$ defined by $\boldsymbol{\ell}$ and is given by

$$m_j(\boldsymbol{\ell}) = \begin{cases} \overline{x}_j(\boldsymbol{\ell}), & \text{if } C_j(\boldsymbol{\ell}) \neq \emptyset, \\ \text{arbitrary, e.g., 0,} & \text{otherwise,} \end{cases} \quad 1 \leq j \leq g. \tag{2.29}$$

After a change of sign so as to avoid minus signs, the classification log-likelihood (2.28) thereby reduces to the (heteroscedastic) **Trimmed Determinant Criterion**

$$r \cdot \mathrm{H}\left(\tfrac{n_1(\boldsymbol{\ell})}{r}, \ldots, \tfrac{n_g(\boldsymbol{\ell})}{r}\right) + \tfrac{1}{2} \sum_{j=1}^{g} n_j(\boldsymbol{\ell})\left[\log \det V_j + \mathrm{tr}\left(V_j^{-1} S_j(\boldsymbol{\ell})\right)\right]. \tag{2.30}$$

It contains the scale parameters $\mathbf{V}$ and the assignment $\boldsymbol{\ell}$. Interesting solutions here are HDBT or size constrained minima w.r.t. V_j and $\boldsymbol{\ell}$ and, most notably, steady solutions. The criterion contains the scatter matrices $S_j(\boldsymbol{\ell})$ of $C_j(\boldsymbol{\ell}) \neq \emptyset$. (Note that the TDC does not need the scatter matrix $S_j(\boldsymbol{\ell})$ when $n_j(\boldsymbol{\ell}) = 0$.)

In the heteroscedastic case, the unconstrained TDC possesses no global minimum w.r.t. the scale parameters when $C_j(\boldsymbol{\ell})$ is deficient. We show next that the HDBT constraints $\mathbf{V} \in \mathcal{V}_c$ guarantee the existence of the minimum w.r.t. the scale parameters $\mathbf{V}$ in criterion (2.30) for all assignments $\boldsymbol{\ell}$; see also Section 1.2.5.

2.19 Proposition *If the data $\mathbf{x}$ is in general position and if $r \geq gd + 1$, then the HDBT constrained minimum of the TDC (2.30) w.r.t. $\mathbf{V} \in \mathcal{V}_c$ and $\boldsymbol{\ell} \in (0 \mathinner{.\,.} g)^{1 \cdots n}$ exists for any constant $0 < c \leq 1$. (Some clusters may be deficient or even empty.)*

PROOF. By finiteness of $(0 \mathinner{.\,.} g)^{1 \cdots n}$ it is sufficient to ensure the minimum w.r.t. the scale parameters. The HDBT constraints imply $\det V_j \geq \det(cV_\ell)$ and $V_j^{-1} \succeq cV_\ell^{-1}$. Hence, we have for any $1 \leq \ell \leq g$

$$\sum_{j=1}^{g} n_j \big[\log \det V_j + \operatorname{tr}(V_j^{-1} S_j(\boldsymbol{\ell})) \big] \geq \sum_{j=1}^{g} n_j \big[\log \det(cV_\ell) + \operatorname{tr}(cV_\ell^{-1} S_j(\boldsymbol{\ell})) \big]$$

$$= r \log \det(cV_\ell) + c \operatorname{tr}(V_\ell^{-1} W(\boldsymbol{\ell})).$$

Here, $W(\boldsymbol{\ell})$ is the pooled SSP matrix specified by $\boldsymbol{\ell}$. By assumption there is some cluster, say ℓ, of size $n_\ell(\boldsymbol{\ell}) \geq d + 1$. By general position, its SSP matrix is positive definite, so $W(\boldsymbol{\ell}) \geq \varepsilon I_d$ with some constant $\varepsilon > 0$ that depends only on the data. Hence

$$\sum_{j=1}^{g} n_j \big[\log \det V_j + \operatorname{tr}(V_j^{-1} S_j(\boldsymbol{\ell})) \big] \geq r \log \det(cV_\ell) + \varepsilon c \operatorname{tr} V_\ell^{-1}.$$

As $\mathbf{V}$ approaches the Alexandrov point of $\mathcal{V}_c$, that is, as some V_j approaches that of PD(d), so does V_ℓ again by the HDBT constraints. That is, either the smallest eigenvalue of V_ℓ tends to zero or the largest one tends to ∞. It is clear and well known that this implies that the right, and hence the left side of the above inequality tends to ∞. This proves the claim. $\qquad\square$

Finally, we will denote the minimizing assignment by $\boldsymbol{\ell}^*$, $R^* = \{i \mid \ell_i \neq 0\}$ will stand for the set of regular elements w.r.t. $\boldsymbol{\ell}^*$, and the partition of R^* associated with $\boldsymbol{\ell}^*$ will be $(C_1^*, \dots, C_g^*)$. The optimal assignment $\boldsymbol{\ell}^*$ induces estimates m_j^* and V_j^* of the location and scale parameters m_j and V_j, which we call the TDC *parameter estimates*. They are $m_j^* = m_j(\boldsymbol{\ell}^*)$ as in (2.29) and the minimizers w.r.t. $\mathbf{V} \in \mathcal{V}_c$ appearing in the TDC, all w.r.t. $\boldsymbol{\ell}^*$.

Although the HDBT constraints are mainly of theoretical interest, it is nevertheless interesting to add here a few words on constrained optimization. In the homoscedastic normal case, $c = 1$, the estimate of the common scale matrix V is the pooled scatter matrix $S(\boldsymbol{\ell})$. Up to an additive constant, the TDC reduces therefore to the trimmed MAP pooled determinant criterion

$$r \cdot \big\{ \mathrm{H}\big(\tfrac{n_1(\boldsymbol{\ell})}{r}, \dots, \tfrac{n_g(\boldsymbol{\ell})}{r}\big) + \tfrac{1}{2} \log \det S(\boldsymbol{\ell}) \big\}. \tag{2.31}$$

A less obvious example of a closed form is presented in the following proposition.

2.20 Proposition *Let $d = 1$, let $g \geq 2$, let $r \geq g + 1$, and let $0 < c \leq 1$. Let $\boldsymbol{\ell}$ be such that the scatter values s_j satisfy $s_2 > 0$ and $cs_\ell \leq s_j \leq s_\ell/c$ for all $3 \leq j \leq g$, $\ell < j$.[2] (In other words, the sample variances satisfy the HDBT constraints except, possibly, for the pair s_1, s_2.) Then partial minimization of the TDC w.r.t. $\mathbf{V} = (v_1, \dots, v_g) \in \mathcal{V}_c$ is solved by*

$$\begin{cases} v_1(\boldsymbol{\ell}) = s_1, & v_2(\boldsymbol{\ell}) = s_2, & \text{if } cs_1 \leq s_2 \leq s_1/c, \\ v_1(\boldsymbol{\ell}) = \frac{w_1 + w_2/c}{n_1 + n_2}, & v_2(\boldsymbol{\ell}) = \frac{cw_1 + w_2}{n_1 + n_2}, & \text{if } s_2 < cs_1, \\ v_1(\boldsymbol{\ell}) = \frac{w_1 + cw_2}{n_1 + n_2}, & v_2(\boldsymbol{\ell}) = \frac{w_1/c + w_2}{n_1 + n_2}, & \text{if } s_1 < cs_2, \end{cases}$$

and $v_j(\boldsymbol{\ell}) = s_j$, $3 \leq j \leq g$.

---

[2] This presupposes that the clusters $2, \dots, g$ contain at least two different elements each.

PROOF. Let us abbreviate $h_j(v) = n_j(\ln v + \frac{s_j}{v})$. In the present case, the partial minimum of the TDC w.r.t. $(v_1, \ldots, v_g)$ can be rewritten in the form (omitting the entropy term)

$$h = \min_{\substack{v_1 > 0 \\ cv_\ell \leq v_j \leq v_\ell/c, \, \ell < j}} \sum_{j=1}^{g} h_j(v_j)$$

$$= \min_{v_1 > 0} \left\{ h_1(v_1) + \min_{cv_1 \leq v_2 \leq v_1/c} \left\{ h_2(v_2) + \min_{\substack{cv_\ell \leq v_j \leq v_\ell/c \\ \ell < j, \, j \geq 3}} \sum_{j \geq 3} h_j(v_j) \right\} \right\}$$

$$\geq \min_{v_1 > 0} \left\{ h_1(v_1) + \min_{cv_1 \leq v_2 \leq v_1/c} \left\{ h_2(v_2) + \sum_{j \geq 3} \min_{v > 0} h_j(v) \right\} \right\}$$

$$= \min_{v_1 > 0} \left\{ h_1(v_1) + \min_{cv_1 \leq v_2 \leq v_1/c} h_2(v_2) \right\} + \sum_{j \geq 3} h_j(s_j),$$

since h_j, $j \geq 3$, assumes its unconstrained minimum at $v = s_j$ (> 0). The constrained minimizer of $h_2(v_2)$ w.r.t. v_2 is

$$\widetilde{v}_2(v_1) = \begin{cases} s_2, & cs_2 < v_1 < s_2/c, \\ cv_1, & v_1 \geq s_2/c, \\ v_1/c, & v_1 \leq cs_2, \end{cases}$$

and we have thus shown

$$h \geq \min_{v_1 > 0} \left\{ h_1(v_1) + h_2(\widetilde{v}_2(v_1)) \right\} + \sum_{j \geq 3} h_j(s_j). \tag{2.32}$$

The function $v_1 \mapsto h_2(\widetilde{v}_2(v_1))$ is differentiable, monotone decreasing in $]0, cs_2]$, constant in $[cs_2, s_2/c]$, and monotone increasing in $[s_2/c, \infty[$. It follows that the sum $v_1 \mapsto h_1(v_1) + h_2(\widetilde{v}_2(v_1))$ has a minimum which is attained in the interval where the minimum of the unimodal function $h_1(v_1)$ is located. The minimizer of the lower bound (2.32) turns out to be the value $v_1(\ell)$ given in the proposition.

We have thus seen that the target function is nowhere less than its value at the parameters stated in the proposition. The proof will be finished after we have shown that these parameters satisfy the HDBT constraints. This is true by assumption for all pairs (j, ℓ), $j, \ell \geq 3$, and was ensured for the pair $(1, 2)$. The remaining pairs $(1, j)$, $(2, j)$, $j \geq 3$, follow from elementary estimates based on the constraints assumed for (s_1, s_j) and (s_2, s_j). The condition $r \geq g + 1$ ensures that the minimum w.r.t. $v_1 > 0$ exists and so $v_j(\ell) > 0$ for all j. □

A second way of ensuring ML estimates uses size constraints on the clusters. Consider again the fully normal case. Assume that the data is in general position, and allow only assignments ℓ with cluster sizes $n_j(\ell) \geq b$ for some $b \geq d + 1$ such that $gb \leq r$. Then minimization of the TDC (2.30) w.r.t. the scale matrices V_j for fixed ℓ may be performed freely cluster by cluster. The minima turn to the scatter matrices $S_j(\ell)$ w.r.t. the clusters $C_j(\ell)$. That is, the estimates of means and covariance matrices become the ML estimates for the g clusters w.r.t. the given assignment. Up to a constant, the TDC turns then to the criterion

$$\tfrac{1}{2} \sum_{j=1}^{g} n_j(\ell) \cdot \log \det S_j(\ell) + r \mathrm{H}\left(\tfrac{n_1(\ell)}{r}, \ldots, \tfrac{n_g(\ell)}{r}\right) \tag{2.33}$$

to be minimized w.r.t. the assignment ℓ under the size constraints. An algorithm will be presented in Section 3.2.3.

The most effective, third, way recurs again to steady solutions. Whereas the first two approaches need HDBT or size constraints, this one faces a multitude of different solutions. How to choose among them will be explained in Section 4.1. The estimation process will be related to optimization.

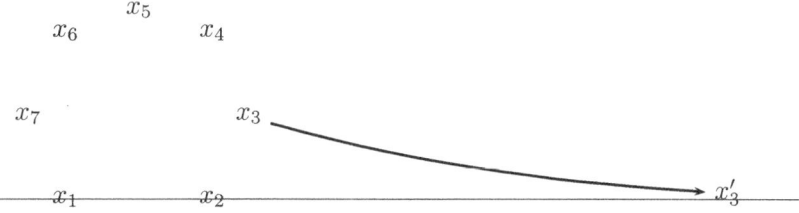

Figure 2.3 *Nonrobustness of criterion (2.33) with fully normal covariance matrices in the* free *heteroscedastic case, minimum cluster size 3.*

2.3.3   Breakdown robustness of the constrained TDC

Although the TDC involves trimming, neither the estimates of the means nor those of the covariance matrices would be robust without scale constraints. In fact, no matter how r is chosen, they would break down under the influence of a single outlier. An example is provided by the data set consisting of seven points $x_1, \ldots, x_7$ shown in Fig. 2.3. We use criterion (2.33) to subdivide it in two groups of minimum cluster size 3 and $r = 6$, that is, we discard one point. There are two equivalent optimal clusterings, $\{\{x_2, x_3, x_4\}, \{x_5, x_6, x_7\}\}$, x_1 discarded, and $\{\{x_3, x_4, x_5\}, \{x_6, x_7, x_1\}\}$, x_2 discarded. We now replace x_3 with a distant outlier x_3' close to the abscissa, say $x_3' = (a, a^{-2})$ for some large a. Although we discard one point, the criterion does not choose the "desired" one, x_3'. In fact, x_3' creates along with x_1 and x_2 a cluster with a small determinant of its scatter matrix which determines the optimal clustering, $\{\{x_1, x_2, x_3'\}, \{x_4, x_5, x_6\}\}$, x_7 discarded. As a consequence, neither do mean and largest eigenvalue of the scatter matrix of the slim cluster remain bounded as $a \to \infty$ nor does the smallest eigenvalue remain bounded away from zero.

Gallegos and Ritter [183] deal with breakdown points of the estimates of the parameters $m_j \in \mathbb{R}^d$ and $V_j \in \mathrm{PD}(d)$ obtained from *constrained* minimization of the TDC w.r.t. $\boldsymbol{\ell}$, $\mathbf{m}$, and $\mathbf{V}$. They show that the HDBT constraints do not only guarantee existence of a solution but also robustness of the TDC. Their main results are the following:

(i) If r is large enough, then the TDC estimates of the covariance matrices resist $n-r+g-1$ arbitrary replacements. However, they break down under $n - r + g$ suitable replacements; see Theorem 2.22.

(ii) There exists a data set such that the TDC estimate of at least one sample mean breaks down under two suitable replacements no matter how many objects are discarded; see Theorem 2.25.

(iii) If the data set bears a clear structure of g substantial clusters and if r is large enough and properly chosen, then the TDC estimates of all means resist $n - r$ arbitrary replacements. On the other hand, it is possible to break down one mean with $n - r + 1$ suitable replacements; see Theorem 2.32.

2.3.4   Universal breakdown point of covariance matrices and means

We will now show that the minimum of the HDBT constrained TDC, with any $0 < c \leq 1$, provides an asymptotically robust estimate of the covariance matrices V_j and compute the universal breakdown point. We first need a lemma. It exploits the pooled SSP matrix $W(\boldsymbol{\ell})$ of an admissible assignment $\boldsymbol{\ell}$.

2.21 Lemma *Let* $\mathbf{V} = (V_1, \ldots, V_g) \in \mathcal{V}_c$, *let* $\mathbf{m} = (m_1, \ldots, m_g) \in \mathbb{R}^{gd}$, *let* $\boldsymbol{\ell}$ *be an admissible labeling, and denote the set of retained objects w.r.t.* $\boldsymbol{\ell}$ *by* R. *We have for all* ℓ, $1 \leq \ell \leq g$,

$$2 \log f(\mathbf{x}_R; \boldsymbol{\ell}, \mathbf{m}, \mathbf{V}) \leq -r \log \det(2\pi c V_\ell) - c \operatorname{tr}(W(\boldsymbol{\ell}) V_\ell^{-1}).$$

PROOF. The HDBT constraints imply

$$2 \log f(\mathbf{x}_R; \boldsymbol{\ell}, \mathbf{m}, \mathbf{V})$$

$$= - \sum_{1 \le j \le g} \left\{ n_j(\boldsymbol{\ell}) \log \det(2\pi V_j) + \sum_{i \in C_j(\boldsymbol{\ell})} (x_i - m_j)^\top V_j^{-1} (x_i - m_j) \right\}$$

$$\le - \sum_{1 \le j \le g} \left\{ n_j(\boldsymbol{\ell}) \log \det(2\pi c V_\ell) + c \operatorname{tr} \sum_{i \in C_j(\boldsymbol{\ell})} (x_i - m_j)(x_i - m_j)^\top V_\ell^{-1} \right\}$$

$$\le -r \log \det(2\pi c V_\ell) - c \operatorname{tr} \sum_{1 \le j \le g} \sum_{i \in C_j(\boldsymbol{\ell})} (x_i - \overline{x}_j(\boldsymbol{\ell}))(x_i - \overline{x}_j(\boldsymbol{\ell}))^\top V_\ell^{-1}$$

$$= -r \log \det(2\pi c V_\ell) - c \operatorname{tr} \left(W(\boldsymbol{\ell}) V_\ell^{-1} \right). \qquad \square$$

The following theorem deals with the universal breakdown point of the HDBT constrained TDC estimates of the covariance matrices.

2.22 Theorem (Universal breakdown point of the covariance matrices for the TDC)
Let the data $\mathbf{x}$ be in general position and assume $r \ge gd + 1$.

(a) If $2r \ge n + g(d+1)$, then the TDC estimates of the covariance matrices remain in a compact subset of $\mathrm{PD}(d)$ that depends only on the original data set $\mathbf{x}$ as at most $n - r + g - 1$ data points of $\mathbf{x}$ are replaced in an arbitrary way.

(b) It is possible to replace $n - r + g$ elements of $\mathbf{x}$ in such a way that the largest eigenvalue of the TDC estimate of some covariance matrix (and hence of all covariance matrices) exceeds any given number.

(c) If $2r \ge n + g(d+1)$, then $\beta_{\mathrm{var}}(n, r, g) = \frac{n-r+g}{n}$.

PROOF. (a) We first note that, no matter what the admissibly modified data M is, the HDBT constrained maximum posterior density and hence the constrained maximum likelihood $f(\mathbf{x}_{R^*}; \boldsymbol{\ell}^*, \mathbf{m}^*, \mathbf{V}^*)$ remain bounded below by a strictly positive constant that depends only on the original data set $\mathbf{x}$. To this end, we compare the optimal solution with one that is constrained irrespective of the constant c. Indeed, let $\boldsymbol{\ell}$ be the labeling that assigns the remaining $r - g + 1$ original points to the first cluster C_1 and $g - 1$ replacements y_j to one-point clusters $C_j = \{y_j\}$, $2 \le j \le g$. Moreover, let $\mathbf{m} = (0, y_2, \ldots, y_g)$, and let $V_j = I_d$ for all $1 \le j \le g$. By optimality, we have

$$-r \mathrm{H}\left(\tfrac{n_1(\boldsymbol{\ell}^*)}{r}, \ldots, \tfrac{n_g(\boldsymbol{\ell}^*)}{r}\right) + \log f(\mathbf{x}_{R^*}; \boldsymbol{\ell}^*, \mathbf{m}(\boldsymbol{\ell}^*), \mathbf{V}(\boldsymbol{\ell}^*))$$

$$\ge -r \mathrm{H}\left(\tfrac{r-g+1}{r}, \tfrac{1}{r}, \ldots, \tfrac{1}{r}\right) + \log f(\mathbf{x}_R; \boldsymbol{\ell}, \mathbf{m}, I_d) = \mathrm{const} - \tfrac{1}{2} \sum_{\ell_i = 1} \|x_i\|^2.$$

The right side of this expression satisfies the HDBT constraints for $c = 1$ and does not depend on the replacements.

Now, by assumption, we replace at most $n - r + g - 1 \le r - (gd + 1)$ (≥ 0) data points of $\mathbf{x}$. Thus, for any assignment, at least one cluster contains at least $d + 1$ original points $T \subseteq \mathbf{x}$. This is in particular true for an optimal assignment $\boldsymbol{\ell}^*$. By general position, it follows $W(\boldsymbol{\ell}^*) \succeq W_T \succeq \varepsilon I_d$ for some $\varepsilon > 0$. Lemma 2.21 and the initial remark imply

$$-r \log \det(2\pi c V_1^*) - c \operatorname{tr} \left(W(\boldsymbol{\ell}^*) V_1^{*-1} \right) \ge 2 \log f(\mathbf{x}_{R^*}; \boldsymbol{\ell}^*, \mathbf{m}^*, \mathbf{V}^*) \ge \mathrm{const} > -\infty.$$

Now, it is well known that the set of matrices V_1^* for which the left side is bounded below is a compact subset of $\mathrm{PD}(d)$. The HDBT constraints finally imply that the associated set of g-tuples $(V_1^*, \ldots, V_g^*)$ is a compact subset of $\mathrm{PD}(d)^d$. This proves claim (a).

(b) Modify $\mathbf{x}$ by $n - r + g$ replacements at a large distance from each other and from all original data points to obtain M. Each r-element subset of M contains at least g replacements. Moreover, there is a cluster C of size at least 2 that contains at least one replacement.

Indeed, if no cluster contains two replacements, then each cluster contains at least one and, by $r \geq gd + 1$, one of them contains another element. Now, let C_ℓ be such a cluster, let $y \in C_\ell$ be a replacement, and let $x \in C_\ell$, $x \neq y$. We have

$$W_\ell(\boldsymbol{\ell}) \geq \left\{ \left(x - \tfrac{x+y}{2}\right)\left(x - \tfrac{x+y}{2}\right)^\top + \left(y - \tfrac{x+y}{2}\right)\left(y - \tfrac{x+y}{2}\right)^\top \right\} = \tfrac{1}{2}(y-x)(y-x)^\top.$$

Now let $(\boldsymbol{\ell}^*, \mathbf{m}^*, \mathbf{V}^*)_j)$ be optimal parameters of the TDC. Comparing them with the inferior parameters $(\boldsymbol{\ell}^*, \mathbf{m}^*, 2\mathbf{V}^*)$ and noting that the entropy terms coincide, we infer

$$0 \leq \sum_j n_j(\boldsymbol{\ell}^*) \left\{ \log \det 2V_j^* + \mathrm{tr}\left((2V_j^*)^{-1} S_j(\boldsymbol{\ell}^*)\right) - \left[\log \det V_j^* + \mathrm{tr}\left(V_j^{*-1} S_j(\boldsymbol{\ell}^*)\right)\right] \right\}$$

$$= \sum_j n_j(\boldsymbol{\ell}^*) \left\{ d \log 2 - \tfrac{1}{2}\mathrm{tr}\left(V_j^{*-1} S_j(\boldsymbol{\ell}^*)\right) \right\} \leq dr \log 2 - \tfrac{1}{2}\mathrm{tr}\left(V_\ell^{*-1} W_\ell(\boldsymbol{\ell}^*)\right)$$

$$\leq dr \log 2 - \tfrac{1}{4}(y-x)^\top V_\ell^{*-1}(y-x).$$

The resulting inequality $(y-x)^\top V_\ell^{*-1}(y-x) \leq 4dr \log 2$ implies that at least one eigenvalue of V_ℓ^* exceeds any positive number as the distance between x and y is chosen large enough. Claim (c) follows from (a) and (b).   □

Like the trimmed, constrained MLE's of Section 2.2.3, the constrained TDC estimates of the covariance matrices withstand $g - 1$ more outliers than there are discards, $n - r$. The reason is the same as before. Outliers that are spread out may be assigned to one-point clusters and outliers located close together are able to form a cluster of their own. In each case the optimal assignment does not completely destroy the estimates.

Recall that the asymptotic breakdown point of an estimator is its limit as $n \to \infty$.

2.23 Corollary *If $r = \lfloor \alpha n \rfloor$ for some $\alpha > 1/2$, then the universal asymptotic breakdown point of the TDC estimates of the covariance matrices is $1 - \alpha$.*

As noted at the beginning of Section 2.3.2, the TDC estimates of the means are the sample means defined by the optimal assignment. In contrast to the covariance matrices, their *universal* breakdown point is low. In order to show this, we need a lemma and denote (univariate) scatter values and sums of squares by the letters s and w, respectively.

2.24 Lemma *Let $F \cup \{z_1, \ldots, z_{g-2}\} \cup \{y_1, y_2\} \subseteq \mathbb{R}$ be a data set of r pairwise distinct elements. If $w_{\{y_1,y_2\}} \leq \frac{2c}{r-2} w_F$, then the constrained normal MLE's $v_j(\boldsymbol{\ell})$ of the variances v_j for the partition $\boldsymbol{\ell} = \{F, \{z_1\}, \ldots, \{z_{g-2}\}, \{y_1, y_2\}\}$ are*

$$v_1(\boldsymbol{\ell}) = \frac{w_F + w_{\{y_1,y_2\}}/c}{r} \quad \text{and} \quad v_j(\boldsymbol{\ell}) = c\, v_1(\boldsymbol{\ell}), \ 2 \leq j \leq g.$$

PROOF. Putting $s_1 = s_F$ and $s_g = s_{\{y_1,y_2\}}$, the TDC requires minimizing the expression

$$h(v_1, \ldots, v_g) = n_1\left(\log v_1 + \frac{s_1}{v_1}\right) + \sum_{2 \leq j \leq g-1} \log v_j + 2\left(\log v_g + \frac{s_g}{v_g}\right)$$

w.r.t. $(v_1, \ldots, v_g) \in \mathcal{V}_c$. We start with the minimum of h on the larger set $\mathcal{V}_c' = \{(v_1, \ldots, v_g) \in \mathbb{R}_>^g \mid cv_1 \leq v_j \leq v_1/c, \ 2 \leq j \leq g\} \supseteq \mathcal{V}_c$. Since $\min_{cv_1 \leq v_j \leq v_1/c} \log v_j = \log cv_1$, dynamic optimization shows

$$\min_{\mathbf{v} \in \mathcal{V}_c'} h(v_1, \ldots, v_g)$$

$$= \min_{cv_1 \leq v_g \leq v_1/c} \left\{ n_1\left(\log v_1 + \frac{s_1}{v_1}\right) + \sum_{2 \leq j \leq g-1} \min_{cv_1 \leq v_j \leq v_1/c} \log v_j + 2\left(\log v_g + \frac{s_g}{v_g}\right) \right\}$$

$$= (g-2)\log c + \min_{cv_1 \leq v_g \leq v_1/c} \left\{ \left((r-2)\log v_1 + \frac{w_1}{v_1}\right) + \left(2\log v_g + \frac{w_g}{v_g}\right) \right\}.$$

This is a virtual two-cluster problem. The second line of the three cases in Proposition 2.20 shows that, under the assumption $\frac{w_g}{2} \leq c\frac{w_1}{r-2}$ stated in the lemma, its solution is indeed given by the values claimed for $v_1(\boldsymbol{\ell})$ and $v_g(\boldsymbol{\ell})$. Finally, the vector $(v_1(\boldsymbol{\ell}), cv_1(\boldsymbol{\ell}) \ldots, cv_1(\boldsymbol{\ell}))$ is even located in $\mathcal{V}_c$, so it is the minimum w.r.t. the smaller parameter set, too.   □

The next theorem deals with the universal breakdown point of the TDC estimates of the means. The result is similar to the mixture case.

2.25 Theorem (Universal breakdown point of the means for the TDC) Let the data $\mathbf{x}$ be in general position and let $g \geq 2$.

(a) If $n \geq r+1$ and $r \geq gd+2$, then the TDC estimates of all means remain bounded by a constant that depends only on the data $\mathbf{x}$ as *one* observation is arbitrarily replaced. (In the case of ties the solution with the largest discard is returned.)

(b) Under the standard assumption $r \geq gd+1$, there is a data set such that the TDC estimate of one sample mean breaks down as *two* particular observations are suitably replaced.

(c) Under the assumptions of (a) we have $\beta_{\mathrm{mean}}(n, r, g) = \frac{2}{n}$.

PROOF. (a) We show by contradiction that an optimal assignment $\boldsymbol{\ell}^*$ discards a remote replacement. Thus, assume that the replacement y lies in cluster ℓ. The cluster must contain a second (original) element x since, by the convention, y would otherwise be swapped with a discarded original element without changing the TDC. Now, by the assumption $r \geq gd+2$, the retained data points contain at least $gd+1$ original elements, so one cluster contains at least $d+1$ of them. Whether this is cluster ℓ or not, general position of $\mathbf{x}$ and this fact imply $\det W(\boldsymbol{\ell}^*) \to \infty$ as $\|y\| \to \infty$. We now use Lemma 2.21. It says that

$$2 \log f(\mathbf{x}_{R^*}; \boldsymbol{\ell}^*, \mathbf{m}^*, \mathbf{V}^*) \leq -r \log \det(2\pi c V_\ell^*) - c \operatorname{tr}\left(W(\boldsymbol{\ell}^*) V_\ell^{*-1}\right).$$

It is well known that, given a positive definite matrix W, the minimum of the function $V \mapsto \log \det V + \operatorname{tr} WV^{-1}$ is $\log \det W + d$. Hence, the right side of the inequality tends to $-\infty$ as $\|y\| \to \infty$ and so does the left side. On the other hand, by the assumption $r < n$, there exists an assignment $\boldsymbol{\ell}$ such that $y \notin R$. Optimality of $\boldsymbol{\ell}^*, \mathbf{m}^*, \mathbf{V}^*$ implies

$$-r\mathrm{H}\left(\tfrac{n_1(\boldsymbol{\ell}^*)}{r}, \ldots, \tfrac{n_g(\boldsymbol{\ell}^*)}{r}\right) + \log f(\mathbf{x}_{R^*}; \boldsymbol{\ell}^*, \mathbf{m}^*, \mathbf{V}^*)$$
$$\geq -r\mathrm{H}\left(\tfrac{n_1(\boldsymbol{\ell})}{r}, \ldots, \tfrac{n_g(\boldsymbol{\ell})}{r}\right) + \log f(\mathbf{x}_R; \boldsymbol{\ell}, \mathbf{0}, I_d).$$

Since the entropies are bounded, this means that $\log f(\mathbf{x}_{R^*}; \boldsymbol{\ell}^*, \mathbf{m}^*, \mathbf{V}^*)$ has a finite lower bound that does not depend on y, a contradiction to what was found before.

(b) A proof in the multivariate case requires a subtle construction of a data set. It must secure that the optimal solution retains at least one outlier. As a main hurdle one has to avoid point patterns that are almost degenerate and mask the desired solution just as in Fig. 1.3. A construction for the case $c = 1$ appears in Gallegos and Ritter [180]. For the sake of illustration, we treat here general c, confining ourselves to the univariate case. Since claim (b) is plainly true if $r \geq n-1$, we assume $r \leq n-2$ and proceed in three steps.

(α) Construction of the modified data set M:
Let x_i, $1 \leq i \leq r-g$, be strictly increasing and put $F = \{x_1, \ldots, x_{r-g}\}$, let $K > 0$, and choose $z_1 < z_2 < \ldots < z_{n-r+g-2}$ such that
    (i) $z_1 - x_{r-g} \geq K$ and $z_{\ell+1} - z_\ell \geq K$ for all $1 \leq \ell < n-r+g-2$.
Let $0 < \varepsilon \leq \sqrt{c\frac{w_F}{r-2}}$, let $y > z_{n-r+g-2} + \varepsilon$, define the replacements $y_{1,2} = y \pm \varepsilon$, and put $M = \{x_1, \ldots, x_{r-g}, z_1, \ldots, z_{n-r+g-2}, y_1, y_2\}$. Plainly, M is in general position.

Let $\widetilde{\boldsymbol{\ell}}$ be the assignment associated with the clustering $\{F, \{z_1\}, \ldots, \{z_{g-2}\}, \{y_1, y_2\}\}$ ($z_{g-1}, \ldots, z_{n-r+g-2}$ discarded).

(β) The maximum posterior density for $\widetilde{\boldsymbol{\ell}}$ does not depend on K and y:

Since $w_{\{y_1,y_2\}} = 2\varepsilon^2 \leq \frac{2c}{r-2}w_F$, Lemma 2.24 shows $v_1(\widetilde{\ell}) = \frac{w_F + w_{\{y_1,y_2\}}/c}{r}$ and $v_2(\widetilde{\ell}) = \ldots = v_g(\widetilde{\ell}) = cv_1(\widetilde{\ell})$. Twice the logarithm of the corresponding posterior density equals

$$2\big((r-g)\log\tfrac{r-g}{r} + 2\log\tfrac{2}{r}\big) - r\log v_1(\widetilde{\ell}) - g\log c - r(1 + \log 2\pi).$$

(γ) If K is large enough, then no assignment ℓ of r points from the set $F \cup \{z_1, \ldots, z_{n-r+g-2}\}$ is optimal:

By $r \leq n-2$, the set contains at least r elements. Since $|F| = r - g$ and since $r > g$, any such assignment ℓ creates a cluster $C_\ell(\ell)$ which contains some z_k and some other point. From (i), it follows

$$w(\ell) \geq w_{C_\ell(\ell)} \xrightarrow[K\to\infty]{} \infty. \tag{2.34}$$

By Lemma 2.21, twice its log-likelihood is bounded above by

$$-r\log(2\pi c\, v_j(\ell)) - c\frac{w(\ell)}{v_j(\ell)} \leq -r\big(\log 2\pi c^2/r + \log w(\ell) + 1\big) \xrightarrow[K\to\infty]{} -\infty, \; 1 \leq j \leq g;$$

here we have used the maximum of the left side as a function of the TDC estimate $v_j(\ell)$ and (2.34). The claim follows from (β) since there are only finitely many ℓ's.

Finally, choose K as in (γ). The optimal solution retains at least one y_h, causing at least one sample mean to break down as $y \to \infty$. This proves part (b) in the special univariate case and part (c) follows from (a) and (b).   □

As a consequence, the asymptotic universal breakdown point of the TDC estimates of the means is zero. More cannot be expected. The reason is that the universal breakdown point makes a statement on any data set for any g, even if the two do not match. On the other hand, García-Escudero and Gordaliza [187] carried out experiments with trimmed k-means, observing that the means of a clear cluster structure are hard to break down with the algorithm. We offer next an analysis of this phenomenon in the present situation.

2.3.5 Restricted breakdown point of the means

We analyze here the HDBT constrained TDC w.r.t. its restricted breakdown point of the means. As in the mixture case (Section 2.2.4) the restriction will again be to a class of data sets with a certain separation property. It defines what a "clear cluster structure" should mean.

The proof of the main result of this section, Theorem 2.32, depends on a series of lemmas which are first stated and proved. Let $\mathcal{P} = \{P_1, \ldots, P_g\}$ be a partition of $\mathbf{x}$ and let $\emptyset \neq T \subseteq \mathbf{x}$. The partition $T \cap \mathcal{P} = \{T \cap P_1, \ldots, T \cap P_g\}$ is the *trace* of $\mathcal{P}$ in T. Let $g' \geq 1$ be a natural number and let $\mathcal{T} = (T_1, \ldots, T_{g'})$ be some partition of T. The *common refinement* of $\mathcal{T}$ and $\mathcal{P}$ is denoted by $\mathcal{T} \sqcap \mathcal{P} = \{T_k \cap P_j \mid k \leq g', j \leq g\}$, a partition of T (some clusters may be empty). The pooled SSP matrix of $\mathcal{T}$ w.r.t. some partition $\mathcal{T}$ is defined by

$$W_{\mathcal{T}} = \sum_{j \leq g'} W_{T_j}.$$

The following proposition states a basic condition which implies robustness of the TDC estimates of the means.

2.26 Proposition *Let the data $\mathbf{x}$ be in general position, let $g \geq 2$ and $gd + 1 < r < n$, and let q be an integer such that $\max\{2r - n, gd + 1\} \leq q < r$. Assume that $\mathbf{x}$ possesses a partition $\mathcal{P}$ in g clusters such that, for all $T \subseteq \mathbf{x}$, $q \leq |T| < r$, and all partitions $\mathcal{T}$ of T in $g - 1$ clusters (some clusters may be empty), the pooled SSP matrix satisfies*

$$\det W_{\mathcal{T}} \geq g^2 \max_{R \in \binom{\mathbf{x}}{r}, R \supseteq T} \det\big(\tfrac{1}{c^2} W_{R \cap \mathcal{P}}\big). \tag{2.35}$$

Then the individual breakdown point of the TDC estimates of the means satisfies

$$\beta_{mean}(n, g, r, \mathbf{x}) \geq \tfrac{1}{n}(r - q + 1).$$

PROOF. Let M be any admissible data set obtained from $\mathbf{x}$ by modifying at most $r - q$ elements and let $(\mathbf{x}_R^*, \boldsymbol{\ell}^*, (m_j^*)_{j=1}^g, (V_j^*)_{j=1}^g)$ be a TDC estimate for M. We will show that its sample means m_j^* are bounded by a number that depends solely on the original data $\mathbf{x}$. Our proof proceeds in several steps.

(α) The matrices V_j^* are bounded above and below by positive definite matrices that depend only on $\mathbf{x}$, not on the replacements:
Let R_j^* be the jth cluster generated by $\boldsymbol{\ell}^*$. Since $|R^*| = r$, $R^* = \bigcup_{j=1}^g R_j^*$ has at least $q \geq gd + 1$ original observations, some R_j^* contains at least $d + 1$ original observations. The proof now finishes in the same way as that of Theorem 2.22(a).

(β) If R_j^* contains some original observation, then m_j^* is bounded by a number that depends only on $\mathbf{x}$:
By (α), $\operatorname{tr} W(\boldsymbol{\ell}^*)$ remains bounded above by a constant which depends solely on the original data $\mathbf{x}$. Now, let $x \in R_j^* \cap \mathbf{x}$. We have $W(\boldsymbol{\ell}^*) \succeq (x - m_j^*)(x - m_j^*)^\top$ and hence $\|x - m_j^*\|^2 \leq \operatorname{tr} W(\boldsymbol{\ell}^*)$ and the claim follows.

(γ) If R_j^* contains some replacement, then $\|m_j^*\| \to \infty$ as the replacement tends to ∞:
This is proved like (β) where x is now the replacement.

From (β) and (γ) it follows: As the replacements tend to ∞, each R_j^*, $1 \leq j \leq g$, consists in the long run solely of original observations or solely of replacements. We next put $c_{d,r} = -\frac{dr}{2}(1 + \log 2\pi)$ and state:

(δ) $-r\mathrm{H}\left(\frac{n_1^*}{r}, \ldots, \frac{n_g^*}{r}\right) + \log f(\mathbf{x}_{R^*}; \boldsymbol{\ell}^*, \mathbf{m}^*, \mathbf{V}^*) < c_{d,r} - dr \log c - \frac{r}{2} \log \det \frac{W(\boldsymbol{\ell}^*)}{r}$, whenever $0 < n_j^* < r$ for some j:
On account of Lemma 2.21 and of the assumption, the left side is strictly bounded above by

$$-dr \log c - \tfrac{dr}{2} \log 2\pi - \tfrac{1}{2}\left[r \log \det(V_1^*/c) + \operatorname{tr}\left(W(\boldsymbol{\ell}^*)(V_1^*/c)^{-1}\right)\right].$$

Part (α) and normal estimation theory now show that the function $A \mapsto r \log \det(A/c) + \operatorname{tr}\left(W(\boldsymbol{\ell}^*)(A/c)^{-1}\right)$, $A \succeq 0$, attains its minimum value $r\left[\log \det\left(\frac{W(\boldsymbol{\ell}^*)}{r}\right) + d\right]$ at $\frac{cW(\boldsymbol{\ell}^*)}{r}$ and the claim follows.

(ϵ) R^* contains no large replacement:
Assume on the contrary that R^* contains some large replacement. In view of the remark right after (γ), some cluster, say R_g^*, consists solely of replacements. Note that $r > |R^* \cap \mathbf{x}| \geq q$. Let $T = R^* \cap \mathbf{x}$ and let $\mathcal{T} = \{R_1^* \cap \mathbf{x}, \ldots, R_{g-1}^* \cap \mathbf{x}\}$. From Steiner's formula we infer the relation $W(\boldsymbol{\ell}^*) \succeq W_\mathcal{T}$ between the pooled SSP matrices and so hypothesis (2.35) implies

$$\det W(\boldsymbol{\ell}^*) \geq \det W_\mathcal{T} \geq g^2 \max_{R \in \binom{\mathbf{x}}{r}, R \supseteq T} \det\left(\tfrac{1}{c^2} W_{R \cap \mathcal{P}}\right).$$

Hence

$$2d \log c + \log \det \frac{W(\boldsymbol{\ell}^*)}{r} \geq 2 \log g + \max_{R \in \binom{\mathbf{x}}{r}, R \supseteq T} \log \det \tfrac{1}{r} W_{R \cap \mathcal{P}}. \qquad (2.36)$$

Now, denoting the assignment associated with $R \cap \mathcal{P}$ by $\boldsymbol{\ell}_{R \cap \mathcal{P}}$, and writing $\mathbf{m}_{R \cap \mathcal{P}} = (m_{R \cap P_1}, \ldots, m_{R \cap P_g})$ and $S_{R \cap \mathcal{P}} = \frac{1}{r} W_{R \cap \mathcal{P}}$, the pooled scatter matrix, we have

$$r \log g + \min_{R \in \binom{M \cap \mathbf{x}}{r}} -\log f(\mathbf{x}_R; \boldsymbol{\ell}_{R \cap \mathcal{P}}, \mathbf{m}_{R \cap \mathcal{P}}, S_{R \cap \mathcal{P}})$$

$$= -c_{d,r} + r \log g + \tfrac{r}{2} \min_{R \in \binom{M \cap \mathbf{x}}{r}} \log \det S_{R \cap \mathcal{P}}$$

$$\leq -c_{d,r} + r \log g + \tfrac{r}{2} \min_{T \subseteq R \in \binom{M \cap \mathbf{x}}{r}} \log \det S_{R \cap \mathcal{P}} \qquad (2.37)$$

$$\leq -c_{d,r} + r \log g + \tfrac{r}{2} \max_{T \subseteq R \in \binom{\mathbf{x}}{r}} \log \det S_{R \cap \mathcal{P}}$$

$$\leq -c_{d,r} + dr \log c + \tfrac{r}{2} \log \det V(\boldsymbol{\ell}^*)$$

$$< r \mathrm{H}\left(\tfrac{n_1^*}{r}, \ldots, \tfrac{n_g^*}{r}\right) - \log f(\mathbf{x}_{R^*}; \boldsymbol{\ell}^*, \mathbf{m}^*, \mathbf{V}^*).$$

Here the last two inequalities follow from (2.36) and (δ), respectively. Note that part (δ) is applicable since $R^* \cap \mathbf{x} \neq \emptyset$ implies $n_j^* > 0$ for some $j < g$ and since $n_g^* > 0$ as well. The last expression above is the constrained minimum of the TDC. It is no larger than its value at the clustering $R \cap \mathcal{P}$ with the parameters $\mathbf{m}_{R \cap \mathcal{P}}$ and $S_{R \cap \mathcal{P}}$ for all $R \in \binom{M \cap \mathbf{x}}{r}$. By an elementary property of the entropy, the latter value is no larger than the first line of (2.37). This contradiction proves claim (ϵ).

Finally, part (β) shows that all sample means m_j^* remain bounded by a number that depends only on $\mathbf{x}$. This proves the proposition. □

The remainder of this section is devoted to elucidating why the hypothesis of Proposition 2.26 actually represents some separation property. We need more notation. Let $g \geq 2$. Given an integer $u \geq 1$ and a real number $0 < \varrho < 1$, we define the number

$$q_{u,\varrho} = \max\left\{2r - n, (g-1)gd + 1, \tfrac{n-u}{1-\varrho}\right\}.$$

If $n > r > (g-1)gd + 1$ and $u \geq n - (1-\varrho)(r-1)$, then $q = \lceil q_{u,\varrho} \rceil$ satisfies the assumption made in Proposition 2.26.

Let $\mathcal{P}$, T, and $\mathcal{T}$ be as in Proposition 2.26. Our next, combinatorial, lemma gives conditions that secure the existence of sufficiently many elements of T in each class P_j and a large intersection $T_k \cap P_j$ for some pair (k, j).

2.27 Lemma *Let* $\mathcal{P} = \{P_1, \ldots, P_g\}$ *be a partition of* $\mathbf{x}$ *in clusters of size* $\geq u$, *let* $T \subseteq \mathbf{x}$ *such that* $q_{u,\varrho} \leq |T| < r$, *and let* $\mathcal{T} = \{T_1, \ldots, T_{g-1}\}$ *be a partition of* T *(some* T_k's *may be empty). Then:*

(a) For all j, *we have* $|T \cap P_j| \geq \varrho|T|$.

(b) At least one T_k *contains elements of two different* P_j's.

(c) There are clusters T_k *and* P_j *such that* $|T_k \cap P_j| \geq \frac{q_{u,\varrho}}{(g-1)g}$ $(> d)$.

PROOF. (a) Assume on the contrary that $|T \cap P_j| < \varrho|T|$ for some j. From $\mathbf{x} \supseteq T \cup P_j$ we infer

$$n \geq |T| + |P_j| - |T \cap P_j| > |T| + u - \varrho|T| = u + (1-\varrho)|T| \geq u + (1-\varrho)q_{u,\varrho} \geq u + n - u$$

by definition of $q_{u,\varrho}$. This is a contradiction.

(b) Since $\varrho|T| > 0$ and since there are more P_j's than T_k's, this follows from the pigeon hole principle along with (a).

(c) The observations in T are spread over the $(g-1)g$ disjoint sets of the form $T_k \cap P_j$. If (c) did not hold, we would have $|T| < q_{u,\varrho}$, contradicting one of the assumptions. □

2.28 Lemma *Let* $g \geq 2$, *let* $0 < \varrho \leq 1/g$, *let* $\mathbf{a} = (a_{kj})_{\substack{1 \leq k < g \\ 1 \leq j \leq g}} \in \mathbb{N}^{(g-1) \times g}$ *be such that* $\|\mathbf{a}\|_1 = \sum_{k,j} a_{kj} > 0$, *let* $\sum_k a_{kj} \geq \varrho\|\mathbf{a}\|_1$ *for all* $1 \leq j \leq g$, *and put* $a_{k+} = \sum_j a_{kj}$. *Then*

$$\sum_{k:a_{k+}>0} \frac{1}{a_{k+}} \sum_{1 \leq j < \ell \leq g} a_{kj}a_{k\ell} \geq \kappa_\varrho\|\mathbf{a}\|_1. \qquad (2.38)$$

PROOF. Write the left hand side of (2.38) as

$$\|\mathbf{a}\|_1 \sum_{k:a_{k+}>0} \frac{a_{k+}}{\|\mathbf{a}\|_1} \sum_{1\leq j<\ell\leq g} \frac{a_{kj}}{a_{k+}} \frac{a_{k\ell}}{a_{k+}} = \|\mathbf{a}\|_1 \sum_{k:a_{k+}>0} \beta_k \sum_{1\leq j<\ell\leq g} A_{k,j} A_{k,\ell}.$$

Since $\beta = (a_{k+}/\|\mathbf{a}\|_1)_{k:a_{k+}>0}$ is a probability vector and since $A = (a_{k,j}/a_{k+})_{k:a_{k+}>0,j}$ is a stochastic matrix such that $\beta A \geq \varrho$ entry by entry, the claim follows from an elementary reasoning. $\square$

In order to define the announced separation property we put

$$\kappa_\varrho = \begin{cases} (1-\varrho)\varrho, & g = 2, \\ \varrho/2, & g \geq 3. \end{cases}$$

2.29 Definition (Separation property for the TDC) Let $u \in \mathbb{N}$ such that $1 \leq u \leq n/g$ and let $0 < \varrho < 1$. We denote by $\mathcal{L}_{u,\varrho,c}$ the system of all d-dimensional admissible data sets $\mathbf{x}$ of size n which have the following **separation property**:

The data $\mathbf{x}$ possesses a partition $\mathcal{P}$ in g subsets of size at least u such that, for all subsets $T \subseteq \mathbf{x}$, $q_{u,\varrho} \leq |T| < r$ and for all partitions $\mathcal{T} = \{T_1, \ldots, T_{g-1}\}$ of T in $g-1$ clusters, we have

$$1 + \kappa_\varrho \cdot \min_{\substack{k,\, j\neq \ell: \\ T_k\cap P_h \neq \emptyset,\ h=j,\ell}} (\overline{x}_{T_k\cap P_j} - \overline{x}_{T_k\cap P_\ell})^\top \left(\frac{W_{\mathcal{T}\sqcap\mathcal{P}}}{|T|}\right)^{-1} (\overline{x}_{T_k\cap P_j} - \overline{x}_{T_k\cap P_\ell})$$

$$\geq g^2 \frac{\max_{R\in\binom{\mathbf{x}}{r},R\supseteq T} \det \frac{1}{c^2} W_{R\sqcap\mathcal{P}}}{\det W_{\mathcal{T}\sqcap\mathcal{P}}}. \tag{2.39}$$

According to Lemma 2.27(b), the minimum extends over at least one triple (k, j, ℓ), $j \neq \ell$, and by Lemma 2.27(c), the pooled scatter matrix $S_{\mathcal{T}\sqcap\mathcal{P}}$ is bounded below by a positive definite matrix which depends only on $\mathbf{x}$. Condition (2.39) is affine invariant. We require the minimum of the Mahalanobis distances of the submeans $\overline{x}_{T_k\cap P_j}$ and $\overline{x}_{T_k\cap P_\ell}$ of P_j and P_ℓ appearing on its left-hand side to be large. Thus, condition (2.39) means that the partition $\mathcal{P}$ subdivides the data set in well-separated clusters; it is the "natural" partition of $\mathbf{x}$. The set $\mathcal{L}_{u,\varrho,c}$ increases with decreasing u and with increasing $\varrho \leq 1/2$.

It is next shown that any data set $\mathbf{x}$ in $\mathcal{L}_{u,\varrho,c}$ satisfies the hypotheses of Proposition 2.26.

2.30 Lemma Let $g \geq 2$, let $n > r > (g-1)gd + 1$, let $u \in \mathbb{N}$ and $0 < \varrho < 1$ satisfy $n - (1-\varrho)(r-1) \leq u \leq n/g$. Let $\mathbf{x} \in \mathcal{L}_{u,\varrho,c}$, let $T \subseteq \mathbf{x}$ be such that $q_{u,\varrho} \leq |T| < r$, and let $\mathcal{T} = \{T_1, \ldots, T_{g-1}\}$ be a partition of T (some T_k's may be empty). We have

$$\det W_{\mathcal{T}} \geq g^2 \max_{R\in\binom{\mathbf{x}}{r},R\supseteq T} \det \tfrac{1}{c^2} W_{R\sqcap\mathcal{P}}.$$

PROOF. Applying Lemma A.12 to each T_k, $k < g$, with partition $\{T_k \cap P_1, \ldots, T_k \cap P_g\}$, we obtain first

$$W_{\mathcal{T}} = \sum_{k=1}^{g-1} W_{T_k}$$

$$= \sum_{k:T_k\neq\emptyset} \left\{ \sum_{j=1}^{g} W_{T_k\cap P_j} + \sum_{1\leq j<\ell\leq g} \frac{a_{kj}a_{k\ell}}{|T_k|} (\overline{x}_{T_k\cap P_j} - \overline{x}_{T_k\cap P_\ell})(\overline{x}_{T_k\cap P_j} - \overline{x}_{T_k\cap P_\ell})^\top \right\},$$

where $a_{kj} = |T_k \cap P_j|$, $1 \leq j \leq g$, $1 \leq k < g$. Now use Lemma A.6(b) and Lemma 2.28 to obtain

$$\det W_{\mathcal{T}}$$

$$\geq \det W_{\mathcal{T}\sqcap\mathcal{P}} \cdot \left\{ 1 + \sum_{k:T_k \neq \emptyset} \sum_{1 \leq j < \ell \leq g} \frac{a_{kj}a_{k\ell}}{|T_k|} (\overline{x}_{T_k \cap P_j} - \overline{x}_{T_k \cap P_\ell})^\top W_{\mathcal{T}\sqcap\mathcal{P}}^{-1} (\overline{x}_{T_k \cap P_j} - \overline{x}_{T_k \cap P_\ell}) \right\}$$

$$\geq \det W_{\mathcal{T}\sqcap\mathcal{P}} \cdot \left\{ 1 + \kappa_\varrho \min_{\substack{k,\, j \neq \ell\,: \\ T_k \cap P_h \neq \emptyset}} (\overline{x}_{T_k \cap P_j} - \overline{x}_{T_k \cap P_\ell})^\top \left(\frac{W_{\mathcal{T}\sqcap\mathcal{P}}}{|T|}\right)^{-1} (\overline{x}_{T_k \cap P_j} - \overline{x}_{T_k \cap P_\ell}) \right\}.$$

The claim follows from the separation property.   □

The conditions on r and u imply that the interval $[q_{u,\varrho}, r[$ contains some integer. It is needed for a set T as in Lemma 2.30 to exist. A simple reasoning shows that the bounds on u imply $\varrho < 1/g$.

The following elementary lemma will be needed.

2.31 Lemma Let $h \geq 0$ and let $k \geq 1$. Let $C = \{x_1, \ldots, x_h, y_1, \ldots, y_k\}$ consist of h original data points and k replacements. Then the norm of the sample mean of C tends to infinity as $\|y_1\| \to \infty$ while the differences $y_i - y_1$, $2 \leq i \leq k$, remain bounded.

PROOF. The sum of C has the representation $\sum_{i=1}^h x_i + ky_1 + \sum_{i=2}^k (y_i - y_1)$ from which the lemma follows.   □

Here is, finally, the main result of this section, the restricted breakdown point of the TDC estimates of the means. If a data set has the separation property, then the TDC estimates of the means are much more robust than predicted by Theorem 2.25.

2.32 Theorem (Restricted breakdown point of the means for the TDC) Let the data $\mathbf{x}$ be in general position, let $g \geq 2$, and let $r < n$.

(a) Assume $r \geq (g-1)gd + 2$ and $n - (1-\varrho)(r-1) \leq u \leq n/g$. Then the restricted breakdown point of the TDC estimates of the means w.r.t. $\mathcal{L}_{u,\varrho,c}$ satisfies

$$\beta_{\text{mean}}(n, g, r, \mathcal{L}_{u,\varrho,c}) \geq \tfrac{1}{n} \min\left\{ n - r + 1, r - (g-1)gd, r + 1 - \tfrac{n-u}{1-\varrho} \right\}.$$

(b) For any data set $\mathbf{x} \in \mathcal{L}_{u,\varrho,c}$, the individual breakdown point of the TDC estimates of the means satisfies

$$\beta_{\text{mean}}(n, g, r, \mathbf{x}) \leq \tfrac{1}{n}(n - r + 1).$$

(c) Let $2r - n \geq (g-1)gd + 1$, let $u \in \mathbb{N}$ such that $2(n-r) < u \leq n/g$, and put $\varrho = \frac{u - 2(n-r)}{2r - n}$. Then

$$\beta_{\text{mean}}(n, g, r, \mathcal{L}_{u,\varrho,c}) = \tfrac{1}{n}(n - r + 1).$$

(A necessary condition for the existence of such a u is the inequality $2(n-r) \leq n/g - 1$.)

(d) Under the assumptions of (a), TDC discards all sufficiently large replacements in any data set that satisfies the separation property (with any parameters).

PROOF. Part (a) is a direct consequence of Proposition 2.26 and Lemma 2.30.

(b) Let M be a data set obtained from $\mathbf{x}$ by replacing $n - r + 1$ of its elements with a tight and distant cluster. The modified data set contains only $r - 1$ original observations, so the optimal set R^* contains some replacement. Then so does $C_j^* = C_j(\ell^*)$ for some j. Lemma 2.31 shows that the norm of m_j^* tends to infinity along with the tight cluster of replacements.

(c) The hypotheses imply $\min\left\{ n - r + 1, r - (g-1)gd, r + 1 - \tfrac{n-u}{1-\varrho} \right\} = n - r + 1$. (Note that ϱ is maximum, so the first term does not exceed the last one for a given u.) Furthermore, the first condition in (a) follows from the first condition, whereas the second condition in (a) follows from the choice of ϱ and from the second condition. Finally, the condition $2(n-r) < u$ implies $\varrho > 0$. The claim now follows from parts (a) and (b).

Claim (d) follows from part (ϵ) of the proof of Proposition 2.26.   □

The inequality $n - (1-\varrho)(r-1) \leq u$ implies $u \geq n-r+2$. That is, in part (a) of Theorem 2.32 the sizes of the natural clusters must exceed the number of discards. Moreover, the assumptions of part (c) imply that these sizes must exceed twice the number of discards.

The following corollary of Theorem 2.32 says that the TDC estimates of the means are asymptotically robust on well-separated, balanced data sets if the parameter g is set to its natural number of clusters.

2.33 Corollary *Let* $g \geq 2$, *let* $0 < \eta < \delta < 1/g$, *let* $r = \lceil n\big(1 - \frac{1}{2g} + \frac{\delta}{2}\big)\rceil$, *let* $u = \lceil n\big(\frac{1}{g} - \eta\big)\rceil$, *and let* $\varrho = \frac{\delta - \eta}{1 - \frac{1}{g} + \delta}$. *Then, asymptotically,*

$$\beta_{mean}(n, g, r, \mathcal{L}_{u,\varrho,c}) \longrightarrow \tfrac{1}{2}\big(\tfrac{1}{g} - \delta\big), \quad as\ n \to \infty.$$

2.34 Remarks (a) (The case $g = 1$) In the case of one component, both the normal trimmed mixture likelihood (2.6) and the TDC, Section 2.3.2, reduce to Rousseeuw's [448] **Minimum Covariance Determinant**, MCD, for robust estimation of location and scatter. If $\alpha < 0.5$, then its asymptotic (universal) breakdown point with parameter $r = \lceil (1-\alpha)n \rceil$ is known to be α; see Rousseeuw [448], p. 291. This is in harmony with the present results on covariance matrices, Theorems 2.5 and 2.22. For $g = 1$, a much smaller number of retained observations, r, still has a positive effect on the universal breakdown of the mean vector. Breakdown needs a much larger number of outliers compared with the case $g > 1$ stated in Theorems 2.9 and 2.25. The reason is that, in the case $g > 1$, the outliers may establish a component of their own if they lie close to each other, thus causing one mean to diverge.

(b) The theory presented so far in Chapters 1 and 2 has shown that the consideration of HDBT ratio and constraints serves four purposes: it guarantees a solution, it reduces the set of local optima, it avoids spurious clusters, and it adds robustness. In Chapter 4 it will be shown that it is also useful to define feasible solutions.

2.3.6 Notes

Trimming goes back to Newcomb [387]. Cuesta-Albertos et al. [106] are the first authors to propose trimming for cluster analysis and to study consistency of the trimmed pooled trace criterion. To the author's knowledge, the first heteroscedastic, normal classification model with full covariance structure and trimming is Rocke and Woodruff's [444] MINO. Besides trimming, MINO also uses constraints on cluster sizes n_j, $1 \leq j \leq g$, in order to enforce maximum likelihood estimates to exist. Gallegos and Ritter [184] extend their method to MAP estimation.

The *Trimmed Determinant Criterion* TDC appears in Gallegos and Ritter [183]. It is the extension of the homonymous homoscedastic ML criterion appearing in Gallegos and Ritter [180] to the heteroscedastic MAP case. The breakdown analysis in Theorems 2.22, 2.25, and 2.32 appears in Gallegos and Ritter [183]. It generalizes their analysis of the homoscedastic ML case, Gallegos and Ritter [180].

An extension of Pollard's [416] theorem to trimming is due to Cuesta-Albertos et al. [106]. García-Escudero et al. [188] present a trimmed, heteroscedastic, normal, classification model investigating its consistency. They require that the ratio of the largest and the smallest eigenvalues of all scale matrices be bounded. These constraints are more restrictive than the HDBT constraints constraining not only the heteroscedasticity but also the asphericity. As a consequence, the criterion lacks equivariance w.r.t. variable scaling. The extension of Pollard's theorem to the HDBT constrained heteroscedastic, elliptical case, Theorem 1.56, and Corollary 4.6 on model selection appear in Gallegos and Ritter [185]. Gallegos and Ritter [181] prove two consistency theorems for classification models valid asymptotically as separation grows. Their paper treats nonstandard data sets that allow multiple observations of the same object, called variants, most or all of them outlying.

The present breakdown points refer to parameter sensitivity. In cluster analysis the most important objects are partitions and their clusters. Hennig [243] therefore proposes robustness criteria for cluster analysis that depend on the clusters themselves. They are based on Jaccard's [266] similarity.

Chapter 3

Algorithms

In Chapters 1 and 2, the HDBT constraints played a major theoretical rôle. They were responsible for the existence of ML and MAP estimates and for their consistency and robustness. Since the HDBT constant is unknown in practical applications and as there is a consistent free *local* optimum, Theorem 1.19, we will focus here on algorithms that compute *free* local optima and steady partitions. They merit the highest attention. Their number may run into the thousands, the great majority being unreasonable. It will, however, be easy to single out the favorite one(s) by the method of scale balance in Section 4.1. Solutions to clustering problems are often not unique.

Likelihood functions and criteria do not themselves suggest methods to actually *compute* solutions. This chapter will deal with the question of how to effectively find local optima to likelihood functions and steady solutions to criteria. This task is in fact computationally difficult depending on heuristics. For both, the mixture and the classification model, dedicated algorithms have been known for some time that serve the purpose. They are iterative and alternate between assignment and parameter estimation. They will be the subject matter of this chapter. Another dedicated clustering method is agglomerative optimization. It yields a hierarchy of partitions, one for each g between $n-1$ and 2. It may be used in the present framework for finding initial solutions for the master algorithms; see Section 3.3.

3.1 EM algorithm for mixtures

A favorite way of estimating a partition is to fit a mixture to the data and to feed its mixing rates π and population parameters γ into MAP discriminant analysis to obtain a partition of the data set. This will be the point of view of this section. These days, the standard way of fitting mixtures to data is the EM algorithm. Dempster, Laird, and Rubin's [115] famous paper on this algorithm stimulated much of the interest and research in mixture models. The basic theory of EM is presented in detail in Appendix G.2. We will now apply it to mixtures, assuming that the basic densities are strictly positive everywhere.

3.1.1 General mixtures

Among other things, Dempster, Laird, and Rubin [115] applied the EM algorithm to estimating the parameters $\vartheta = (\pi, \gamma)$ of a mixture (1.9) with strictly positive density $f_{\pi,\gamma}(x) = \sum_j \pi_j f_{\gamma_j}(x)$, mixing rates $\pi = (\pi_1, \ldots, \pi_g) \in \Delta_{g-1}$, and population parameters $\gamma = (\gamma_1, \ldots, \gamma_g) \in \Gamma^g$ where Γ is the parameter space of some basic model on a suitable sample space E. The observed joint log-likelihood function is thus $\sum_i \log \sum_j \pi_j f_{\gamma_j}(x_i)$ with $\mathbf{x} = (x_1, \ldots, x_n)$ being again realizations of the (independent) observed random variables $X = (X_1, \ldots, X_n)$. They considered the random labels $L_i : \Omega \to 1..g$ of the observations $i \in 1..n$ as hidden variables; see Appendix G.2.1. With $L = (L_1, \ldots, L_n)$, the complete variable becomes (X, L). The application of the general framework of the EM algorithm begins with formulæ (G.9), (G.10) and counting measure ϱ_x on $1..g$. The representation

$X = Y^{(L)}$, $Y^{(j)} \sim \gamma_j$, from Lemma 1.12 and the product formula yield the complete likelihood

$$g_{\pi,\gamma}(x,j) = f_{(X,L)}(x,j) = f_{(Y^{(j)},L)}(x,j) = f_{Y^{(j)}}[x \mid L = j]P[L = j] = \pi_j f_{\gamma_j}(x),$$

and so the conditional probabilities (G.9) turn to

$$w_j^{\pi,\gamma}(x) = \frac{\pi_j f_{\gamma_j}(x)}{\sum_\ell \pi_\ell f_{\gamma_\ell}(x)}. \tag{3.1}$$

A further simple computation shows that, with another pair of parameters $\theta = (\tau, \eta) \in \Delta_{g-1} \times \Gamma^g$, the Q-functional (G.5) becomes

$$Q((\pi,\gamma),(\tau,\eta)) = \sum_i \sum_j w_j^{\pi,\gamma}(x_i) \log g_{\tau,\eta}(x_i,j) \tag{3.2}$$

$$= \sum_j \left(\sum_i w_j^{\pi,\gamma}(x_i) \right) \log \tau_j + \sum_j \sum_i w_j^{\pi,\gamma}(x_i) \log f_{\eta_j}(x_i),$$

with the weights $w_j^{\pi,\gamma}(x)$ defined in (3.1) for π and γ. This representation separates the variables τ_j and η_j and the two terms on the right of (3.2) can be maximized separately. By the entropy inequality, the maximum w.r.t. τ is π_{new} defined by

$$\pi_{\text{new},j} = \frac{1}{n} \sum_i w_j^{\pi,\gamma}(x_i). \tag{3.3}$$

The EM step G.11 thus turns into the following EM step for general mixtures.

3.1 Procedure (EM step for general mixtures)
// Input: Parameters π and γ.
// Output: Parameters π_{new} and γ_{new} with larger observed likelihood.

1. *(E-step)* Compute the weights $w_\ell^{\pi,\gamma}(x_i)$ from the current parameters π and γ; see (3.1).
2. *(M-step)* Define π_{new} by (3.3) and maximize $\sum_i w_j^{\pi,\gamma}(x_i) \log f_{\eta_j}(x_i)$ w.r.t. $\eta_j \in \Gamma$ for all j to obtain the new parameter $\gamma_{\text{new}} = (\eta_1^*, \ldots, \eta_g^*)$ (if the maximum exists).

Of course, the EM step can only be applied to models f_γ that actually allow maximization in the M-step. This is, for instance, the case when Γ is compact and the likelihood function is continuous. If Γ is locally compact and noncompact and if the likelihood function $\eta \mapsto f_\eta(x_i)$ vanishes for all i as η approaches the Alexandrov point of Γ, then the same is true for the sum $\sum_{1 \le i \le n} w_j^{\pi,\gamma}(x_i) \log f_{\eta_j}(x_i)$ and it is again plain that the maximum exists. If components are normal, then the solution even exists in closed form; see Section 3.1.2.

The **EM algorithm for general mixtures** is the iteration of EM steps. It is iterative and alternating, proceeding as follows:

$$(\pi^{(0)}, \gamma^{(0)}) \to w^{(1)} \to (\pi^{(1)}, \gamma^{(1)}) \to w^{(2)} \to (\pi^{(2)}, \gamma^{(2)}) \to w^{(3)} \to \cdots$$

Its convergence depends on the initial parameters and on the model. For the sake of completeness the main properties of the EM step and the EM algorithm will be formulated for general mixtures. They follow from Theorem G.10.

3.2 Proposition *(a) The EM step improves the observed log-likelihood function* $\sum_i \log \sum_j \pi_j f_{\gamma_j}(x_i)$.
(b) If the EM algorithm converges, then the limit is a fixed point of Q.
(c) If Θ is an open subset of some Euclidean space and if the likelihood function is differentiable there, then the limit is a critical point of the likelihood function.

3.1.2   Normal mixtures

Unconstrained heteroscedastic as well as *homoscedastic* normal mixtures allow analytic optimization of the Q-functional in the M-step of Procedure 3.1. The optimal parameters are representable in closed form. These facts much facilitate and accelerate the algorithm.

3.3 Proposition *Let $w_j(x_i)$ be the weight matrix (3.1) output from an E-step and let $n_j = \sum_i w_j(x_i)$.*

(a) In the unconstrained, heteroscedastic normal case, the optimal mixing rates, means, and covariance matrices in the subsequent M-step are, respectively,

$$
\begin{cases}
\pi_j^* = \frac{1}{n} n_j, \\
m_j^* = \bar{x}(w_j) = \frac{1}{n_j} \sum_i w_j(x_i) x_i, & 1 \le j \le g, \\
V_j^* = S(w_j) = \frac{1}{n_j} \sum_i w_j(x_i)(x_i - m_j^*)(x_i - m_j^*)^\top,
\end{cases}
$$

provided that the weighted scatter matrix $S(w_j)$ is positive definite. The maximum of the double sum in the M-step is $-\frac{1}{2}nd(1 + \log 2\pi) - \frac{1}{2}\sum_j n_j \log \det S(w_j)$.

(b) In the homoscedastic, normal case, the optimal mixing rates and means in the subsequent M-step are as in part (a) and the optimal common covariance matrix is

$$
V^* = \frac{1}{n} \sum_j \sum_i w_j(x_i)(x_i - m_j^*)(x_i - m_j^*)^\top = \sum_{j=1}^g \pi_j V_j^*.
$$

(c) If

$$
w_j(x_i) = \frac{\pi_j N_{m_j^*, V_j^*}(x_i)}{\sum_\ell \pi_\ell N_{m_\ell^*, V_\ell^*}(x_i)} \qquad \left(\text{respectively } w_j(x_i) = \frac{\pi_j N_{m_j^*, V^*}(x_i)}{\sum_\ell \pi_\ell N_{m_\ell^*, V^*}(x_i)} \right),
$$

then $(\boldsymbol{\pi}^*, \mathbf{m}^*, \mathbf{V}^*)$ is a fixed point of the EM algorithm.

PROOF. (a) In this case $\eta_j = (m_j, V_j)$. Let $\bar{x}_j = \bar{x}(w_j)$ be the weighted mean of $x_1, \ldots, x_n$ w.r.t. the weights $w_j(x_i)$. The target function maximized in the heteroscedastic M-step is the sum over j of

$$
\sum_i w_j(x_i) \log N_{m_j, V_j}(x_i) = -\frac{1}{2} \sum_i w_j(x_i) \left[\log \det 2\pi V_j + (x_i - m_j)^\top V_j^{-1}(x_i - m_j) \right]
$$

$$
= -\frac{n_j}{2} \log \det 2\pi V_j - \frac{1}{2} \sum_i w_j(x_i)(x_i - m_j)^\top V_j^{-1}(x_i - m_j)
$$

$$
= -\frac{n_j}{2} \log \det 2\pi V_j - \frac{1}{2} \sum_i w_j(x_i) \left[(x_i - \bar{x}_j)^\top V_j^{-1}(x_i - \bar{x}_j) + (\bar{x}_j - m_j)^\top V_j^{-1}(\bar{x}_j - m_j) \right].
$$

The last line follows from the weighted Steiner formula A.12. Hence, the optimal location parameter is $m_j = \bar{x}_j$. It remains to minimize

$$
n_j \log \det V_j + \sum_i w_j(x_i)(x_i - \bar{x}_j)^\top V_j^{-1}(x_i - \bar{x}_j)
$$

$$
= n_j \log \det V_j + \operatorname{tr} V_j^{-1} \sum_i w_j(x_i)(x_i - \bar{x}_j)(x_i - \bar{x}_j)^\top.
$$

If the last sum is positive definite, then the optimal V_j follows from Lemma F.10. The remaining claim follows from inserting the optimal parameters.

(b) With $\eta_j = (m_j, V)$, the target function in the homoscedastic M-step is

$$
\sum_j \sum_i w_j(x_i) \log f_{\eta_j}(x_i) = -\frac{1}{2} \sum_j \sum_i w_j(x_i) \left[\log \det 2\pi V + (x_i - m_j)^\top V^{-1}(x_i - m_j) \right]
$$

$$
= -\frac{n}{2} \log \det 2\pi V - \frac{1}{2} \sum_j \sum_i w_j(x_i)(x_i - m_j)^\top V^{-1}(x_i - m_j)
$$

$$
= -\frac{n}{2} \log \det 2\pi V - \frac{1}{2} \sum_j \sum_i w_j(x_i) \left[(x_i - \bar{x}_j)^\top V^{-1}(x_i - \bar{x}_j) + (\bar{x}_j - m_j)^\top V^{-1}(\bar{x}_j - m_j) \right],
$$

again by the weighted Steiner formula A.12. The maximum w.r.t. m_j is again $m_j = \overline{x}_j$ and it remains to minimize

$$n \log \det V + \sum_j \sum_i w_j(x_i) [(x_i - \overline{x}_j)^\top V^{-1} (x_i - \overline{x}_j)$$

$$= n \log \det V + \operatorname{tr} V^{-1} \sum_j \sum_i w_j(x_i)(x_i - \overline{x}_j)(x_i - \overline{x}_j)^\top$$

w.r.t. V. This follows again from Lemma F.10.

(c) The hypothesis means that the subsequent E-step reproduces the weight matrix w. $\square$

3.4 Remarks (a) Dempster, Laird, and Rubin [115] note that the EM algorithm applied to mixtures comes down to an iterative solution of the maximum likelihood equations. Hathaway [233] derives it by way of coordinate descent on Eq. (2.4) with $\widetilde{w} = w$.

(b) Note that it is the updated mean vectors that are inserted into the update formula of the scatter matrices in Proposition 3.3(a).

(c) The parameters π_j, m_j, V_j, and V in Proposition 3.3(a) and (b) are uniquely defined by the stochastic weight matrix $(w_j(x_i))$. The converse is not true. That is, there is a whole continuum of stochastic weight matrices $(w_{i,j})$ that generate the same parameters. Consider the homoscedastic case. The system of linear equations in the variables $w_{i,j}$

$$\begin{aligned}
\sum_j w_{i,j} &= 1, & i \in 1..n, \\
\sum_i w_{i,j} &= n\pi_j, & j \in 1..g, \\
\sum_i w_{i,j}(x_i - m_j) &= 0, & j \in 1..g, \\
\sum_j \sum_i w_{i,j}(x_i - m_j)(x_i - m_j)^\top &= nV,
\end{aligned}$$

has at most $n + (g-1) + g + \binom{d+1}{2}$ linearly independent rows (the first two are linearly dependent because of $\sum \pi_j = 1$) but there are gn variables. Since $(w_j(x_i))$ is a solution, the dimension of the affine solution manifold is $\geq gn - n - 2g + 1 - \binom{d+1}{2} = (g-1)n - 2g + 1 - \binom{d+1}{2}$. For instance, for $d = 1$, $g = 2$ and $n \geq 5$ dimension is ≥ 1. If the numbers $w_j(x_i)$ are all strictly positive, then the solutions close to $(w_j(x_i))$ are strictly positive, too, thus representing stochastic matrices.

3.1.3   Mixtures of multivariate t-distributions

Peel and McLachlan [407] (see also McLachlan and Peel [367]) applied the EM algorithm to the ML estimation of multivariate t-mixtures. The following derivation of their results refers to Appendix G.2.2, which treats pure parameter estimation, that is, the case of a single component. We fix the degrees of freedom $\lambda \geq 2$. For each i there is the representation $X_i = Y_i^{(L_i)}$ with $L_i \sim \boldsymbol{\pi}$ and $Y_i^{(\ell)} \sim E_{\lambda, m_\ell, V_\ell}$, $1 \leq \ell \leq g$; see Lemma 1.12. Dropping the index i, Example E.25 yields the representation $Y^{(\ell)} = m_\ell + \frac{N^{(\ell)}}{\sqrt{Z}}$, $1 \leq \ell \leq g$, with $m_\ell \in \mathbb{R}^d$, $N^{(\ell)} \sim N_{0, V_\ell}$, and $Z = \chi_\lambda^2 / \lambda$ independent of $N^{(\ell)}$. The variables L and Z are missing and so the complete variable is (X, L, Z). From $X = m_L + \frac{N^{(L)}}{\sqrt{Z}}$ and the product formula we obtain in a similar way as in Appendix G.2.2 the complete likelihood

$$g_{\boldsymbol{\pi}, \mathbf{m}, \mathbf{V}}(x, \ell, z) = f_{(X, L, Z)}(x, \ell, z) = f_{(Y^{(\ell)}, L, Z)}(x, \ell, z)$$

$$= P[L = \ell] f_{(Y^{(\ell)}, Z)}(x, z) = \pi_\ell z^{d/2} f_{(N^{(\ell)}, Z)}(\sqrt{z}(x - m_\ell), z)$$

$$= \pi_\ell z^{d/2} f_{N^{(\ell)}}(\sqrt{z}(x - m_\ell)) f_Z(z).$$

Up to additive terms that do not depend on $\boldsymbol{\pi}'$, $\mathbf{m}'$, and $\mathbf{V}'$, its logarithm is

$$\log g_{\boldsymbol{\pi}', \mathbf{m}', \mathbf{V}'}(x, \ell, z) \sim \log \pi_\ell' - \tfrac{1}{2} \log \det V_\ell' - \tfrac{1}{2} z(x - m_\ell')^\top V_\ell'^{-1}(x - m_\ell').$$

Following (G.9), we define the weights

$$w_{\boldsymbol{\pi},\mathbf{m},\mathbf{V}}(x,\ell,z) = \frac{g_{\boldsymbol{\pi},\mathbf{m},\mathbf{V}}(x,\ell,z)}{\sum_j \int_0^\infty g_{\boldsymbol{\pi},\mathbf{m},\mathbf{V}}(x,j,u)\mathrm{d}u}.$$

The part of the partial Q-functional in formula (G.10) that depends on $\boldsymbol{\pi}'$, $\mathbf{m}'$, and $\mathbf{V}'$ is

$$Q_i((\boldsymbol{\pi}',\mathbf{m}',\mathbf{V}'),(\boldsymbol{\pi},\mathbf{m},\mathbf{V})) = \sum_\ell \int_0^\infty w_{\boldsymbol{\pi},\mathbf{m},\mathbf{V}}(x_i,\ell,z) \log g_{\boldsymbol{\pi}',\mathbf{m}',\mathbf{V}'}(x_i,\ell,z)\mathrm{d}z$$

$$\sim \sum_\ell \Big\{ p_{\boldsymbol{\pi},\mathbf{m},\mathbf{V}}(x_i,\ell) \log \pi'_\ell - \tfrac{1}{2} p_{\boldsymbol{\pi},\mathbf{m},\mathbf{V}}(x_i,\ell) \log \det V'_\ell$$

$$-\tfrac{1}{2} e_{\boldsymbol{\pi},\mathbf{m},\mathbf{V}}(x_i,\ell)(x_i - m'_\ell)^\top V_\ell'^{-1}(x_i - m'_\ell)) \Big\},$$

where

$$p_{\boldsymbol{\pi},\mathbf{m},\mathbf{V}}(x,\ell) = \frac{\int_0^\infty g_{\boldsymbol{\pi},\mathbf{m},\mathbf{V}}(x,\ell,z)\mathrm{d}z}{\sum_j \int_0^\infty g_{\boldsymbol{\pi},\mathbf{m},\mathbf{V}}(x,j,z)\mathrm{d}z},$$

$$e_{\boldsymbol{\pi},\mathbf{m},\mathbf{V}}(x,\ell) = \frac{\int_0^\infty z g_{\boldsymbol{\pi},\mathbf{m},\mathbf{V}}(x,\ell,z)\mathrm{d}z}{\sum_j \int_0^\infty g_{\boldsymbol{\pi},\mathbf{m},\mathbf{V}}(x,j,z)\mathrm{d}z}.$$

From the entropy inequality, the weighted Steiner formula A.12, and from Lemma F.10, it follows that the optimal mixing rates π'_ℓ, means m'_ℓ, and scale matrices V'_ℓ in the Q-functional $\sum_i Q_i$ are

$$\pi^*_\ell = \frac{1}{n} \sum_i p_{\boldsymbol{\pi},\mathbf{m},\mathbf{V}}(x_i,\ell), \quad m^*_\ell = \frac{\sum_i e_{\boldsymbol{\pi},\mathbf{m},\mathbf{V}}(x_i,\ell)x_i}{\sum_i e_{\boldsymbol{\pi},\mathbf{m},\mathbf{V}}(x_i,\ell)}, \quad \text{and} \tag{3.4}$$

$$V^*_\ell = \frac{\sum_i e_{\boldsymbol{\pi},\mathbf{m},\mathbf{V}}(x_i,\ell)(x_i - m^*_\ell)(x_i - m^*_\ell)^\top}{\sum_i p_{\boldsymbol{\pi},\mathbf{m},\mathbf{V}}(x_i,\ell)}, \tag{3.5}$$

respectively, whenever V^*_ℓ is positive definite.

It remains to compute the conditional probabilities $p_{\boldsymbol{\pi},\mathbf{m},\mathbf{V}}(x,\ell)$ and expectations $e_{\boldsymbol{\pi},\mathbf{m},\mathbf{V}}(x,\ell)$. Putting $\kappa = \lambda + d$ (a constant) and $a_\ell(x) = \lambda + (x - m_\ell)^\top V_\ell^{-1}(x - m_\ell)$ and using the corresponding results in Appendix G.2.2, we find up to the same multiplicative constants

$$\int_0^\infty g_{\boldsymbol{\pi},\mathbf{m},\mathbf{V}}(x,\ell,z)\mathrm{d}z \sim \frac{\pi_\ell}{\sqrt{\det V_\ell}} \frac{\Gamma(\kappa/2)}{(a_\ell(x)/2)^{\kappa/2}},$$

$$\int_0^\infty z g_{\boldsymbol{\pi},\mathbf{m},\mathbf{V}}(x,\ell,z)\mathrm{d}z \sim \frac{\pi_\ell}{\sqrt{\det V_\ell}} \frac{\Gamma(\kappa/2+1)}{(a_\ell(x)/2)^{\kappa/2+1}}.$$

Hence

$$p_{\boldsymbol{\pi},\mathbf{m},\mathbf{V}}(x,\ell) = \frac{\pi_\ell(\det V_\ell)^{-1/2} a_\ell(x)^{-\kappa/2}}{\sum_j \pi_j(\det V_j)^{-1/2} a_j(x)^{-\kappa/2}}, \tag{3.6}$$

$$e_{\boldsymbol{\pi},\mathbf{m},\mathbf{V}}(x,\ell) = \kappa \frac{\pi_\ell(\det V_\ell)^{-1/2} a_\ell(x)^{-(\kappa/2+1)}}{\sum_j \pi_j(\det V_j)^{-1/2} a_j(x)^{-\kappa/2}} = \frac{\kappa}{a_\ell(x)} p_{\boldsymbol{\pi},\mathbf{m},\mathbf{V}}(x,\ell). \tag{3.7}$$

We have thus derived the EM step for t-mixtures: Use (3.6) and (3.7) in order to compute $p_{\boldsymbol{\pi},\mathbf{m},\mathbf{V}}(x_i,\ell)$ and $e_{\boldsymbol{\pi},\mathbf{m},\mathbf{V}}(x_i,\ell)$ in the E-step and compute the parameters (3.4) and (3.5) in the M-step. Again, the resulting EM algorithm does not necessarily converge. Following an idea of Kent et al. [288] in the case $g = 1$, we modify the M-step, replacing the denominator in (3.5) with $\sum_i e_{\boldsymbol{\pi},\mathbf{m},\mathbf{V}}(x_i,\ell)$ in order to gain more speed. The next lemma implies that a limit of the EM algorithm is a fixed point of the modified algorithm.

3.5 Lemma *If the EM algorithm converges, then the limits of the sums $\sum_i p_{\boldsymbol{\pi},\mathbf{m},\mathbf{V}}(x_i,\ell)$ and $\sum_i e_{\boldsymbol{\pi},\mathbf{m},\mathbf{V}}(x_i,\ell)$ are equal.*

PROOF. Set $s^{(p)}(\ell) = \sum_i p_{\boldsymbol{\pi},\mathbf{m},\mathbf{V}}(x,\ell)$ and $s^{(e)}(\ell) = \sum_i e_{\boldsymbol{\pi},\mathbf{m},\mathbf{V}}(x,\ell)$. In the limit we have $m_\ell^* = m_\ell$ and $V_\ell^* = V_\ell$ and hence (3.5), the definition of $a_\ell(x)$, and (3.7) combine to show

$$s^{(p)}(\ell)d = s^{(p)}(\ell)\mathrm{tr}\left(V_\ell^* V_\ell^{-1}\right) = \mathrm{tr}\left(\sum_i e_{\boldsymbol{\pi},\mathbf{m},\mathbf{V}}(x_i)(x_i - m_\ell^*)^\top V_\ell^{-1}(x_i - m_\ell^*)\right)$$

$$= \mathrm{tr}\left(\sum_i e_{\boldsymbol{\pi},\mathbf{m},\mathbf{V}}(x_i)(x_i - m_\ell)^\top V_\ell^{-1}(x_i - m_\ell)\right)$$

$$= \sum_i e_{\boldsymbol{\pi},\mathbf{m},\mathbf{V}}(x_i)(a_\ell(x_i) - \lambda) = \kappa s^{(p)}(\ell) - \lambda s^{(e)}(\ell).$$

The claim follows from $\kappa - d = \lambda$. $\square$

We have the following modified EM step for mixtures of multivariate t-distributions.

3.6 Procedure (Modified EM step for mixtures of multivariate t-distributions)
// Input: Parameters $\boldsymbol{\pi}$, $\mathbf{m}$, and $\mathbf{V}$.
// Output: Parameters $\boldsymbol{\pi}_{\text{new}}$, $\mathbf{m}_{\text{new}}$, and $\mathbf{V}_{\text{new}}$ with larger observed likelihood.

1. *(E-step)* Compute the weights $p_{\boldsymbol{\pi},\mathbf{m},\mathbf{V}}(x_i,\ell)$ and conditional expectations $e_{\boldsymbol{\pi},\mathbf{m},\mathbf{V}}(x_i,\ell)$ from the current parameters $\boldsymbol{\pi}$, $\mathbf{m}$, and $\mathbf{V}$; see (3.6) and (3.7).

2. *(M-step)* Return the new parameters $\boldsymbol{\pi}_{\text{new}} = \boldsymbol{\pi}^*$ and $\mathbf{m}_{\text{new}} = \mathbf{m}^*$ according to Eq. (3.4) and $\mathbf{V}_{\text{new},\ell} = \frac{\sum_i e_{\boldsymbol{\pi},\mathbf{m},\mathbf{V}}(x_i,\ell)(x_i - m_\ell^*)(x_i - m_\ell^*)^\top}{\sum_i e_{\boldsymbol{\pi},\mathbf{m},\mathbf{V}}(x_i,\ell)}$.

A nice feature of the multivariate t-distribution is that the parameters $\boldsymbol{\pi}_{\text{new}} = \boldsymbol{\pi}^*$, $\mathbf{m}_{\text{new}} = \mathbf{m}^*$, and $\mathbf{V}_{\text{new},\ell}$ can again be represented in closed form in the M-step. Peel and McLachlan [407] also estimate the degrees of freedom λ, but they do not admit a closed form.

3.1.4   Trimming – the EMT algorithm

We will next compute (local) maxima of the *trimmed* likelihood function (2.3) w.r.t. $\boldsymbol{\pi}$ and $\boldsymbol{\gamma}$. Gallegos and Ritter [182] propose extending the EM step 3.1 to trimming by a subsequent T-step. The resulting robust **EMT step** is thus the suite of an E-, an M-, and a T-step. The EM step is carried out w.r.t. an r-element subset of the data set, leading to new parameters, while the trimming step retains the r elements that best conform to the new parameters. Properties of the EMT step will also be proved.

3.7 Procedure (EMT step for general mixtures)
// Input: An initial subset $R \subseteq \mathbf{x}$ of r elements, mixing rates $\pi_j > 0$, and
             initial population parameters γ_j, $1 \leq j \leq g$.
// Output: A subset, mixing rates, and population parameters with improved
             trimmed likelihood (2.3); see Proposition 3.8.

1. *(E-step)* Compute the weights $w_j(x) = \frac{\pi_j f_{\gamma_j}(x)}{\sum_\ell \pi_\ell f_{\gamma_\ell}(x)}$, $x \in R$, $j \in 1..g$.

2. *(M-step)* Set $\pi_{\text{new},j} = \frac{1}{r}\sum_{x \in R} w_j(x)$, and maximize $\sum_{x \in R} w_j(x) \log f_{\eta_j}(x)$ w.r.t. $\eta_j \in \Gamma$, $1 \leq j \leq g$, to obtain $\gamma_{\text{new}} = (\eta_1^*, \ldots, \eta_g^*)$ (if the maximum exists).

3. *(T-step)* Define R_{new} to be the set of data points $x \in \mathbf{x}$ with the r largest values of $f_{\boldsymbol{\pi}_{\text{new}},\gamma_{\text{new}}}(x) = \sum_j \pi_{\text{new},j} f_{\gamma_{\text{new}},j}(x)$.

If no parameter constraints are applied in the M-step, that is, if the parameter space is the g-fold Cartesian product Γ^g, then optimization of $\sum_j \sum_{x \in R} w_j(x) \log f_{\gamma_j'}(x)$ reduces to separate optimizations of $\sum_{x \in R} w_j(x) \log f_{\gamma_j'}(x)$ for $1 \leq j \leq g$.

The **EMT algorithm** is the iteration of EMT steps. Like the EM algorithm, it is iterative and alternating, proceeding as follows:

$$(R^{(0)}, \mathbf{w}^{(0)}) \xrightarrow{\text{M-step}} (\boldsymbol{\pi}^{(1)}, \boldsymbol{\gamma}^{(1)}) \xrightarrow{\text{T-step}} (R^{(1)}, \boldsymbol{\pi}^{(1)}, \boldsymbol{\gamma}^{(1)}) \xrightarrow{\text{E-step}} (R^{(1)}, \mathbf{w}^{(1)}) \xrightarrow{\text{M-step}} \ldots$$

It may be started from the M-step with a randomly or expediently chosen r-element subset $R^{(0)}$ and a stochastic matrix $\mathbf{w}^{(0)}$ as initial quantities. An elegant procedure for uniform generation of an r-element subset $R^{(0)}$ appears in Knuth [295], p. 136 ff. The rows of the initial weight matrix $w^{(0)}$ may be chosen uniformly from the $g{-}1$-dimensional unit simplex Δ_{g-1}. An efficient procedure is OSIMP; see Fishman [157]. An alternative is a set of randomly sampled unit vectors or the output obtained from some clustering algorithm; see Sections 3.2 and 3.3. If components are sufficiently separated, the algorithm may also be started from the E-step with initial parameters $(R^{(0)}\boldsymbol{\pi}^{(0)}, \boldsymbol{\gamma}^{(0)})$. If all initial mixing rates are strictly positive, then they preserve this property during iteration. The iteration is successfully stopped as soon as the trimmed likelihood (2.3) is close to convergence, or with a failure, as there is indication that convergence will not take place. If the MLE exists (see Proposition 1.24), then there is always convergence.

The remarks after the statement of the EM step for mixtures also apply to the EMT algorithm. Like the likelihood function, its trimmed version, too, has many local maxima. Let us apply Sections G.1.1 and G.2 to analyze the behavior of the algorithm in this regard. Call $(R, \boldsymbol{\pi}, \boldsymbol{\gamma})$ a *halting point* of the EMT step if the ML criterion (2.3) remains unchanged after an application of an EMT step. The EMT algorithm starting from a halting point has this point as a possible output and the algorithm stops. A *limit point* $(R, \boldsymbol{\pi}, \boldsymbol{\gamma})$ is a point of convergence of the EMT algorithm starting from some initial parameters. A *critical* point of a differential function is a point where its gradient vanishes. There are relationships between fixed, halting, limit, critical, and optimal points. Let us first see why the successive values of the target function are monotone increasing.

3.8 Proposition *Let the statistical model be as described at the beginning of Section 3.1.1.*
(a) The EMT step improves the trimmed likelihood $f(\mathbf{x}_R; \boldsymbol{\pi}, \boldsymbol{\gamma})$ in the sense of $\geq$.
(b) If $(R, \boldsymbol{\pi}, \boldsymbol{\gamma})$ is optimal, then so is $(R_{\text{new}}, \boldsymbol{\pi}_{\text{new}}, \boldsymbol{\gamma}_{\text{new}})$, and $(R, \boldsymbol{\pi}, \boldsymbol{\gamma})$ is a halting point.

PROOF. (a) The inequality $f(\mathbf{x}_R; \boldsymbol{\pi}, \boldsymbol{\gamma}) \leq f(\mathbf{x}_R; \boldsymbol{\pi}_{\text{new}}, \boldsymbol{\gamma}_{\text{new}})$ is the well-known fact that the EM algorithm is monotone, here applied to the data set R; see Theorem G.10 and [115], p. 8. Moreover,

$$\log f(\mathbf{x}_R; \boldsymbol{\pi}_{\text{new}}, \boldsymbol{\gamma}_{\text{new}}) = \sum_{x \in R} \log \sum_{\ell} \pi_{\text{new},\ell} f_{\gamma_{\text{new},\ell}}(x)$$

$$\leq \sum_{x \in R_{\text{new}}} \log \sum_{\ell} \pi_{\text{new},\ell} f_{\gamma_{\text{new},\ell}}(x) = \log f(\mathbf{x}_{R_{\text{new}}}; \boldsymbol{\pi}_{\text{new}}, \boldsymbol{\gamma}_{\text{new}})$$

by maximality of the data points in R_{new}.
(b) follows from the increasing property (a). $\square$

We will need the H-functional (G.3) w.r.t. an r-element subset $R \subseteq \mathbf{x}$,

$$H_R((\boldsymbol{\pi}, \boldsymbol{\gamma}), (\boldsymbol{\tau}, \boldsymbol{\eta})) = \log f(\mathbf{x}_R; \boldsymbol{\tau}, \boldsymbol{\eta}) - D_R((\boldsymbol{\pi}, \boldsymbol{\gamma}), (\boldsymbol{\tau}, \boldsymbol{\eta})),$$

where $D_R((\boldsymbol{\pi}, \boldsymbol{\gamma}), (\boldsymbol{\tau}, \boldsymbol{\eta}))$ is the Kullback–Leibler divergence of the complete model w.r.t. R conditional on $[\Phi = \mathbf{x}_R]$. (The letter Φ designates the map from the complete variable to the observed variable; see Appendix G.2.1.)

3.9 Proposition *(a) If $(R^*, \boldsymbol{\pi}^*, \boldsymbol{\gamma}^*)$ is a halting point of the EMT step, then $(\boldsymbol{\pi}^*, \boldsymbol{\gamma}^*)$ is a fixed point w.r.t. R^*; see Section G.2.*
(b) If $(\boldsymbol{\pi}^*, \boldsymbol{\gamma}^*)$ is the unique fixed point w.r.t. R^*, then $(R^*, \boldsymbol{\pi}^*, \boldsymbol{\gamma}^*)$ is a halting point.

PROOF. Let us put $\vartheta^* = (\boldsymbol{\pi}^*, \boldsymbol{\gamma}^*)$ and $\vartheta_{\text{new}} = (\boldsymbol{\pi}_{\text{new}}, \boldsymbol{\gamma}_{\text{new}})$, the output of the EM step starting from (R^*, ϑ^*).

(a) If (R^*, ϑ^*) is a halting point of the EMT step, then

$$H_{R^*}(\vartheta^*, \vartheta_{\text{new}}) = \log f(\mathbf{x}_{R^*}; \vartheta_{\text{new}}) - D_{R^*}(\vartheta^*, \vartheta_{\text{new}}) \le \log f(\mathbf{x}_{R^*}; \vartheta_{\text{new}})$$
$$\le \log f(\mathbf{x}_{R_{\text{new}}}; \vartheta_{\text{new}}) = \log f(\mathbf{x}_{R^*}; \vartheta^*) = H_{R^*}(\vartheta^*, \vartheta^*),$$

that is, ϑ^* is a fixed point w.r.t. R^*.

(b) By assumption, we find $\vartheta_{\text{new}} = \vartheta^*$ as the output of the EM step starting from (R^*, ϑ^*) and the claim follows from the definition of the T-step. $\square$

3.10 Proposition *Limit and halting points are the same.*

PROOF. Let the sequence $(R_t, \boldsymbol{\pi}_t, \boldsymbol{\gamma}_t) = (R_t, \vartheta_t)$ converge to $(R^*, \boldsymbol{\pi}^*, \boldsymbol{\gamma}^*) = (R^*, \vartheta^*)$. Since we have $R_t = R^*$ for eventually all t, it is sufficient to fix R_t. Abbreviate $\theta = (\boldsymbol{\tau}, \boldsymbol{\eta})$ and $\vartheta_t = (\boldsymbol{\pi}_t, \boldsymbol{\gamma}_t)$. From $H_{R_t}(\vartheta_t, \theta) \le H_{R_t}(\vartheta_t, \vartheta_{t+1})$ for all θ we infer

$$H_{R_t}(\vartheta^*, \theta) = \lim_{t \to \infty} H_{R_t}(\vartheta_t, \theta) \le \lim_{t \to \infty} H_{R_t}(\vartheta_t, \vartheta_{t+1}) = H_{R_t}(\vartheta^*, \vartheta^*).$$

This shows that limit points are halting points.

Conversely, if (R^*, ϑ^*) is a halting point, then ϑ^* is a fixed point w.r.t. R^*. We may, therefore, choose $\vartheta_{\text{new}} = \vartheta^*$ in the EM step w.r.t. R^* and $R_{\text{new}} = R^*$ in the subsequent T-step. This proves that (R^*, ϑ^*) is a limit point. $\square$

The following proposition investigates optimality of the mixing rates. It applies in particular to fixed points. We will need the concept of a face of the simplex Δ_{g-1}. Refer also to Section C.1, where faces of general convex sets are described. In this special case, it is just the convex hull of a nonempty set of unit vectors in $\mathbb{R}^g$. A subset $F \subseteq \Delta_{g-1}$ is a face if it is the nonempty intersection of Δ_{g-1} with some hyperplane H of $\mathbb{R}^g$ such that $\Delta_{g-1} \backslash H$ is convex or again if it is the set of points in Δ_{g-1} where the restriction to Δ_{g-1} of some linear form on $\mathbb{R}^g$ assumes its minimum. To each nonempty subset $T \subseteq \Delta_{g-1}$ there is a smallest face that contains it, the face *generated* by T. The face generated by a subset that contains an interior point of the simplex is the whole simplex. The face generated by one point contains this point in its interior. (This is also true if the point is extremal.)

3.11 Proposition *Let $R \subseteq \mathbf{x}$, $|R| = r$, let $\boldsymbol{\gamma} \in \Gamma^g$, and let $\widetilde{\boldsymbol{\pi}} \in \Delta_{g-1}$.*
(a) The following statements are equivalent.
(i) The EM step with input $(R, \widetilde{\boldsymbol{\pi}}, \boldsymbol{\gamma})$ retrieves $\widetilde{\boldsymbol{\pi}}$;
(ii) the vector $\widetilde{\boldsymbol{\pi}}$ is an extreme point of the simplex or a critical point of the function $\boldsymbol{\pi} \mapsto f(\mathbf{x}_R; \boldsymbol{\pi}, \boldsymbol{\gamma})$ restricted to the face generated by it;
(iii) the function $\boldsymbol{\pi} \mapsto f(\mathbf{x}_R; \boldsymbol{\pi}, \boldsymbol{\gamma})$ restricted to the face generated by $\widetilde{\boldsymbol{\pi}}$ is maximal at $\widetilde{\boldsymbol{\pi}}$.
(b) Let the equivalent conditions (i)–(iii) be satisfied. If $\widetilde{\boldsymbol{\pi}}$ is an interior point of the simplex, then it is a maximum of the function $\boldsymbol{\pi} \mapsto f(\mathbf{x}_R; \boldsymbol{\pi}, \boldsymbol{\gamma})$. If, moreover, the g vectors

$$\left(f_{\gamma_1}(x)\right)_{x \in R}, \dots, \left(f_{\gamma_g}(x)\right)_{x \in R}$$

are affine independent, then it is the only maximum.
(c) Any fixed point $(\widetilde{\boldsymbol{\pi}}, \boldsymbol{\gamma})$ w.r.t. R satisfies the equivalent conditions (i)–(iii) in (a).

PROOF. Part (a) is immediate if $\widetilde{\boldsymbol{\pi}}$ is extremal. Otherwise, assume without loss of generality $\widetilde{\pi}_g \ne 0$, let F be the face generated by $\widetilde{\boldsymbol{\pi}}$, and let $\widetilde{\mathbf{w}}$ be returned from the E-step with input $(R, \widetilde{\boldsymbol{\pi}}, \boldsymbol{\gamma})$. Since $f_{\gamma_j}(x) > 0$ by general assumption, the partial derivative of the function

$$F \to \mathbb{R}, \quad \boldsymbol{\pi} \mapsto \log f(\mathbf{x}_R; \boldsymbol{\pi}, \boldsymbol{\gamma}) = \sum_{x \in R} \log \left[\sum_{j \ne g} \pi_j f_{\gamma_j}(x) + \left(1 - \sum_{j \ne g} \pi_j\right) f_{\gamma_g}(x) \right]$$

w.r.t. $j \ne g$ such that $\widetilde{\pi}_j \ne 0$ shows that $\widetilde{\boldsymbol{\pi}}$ is critical if and only if $\widetilde{w}_j(R) = \text{const} \cdot \widetilde{\pi}_j$ for all j. The equivalence of (i) and (ii) now follows from $\pi_{\text{new},j} = \widetilde{w}_j(R)/r$.

In view of the implication from (ii) to (iii) note that

$$\log f(\mathbf{x}_R; \theta) = \sum_{x \in R} \log \sum_j \pi_j f_{\gamma_j}(x)$$

is of the form $\sum_{x \in R} \log(A\pi)_x$ with $A_{x,j} = f_{\gamma_j}(x)$. Assertion (iii) now follows from concavity C.7(a) of this function restricted to the mixing parameters and from (ii). The implication from (iii) to (ii) is plain.

(b) If $\widetilde{\pi}$ is an interior point, then the face it generates is the whole simplex and the first claim follows from (a). The second claim follows from Lemma C.7(b).

(c) Let $\widetilde{\mathbf{w}}$ be defined by $(\widetilde{\pi}, \gamma)$ and let $(\pi_{\text{new}}, \gamma_{\text{new}})$ be the parameters after an EM step starting from $(\widetilde{\pi}, \gamma)$. Since both pairs maximize the Q-functional w.r.t. R, $Q_R(\cdot, (\widetilde{\pi}, \gamma))$, and since $\widetilde{w}_j(R) = r\pi_{\text{new},j}$, Eq. (3.2) shows

$$r \sum_j \pi_{\text{new},j} \log \pi_{\text{new},j} + \sum_j \sum_{x \in R} \widetilde{w}_j(x) \log f_{\gamma_{\text{new},j}}(x)$$

$$= r \sum_j \pi_{\text{new},j} \log \widetilde{\pi}_j + \sum_j \sum_{x \in R} \widetilde{w}_j(x) \log f_{\gamma_j}(x).$$

The weighted log-likelihood grows with the EM step; hence $\sum_j \pi_{\text{new},j} \log \pi_{\text{new},j} \leq \sum_j \pi_{\text{new},j} \log \widetilde{\pi}_j$ and the entropy inequality shows $\pi_{\text{new}} = \widetilde{\pi}$, that is, (a)(i). $\square$

The next proposition discusses the population parameters and the set of retained observations of a halting point.

3.12 Proposition *Let (R^*, π^*, γ^*) be a halting point of the EMT step.*
(a) Assume that Γ^g is an open subset of some Euclidean space and that $f_\gamma(x)$ is differentiable w.r.t. $\gamma \in \Gamma^g$ for all x. Then γ^* is a critical point of the function $\gamma \mapsto f(\mathbf{x}_{R^*}; \pi^*, \gamma)$.
(b) R^* is consistent with the output $(\pi_{\text{new}}, \gamma_{\text{new}})$ of the EM step starting from (R^*, π^*, γ^*).

PROOF. By Proposition 3.9, (R^*, π^*, γ^*) maximizes the map $\gamma \to Q_{R^*}((\pi^*, \gamma), (\pi^*, \gamma^*))$, $\gamma \in \Gamma^g$. Since Γ^g is open, this fact and (3.2) show that the gradient of

$$\gamma \to \sum_{i \in R^*} \sum_j w_j^{\pi^*, \gamma^*}(x_i) \log f_{\gamma_j}(x_i)$$

vanishes at $\gamma = \gamma^*$. The gradient is

$$\sum_{i \in R^*} \frac{w_j^{\pi^*, \gamma^*}(x_i)}{f_{\gamma_j}(x_i)} \frac{\partial f_{\gamma_j}(x_i)}{\partial \gamma_j}.$$

On the other hand, the gradient of the mapping $\gamma \to \log f(x_{R^*}; \pi^*, \gamma)$ is

$$\sum_{i \in R^*} \frac{\pi_j^*}{f(x_i; \pi^*, \gamma)} \frac{\partial f_{\gamma_j}(x_i)}{\partial \gamma_j}$$

and claim (a) follows from $w_j^{\pi^*, \gamma^*}(x_i) = \frac{\pi_j^* f_{\gamma_j^*}(x_i)}{f(x_i; \pi^*, \gamma^*)}$. Claim (b) follows immediately from the estimate $f(\mathbf{x}_{R_{\text{new}}}; \vartheta_{\text{new}}) = f(\mathbf{x}_{R^*}; \vartheta^*) \leq f(\mathbf{x}_{R^*}; \vartheta_{\text{new}})$. $\square$

It rarely happens that the EMT algorithm does not converge. The following corollary, a consequence of Propositions 3.10–3.12, summarizes properties of the limit.

3.13 Corollary *Let the assumptions of Proposition 3.12(a) hold and assume that the sequence of successive outputs of the EMT algorithm converges with limit (R^*, π^*, γ^*). Then Propositions 3.11(b),(c) and 3.12(a),(b) apply to (R^*, π^*, γ^*).*

3.14 Remarks (a) In general, the EM algorithm has several or even many halting points.

Table 3.1 *Halting points for the data set* $-4, -2, 2, 4$ *and the homoscedastic normal model with three components.*

log-likelihood	π	m	v
-8.39821	$0.5, 0.25, 0.25$	$-3.00000, 2.07741, 3.92258$	0.57442
-8.44833	$0.5, 0.5 - \alpha, \alpha$	$-2.99999, 2.99999, 2.99999$	1.00007
-10.2809	$1 - \alpha - \beta, \alpha, \beta$	$0, 0, 0$	10

A simple, one-dimensional, normal example where there is even a continuum of them is as follows. The data set consists of the four points $-4, -2, 2, 4$. It has two obvious clusters. Running the *homo*scedastic version of the EM algorithm (no trimming) with $g = 3$ we find the halting points shown in the rows of Table 3.1. The first is the global maximum. It essentially uses the two negative observations for one component and each of the two positive ones for the remaining components. The second line represents a continuum of halting points corresponding to the obvious solution with two components and means close to -3 and 3. One of the components is split into two very similar parts. These halting points lie in a region where the likelihood is very flat in two directions, two eigenvalues of the Hessian being close to zero. The last line describes a two-dimensional manifold of halting points with equal log-likelihoods. The positive semi-definite Hessian is the same at each point and has four vanishing eigenvalues. In the first line, the mixing rates are unique by Proposition 3.11(b). Each of the first two lines induces a number of symmetrical, equivalent solutions.

(b) There are the following relationships between the various interesting parameters:

$$\text{optimal} \xrightarrow{\;3.8\;} \text{halting} \underset{3.9}{\rightleftarrows} \text{fixed}$$
$$\text{limit} \xleftarrow{\;3.10\;} \qquad \underset{3.11,\ 3.12}{\longrightarrow} \text{critical}$$

(c) Modifications to the M-step are possible. It is not necessary to go to the maximum in the M-step. Each improvement in the M-step or in the T-step improves the observed likelihood.

(d) If Γ is not open as required in Proposition 3.12(a) and if γ^* is a boundary point of Γ, then the *directional* derivatives of $\gamma \mapsto f(x_{R^*}; \pi^*, \gamma)$ at γ^* must be ≤ 0 in all interior directions.

In contrast to algorithms without trimming, it suffices to use the EMT algorithm with a light-tailed model. The trimming will take care of potential heavy tails or even gross outliers contained in the data. It is, therefore, reasonable and convenient to choose the *normal family*. This much facilitates the M-step. The related algorithm for full covariance matrices is this.

3.15 Procedure (EMT step for normal mixtures)
// Input: A subset $R \subseteq 1 .. n$ of r indices, mixing rates $\pi_j > 0$, and initial population parameters m_j and V_j, $1 \leq j \leq g$.
// Output: A subset, mixing rates, and population parameters with improved trimmed likelihood (2.3); see Proposition 3.8.

1. *(E-step)* Compute the weights $w_j(x_i) = \frac{\pi_j N_{m_j, V_j}(x_i)}{\sum_\ell \pi_\ell N_{m_\ell, V_\ell}(x_i)}$, $i \in R$, and the numbers $n_j = \sum_{i \in R} w_j(x_i)$, $j \in 1 .. g$.

2. *(M-step)* Return $\pi_{\text{new}, j} = \frac{n_j}{r}$, $m_{\text{new}, j} = \frac{1}{n_j} \sum_{i \in R} w_j(x_i) x_i$, and
$V_{\text{new}, j} = \frac{1}{n_j} \sum_{i \in R} w_j(x_i)(x_i - m_{\text{new}, j})(x_i - m_{\text{new}, j})^\top$, $1 \leq j \leq g$.

3. *(T-step)* Define R_{new} to be the set of data points $i \in 1 .. n$ with the r largest values of $f_{\pi_{\text{new}}, m_{\text{new}}, V_{\text{new}}}(x_i) = \sum_j \pi_{\text{new}, j} N_{m_{\text{new}, j}, V_{\text{new}, j}}(x_i)$.

The algorithm can also be defined for normal submodels. The estimates of the *heteroscedas-tic normal* scale parameters *given* the weights w_j are known to be

$$\begin{cases} V_j^* = S_R(w_j) & \text{(full)}, \\ v_{j,k}^* = V_j^*(k,k) & \text{(diagonal)}, \\ v_j^* = \frac{1}{d} \sum_k v_{j,k}^* & \text{(spherical)}, \end{cases} \tag{3.8}$$

$j \in 1..g$, $k \in 1..d$.

For $g = 1$, not only the criterion but also the algorithm was known earlier. The weights are all 1, so the E-step is trivial. The M- and T-steps of EMT reduce to a procedure for computing Rousseeuw's MCD. It was simultaneously designed by Pesch [410] and Rousseeuw and Van Driessen [450].

3.1.5   Order of convergence

If the Hessian is negative definite at a local maximum of a $\mathcal{C}^2$ target function, then it is so in a neighborhood and the function is strictly concave there. The time till convergence of iterative optimization algorithms to a local maximum is composed of the time to reach the concave part and the time for the final, local, convergence. Let us next analyze the latter in the case of the EM. The analysis needs some preliminaries. The following concepts are found in Polak [414].

3.16 Definition *(a) The convergence $\vartheta_t \to \vartheta^*$ of a sequence $(\vartheta_t)_t$ of elements in some normed space with norm $\|\cdot\|$ is called*

(i) **Q-linear** *(or of **first order**) if $\|\vartheta_{t+1} - \vartheta^*\| \le r\|\vartheta_t - \vartheta^*\|$ for some $r < 1$ and eventually all t;*

(ii) **Q-superlinear** *if $\|\vartheta_{t+1} - \vartheta^*\| \le r_t\|\vartheta_t - \vartheta^*\|$ for some sequence $r_t \to 0$ and all t;*

(iii) *of **Q-order** $p > 1$ if $\|\vartheta_{t+1} - \vartheta^*\| \le \text{const} \cdot \|\vartheta_t - \vartheta^*\|^p$ for some constant and (eventually) all t.*

(b) The greatest lower bound of the numbers r appearing in (i) for any norm is called the **rate of Q-linear convergence**. *Thus, superlinear convergence is characterized by the rate zero.*

These concepts are adequate for describing the speed of convergence of iterative algorithms. The "Q" concepts depend on the norm chosen. There is also the concept of R-linear[1] convergence. It says $\|\vartheta_t - \vartheta^*\| \le \text{const} \cdot r^t$ for some $r < 1$ and (eventually) all t. Since any two norms on Euclidean space are compatible (see Section B.6.2), it is here irrelevant, which norm is chosen. Q-linearity implies R-linearity with the same rate but not vice versa. Consider the sequence $\frac{1}{4}, \frac{1}{4}, \frac{1}{16}, \frac{1}{16}, \frac{1}{64}, \frac{1}{64}, \cdots$. It converges to zero R-linearly with the rate $1/2$ but not Q-linearly with any rate $r < 1$.

A linearly convergent sequence follows essentially a geometric progression. The speed of linear convergence is controlled by the rate r. The smaller r is, the more rapid the convergence. For instance, the fairly fast converging sequence 2^{-t} converges just linearly, not even superlinearly. If r has to be chosen close to 1, then the felt convergence is quite slow. Superlinear convergence is controlled by the decay of the numbers r_t. It is rapid, unless the sequence r_t must be chosen large for a long time. Note that convergence means the behavior for t close to ∞, not at its beginning. It cannot be fathomed out after a finite number of steps. Nevertheless, the convergence orders are useful concepts. There are the following obvious relationships:

$$\text{Order } q > p \; (> 1) \quad \Rightarrow \quad \text{order } p \quad \Rightarrow \quad \text{superlinear} \quad \Rightarrow \quad \text{linear}.$$

---

[1]The letter Q stands for *quotient*, R for *rate*.

The map M that maps each parameter ϑ to the output of the EM step started from ϑ is called **EM operator**. Insight into the convergence of the EM iteration to a local maximum ϑ^* follows from a general lemma on the convergence of iterates of differential maps to fixed points and from an analysis of the Jacobian of the EM operator at ϑ^*. It turns out that the largest eigenvalue of the latter governs the worst-case convergence.

3.17 Lemma *Let $U \subseteq \mathbb{R}^p$ be open, let $g: U \to U$ be continuously differentiable, and let $\vartheta^* \in U$ be a fixed point of g such that the spectral radius of the Jacobian $\mathrm{D}g(\vartheta^*)$ is strictly less than 1. Then there is a norm on $\mathbb{R}^p$ such that the iteration $g^t(\vartheta_0)$, $t \geq 0$, converges Q-linearly to ϑ^* for all ϑ_0 in some neighborhood of ϑ^*. The rate of convergence is less than or equal to the spectral radius of $\mathrm{D}g(\vartheta^*)$.*

PROOF.   Lemma B.23 provides a norm $\||\cdot\||$ on $\mathbb{R}^p$ such that $\||\mathrm{D}g(\vartheta^*)\|| < 1$. Here, the associated matrix norm on the $p \times p$ matrices is again denoted by $\||\cdot\||$. Define $\vartheta_t = g^t(\vartheta_0)$. A version of the mean value theorem for differentiable maps (see, for instance, Dieudonné [126], 8.5.4) implies

$$\||\vartheta_{t+1} - \vartheta^*\|| = \||g(\vartheta_t) - g(\vartheta^*)\|| \leq \sup_{0 \leq s \leq 1} \||\mathrm{D}g(\vartheta^* + s(\vartheta_t - \vartheta^*))\|| \, \||\vartheta_t - \vartheta^*\||. \quad (3.9)$$

Denote the spectral radius of $\mathrm{D}g(\vartheta^*)$ by ϱ and let $0 < \varepsilon < 1 - \varrho$. Since $\varrho < 1$ it is now sufficient to choose $\delta > 0$ so that $\||\mathrm{D}g(\vartheta)\|| < \varrho + \varepsilon$ for all $\vartheta \in B_\delta(\vartheta^*)$. Equation 3.9 shows that the sequence ϑ_t converges to ϑ^* Q-linearly with rate $\leq \varrho + \varepsilon < 1$ from any initial point $\vartheta_0 \in B_\delta(\vartheta^*)$.   □

3.18 Theorem (Order of convergence of the EM algorithm)

(a) *If the EM algorithm converges to a local maximum with a negative definite Hessian, then it does so Q-linearly w.r.t. some norm and R-linearly w.r.t. any norm. For any initial point, the rate of convergence is less than or equal to the spectral radius of $\mathrm{DM}\vartheta^*$.*

(b) *Let $g \geq 2$ and consider an EM step starting from the normal parameters $(\boldsymbol{\pi}, \mathbf{m}, \mathbf{V})$, $\boldsymbol{\pi} \in \overset{\circ}{\Delta}_{g-1}$, and let $j, \ell \geq 2$. We have*

$$\frac{\partial (\mathbf{M}\boldsymbol{\pi})_j}{\partial \pi_\ell} = \begin{cases} \frac{1}{n} \sum_i \frac{\pi_j N(x_i; m_j, V_j)}{f^2(x_i; \mathbf{m}, \mathbf{V})} (N(x_i; m_1, V_1) - N(x_i; m_\ell, V_\ell)), & \ell \neq j, \\ \frac{1}{n} \sum_i \frac{N(x_i; m_j, V_j)}{f^2(x_i; \mathbf{m}, \mathbf{V})} (f(x_i; \mathbf{m}, \mathbf{V}) + \pi_j (N(x_i; m_1, V_1) - N(x_i; m_j, V_j))), & \ell = j. \end{cases}$$

(c) *The convergence of the EM algorithm is not superlinear in general.*

PROOF. (a) We apply Lemma 3.17. Denote the local maximum of the log-likelihood function by ϑ^* and let $\alpha < \log f(\mathbf{x}; \vartheta^*)$ and $U = [\log f(\mathbf{x}; \cdot) > \alpha]$. Once entered, the iteration will stay in U. This follows from the ascending property of the EM algorithm. The Hessian $\mathrm{D}_\vartheta^2 f(\mathbf{x}; \vartheta^*)$ being negative definite, Dempster et al. [115], pp. 9 and 10, assert that all eigenvalues of $\mathrm{DM}\vartheta^*$ lie in the half-open interval $[0, 1[$. The claim on Q-linearity now follows from Lemma 3.17. Finally, Q-linear implies R-linear convergence for any norm.

(b) This follows from

$$(\mathbf{M}\boldsymbol{\pi})_j = \frac{1}{n} \sum_i \frac{\pi_j N(x_i; m_j, V_j)}{f(x_i; \mathbf{m}, \mathbf{V})}$$

$$= \frac{1}{n} \sum_i \frac{\pi_j N(x_i; m_j, V_j)}{N(x_i; m_1, V_1) + \sum_{t=2}^g \pi_t (N(x_i; m_t, V_t) - N(x_i; m_1, V_1))}.$$

(c) The mean value theorem applied to the function $\pi_2 \to (\mathbf{M}\boldsymbol{\pi})_2$ says

$$(\mathbf{M}\boldsymbol{\pi})_2 - \pi_2^* = (\mathbf{M}\boldsymbol{\pi})_2 - (\mathbf{M}\boldsymbol{\pi}^*)_2 = \frac{\partial \mathbf{M}(\boldsymbol{\pi}^* + s(\boldsymbol{\pi} - \boldsymbol{\pi}^*))_2}{\partial \pi_2} (\pi_2 - \pi_2^*)$$

Table 3.2 *Eigenvalues of the Jacobian of* M *at* $\vartheta^* = (\pi_2^*, m_1^*, m_2^*, v_1^*, v_2^*)$ *for two normal components.*

data	π_2^*	m_1^*	m_2^*	v_1^*, v_2^*	eigenvalues of DMϑ^*				
0..5	0.5	1.0554	3.9447	0.8297	0.95590	0.43375	0.01651	0	0
0..7	0.5	1.5767	5.4234	1.5507	0.98840	0.40987	0.02429	0	0
0..9	0.5	2.0944	6.9056	2.4632	0.99199	0.40869	0.02639	0	0
0..11	0.5	2.6119	8.3881	3.5756	0.99204	0.41075	0.02714	0	0

for some $s \in [0,1]$. By the second case in (b), the partial derivative is strictly positive. Therefore, the convergence cannot be superlinear for all initial points. $\qquad\square$

Part (a) of the theorem says actually more than just Q-linear convergence. As Eq. (3.9) shows, the inequality in Definition 3.16(i) holds as soon as the algorithm has entered the concave neighborhood of the local maximum. Whether part (a) means rapid (local) convergence from any initial point depends on how much smaller than 1 the largest eigenvalue of the Jacobian DMϑ^* is. The poorer the separation is, the larger the rate and the slower the convergence. The data sets 0..5, 0..7, 0..9, and 0..11 of equidistant points all have a symmetric local maximum ϑ^* w.r.t. the two-component normal mixture. Table 3.2 shows the four solutions along with all eigenvalues of the Jacobian of M, DM, at their parameters. The larger the data set is, the poorer the separation and the larger the greatest eigenvalue. Part (c) of the theorem describes the worst case. Of course, there are initial points with superlinear convergence, for instance, ϑ^* itself. Besides large eigenvalues, the Jacobian DMϑ^* may also have vanishing ones; see again Table 3.2. This means that convergence from special initial points may be superlinear.

3.1.6 Acceleration of the mixture EM

Since it is not possible to solve the mixture likelihood equations analytically, *iterative* algorithms such as EM are used to create increasing orbits that converge to local maxima of the likelihood function. They must meet several requirements. Except in special situations, where many similar data sets are analyzed in real time, such an algorithm should provide us with a survey of all or at least a *good number* of local likelihood maxima. They are the raw material from which to select the favorite solution(s). Therefore, it must be possible to start the algorithm from *arbitrary* initial parameters. Moreover, the algorithm should work on all sufficiently separated data sets, not just special ones, and it should apply to models with *arbitrary, full scales*. Likelihood functions of mixture models are complex "landscapes" with local maxima, minima, and saddle points, of which there are plenty. They even bear complex singularities where a search algorithm may get lost. The algorithm should avoid all critical points other than the desired local maxima.

Appealing properties of the EM algorithm are its nice probabilistic interpretation, its monotone convergence, its automatic adherence to the domain of definition of the likelihood function (constraint satisfaction), its almost global convergence,[2] and its simplicity. It complies with many of the requirements stated above. However, application of the EM to a mixture likelihood is often reported to be slow. Common attributes are "painful" or "excruciating." This is no surprise and not special to mixture analysis. All complex optimization problems consume more time than we may hope. Just talk, for instance, to a biochemist involved in protein folding. One reason is the rate of linear convergence, which is large when separation is poor; see Section 3.1.5. However, this describes only the performance toward the end of the program run. We next try to speed up the overall convergence of the algorithm.

---

[2]It converges except when started from a point close to a singularity.

It is first useful to investigate where the time goes. When sample space dimension is small and data size is not excessive, the EM algorithm does a reasonable job. The problem grows mainly with dimension. An explanation of this fact is easy. Let us analyze in more detail the time complexity of the E-, M-, and T-steps. We restrict attention to operations that involve matrices. This is justified when sample space dimension is large. Here the need for saving time is most urgent. A glance at Procedure 3.15 shows that it needs the ng values $f_{m_j, V_j}(x_i)$ for the E- and T-steps and for monitoring the likelihood values. The main ingredients here are the Mahalanobis distances $(x_i - m_j)^\top \Lambda_j (x_i - m_j)$. A somewhat sophisticated implementation uses the identity

$$(x_i - m_j)^\top \Lambda_j (x_i - m_j) = \operatorname{tr}(x_i x_i^\top \Lambda_j) - 2 x_i^\top (\Lambda_j m_j) + m_j^\top \Lambda_j m_j.$$

The time needed for computing the g products $\Lambda_j m_j$ and $m_j^\top \Lambda_j m_j$ on the right is negligible and the second term on the right does not involve a further matrix multiplication. The symmetric matrices $x_i x_i^\top$ can be prepared at the beginning of the program run. It remains the trace of the ng products of the symmetric matrices $x_i x_i^\top$ and Λ_j. They need d^2 elementary operations, each.[3] Therefore, the overall cost of the monitoring and the E- and T-steps amounts to ngd^2 elementary operations. Let us call this number *one time unit*. By symmetry, the cost of computing the g (symmetric) scatter matrices in the M-step is again one time unit. Hence, an EM or an EMT step amounts to about two time units. The algorithm is quadratic in d while it is only linear in n. Note that the diagonal and spherical cases are less expensive, their complexities being linear in d.

Some methods for acceleration use the gradient of the joint mixture log-likelihood at the current point $(\pi, \mathbf{m}, \mathbf{V})$. In the normal case it is

$$\log f(\mathbf{x}; \pi, \mathbf{m}, \mathbf{V}) = \sum_i \log \sum_{j=1}^{g} \pi_j N_{m_j, V_j}(x_i).$$

Its gradient follows from that of the log-likelihood $\log \sum_{j=1}^{g} \pi_j N_{m_j, V_j}$ and is easily computed. Since the mixing rates π_j sum up to 1, only $g - 1$ of them are free. The partial derivatives of

$$\sum_{j=1}^{g} \pi_j N_{m_j, V_j}(x_i) = N_{m_1, V_1}(x) + \sum_{j=2}^{g} \pi_j \big(N_{m_j, V_j}(x) - N_{m_1, V_1}(x) \big)$$

w.r.t. π_ℓ, $\ell \geq 2$, and the gradients w.r.t. m_ℓ and the inverse covariance matrices Λ_ℓ, $\ell \geq 1$, are, respectively,

$$\frac{\partial}{\partial \pi_\ell} \sum_{j=1}^{g} \pi_j N_{m_j, V_j}(x) = N_{m_\ell, V_\ell}(x) - N_{m_1, V_1}(x),$$

$$\mathrm{D}_{m_\ell} \sum_{j=1}^{g} \pi_j N_{m_j, V_j}(x) = \pi_\ell \mathrm{D}_{m_\ell} N_{m_\ell, V_\ell}(x) = \pi_\ell N_{m_\ell, V_\ell}(x) \cdot (x - m_\ell)^\top \Lambda_\ell, \qquad (3.10)$$

$$\mathrm{D}_{\Lambda_\ell} \sum_{j=1}^{g} \pi_j N_{m_j, V_j}(x) = \frac{\pi_\ell}{2} N_{m_\ell, V_\ell}(x) \cdot \big(V_\ell - (x - m_\ell)(x - m_\ell)^\top \big). \qquad (3.11)$$

The proof of Eq. (3.11) uses Example C.9(d). These formulæ define the gradient of the joint normal mixture likelihood w.r.t. $\pi_2, \ldots, \pi_g$, $\mathbf{m}$, and Λ. The additional cost caused by Eq. (3.11) given the values N_{m_j, V_j} is again one time unit.

Fortunately, the gradient of the joint log-likelihood at $(\pi, \mathbf{m}, \mathbf{V})$ can be computed with negligible cost from the output parameters $M\vartheta = (M\pi, M\mathbf{m}, M\mathbf{V})$ of the EM step starting

---

[3] Elementary operations are additions, subtractions, multiplications, and divisions of floating point numbers. The more expensive exponentials and logarithms are dominant only in low dimensions and are omitted.

from $\vartheta = (\boldsymbol{\pi}, \mathbf{m}, \mathbf{V})$ already at hand. (Note that $\mathrm{M}\boldsymbol{\pi}$, $\mathrm{M}\mathbf{m}$, and $\mathrm{M}\mathbf{V}$ depend on all of ϑ.) We will write $\mathrm{M}\boldsymbol{\pi}_\ell$ for $(\mathrm{M}\boldsymbol{\pi})_\ell$. The proof for Λ_ℓ uses the weighted Steiner formula A.12.

3.19 Lemma (Gradient in terms of EM) *The gradient of the joint log-likelihood function at the current point $(\boldsymbol{\pi}, \mathbf{m}, \mathbf{V})$ has the following representation.*

(a) The partial derivatives w.r.t. π_ℓ, $\ell \geq 2$, are

$$\frac{\partial}{\partial \pi_\ell} \log f(\mathbf{x}; \boldsymbol{\pi}, \mathbf{m}, \mathbf{V}) = n\left(\frac{\mathrm{M}\boldsymbol{\pi}_\ell}{\pi_\ell} - \frac{\mathrm{M}\boldsymbol{\pi}_1}{\pi_1}\right).$$

(b) For all ℓ, the gradients w.r.t. m_ℓ and Λ_ℓ are, respectively,

$$\mathrm{D}_{m_\ell} \log f(\mathbf{x}; \boldsymbol{\pi}, \mathbf{m}, \mathbf{V}) = n\mathrm{M}\boldsymbol{\pi}_\ell (\mathrm{M}\mathbf{m}_\ell - m_\ell)^\top \Lambda_\ell,$$

$$\mathrm{D}_{\Lambda_\ell} \log f(\mathbf{x}; \boldsymbol{\pi}, \mathbf{m}, \mathbf{V}) = -n\frac{\mathrm{M}\boldsymbol{\pi}_\ell}{2}\left(\mathrm{M}\mathbf{V}_\ell - V_\ell + (\mathrm{M}\mathbf{m}_\ell - m_\ell)(\mathrm{M}\mathbf{m}_\ell - m_\ell)^\top\right).$$

It is also possible to invert the method and to write the output of the EM step in terms of the gradient. The lemma implies that the EM step departs from its initial point in an ascent direction of the joint mixture likelihood as will now be shown.

3.20 Proposition *The vectors $\mathrm{M}\boldsymbol{\pi} - \boldsymbol{\pi}$ and $\mathrm{M}\mathbf{m} - \mathbf{m}$, the joint vector $(\mathrm{M}\mathbf{m} - \mathbf{m}, \mathrm{M}\mathbf{V} - \mathbf{V})$, and the EM direction $\mathrm{M}\vartheta - \vartheta$ all have positive projections on the gradient of the joint mixture likelihood at $\vartheta = (\boldsymbol{\pi}, \mathbf{m}, \mathbf{V})$.*

PROOF. We use the representation of the gradients given in Lemma 3.19. An elementary computation shows

$$\sum_{j=2}^g (\mathrm{M}\boldsymbol{\pi}_j - \pi_j)^\top \left(\frac{\mathrm{M}\boldsymbol{\pi}_j}{\pi_j} - \frac{\mathrm{M}\boldsymbol{\pi}_1}{\pi_1}\right) = \sum_{j=1}^g \mathrm{M}\boldsymbol{\pi}_j^2/\pi_j - 1.$$

Now, for any two probabilities $\boldsymbol{\pi}$ and $\mathbf{u}$ on $1 \mathinner{.\,.} g$, we have $\sum_{j=1}^g \frac{u_j^2}{\pi_j} - 1 = \sum_{j=1}^g \frac{(u_j - \pi_j)^2}{\pi_j} \geq 0$, with equality if and only if $\boldsymbol{\pi} = \mathbf{u}$. This is the first claim. The second claim is the obvious positivity of $\sum_{j=1}^g \mathrm{M}\boldsymbol{\pi}_j (\mathrm{M}\mathbf{m}_j - m_j)^\top \Lambda_j (\mathrm{M}\mathbf{m}_j - m_j)$. For the third claim we show that each sum

$$\mathrm{D}_{m_\ell} \log f(\mathbf{x}; \boldsymbol{\pi}, \mathbf{m}, \mathbf{V})(\mathrm{M}\mathbf{m}_\ell - m_\ell) + \operatorname{tr} \mathrm{D}_{V_\ell} \log f(\mathbf{x}; \boldsymbol{\pi}, \mathbf{m}, \mathbf{V})(\mathrm{M}\mathbf{V}_\ell - V_\ell), \qquad (3.12)$$

$1 \leq \ell \leq g$, is positive. By Lemma 3.19(b) and Example C.9(b), we have

$$\mathrm{D}_{V_\ell} \log f(\mathbf{x}; \boldsymbol{\pi}, \mathbf{m}, \mathbf{V}) = -\Lambda_\ell \mathrm{D}_{\Lambda_\ell} \log f(\mathbf{x}; \boldsymbol{\pi}, \mathbf{m}, \mathbf{V})\Lambda_\ell$$

$$= n\frac{\mathrm{M}\boldsymbol{\pi}_\ell}{2}\Lambda_\ell\left(\mathrm{M}\mathbf{V}_\ell - V_\ell + (\mathrm{M}\mathbf{m}_\ell - m_\ell)(\mathrm{M}\mathbf{m}_\ell - m_\ell)^\top\right)\Lambda_\ell.$$

The second term in (3.12) is therefore

$$\frac{n\mathrm{M}\boldsymbol{\pi}_\ell}{2}\operatorname{tr}\left[(\Lambda_\ell \mathrm{M}\mathbf{V}_\ell - I_d)^2 + \Lambda_\ell(\mathrm{M}\mathbf{m}_\ell - m_\ell)(\mathrm{M}\mathbf{m}_\ell - m_\ell)^\top \Lambda_\ell \mathrm{M}\mathbf{V}_\ell\right]$$
$$- \frac{n\mathrm{M}\boldsymbol{\pi}_\ell}{2}\operatorname{tr}\Lambda_\ell(\mathrm{M}\mathbf{m}_\ell - m_\ell)(\mathrm{M}\mathbf{m}_\ell - m_\ell)^\top.$$

Adding here the first term in (3.12) and observing the first half of Lemma 3.19(b) makes the minus sign in front of the last term a plus. This is the third claim. The last claim follows from this one and the first. □

The last claim says that the angle between the EM direction and the gradient is acute. This is equivalent to saying that each one of the two vectors is obtained from the other by a (nonunique) positive definite, linear map. It is not generally true that the mixture likelihood increases in the direction $\mathrm{M}\mathbf{V} - \mathbf{V}$. A simple univariate counterexample with two components is given by $\mathbf{x} = (-1, 0, 1, 10, 11, 12)$, $\pi_1 = \pi_2 = 0.5$, $m_1 = 0$, $m_2 = 12$, and $v_1 = v_2 = 1$. Because of the separation of the two natural clusters, the weights are almost trivial. Therefore, $(\mathrm{M}v)_2$ is close to $2/3$. The derivative at $(\boldsymbol{\pi}, \mathbf{m}, \mathbf{v})$ in the direction

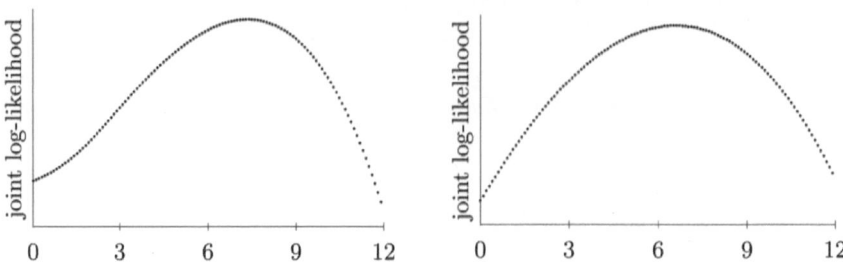

Figure 3.1: *Two courses of the log-likelihood along the EM direction.*

of $(Mv)_2 - v_2$ turns out to be ≈ -0.67. After these preliminaries, let us now discuss some acceleration methods.

(a) When the concave domain of attraction of a local maximum has been reached, EM will easily complete the job. The problem is just how to get there. A common proposal is to use special *initial solutions*. One way to generate them is the application of a probabilistic cluster criterion along with the k-parameters algorithm or agglomerative hierarchical optimization. They will be discussed in Sections 3.2 and 3.3 below. Both are much faster than the EM, their only disadvantage being the bias of the classification model. This does not lead to severely distorted solutions when mixture components are sufficiently separated. Once the target is in sight, it is in general fairly easy to hit it. However, when separation is poor, it is in the end safer to leave this main task to the mixture model itself.

(b) On data sets with many observations, initial parameters may be estimated with *subsets* of the data. Variants of this method have been proposed in computer science by various authors. Please see the discussion in McLachlan and Peel [366], Chapter 12, and Ng and McLachlan [393].

(c) This method tries to reduce the computational load of each step. It refers to the analysis of the computation time above. The Mahalanobis distances appear as negative exponents. Those that turn out to be large relative to others can therefore be neglected. This needs early, inexpensive lower estimates on the Mahalanobis distances. The large ones may be omitted.

(d) The number of iterations of the EMT step needed for one run of the algorithm varies from ten in toy examples to hundreds and even thousands in larger applications. The next two measures try to reduce this number. In each step, iterative optimization algorithms face two decisions: the search direction and the step length. Let us next discuss step lengthening along the EM direction. Figure 3.1 shows two instances where the step size proposed by EM is far from optimal. It takes a step of one unit along the abscissa. This is guaranteed to be admissible and to improve the likelihood. However, seven units would be optimal in both cases. Of course, the optimal step size is not always this large. On the other hand, it can increase far beyond ten. It would thus be nice to guess, estimate, or compute the optimal step size. However, this is easier said than done. It is of course possible to determine the maximum of the target function along the search direction. This is called an **exact line search** and was used by numerical analysts up to 1970. Unfortunately, it needs a good number of function evaluations. The overall effect of this overhead would result in a clear slowdown instead of an acceleration.

The other extreme would be to just guess the step length. This avoids any regular overhead but bears the risk of finding an inferior point or even leaving the parameter space. So our guess must be well founded and conservative. A step size of 1.5 is in most cases quite safe. When it (seldom) fails, we just fall back on a plain EM step. This is extremely easily implemented and results in a modest acceleration of about fifteen percent. Other ideas attempt to predict the step size based on the past orbit. For instance, if two consecutive EM steps have taken an almost equal direction, there is reason to conjecture that matters will continue this way for a while.

Also a compromise between computing and guessing is possible: estimating. The **parabolic line search** tries to fit a parabola to the likelihood along the search direction in order to estimate the maximum. A proposal for general optimization algorithms is contained in Press et al. [423], Section 9.7. If the EM direction is used, it can be simplified because the EM step offers a guaranteed step length. Let us discuss this method in more detail. Recall that $M\vartheta$ denotes the output of the EM step starting from ϑ. Endow the straight (search) line through ϑ and $M\vartheta$ with a linear scale so that ϑ becomes the origin and $M\vartheta$ is located at 1. Let the domain of definition of the normal mixture likelihood be denoted by U. Let h denote the log-likelihood function along the search line. Define a tentative next base by the auxiliary point $s > 1$ on the line. The procedure needs two parameters $0 < \gamma < 1$ and $\mu > 1$.

3.21 Procedure (Lengthened EM step)
Input: A base point $\vartheta \in U$ and a tentative step length $s \geq 1$.
Output: A new base point $\vartheta_{\text{new}} \in U$ and a new tentative step length $s_{\text{new}} \geq 1$.

1. Compute $M\vartheta$ and set $\vartheta_{\text{tent}} = \vartheta + s(M\vartheta - \vartheta)$.

2. *If* $\vartheta_{\text{tent}} \notin U$, then return $\vartheta_{\text{new}} = M\vartheta$ and $s_{\text{new}} = 1$;

3. *else* compute $h(0)$, $h'(0)$ (use Lemma 3.19), $h(s)$, and the parabola $p(t) = at^2 + bt + c$ determined by these quantities (that is, $c = h(0)$, $b = h'(0)$, $a = (h(s) - bs - c)/s^2$).

    (a) *If* $a \geq 0$, then return $\vartheta_{\text{new}} = \vartheta_{\text{tent}}$ and $s_{\text{new}} = \mu$;

    (b) *else* compute the maximizer $t_{\max} = \frac{1}{2}h'(0)s^2/(h(0)s + h'(0) - h(s))$ of the parabola.

        (i) *If* $t_{\max} \geq \gamma s$, then return $\vartheta_{\text{new}} = \vartheta_{\text{tent}}$ and $s_{\text{new}} = t_{\max} \wedge \mu$;
        (ii) *else* return $\vartheta_{\text{new}} = \vartheta + (t_{\max} \vee 1)(M\vartheta - \vartheta)$ and $s_{\text{new}} = 1$.

The procedure attempts to compute the likelihood values $f(x_i; \vartheta)$ only once at each call to the procedure. This happens in computing $h(s)$ in step 3. If ϑ_{tent} is actually returned as the new base, then these values can be reused and the E-step in step 1 and $h(0)$ and $h'(0)$ at the next call are free lunches. The time for each call to the procedure is then similar to the EM. Of course, ϑ_{tent} should be returned only when its likelihood is large, at least as large as an EM step would provide. The procedure uses the parabola p as an approximation to h in order to estimate the likelihood for $t \in [1, s]$. The best case is 3(a) with the parabola open upwards or degenerate. This indicates that the tentative step could have been larger. Here, ϑ_{tent} is made the next base and the maximum tentative step length μ is proposed for the next round. Otherwise, 3(b), the maximizer of p approximates that of h. If it is larger or at least not much smaller than s, 3(b)(i), we still use ϑ_{tent} for the next base but possibly a more conservative tentative (next) step length. If it is much smaller, 3(b)(ii), we make the maximizer the next base unless it is less than 1. In this case and when the likelihood is not even defined at ϑ_{tent}, step 2, the procedure is maneuvering in difficult waters. We resort in this and the next step to the safe port $s_{\text{new}} = 1$, a pure EM step. It is only in these cases that the time unit for computing $h(s)$ is lost. The procedure is trimmed so that this does not often happen. The parameters $\gamma \approx 0.5$, μ beyond 10, and an initial step length $s \approx 2$ are often practicable. In the presence of a data set requiring excessive computing time, it is reasonable to calibrate these parameters with short program runs. This describes the dynamic line search. The procedure increases the speed on moderately separated data sets, as they typically appear in cluster analysis, by a factor of two to three. If $s = 1$ is chosen, then it collapses to a pure EM step. Substituting a **cubic** for the parabola renders the approximation more accurate. However, the higher accuracy does not pay since it needs either the slope at the auxiliary point or the likelihood value at another point.

(e) Modifying the *search direction* leads us to the realm of numerical optimization, Newton-Raphson optimization and the Conjugate Gradient and quasi-Newton methods. The former is impractical here since it heads toward any critical point close by, also local minima and saddle points. Another algorithm has to be employed first to get close to a local maximum. As said before, this is actually the main work. The other optimization algorithms are more expedient. An example of a quasi-Newton method is the double-rank BFGS algorithm G.8 cited in Section G.1.2. The target function G appearing there is here the joint log-likelihood function. Hence, $s = \mathrm{D}G$ is the score functional, and the search direction taken by the BFGS step is $Bs(\vartheta)$. The Conjugate Gradient and quasi-Newton methods have been designed primarily for functions defined globally on $\mathbb{R}^p$. Their application to partially defined functions, such as mixture likelihoods, is possible but care must be taken so that the search path stays in the domain of definition. This creates computational overhead.

Jamshidian and Jennrich [269] proposed two quasi-Newton accelerations of the EM algorithm. The first one uses Broyden's [71] root finding algorithm. It tries to find fixed points of the EM operator M. However, root finding algorithms may get caught in regions without ever hitting a zero. Just consider the function $x^2 + 1$. This is a major shortcoming. The second one is based on the BFGS rank-two optimization method covered in Section G.1.2. Jamshidian and Jennrich [269] modify Procedure G.8 by searching along a combination $As(\vartheta) + \widetilde{s}(\vartheta)$ of the gradient s and the EM direction $\widetilde{s}$. They correspondingly replace the term $Bv = B(s(\vartheta_{\mathrm{new}}) - s(\vartheta))$ with $A(s(\vartheta_{\mathrm{new}}) - s(\vartheta)) + (\widetilde{s}(\vartheta_{\mathrm{new}}) - \widetilde{s}(\vartheta))$ everywhere in the update formula G.2 for B. They report substantial accelerations compared with EM on heavily overlapped, homoscedastic data sets, starting the algorithm always from the same initial point. However, a comparison of their method with EM on better separated, heteroscedastic data and random starting points is rather sobering. The direction the update formula proposes is sometimes not even ascending. Yet, the method performs on these data sets like the line search above. Maybe the progress is due to the line search, which is part of the algorithm, and not to the special search direction.

The aim is a method to determine a reasonable search direction from any initial point using only local information at the points visited. The method is supposed to work well on all data sets. This is not easy. The joint normal likelihood function is analytic. Thus, the behavior on its entire domain of definition is coded locally at the current point. However, value and gradient there are by no means sufficient. There are many analytic functions with the same value and gradient and yet very different local maxima. The optimal search direction and step size would hit the target in one step, but this pair is unknown. In the end, a most effective method will be a combination of all methods above and more. In optimization any "dirty" method is welcome as long as it works. The ultimate value of an optimization method does not lie in its theory but in the time it saves us on a wide variety of instances. This must be verified by experiment.

3.1.7 Notes

The special case of the EM algorithm for normal mixtures goes back to Hasselblad [230] and Day [113]. These were actually two of the papers that triggered the work of Dempster et al. [115], who coined the term EM algorithm. On p. 10, they also analyze the speed of convergence. Meng and Rubin [374] show Q-linear convergence of the normal EM w.r.t. the Euclidean norm. The quotient $\|\vartheta_{t+1} - \vartheta^*\|_2 / \|\vartheta_t - \vartheta^*\|_2$ converges to some eigenvalue of $\mathrm{DM}(\vartheta^*)$, depending on the initial and the limit parameters. An application of EM to multivariate t-mixtures is due to Peel and McLachlan [407]. Besides symmetric distributions, skewed models such as the skew normal (see Azzalini and Dalla Valle [18] and the survey article by Kotz and Vicari [298]) are also of interest. Applications of skewed t-distributions to mixture modeling appear in McLachlan et al. [361], Lin et al. [322], Wang et al. [521], Lin [321], and Lee and McLachlan [313, 314]. Skewed distributions may carry uncertainties into the analysis if clusters are clinging together. It is often difficult to distinguish between data sampled from a skewed distribution and from two overlapping symmetric ones.

Ingrassia and Rocci [264, 265] consider scale constraints of the form $\lambda_{\min}(V_j)/\lambda_{\max}(V_\ell) \geq c$ in normal mixtures, where $\lambda_{\min}$ ($\lambda_{\max}$) stands for the minimal (maximal) eigenvalue. The constraints are more restrictive than the HDBT constraints and not invariant w.r.t. variable scaling. The authors design various strategies to enforce the constraints on the EM algorithm. The EMT algorithm and its properties go back to Gallegos and Ritter [182].

General-purpose optimization tools, too, can and are being used for mixture analysis. Examples are local search, the Metropolis-Hastings algorithm [231, 193, 284], and the Gibbs sampler [195, 193, 284]. The latter are based on Markov chains. They are tailored to visiting high values of the target function and *global* optima. This is not exactly what is needed in mixture analysis. Refer to the experience reported in McLachlan and Peel [366], Section 3.11.

3.2 k-**Parameters algorithms**

We will next compute steady solutions to the probabilistic cluster criteria designed in Sections 1.3.1, 1.3.4, 1.3.5, 2.3.1, and 2.3.2. Estimating an element of the combinatorial structure of all admissible assignments is not a simple task either. In fact, clustering is NP-complete; see Garey and Johnson [190], [191], Papadimitriou in Shi and Malik [471], Czumaj and Sohler [107], and Brandes et al. [64]. Only for Rousseeuw's [448] MCD, the special case of the TDC for $g = 1$, has it been shown that optimization can be carried out in polynomial time. Bernholt and Fischer [37] designed an algorithm of complexity $\mathcal{O}(n^v)$, $v = d(d+3)/2+1$, based on elliptical separability of retained and discarded data points. In principle, one has recourse to general optimization schemes such as local descent methods combined with multistarting. They tentatively modify the assignment, compute the new parameters and the new value of the criterion, and decide finally whether to accept the move. Their shortcoming is the need to update the parameters with each move, even the ones that will turn out to be unsuccessful. More efficient algorithms tailored to the criteria detect whether a move will be successful *before* updating the parameters and recomputing the criterion. They consist of the iteration of so-called *reduction steps*. A reduction step admissibly removes some or all misfits from a given admissible assignment w.r.t. its own parameters. This leads to iterative relocation algorithms more efficient than the EM. Their only disadvantage is the bias in the estimate; see Remark 1.57(c).

In some cases the ML estimates of the in general continuous parameters γ_j, given the admissible assignment ℓ, can be computed analytically, a circumstance that much speeds up the resulting algorithms. Popular examples are the normal and coin tossing models where ML estimation reduces to simple summation; see, for instance, the trace and determinant criteria in Section 1.3.5 and the TDC, Section 2.3.2.

Reduction steps will be subsequently specialized to elliptical and normal trace and determinant criteria; see Corollaries 1.37, 1.40, and 1.41. Simplifications are possible for the (normal) determinant and trace criteria, Examples 3.26. In the mixture case the focus has been on unconstrained local optima. Here we will concentrate on *unconstrained* steady partitions. The reason is again that the HDBT constant is unknown and is favorably estimated along with the solution. It is therefore *steady, admissible* partitions for a full product model Γ^g that we seek. Small data sets will make it necessary to put lower constraints on cluster sizes and to recur to the methods of combinatorial optimization; see Section 3.2.3.

3.2.1 *General and elliptically symmetric models*

The algorithms to be discussed here are again iterative and alternate between parameter estimation and relocation. Their theoretical basis is the following general proposition that states a practicable condition for a new assignment to improve the criterion. It goes back to Diday and Schroeder [124] and Schroeder [459].

3.22 Proposition *Let ℓ and ℓ_{new} be two Γ^g-admissible assignments and let $\mathcal{C} = (C_1, \ldots, C_g)$ and $\mathcal{D} = (D_1, \ldots, D_g)$ be their induced partitions. Let $\gamma_j(\ell)$ be the MLE w.r.t. C_j. Assume that the following inequality holds:*

$$\sum_i \big\{ \log f_{\gamma_{\ell_{\mathrm{new},i}}(\ell)}(x_i) + \log |C_{\ell_{\mathrm{new},i}}| \big\} \geq \sum_i \big\{ \log f_{\gamma_{\ell_i}(\ell)}(x_i) + \log |C_{\ell_i}| \big\}. \tag{3.13}$$

(a) The assignment ℓ_{new} improves the MAP classification likelihood (1.38) (meaning "$\geq$").

(b) If the inequality (3.13) is strict, then the criterion is strictly improved.

(c) Assume that the MLE's are unique. If the criterion is not improved, then the MLE's of the parameters w.r.t. D_j and C_j are equal and $|D_j| = |C_j|$ for all j.

PROOF. (a) Note that $\ell_{\mathrm{new},i} = j$ is equivalent with $i \in D_j$. Using (3.13) we estimate

$$\sum_{j=1}^g \sum_{i \in C_j} \log f_{\gamma_j(\ell)}(x_i) - n \cdot \mathrm{H}\Big(\frac{|C_1|}{n}, \ldots, \frac{|C_g|}{n} \Big)$$

$$= \sum_j \sum_{i \in C_j} \Big\{ \log f_{\gamma_j(\ell)}(x_i) + \log \frac{|C_j|}{n} \Big\} = \sum_{i=1}^n \Big\{ \log f_{\gamma_{\ell_i}(\ell)}(x_i) + \log \frac{|C_{\ell_i}|}{n} \Big\}$$

$$\leq \sum_{i=1}^n \Big\{ \log f_{\gamma_{\ell_{\mathrm{new},i}}(\ell)}(x_i) + \log \frac{|C_{\ell_{\mathrm{new},i}}|}{n} \Big\} = \sum_j \sum_{i \in D_j} \Big\{ \log f_{\gamma_j(\ell)}(x_i) + \log \frac{|C_j|}{n} \Big\}$$

$$= \sum_j \sum_{i \in D_j} \log f_{\gamma_j(\ell)}(x_i) + \sum_j |D_j| \log \frac{|C_j|}{n}$$

$$\leq \sum_j \sum_{i \in D_j} \log f_{\gamma_j(\ell_{\mathrm{new}})}(x_i) + \sum_j |D_j| \log \frac{|D_j|}{n}.$$

In the last line both the likelihood term and the entropy term increase, the latter due to the entropy inequality D.17(a).

(b) Under the assumption of (b) the first inequality in the chain above is strict.

(c) In this case, all inequalities in the chain above must be equalities. In particular, the two likelihoods and the two entropy terms in the last two lines are equal. From the uniqueness assumption it follows $\gamma_j(\ell_{\mathrm{new}}) = \gamma_j(\ell)$ and equality in the entropy inequality implies $|D_j| = |C_j|$. $\square$

Note that it is only parameters w.r.t. $\mathcal{C}$, and not w.r.t. $\mathcal{D}$, that appear in (3.13). It is just the index $\ell_{\mathrm{new},i}$ that changes on the left-hand side. There are many ways of exploiting the previous proposition for the design of algorithms. Note that (3.13) is satisfied if it is satisfied for each index i separately. If $|D_j| f_{\gamma_{\ell_{\mathrm{new},i}}(\ell)}(x_i) > |C_j| f_{\gamma_{\ell_i}(\ell)}(x_i)$ for some i, then this index is a misfit in ℓ; see Definition 1.34. Therefore Proposition 3.22(b) has the following corollary.

3.23 Corollary

(a) The admissible removal of misfits strictly improves the MAP classification likelihood (1.38).

(b) Any optimal partition is steady.

3.24 Remarks (a) If the mixing rates π_j are a priori known, then the sizes appearing in (3.13) may be replaced with the values $n\pi_{\ell_i}$ and $n\pi_{\ell_{\mathrm{new},i}}$. If mixing rates are known to be equal, then they may be omitted. This corresponds to the ML classification likelihood (1.39).

(b) The converse of part (b) of the corollary is wrong. That is, there are suboptimal steady

Figure 3.2: *Inadmissible removal of a misfit.*

partitions. This was discussed after Proposition 1.35.

(c) The application of Proposition 3.22 and its corollary to elliptical models $E_{\varphi,m,V}$ is straightforward. If the density generator φ is chosen in advance, then put $\gamma_j = (m_j, V_j)$. The ML estimation of elliptical parameters is the subject of Section G.2.2.

(d) The criterion may improve, although there is equality in (3.13). An example with a univariate, homoscedastic, isobaric normal model is this: $\mathbf{x} = (-2, -1, 1, 2)$, $\mathcal{C} = (\{-2, 2\}, \{-1, 1\})$, $\mathcal{D} = (\{-2, -1\}, \{1, 2\})$. We have $W_{\mathcal{C}} = 10$ und $W_{\mathcal{D}} = 1$, but $\overline{x}_{C_i} = 0$ implies equality in (3.13).

(e) It may happen that a misfit cannot be admissibly removed: Let the model be homoscedastic, normal and let the data be as shown in Fig. 3.2. They are in general position. The point x_3 is a misfit in the partition $\{\{x_1, x_2, x_3\}, \{x_4\}\}$ and cannot be admissibly removed if the scale is assumed to be full or diagonal. It is, however, possible in the spherical case.

Corollary 3.23 says that the removal of misfits improves the criterion. It is possible to relocate one or more misfits at the same time but the new labeling must remain admissible. This can in many cases be checked by just monitoring cluster sizes. A number of useful algorithms iterate **reduction steps** of an alternating character:

ML parameter estimation – discovery of misfits – relocation.

The simplest useful one reads as follows.

3.25 Procedure (Elementary multipoint reduction step)
// Input: An admissible assignment ℓ.
// Output: An admissible assignment with improved criterion *or* "steady."

1. *(Estimation)* Compute the MLE's of the parameters for all clusters w.r.t. ℓ.

2. *(Relocation)* Explore all $i \in 1..n$ for misfits; see Definition 1.34;
   after each detected misfit, update the assignment if it remains admissible.

3. *If* some misfit was relocated, output the current assignment;
   else "steady."

The multipoint reduction step uses *ML parameter estimation* and has to recognize *misfits* and *admissible partitions* (see Lemma 1.39), all w.r.t. the assumed statistical model. For elliptical and normal data in general position it is sufficient to monitor cluster sizes. In practice it would be too computationally expensive to verify general position of the data. Instead, it is sufficient to monitor the eigenvalues of the scale matrices during parameter estimation. The removal of a misfit amounts to applying the Bayesian discriminant rule; see Definition 1.34. Therefore, the multipoint reduction step is the combination of ML parameter estimation with the Bayes discriminant rule. It is clear that both improve the MAP classification likelihood (1.38).

The relocation step of 3.25 depends on the order of the data. By definition, it does not produce inadmissible assignments. It is rendered more efficient by exploring the largest clusters first. This facilitates admissibility, since elements are shifted to smaller clusters. It is not always possible to simply reassign all misfits at the same time – admissibility would not be guaranteed. Step 2 updates only the assignment, *not* the parameters. The reduction step is "simple" in the sense that it does not output the optimal partition w.r.t. the parameters computed in step 1. This is, however, achieved with the exact version, Procedure 3.29.

The iteration of multipoint reduction steps will be called the k-**parameters** algorithm.[4] It is started from some initial partition $\mathcal{C}_0$. As long as no steady partition is attained, the criterion improves. Since there are only finitely many partitions, the iteration must terminate after a finite number of steps. The steady partition starting from $\mathcal{C}_0$ has been reached. It is not necessarily the desired or just an interesting one (in complex cases it rarely will be). It is rather necessary to replicate the whole process many times starting from possibly many different, randomly or expediently chosen, initial assignments or parameters in order to generate many steady partitions. Hopefully the desired solution will be among them.

Recognizing misfits plays an important role in the reduction step. The characterization of a misfit for an elliptical basic model $E_{\varphi,m,V}$ with radial function $\phi = -\log\varphi$ is as follows: An index i is a misfit in $\boldsymbol{\ell}$ if there is $j \neq \ell_i$ such that

$$\tfrac{1}{2}\log\det V_j(\boldsymbol{\ell}) + \phi\big((x_i - m_j(\boldsymbol{\ell}))^\top V_j(\boldsymbol{\ell})^{-1}(x_i - m_j(\boldsymbol{\ell}))\big) - \log n_j(\boldsymbol{\ell})$$
$$< \tfrac{1}{2}\log\det V_{\ell_i}(\boldsymbol{\ell}) + \phi\big((x_i - m_{\ell_i}(\boldsymbol{\ell}))^\top V_{\ell_i}(\boldsymbol{\ell})^{-1}(x_i - m_{\ell_i}(\boldsymbol{\ell}))\big) - \log n_{\ell_i}(\boldsymbol{\ell}).$$

Since ML parameter estimation can be carried out analytically in the normal cases, simplifications are possible. These relations will also be needed for the trimming algorithms in Section 3.2.2. Let $n_j = n_j(\boldsymbol{\ell})$, $\overline{x}_j = \overline{x}_j(\boldsymbol{\ell})$, and $S_j = S_j(\boldsymbol{\ell})$ be the size, the sample mean, and the scatter matrix, respectively, of cluster $C_j(\boldsymbol{\ell})$, let $s_j = s_j(\boldsymbol{\ell})$ be the mean value of the diagonal of S_j, let $S = S(\boldsymbol{\ell})$ be the pooled scatter matrix, and let s be the mean value of the diagonal of S. Here are characterizations of a misfit i in the six normal cases:

3.26 Examples (a) (Unconstrained heteroscedastic normal models) The index i is a misfit in $\boldsymbol{\ell}$ if there exists $j \neq \ell_i$ such that

$$\tfrac{1}{2}\log\det S_j - \tfrac{1}{2}(x_i - \overline{x}_j)^\top S_j^{-1}(x_i - \overline{x}_j) - \log n_j$$
$$< \tfrac{1}{2}\log\det S_{\ell_i} - \tfrac{1}{2}(x_i - \overline{x}_{\ell_i})^\top S_{\ell_i}^{-1}(x_i - \overline{x}_{\ell_i}) - \log n_{\ell_i}; \tag{full}$$

$$\tfrac{1}{2}\sum_k \big\{\log S_j(k,k) + (x_{i,k} - \overline{x}_{j,k})^2/S_j(k,k)\big\} - \log n_j$$
$$< \tfrac{1}{2}\sum_k \big\{\log S_{\ell_i}(k,k) + (x_{i,k} - \overline{x}_{\ell_i,k})^2/S_{\ell_i}(k,k)\big\} - \log n_{\ell_i}; \tag{diagonal}$$

$$\tfrac{1}{2}\big\{d\log s_j + \tfrac{1}{s_j}\|x_i - \overline{x}_j\|^2\big\} - \log n_j < \tfrac{1}{2}\big\{d\log s_{\ell_i} + \tfrac{1}{s_{\ell_i}}\|x_i - \overline{x}_{\ell_i}\|^2\big\} - \log n_{\ell_i}. \tag{spherical}$$

(b) (Homoscedastic normal models) The index i is a misfit in $\boldsymbol{\ell}$ if there exists $j \neq \ell_i$ with

$$\tfrac{1}{2}(x_i - \overline{x}_j)^\top S^{-1}(x_i - \overline{x}_j) - \log n_j < \tfrac{1}{2}(x_i - \overline{x}_{\ell_i})^\top S^{-1}(x_i - \overline{x}_{\ell_i}) - \log n_{\ell_i}; \tag{full}$$

$$\tfrac{1}{2}\sum_k \frac{(x_{i,k} - \overline{x}_{j,k})^2}{S(k,k)} - \log n_j < \tfrac{1}{2}\sum_k \frac{(x_{i,k} - \overline{x}_{\ell_i,k})^2}{S(k,k)} - \log n_{\ell_i}; \tag{diagonal}$$

$$\tfrac{1}{2s}\|x_i - \overline{x}_j\|^2 - \log n_j < \tfrac{1}{2s}\|x_i - \overline{x}_{\ell_i}\|^2 - \log n_{\ell_i}. \tag{spherical}$$

(c) (The isobaric cases) If it is a priori known that clusters are of about the same size, then we may assume an isobaric model and the logarithms of the cluster sizes are everywhere dropped.

(d) (The homoscedastic, isobaric, spherical, normal model) This is the simplest special case. It is often used and arises from the spherical case in (b) taking account of (c). Here, parameter estimation in item 1 of the multipoint reduction step means computing the sample means of all clusters. Therefore, the k-parameters algorithm for this model is called the k-**means algorithm**. The condition for a misfit i is just

$$\|x_i - \overline{x}_j\|^2 < \|x_i - \overline{x}_{\ell_i}\|^2 \quad \text{for some } j.$$

---

[4]Its step 1 is ML parameter estimation in the assumed statistical model. The name is adapted from k-means (see Example 3.26(c) below), where the parameters to be estimated in step 1 are just the means of a homoscedastic, isobaric, spherical classification model.

That is, the removal of a misfit in step 2 of the multipoint reduction step of 3.25 comes down to relocating each misfit to the closest cluster center (w.r.t. the Euclidean metric) computed in step 1. Therefore, step 2 creates the Voronoi decomposition w.r.t. the means computed in step 1. Admissibility means that at least one cluster contains two points. If all data points are distinct, then it is sufficient to have $n \geq g + 1$.

The k-means algorithm is most popular in (optimal) vector quantization; see Remark 1.57. It is known there under the name **Lloyd's algorithm**, Lloyd [333]. While the data analyst is concerned with establishing a mixture model of the data and a corresponding partition in distinguishable clusters, no matter what their sizes, scales, or extents may be, the engineer wishes to extract a "uniform" sample from the data called **principal points**. Each data point is supposed to lie close to one of them w.r.t. the Euclidean (or some other given) metric. These goals are different from cluster analysis. The result obtained in Fig. 1.11 may be acceptable in optimal vector quantization. It is not in cluster analysis.

3.2.2 Steady solutions and trimming

The k-parameters algorithm can again be combined with trimming. This allows us to compute steady solutions of the unconstrained TDC; see Section 2.3.2. The following proposition extends Proposition 3.22 to the trimmed MAP classification likelihood (2.28).

3.27 Proposition *Let $\boldsymbol{\ell}$ and $\boldsymbol{\ell}_{\text{new}}$ be two Γ^g-admissible labelings and let $\mathcal{C} = (C_1, \ldots, C_g)$ be the partition of the set of retained elements w.r.t. $\boldsymbol{\ell}$. Let $\gamma_j(\boldsymbol{\ell})$ be the MLE w.r.t. C_j and assume the inequality*

$$\sum_{i:\ell_{\text{new},i}\neq 0} \left(\log f_{\gamma_{\ell_{\text{new},i}}(\boldsymbol{\ell})}(x_i) + \log |C_{\ell_{\text{new},i}}| \right) > \sum_{i:\ell_i \neq 0} \left(\log f_{\gamma_{\ell_i}(\boldsymbol{\ell})}(x_i) + \log |C_{\ell_i}| \right). \quad (3.14)$$

(a) Then $\boldsymbol{\ell}_{\text{new}}$ strictly improves the trimmed MAP classification likelihood (2.28) w.r.t. $\boldsymbol{\ell}$.
(b) The same holds for the trimmed ML classification likelihood after dropping the summand $\log |C|$ on both sides of estimate (3.14).

PROOF. Let $\mathcal{D} = (D_1, \ldots, D_g)$ be the partition of the set of retained elements w.r.t. $\boldsymbol{\ell}_{\text{new}}$. Note that $\ell_i = j$ ($\ell_{\text{new},i} = j$) is equivalent with $i \in C_j$ ($i \in D_j$). Applying (3.14), ML estimation, and the entropy inequality in this order, we have the following chain of estimates:

$$\sum_{j=1}^{g} \sum_{i \in C_j} \log f_{\gamma_j(\boldsymbol{\ell})}(x_i) - r \cdot \mathrm{H}\left(\frac{|C_1|}{r}, \ldots, \frac{|C_g|}{r} \right)$$

$$= \sum_{j=1}^{g} \sum_{i \in C_j} \left\{ \log f_{\gamma_j(\boldsymbol{\ell})}(x_i) + \log \frac{|C_j|}{r} \right\} < \sum_{j} \sum_{i \in D_j} \left\{ \log f_{\gamma_j(\boldsymbol{\ell})}(x_i) + \log \frac{|C_j|}{r} \right\}$$

$$= \sum_{j} \sum_{i \in D_j} \log f_{\gamma_j(\boldsymbol{\ell})}(x_i) + \sum_{j} |D_j| \log \frac{|C_j|}{r} \leq \sum_{j} \sum_{i \in D_j} \log f_{\gamma_j(\boldsymbol{\ell}_{\text{new}})}(x_i) + \sum_{j} |D_j| \log \frac{|D_j|}{r}.$$

This proves case (a); case (b) is similar.   □

The fact that both sides in the hypothesis of Proposition 3.14 contain the *current* population parameters $\gamma_j(\boldsymbol{\ell})$ substantially reduces again the complexity of the optimization. The proposition may be exploited to design several reduction steps, depending on the transition from $\boldsymbol{\ell}$ to $\boldsymbol{\ell}_{\text{new}}$ employed. One of them is the following simple reduction step for trimmed clustering. Denote the posterior probabilities (weights) by

$$u_{i,j} = \log n_j + \log f_{\gamma_j}(x_i), \quad i \in 1..n, \; j \in 1..g. \quad (3.15)$$

3.28 Procedure (Elementary trimmed multipoint reduction step)
// <u>Input:</u> An admissible labeling $\boldsymbol{\ell}$.
// <u>Output:</u> An admissible labeling with larger criterion *or* "fail" *or* "stop."

1. *(Estimation)* Compute the MLE's $\gamma_j = \gamma_j(\boldsymbol{\ell})$ for all clusters j w.r.t. $\boldsymbol{\ell}$.

2. *(Relocation)* Let $n_j = n_j(\boldsymbol{\ell})$. Assign each element i to the index j with maximum weight $u_{i,j}$ to obtain a labeling $\boldsymbol{\ell}'$.

3. *(Trimming)* Discard the $n - r$ data points with smallest weights u_{i,ℓ'_i} from $\boldsymbol{\ell}'$ to obtain $\boldsymbol{\ell}_{\text{new}}$.

4. *If* $\boldsymbol{\ell}_{\text{new}}$ is inadmissible, then respond "fail;"
   else if $\displaystyle\sum_{i:\ell_{\text{new},i}\neq 0} u_{i,\ell_{\text{new},i}} = \sum_{i:\ell_i\neq 0} u_{i,\ell_i}$, then respond "stop;"
   else return the new labeling (it is superior).

This reduction step consists essentially of three successive steps: ML estimation of parameters, the MAP discriminant rule, and trimming. Note that parameters are estimated only once at the beginning of Procedures 3.25 and 3.28. In particular, step 4 of 3.28 needs the value $f_{\gamma_j}(x_i)$ for $j = \ell_{\text{new},i}$, but not the new parameters $\gamma_{\text{new},j}$. The relocation step does not depend on the order of the data. However, its relocation and trimming steps may produce deficient clusters in which case the procedure responds with a failure. Therefore, it should not be applied when small clusters are expected, for instance, because of small n or large g. There is a more complete reduction step that removes misfits more comprehensively, avoiding also the dependence of the relocation step of Procedure 3.25 on the order of the data as well as the possible failures of Procedure 3.28: the *exact reduction step*. Its description needs some preliminaries and leads into the realm of combinatorial optimization; see Section G.3.

3.2.3   Using combinatorial optimization

In order to attain the maximum on the left-hand side of (3.14), we have to optimize the sum

$$\sum_{j=1}^{g} \sum_{i:\ell_{\text{new},i}=j} \left(\log n_j + \log f_{\gamma_j}(x_i) \right) = \sum_{j=1}^{g} \sum_{i:\ell_{\text{new},i}=j} u_{i,j} = \sum_{\ell_{\text{new},i}\neq 0} u_{i,\ell_{\text{new},i}}$$

w.r.t. admissible labelings $\boldsymbol{\ell}_{\text{new}}$ given the numbers $n_j = n_j(\boldsymbol{\ell})$ and parameters $\gamma_j = \gamma_j(\boldsymbol{\ell})$. We will restrict matters here to labelings with minimum cluster sizes $n_j \geq b_j$ assuming that they are admissible. Sometimes, admissibility is even characterized by size constraints. This includes multivariate t-distributions and normal models for $b_j \geq d+1$. A lower bound b_j can also be used to exclude solutions with small, spurious clusters. We arrive at the multipoint optimization problem

(MPO) $\displaystyle\sum_{\ell_i\neq 0} u_{i,\ell_i}$ maximal over all labelings $\boldsymbol{\ell}$ subject to the constraints

$$\begin{cases} |\{i \in 1..n \mid \ell_i = j\}| \geq b_j, & j \in 1..g, \\ |\{i \in 1..n \mid \ell_i = 0\}| = n - r. \end{cases}$$

The parameters of this problem are the number of retained elements, r, the bounds, b_j, and the *weights* $u_{i,j}$ defined in (3.15). This looks at first sight like a difficult problem. Yet it belongs to a class of problems known to be tractable in discrete mathematics and theoretical computer science. Introduce an artificial $(g + 1)$th group with associated weights

$$u_{i,g+1} = \max_{j\in 1..g} u_{i,j}.$$

This group serves to accommodate the excess members w.r.t. the constraints $n_j \geq b_j$ of the g proper groups. The discards do not need weights $u_{i,0}$; they could be assigned any constant value if necessary, e.g., 0. Now, Section G.3 shows that problem (MPO) can be transformed to a special linear optimization problem, the λ-assignment problem

$$(\lambda A)_{r,\mathbf{b},\mathbf{u}} \quad \sum_{i=1}^{n} \sum_{j=1}^{g+1} u_{i,j} z_{i,j} \text{ maximal over all matrices } \mathbf{z} \in \mathbb{R}^{n \times (g+2)} \text{ subject to}$$

$$\begin{cases} \sum_{j=0}^{g+1} z_{i,j} = 1, & i \in 1..n, \\ \sum_i z_{i,0} = n - r, \\ \sum_i z_{i,j} = b_j, & j \in 1..g, \\ \sum_i z_{i,g+1} = r - \sum_j b_j, \\ z_{i,j} \geq 0, & i \in 1..n, \ j \in 0..(g+1). \end{cases}$$

It depends again on the parameters r, $\mathbf{b} = (b_j)_j$, and $\mathbf{u} = (u_{i,j})$. The matrix $z_{i,j}$ has real-valued entries. However, according to Section G.3, (λA) has a binary solution $\mathbf{z}^*$ with exactly one entry 1 in each row, a fact that justifies the name *assignment matrix*. A matrix $\mathbf{z}$ that satisfies the *equality* constraints of (λA) will be called *feasible*. By contrast, an admissible solution satisfies the *inequality* constraints of (MPO). Section G.3 shows that the assignment matrix $\mathbf{z}^*$ induces a solution $\boldsymbol{\ell}_{\text{new}}$ to the multipoint optimization problem (MPO), namely,

$$\ell_{\text{new},i} = \begin{cases} j, & \text{if } \mathbf{z}^*_{i,j} = 1 \text{ and } 0 \leq j \leq g, \\ \text{the group of } i, & \text{if } \mathbf{z}^*_{i,g+1} = 1; \end{cases} \tag{3.16}$$

here, the "group" of object i is $\operatorname{argmax}_{j \in 1..g} u_{i,j}$.

The λ-assignment problem would also allow simultaneous lower and upper bounds $b_j \leq n_j \leq c_j$, and hence also equality constraints. However, in order to reduce the number of computational parameters, we will set all b_j's to the same value, b. We obtain the following exact multipoint reduction step with trimming.

3.29 Procedure (Exact multipoint reduction step with trimming)
// Input: An admissible labeling $\boldsymbol{\ell}$.
// Output: An admissible labeling $\boldsymbol{\ell}_{\text{new}}$ with larger criterion
        or the response "stop."

1. *(Estimation)* Compute the MLE's $\gamma_j(\boldsymbol{\ell})$ for all clusters j w.r.t. $\boldsymbol{\ell}$.

2. Solve the λ-assignment problem $(\lambda A)_{r,b,\mathbf{u}}$ with weights

$$\begin{cases} u_{i,j} = \log n_j(\boldsymbol{\ell}) + \log f_{\gamma_j(\boldsymbol{\ell})}(x_i), & j \in 1..g, \\ u_{i,g+1} = \max_{j \in 1..g} u_{i,j}, \end{cases} \quad i \in 1..n.$$

3. *(Relocation)* Determine the optimal assignment $\boldsymbol{\ell}_{\text{new}}$ according to Eq. (3.16).

4. *If* $\displaystyle\sum_{i:1 \leq \ell_{\text{new},i} \leq g} u_{i,\ell_{\text{new},i}} > \sum_{i:1 \leq \ell_i \leq g} u_{i,\ell_i}$, then return $\boldsymbol{\ell}_{\text{new}}$ and its parameters and criterion;
   else return "stop."

The reduction step may also be started from a given weight matrix $(u_{i,j})$. Any reduction step is the combination of parameter estimation and a discriminant rule. In this sense, the multipoint reduction step 3.29 may be viewed as a combination of ML parameter estimation and a *constrained* discriminant rule.

The value for the bound b can be detected by experiment. Starting with a low value, say $d + 1$, raise b until the size of the smallest cluster is substantially larger than b. Solutions $\boldsymbol{\ell}$ satisfying $\min_j n_j(\boldsymbol{\ell}) = b$ are in most cases forced and undesirable. This method will, of course, not detect small clusters, which are then found among the discards.

Section G.3.3 contains (primal) heuristics for producing *feasible* initial partitions. They can also be used on their own for approximately solving the λ-assignment problem in Procedure 3.29. However, since they need not improve the criterion, they do not necessarily terminate with a steady partition. Iteration of exact or heuristic multipoint reduction steps thus gives rise to several stable optimization methods for *fixed* numbers of clusters and discards. They produce reasonable solutions even if a cluster tends toward becoming too small by the end of an iteration, where the simple versions run into problems. Section G.3.2 is devoted to a survey on efficient algorithms for solving (λA).

3.2.4 Overall algorithms

Reduction steps receive a labeling and improve it according to Propositions 3.22 and 3.27. Their iteration thus gradually increases the criterion. Since there are only finitely many labelings, the iteration must stall after a finite number of steps with the "stop" signal. Each suite of reduction steps to termination is a run of the k-parameters algorithm. This process takes typically a few or a few tens of reduction steps. The partition returned is steady, that is, self-consistent in the sense that it reproduces its parental parameters. Finding the HDBT constant along with the desired solution needs in general a survey of all steady clusterings, as will be explained in Section 4.1. This needs in general an application of the multistart method, for instance, starting from many random initial labelings. Since clustering is known to be NP hard (see the beginning of Section 3.2), we cannot expect to find all steady solutions with certainty in the time available except in simple cases. This will depend heavily on size and structure of the data set, on the initial labelings, on the parameters g and r, and on the statistical model chosen.

Each run of the algorithm needs an initial solution to begin with. It can be chosen deliberately or randomly. First pick a subset of size r uniformly at random, the retained elements. There exists a simple yet sophisticated algorithm that accomplishes just this in one sweep through the data; see Knuth [295], p. 136 ff. Next, assign the retained elements of this set to g clusters again uniformly at random. The clusters must be large enough to allow estimation of their parameters. In general, this requirement does not pose a problem unless g is large. In this way, the clusters will be of about equal size with very high probability. It follows that the entropy of the mixing rates is large at the beginning making it easier to fulfill assumption (3.14) of Proposition 3.27. The algorithm may also be started from initial parameters $\gamma^{(0)}$ and the Bayes discriminant rule. Coleman et al. [98] compare the effectiveness of some initial solutions.

3.2.5 Notes

The archetype of k-parameters is k-means. It has a long history and has been rediscovered many times in different disciplines. First versions appear in Thorndike [502] (in a statistical context), Steinhaus [486], Jancey [270], and Maranzana [343] (in Operations Research). MacQueen [340] coined the name "k-means." In the engineering literature it is known as Lloyd's [333] algorithm. The extension to more general models, in particular Proposition 3.22 and the multipoint reduction step 3.25, go back to Diday and Schroeder [124] and Schroeder [459]. A formulation of their rather abstract algorithm in terms of the normal classification model appears in Celeux and Govaert [83]. Bock [52] offers a historic view of iterative replication algorithms.

Rocke and Woodruff [444] and Woodruff and Reiners [534] applied local search to a heteroscedastic cluster criterion with cluster size constraints and trimming, called MINO. The use of sorting for trimming in the trimmed multipoint algorithm 3.28 was proposed by Gallegos and Ritter [180, 183]. Gallegos and Ritter [183] contains a method that enforces the HDBT constraints in the univariate case with two clusters. With this exception and apart from homoscedastic models, the algorithms described here are based on unconstrained heteroscedastic models. Fritz et al. [175] design an algorithm that optimizes the related MAP criterion under these constraints. The implementation as an R package under the name TCLUST is publicly accessible. A earlier version appears in García-Escudero et al. [188].

The transportation problem in connection with constrained *discriminant analysis* goes back to Tso et al. [511]. These authors deal with automatic chromosome classification using the constraints to ensure the correct number of chromosomes of each class in a biological cell. An application of λ-assignment to a constrained, least-squares clustering problem with *fixed* cluster centers and sizes is found in Aurenhammer et al. [17]. Versions of the exact multipoint reduction step with cluster size constraints for k-means appear in Demiriz et al. [114] and Banerjee and Ghosh [23]. The present exact multipoint reduction step with cluster size constraints and trimming based on λ-assignment (Procedure 3.29) appears in Gallegos and Ritter [184].

3.3 Hierarchical methods for initial solutions

Producing the desired local likelihood maximum with the EM and the EMT algorithms may consume much time. The algorithm has to be restarted from (very) many different initial values in order to find many local solutions and the interesting one(s). This shortcoming may be counteracted by using proper initial solutions. Literature often proposes a hierarchical method for initialization; see, for instance, Fraley [169]. It may be used with the EM and EMT algorithm and also with cluster criteria. The result is a hierarchical method with an EM booster. It produces just one solution and is applicable to data sets with a clear cluster structure. Note, however, that the EM algorithm may not be able to liberate itself from any weakness received from the initial solution. The weakness may (and usually will) appear in the final result. In the end we'd better use the EM right away. Speak to the organ grinder and not to the monkey.

The agglomerative hierarchical method (Florek et al. [163]) is a fast "greedy" optimization method for general target functions on partition spaces, here cluster criteria. It starts from a partition with a large number of clusters and iteratively merges two of them with minimum loss of the target function until a reasonable partition with the desired number of clusters is found. The method is fast and can be used on its own or it can be employed to generate initial configurations for other clustering methods. Classical target functions are the *complete* linkage (Sørensen [480]) and *single* linkage (Florek et al. [163]) cluster criteria. Complete linkage minimizes

$$\max_C d(C),$$

that is, the maximum diameter of all clusters. It strives for *cluster cohesion*. It is related to the clique problem in graph theory which is known to be NP complete. Single linkage maximizes

$$\min_{C_1 \neq C_2} d(C_1, C_2),$$

that is, the minimum distance between any two clusters. It strives for *cluster isolation*. Two efficient algorithms are by Jarník [272] (often attributed to Prim [424]) and by Kruskal [300]. They construct a minimum spanning tree. Later, Ward [522] used the hierarchical method to optimize his sum-of-squares criterion, the ML pooled trace criterion in the taxonomy of this text. An overview of classical criteria used in hierarchical optimization is found, for instance, in Johnson [280], Murtagh [385], Baker and Lawrence [22], Ling [326], and Gordon [209]. However, the agglomerative method may in principle be applied to any cluster criterion.

Returning to the probabilistic cluster criteria treated in this text, let $\mathcal{P}$ be the set of all partitions of a (finite) data set $\mathbf{x}$ in nonempty clusters. In general, the criteria are not defined on all of $\mathcal{P}$ but only on the system of admissible partitions; see Definitions 1.31 and 1.33. Examples are the cluster criteria of Sections 1.3.1, 1.3.4, and 1.3.5. Friedman and Rubin's [172] pooled determinant criterion (Theorem 1.41(a)), for instance, needs at least *one* cluster of size $\geq d+1$. Scott and Symons' [461] heteroscedastic determinant criterion requires that *all* clusters be of size $\geq d+1$. We therefore consider a target function $t : \mathcal{D}(t) \rightarrow$

$\mathbb{R}$ that is defined partially on $\mathcal{P}$ with domain of definition $\mathcal{D}(t) \subseteq \mathcal{P}$. The different parameter constraints determine the set $\mathcal{D}(t)$. Assume that $\mathcal{D}(t)$ is closed w.r.t. the formation of unions of clusters: If two clusters of a partition from $\mathcal{D}(t)$ are merged, then the resulting partition is again a member of $\mathcal{D}(t)$. A coarsening of a partition $\mathcal{C}$ is a partition that consists of unions of clusters in $\mathcal{C}$. In this way $\mathcal{D}(t)$ contains all its coarsenings. The agglomerative method is particularly suited for target functions that are additively composed of terms that depend on single clusters: If $\mathcal{C} = \{C_1, \ldots, C_{g_0}\}$, then $t(\mathcal{C}) = \sum_j a(C_j)$ for some function a defined for the clusters C_j. This has the following consequence: If two clusters C_1, C_2 in a partition are merged, then the difference between the two target values is just $a(C_1 \cup C_2) - (a(C_1) + a(C_2))$. Although the agglomerative method is applicable to any target function, the optimization step will be formulated here for this convenient case only. It excludes some homoscedastic cases. The following procedure applies to *minimization* of t.

3.30 Procedure (Agglomerative optimization step)

// <u>Input:</u>   A member $\mathcal{C} \in \mathcal{D}(t)$ with $g_0 \geq 2$ clusters and the values $a(C)$ for
                all $C \in \mathcal{C}$.

// <u>Output:</u> A member of $\mathcal{D}(t)$ with $g_0 - 1$ clusters along with all its values.

1. Compute the values $a(C_1 \cup C_2)$ for all pairs $C_1 \neq C_2$ of clusters in $\mathcal{C}$.

2. Merge the two clusters C_1 and C_2 with minimum difference $a(C_1 \cup C_2) - (a(C_1) + a(C_2))$ and return the resulting partition along with its values.

If a partition in g clusters is eventually requested, then one starts from some partition with $g_0 > g$ clusters, iterating agglomerative optimization steps until g is reached. Of course, the result is not the optimal partition with g clusters but there is a chance of obtaining a reasonable solution. An advantage of this method is the fact that it yields solutions with g clusters for all $g < g_0$. That is, we obtain a whole hierarchy of solutions at once.

In the following examples we have to keep track of the sizes $|C|$, the mean values $\overline{x}_C$ (in order to compute the scatter matrices and new means), and the scatter matrices S_C of all unions C of two clusters. Of course, $|C_1 \cup C_2| = |C_1| + |C_2|$ and $\overline{x}_{C_1 \cup C_2} = \frac{1}{|C_1| + |C_2|}(|C_1| \overline{x}_{C_1} + |C_2| \overline{x}_{C_2})$. Moreover, by Lemma A.14, the SSP matrix $W_{C_1 \cup C_2}$ is a simple rank-one update,

$$W_{C_1 \cup C_2} = W_{C_1} + W_{C_2} + \frac{|C_1| \cdot |C_2|}{|C_1| + |C_2|}(\overline{x}_{C_1} - \overline{x}_{C_2})(\overline{x}_{C_1} - \overline{x}_{C_2})^{\top},$$

and the update of the scatter matrix $S_{C_1 \cup C_2}$ follows.

3.31 Examples In order to apply the agglomerative optimization step, we have to specify the domain of definition $\mathcal{D}(t)$ and the function a.

(a) (Heteroscedastic MAP determinant criterion; see Theorem 1.40) Since the log-determinants of scatter matrices have to be computed, all clusters of all partitions in $\mathcal{D}(t)$ must be of size $\geq d + 1$ – assuming general position of the data. A small computation shows that the function a reads

$$a(C) = |C|\left(-\tfrac{1}{2}\log \det S_C + \log \tfrac{|C|}{n}\right).$$

(b) (Heteroscedastic MAP trace criterion; see again Theorem 1.40) Here the log-traces of scatter matrices must be defined. This is possible when all clusters are of size at least two – assuming that data points are pairwise distinct. In this case, we have

$$a(C) = |C|\left(-\tfrac{d}{2}\log \tfrac{1}{d}\operatorname{tr} S_C + \log \tfrac{|C|}{n}\right).$$

(c) (ML pooled trace (Ward's) criterion; cf. Theorem 1.41) Let W_C be the SSP matrix of a cluster C and let $W_{\mathcal{C}}$ be the pooled SSP matrix of a partition $\mathcal{C}$. We have to minimize $\operatorname{tr} W_{\mathcal{C}} = \sum_{C \in \mathcal{C}} \operatorname{tr} W_C$. Therefore, we define

$$a(C) = \operatorname{tr} W_C.$$

Since this function is defined for all clusters, even singletons, $\mathcal{D}(t)$ is the collection of *all* partitions. Of course, $a(\{x\}) = 0$. Moreover, the difference $a(C_1 \cup C_2) - (a(C_1) + a(C_2))$ has the simple representation

$$a(C_1 \cup C_2) - (a(C_1) + a(C_2)) = \operatorname{tr}(W_{C_1 \cup C_2} - W_{C_1} - W_{C_2}) = \frac{|C_1| \cdot |C_2|}{|C_1| + |C_2|} \|\bar{x}_{C_1} - \bar{x}_{C_2}\|^2.$$

The related iteration needs only the sizes and mean vectors of all clusters.

The pooled determinant criterion is less obvious, since the change in criterion due to the merging of two clusters is influenced by all clusters.

A word is in order concerning initial partitions for the agglomerative hierarchical method. No problem arises in Example 3.31(c), since the criterion is defined even for the finest partition in n singletons. In other cases, we must ensure that the criterion for the union of two clusters makes sense. This needs initial clusters of size $\geq b$, depending on the case. For instance, for Scott and Symons' [461] heteroscedastic ML determinant criterion (see Section 1.40), we need $b = \lceil (d+1)/2 \rceil$ if data is in general position. We begin with decomposing the data set in a mosaic of small spherical clusters of size at least b. An elementary version with k-means was proposed by Ghosh and Chinnaiyan [198], p. 279. A more refined initial algorithm that accomplishes this is the exact multipoint reduction step with or without trimming (Procedure 3.29) with the initial number of clusters g_0. The statistical model to be used is the homoscedastic spherical normal model. In order to obtain many small clusters, the number g_0 should be large. We should, however, allow the algorithm to produce small clusters that do not transgress the boundaries of the g desired clusters. The agglomerative method could not correct this error later. For instance, if a data set consists of 27 points and we decompose it in nine small clusters of three points each, then the condition will certainly be violated if there are two true clusters of size 13 and 14. Finding a favorable number g_0 needs a small reasoning. Each natural cluster C_ℓ can be decomposed in k_ℓ small clusters of size b with $r_\ell < b$ elements remaining,

$$|C_\ell| = k_\ell b + r_\ell.$$

Summing up we obtain $r = b \sum k_\ell + \sum r_\ell \leq b \sum k_\ell + g(b-1)$ and hence we can always achieve $\sum k_\ell \geq \frac{r - g(b-1)}{b}$. That is, the present task can be solved with $\lceil \frac{r - g(b-1)}{b} \rceil$ small clusters no matter what the sizes of the true clusters are. This is the maximum number g_0 of small clusters that should be used. The remaining $\sum r_\ell$ elements are accommodated in the k_ℓ clusters by the initial algorithm. For a solution with $g_0 \geq g$ to exist in all cases, it is necessary and sufficient that $(r - g(b-1))/b > g - 1$, that is, $r > 2gb - g - b$.

More precisely, there is the following procedure. It provides the agglomerative optimization step with an initial configuration consisting of small *spherical* clusters. This also avoids spurious final solutions.

3.32 Procedure (Initial partition for agglomerative optimization)
// <u>Input:</u>   A data set $(x_1 \ldots, x_n)$, the number r of retained elements, and a
           size constraint $b \geq 2$.
// <u>Output:</u> A subset of r retained elements and its partition in a large number
           of clusters of size $\geq b$.

1. Sample $g_0 = \lceil \frac{r - g(b-1)}{b} \rceil$ indices i_j, $1 \leq j \leq g_0$, uniformly from $1 .. n$ and put $m_j = x_{i_j}$.
2. Run the exact multipoint algorithm based on the ML spherical normal model with unitary covariance matrix and parameters r and b until convergence, starting from the weight matrix
$$u_{i,j} = \|x_i - m_j\|^2, \quad i \in 1 .. n, \ 1 \leq j \leq g_0.$$

In step 1, g_0 cluster centers (or "seeds") m_j are determined to begin with. They should be spread uniformly over the data set. A deterministic alternative is the following iteration due to DeSarbo et al. [117], (A-4): The first two seeds are the two points with maximum distance. Let now $\mathcal{S}$ be the current set of seeds already constructed. The next seed maximizes the Euclidean distance to $\mathcal{S}$, that is, it is a solution to the max-min criterion

$$\operatorname*{argmax}_{y \notin \mathcal{S}} \min_{x \in \mathcal{S}} \|y - x\|.$$

In step 2, the first exact multipoint reduction step is not started from a labeling as in Procedure 3.29 but from a weight matrix. In the following iterations, the weight matrix w.r.t. the model is

$$\begin{cases} u_{i,j} = \|x_i - \overline{x}_j(\boldsymbol{\ell})\|^2, & 1 \le j \le g_0, \\ u_{i,g_0+1} = \max_{1 \le j \le g_0} u_{i,j}, \end{cases} \quad i \in 1..n,$$

where $\boldsymbol{\ell}$ is the current assignment. Using the unitary covariance matrix is tantamount to applying the ML pooled trace criterion (cf. Theorem 1.41). It is not necessary to compute the pooled variance since it appears as a common factor on the weights.

Chapter 4

Favorite solutions and cluster validation

The criteria and algorithms in the previous chapters depend on computational parameters that were essentially arbitrary but fixed. For any number of clusters, g, and of discards, $n - r$, we have obtained a finite and moderate number of locally optimal (mixture model) and steady (classification model) solutions. Although this number may still be fairly large, this means a substantial reduction from the initially astronomical number of all possible partitions to a much smaller set. If there is a true solution, we hope that it is still contained there. The final task is to further reduce this number to a handful of acceptable solutions or, in the best case, even to a single one.

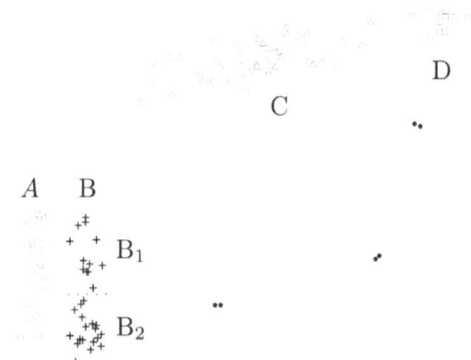

Figure 4.1 *The sample data set of four clusters and six outliers used for illustration.*

This is not easy. One reason is that complex data does not know its true structure, even if there is one. Uniqueness of the solution belongs to classical inferential thinking. This is not appropriate here. Even if the number of components is known, the solution may yet be ambiguous. In fact, one of the more complex problems of cluster analysis is overcoming the nonuniqueness of the solution given the number of clusters. For instance, the data set shown in Fig. 4.1 without the six outliers has been sampled from four normal components. However, there are small gaps in the lower part of cluster A and about the middle of cluster B. They could give rise to two different five-component solutions. Selecting solutions, estimating the number of clusters, as that of outliers, and cluster validation are therefore sometimes considered the realm of exploratory and heuristic methods such as visualization. Nevertheless, there are some statistical tools such as tests that are helpful.

Besides undesired solutions, there may even be multiple *sensible* ones. In fact, the nonuniqueness of the solution in Fig. 4.1 is not by coincidence. A data set may allow more than just one interesting solution, a fact that has been well known to practitioners. Gondek [207] writes on p. 245 *"Data contains many plausible clusterings"* and on p. 249 *"It is often the case that there exist multiple clusterings which are of high quality, i.e., obtain high values of the objective function. These may consist of minor variations on a single clustering or may include clusterings which are substantially dissimilar."* Jain et al. [268] express in the same spirit *"Clustering is a subjective process; the same set of data items often needs to be partitioned differently for different applications."* These quotations support the fact that, except in simple situations, the clustering problem is ambiguous. It is eventually the investigator who must decide among the sensible solutions. Attempting to design systems that always produce a *single* solution in one sweep is, therefore, not advisable. Too often the outputs of computer programs are taken for granted.

This chapter deals with estimating the number of groups and outliers and valid solutions.

As a byproduct, we will also obtain the HDBT ratio. Section 4.1 will show how to obtain acceptable solutions given the parameters r and g. It will turn out that the likelihood principle does not apply to mixture and cluster analysis. Classical inference rather needs to be complemented here by an additional principle, that of *scale balance*. This will follow from the consistency theorems of Chapter 1. Estimating the number of components and outliers will be the subject matter of Sections 4.2 and 4.3. Partitioning methods produce solutions also when the data is devoid of structure. Therefore, Section 4.4 will deal with validation methods.

The methods presented here will be illustrated with the simple data set plotted in Fig. 4.1. Its four true clusters A, B, C, and D of size 25, 30, 50, and 15, respectively, were sampled from four normal distributions. Cluster A was manually alienated so as to observe what happens when a cluster is not sampled from a normal. The additional three pairs of outliers were manually placed so that they can hardly be interpreted as clusters. A deficient cluster of two points in the plane appears as composed of outliers in the context of a fully normal or MV Student-t model. These illustrations are, of course, far from being a moderate, let alone comprehensive, study. However, given that there are four clusters, methods that fail on this data set would be of restricted utility only.

4.1   Scale balance and Pareto solutions

Even when the key parameters r (number of outliers) and g (number of clusters) are fixed, one still faces the problem of finding the consistent solution. The trimmed likelihood function does not in general possess a unique local maximum. Some authors advise choosing the largest one as the estimate. However, this advice is often, although not always, misleading, as Fig. 1.3 shows. This section will derive the *HDBT ratio* (1.34) as a complement to likelihood. It is the combination of both that solves the problem for a finite data set.

The following description refers to the mixture model (1.19). It can also be applied to classification models (1.38) and their criteria. Theorem 1.19 and its corollary state that the likelihood function (2.3) possesses a consistent, *free local* maximum. Unfortunately, the theorem does not uncover which one it is. The (trimmed) likelihood function given the parameters r and g has in general a large number of local maxima, at least when different, full scales are estimated. Therefore, maximum likelihood parameter estimation in mixture models has been considered unreliable, even in the event of an uncontaminated data set; see, for instance, Milligan [376], p. 358. This is often due to a large number of local maxima it admits already for fixed parameters g and $r = n$. The problem is less urgent when cross cluster constraints or scale constraints such as diagonal or spherical scale matrices are used. But these models risk returning false solutions unless the parent populations actually satisfy these assumptions. Therefore, an algorithm based on the general heteroscedastic model is in general preferable.

On the other hand, by Hathaway's [232] consistency theorem for HDBT constrained mixture models and its extensions (Theorems 1.22 and 1.27), the *global* maximum of the properly *HDBT constrained* likelihood function, too, is consistent. However, this theorem leaves us again in the dark about the proper scale constraint. An arbitrarily small constant c is applicable only asymptotically. In the presence of a *finite* data set, the constrained maximizer of the likelihood function depends on the HDBT constant c; the set of possible solutions increases as c decreases. Choosing the constraint too large may exclude the desired, consistent solution. If it is chosen too small, undesirable solutions intervene, hindering the analysis of real and simulated data sets; see Section 6.1. Optimizing the target function under a *fixed* constraint c is therefore not advisable; this parameter should rather be estimated. What is more, HDBT constrained optima are not easily computed. The crux is the estimation step. We will now see that constrained optimization is neither needed nor useful. The presentation here will be for normal mixtures. Extensions to other populations are straightforward as long as a theorem of Hathaway's type is available. The key is a sharpening of Theorem 1.27. We first need a simple lemma.

4.1 Lemma *Let $0 < c < 1$ and let $V_1, \ldots, V_g \in \mathrm{PD}(d)$ be such that $cV_j \prec V_\ell$ for all $j \neq \ell$. Then $(V_1, \ldots, V_g)$ is an interior point of $\mathcal{V}_c$ w.r.t. $\mathrm{PD}(d)^g$ (and w.r.t. $\mathrm{SYM}(d)^g$). (See also the discussion of interiority on p. 252.)*

PROOF. Let $\varepsilon > 0$ be such that $cV_j + 2\varepsilon I_d \preceq V_\ell$ for all j, ℓ. By Section B.6.2, the Cartesian product $\prod_{j=1}^g [V_j - \varepsilon I_d, V_j + \varepsilon I_d]$ of Löwner intervals is a neighborhood of $(V_1, \ldots, V_g)$ w.r.t. $\mathrm{SYM}(d)^g$. Moreover, if $A_j \in [V_j - \varepsilon I_d, V_j + \varepsilon I_d]$, then, for all j, ℓ,

$$A_\ell - cA_j \succeq V_\ell - \varepsilon I_d - c(V_j + \varepsilon I_d) = V_\ell - cV_j - (1+c)\varepsilon I_d \succ V_\ell - (cV_j + 2\varepsilon I_d) \succeq 0,$$

that is, $\prod_{j=1}^g [V_j - \varepsilon I_d, V_j + \varepsilon I_d] \subseteq \mathcal{V}_c$. This proves the lemma. Another proof uses the continuity of $\mathbf{V} \to \min_{j,\ell,k} \lambda_k(V_\ell - cV_j)$ on $\mathrm{PD}(d)^g$ (or on $\mathrm{SYM}(d)^g$). □

Now recall the parameter space

$$\Theta_g = \overset{\circ}{\Delta}_{g-1} \times \{(\mathbf{m}, \mathbf{V}) \,|\, (m_j, V_j) \in \mathbb{R}^d \times \mathrm{PD}(d) \text{ pairwise distinct}\}$$

of the normal mixture model and the HDBT constrained parameter space

$$\Theta_{g,c} = \overset{\circ}{\Delta}_{g-1} \times \{(\mathbf{m}, \mathbf{V}) \in \mathbb{R}^{gd} \times \mathcal{V}_c \,|\, (m_j, V_j) \text{ pairwise distinct}\}$$

of the parametric normal mixture model with g components; see Eqs. (1.16) and (1.33). Like Theorem 1.19, the following proposition guarantees the existence of a local MLE. Whereas the former hinges on differentiability, this one does not.

4.2 Proposition *Let $(\boldsymbol{\pi}_0, \mathbf{m}_0, \mathbf{V}_0) \in \Theta_g$ be the parameter of the parental normal mixture and let $0 < c < r_{\mathrm{HDBT}}(\mathbf{V}_0)$.*

(a) P-a.s. and for eventually all n, the consistent, constrained MLE T_n according to Theorem 1.27 is interior to $\Theta_{g,c}$ as a subset of Θ_g.

(b) The maximum $T_n \in \Theta_{g,c}$ is a local maximum in Θ_g.

PROOF. Write $T_n = (\boldsymbol{\pi}^{(n)}, \mathbf{m}^{(n)}, \mathbf{V}^{(n)})$. Because of the assumption on c, Theorem 1.27 teaches us that, after label switching, $\lim_{n\to\infty}(\boldsymbol{\pi}^{(n)}, \mathbf{m}^{(n)}, \mathbf{V}^{(n)}) = (\boldsymbol{\pi}_0, \mathbf{m}_0, \mathbf{V}_0)$, P-a.s. In particular, $\mathbf{V}^{(n)} \to \mathbf{V}_0$; therefore, $r_{\mathrm{HDBT}}(\mathbf{V}^{(n)}) > c$ for large n. By Lemma 4.1, $\mathbf{V}^{(n)}$ is interior to the subset $\mathcal{V}_{g,c}$ of the g-fold Cartesian product $\mathrm{PD}(d)^g$. Hence, P-a.s. for large n, $T_n = (\boldsymbol{\pi}^{(n)}, \mathbf{m}^{(n)}, \mathbf{V}^{(n)})$ is interior to $\Theta_{g,c}$ as a subset of Θ_g. This is claim (a) and (b) is an immediate consequence. □

It has thus turned out that the HDBT constrained solution promised by Theorem 1.27 is actually a *free* solution if $c < r_{\mathrm{HDBT}}(\mathbf{V}_0)$ and if n is large. It can be generated with the EM algorithm. It seems at first that finding this solution, too, needs a scale constraint. But this is not the case, as will now become clear. The solution T_n above is a *free* local maximum of the likelihood function optimal w.r.t. *some* HDBT constraint. Let us focus on the collection of all such solutions. They are candidates for the consistent solution. In order to characterize them, we need the product order of the plane $\mathbb{R}^2$. It is defined by $(s_0, t_0) \leq (s, t)$ if and only if $s_0 \leq s$ and $t_0 \leq t$. We also define $(s_0, t_0) < (s, t)$ if $(s_0, t_0) \leq (s, t)$ and $(s_0, t_0) \neq (s, t)$, that is, if $s_0 \leq s$ and $t_0 < t$ or if $s_0 < s$ and $t_0 \leq t$. We say in this case that (s, t) *dominates* (s_0, t_0). Domination endows the plane with a partial order. It is not total because neither of the two points $(1, 0)$ and $(0, 1)$, for instance, dominates the other. Given a nonempty subset $\mathscr{L} \subseteq \mathbb{R}^2$, a point is maximal in $\mathscr{L}$ if it is dominated by no other point in $\mathscr{L}$. Any finite, nonempty subset possesses a maximal point but, in contrast to the usual order on the real line, there may be more than one. Computing maximal points in product orders belongs to the realm of *biobjective optimization*, where they are called **Pareto points**. Pareto [399] introduced them to economics in 1906. They describe optimal solutions in the presence of antagonistic objectives. This calls usually for a compromise. In economy, the variables are, for instance, quality and inverse price. The only products competitive on the market belong to Pareto points. No customer wants to buy lower quality at a higher price. The set of Pareto points of a finite subset of the plane is also characterized as the *smallest* subset such that the left-lower quadrants defined by them cover the whole set and at the same time the set of all points with an empty right-upper quadrant. The reader interested learning more about multiobjective optimization is referred, for instance, to Steuer [488], Serafini [464], and Erfani and Utyuzhnikov [144].

The search for a feasible, consistent estimate uses the following simple abstract lemma. It will in fact lead to a very simple and effective estimation procedure devoid of parameters.

4.3 Lemma *Let $\mathscr{L} \subseteq \mathbb{R}^2$ be a finite (nonempty) set of points the first entries of which are pairwise different. The following statements on a point $(l_0, r_0) \in \mathscr{L}$ are equivalent.*

(a) There exists $c \leq r_0$ such that $l_0 \geq \max\limits_{(l,r)\in\mathscr{L}, r\geq c} l$;

(b) $l_0 = \max\limits_{(l,r)\in\mathscr{L}, r\geq r_0} l$;

(c) (l_0, r_0) is a Pareto point (that is, maximal in $\mathscr{L}$ w.r.t. the product order).

PROOF. Assume (c). By maximality, $l_0 \geq l$ for all $(l,r) \in \mathscr{L}$ such that $r \geq r_0$, that is, (b). In order to see that (b) implies (a), just put $c = r_0$. Finally, assume (a). We wish to prove that (l_0, r_0) is maximal. Because of $c \leq r_0$, we have in particular $l_0 \geq l$ for all $(l,r) \in \mathscr{L}$ such that $r \geq r_0$. By assumption on pairwise difference, we even have $l_0 > l$. This is maximality (c) of (l_0, r_0). □

The assumption of pairwise difference is no loss of generality. Without the assumption, a pair (l_0, r_0) satisfying (a) might not be maximal but there is a maximal one with the same l_0 and a larger t_0.

Returning to mixture decomposition, we apply Lemma 4.3 to the pairs (l,r) formed by the likelihood values and the HDBT ratios (1.34) of *free* local maxima of the mixture likelihood. Part (c) of Lemma 4.3 says that the solutions maximal w.r.t. *some* HDBT constraint are just the Pareto points. By Proposition 4.2, they are the only ones to be further considered. They will be called **Pareto solutions**. In most cases, there will be more than one. In fact, several Pareto solutions could be the desired estimate of the parent. The nonuniqueness of Pareto points is not a weakness but rather a strength. Note that the data set of Fig. 1.3 could also have been sampled from three populations, two spherical and one with a very slim covariance matrix and a small mixing rate. It should thus be no surprise that we often retrieve more than one solution. However, some of them are less likely to represent the data-generating mechanism sought than others as will be explained below. Further reduction of the set of solutions will need validation methods or additional assumptions, for instance, further characteristic variables or exterior information. The appearance of nonunique Pareto solutions in mixture analysis also explains the ambiguity observed by various authors; see the introduction to this chapter.

The HDBT ratio is a measure of **scale balance**. A solution with a large scale balance, that is, a large HDBT ratio, does not deviate dramatically from homoscedasticity. This, in turn, means that there is some (unknown) linear transformation on the sample space that transfers the components of the parent *simultaneously* not too far away from common sphericity. Proposition 4.2 and Lemma 4.3 have led to a trade-off between the two antagonistic objectives "fit" and "scale balance." Both are important in mixture and cluster analysis. Fit is not everything here; scale balance matters as well. Figure 1.3 illustrates this. Proposition 4.2 and Lemma 4.3 show that likelihood and HDBT ratio play a dual rôle in mixture decomposition. Therefore, large scale balance is as valuable an asset as a large likelihood is. Focusing on fit alone, that is, returning the solution with the largest local maximum of the likelihood function may be as misleading as focusing on scale balance alone, that is, using a homoscedastic model.[1] The best-fitting solution may be spurious (see Section 1.2.4), and a homoscedastic model may fail to detect the right cluster scales. Of course, we may be lucky with either one, but there is no guarantee. We wish to decompose *all* elliptical mixtures in a reasonable way. The present method does this unless the overlap is too heavy or the data set is too small, in which case appealing to consistency theorems would not make sense. When the largest local maximum or the sum-of-squares criterion is successful, the present method recovers just this solution.

The likelihood principle says that the likelihood function contains the full information on the estimate; see Section F.2.1. It fails in the context of mixture analysis. The desired estimate cannot in general be derived from the likelihood function alone. We have already been warned by the fact that the likelihood has no maximum. The inference in mixture and cluster analysis is different from the unimodal case.

The set of Pareto solutions produced by a series of runs of the EM(T) algorithm is easily computed online by following each run with the procedure below. A number of solutions are *independent* if none dominates any other. We have to administer a list $\mathscr{L}$ of independent solutions, deleting current solutions or entering a new one as necessary after each run.

4.4 Procedure (List updating)
// Input: A list $\mathscr{L}$ of independent solutions, a new solution $\mathcal{S}_{\text{new}}$.
// Output: An updated list of independent solutions.

1. For each $\mathcal{S}$ in $\mathscr{L}$ do
   if $\mathcal{S}$ dominates $\mathcal{S}_{\text{new}}$, return $\mathscr{L}$;
   else if $\mathcal{S}_{\text{new}}$ dominates $\mathcal{S}$, delete $\mathcal{S}$ from $\mathscr{L}$;
2. return $\mathscr{L} \cup \{\mathcal{S}_{\text{new}}\}$.

If the "if" case occurs in step 1, then $\mathcal{S}_{\text{new}}$ is no Pareto solution and the procedure stops. If the "else" case occurs, then the "if" case will never occur later for the same $\mathcal{S}_{\text{new}}$ but all other solutions $\mathcal{S}$ in the list will still have to be explored. In this case, the list is updated with a stronger Pareto solution. It may happen that neither of the two cases occurs. Then $\mathcal{S}_{\text{new}}$ is a solution independent of all others in the list. The overall algorithm is started from the empty list. Note that the algorithm does not use any constraints. A further advantage of the method: It also offers some guidance about the number of runs needed – run the EMT (or k-parameters) algorithm until the list has not changed for some time. As a rule of thumb, begin with a minimum number of runs, say 300, dynamically extending this number to at least ten times the number of the last update. So, if the last update among the first 300 runs was at run 10, stop at 300, hoping that no new update would occur later. If it was at 200, run the algorithm at least 2000 times, and so on. If this procedure does not stop in the time available, then use the last list. At the end, the list contains all Pareto solutions found during the program run.

Apart from the trade-off between fit and scale balance discussed above, there is also a trade-off between cluster qualities. Solutions often consist of clusters of different quality. A single mixture component that fits its cluster very well raises the overall likelihood. It may even lead to the largest local maximum, although the other clusters are of low quality. Examples are spurious clusters. Partitions with balanced cluster qualities might be preferred to others with very unequal qualities. Therefore, some Pareto solutions are less likely to be estimates of the parent. We further reduce their number by removing any solution with at least one of the following properties.

(i) The solution is a halting point of the EM algorithm, but no local maximum. See the discussion in Remark 3.14(a).

---

[1]A special case is the sum-of-squares criterion.

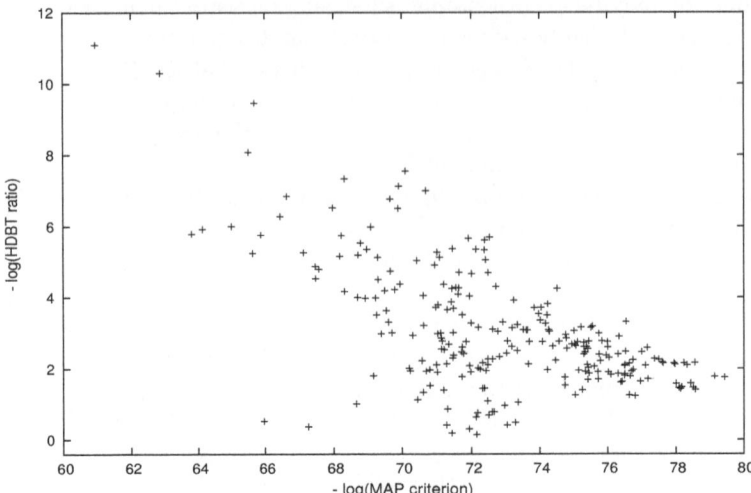

Figure 4.2 *Simulated data set of Fig. 1.3: SBF plot for a large number of steady partitions with two clusters without trimming. There are eight Pareto points.*

(ii) The HDBT ratio of the solution is too small. Roughly, an HDBT ratio of 10^{-2} or a little less is not unusual, but 10^{-3} or less usually indicates a spurious solution in a real data set. There is, however, no logical foundation of this rule and it is easy to create artificial data sets with arbitrary HDBT ratios.

(iii) All points of some cluster fit well in other clusters.

(iv) There are clusters of poor quality; use a validation method of Section 4.4 below.

Of course, solutions which contradict external knowledge should also be excluded. After this phase, only very few solutions, maybe even one, usually remain.

In order to visualize the biobjective optimization problem, Gallegos and Ritter [182, 183] proposed plotting the negative logarithms of the HDBT ratios vs. the negative log-likelihoods of all solutions;[2] see Fig. 4.2. This is a plot of *scale balance vs. fit* (**SBF plot**). Since the logarithm is strictly increasing, the Pareto points (here minimal) are not changed and appear on the left and lower borders of the SBF plot. A point is Pareto if the closed, southwest quadrant it specifies contains no other point. Of particular interest are the southwest extreme points of the convex hull of all points in the plot. They are Pareto points, touched from below by a supporting line of negative slope such that all points lie in northeastern direction. The slope is not unique and depends on the point. The related solutions will be called **extreme Pareto solutions**. They depend on the special (logarithmic) transformation. There may exist nonextreme Pareto points and one of them, too, may be the correct solution. In clear cases, all points in the plot are supported from below by an almost horizontal line segment and the desired Pareto point is unique and found close to its left end as in Fig. 4.2. In unclear cases, the southwest border of the points describes a quarter circle; see, for instance, the right lower plot in Fig. 6.1 on page 221. In these cases, there is usually more than one sensible solution for this pair (r, g).

The extreme Pareto points can be described with the aid of a criterion that depends, however, on a parameter. A suitable slope as above depends on the data set. Let $(1, \sigma)$, $\sigma \geq 0$, be a vector perpendicular to the supporting line and pointing northeast. The extreme Pareto point is the minimum of the **aggregate objective function**

$$-\log f[\mathbf{x}_{R^*} \mid \mathbf{u}^*, \mathbf{m}^*, \mathbf{V}^*] - \sigma \log r_{\text{HDBT}}(\mathbf{V}^*) \qquad (4.1)$$

---

[2]Negative logarithms are used in order to avoid negative values. This is, of course, not essential.

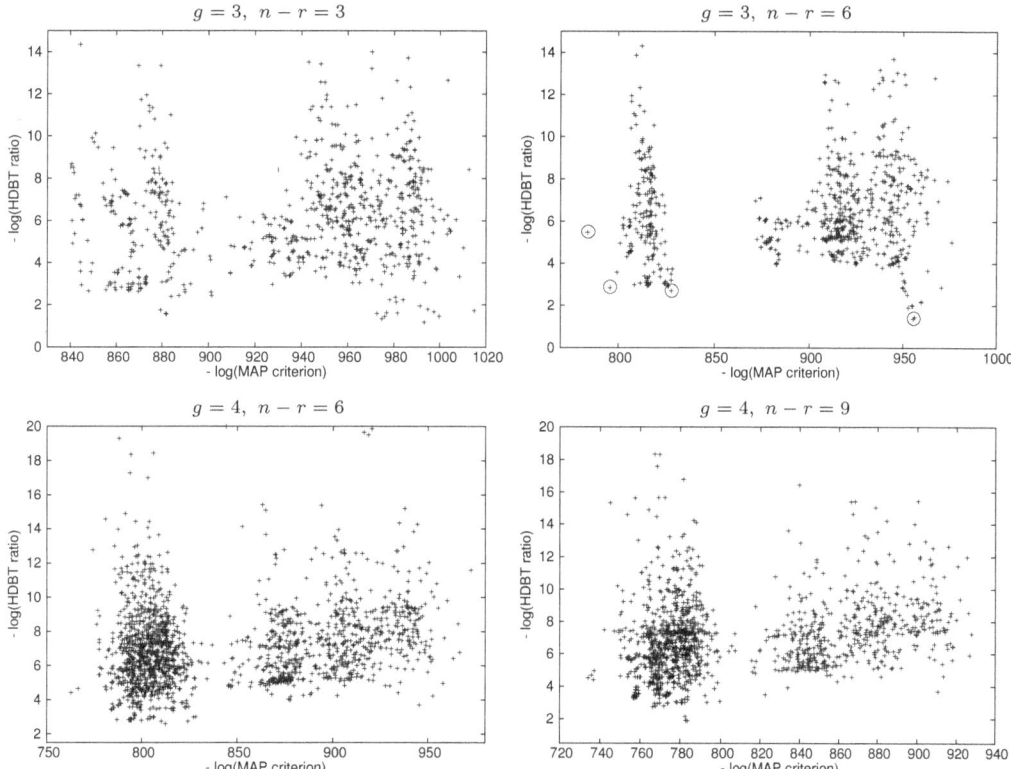

Figure 4.3 *SBF plots for the sample data set of Fig. 4.1 and various numbers of components and discards. The Pareto solutions of the right upper plot are circled. Three of them are extreme.*

w.r.t. all limit points $(R^*, \mathbf{u}^*, \mathbf{m}^*, \mathbf{V}^*)$ of the EMT algorithm.[3] By consistency, the log-likelihood is asymptotically linear with r and the HDBT ratio converges. Hence, σ/r does not depend much on the size of the data. In the homoscedastic case, the theoretical HDBT ratio is 1 and the negative log-likelihood is omitted.

All methods apply as well to the classification model and estimation of steady solutions of the heteroscedastic determinant criterion, Theorem 1.40(a), Sections 2.3 and 3.2.2. Here, the criterion replaces the log-likelihood and free steady solutions substitute for local maxima. By way of illustration, consider again the data set of 120 points sampled from a normal mixture with four well separated components and six outliers shown in Fig. 4.1. SBF plots for several choices of g and $n - r$ are shown in Fig. 4.3. In Table 4.1 some extreme Pareto solutions obtained with the unconstrained TDC (Section 2.3.2) for three to five clusters, zero to nine outliers, and various values of σ are described. The table shows that the criterion merges two clusters if $g = 3$ is assumed. For $g = 5$, it either splits a small part off some cluster to create a new cluster or it creates a cluster of outliers. If the number of discarded elements is smaller than the number of outliers, that is, if $n - r < 6$, then the clusters have to accommodate some outliers. This grossly distorts the estimated parameters. On the other hand, no dramatic effect is observed if it is assumed to be larger – just a few extreme elements are removed from the clusters. Therefore, the number of discards should not be chosen too small. For the original number of clusters (four) and outliers (six), the TDC recovers exactly the correct solution.

---

[3]There is a resemblance to the penalized log-likelihood functions proposed by Ciuperka et al. [96], Chen and Tan [89], and Tanaka [495]. The additive penalty terms depend on the scale matrices and on n. They have also a Bayesian interpretation. (Alexandrovich [7] detected and corrected a logical inconsistency in Chen and Tan's proof.)

Table 4.1 *Data set of Fig. 4.1: Table of extreme Pareto solutions w.r.t. free TDC for various numbers of clusters and discards.* E = *the set of outliers,* $o_1, \ldots, o_6$ = *the outliers from left to right,* F = *two points in the upper left part of* C, a = *the left lower point of* A, c = *the rightmost point in* C, d = *the leftmost point in* C.

g	$n-r$	σ	$-\log r_{\text{HDBT}}$	clusters	discarded
3	3	2	2.94	A∪B, C, D∪$\{o_4, o_5, o_6\}$	$\{o_1, o_2, o_3\}$
3	3	25	2.66	A∪B∪$\{o_1\}$, C, D∪$\{o_5, o_6\}$	$\{o_2, o_3, o_4\}$
3	6	0	5.47	A, B, C∪D	E
3	6	20	2.84	A∪B, C, D	E
3	9	2	5.73	A\$\{a\}$, B,(C∪D) \F	E∪F∪$\{a\}$
3	9	6	2.80	(A\$\{a\}$)∪B, C\$\{c,d\}$, D	E∪$\{a, c, d\}$
4	3	4	4.41	A, B, C, D∪$\{o_4, o_5, o_6\}$	$\{o_1, o_2, o_3\}$
4	6	4	4.41	A, B, C, D	E
4	9	5	4.51	A\$\{a\}$, B, C\$\{c,d\}$, D	E∪$\{a, c, d\}$
5	0	0.3	7.64	A, B, C, D∪$\{o_5, o_6\}$, $\{o_1, \ldots, o_4\}$	none
5	0	6	5.36	A, B, C, D, E	none
5	6	6	4.41	A, B_1, B_2, C, D	E

Figure 4.4 resumes the simulated, clean data set of Fig. 1.3 with two clusters of ten points each, randomly sampled from the normal distributions N_{-2e_1, I_2} and N_{2e_1, I_2}, respectively (separated by the dashed line). Shown are all nine almost collinear "spurious clusters." The partitions defined by them all mask the original partition in two clusters, their negative posterior log-densities (2.33) falling below its value 65.96. However, the HDBT ratio (1.34) of the original partition is 1/1.69, whereas the largest of the spurious ones shown is 1/2 757 (the spurious cluster of five points). The optimal unconstrained solution uses the uppermost horizontal cluster and has a negative posterior log-density (2.33) of 60.95 but an HDBT ratio of 1/66 244.

Besides free steady solutions, also lower constraints on *cluster sizes* are sometimes useful for finding reasonable solutions in the context of the classification model; see Rocke and Woodruff [444]. An example appears in Section 6.1.1. As shown by Gallegos and Ritter [184], the constraints lead to the subject of Section 3.2.3: combinatorial optimization and λ-assignment. The method is effective but has the disadvantage of being unable to detect clusters smaller than the size constraint. Their elements will then appear in other clusters or among the outliers which must then receive special attention.

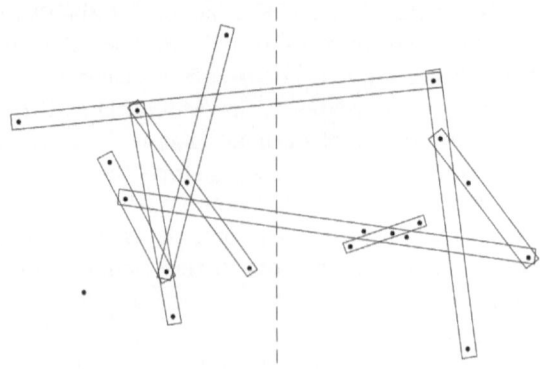

Figure 4.4: *The randomly sampled data set of Fig. 1.3 with all spurious clusters.*

4.2 Number of components of uncontaminated data

All previous partitioning methods need the, in general unknown, numbers r and g as computational parameters. Their estimation is therefore an important part of mixture and cluster analysis. We begin with the number of components. A general strategy for its estimation is to produce solutions for all reasonable numbers and to assess their quality. This requires an upper bound on the number since, otherwise, the preferred solution with the best fit and scale balance according to Section 4.1 above would consist of n singletons. The strategy of parameter variation is computationally expensive, but running the algorithm for various parameter settings is "embarrassingly parallel." It can be easily distributed on different processors, which greatly reduces computing time.

A crude indicator of the number of components is the number of modes, which is, however, not easily determined in a finite data set. Moreover, the number of modes does not have to equal the number of components even if clear modes can be determined; see Section 4.2.4. There are several approaches to the number of components of uncontaminated data, *statistical tests* such as the likelihood ratio test (Section 4.2.1) and the gap statistic (Section 4.2.2). Comparing the maximum likelihoods between solutions for different numbers of components does not make sense, since each additional component allows better fit and these values increase with g. A *model selection* criterion counteracts this tendency by subtracting from the maximum of the log-likelihood or from the posterior log-density a penalty term that increases with g (Section 4.2.3). A method with much authority is *cluster validation* (Section 4.4). A valid solution yields at the same time a number of components. One must be aware of the fact that the number of mixture components is sometimes ambiguous.

4.2.1 Likelihood-ratio tests

Wolfe [531] proposed a χ^2 likelihood-ratio test for testing a g-component mixture against $g' > g$ components. The distribution of the test statistic is not easily determined, since it involves optimal decompositions. Therefore, he used Aitchison and Silvey's [4] χ^2-approximation; see Section F.3.1. This is, however, not permissible, since one of the key assumptions of their theorem does not hold true – the parameter space of the smaller model is not a $\mathcal{C}^2$-submanifold of that of the larger one. This has often been criticized; see, for instance, Wolfe [532], Binder [43], Titterington et al. [505], Mclachlan [360], and McLachlan and Peel [366], p. 185. However, the distribution of the test statistic can be approximately estimated with a Monte Carlo method; see again McLachlan and Peel [364, 366]. A number of q data sets are sampled from the parameters of the estimated g-component mixture. The q values of the likelihood ratio test statistic are computed after fitting a g-component and a g'-component mixture to each of them. If q is large enough, this gives a fair impression of the null distribution. Smyth [478] notes that computation time can be substantially reduced by using q processors in parallel.

The task can be viewed as a nonparametric hypothesis testing problem in a statistical model not identifiable at the null distribution. Liu and Shao [332] propose a general approach to this type of problem. Their method establishes a quadratic approximation to general likelihood ratios in a Hellinger neighborhood of the true density function. In the case of (finite) normal mixtures, the asymptotic distribution of the likelihood ratio test statistic under the null is obtained by maximizing the quadratic form. Unfortunately, the authors do not present their method in implementable form.

Also, Wolfe's original idea of a χ^2-test can be rescued with a trick. Consider $g' = g+1$. If in doubt, we prefer g components to $g+1$; therefore the null hypothesis is g. Complete some point(s) close to the center of the largest cluster of the g-component solution to a small d-dimensional simplex (altogether $d+1$ points). This should be accomplished so as to create a solution with $g+1$ components close to the g-component solution up to an additional component with a small mixing rate created by the simplex. Call this solution a "fake g-component" mixture. Such a solution with $g+1$ components exists when the simplex is small enough. It is often not even necessary to add points. Thus, the extended data set yields two mixtures: the fake g-component mixture and a new estimated $g+1$-component mixture close to the original. They are now compared in the test. The parameter space with $g+1$ components but with the small component *fixed* becomes the null model. This approach makes it a differentiable submanifold of the $g+1$-component model, its dimension being that of the model with g components. Moreover, the fake solution is a local maximum in the null model, too. Thus, Aitchison and Silvey's [4] classical χ^2-approximation is applicable. In the normal case, the difference of the dimensions amounts to 1 (mixing rate) plus d (mean vector) plus $\binom{d+1}{2}$ (scale matrix), which is $\binom{d+2}{2}$. This number is independent of g. In short, the test statistic is

$2 \cdot$(local maximum for $g+1$ components $-$ local maximum for fake g components).

It is approximately distributed as $\chi^2_{\binom{d+2}{2}}$. This leads to a p-value. The estimated number of components is the smallest number g which is not rejected in favor of $g+1$. When fake solutions with large p-values only are found, there is evidence for g components. As a heuristic, the test is also applicable to the MAP version of the classification models, since criteria of steady solutions and local maxima of likelihood functions are similar; see Section 1.3.1. The fake solution is here an admissible, steady solution with a cluster of small size. It can be created with the TDC and combinatorial optimization as described in Section 3.2.3.

Table 4.2 offers an application of the test to the sample data set of Fig. 4.1. Up to $g=3$ it rejects g components in favor of $g+1$ with very small p-values. A fake solution which would reject four components in favor of five was not found, that is, there is no evidence for five components. The table contains in its last row a fake four-component solution close

Table 4.2: *The negative maximum log-likelihoods and p-values of the modified Wolfe test.*

g	negative log-likelihoods		2*difference	p-value
	fake g	$g+1$		
1	957.9	817.1	281.6	$7.14 \; 10^{-58}$
2	810.5	784.1	52.8	$1.29 \; 10^{-9}$
3	791.5	762.8	57.4	$1.52 \; 10^{-10}$
4	755.9	756.0	-0.2	1

to the true one in the original data set. The fact that its difference is negative has a simple explanation. Its small component is based on three elements taken from the big cluster, the mixing rate being 0.02439, and the solution does not satisfy the HDBT constraints. The likelihood is larger than that of the genuine, HDBT constrained five-component solution indicated in Fig. 4.1.

Lo et al. [334] proposed a different approach via the likelihood-ratio statistic based on the Kullback–Leibler divergence. Their method is based on White's [526] consistency for unspecified models. It uses the test statistic

$$\mathrm{LR}(\mathbf{x}; \vartheta_0, \vartheta_1) = \sum_{i=1}^{n} \log \frac{f(x_i; \vartheta_1)}{f(x_i; \vartheta_0)}.$$

Under compactness assumptions on the parameter spaces and differentiability assumptions on the likelihood functions, Vuong [517], Theorem 3.3(i), showed that $2 \cdot \mathrm{LR}$ is equivalent to a certain weighted sum of $q_0 + q_1$ independent χ^2_1 random variables, q_b being the dimension of ϑ_b. In the univariate case, the authors provide analytical formulæ for its tail distribution based on the weights. The expressions contain integrals to be evaluated numerically. Simulation studies suggest that the test performs well in the case of homoscedastic, univariate, normal mixtures.

4.2.2 Using cluster criteria as test statistics

Bock [50] proposes using cluster criteria as test statistics. A criterion for g clusters is expected to be smaller when data is actually sampled from a mixture of g clusters instead of a unimodal population. Therefore, the latter can serve as the null hypothesis for a test of unimodality vs. the alternative of heterogeneity. The p-value is the probability that the g-cluster criterion on unimodal data with the parameters of the present data is at least as small as the criterion on the data themselves. Using Pollard's [417, 418] asymptotic distribution of the means in k-means clustering, Bock applies the idea to Ward's minimum within-clusters sum-of-squares criterion. He computes the asymptotic distribution of the test statistic under the hypothesis of one component. In this context it is important that Pollard's result deals with *unspecified* parent distributions, since, otherwise, it could not be applied with $g > 1$ to homogeneous or unimodal data.

Bock's analytic approach is mainly applicable when clusters actually satisfy the assumptions underlying Ward's criterion, in particular equal sphericity; see Remark 1.42(b). The method is not equivariant w.r.t. variable scaling, since the criterion lacks this property. For criteria related to more general cluster scales, the distribution of the test statistic under the null model has not been computed yet. In a significance test of unimodality vs. heterogeneity, Monte Carlo simulation, too, can be used. The method has also the potential to distinguish between several numbers of clusters. The number g in some range $1..g_{\max}$ with the smallest p-value against unimodality is also a reasonable estimate of the number of clusters. Unfortunately, simulation does not help us distinguish between different g's. The p-values are in general much too small to be accessible by simulation.

This problem does not occur when the *expected* criterion under a null model is considered instead of a p-value. This was proposed by Tibshirani et al. [503]. They discuss a least favorable, single-component, log-concave null distribution, showing that it is uniform if the sample space is a (one-dimensional) bounded interval. In the multivariate case, such a distribution must be degenerate and is therefore not useful. For this reason they take a uniform (or Poisson) model ϱ on a small box that contains the data. Such a box is most easily constructed by rotating the data so that its principal axes are aligned with the coordinates. The aligned box is determined by the smallest and largest entries of the rotated data in each coordinate. The proposed null distribution is the back-rotation of the uniform on this box. Sampling a vector is done by sampling a vector in the aligned box and rotating back. If the cluster criterion is invariant w.r.t. rotations, then rotating back is unnecessary. The method is not affine equivariant in general, since linear transformations do not generally commute with the construction of the box. A two-dimensional example is provided by a data set with a diamond shaped convex hull (a square rotated by 45 degrees) and the linear transformation $\mathrm{Diag}(1, 2)$. Therefore, the normal distribution estimated from the data or the uniform distribution on an ellipsoid adapted to the data may be preferable.

Tibshirani et al. [503] reject their null model in favor of a g-component model if the evidence for g components is strongest, thereby protecting against erroneous rejection of the one-component model; see also Roeder [445]. Specifically, denote the criterion of the favorite partition of the data set $\mathbf{x}$ in g clusters by $K_g^*(\mathbf{x})$. The authors define the **gap statistic**

$$\mathrm{Gap}(g) = \mathrm{E}K_g^*(Y) - K_g^*(\mathbf{x}).$$

Here, $Y = (Y_1, \ldots, Y_n)$ is the random data set sampled from the null distribution ϱ. Thus, the gap statistic compares the criterion with its expectation under the null distribution ϱ.

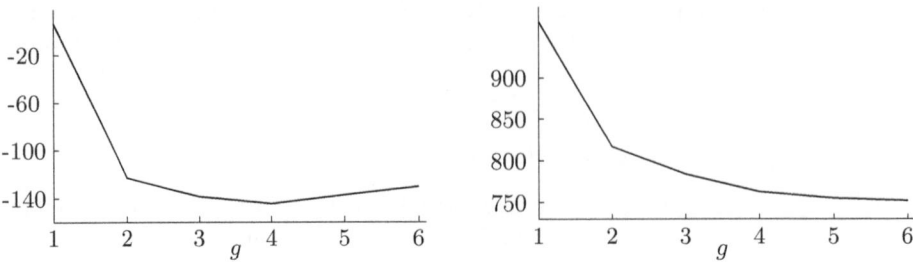

Figure 4.5: *The reduced gap statistic (left) and the elbow criterion.*

It is a Monte Carlo method applicable, for instance, to logarithmic criteria of classification models (Section 1.3) and to log-likelihoods of mixture models (Section 1.2).

4.5 Procedure (Gap statistic)
// Input: A data set $\mathbf{x}$ of length n, a number $g_{\max}$.
// Output: An estimate of the number of clusters of $\mathbf{x}$ up to $g_{\max}$.

1. Cluster the given data in $g \in 1..g_{\max}$ clusters and determine the criteria $K_g^*(\mathbf{x})$;

2. generate q reference data sets $\mathbf{y}_b$, $1 \leq b \leq q$, from ϱ;

3. cluster all reference data sets in $g \in 1..g_{\max}$ clusters and determine their criteria $K_g^*(\mathbf{y}_b)$, $1 \leq b \leq q$;

4. for all g, compute the standard deviations SD_g of the criteria $(K_g^*(\mathbf{y}_b))_{b=1}^q$ and define
$$s_g = \sqrt{1 + \tfrac{1}{q}}\,\mathrm{SD}_g;$$

5. approximate the expectations $EK_g^*(Y)$ in the definition of $\mathrm{Gap}(g)$ by the mean values $\frac{1}{q}\sum_{b=1}^q K_g^*(\mathbf{y}_b)$ to obtain the estimates $\mathrm{Gap}'(g)$;

6. finally, return the smallest g such that $\mathrm{Gap}'(g) \leq \mathrm{Gap}'(g+1) + s_{g+1}$ as the estimated number of clusters.

The numbers s_{g+1} account for the simulation error in the approximation $\frac{1}{q}\sum_{b=1}^q K_g^*(\mathbf{y}_b)$. The **reduced gap statistic** $\mathrm{Gap}'(g) + \sum_{k=2}^g s_k$ exhibits the location of its first local minimum as the estimated number of clusters.

A primitive version of the gap statistic is the traditional **elbow criterion**. This is a so-called **scree test** widely applied to estimate the number of clusters. It uses a plot of the criterion vs. the number g. The estimate is the number at the "elbow" of the curve, the location where it flattens "markedly."

Figure 4.5 illustrates the gap statistic and the elbow criterion with the sample data set of Fig. 4.1. The outliers are omitted. It shows the two curves for the favorite Pareto solutions obtained from the mixture model. The solutions merge the cluster pairs A, B and C, D ($g = 2$) and C, D ($g = 3$). For $g = 4$, the solution is the original structure. For $g = 5$ it essentially divides cluster B into an upper and a lower half and for $g = 6$ it additionally splits the subset F described in the caption of Table 4.1 off cluster C. All solutions are acceptable given their numbers of clusters. The gap statistic supports again the original number of clusters, four. The point where the elbow criterion flattens markedly is unclear, a major disadvantage of the method.

4.2.3   Model selection criteria

Model selection criteria are penalized MLE's designed to decide between one of several proposed statistical models. The penalty term is subtracted from the log-likelihood. It contains the model dimension, $\dim \Theta$, that is, the number of free model parameters. An early criterion was Akaike's [6] AIC, defined by

$$\mathrm{AIC}(\mathbf{x}) = \log f(\mathbf{x}; \vartheta^*) - \dim \Theta.$$

This is a maximum likelihood estimate with the additional term $-\dim \Theta$ to penalize large model dimensions. It requests a parsimonious use of parameters without losing too much fit. Applying the Bayesian paradigm, Schwarz [460] derived a selection criterion for the number of parameters in nested exponential models. His **Bayesian information criterion** (BIC) is an approximation to the logarithm of the likelihood integrated over the model parameters. It has again the form of a penalized maximum likelihood and reads

$$\mathrm{BIC}(\mathbf{x}) = \log f(\mathbf{x}; \vartheta^*) - \tfrac{\log n}{2} \dim \Theta. \qquad (4.2)$$

A similar earlier approximation goes back to Jeffreys [274], Chap. 5. Besides model dimension, the penalty term of BIC also depends on data size. Although it is Bayesian in nature, it has the advantage of not depending on a prior probability (although there is one in the background[4]). It is a "simple and convenient, albeit crude, approximation" (Raftery [425]) to the integrated likelihood. We should not rely solely on it when a wrong decision causes high cost. Since $\tfrac{\log n}{2} > 1$ for $n \geq 8$, BIC tends to yield smaller estimates than AIC does.

Schwarz showed that BIC consistently estimates the dimension of exponential models. Kass and Wasserman [286] showed that, under special assumptions on the prior probabilities, the rate of approximation is $\mathcal{O}(n^{-1/2})$. It later turned out that the precise form of the penalty term is not important, asymptotically. In fact, Nishi [394] proved for more general models under certain regularity conditions that the more general model selection criterion

$$\mathrm{MSC}(\mathbf{x}) = \log f(\mathbf{x}; \vartheta^*) - c_n \dim \Theta$$

leads to a strongly consistent model selector for a wide variety of sequences c_n. It is sufficient that $\lim_{n\to\infty} c_n / \log \log n = \infty$ and $\lim_{n\to\infty} c_n / n = 0$. A special example is BIC, whereas AIC is not. All criteria have the drawback of failing to yield a score such as a p-value that quantifies the confidence in the result. A difference of ten or more is in general considered significant for BIC.

Model selection has been shown to be applicable to the estimation of the number of components in mixture analysis, the different models being determined by their numbers of components. Under certain regularity conditions on class-conditional populations and compactness constraints on the parameters, Keribin [289, 290] proved consistency for mixtures with a penalty term similar to Nishi's. Her result is applicable, for instance, to Gaussian families if the mean values are bounded and if the covariance matrices are Löwner bounded away from the zero matrix. Her criterion assumes the form

$$\mathrm{MSC}(\mathbf{x}, g) = \log f(\mathbf{x}; \vartheta^*) - c_n s_g,$$

where $\lim_{n\to\infty} c_n = \infty$, $\lim_{n\to\infty} c_n / n = 0$, and where s_g strictly increases with g.[5] However, when clusters are not well separated, it needs a large data set. As in Nishi's result above, the conditions to be met by c_n and s_g are quite unspecific. The special case with $c_n = (\log n)/2$ and $s_g = \dim \Theta_g$ is again denoted by $\mathrm{BIC}(\mathbf{x}, g)$. Model dimension $\dim \Theta = \dim \Theta_g$ equals here $g-1$ (for the mixing rates) plus the total number of (independent, real) parameters of the g components. Also $\mathrm{AIC}(\mathbf{x}, g)$ with $c_n = 1$ and the same s_g has been proposed for this purpose, but $c_n = 1$ does not satisfy Keribin's first condition. Both model selection criteria AIC and BIC have been extensively studied with simulated data. BIC (AIC) is reported to be negatively (positively) biased; see McLachlan and Peel [366], Ch. 6, and Yang and Yang [536], who conclude that BIC performs well unless sample size is small or separation is poor. Unfortunately, these are the circumstances when a formal decision would be most welcome. Steele and Raftery [485] report on the basis of extensive simulations that BIC,

---

[4]Readers interested in prior distributions are invited to consult Steele and Raftery [485].
[5]Instead of a product $c_n s_g$ she actually uses a coefficient $a_{n,g}$. But this generality is hardly needed.

with the fully Bayesian method of Stephens [487], outperforms five other model selection criteria, among them AIC.

Model selection is applicable to classification models as well if there is sufficient separation. A prototype can be derived from Theorem 1.56. We return to the notation and the drop points defined in Section 1.3.7. Examples 1.59(a),(c) indicate that the number of components of a clearly separated, normal mixture is usually a drop point. But the last paragraph of Example 1.59(b) shows that not every drop point can be considered a valid number of components. Given an upper bound $g_{\max} \geq 1$, the largest (HDBT constrained) drop point g^* up to $g_{\max}$ may often be taken to mean the "number of components" that make up μ. It is the size of the minimizer of the population criterion (1.47), Φ, on $\Theta_{\leq g_{\max}, c}$ defined in (1.55). In this sense, the task is to recognize this drop point from the given sample. Drop points are accompanied by **sample drop points** g of the sampling criterion (1.46), Φ_n, defined by

$$\inf_{A \in \Theta_{\leq g, c}} \Phi_n(A) < \inf_{A \in \Theta_{\leq g-1, c}} \Phi_n(A).$$

Also $g = 1$ is a sample drop point. Because of random fluctuation of the data, a population drop point may not be a sample drop point and vice versa. The following corollary of Theorem 1.56 says that a small (but unknown) distortion of Φ_n converts the largest sample drop point up to $g_{\max}$ to the largest drop point g^* there, at least asymptotically. Denote the sequence of (population) drop points by $1 = g_1 < g_2 < \cdots$.

4.6 Corollary *Let $g_{\max} \geq 2$ and let $(s_1, \ldots, s_{g_{\max}})$ be any strictly increasing $g_{\max}$-tuple of real numbers. If the assumptions of Theorem 1.56 are satisfied, then there exists $\varepsilon_0 > 0$ such that, for any $0 < \varepsilon \leq \varepsilon_0$, the maximum (population) drop point up to $g_{\max}$ is given by*

$$\operatorname*{argmin}_{1 \leq g \leq g_{\max}} \left(\min_{A \in \Theta_{\leq g, c}} \Phi_n(A) + \varepsilon s_g \right),$$

for eventually all n.

PROOF. Write $g^* = \max_{g_j \leq g_{\max}} g_j$ and $h_g = \min_{A \in \Theta_{\leq g, c}} \Phi(A)$. We have $h_1 \geq \cdots \geq h_{g^*-1} > h_{g^*} = h_{g^*+1} = \cdots = h_{g_{\max}}$. If $g^* = 1$, we choose an arbitrary $\varepsilon_0 > 0$. If $g^* > 1$, then there is $\varepsilon_0 > 0$ such that

$$h_g + \varepsilon_0 s_g > h_{g^*} + \varepsilon_0 s_{g^*}$$

for all $1 \leq g < g^*$. This relation continues to hold for all strictly positive $\varepsilon \leq \varepsilon_0$ and, of course, for $g > g^*$, too. The corollary now follows from $\min_{A \in \Theta_{\leq g, c}} \Phi_n(A) \to h_g$ as $n \to \infty$; see Lemma 1.54(b).   □

We have again obtained a *penalized* (MAP) sampling criterion. The proof shows that the penalty term is needed because of the random fluctuation of the nth minimum $\min_{A \in \Theta_{\leq g, c}} \Phi_n(A)$ for g between g^* and $g_{\max}$. Being defined by the population criterion Φ, the point g^* is an asymptotic quantity that also depends on the choice of the HDBT constant c. If the HDBT constraints were replaced with minimal size constraints (see Section 3.2.3, then, beyond g^*, the sampling criterion would usually split some cluster or split off clusters of small or even deficient size, producing spurious solutions.

Sample drop points are related to the elbow criterion (Section 4.2.2). According to the corollary, adding a penalty term εs_g with a suitable small ε to the elbow depicted in Fig. 4.5 causes the curve to increase again from g^* on. In the notation of MSC,

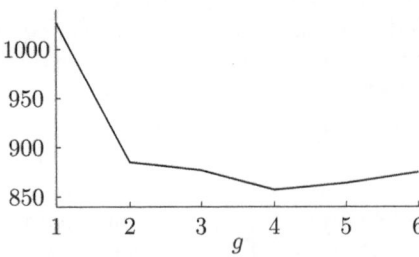

Figure 4.6: *The BIC curve.*

where $\log f(\mathbf{x}; \vartheta^*)$ is not divided by n, we have obtained $c_n = \varepsilon n$. In classification, the decrease to zero of c_n/n is therefore not needed; it has just to be small enough.

Figure 4.6 shows the BIC curve obtained with the trimmed classification model (TDC) for the sample data set of Fig. 4.1. Its minimum is indeed located at the original number of components, four.

4.2.4 Ridgeline manifold

Let g^* be the number of clusters of the parent population. The algorithms of Chapter 3 with a number smaller than g^* usually merge some close clusters. If we prescribe a number larger than g^*, then true clusters are usually split; often small clusters are split off. This section deals with, among others, the question of how to detect split and merged clusters.

Ray and Lindsay [432] associate with any normal mixture a low-dimensional (nonlinear) manifold that provides insight into its characteristics. The manifold facilitates the analysis of major characteristics such as the modal structure. Let $\boldsymbol{\pi} \in \Delta_{g-1}$, $m_j \in \mathbb{R}^d$, $V_j \in \mathrm{PD}(d)$ be fixed, abbreviate $\beta_j(x) = \sqrt{\det(\Lambda_j/2\pi)}e^{-\frac{1}{2}(x-m_j)^\top \Lambda_j(x-m_j)}$, and let

$$f(x) = f(x; \boldsymbol{\pi}, \mathbf{m}, \mathbf{V}) = \sum_{j=1}^{g} \pi_j \beta_j(x)$$

be the normal mixture with parameters $\boldsymbol{\pi} \in \Delta_{g-1}$, $\mathbf{m}$, and $\mathbf{V}$. Ray and Lindsay study the "topography" of the mixture by its (differentiable) **ridgeline function** $h : \Delta_{g-1} \to \mathbb{R}^d$,

$$h(\boldsymbol{\alpha}) = \Big(\sum_{j=1}^{g} \alpha_j \Lambda_j\Big)^{-1}\Big(\sum_{j=1}^{g} \alpha_j \Lambda_j m_j\Big), \quad \boldsymbol{\alpha} \in \Delta_{g-1}. \tag{4.3}$$

It depends on the parameters $\mathbf{m}$ and $\mathbf{V}$ of the mixture, but not on $\boldsymbol{\pi}$. Plainly, the vertices of the simplex are mapped to the mean values m_j. The (compact) image $\mathcal{M}$ of h is called the **ridgeline manifold**.[6] For $g = 2$ it is a curve connecting m_1 to m_2.

Recall that a critical point of a differentiable function is a point where the gradient vanishes. Local maxima, local minima, and saddle points in $\mathbb{R}^d$ are all critical. Lagrange interpolation theory teaches that any finite number of points in $\mathbb{R}^d$ can be joined with a polynomial. It could be applied to the critical points of a normal mixture, however, it would need knowledge of these points. The following theorem due to Ray and Lindsay [432] shows that the ridgeline manifold, which does not need this knowledge, also interpolates all critical points.

4.7 Theorem *The ridgeline manifold contains all critical points of f.*

PROOF. Let x^* be critical, that is, $\mathrm{D}f(x^*) = 0$ and define

$$\alpha_j^* = \frac{\pi_j \beta_j(x^*)}{f(x^*)}.$$

It is clear that $\boldsymbol{\alpha}^* \in \Delta_{g-1}$. We show that $h(\boldsymbol{\alpha}^*) = x^*$. Indeed,

$$0 = \mathrm{D}f(x^*)^\top = \sum_j \pi_j \beta_j(x^*)\Lambda_j(x^* - m_j) = f(x^*)\sum_j \alpha_j^* \Lambda_j(x^* - m_j)$$

$$= f(x^*)\Big(\sum_j \alpha_j^* \Lambda_j x^* - \sum_j \alpha_j^* \Lambda_j m_j\Big) = f(x^*)\Big(\sum_j \alpha_j^* \Lambda_j x^* - \sum_j \alpha_j^* \Lambda_j h(\boldsymbol{\alpha}^*)\Big).$$

This is the claim. $\qquad\square$

If dimension d is large and g is small, then the ridgeline manifold facilitates the search of critical points and of local maxima. If $g = 2$, then it even reduces the problem of analyzing local maxima from d dimensions down to one. In the homoscedastic case we have $h(\alpha) = \sum_j \alpha_j m_j$, that is, h is linear. Thus, the ridgeline manifold is just the simplex in $\mathbb{R}^d$ with vertices m_j, the convex hull of the set of mean vectors $\{m_1, \ldots, m_g\}$.

---

[6]The term "ridgeline" is adopted from the language of the earth's topography.

4.8 Corollary *In the homoscedastic case, the convex hull of the mean vectors m_j contains all critical points of f.*

Theorem and corollary may be extended to elliptical mixtures with differentiable density generators φ such that $\varphi' < 0$. Just put $\beta_j(x) = \sqrt{\det \Lambda_j}\, \varphi((x - m_j)^\top \Lambda_j (x - m_j))$ and define α_j^* properly.

If some means of a normal mixture of $g \geq 2$ components are close, then the mixture may have less than g modes. A *univariate* normal mixture of $g \geq 2$ components always has at most g modes. Carreira-Perpiñán and Williams [81] noticed that this statement loses its validity in the multivariate context. A *multivariate* mixture of $g \geq 2$ normals may have more than g modes. It is sufficient to superpose two elongated, perpendicular normal distributions on their long flanks. There will be three modes, two near the distribution centers and one near the center of the superposition; see Fig. 4.7. Similarly, a g-component mixture with $2g-1$ modes can be constructed. Hence, there is no one-to-one correspondence of modes and normal components (causes). Ray and Ren [433] show that a mixture of two d-dimensional normals can have at most $d+1$ modes. A study of the modes of elliptical mixtures appears in Alexandrovich et al. [8].

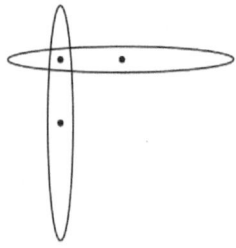

Figure 4.7 *A normal, two-component mixture with three modes.*

In their Discussion, Ray and Lindsay [432] indicate some potential applications of the ridgeline. It is known that *any* (absolutely continuous, say) distribution can be approximated arbitrarily closely by a finite Gaussian mixture. When some of the parent components are not close to normal, it will happen that the estimated mixture tries to fit each nonnormal component with normal ones. This results in too wmany estimated components, clusters, and causes. In this case, two or more of the estimated components will be so close that the associated subsets created by the Bayesian discriminant rule, $\ell_x = \mathrm{argmax}_j\, \pi_j N_{m_j, V_j}(x)$, are not sufficiently isolated to be perceived as distinct clusters. The subsets should rather be merged to form a single cluster. Hennig [244] picked up the idea, proposing several methods for this purpose based on local maxima. His *ridgeline ratio method* postulates that the clusters of two components should be merged if either the mixture density is unimodal or if the ratio between the minimum of the density along the ridgeline and its second local maximum exceeds some value to be specified in advance; Hennig proposes 0.2. This can be easily tested by inserting the ridgeline function (4.3) into the mixture density.

Let us apply Hennig's method to the sample data set of Fig. 4.1. Trying to fit a mixture of five components, one obtains three components corresponding to the original clusters A, C, and D. Cluster B is divided in two parts, an upper, B_1, and a lower, B_2. The reason is that cluster B is unusually thin about its center. Figure 4.8 shows the density of the mixture of the two components along the ridgeline. It has two local maxima close to $\alpha = 0$ and $\alpha = 1$, respectively. The ridgeline ratio is 0.56 and so the method favors merging the subsets defined by the two components, here a sensible decision.

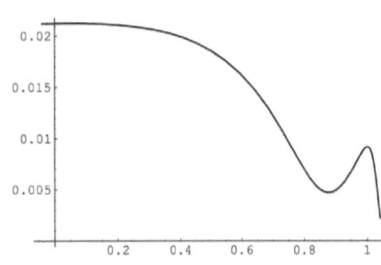

Figure 4.8 *The mixture density along the ridgeline.*

It is often very difficult, if not impossible, to tell a broken or bent cluster from data sampled from two overlapping unimodal populations. A similar problem arises when two clusters heavily overlap in the form of an X. The method may have the unintended effect of merging two clusters from different sources. Therefore a closer analysis of the situation with additional information is necessary in doubtful cases.

4.3 Number of components and outliers

This section presents some ad hoc methods for simultaneously selecting the numbers of components and outliers. In contrast to Section 4.2, we will now vary not only the number of groups, g, but also the number of retained elements, r. The methods presented here are heuristic, such as the classification trimmed likelihood and the trimmed BIC. They work when the true clusters are well separated but fail in more complex situations, where they yield unclear results.

4.3.1 Classification trimmed likelihood curves

A very simple approach to the number of clusters and outliers is due to García-Escudero et al. [189], the **classification trimmed likelihood** (CTL). Assume that the true number of clusters is g^* and that the data set contains a^* outliers. As long as $n - r < a^*$ and $g \leq g^*$,

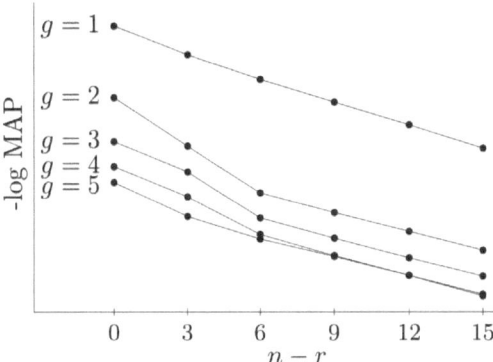

the solution has to accommodate some outliers in the clusters *or* to merge clusters and to create clusters of outliers. Therefore, the fit cannot be good, and the criterion has a poor value. Choosing $g > g^*$, say $g = g^* + 1$, allows the surplus outliers to create a cluster of their own while keeping the true ones. This means better fit and a larger value of the criterion. Therefore, the difference between the two criteria is large. As we increase $n - r$ to a^*, the outliers will be discarded but, if $g = g^* + 1$, the solution will essentially be formed of the g^* true clusters except for a cluster that is divided in two; sometimes a small cluster is split off. The criterion is not far away from

Figure 4.9 *Sample data set of Fig. 4.1: Classification trimmed likelihood curves.*

that of the true solution. That is, in a plot of the criterion against $n - r$ for various values of g, the two curves for $g = g^*$ and $g = g^* + 1$ almost flow together at the value a^* and the point of confluence indicates both values g^* and a^*.

The method is illustrated in Fig. 4.9 with the sample data set of Fig. 4.1. The favorite solution for $g = 4$ and $n - r = 3$ has a negative log-MAP criterion of 812; for $n - r = 6$ it is 763. The values for $g = 5$ are 786 and 760, respectively. The curves for $g = g^* = 4$ and $g = 5$ are thus almost confluent at the value $n - r = 6$ and the CTL favors the correct solution with four clusters and six outliers.

4.3.2 Trimmed BIC

Another way of estimating both numbers at the same time was proposed by Neykov et al. [389], the **trimmed BIC**. The method uses a table of BIC values for solutions indexed by g and $n - r$ and returns the parameter values where the minima w.r.t. g stabilize. In order to simplify the procedure, it is sufficient to consider a lacunary set of values $n - r$. The idea underlying the method is that the positions of the minimizers w.r.t. g in each column stabilize as more elements are discarded as soon as the number of discarded elements equals that of the outliers in the data set. As soon as the outliers are removed from the data the number of components is expected to remain stable; the clusters just lose some of their extreme elements. The solutions represented by the entries of the table must be carefully chosen, for instance, by the method of Section 4.1. Neykov et al. [389] use the median w.r.t. many runs of the algorithm. This may be problematic, since the influence of unreasonable, spurious, solutions remains.

In Table 4.3 shows the trimmed BIC values of the favorite Pareto solutions obtained

Table 4.3 *Sample data set of Fig. 4.1: Table of trimmed BIC values for various numbers of clusters and discards (TDC). The boldface numbers mark the columnwise minima. They stabilize at the values $g = 4$ and $n - r = 6$.*

$n - r$	0	3	6	9	12	15
g						
1	1050	1012	979	949	919	888
2	970	906	843	817	792	767
3	926	886	825	797	772	747
4	908	867	**818**	**789**	**762**	**738**
5	**901**	**860**	825	796	771	746

with the TDC for the data set of Fig. 4.1. The heights of the (boldface) minima w.r.t. g stabilize at $g = 4$ and the point where this occurs is at $n - r = 6$. The method therefore favors the correct solution.

4.3.3   Adjusted BIC

The **adjusted BIC** (Gallegos and Ritter [183, 184]) is another method for simultaneously estimating the numbers of groups and outliers based on the model selection criterion BIC. It combines an approach to estimating the number of clusters with a validation method; see Section 4.4. It is presented here for the classification model but it can as well be applied to the mixture model. In a first step, establish a table of the optimal clusterings for all (reasonable) numbers of clusters, g, and all numbers of discards, $n - r$. Again a lacunary set of values of $n - r$ is sufficient. Next, reduce the number of possible solutions by validating their regular parts; see Section 4.4 below. If g admits an acceptable pair $(g, n - r)$, keep the one with maximum r $(=: r_g)$ as a candidate. After having run through all values of g up to a maximum, at most one pair is left in each row. It remains to choose the favorite g. Since the estimated numbers r_g of regular observations depend on g, they have to be normalized, for instance, to n. By consistency of parameter estimation, Chapter 1, the value of the MAP criterion (1.38) increases approximately linearly with the number r, asymptotically, at least if there is sufficient separation. Therefore, it is reasonable to adapt the MAP part in the BIC with the factor n/r_g. The estimated number of clusters is the maximizer w.r.t. g of the *adjusted BIC*,

$$-n\,\mathrm{H}\!\left(\frac{n_1(\boldsymbol{\ell}^*)}{r_g}, \dots, \frac{n_g(\boldsymbol{\ell}^*)}{r_g}\right) + \frac{n}{r_g}\sum_{j=1}^{g}\sum_{\ell_i^*=j} \log f_{\gamma_j(\boldsymbol{\ell}^*)}(x_i) - \frac{q(g)}{2}\log n. \qquad (4.4)$$

The number $q(g)$ is the dimension of the associated mixture model with g components.

An application of the method to the sample data set of Fig. 4.1 is shown in Table 4.4. The entries "yes" indicate that a normality test applied to the clusters has accepted the solution, whereas "no" means rejection. It turns out that the rows for $g = 4$ and $g = 5$

Table 4.4: *Sample data set of Fig. 4.1: Adjusted BIC.*

$n - r$	0	3	6	9	12	15
g						
1	no	no	no	no	no	no
2	no	no	no	no	no	no
3	no	no	no	no	no	no
4	no	no	yes	yes	yes	yes
5	no	no	yes	yes	yes	yes

contain acceptable solutions. The other values of g have to merge different clusters which the validation method does not accept. The leftmost acceptable solutions are $(g, n - r) = (4, 6)$ and $(g, n - r) = (5, 6)$, the original clustering and a solution which discards the six outliers and splits one cluster into two parts. The values of the adjusted BIC for these two solutions are -856.4 and -863.0, respectively. Thus, the adjusted BIC pleads for the solution with four clusters and six outliers. Here, $r_4 = r_5$; examples more characteristic have been treated in Gallegos and Ritter [183].

4.4 Cluster validation

In clear cases, the methods of the previous sections allow us to decide for a single solution, including the number of components and outliers. In others there remain several solutions, possibly with different numbers of clusters and outliers. This uncertainty is intrinsic to mixture and cluster analysis and only partly a consequence of insufficiencies of the data set. Clustering algorithms have a tendency to partition data sets even if they are samples from a unimodal, for instance, normal or uniform, population. They can reveal true clusters but also create artificial structure. They are just eager to find fractures in data which they can exploit for decomposition. They may split a cluster or also merge two clusters, in particular in subtle situations. The methods of Section 4.1 provide first insights into these issues. However, blind faith in the results of clustering methods is not advisable. It is eventually safer to examine a proposed partition for validity.

Unless the components of a mixture heavily overlap, we expect a sample to be composed of subsets that meet the heuristic requirements of external *isolation* and internal *cohesion*; see Cormack [101]. A paraphrase of isolation is *separation* between the subset and its complement. These subsets are the witnesses of the mixture components. It is the idea underlying cluster analysis to retrieve them as the clusters of the estimated partition. The tools are criteria that rate all admissible partitions in view of this goal. Unfortunately, there is no generally accepted, formal and mathematical definition of what "cohesion" and "separation" should mean.

There is a family resemblance to the concept of connectedness in mathematical topology but this is not directly applicable here. McQuitty [368, 369, 370] defined a cluster ("type" in his terminology) as a subset $C \subseteq \mathbf{x}$ such that the dissimilarity of each $x \in C$ to any point in C is less than its dissimilarity to any point in the complement of C. A similar definition is due to van Rijsbergen [436]. In his terminology a subset satisfies the L_1-condition if its diameter is less than the distance to its complement. From a formal point of view these definitions do probably not correspond to what the authors had in mind. Consider the three sets $A = \{0, 1, 2, 3, 4\}$, $B = \{10, 11, 12, 13, 14\}$, and $C = \{30, 31, 32, 33, 34\}$ with the usual distance as dissimilarity. In the strict sense of their definitions, $A \cup B$ would be a cluster in the union $A \cup B \cup C$. On the other hand, both A and B would be clusters. Hence, A and B would at the same time be clusters and subsets of a cluster. What their definition lacks is the property of cohesion. It could be added to their definitions that a cluster should be minimal with the stated property. But then we would get singletons, since they, too, satisfy their definitions. Also adding the requirement that there should be at least *two* points in each cluster gets us nowhere, since the two points with minimal dissimilarity would then always be a cluster. Again, this is not always desired. It does not make sense to continue this discussion any further. These are attempts to define the notion of an *ideal* cluster. Unfortunately, such attempts fall short of real-world examples. The definitions are too restrictive, since clusters are rarely so well separated and often even overlap.

The problem of assessing the validity of the few potential solutions remaining after the methods of Section 4.1 is more demanding than that of reducing the 10^{50}, say, potential clusterings present at the beginning of an analysis to a few. The task is in fact often frustrating, like removing the last typos from a long text. A myriad of proposals is found in the literature. In order to avoid boundless discussions, matters are here restricted to methods that are at least invariant w.r.t. variable scaling, that is, yield the same result before and after the transformation. This excludes all methods based on ad hoc metrics such as the Euclidean or the Manhattan metric. These work well when they happen to be in harmony with the scales of the parent. They mainly say that near data points should belong to the same cluster while distant ones should not. This is rather the perspective of *optimal vector quantization*; see Graf and Luschgy [212]. However, real data bears intrinsic metrics of its own. It does not have to stick to any prior metrics.

There are three different types of validation methods: *external*, *internal*, and *relative*; see Jain and Dubes [267], Chapter 4. *External* methods compare a classification with information not used in constructing the partition. For instance, comparing the result of a clustering algorithm with the true labeling known in a simulation study or obtained via a "gold standard" will provide insight into its *performance*. Internal methods evaluate a mixture or partition in relation to the given data, thus assessing its quality. Relative methods compare the results of different classification methods on the same data and are useful in both respects.

Estimates in mixture analysis are usually point estimates. Such estimates must be accompanied by confidence regions which quantify their accuracy; see, for instance, Barndorff-Nielsen and Cox [25], p. 301. In cluster analysis, confidence regions are usually replaced by cluster validation. This section describes mainly internal ways of assessing the validity of estimated (or otherwise given) mixtures, clusters, and partitions. They are based on clusters and not on mixtures. Some methods appearing in the literature concern the validity of a single cluster, others the validity of a whole partition. In designing clustering algorithms, it is reasonable to deal with *all* data at once because the assignment of one data point may affect all others. Validation is different. Here we are given a specific partition and ask whether it is reasonable. Validating an entire partition may not give a fair assessment of its quality, since it may contain some very distinct clusters and others of lower quality. It therefore makes sense to approach the cohesion problem *clusterwise* and the isolation problem *pairwise*; see, for instance, Dubes and Jain [132], Bertrand and Bel Mufti [38], and Hennig [242]. The assumptions that underlie an assessment should be independent of the partitioning method. Some interesting methods that comply with this requirement have been proposed in the past decades. They rest mainly on counting and combinatorial arguments.

Internal validation methods can be classified as *statistical tests* or *heuristic*. Although cohesion is different from normality, normality tests may sometimes be used in testing cohesion. This will be discussed in Section 4.4.2. Heuristic methods use test statistics based on cohesion and isolation but abstain from formulating a null hypothesis or distribution. An example is the separation index of Section 4.4.1. Visualization, too, uses test statistics but these are subjective and not even clearly specified; see Section 4.4.3. Stability analysis in Section 4.4.5 compares the clustering results of different subsets of the data.

Since validation methods are higher authorities used to assess the results of clustering algorithms, the reader may wonder why they are not themselves used for the purpose of clustering. The reason is simple: They do not lead to computationally efficient algorithms.

4.4.1 Separation indices

The dissimilarity of the sampling distributions of two clusters w.r.t. some metric or other distance measure is useful in order to assess their separation. Examples are the symmetrized Kullback–Leibler divergence D_{KL}, the negative logarithm of the Hellinger integral, and the Hellinger distance d_H; see Appendix D.5.2. Although they do not provide p-values, they can be employed to compare different partitions.

The dissimilarities of two normal populations with densities $f = N_{m_1, V_1}$ and $g = N_{m_2, V_2}$

can be computed from their parameters. Their symmetrized Kullback–Leibler divergence is

$$D_{\text{sym}}(f,g) = \tfrac{1}{2}\text{tr}\left(V_1^{-1} - V_2^{-1}\right)(V_2 - V_1) + \tfrac{1}{2}(m_2 - m_1)^\top\left(V_1^{-1} + V_2^{-1}\right)(m_2 - m_1),$$

the negative logarithm of the Hellinger integral becomes

$$\tfrac{1}{2}\log\det\tfrac{1}{2}(V_1 + V_2) - \tfrac{1}{4}\log\det(V_1 V_2) + \tfrac{1}{4}(m_2 - m_1)^\top(V_1 + V_2)^{-1}(m_2 - m_1),$$

and their Hellinger distance square has the representation

$$d_H^2(f,g) = 2\left(1 - \sqrt{\frac{\sqrt{\det(V_1 V_2)}}{\det\tfrac{1}{2}(V_1 + V_2)}}\, e^{-\frac{1}{4}(m_2 - m_1)^\top(V_1 + V_2)^{-1}(m_2 - m_1)}\right).$$

The above dissimilarities of two normal populations do not vanish when the two mean values are equal as long as the scale matrices are different. It is often desirable that the components of a mixture be separated merely by *location*. This can be verified by projecting the data to an appropriate straight line. The first idea that comes to mind is Fisher's [156] linear discriminant rule; see also Mardia et al. [345]. Passing from populations to data, let $\overline{x}_1$ and $\overline{x}_2$ be the mean values of two clusters C_1 and C_2, respectively, and let

$$W = \sum_{x \in C_1}(x - \overline{x}_1)(x - \overline{x}_1)^\top + \sum_{x \in C_2}(x - \overline{x}_2)(x - \overline{x}_2)^\top$$

be their pooled within-groups SSP matrix. Fisher's linear discriminant rule assigns an observation x to class 1 if and only if

$$(\overline{x}_2 - \overline{x}_1)^\top W^{-1}\left(x - \tfrac{1}{2}(\overline{x}_1 + \overline{x}_2)\right) \leq 0.$$

When data is drawn from two normal populations $N_{m_1,V}$ and $N_{m_2,V}$ with *common* covariance matrix V, it is just the ML discriminant rule. The related MAP discriminant rule reads in this case

$$(\overline{x}_2 - \overline{x}_1)^\top W^{-1}\left(x - \tfrac{1}{2}(\overline{x}_1 + \overline{x}_2)\right) \leq \log\tfrac{p_1}{p_2}.$$

Projection to a line in the direction of the vector $W^{-1}(\overline{x}_2 - \overline{x}_1)$ exhibits the separation between the two clusters.

When the covariance matrices are different, the rule can still be applied but is no longer optimal. Fisher did not primarily design his method in view of a separation index and the direction $W^{-1}(\overline{x}_2 - \overline{x}_1)$ can be improved in this case. The rest of this section is devoted to the definition of an affine invariant index of separation by location of two normal distributions independent of prior probabilities. It will also be used for the purpose of visualization in the next section. Before introducing the index, we first take a look at *elliptical* separation – for algorithmic reasons. Given a normal distribution $N_{m,V}$, let us call its p-**ellipsoid** the ellipsoid of the form

$$(x - m)^\top V^{-1}(x - m) = t \tag{4.5}$$

which encloses $N_{m,V}$-probability $p < 1$. By the transformation theorem, the p-ellipsoid of the affine transform of a normal distribution is the transform of its p-ellipsoid. We are interested in the largest p such that the p-ellipsoids of two normal distributions just touch (at one point) from their exterior. It exists whenever the centers of the two ellipsoids are distinct. In this case, the p-ellipsoid of each distribution is contained in the complement of the p-ellipsoid of the other and p is maximal with this property. The value of p does not change with affine transformations. It will be shown below that the normal vector to the two ellipsoids at the touching point defines a hyperplane that optimally displays the linear separation of the two populations.

4.9 Lemma (Elliptical separation) *Consider two normal distributions N_{m_1,V_1}, N_{m_2,V_2} on $\mathbb{R}^d$ with mean vectors $m_1 \neq m_2$. Let $m = V_1^{-1/2}(m_2 - m_1)$ and $V = V_1^{-1/2}V_2 V_1^{-1/2}$.*

(a) *The equation*

$$\|(I_d + \gamma V)^{-1}m\|^2 = \|V^{1/2}(I_d/\gamma + V)^{-1}m\|^2 \tag{4.6}$$

for $\gamma > 0$ *has a unique solution* γ^*.

(b) *Let* t^* *be the value of* (4.6) *for* $\gamma = \gamma^*$. *The common probability of the touching p-ellipsoids is* $p^* = P[\chi_d^2 \le t^*]$.

(c) *The touching point of the* p^*-*ellipsoids is*

$$\left(\gamma^* V_1^{-1} + V_2^{-1}\right)^{-1}\left(\gamma^* V_1^{-1}m_1 + V_2^{-1}m_2\right).$$

(d) *The normal vector to the* p^*-*ellipsoids at their touching point is the normalization of* $(V_1 + \gamma^* V_2)^{-1}(m_2 - m_1)$.

PROOF. It is comfortable to first sphere one of the distributions. We take the first, that is, we perform the transformation $x = \sqrt{V_1}y + m_1$. The second distribution becomes the normal with parameters m and V and it is sufficient to deal with the p-ellipsoids of the standard multivariate normal and of $N_{m,V}$. If they touch from the exterior at a point y_0, the normal vectors to the sphere and to the ellipsoid at y_0 point in opposite directions. The outward normal vector to the ellipsoid (4.5) at y is $V^{-1}(y - m)$. We infer $-\gamma y_0 = V^{-1}(y_0 - m)$ and $y_0 = (I_d + \gamma V)^{-1}m$ for some $\gamma > 0$; hence

$$m - y_0 = (I_d - (I_d + \gamma V)^{-1})m \quad \text{and} \quad m - y_0 = \gamma V y_0 = V(I_d/\gamma + V)^{-1}m.$$

Since the two probabilities are supposed to be equal, this implies

$$P[\chi_d^2 \le \|(I_d + \gamma V)^{-1}m\|^2] = P[\chi_d^2 \le \|y_0\|^2] = P[\chi_d^2 \le \|V^{-1/2}(m - y_0)\|^2]$$
$$= P[\chi_d^2 \le \|V^{1/2}(I_d/\gamma + V)^{-1}m\|^2],$$

that is, (4.6). The solution to this equation is unique, since, by $m \ne 0$, its left side decreases to zero as $\gamma \to \infty$ and its right side decreases to zero as $\gamma \to 0$. This proves parts (a) and (b).

Transforming back shows that the touching point is

$$x_0 = \sqrt{V_1}y_0 + m_1 = \sqrt{V_1}(I_d + \gamma^* V)m + m_1$$
$$= \sqrt{V_1}\left(I_d + \gamma^* \sqrt{\Lambda_1}V_2\sqrt{\Lambda_1}\right)^{-1}\sqrt{\Lambda_1}(m_2 - m_1) + m_1$$
$$= (I_d + \gamma^* V_2\Lambda_1)^{-1}(m_2 - m_1) + m_1 = (I_d + \gamma^* V_2\Lambda_1)^{-1}(m_2 + \gamma^* V_2\Lambda_1 m_1).$$

This is the expression (c). The normal vector at this point requested in (d) is, up to normalization,

$$n_x = V_1^{-1}(x_0 - m_1) = V_1^{-1/2}y_0 = V_1^{-1/2}(I_d + \gamma^* V)^{-1}m$$
$$= V_1^{-1/2}(I_d + \gamma^* V_1^{-1/2}V_2 V_1^{-1/2})^{-1}V_1^{-1/2}(m_2 - m_1)$$
$$= (V_1 + \gamma^* V_2)^{-1}(m_2 - m_1). \qquad \square$$

In the homoscedastic case, $V_1 = V_2$, we have $V = I_d$ and Eq. (4.6) has the unique solution $\gamma^* = 1$. Moreover, $t^* = \|m\|^2/4$ and $p^* = P[\chi_d^2 \le \|m\|^2/4]$, an increasing function of $\|m\|$. In this case, the two probability densities agree at the touching point and the normal vector is the normalization of $(2V_1)^{-1}(m_2 - m_1)$. In the general univariate case with variance v, Eq. (4.6) is solved by $\gamma^* = 1/\sqrt{v}$ and $t^* = m^2/(1 + \sqrt{v})^2$. In general, Eq. (4.6) can be solved for γ numerically. It can be rewritten as $\|(I_d + \gamma V)^{-1}m\|^2 = \|\gamma\sqrt{V}(I_d + \gamma V)^{-1}m\|^2$.

The normal vector to the p^*-ellipsoids, $(V_1 + \gamma^* V_2)^{-1}(m_2 - m_1)$, bears a similarity to but is different from the eigenvector $\left(V_1 + \frac{p}{1-p}V_2\right)^{-1}(m_2 - m_1)$ obtained by Fisher [156] in his linear discriminant analysis of a heteroscedastic mixture $(1 - p)N_{m_1,V_1} + pN_{m_2,V_2}$. Note, however, that $\frac{p}{1-p}$ is a weight ratio, whereas γ^* depends only on the two populations.

We next divide the Euclidean space by a hyperplane in two halves, each bearing equal and maximal probability w.r.t. the population whose center it contains. The probability is the index of linear separation.

4.10 Proposition (Linear separation) *Let $Y^{(j)} \sim N_{m_j, V_j}$, $j = 1, 2$, with $V_1, V_2 \in \mathrm{PD}(d)$ and distinct vectors $m_1, m_2 \in \mathbb{R}^d$.*

(a) Let $v \in \mathbb{R}^d$ be of unit length and assume without loss of generality $v^\top (m_2 - m_1) \geq 0$. Define the convex combination

$$m^* = \frac{m_1 \sqrt{v^\top V_2 v} + m_2 \sqrt{v^\top V_1 v}}{\sqrt{v^\top V_2 v} + \sqrt{v^\top V_1 v}}$$

of m_1 and m_2 and let $\mathcal{H}_1$ and $\mathcal{H}_2$ be the half-spaces defined by m^* and v:

$$\mathcal{H}_1 : v^\top (x - m^*) \leq 0, \quad \mathcal{H}_2 : v^\top (x - m^*) > 0.$$

*Then, we have $P[Y^{(1)} \in \mathcal{H}_1] = P[Y^{(2)} \in \mathcal{H}_2] \geq 1/2$ (**index of linear separation**).*

(b) The probabilities are maximal for the unit vector v^* that maximizes the expression

$$q(v) = \frac{(m_2 - m_1)^\top v}{\sqrt{v^\top V_2 v} + \sqrt{v^\top V_1 v}}$$

and, in this case, both equal $N_{0,1}\big(]-\infty, q(v^*)]\big)$.

(c) The vector v^* equals the normal vector to the p^*-ellipsoids of Lemma 4.9(d) and the half-spaces $\mathcal{H}_1$ and $\mathcal{H}_2$ are separated by the common tangent hyperspace through the touching point there.

PROOF. (a) Let Φ denote the standard normal c.d.f. The unique number t that solves the equation

$$\Phi\left(\frac{t - v^\top m_1}{\sqrt{v^\top V_1 v}}\right) + \Phi\left(\frac{t - v^\top m_2}{\sqrt{v^\top V_2 v}}\right) = 1 \tag{4.7}$$

is $t^* = v^\top m^*$. Indeed, the equation is equivalent to

$$\frac{t - v^\top m_1}{\sqrt{v^\top V_1 v}} = -\frac{t - v^\top m_2}{\sqrt{v^\top V_2 v}}$$

and the claim follows from simple algebraic manipulations.

Because of $v^\top Y^{(j)} \sim N_{v^\top m_j, v^\top V_j v}$ and (4.7), we have with $Z \sim N_{0,1}$ and $t^* = v^\top m^*$

$$P[Y^{(1)} \in \mathcal{H}_1] = P[v^\top Y^{(1)} \leq t^*] = P\left[Z \leq \frac{t^* - v^\top m_1}{\sqrt{v^\top V_1 v}}\right] = \Phi\left(\frac{t^* - v^\top m_1}{\sqrt{v^\top V_1 v}}\right)$$

$$= 1 - \Phi\left(\frac{t^* - v^\top m_2}{\sqrt{v^\top V_2 v}}\right) = P\left[Z > \frac{t^* - v^\top m_2}{\sqrt{v^\top V_2 v}}\right] = P[v^\top Y^{(2)} > t^*] = P[Y^{(2)} \in \mathcal{H}_2].$$

This is the desired equality of the probabilities. They are at least $1/2$ because of $m_1 \in \mathcal{H}_1$ (and $m_2 \in \mathcal{H}_2$).

(b) The computation a few lines above shows that we have to maximize the expression $\frac{t^* - v^\top m_1}{\sqrt{v^\top V_1 v}} = \frac{v^\top (m^* - m_1)}{\sqrt{v^\top V_1 v}}$. The claim follows from the representation of m^* given in (a).

(c) The $N_{m,V}$-probability of the ellipsoid (4.5) depends only on t and not on (m, V). The same is true for the probability of any half-space touching the ellipsoid. Thus, each of the two probabilities is a function of the other, independent of the parameters of the ellipsoid. Now consider the two ellipsoids w.r.t. (m_1, V_1) and (m_2, V_2), respectively, that touch the hyperplane separating $\mathcal{H}_1$ and $\mathcal{H}_2$. Since the probabilities of the half-spaces are equal, so are their N_{m_1,V_1}- and N_{m_2,V_2}-probabilities. If the ellipsoids touched the separating hyperplane at different points, their distance would be strictly positive and the touching p-ellipsoids would be strictly larger, leading to two half-spaces with larger N_{m_1,V_1}- and N_{m_2,V_2}-probabilities, a contradiction. Thus, the two ellipsoids touch the hyperplane at the same point and they are in fact the ellipse found in Lemma 4.9. This proves part (c). $\square$

The first part of Proposition 4.10 hinges on a given vector v that defines the separating hyperplane. Any vector (or its negative) could be used, for instance, Fisher's vector $W^{-1}(m_2 - m_1)$ mentioned above. The second part exhibits the optimal choice of v, disregarding population weights. In general, this does not mean optimality in the sense of minimal error rates which would be provided by the MAP discriminant rule and needs *weighted* populations. It just means that each half-space contains a maximum and equal part of the population whose center it contains. We have thus found that the optimal linear and elliptical separations are closely related.

Assessing separation with the value of p^* in Lemma 4.9(b) would be difficult, since p^* tends to disappear in higher dimensions even when populations are well separated. Therefore, the index of linear separation is preferable. The optimal vector v^* needed to compute it via $q(v^*)$ could be computed by some iterative descent algorithm applied to the maximization problem of Proposition 4.10(b). However, part (d) of the proposition shows that the vector v^* can be computed with the more convenient univariate method provided by *elliptical* separation, Lemma 4.9(a).

4.4.2 Normality and related tests

Cluster validation fits nicely into the framework of hypothesis testing. Statistical tests have the most authority but the known ones in this field suffer from insufficient power or special distributional assumptions. To date, there are no powerful tests specifically designed for "cohesion" and "isolation." Although unimodality and cohesion are different from normality, normality implies cohesion and normality tests may therefore sometimes be used for testing cohesion. This holds in particular when the number of observations is not large. Note that data from normal populations may look very different in this case. Thode [501] states that *"a test of normality might have some promise as a tool in cluster analysis."*

Univariate normality tests such as Shapiro and Wilk's [470] impressive omnibus test (see also Royston [452, 453]) may be applied along the favorable directions v proposed in Section 4.4.1 to display separation. When two clusters arise actually from dissecting a data set drawn from a normal, the orthogonal projection of their union to any straight line is again theoretically normal. This can be tested with a univariate normality test. It provides a p-value. This, however, poses a problem in higher dimensions: From a continuum of directions, we pick an optimal one to decompose the data set. The projection of the data to this direction cannot be expected to look very normal. Indeed, using a result of Geman's [194], Diaconis and Freedman [122], Remark 3, show that the least normal projection of *normal* data will deviate from normality unless n/d is very large. Therefore, in particular in higher dimensions, normality should be rejected only if the p-value is very small, the smaller, the higher the dimension is. A p-value of 0.01 does not generally reject normality unless the dimension is low.

Multivariate normality tests such as the multivariate extensions of Shapiro and Wilk's test by Srivastava and Hui [484] and Liang et al. [320], too, are useful in this respect. See also the extensive reviews by Mecklin and Mundfrom [371] and Thode [501]. Note, however, that normality tests register any deviation from normality and not just missing cohesion. So they should be applied with care. Also tests for elliptical symmetry (Eaton and Kariya [138], Romano [447], Gupta and Kabe [215], Schott [458], Manzotti et al. [342], Serfling [465], Huffer and Park [258]) may be consulted. Goodness-of-fit tests, too, offer themselves. Does

Table 4.5: *Separation indices and p-values for the sample data set of Fig. 4.1.*

Cluster pair	Fisher	linear separ.	Sym. diverg.	−log Helling.	p-value MV S.-W.	p-value lin. separ.
(A, B_1)	1.000	1.000	61.96	6.5	0.000	0.000
(A, B_2)	1.000	1.000	66.12	6.3	0.000	0.000
(A, C)	1.000	1.000	795.7	8.6	0.000	0.000
(A, D)	1.000	1.000	3091	181	0.000	0.000
(B_1, B_2)	0.984	0.984	20.40	2.3	0.073	0.029
(B_1, C)	1.000	1.000	471.7	12.6	0.000	0.000
(B_1, D)	1.000	1.000	2011	173	0.000	0.000
(B_2, C)	1.000	1.000	525.1	34.9	0.000	0.000
(B_2, D)	1.000	1.000	1876	202	0.000	0.000
(C, D)	0.997	0.997	117.3	3.3	0.001	0.001

the estimated mixture fit the data? However, spurious clusters fit the data particularly well, so again use with care. Finally, common outlier tests should be applied. Refer, for instance, to Hawkins [235], Rousseeuw and Leroy [449], Davies and Gather [111], and Barnett and Lewis [26].

Table 4.5 shows some separation indices and p-values for cluster pairs as described in Section 4.4.1. The meanings of the six columns are, from left to right: the index of separation by Fisher's vector, the index of linear separation, Proposition 4.10(a), the symmetrized Kullback–Leibler divergence, the negative logarithm of Hellinger's integral, and the p-values of the "sum" version of Srivastava and Hui's [484] normality test and of Shapiro and Wilk's [470] normality test applied to the direction of elliptical and linear separation, v, defined in Proposition 4.9(d) and 4.10(c). Each row shows the results for one cluster pair of the favorite steady solution of the sample data set of Fig. 4.1 with five clusters and the pairs are sorted in lexical order. The solution splits Cluster B into an upper (B_1) and a lower (B_2) part. The different assessments are unanimous. As expected, the separation indices of the two subsets B_1 and B_2 are poorest and the p-value is largest, followed by the pairs (A, B_1), (A, B_2), and (C, D). Although the p-values of the tests for the union $B_1 \cup B_2$ in the last two columns are not large, both indicate acceptance of normality of the union. In particular, 0.029 in the last column does not reject normality because it refers to the direction of best separation!

4.4.3 Visualization

Visualization is a very powerful validation method. It is no coincidence that, despite much progress in automatic image processing in the last decades, physicians still rely on their eyes when analyzing X-ray, NMR, ultrasound, and other image data. The only disadvantage is its subjectiveness. On the other hand, the *trained* human eye is able to explicitly or implicitly capture characteristics of a geometric object such as a data set that are not easily conveyed to a test statistic.

Histograms or scatter plots are used to visually confirm an alleged cluster as cohesive and isolated. As to *isolation*, it is sufficient to show that any two clusters of a given partition are indeed separated. Several approaches are described below. Unless the dimension is one or two, *cohesion* is more difficult, since two-dimensional plots would have to be produced for *many* data projections. Otherwise, one cluster might hide the other, leading us to believe in cohesion. Verifying cohesion can, however, be reduced to separation. Remember that a set is cohesive if and only if none of its subsets is isolated. This can be checked by using a clustering algorithm to optimally split the set in two clusters and by examining their

Figure 4.10 *Unfavorable orthogonal projections (top) and optimal 2-D projections of two 3-D data sets.*

separation. If separation fails, there is a good chance of cohesion. It is therefore sufficient to restrict matters to verifying pairwise separation of all pairs of alleged clusters if the dimension is ≥ 3.

Two subsets C_1 and C_2 display their degree of linear separation when projected to a straight line between them. For this purpose, a line in the direction of Fisher's vector $W^{-1}(\overline{x}_2 - \overline{x}_1)$ or the vector v of elliptical and linear separation introduced in Propositions 4.9(d) and 4.10(c) suggests itself. Both are based on the population parameters estimated from the subsets. They will be successful when separation is actually linear. In Section 4.4.1 they have been used for separation indices. They can also be used to visualize the separation between the two subsets. Distinct bimodality of the histogram of the projected data implies that C_1 and C_2 are actually clusters.

Screens offer two directions, and a second direction u perpendicular to v is welcome to reveal additional information useful for a decision in subtle cases. One such direction is the largest eigenvector of the scatter matrix of $C_1 \cup C_2$ projected to a hyperplane orthogonal to v. To define another one, project the scatter matrices of C_1 and C_2 to the hyperplane perpendicular to v.[7] The largest eigenvectors w_1 and w_2 of the two projections contain again information about the two subsets. A second direction u is therefore chosen such that both scalar products $|(u, w_1)|$ and $|(u, w_2)|$ are simultaneously large. One way is to maximize the sum of squares $(u, w_1)^2 + (u, w_2)^2 = u^\top (w_1 w_1^\top + w_2 w_2^\top) u$, that is, to compute the largest eigenvector of the symmetric rank-2 matrix $w_1 w_1^\top + w_2 w_2^\top$. It is the normalization of $w_1 + w_2$ or $w_1 - w_2$ according to the sign of the scalar product (w_1, w_2). The same result is obtained if the modulus of the product $2(u, w_1)(u, w_2) = 2u^\top w_1 w_2^\top u = u^\top (w_1 w_2^\top + w_2 w_1^\top) u$ is maximized. The left side of Fig. 4.10 presents two views of a 3-D data set made up of two clusters. The projection on top is to the xy-plane, that on the bottom to the xz-plane. The bottom view exhibits the two-cluster structure much better.

The vectors v and u are not adequate when the proposed clusters are separated only by scale, that is, when their mean vectors are close relative to their scales. Instances are two clusters forming a cross, a V, or a T. In such a case, project the data to the plane spanned by the largest eigenvectors of the two proposed clusters. The right-hand side of Fig. 4.10 shows two projections of a V-shaped 3-D data set composed of two clusters. Again, the structure is visible in the lower plot only.

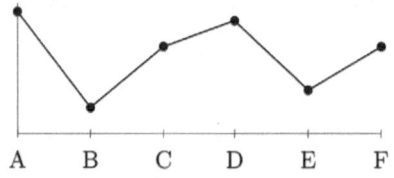

Figure 4.11 *A data point with the variables A–F represented as a profile.*

In the event of very many data points, a large region in the scatter plot will be just black. In order to

---

[7]They are the scatter matrices of the data projections.

detect structure here, plot a random sub-sample. Histograms and scatter plots are possible in at most three dimensions. Plots in the form of *profiles* are able to represent higher dimensional data; see Fig. 4.11. An example is the graph of a time series such as the time course of a share price. But profiles are applicable to general numerical or ordinal data. Although the dimension may be arbitrary, their interpretation is often not so easy.

4.4.4 Measures of agreement of partitions

Several measures of the agreement of two partitions are known. The **Rand index** (Rand [427]) is the proportion of (unordered) pairs of data points that agree w.r.t. the two cluster structures, that is, lie jointly either within or between clusters. More precisely, let $\mathcal{C}$ and $\mathcal{D}$ be any two partitions of the same data set and let the numbers of edges within clusters in both partitions, within clusters in $\mathcal{C}$ and between clusters in $\mathcal{D}$, within clusters in $\mathcal{D}$ and between clusters in $\mathcal{C}$, and between clusters in both partitions be denoted by a, b, c, and d, respectively. The Rand index is defined as

$$\mathrm{RI}(\mathcal{C}, \mathcal{D}) = \frac{a+d}{\binom{n}{2}} = \frac{a+d}{a+b+c+d}.$$

It lies in the closed unit interval; both limits are attained. Another agreement measure based on the four numbers is **Jaccard's** [266] **index** $a/(a+b+c)$.

We often wish to know whether an agreement is significant or just by chance. The message of the Rand index is not straightforward in this respect. Is a Rand index of 0.7 haphazard or significant? That is, if the data set were endowed with a random partition subject to the cluster sizes of $\mathcal{D}$, would we expect a value larger or smaller than 0.7? Hubert and Arabie's [257] **adjusted Rand index** (ARI) answers this question, relating the index to the expected chance agreement subject to the original cluster sizes; see [257], p. 197. Except in trivial cases,[8] their adjusted index has the representation

$$\mathrm{ARI}(\mathcal{C}, \mathcal{D}) = \frac{\binom{n}{2}(a+d) - \big((a+b)(a+c) + (d+b)(d+c)\big)}{\binom{n}{2}^2 - \big((a+b)(a+c) + (d+b)(d+c)\big)}.$$

It lies in the interval $]-1, 1]$, is close to zero in the random case, and may be negative, an event realized, for instance, by the worst-case agreement.

Milligan and Cooper [380] undertook an extensive study about the interpretability of several agreement measures of partitions, among them the original and the Hubert and Arabie adjusted Rand index. They uniformly sampled various data sets from hypercubes of different dimensions, imposing on them artificial cluster structures of two to five clusters of various sizes. They then used several criteria standard in hierarchical clustering to create partitions of between one and fifty clusters, thus recovering weak, accidental cluster structures present in the uniform data. The plotted adjusted Rand indices between the imposed and the accidental partitions turned out to be flat and close to zero in the mean, reflecting the arbitrariness of the imposed structure. Moreover, only small fluctuations about the mean are observed. On the other hand, the message of the original Rand index is not clear. It exhibits mean values between 0.4 and 0.7, depending on the parameters, and heavily fluctuates about them. The interpretation of the adjusted Rand index is much easier. They concluded that the Hubert and Arabie adjusted Rand index was superior to the plain Rand index as a measure of agreement in cluster analysis.

Mutual information (Appendix D.5.1) offers another measure of agreement of two partitions. Let $\big(L_i^{(1)}, L_i^{(2)}\big) : \Omega \to (1 \ldots g_1) \times (1 \ldots g_2)$, $1 \le i \le n$, be n i.i.d. pairs of random labels with joint probabilities $\pi_{j,\ell} = P[L_1^{(1)} = j, L_1^{(2)} = \ell]$. Abbreviate $\pi_{j,+} = \sum_\ell \pi_{j,\ell}$ and

---

[8]That is, both partitions consist of one or of n clusters and are hence equal.

$\pi_{+,\ell} = \sum_j \pi_{j,\ell}$. The mutual information

$$\mathrm{MI}(L_1^{(1)}, L_1^{(2)}) = \sum_{j,\ell} \pi_{j,\ell} \log \frac{\pi_{j,\ell}}{\pi_{j,+} \cdot \pi_{+,\ell}}$$

of $L_1^{(1)}$ and $L_1^{(2)}$ is a measure of agreement of the associated random partitions $\mathcal{C}^1 = \{C_j^1\}_{j=1}^{g_1}$ and $\mathcal{C}^2 = \{C_\ell^2\}_{\ell=1}^{g_2}$ of $1..n$. It is symmetric and positive, zero if and only if $L_1^{(1)}$ and $L_1^{(2)}$ are independent, and equal to the entropy if $L_1^{(1)} = L_1^{(2)}$. Lacking a fixed upper bound, it becomes more easily accessible to interpretation and comparison after normalization (D.11), for instance,

$$\mathrm{NMI}(L_1^{(1)}, L_1^{(2)}) = \frac{\mathrm{MI}(L_1^{(1)}, L_1^{(2)})}{\sqrt{\mathrm{HL}_1^{(1)} \cdot \mathrm{HL}_1^{(2)}}} = \frac{\sum_{j,\ell} \pi_{j,\ell} \log \frac{\pi_{j,\ell}}{\pi_{j,+}\cdot\pi_{+,\ell}}}{\sqrt{(\sum_j \pi_{j,+} \log \pi_{j,+})(\sum_\ell \pi_{+,\ell} \log \pi_{+,\ell})}}.$$

NMI is well defined except in trivial cases. Given partitions $\mathcal{C}^1 = \{C_j^1\}$ and $\mathcal{C}^2 = \{C_\ell^2\}$ of $1..n$, the ML estimates of the joint probabilities $\pi_{j,\ell}$,

$$\pi_{j,\ell}(\mathcal{C}^1, \mathcal{C}^2) = \frac{n_{j,\ell}}{n} = \frac{|C_j^1 \cap C_\ell^2|}{n},$$

are inserted for $\pi_{j,\ell}$. Depending on the contingency matrix $(n_{j,\ell})_{j,\ell}$ of $\mathcal{C}^1$ and $\mathcal{C}^2$, the estimate of NMI is based on counting points instead of pairs.

4.4.5 Stability

The consistency theorem 1.56 says that the estimates obtained with the TDC under HDBT constraints converge along n if, besides the regularity conditions (i)–(iv), two conditions are met: The minimizer of the population criterion must be unique and the assumed number of clusters, g, must be a drop point; see Definition 1.55. Let us accept the uniqueness. If g is a drop point, then the estimates obtained for two samples large enough for the theorem to take hold are expected to replicate. In particular, assuming that the number of observations is large enough, the solutions for different samples randomly drawn from a given sample fluctuate strongly only if g fails to be a drop point. Now, drop points are related to numbers of components; see Remark 1.57, Examples 1.59, in particular (c), and Section 4.2.3. That is, instability of random sub-samples casts doubt on the assumed number of components and, more generally, on the assumed model. Roughly speaking, model *validity leads to stability* or equivalently, *instability questions validity* of the model. For instance, a spurious, small cluster is extremely unstable, since it is sufficient to remove (or to slightly modify) a single, selected data point in order to have it disappear. Therefore, it is likely to be invalid.

The implication may not be reversed and stability must not be confounded with validity as McIntyre and Blashfield [358], Morey et al. [384], and Hennig [242] emphasize. While a distinct, valid solution is expected to be stable, a stable solution may well be invalid. Just think of the extreme case $g = 1$. There is only one solution which is therefore maximally stable. But if the data bears a cluster structure, then the solution for $g = 1$ is invalid. This is also exemplified by the two wrong clusters obtained with the k-means algorithm shown in Fig. 1.11. Its application to subsets results in about the same solutions. The clustering is, therefore, stable w.r.t. k-means. But it is not valid. The discrepancy is caused by the fact that stability depends not only on data and clustering but also on the, in this case insensitive, criterion used. Had we used the determinant criterion, the improper nature of the partition would have been revealed.

The general idea behind stability is that distinct partitions and clusters should not change much after an *inessential* modification of the data set or of the algorithmic parameters. In practice, the parent populations are not at our disposal. Therefore, we draw sub-samples from the data at hand and analyze them. Stability is measured by the degree of agreement of the partitions obtained. Rogers and Linden [446] introduced cross-validation for this purpose. Their method is more formally described in McIntyre and Blashfield [358] and Morey et al. [384]. An abstract version of their **replication analysis** adapted to trimming reads as follows.

4.11 Procedure (Replication analysis)
// <u>Input</u>: A clustering method, a data set $\mathbf{x}$ of n points, and natural
           numbers $g \geq 2$ and $r \leq n$.
// <u>Output</u>: A measure of stability of the clustering method's action on $\mathbf{x}$.

1. Divide the data set randomly into two parts $\mathbf{x}_1$ and $\mathbf{x}_2$.

2. Decompose the subset $\mathbf{x}_1$ with the given method in g clusters, discarding $n - r$ elements, and use the result to establish a discriminant rule for g components on the sample space.

3. Use the discriminant rule to decompose the data set $\mathbf{x}_2$ in the same number of clusters and discards.

4. Decompose the subset $\mathbf{x}_2$ with the given method in g clusters, again discarding $n - r$ elements.

5. Determine the level of agreement of the two clusterings obtained in steps 3 and 4. (The outlier set is treated here as a cluster.)

6. Repeat steps 1–5 with different subdivisions and return the mean value of the levels obtained in step 5.

Since the outliers and thus their portions in the random subsets $\mathbf{x}_1$ and $\mathbf{x}_2$ drawn in Procedure 4.11 are unknown and since the clustering method is sensitive to discarding too few elements, $n - r$ elements are discarded in steps 2–4. Under a valid model, both the discriminant rule obtained for $\mathbf{x}_1$ in step 2 and the cluster structure obtained for $\mathbf{x}_2$ in step 4 are expected to conform to the estimated structure of $\mathbf{x}$ and so will yield high degrees of agreement in step 5. There is no evidence of instability. If the discriminant rule and the clustering often disagree, the mean value in step 6 will be lower and we may infer invalidity of the model. Hubert and Arabie's adjusted Rand index and the normalized mutual information are adequate measures in step 5, since they allow us to compare the levels of agreement; see Section 4.4.4.

Replication analysis can be applied with the clustering algorithms described in Chapter 3. Both the mixture and the classification approaches establish mixture models of the data set and their estimated parameters can be used for a Bayesian discriminant rule in step 2. Estimation errors affect the accuracy of the method.

Milligan and Cooper [380] note as a side remark that the adjusted Rand indices of a given partition w.r.t. other cluster solutions of the same data set might offer some insight into its validity. Instability may, for instance, be used to exclude certain values as the number of clusters. Breckenridge's [66] study of replication analysis, too, showed that the method could be useful in determining the number of clusters, although he observes that "*choosing the number of clusters by maximum replication results in a negatively biased estimate.*"

Procedure 4.11 can be modified to evaluate the stability of a *given* clustering. It is sufficient to drop step 2 and to replace the output of step 3 with the trace of the clustering given in $\mathbf{x}_2$.

Table 4.6 presents a replication analysis of the sample data set of Fig. 4.1. The data set is cluster analyzed with the TDC in $g = 2, \ldots, 6$ clusters, discarding a fixed number of six elements. Both Rand indices seem to exclude three and six as numbers of clusters, the Hubert and Arabie adjusted Rand index also five. As to the remaining numbers, two and four, the indices do not give a convictive answer.

Stability indices assess cohesion and isolation of *single* clusters instead of validity of whole partitions. Two stability indices are due to Bertrand and Bel Mufti [38] and Hennig [242]. Both papers draw sub-samples from the data set. Hennig also investigates other ways of data modification (bootstrapping, jittering, and adding noise), finding that this is among the best. It also avoids multiple points, necessary for applying the mixture analysis presented in Section 2.2, and the TDC, Section 2.3. Their papers use a common scheme for measuring the stability of a *single* cluster. It reads as follows.

4.12 Procedure (Cluster stability index)
// Input: A data set $\mathbf{x}$ of n pairwise different points and a subset $C \subseteq \mathbf{x}$.
// Output: Stability index of C.

1. Draw a random subset $S \subseteq \mathbf{x}$ of size $|S| = \alpha|\mathbf{x}|$, $0 < \alpha < 1$, such that $C \cap S \neq \emptyset$.

2. Cluster analyze the subset S with the given method to obtain a partition $\mathcal{D}$ of S.

3. Determine the degree of similarity between $C \cap S$ and the clusters in $\mathcal{D}$.

4. Repeat steps 1–3 m times with different subsets S and return the mean value of the degrees obtained in step 3.

The approaches of Hennig [242] and Bertrand and Bel Mufti [38] differ in the details. The first concerns the subset S in step 1. Hennig draws a portion α uniformly from $\mathbf{x}$, recommending $\alpha = 1/2$. Bertrand and Bel Mufti recommend $\alpha > 0.7$ and use *proportionate stratified sampling*. That is, they randomly draw equal proportions α from both sets C and $\mathbf{x} \backslash C$. After S has been clustered in step 2, the new clustering $\mathcal{D}$ of S and the subset $C \cap S \subseteq S$ induced by C are available. In step 3, both methods use mainly the indices $\in 1..n$ of the data points (and not their values) and do not have to assume a particular data model. The main difference between the two methods is the way the degree of similarity is determined in step 3. Hennig uses Jaccard's [266] similarity of two subsets $C, D \subseteq \mathbf{x}$, $C \cup D \neq \emptyset$,

$$\gamma(C, D) = \frac{|C \cap D|}{|C \cup D|}.$$

This number lies in the unit interval representing the proportion of points belonging to both sets relative to their union. The dissimilarity $1 - \gamma$ is even a metric on the collection of all nonempty subsets of $\mathbf{x}$; see Gower and Legendre [211]. Hennig selects in step 3 the set in the new clustering $\mathcal{D}$ most similar to the induced cluster $C \cap S$ in the sense of Jaccard's similarity. That is, he proposes the maximum similarity $\gamma^* = \max_{D \in \mathcal{D}} \gamma(C \cap S, D)$ between $C \cap S$ and the clusters $D \in \mathcal{D}$.

Whereas the Jaccard similarity counts data points, Bertrand and Bel Mufti [38] count links between points. This is a detail their method shares with the Rand indices of Section 4.4.4. More precisely, they use a measure of rule satisfaction due to Loevinger [335]. Given two events E and F in some probability space such that $P(E) > 0$ and $P(F) < 1$, he defines the measure of satisfaction of the inclusion $E \subseteq F$ as

$$1 - \frac{P(\complement F \mid E)}{P(\complement F)} = 1 - \frac{P(E \cap \complement F)}{P(E)P(\complement F)}.$$

It is one minus a weighted probability of violation of the inclusion $E \subseteq F$. The measure is 1 if $E \subseteq F$, it vanishes if E and F are independent. It may become negative, for instance, when $E = \complement F$ and $0 < P(E) < 1$. The authors use this measure for stability indices of

Table 4.6 *Replication analysis of partitions of the sample data set in two to six clusters and six outliers with the TDC.*

g	2	3	4	5	6
Rand	0.92	0.82	0.93	0.90	0.87
H&A adj. Rand	0.83	0.59	0.81	0.70	0.60

Table 4.7 *Cluster stability indices for decomposition of the sample data set in five clusters with two cluster methods.*

Stability method	Cluster method	A	B_1	cluster B_2	C	D
Jaccard	EMT	0.735	0.501	0.512	0.834	0.708
	TDC	0.966	0.706	0.735	0.922	0.973
Loevinger isolation	EMT	0.824	0.613	0.550	0.885	0.752
	TDC	0.992	0.651	0.744	0.994	0.987
Loevinger cohesion	EMT	0.865	0.798	0.962	0.855	0.983
	TDC	0.853	0.934	0.917	0.903	0.979

isolation and cohesion. If C is well separated from its complement, then we expect any two points $x, y \in S$ that are not in the same set C or $\complement C$ to lie in different clusters of $\mathcal{D}$. The degree to which this is actually true is Loevinger's measure for

$$E = \text{all pairs of points in } S \text{ not in the same set } C \text{ or } \complement C \text{ and}$$

$$F = \text{all pairs of points in } S \text{ that are in different clusters of } \mathcal{D}.$$

Here, the discarded elements are to be treated as a cluster, say with index 0. Putting $n' = |S|$, $n'_C = |C \cap S|$, and $n'_j = |D_j|$, we have $|E| = n'_C(n' - n'_C)$ and $|\complement F| = \sum_{j=0}^{g} \binom{n'_j}{2}$. Assuming the uniform probability on all edges of the complete graph on S, Loevinger's measure applied to isolation reads

$$1 - \frac{\binom{n'}{2}|E \cap \complement F|}{|E||\complement F|} = 1 - \frac{\binom{n'}{2}|E \cap \complement F|}{n'_C(n' - n'_C)\sum_{j=0}^{g}\binom{n'_j}{2}}.$$

On the other hand, if C is cohesive, then we expect any two points $x, y \in C \cap S$ to lie in the same cluster of $\mathcal{D}$. This time, we put

$$E = \text{all pairs of points in } C \cap S \quad \text{and} \quad F = \bigcup_{j=0}^{g} \binom{D_j}{2}$$

and Loevinger's measure applied to cohesion becomes

$$1 - \frac{\binom{n'}{2}|E \cap \complement F|}{|E||\complement F|} = 1 - \frac{\binom{n'}{2}|E \cap \complement F|}{\binom{n'_C}{2}\left(\binom{n'}{2} - \sum_{0 \le j \le g}\binom{n'_j}{2}\right)}.$$

The subsets are drawn at random in step 1 of Procedure 4.12. This allows application of the CLT to estimate the length of the p-confidence interval after m repetitions in step 4. It is $2q_{m-1}\sigma_m$, where q_{m-1} denotes the $1 - p/2$ quantile of the Student t-distribution and σ_m is the sample standard deviation of the first m observed indices. Bertrand and Bel Mufti [38] proceed to combine the two stability indices into a single validity index.

As in the case of the replication analysis above, validity of a cluster implies stability but not vice versa. A single cluster may well be stable without being valid. A witness is again the improper partition shown in Fig. 1.11.

The three methods are applied to the five subsets and six outliers of the sample data set of Fig. 4.1 with the normal mixture and the normal classification model with full scales. The favorite Pareto solution was selected by the method presented in Section 4.1. It decomposes cluster B in the two subsets B_1 and B_2. The result is shown in Table 4.7. Most figures show that clusters A, C, and D are superior in stability to B_1 and B_2. This conforms to the fact that the data set was sampled from a mixture of only four components. Exceptions are the high cohesion scores of B_2 and also B_1. They are justified, as a look at Fig. 4.1 shows.

4.5   Notes

The idea of using the SBF plot is due to Gallegos and Ritter [182, 183]. They also proposed the extreme Pareto solutions. The general Pareto solutions are new here. Fang and Wang [149] define the instability of partitions of bootstrap samples whence they derive an estimate of the number of clusters.

A large number of validity measures based on prespecified distances are found in Bezdek et al. [39]. They are most useful when cluster scales are consistent with the prior metric. In the case of (almost) spherical clusters, the total within-clusters sum of squared distances about the centroids is used as a measure of cohesion and the total between-clusters sum of squared distances for isolation. However, Milligan [378] noted that, in general, "such traditional multivariate measures as $\operatorname{tr} W$ and $\det W$ performed poorly as measures of internal validity." The reader interested in distance-based validation methods is referred to Milligan and Cooper's [379] critical survey; see also the abridged version of their work by Gordon [210]. They compare a large number of procedures for estimating the number of clusters. The methods discussed by them provide a ranking of the solutions for different values of g. Most of them do not allow a comparison with $g = 1$; see Tibshirani et al. [503].

Milligan's [376] critical survey contains material on variable selection, variable standardization, determining the number of clusters, validation, and on replication analysis. Chen et al. [87] propose a modified likelihood ratio test for a mixture of two components vs. $g \geq 3$. The idea of cluster stability goes back to Rand [427].

Attempts have been made in the past to establish significance tests for cohesion and isolation with test statistics based on distances between data points, either all or just nearest neighbors; see Hubert [256], Baker and Hubert [21], Bailey and Dubes [20], and Bock [50]. It seems that such tests have not become widespread; Bock even raises doubts about the power of distance-based tests.

The archetype of the visualization methods presented in Section 4.4.3 is Fisher's discriminant rule. The standard extension to $g \geq 2$ clusters is found in Rao [430]; see also Mardia et al. [345]. Rao uses the within- and between-groups SSP matrices W and B, computing vectors $c_1, \ldots, c_{g-1}$ that iteratively maximize the quotient $\frac{c^\top Bc}{c^\top Wc}$ on the orthogonal complement w.r.t. W of the set of vectors already computed. These **canonical variates**[9] are just the eigenvectors of the matrix $W^{-1}B$. Various reinterpretations of "betweenness" and "withinness" lead to different variants of canonical variates; see Gnanadesikan [200], Sect. 4.2, and Fukunaga [176], Chapter 10. In pattern recognition, canonical variates serve for extracting features that optimally describe a data structure; see again Fukunaga [176]. Hennig [240] also studies visualization of a cluster versus a subset of unspecified structure. Scrucca [462] discusses a dimension reduction method based on class means and covariances for visualizing normal mixtures.

---

[9]Gnanadesikan [200] calls them **discriminant coordinates** in order to avoid a terminological collision with the "canonical variates" appearing in canonical correlation analysis.

Chapter 5

Variable selection in clustering

There is no reason why *all* variables of a data set should be useful in view of a cluster structure. Often it is hidden in a subset of all variables. In these cases, only the variables that determine the clustering should be included in the analysis. This is in particular true when there are many variables. Often, some of them are just *noise*; others are *redundant*. This refers to variables that contain *no* or *repeated* information on the structure to be detected. Adding noninformative variables to a clustered data set may strongly hamper the performance of clustering algorithms; see, for instance, Milligan [377]. Fowlkes et al. [168] analyzed a plane data set of five well separated clusters of 15 data points each. They extended the dimension to five by taking the product with three independent standard normal variables. Although the first two variables show a clear cluster structure that can be detected by any reasonable cluster method, they found that the structure was completely upset when all five variables were used. Their context was hierarchical clustering with single and complete linkage and with Ward's criterion.

The question arises whether likelihood-based algorithms perform better in this respect. The answer is yes and no. Let us follow Fowlkes et al. [168] and sample five clusters of 15 points each from $N_{(0,0),I_2} \otimes \nu$, $N_{(5,5),I_2} \otimes \nu$, $N_{(-5,5),I_2} \otimes \nu$, $N_{(-5,-5),I_2} \otimes \nu$, and $N_{(5,-5),I_2} \otimes \nu$, respectively, where $\nu = N_{(0,0,0),100I_3}$. The last three variables are irrelevant and just noise. The ML version of the heteroscedastic, *diagonally* normal criterion, that is, Theorem 1.40(b) without the entropy term, yields the correct solution. This is due to the fact that the ML criterion favors equal cluster sizes and that the assumed diagonal model is correct here. The same is true if the MAP version of the pooled determinant criterion (Theorem 1.41(b)) is employed. It exploits the fact that the sampling mixture is indeed homoscedastic. However, as soon as we deviate from these assumptions, MAP instead of ML or the full heteroscedastic model instead of the diagonal one, the algorithm finds many unsuccessful steady solutions, although it was told the correct number of clusters. None of them is convincing. The reason is that the algorithm finds haphazard fractures in the (small) data set that it utilizes for dissection. However, the correct clusters are recovered as cluster size is increased to 150, since the haphazard fractures disappear. (Interestingly, it does not help much to increase cluster separation in the first two variables, even to 10,000!) The EM algorithm behaves similarly. The conclusion is again that it may be favorable to base clustering on a *selected subset* of variables only. Variable selection is thus needed to reduce the dimension of the sample space by removing irrelevant and noisy variables. When done effectively, it can greatly improve the performance of clustering; see Gnanadesikan et al. [201]. Also useful but redundant variables may be dispensed with. In the image on the left of Fig. 5.1, only the first variable is relevant and the second is just noise, whereas on the right each of the two variables is redundant w.r.t. the other.

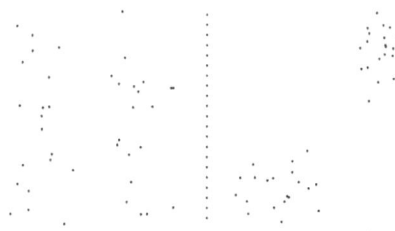

Figure 5.1 *An irrelevant variable (left) and redundant variables (right).*

There are more reasons for selecting variables. The heteroscedastic clustering methods with full scales presented in Chapters 1–3 require that the number of objects (cases) n exceed the number of variables, d. They typically need $n \geq gd+1$ when a full-scale model is to be employed. For the methods to work reliably, n should even be much larger than this number. This calls for a small d. Some more recent applications of data analysis to text processing, genetics and proteomics, systems biology, and combinatorial chemistry grossly violate this requirement. The number of variables exceeds by far the number of cases and the algorithms are not even applicable without drastic reduction of data dimension. Finally, variable selection does not only have statistical and computational advantages. In many applications the selected variables have an intrinsic meaning of their own. They are as important as the partition induced by them.

Variable selection can be classified w.r.t. three aspects. *First*, it can be applied to both *discriminant analysis* and *clustering* (see, for instance, Guyon and Elisseeff [219]), as well as to many other statistical tasks, for instance, regression; see Fowlkes et al. [167]. Of course, variable selection in discriminant analysis is easier than in clustering since the groups are a priori given and can be used for selection. There is a large body of literature about variable selection when the target variable is known; see, for instance, Liu and Yu [331], Saeys et al. [456], Guyon et al. [218], and Yu [540]. In contrast, publications on variable selection in clustering are rarer and often less specific. Besides statistics, variable selection has also been detected by the machine learning community. A source of information is the collection edited by Liu and Motoda [330] and the literature cited there.

Second, variable selection comes in two modes, filters and wrappers, two concepts named by John et al. [278]. A *filter* discards variables that appear noninformative just by analyzing their distributions. For instance, a variable with a constant value across all observations is useless. It can be removed. In contrast, a *wrapper* uses the clustering algorithm itself in order to assess the quality of a selected subset of variables. Here, feature selection is controlled by classification; it is "wrapped" around the clustering method. The wrapper strives for the subset that leads to the most convincing partition of the data. Wrappers are more accurate but also more complex, consuming more computing time. They can be applied, and are preferable, in the presence of moderately many variables. When the number of variables is excessive, it is in general necessary to first reduce the dimension with a filter so that the data can then be processed with a wrapper. *Third*, a soft version of variable selection is *variable weighting*. Variables are not discarded or retained in a crisp way but weighted according to their importance in view of the goal.

After introducing the concept of irrelevance in Section 5.1, this chapter will deal with both filters and wrappers, Sections 5.2 and 5.3. The filters will be univariate (Section 5.2.1) as well as multivariate (Section 5.2.2). Some will select subsets of the given variables, others linear combinations. The wrappers in Sections 5.3.1 and 5.3.2 will be based on the LRT and the Bayesian testing paradigms, those in Section 5.3.3 on likelihood-based clustering algorithms. The chapter will terminate with the author's guidelines of how to proceed in a mixture or cluster analysis.

It is interesting to note that variable selection is in some sense dual to clustering: Indeed, variable selection looks for a *one-to-one* map from a small set to the set of all variables. A labeling, on the other hand, is a map from the set of all objects *onto* a small set. It turns out that the combination of the two is needed for solving important practical problems.

5.1   Irrelevance

The main focus will here be on irrelevance of subsets of variables in mixture analysis on $\mathbb{R}^D$. Two theoretical results have to be noted. First, the set of variables decomposes in a unique way in a subset devoid of irrelevant components and the irrelevant complement; see Theorem 5.5. It is the aim to find this subset. Second, when irrelevant variables are used as the response variables in a normal linear regression, the regression coefficients do not depend on the classes; see Theorem 5.7. This property will prove crucial in the design of likelihood-based wrappers in Section 5.3.3.

The following notation will be used. The restriction of a vector $a \in \mathbb{R}^D$ to a nonempty subset $F \subseteq 1..D$ is denoted by a_F. The given data set contains n cases of D real variables, each, that is, case i is $x_i = (x_{i,1}, \ldots, x_{i,D}) \in \mathbb{R}^D$, $1 \le i \le n$, its kth entry being $x_{i,k}$. The restriction of the data set to the indices in F will consequently be denoted by $\mathbf{x}_F$. Its dimension is $|F| = d$. The ith observation in $\mathbf{x}_F$ (or x_i restricted to F) is $x_{i,F} \in \mathbb{R}^F$. A parameter γ of the basic statistical model acts on $x \in \mathbb{R}^D$ via the likelihood function, $f(x; \gamma)$. But it also acts on $x_F \in \mathbb{R}^F$: Since the projection $P[X_F \in \mathrm{d}x_F \,|\, \gamma]$ onto $\mathbb{R}^F$ is well defined, so is $f(x_F; \gamma)$.

5.1.1 Definition and general properties

The following definitions depend on a given mixture with a fixed number of components g. We use again the stochastic representation $X_i = Y_i^{(L)}$ with random variables $Y_i^{(j)} : \Omega \to \mathbb{R}^D$, $1 \le j \le g$, in the sense of Lemma 1.12 and assume throughout that the support of L is $1..g$.

5.1 Definition Let $F_1, F_2 \subseteq 1..D$ be disjoint subsets.

(a) The subset F_2 is **irrelevant** w.r.t. F_1 if L is conditionally independent of X_{F_2} given X_{F_1}, that is, P-a.s. for all j,
$$P[L = j \,|\, X_{F_1}, X_{F_2}] = P[L = j \,|\, X_{F_1}].$$
If X_{F_1} is given, then X_{F_2} provides no additional information on the class labels.

(b) The subset F_2 is **irrelevant** if it is irrelevant w.r.t. its complement.

(c) A subset $F \subseteq 1..D$ is **structural** if no subset $\emptyset \subset C \subseteq F$ is irrelevant w.r.t. $F \backslash C$.

Intuitively, a subset is *redundant* if the information on the class labels is also contained in its complement; it is *noise* if it contains no information on the class labels. Irrelevance is a combination of both. There are the following equivalent statements for irrelevance.

5.2 Lemma *The following statements on two nonempty, disjoint subsets $F_1, F_2 \subseteq 1..D$ are equivalent.*

(i) *The subset F_2 is irrelevant w.r.t. F_1.*

(ii) *P-a.s., the conditional probability $P[L = j \,|\, X_{F_1}, X_{F_2}]$ does not depend on X_{F_2}.*

(iii) *P-a.s. and for all $j \in 1..g$, $P[X_{F_2} \in \mathrm{d}x_{F_2} \,|\, X_{F_1}] = P\big[Y_{F_2}^{(j)} \in \mathrm{d}x_{F_2} \,|\, Y_{F_1}^{(j)}\big]$.*

(iv) *The conditional probabilities $P\big[Y_{F_2}^{(j)} \in \mathrm{d}x_{F_2} \,|\, Y_{F_1}^{(j)}\big]$ do not depend on $j \in 1..g$.*

PROOF. Plainly, (i) implies (ii). If (ii) is true, then the formula of total probability implies, for all $y \in \mathbb{R}^{F_2}$,
$$P[L = j \,|\, X_{F_1} = x_{F_1}]$$
$$= \int P[L = j \,|\, X_{F_1} = x_{F_1}, X_{F_2} = x_{F_2}] P[X_{F_2} \in \mathrm{d}x_{F_2} \,|\, X_{F_1} = x_{F_1}]$$
$$= \int P[L = j \,|\, X_{F_1} = x_{F_1}, X_{F_2} = y] P[X_{F_2} \in \mathrm{d}x_{F_2} \,|\, X_{F_1} = x_{F_1}]$$
$$= P[L = j \,|\, X_{F_1} = x_{F_1}, X_{F_2} = y],$$
that is, (i). Now, by Bayes' formula and independence of Y and L,
$$P[L = j \,|\, X_{F_1} = x_{F_1}, X_{F_2} = x_{F_2}] = \frac{f_{X_{F_1}, X_{F_2}}[x_{F_1}, x_{F_2} \,|\, L = j] P(L = j)}{f_{X_{F_1}, X_{F_2}}(x_{F_1}, x_{F_2})}$$
$$= \frac{f_{Y_{F_1}^{(j)}, Y_{F_2}^{(j)}}(x_{F_1}, x_{F_2}) P(L = j)}{f_{X_{F_1}, X_{F_2}}(x_{F_1}, x_{F_2})} = \frac{f_{Y_{F_2}^{(j)}}[x_{F_2} \,|\, Y_{F_1}^{(j)} = x_{F_1}] f_{Y_{F_1}^{(j)}}(x_{F_1}) P(L = j)}{f_{X_{F_1}, X_{F_2}}(x_{F_1}, x_{F_2})}.$$

Similarly,

$$P[L = j \mid X_{F_1} = x_{F_1}] = \frac{f_{X_{F_1}}[x_{F_1} \mid L = j]P(L = j)}{f_{X_{F_1}}(x_{F_1})} = \frac{f_{Y_{F_1}^{(j)}}(x_{F_1})P(L = j)}{f_{X_{F_1}}(x_{F_1})}.$$

Therefore, (iii) and (i) are equivalent. Plainly, (iii) implies (iv) and the converse follows from the formula of total probability. □

5.3 Properties

(a) A subset F is irrelevant w.r.t. the empty set if and only if L and X_F are independent.

(b) It follows from Lemma 5.2(iv) that a nonempty subset F is irrelevant w.r.t. the empty set if $Y_F^{(j)} \sim Y_F^{(\ell)}$ for all j, ℓ.

(c) If F_2 is irrelevant w.r.t. F_1, then any subset of F_2 is again irrelevant w.r.t. F_1.

(d) Let F_1, F_2, and F_3 be pairwise disjoint subsets of $1..D$. Any two of the following statements imply the third:

   (i) F_2 is irrelevant w.r.t. F_1;
   (ii) $F_2 \cup F_3$ is irrelevant w.r.t. F_1;
   (iii) F_3 is irrelevant w.r.t. $F_1 \cup F_2$.

   This follows from the product formula

$$f_{Y_{F_2 \cup F_3}^{(j)}}[x_{F_2 \cup F_3} \mid Y_{F_1}^{(j)} = x_{F_1}] = f_{Y_{F_3}^{(j)}}[x_{F_3} \mid Y_{F_1 \cup F_2}^{(j)} = x_{F_1 \cup F_2}]P_{Y_{F_2}^{(j)}}[x_{F_2} \mid Y_{F_1}^{(j)} = x_{F_1}].$$

(e) If the random variables $Y_k^{(j)}$, $1 \le k \le D$, are independent for all j, then F is irrelevant if and only if, for all $k \in F$, the distribution of $Y_k^{(j)}$ does not depend on j. This follows from Lemma 5.2(iv).

(f) If a subset F is structural, then there exist j, ℓ such that $Y_F^{(j)} \not\sim Y_F^{(\ell)}$. This follows from the fact that F is not irrelevant w.r.t. the empty set.

(g) Let $F \subset 1..D$ be structural and let $k \in \complement F$ be relevant w.r.t. F. Then $F \cup \{k\}$ is structural if and only if each $\emptyset \subset C \subseteq F$ is relevant w.r.t. $(F \cup \{k\}) \backslash C$.
   We have to show that each $\emptyset \subset C \subseteq F \cup \{k\}$, $k \in C$, is relevant w.r.t. $F \backslash C$. By assumption this is true if $C = \{k\}$. If $\emptyset \subset C \subseteq F \cup \{k\}$, $C \supset \{k\}$, were irrelevant w.r.t. $F \backslash C$, then $C \backslash \{k\}$ would be irrelevant w.r.t. $F \backslash C$ by (b), a contradiction since F is structural.

The following well-known technical lemma shows that, under mild assumptions, a distribution is specified by its conditionals; see Patil [402] and Arnold et al. [16].

5.4 Lemma *If the joint Lebesgue density of $X : \Omega \to \mathbb{R}^d$ and $Y : \Omega \to \mathbb{R}^e$ is continuous and strictly positive, then it is specified by the conditional densities $f_X[x \mid Y = y]$ and $f_Y[y \mid X = x]$ and we have*

$$f_{(X,Y)}(x, y) = c \cdot \frac{f_Y[y \mid X = 0]}{f_X[0 \mid Y = y]} f_X[x \mid Y = y]$$

with $\frac{1}{c} = \int \frac{f_Y[y \mid X=0]}{f_X[0 \mid Y=y]} dy$.

PROOF. By conditioning (Example D.16(b)), there exists a number c_y not depending on x such that $f_{(X,Y)}(x, y) = c_y f_X[x \mid Y = y]$. In particular, $f_{(X,Y)}(0, y) = c_y f_X[0 \mid Y = y]$. Hence

$$f_{(X,Y)}(x, y) = \frac{f_{(X,Y)}(0, y)}{f_X[0 \mid Y = y]} f_X[x \mid Y = y].$$

The claim now follows from $f_{(X,Y)}(0, y) = c \cdot f_Y[y \mid X = 0]$. □

Without strict positivity of the joint density the lemma is not necessarily true. Just consider the two joint densities $\frac{1}{2}\mathbf{1}_{[0,1]^2} + \frac{1}{2}\mathbf{1}_{[2,3]^2}$ and $\frac{2}{3}\mathbf{1}_{[0,1]^2} + \frac{1}{3}\mathbf{1}_{[2,3]^2}$. Their conditional densities coincide. The following theorem shows that the index set $1..D$ has a unique decomposition in a structural and an irrelevant subset.

5.5 Theorem (Relevance decomposition) *Let the real random variables* X_i, $i \in 1..D$, *have a strictly positive and continuous joint Lebesgue density* $f_{(X_1,\ldots,X_D)}$. *There exists exactly one structural subset* $F \subseteq 1..D$ *with an irrelevant complement.*

PROOF. Let us first show uniqueness and assume that F and F' both share the two properties. Put $A = F\backslash F'$, $B = F'\backslash F$, and $C = F \cap F'$. By Property 5.3(c), A is irrelevant w.r.t. $F' = B \cup C$, and B is irrelevant w.r.t. $F = A \cup C$. By Lemma 5.2, the conditional densities $f_{Y_A^{(j)}}[x_A \,|\, Y_B^{(j)} = x_B, Y_C^{(j)} = x_C]$ and $f_{Y_B^{(j)}}[x_B \,|\, Y_A^{(j)} = x_A, Y_C^{(j)} = x_C]$ do not depend on j. Being a function of the two densities by Lemma 5.4, the conditional density $f_{Y_{A\cup B}^{(j)}}[x_{A\cup B} \,|\, Y_C^{(j)} = x_C]$ does not depend on j. This shows that $A \cup B$ and hence A and B are irrelevant w.r.t. C. Since F and F' are structural, we conclude that $A = B = \emptyset$, that is, $F = F'$. This is uniqueness.

Let us then show existence. Since the empty set is irrelevant, there exists a maximally irrelevant subset of $1..D$. Its complement F is structural. Indeed, let $C \subseteq F$ be irrelevant w.r.t. $F\backslash C$. Then

$$f_{Y_{\complement F \cup C}^{(j)}}\left[x_{\complement F \cup C} \,|\, Y_{F\backslash C}^{(j)} = x_{F\backslash C}\right] = f_{Y_{\complement F}^{(j)}}\left[x_{\complement F} \,|\, Y_F^{(j)} = x_F\right] \cdot f_{Y_C^{(j)}}\left[x_C \,|\, Y_{F\backslash C}^{(j)} = x_{F\backslash C}\right]$$

does not depend on j, that is, $\complement F \cup C$ is irrelevant. By maximality of $\complement F$, we have $C = \emptyset$, that is, F is structural. $\quad\square$

The following examples illustrate Definition 5.1 and Theorem 5.5.

5.6 Examples (a) Let (X_1, X_2, X_3) be uniform on the Boolean space $\{0,1\}^3$ and let $L = X_1 \oplus X_2$, the exclusive "or." It is no surprise that the structural subset according to Theorem 5.5 is $\{1,2\}$. Indeed, $\{3\}$ is irrelevant since

$$P[L = 0 \,|\, X_{\{1,2\}} = (y_1, y_2), X_3 = x] = \begin{cases} 0, & y_1 \neq y_2, \\ 1, & y_1 = y_2, \end{cases}$$

does not depend on x, $\{1\}$ is relevant w.r.t. $\{2\}$, and $\{2\}$ is relevant w.r.t. $\{1\}$ since

$$P[L = 0 \,|\, X_2 = 0, X_1 = x] = P[L = 0 \,|\, X_1 = 0, X_2 = x] = 1 - x,$$

and $\{1,2\}$ is relevant w.r.t $\emptyset$ since

$$P[L = 0 \,|\, X_{\{1,2\}} = (y_1, 0)] = 1 - y_1.$$

(b) If the random variables $Y_k^{(j)}$, $1 \leq k \leq D$, are independent for all j, then the unique structural subset promised by Theorem 5.5 is $F = \{k \,|\, \text{there exist } j, \ell \text{ such that } Y_k^{(j)} \not\sim Y_k^{(\ell)}\}$. Its complement is $\complement F = \{k \,|\, Y_k^{(j)} \sim Y_k^{(\ell)} \text{ for all } j, \ell\}$. An example with dependent variables is the following.

(c) Let $g = D = 2$ and let $(Y_1^{(j)}, Y_2^{(j)})$, $j = 1, 2$, be bivariate, normal with mean $(m_{j,1}, m_{j,2})$ and covariance matrix $\begin{pmatrix} v_{j,1} & \\ v_{j,21} & v_{j,2} \end{pmatrix}$. Then $\{1\}$ is structural and $\{2\}$ is irrelevant if and only if $(m_{1,1}, v_{1,1}) \neq (m_{2,1}, v_{2,1})$ and

$$m_{2,2} - m_{1,2} = \frac{v_{1,21}}{v_{1,1}}(m_{2,1} - m_{1,1}), \quad \frac{v_{1,21}}{v_{1,1}} = \frac{v_{2,21}}{v_{2,1}}, \quad \text{and} \quad v_{1,2} - \frac{v_{1,21}^2}{v_{1,1}} = v_{2,2} - \frac{v_{2,21}^2}{v_{2,1}};$$

use Theorem F.19. In this case, $\{1\}$ is the structural subset F of Theorem 5.5. In the homoscedastic case, the conditions reduce to $m_{1,1} \neq m_{2,1}$ and

$$m_{2,2} - m_{1,2} = \frac{v_{21}}{v_1}(m_{2,1} - m_{1,1}).$$

An example is $Y^{(1)} \sim N_{0,V}$, $Y^{(2)} \sim N_{(2,1),V}$ with $V = \begin{pmatrix} 2 & 1 \\ 1 & 2 \end{pmatrix}$.

The variable 1 provides indeed better separation than 2 does. If the model is homoscedastic, a standard measure of separation between the two normal components is the Mahalanobis distance between their means w.r.t. the common covariance matrix V. Let $u_k = m_{2,k} - m_{1,k}$, $k = 1, 2$. Because of $v_1 u_2 = v_{21} u_1$, the distance square between the two populations equals

$$(u_1, u_2)V^{-1}(u_1, u_2)^\top = \frac{v_2 u_1^2 + v_1 u_2^2 - 2v_{21} u_1 u_2}{\det V}$$

$$= \frac{u_1^2}{v_1}\frac{v_1 v_2 + v_{21}^2 - 2v_{21}^2}{\det V} = \frac{u_1^2}{v_1}.$$

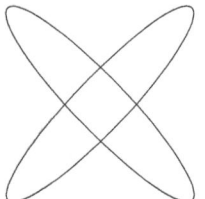

Figure 5.2 *The set $\{1,2\}$ is structural although both variables are independent of L.*

Thus, the presence of the second variable does not increase the distance. The squared distance w.r.t. the first variable alone is again u_1^2/v_1, whereas that w.r.t. the second variable is u_2^2/v_2. Because of $0 < \det V = v_1 v_2 - v_{21}^2$ we have $v_2/v_1 > v_{21}^2/v_1^2 = u_2^2/u_1^2$, that is, $u_1^2/v_1 > u_2^2/v_2$ and s, variable 1 separates better.

(d) A subset can be structural although all its variables are independent of L. An example is $g = D = 2$, $Y^{(1)}$ centered, normal with covariance matrix $V_1 = UBU^\top$, $Y^{(2)}$ centered normal with covariance matrix $V_2 = U^\top BU$, where $U = \frac{1}{\sqrt{2}}\begin{pmatrix} 1 & -1 \\ 1 & 1 \end{pmatrix}$, B diagonal with different entries; see Fig. 5.2.

5.1.2   Normal case

It is well known that normal mixtures can be reparameterized by regressing some of the variables on the rest. In connection with irrelevance, this results is a much simplified model. The restriction of a matrix $V \in \mathbb{R}^{D \times D}$ to the variables in $F \subseteq 1..D$ will be denoted by V_F. If $|F| = d$, the vectors a_F and matrices V_F will also be considered as vectors in $\mathbb{R}^d$ and matrices in $\mathbb{R}^{d \times d}$ via the natural order on F. The canonical projection of a normal distribution to a subset F of variables will be described by the restrictions m_F and V_F of its parameters and by the equality $N_{m,V}(x_F) = N_{m_F,V_F}(x_F)$. In *normal* mixtures, the information on the irrelevance of F_2 w.r.t. F_1 is contained in the parameters of the random vectors $Y^{(j)}_{F_1 \cup F_2}$, $1 \leq j \leq g$, from Lemma 1.12, that is, the mean vectors $m_{j,F_1 \cup F_2} = (m_{j,F_1}, m_{j,F_2})$ and covariance matrices

$$V_{j,F_1 \cup F_2} = \begin{pmatrix} V(Y^{(j)}_{F_1}) & \mathrm{Cov}(Y^{(j)}_{F_1}, Y^{(j)}_{F_2}) \\ \mathrm{Cov}(Y^{(j)}_{F_2}, Y^{(j)}_{F_1}) & V(Y^{(j)}_{F_2}) \end{pmatrix} = \begin{pmatrix} V_{j,F_1} & V_{j,F_1,F_2} \\ V_{j,F_2,F_1} & V_{j,F_2} \end{pmatrix},$$

where $V_{j,F_2,F_1} = V_{j,F_1,F_2}^\top$. By Theorem F.19(a), the variable $Y^{(j)}_{F_2}$ is the linear regression

$$Y^{(j)}_{F_2} = m_{j,F_2|F_1} + G_{j,F_2|F_1} Y^{(j)}_{F_1} + U^{(j)}_{F_2|F_1} \qquad (5.1)$$

of $Y^{(j)}_{F_1}$. Here, $G_{j,F_2|F_1} = V_{j,F_2,F_1} V_{j,F_1}^{-1} \in \mathbb{R}^{F_2 \times F_1}$, $m_{j,F_2|F_1} = m_{j,F_2} - G_{j,F_2|F_1} m_{j,F_1} \in \mathbb{R}^{F_2}$, and the residual $U^{(j)}_{F_2|F_1}$ is an $\mathbb{R}^{F_2}$-valued, centered normal error that explains $Y^{(j)}_{F_2}$ as an affine transform of $Y^{(j)}_{F_1}$. Its covariance matrix is $V_{j,F_2|F_1} = V_{j,F_2} - V_{j,F_2,F_1} V_{j,F_1}^{-1} V_{j,F_1,F_2} = V_{j,F_2} - G_{j,F_2|F_1} V_{j,F_1,F_2} \in \mathbb{R}^{F_2 \times F_2}$. Lemma F.18(b) implies

$$\det V_{j,F_1 \cup F_2} = \det V_{j,F_1} \det V_{j,F_2|F_1}. \qquad (5.2)$$

The residual $U^{(j)}_{F_2|F_1}$ is statistically independent of the random variable $Y^{(j)}_{F_1}$. There are substantial simplifications when F_2 is irrelevant w.r.t. F_1.

5.7 Theorem *Let F_1 and F_2 be nonempty disjoint subsets of $1..D$ such that the covariance matrix $\mathrm{V}X_{F_1}$ is invertible.*

(a) If X is a normal mixture, then the following statements are equivalent.

    (i) The subset F_2 is irrelevant w.r.t. F_1;

    (ii) for all $x_{F_1} \in \mathbb{R}^{F_1}$, the distribution of $m_{j,F_2|F_1} + G_{j,F_2|F_1}x_{F_1} + U^{(j)}_{F_2|F_1}$ does not depend on j;

    (iii) the parameters $G_{j,F_2|F_1}$, $m_{j,F_2|F_1}$, and $V_{j,F_2|F_1}$ do not depend on j.

(b) In this case the common regression matrix $G_{F_2|F_1} = G_{j,F_2|F_1}$ can be computed from X, namely $G_{F_2|F_1} = \mathrm{Cov}(X_{F_2}, X_{F_1})(\mathrm{V}X_{F_1})^{-1}$.

PROOF. (a) By (5.1) and independence, we have for all x_{F_1}

$$P[Y^{(j)}_{F_2} \in \mathrm{d}x_{F_2} \,|\, Y^{(j)}_{F_1} = x_{F_1}]$$
$$= P[m_{j,F_2|F_1} + G_{j,F_2|F_1}Y^{(j)}_{F_1} + U^{(j)}_{F_2|F_1} \in \mathrm{d}x_{F_2} \,|\, Y^{(j)}_{F_1} = x_{F_1}]$$
$$= P[m_{j,F_2|F_1} + G_{j,F_2|F_1}x_{F_1} + U^{(j)}_{F_2|F_1} \in \mathrm{d}x_{F_2} \,|\, Y^{(j)}_{F_1} = x_{F_1}]$$
$$= P[m_{j,F_2|F_1} + G_{j,F_2|F_1}x_{F_1} + U^{(j)}_{F_2|F_1} \in \mathrm{d}x_{F_2}].$$

If (i) holds, then Definition 5.2(iv) says that the first and hence the last line do not depend on j for all x_{F_1}, that is, (ii). If (ii) is true, then identifiability of the normal model asserts that the mean value $m_{j,F_2|F_1} + G_{j,F_2|F_1}x_{F_1}$ and the covariance matrix $V_{j,F_2|F_1}$ of $m_{j,F_2|F_1} + G_{j,F_2|F_1}x_{F_1} + U^{(j)}_{F_2|F_1}$ do not depend on j for all x_{F_1} and (iii) follows. The implication from (iii) to (i) is an obvious consequence of the first and last lines of the above chain of equalities.

(b) By independence of L and $Y^{(j)}$ and of $U^{(j)}_{F_2|F_1}$ and $Y^{(j)}_{F_1}$ and by the formula of total expectation, we first have

$$\mathrm{Cov}(U^{(L)}_{F_2|F_1}, X_{F_1}) = \mathrm{E}\big(U^{(L)}_{F_2|F_1}\widehat{Y}^{(L)\,\top}_{F_1}\big)$$
$$= \sum_j \mathrm{E}\big[U^{(j)}_{F_2|F_1}\widehat{Y}^{(j)\,\top}_{F_1} \,|\, L = j\big]P[L = j] = \sum_j \mathrm{E}\big[U^{(j)}_{F_2|F_1}\widehat{Y}^{(j)\,\top}_{F_1}\big]P[L = j] = 0.$$

The equality $\widehat{X}_{F_2} = G_{F_2|F_1}\widehat{X}_{F_1} + U^{(L)}_{F_2|F_1}$ therefore implies $\mathrm{Cov}(X_{F_2}, X_{F_1}) = G_{F_2|F_1}\mathrm{V}X_{F_1}$ and claim (b). $\qquad\square$

We have obtained a *reparameterization* of the density of $Y^{(j)}_{F_2}$ conditional on $Y^{(j)}_{F_1}$,

$$\Gamma_{F_2|F_1} = \{(G, m, V) \,|\, G \in \mathbb{R}^{F_2 \times F_1}, m \in \mathbb{R}^{F_2}, V \in \mathrm{PD}(F_2)\}.$$

By the product formula, the joint density of $X_{F_1 \cup F_2} = Y^{(L)}_{F_1 \cup F_2}$ is

$$f_{X_{F_1 \cup F_2}}(x_{F_1 \cup F_2}) = f_{X_{F_1}}(x_{F_1})f_{X_{F_2}}[x_{F_2} \,|\, X_{F_1} = x_{F_1}]$$
$$= \Big(\sum_{j=1}^{g} \pi_j N_{m_j, V_j}(x_{F_1})\Big)N_{m, V}(x_{F_2} - Gx_{F_1}). \tag{5.3}$$

Therefore, denoting the parameter space of the predictor variables X_{F_1} by

$$\Gamma_{F_1} = \overset{\circ}{\Delta}_{g-1} \times \{(\mathbf{m}, \mathbf{V}) \,|\, \mathbf{m} \in (\mathbb{R}^{F_1})^g, \mathbf{V} \in \mathcal{V}_c\},$$

the new parameter space becomes

$$\Gamma_{F_1, F_2|F_1} = \Gamma_{F_1} \times \Gamma_{F_2|F_1}.$$

If F_2 is irrelevant w.r.t. F_1, then $U_{F_2|F_1}^{(j)}$ can be taken to be the same random variable for all j within the data model $(Y_i^{(L_i)})_i$. We obtain the representation

$$Y_{i,F_2}^{(j)} = m_{F_2|F_1} + G_{F_2|F_1} Y_{i,F_1}^{(j)} + U_{i,F_2|F_1} \tag{5.4}$$

with the parameters $G_{F_2|F_1}$, $m_{F_2|F_1}$, and $V_{F_2|F_1}$. The parameters $G_{F_2|F_1}$ of diagonal and spherical models vanish and the random variables $Y_{i,F_2}^{(j)}$ do not depend on j.

The next lemma focuses on the homoscedastic model and shows in particular that the Mahalanobis distance between groups does not change with D as long as the subset $(d+1)..D$ is irrelevant.

5.8 Lemma *Let* $Y^{(j)} : \Omega \to \mathbb{R}^D$ *be distributed according to* $N_{m_j, V}$ *with the same* V *for all* $j \in 1..g$. *Let* $F \subseteq 1..D$ *such that* $\mathsf{C}F$ *is irrelevant. Then the Mahalanobis distance of* $EY^{(j)}$ *and* $EY^{(\ell)}$ *equals that of* $EY_F^{(j)}$ *and* $EY_F^{(\ell)}$.

PROOF. The proof rests on the representation

$$V^{-1} = \begin{pmatrix} V_F^{-1} & 0 \\ 0 & 0 \end{pmatrix} + \begin{pmatrix} -G_{\mathsf{C}F|F}^\top \\ I_{\mathsf{C}F} \end{pmatrix} \left(V_{\mathsf{C}F}^{-1} - G_{\mathsf{C}F|F} V_{\mathsf{C}F,F}^\top \right)^{-1} \left(-G_{\mathsf{C}F|F} \quad I_{\mathsf{C}F} \right);$$

see Lemma F.18(c). By irrelevance and by (5.4), we have $m_{\mathsf{C}F} = m_{\mathsf{C}F|F} + G_{\mathsf{C}F|F} m_F$ and, writing $m_j = \binom{m_{j,F}}{m_{j,\mathsf{C}F}}$ and $m_\ell = \binom{m_{\ell,F}}{m_{\ell,\mathsf{C}F}}$,

$$\left(-G_{\mathsf{C}F|F} \quad I_{\mathsf{C}F} \right) m_j = m_{\mathsf{C}F|F} = \left(-G_{\mathsf{C}F|F} \quad I_{\mathsf{C}F} \right) m_\ell,$$

that is, $\left(-G_{\mathsf{C}F|F} \quad I_{\mathsf{C}F} \right)(m_\ell - m_j) = 0$. The claim

$$(m_\ell - m_j)^\top V^{-1}(m_\ell - m_j) = (m_{\ell,F} - m_{j,F})^\top V_F^{-1}(m_{\ell,F} - m_{j,F})$$

follows. □

5.2 Filters

Filter methods use statistics of the data as scores for removing irrelevant and redundant variables. Being simpler than wrappers, they are mainly employed for preselection in analyses of data sets with excessively many variables. In fact, an important application of filters is reducing dimension to a size that can be processed by a wrapper. Often eigenvalue decompositions such as Principal Components Analysis are used for this purpose. However, this approach is naïve. The interesting directions, for instance, those that decompose clustered data, do not have to have anything to do with the largest or smallest eigenvector; see also Sneath [479], Chang [85], Green and Krieger [213], Yeung and Ruzzo [539], and the Introduction of Law et al. [308]. The principal components depend on scaling, which the selected variables should not if they are supposed to be invariant w.r.t. variable rescaling. Many known filter methods are univariate and do not take into account dependencies between variables; see Section 5.2.1. Three more refined multivariate methods are covered in Section 5.2.2. The first is a selector of irrelevant variables applicable to normal mixtures. The other two are generally applicable and select also linear combinations oblique in space. They are the "invariant coordinate system" (ICS) of Tyler et al. [514] and Hui and Lindsay's [259] "white noise matrix."

5.2.1 Univariate filters

Univariate filters are unable to detect dependencies between variables. However, in the presence of a data sets of very high dimension, there is often no other choice but to use them. Many such filters originate from Projection Pursuit. The first filter is somewhat different. Sometimes, but not always, independence indicates irrelevance. The following lemma substantiates a filter for detecting irrelevance in clustering.

5.9 Lemma *Assume that there exists a subset of indices F, $|F| \geq 2$, such that X_F determines the class L and that no pair of random variables with indices in F is independent. If X_k is independent of $X_{\mathsf{C}\{k\}}$, then $\{k\}$ is irrelevant.*

PROOF. The assumptions imply $k \notin F$. Hence, the class L is $X_{\mathsf{C}\{k\}}$-measurable. Therefore, by independence,

$$P[L = j \mid X = x] = \frac{P[L = j, X_{\mathsf{C}\{k\}} \in \mathrm{d}x_{\mathsf{C}\{k\}}, X_k \in \mathrm{d}x_k]}{P[X_{\mathsf{C}\{k\}} \in \mathrm{d}x_{\mathsf{C}\{k\}}, X_k \in \mathrm{d}x_k]}$$

$$= \frac{P[L = j, X_{\mathsf{C}\{k\}} \in \mathrm{d}x_{\mathsf{C}\{k\}}]}{P[X_{\mathsf{C}\{k\}} \in \mathrm{d}x_{\mathsf{C}\{k\}}]} = P[L = j \mid X_{\mathsf{C}\{k\}} = x_{\mathsf{C}\{k\}}].$$

This is the irrelevance of $\{k\}$. □

The assumption of dependence of the variables with indices in F is often satisfied in practice – just think of redundant variables. It is necessary for the lemma to be true. As a counterexample where it fails, let $f = \frac{1}{2}(\mathbf{1}_{[0,1]} + \mathbf{1}_{[2,3]})$ and let X be distributed according to the tensor product $f \otimes f \otimes N_{0,1}$. There are four classes and $F = \{1,2\}$ satisfies all assumptions except that of dependence. Nevertheless, the first two variables are relevant.

Projection pursuit in high-dimensional Euclidean space (Friedman and Tukey [174]) looks for "interesting" low-dimensional subspaces that elucidate structure such as clusters or outliers in a data set. It ignores noisy and information poor variables. From a practical point of view, relevant variables may often be equated with interesting ones. The normal distribution has maximum entropy among all distributions with given mean and variance. Thus, the closer to Gaussian a variable is the less interesting it is. Therefore, projection pursuit considers normal directions least interesting containing no or only little information; see also Diaconis and Freedman [122], Huber [255], Friedman [173], Jones and Sibson [282], and Sun [490].

Therefore, test statistics for normal variables are fertile sources of filters characterizing interestingness quantitatively. Refer to Thode's [501] monograph and Mecklin and Mundfrom's [371] review article on normality testing. Location and scale invariant statistics are particularly appealing. Such scores arise as monotone functions of quotients $S_1(X)/S_2(X)$, where both statistics S_1 and S_2 are location invariant and scale equivariant in the sense of $S_i(sX + t) = sS_i(X)$, $s \geq 0$, $t \in \mathbb{R}$. Some examples of such (univariate) scores $Q(X)$ for nonnormality are as follows. Although projection pursuit refers to arbitrary directions in space, it may be applied to advantage to coordinate directions as well.

5.10 Examples (a) (Moment projection indices) If the mth moment exists, the mth **cumulant** of X, $m \geq 1$, has the representation

$$c_m(X) = \frac{\mathrm{d}^m}{\mathrm{d}z^m} \log \mathrm{E}e^{zX}_{\,|z=0},$$

where the derivative is to be taken along the imaginary axis $i\mathbb{R}$ unless the distribution of X decays fast enough at infinity. Expanding both log and e^{zX} in their power series, we find the first cumulants of X,

$$c_1(X) = \mathrm{E}X,$$
$$c_2(X) = \mathrm{E}X^2 - \mathrm{E}^2X = \mathrm{E}(X - \mathrm{E}X)^2 = \mathrm{V}X,$$
$$c_3(X) = \mathrm{E}X^3 - 3\mathrm{E}X\mathrm{E}X^2 + 2\mathrm{E}^3X = \mathrm{E}(X - \mathrm{E}X)^3,$$
$$c_4(X) = \mathrm{E}X^4 - 4\mathrm{E}X\mathrm{E}X^3 - 3\mathrm{E}^2X^2 + 12\mathrm{E}X^2\mathrm{E}^2X - 6\mathrm{E}^4X.$$

The mth roots of the cumulants are location and scale equivariant in the sense above. It follows that the **standardized absolute cumulants**

$$Q(X) = \frac{|c_m(X)|}{\mathrm{V}^{m/2}X}, \quad m > 2,$$

are location and scale invariant indices. For $m = 3$ and $m = 4$ we obtain the absolute **skewness** and the absolute **kurtosis**, respectively. We have, of course, $Q \geq 0$ with equality if X is normal, since the cumulants for $m \neq 2$ vanish in this case. This follows from the fact that the characteristic function of $N_{0,v}$ is e^{ivt^2}. Brys et al. [74] discuss robust measures of skewness. Ferguson [151] asserts that skewness and kurtosis are most powerful for testing normality against the presence of outliers.

(b) Let X possess a finite second moment and let f_X be continuously differentiable such that xf_X vanishes at $\pm\infty$. The **standardized Fisher information**

$$Q(X) = \mathrm{V}X \int \left(\frac{f_X'(x)}{f_X(x)}\right)^2 f_X(x)\mathrm{d}x - 1$$

arises from $S_1(X) = \sqrt{\mathrm{V}X}$ and $S_2(X) = \left(\int \left(\frac{f_X'(x)}{f_X(x)}\right)^2 f_X(x)\mathrm{d}x\right)^{-1/2}$. Let ϕ be the Lebesgue density of the normal distribution with the same mean value m and variance v as X. From $(\log \phi)'(x) = -\frac{x-m}{v}$, the assumptions, and from integration by parts (D.3), it follows

$$\int \left((\log f_X)'(x) - (\log \phi)'(x)\right)^2 f_X(x)\mathrm{d}x$$

$$= \int (\log f_X)'^2(x)f_X(x)\mathrm{d}x + 2\int f_X'(x)\frac{x-m}{v}\mathrm{d}x + \int \left(\frac{x-m}{v}\right)^2 f_X(x)\mathrm{d}x$$

$$= \int (\log f_X)'^2(x)f_X(x)\mathrm{d}x - \frac{1}{v} = Q(X)/v.$$

Hence, $Q(X) \geq 0$ with equality if and only if X is normal.

(c) Let X possess a finite second moment. The statistics $S_1(X) = \sqrt{2\pi e \mathrm{V}X}$ and $S_2(X) = \exp\left(-\int f_X \log f_X \mathrm{d}x\right)$ and the logarithm as the monotone function yield the **standardized negative Shannon entropy**

$$Q(X) = \int f_X \log f_X \mathrm{d}x + \tfrac{1}{2}\log(2\pi e \mathrm{V}X).$$

With the standard normal Lebesgue density ϕ, we have

$$\mathrm{D}_{\mathrm{KL}}(f_X, \phi) = \int \log \frac{f_X(x)}{\phi(x)} f_X(x)\mathrm{d}x = \int f_X(x)\log f_X(x)\mathrm{d}x - \int f_X(x)\log \phi(x)\,\mathrm{d}x$$

$$= \int f_X(x)\log f_X(x)\mathrm{d}x + \tfrac{1}{2}\int f_X(x)\left(\log 2\pi v + \frac{(x-m)^2}{v}\right)\mathrm{d}x$$

$$= \int f_X(x)\log f_X(x)\mathrm{d}x + \tfrac{1}{2}\log 2\pi v + \tfrac{1}{2} = Q(X).$$

Thus, $Q(X)$ is the Kullback–Leibler divergence $\mathrm{D}_{\mathrm{KL}}(f_X, \phi)$. Again $Q(X) \geq 0$ with equality if and only if X is normal.

Other filters are based on normality or symmetry. Here are some examples.

5.11 Examples (a) Normality test statistics are useful filter indices. Examples are χ^2 goodness-of-fit normality tests, the univariate Shapiro-Wilk [470] test. See also the monograph of Thode [501] and Mecklin and Mundfrom's [371] survey.

(b) (**Friedman's index** [173]) Let Φ be the standard normal c.d.f. If X is standard normal, then $\Phi(X)$ is uniform on $[0,1]$. Therefore, for an arbitrary X with standardization X_0, the integral

$$\int_0^1 (f_{\Phi(X_0)}(x) - 1)^2 \mathrm{d}x$$

measures the deviation of X from the normal distribution with the same mean value and covariance. Friedman [173] also presents a sample version of his index. Hall [223] proposes a similar index that uses the normal density rather than the distribution function.

(c) Asymmetry is a simple sign of nonnormality. A classical measure of asymmetry is the skewness; see Example 5.10(a). If some quantile of X is located far away from what is expected under normality, then there is indication that X is interesting. Given a numeric data set $\mathbf{x} = (x_1, \ldots, x_n)$, the simple statistic

$$s(\mathbf{x}) = \frac{\sum_i (x_i - \operatorname{med} \mathbf{x})}{\operatorname{med}_i |x_i - \operatorname{med} \mathbf{x}|}, \tag{5.5}$$

where $\operatorname{med} \mathbf{x} = \operatorname{med}_i x_i$ denotes the median of the data $\{x_i\}$, is close to zero for symmetric data while it is positive for data skewed to the right. The denominator is the **median absolute deviation from the median**, $\operatorname{MAD}(\mathbf{x})$.

(d) Sometimes the sample space is not a high-dimensional Euclidean space but a high-dimensional *hyperbox* $\prod_k R_k$. Thus, assume now that X_k is bounded with range R_k, the smallest real interval that contains all values of X_k. Yao [537] considers a variable X_k interesting if it is structured, that is, if its distribution is far away from the uniform distribution on R_k. This has maximum entropy among all distributions on R_k. A measure of structuredness is again the Kullback–Leibler divergence $\mathrm{D_{KL}}(X_k, U) = \int f_{X_k} \log \frac{f_{X_k}}{1/|R_k|} = \log |R_k| - \mathrm{H}X_k$, where $|R_k|$ is the length of R_k. After normalization, it turns into a normalized measure of relevance of X_k,

$$W(X_k) = 1 - \frac{\mathrm{H}X_k}{\log |R_k|} \in [0, 1];$$

see also Hwang and Yoon [262].

Approaches to *redundancy* are correlation and stochastic dependence. Some examples follow.

5.12 Examples (a) The classical measure for correlation between the variables k and m is Pearson's **correlation coefficient**

$$r_{k,m}(X) = \frac{\operatorname{Cov}(X_k, X_m)}{\sqrt{\mathrm{V}X_k \mathrm{V}X_m}} = \frac{\mathrm{E}\big(\widehat{X}_k \widehat{X}_m\big)}{\sqrt{\mathrm{E}\widehat{X}_k^2 \, \mathrm{E}\widehat{X}_m^2}}.$$

By the Cauchy-Schwarz inequality, it lies between -1 and 1. A value of $|r_{k,m}|$ near 1 indicates high correlation. The matrix $(|r_{k,m}|)_{k,m}$ can be considered a similarity matrix on the variables. In general, the set of variables is not clustered in a natural way w.r.t. this matrix but vector quantization can be used to partition it in subsets of highly correlated variables. If a k-centers algorithm is used, then the centers act as the selected variables.

(b) A measure of dependence of two random variables X_k and X_t is the normalized mutual information (D.11)

$$\mathrm{NMI}(X_k | X_t) = \frac{\mathrm{MI}(X_k | X_t)}{\mathrm{H}X_k \wedge \mathrm{H}X_t} \in [0, 1].$$

Again, the matrix $\big(\mathrm{NMI}(X_k | X_t)\big)_{k,t}$ represents a similarity matrix and is partitioned by vector quantization as before. Since the densities are not at our disposal, a sample version of the mutual information is used. Yu [540] reports that the mutual information is more effective than the sample correlation.

(c) For the sake of simplicity, let the variables now be discrete, $(1 \ldots a)$-valued, and let K be the stochastic matrix defined by $K(x, y) = P[X_t = y \mid X_k = x]$. A measure of the uncertainty about X_t given $X_k = x$ is the entropy $\mathrm{H}K(x, \cdot)$. Therefore, the sum

$$\sum_x P(X_k = x)\mathrm{H}K(x, \cdot) = -\sum_{x,y} P(X_k = x)K(x, y) \log K(x, y)$$

$$= -\sum_{x,y} P(X_k = x, X_t = y) \log P[X_t = y \mid X_k = x] \in [0, \log a]$$

is a measure of the average uncertainty about X_t given X_k. A small value indicates that X_k predicts X_t well and so X_t may be taken to be redundant given X_k. Talavera [494] uses the square $K(x,y)^2$ instead of $-K(x,y)\log K(x,y)$, obtaining the expression

$$\sum_{x,y} P(X_k = x)K(x,y)^2 = \sum_{x,y} P(X_k = x, X_t = y)P[X_t = y \mid X_k = x] \in [\tfrac{1}{a}, 1]$$

as a measure of predictability of X_t by X_k. It is 1 if X_t is a function of X_k.

Filter methods lead to multiple testing problems which are known to produce many false positive errors, for instance, false rejection of normality. When this is problematic, Benjamini and Hochberg's [35] adjustment of multiple testing should be applied as a remedy; see Appendix F.3.2. A survey of univariate feature selection filters for both supervised and unsupervised classification appears in Ferreira and Figueiredo [152]. They also propose two new dispersion measures for unsupervised classification.

Of course, univariate projections may fail to exhibit structure in the data set. The two coordinate projections in Fig. 5.3 look fairly normal, although the data exhibits a nice structure of two clusters. Therefore, multidimensional projections are preferable in detecting structure. This will be the subject matter of the following section. The interested reader is also referred to Posse [421].

Figure 5.3 *A two-dimensional structured data set with (almost) normal coordinate projections.*

5.2.2 Multivariate filters

Multivariate filters are more recent than univariate ones. The first multivariate filter describes a criterion for relevance. It is applicable when X is a normal mixture and flows from Theorem 5.7.

5.13 Proposition *Let the random variable X be a mixture of g normals and let F_1 and F_2 be nonempty, disjoint subsets of $1..D$. Assume that the covariance matrix $\mathrm{V}X_{F_1}$ is regular and let $G_{F_2|F_1} = \mathrm{Cov}(X_{F_2}, X_{F_1})(\mathrm{V}X_{F_1})^{-1}$ be the regression matrix. The subset F_2 is irrelevant w.r.t. F_1 if and only if the residual $X_{F_2} - G_{F_2|F_1}X_{F_1}$ is normal and independent of X_{F_1}.*

PROOF. Write $X = Y^{(L)}$. For the proof of the direct sense, we know from Theorem 5.7 that $m_{j,F_2|F_1} = m_{F_2|F_1}$ and $G_{j,F_2|F_1} = G_{F_2|F_1}$ are independent of $j \in 1..g$ and that $\widetilde{U}^{(j)}_{F_2|F_1} = Y^{(j)}_{F_2} - G_{F_2|F_1}Y^{(j)}_{F_1}$ is normal $N_{m_{F_2|F_1},V_{F_2,F_1}}$. Let $\widetilde{U}_{F_2|F_1} = \widetilde{U}^{(L)}_{F_2|F_1} = X_{F_2} - G_{F_2|F_1}X_{F_1}$. By the Law of Total Probability and by independence of $Y^{(j)}$ and L and of $\widetilde{U}^{(j)}_{F_2|F_1}$ and $Y^{(j)}_{F_1}$, we have

$$f_{(\widetilde{U}_{F_2|F_1},X_{F_1})}(u,x_{F_1}) = \sum_j f_{(\widetilde{U}^{(j)}_{F_2|F_1},Y^{(j)}_{F_1})}(u,x_{F_1})P[L=j]$$

$$= \sum_j f_{\widetilde{U}^{(j)}_{F_2|F_1}}(u)f_{Y^{(j)}_{F_1}}(x_{F_1})P[L=j] = \sum_j N_{m_{F_2|F_1},V_{F_2,F_1}}(u)f_{Y^{(j)}_{F_1}}(x_{F_1})P[L=j]$$

$$= N_{m_{F_2|F_1},V_{F_2,F_1}}(u)f_{X_{F_1}}(x_{F_1})$$

and the two claims.

In the opposite sense, let $X_{F_2} - G_{F_2|F_1}X_{F_1} \sim N_{m,V}$. From independence of $Y^{(j)}$ and L

and of $\widetilde{U}_{F_2|F_1}^{(j)}$ and $Y_{F_1}^{(j)}$, we infer for all $x_{F_1} \in \mathbb{R}^{F_1}$

$$N_{m,V}(\mathrm{d}u) = P[X_{F_2} - G_{F_2|F_1}X_{F_1} \in \mathrm{d}u] = P[\widetilde{U}_{F_2|F_1} \in \mathrm{d}u \mid X_{F_1} = x_{F_1}]$$

$$= \sum_j P[L = j \mid X_{F_1} = x_{F_1}]P[\widetilde{U}_{F_2|F_1} \in \mathrm{d}u \mid L = j, X_{F_1} = x_{F_1}]$$

$$= \sum_j P[L = j \mid X_{F_1} = x_{F_1}]P[\widetilde{U}_{F_2|F_1}^{(j)} \in \mathrm{d}u \mid L = j, Y_{F_1}^{(j)} = x_{F_1}]$$

$$= \sum_j P[L = j \mid X_{F_1} = x_{F_1}]P[\widetilde{U}_{F_2|F_1}^{(j)} \in \mathrm{d}u].$$

By identifiability of finite normal mixtures, Proposition 1.16, the normal residuals $\widetilde{U}_{F_2|F_1}^{(j)}$ are identically distributed for all j (let $x_{F_1} = 0$, say). It first follows that the mean values $m_{j,F_2|F_1}$ and covariance matrices V_{j,F_2,F_1} of $\widetilde{U}_{F_2|F_1}^{(j)}$ equal m and V, respectively, for all j. Second, the chain of equalities also shows

$$m = \mathrm{E}N_{m,V}(\mathrm{d}u) = \sum_j P[L = j \mid X_{F_1} = x_{F_1}]\mathrm{E}[\widetilde{U}_{F_2|F_1}^{(j)} \mid Y_{F_1}^{(j)} = x_{F_1}]$$

$$= \sum_j P[L = j \mid X_{F_1} = x_{F_1}]\mathrm{E}[Y_{F_2}^{(j)} - G_{j,F_2|F_1}x_{F_1} \mid L = j, Y_{F_1}^{(j)} = x_{F_1}]$$

$$= \sum_j P[L = j \mid X_{F_1} = x_{F_1}]\mathrm{E}[Y_{F_2}^{(j)} - G_{j,F_2|F_1}x_{F_1} \mid Y_{F_1}^{(j)} = x_{F_1}]$$

$$= \sum_j P[L = j \mid X_{F_1} = x_{F_1}]\mathrm{E}[Y_{F_2}^{(j)} \mid Y_{F_1}^{(j)} = x_{F_1}] - G_{j,F_2|F_1}x_{F_1}.$$

Since this is true for all $x_{F_1} \in \mathbb{R}^{F_1}$, $G_{j,F_2|F_1}$, too, is independent of j. The claim finally follows from Theorem 5.7. □

The matrix $G_{F_2|F_1}$ minimizes the trace of the covariance matrix of $X_{F_2} - G_0 X_{F_1}$ w.r.t. G_0. Proposition 5.13 suggests a filter in the form of two statistical tests. Assume that X is a mixture of normals. If a test rejects normality of $X_{F_2} - G_{F_2|F_1}X_{F_1}$ or another test rejects independence of $X_{F_2} - G_{F_2|F_1}X_{F_1}$ and X_{F_1}, then there is evidence for relevance of F_2 w.r.t. F_1. Appropriate normality tests are the multivariate extensions of Shapiro and Wilk's [470] univariate test by Srivastava and Hui [484] and Liang et al. [320].

A second proposal is due to Tyler et al. [514]. Their method is general and independent of mixtures. It also detects relevant directions *oblique* in space. Their starting point is that comparing *different* estimates of multivariate scatter reveals departures from elliptical symmetry. It helps to choose appropriate scatter statistics depending on whether outliers or cluster structures, for instance, are of interest. When outliers are an issue, then the comparison of the scatter matrix with a robust covariance estimate such as a robust M-estimate (Maronna [349], Huber [254]) or Rousseeuw's [448] Minimum Covariance Determinant (MCD) contains the relevant information. They all are easily computed.

The authors propose detecting directions in space responsible for a departure from ellipticity by the spectral decomposition of one matrix w.r.t. the other. Thus, let $S_{\mathbf{x}}$ and $\widetilde{S}_{\mathbf{x}}$ be two scatter estimates of the data $\mathbf{x}$. Assume that both satisfy the usual equivariance w.r.t. affine transformations $x \mapsto Ax + b$: $S_{Ax+b} = \varrho \cdot AS_{\mathbf{x}}A^\top$ and $\widetilde{S}_{Ax+b} = \widetilde{\varrho} \cdot A\widetilde{S}_{\mathbf{x}}A^\top$ with strictly positive numbers ϱ and $\widetilde{\varrho}$. Let $\lambda_{\mathbf{x},k}$, $k \in 1..d$, be the d eigenvalues of $S_{\mathbf{x}}^{-1}\widetilde{S}_{\mathbf{x}}$ with associated eigenvectors collected in the columns of $U_{\mathbf{x}} \in \mathbb{R}^{d \times d}$. Writing $\boldsymbol{\lambda}_{\mathbf{x}} = (\lambda_{\mathbf{x},1}, \ldots, \lambda_{\mathbf{x},d})$, there is the eigenvalue-eigenvector decomposition $S_{\mathbf{x}}^{-1}\widetilde{S}_{\mathbf{x}} = U_{\mathbf{x}}\mathrm{Diag}(\boldsymbol{\lambda}_{\mathbf{x}})U_{\mathbf{x}}^{-1}$, or

$$\widetilde{S}_{\mathbf{x}}U_{\mathbf{x}} = S_{\mathbf{x}}U_{\mathbf{x}}\mathrm{Diag}(\boldsymbol{\lambda}_{\mathbf{x}}) \tag{5.6}$$

of one scatter matrix relative to the other; see also Section A.7(a). Projections of the data to planes spanned by pairs of eigenvectors are expected to reveal departures of the data from ellipticity. They are contained in the matrix $U_{\mathbf{x}}^{\top}\mathbf{x}$. Particularly interesting directions are defined by "small" and "large" eigenvectors. An important property of the decomposition is the affine invariance of the directions defined by the eigenvectors.

5.14 Proposition *Let the eigenvalues of $S_{\mathbf{x}}^{-1}\widetilde{S}_{\mathbf{x}}$ be pairwise distinct. The effect of a (non-singular) affine transformation $x \mapsto Ax + b$ of $\mathbb{R}^d$ is as follows.*

(a) *The eigenvalues of $S_{A\mathbf{x}+b}^{-1}\widetilde{S}_{A\mathbf{x}+b}$ are a (common) positive multiple of those of $S_{\mathbf{x}}^{-1}\widetilde{S}_{\mathbf{x}}$, namely, $\lambda_{A\mathbf{x}+b,k} = \frac{\widetilde{\varrho}}{\varrho}\lambda_{\mathbf{x},k}$.*

(b) *The data set $U_{A\mathbf{x}+b}^{\top}(A\mathbf{x} + b)$ is a coordinatewise nonsingular, affine transform of $U_{\mathbf{x}}^{\top}\mathbf{x}$.*

PROOF. Equation (5.6) applied to $A\mathbf{x}$ reads $\widetilde{S}_{A\mathbf{x}}U_{A\mathbf{x}} = S_{A\mathbf{x}}U_{A\mathbf{x}}\mathrm{Diag}(\boldsymbol{\lambda}_{A\mathbf{x}})$. Inserting here the two equivariance assumptions above yields

$$S_{\mathbf{x}}^{-1}\widetilde{S}_{\mathbf{x}}\big(A^{\top}U_{A\mathbf{x}}\big) = \frac{\widetilde{\varrho}}{\varrho}\big(A^{\top}U_{A\mathbf{x}}\big)\mathrm{Diag}(\boldsymbol{\lambda}_{A\mathbf{x}}).$$

That is, the vector $\boldsymbol{\lambda}_{\mathbf{x}}$ of eigenvalues of $S_{\mathbf{x}}^{-1}\widetilde{S}_{\mathbf{x}}$ is $\frac{\widetilde{\varrho}}{\varrho}\boldsymbol{\lambda}_{A\mathbf{x}}$. Since the eigenvectors are unique up to scalar multiples, we also have $A^{\top}U_{A\mathbf{x}} = U_{\mathbf{x}}B$ for some diagonal matrix B. By $U_{A\mathbf{x}+b} = U_{A\mathbf{x}}$, the first conclusion is claim (a). The second conclusion implies

$$U_{A\mathbf{x}}^{\top}(A\mathbf{x} + b) = U_{A\mathbf{x}}^{\top}A\mathbf{x} + U_{A\mathbf{x}}^{\top}b = BU_{\mathbf{x}}^{\top}\mathbf{x} + U_{A\mathbf{x}}^{\top}b,$$

that is, claim (b). □

The proposition is readily extended to multiple eigenvalues. The directions of the eigenvectors are just no longer unique and it is the eigenspaces that assume their role. On account of part (b) of the proposition, the authors call their method "invariant coordinate selection" (**ICS**). The data set $U_{\mathbf{x}}^{T}\mathbf{x}$ will be called the **ICS transform** of $\mathbf{x}$ (related to $S_{\mathbf{x}}$ and $\widetilde{S}_{\mathbf{x}}$). Although $U_{\mathbf{x}}^{T}\mathbf{x}$ is a linear transform of $\mathbf{x}$, the ICS transformation $\mathbf{x} \mapsto U_{\mathbf{x}}^{T}\mathbf{x}$ is nonlinear since $U_{\mathbf{x}}$ depends on $\mathbf{x}$. An R package entitled ICS is freely available.

Theorem 4 of Tyler et al. [514] deals with mixtures of elliptical distributions with different density generators but a common scale matrix. It says that the span of some subset of invariant coordinates corresponds to Fisher's [156] linear discriminant subspace; see also Mardia et al. [345].

Finally, Hui and Lindsay [259] take an algorithmic approach to projection pursuit via "white noise matrices," again detecting interesting directions oblique in space. Rejecting the largest white noise space, they use its orthogonal complement for the analysis. The starting point is again the general consensus that irrelevance may be equated with uninterestingness and normality. The method uses the spectral decomposition of an easily computed symmetric matrix. Let X be a random vector in $\mathbb{R}^d$ with continuously differentiable Lebesgue density $f = f_X$. We will assume in the rest of this section that X and $\mathrm{D}\log f(X)$ are square integrable.

5.15 Definition *(a) The **density information matrix** of f is defined by*

$$\widetilde{J}_f = \mathrm{E}(\mathrm{D}\log f(X))(\mathrm{D}\log f(X))^{\top} = \int (\mathrm{D}f)(\mathrm{D}f)^{\top}/f\,\mathrm{d}\lambda^d.$$

*(b) The **standardized density information matrix** of f is*

$$J_f = \sqrt{VX}\,\mathrm{E}(\mathrm{D}\log f(X))(\mathrm{D}\log f(X))^{\top}\sqrt{VX} = \sqrt{VX}\int (\mathrm{D}f)(\mathrm{D}f)^{\top}/f\,\mathrm{d}\lambda^d\sqrt{VX}.$$

Both matrices are positive definite. Despite their superficial similarity to the Fisher information, they much differ from it. The Fisher matrix contains *parameter* information, the derivatives being w.r.t. parameters. The density information refers to a single density and differentiation is in sample space. Nevertheless, they share some common characteristics due to the differential calculus applicable to both cases. The following lemma is a consequence of the transformation theorem E.7 for Lebesgue densities. The random vector X and its Lebesgue density are standardized if $EX = 0$ and $VX = I_d$.

5.16 Lemma *The standardized density information matrix of f is the density information matrix of the standardization $x \mapsto \sqrt{\det VX} \cdot f(\sqrt{VX}x + EX)$ of f.*

Hui and Lindsay [259] use the information matrices to develop a goodness-of-fit methodology for the multivariate normal. Recall that the normal distribution has the largest entropy among all distributions with the same mean vector and covariance matrix; see Example D.19. The density information enjoys a similar property.

5.17 Lemma *Let $f > 0$ and assume that $x D f(x)^\top$ is λ^d-integrable and that $x f(x)$ vanishes as $\|x\| \to \infty$. Then $J_f \succeq I_d$ and we have $J_f = I_d$ if and only if X is normal.*

PROOF. Let X be a standardized random vector with Lebesgue density $f > 0$. Writing $f = h N_{0,I_d}$, we have $D \log f(x) = D \log h(x) - x$ and

$$(D \log f)(D \log f)^\top = (D \log h)(D \log h)^\top - (D \log h)x^\top - x(D \log h)^\top + xx^\top.$$

Hence

$$J_f = E(D \log h(X))(D \log h(X))^\top - E(D \log h(X))X^\top - EX(D \log h(X))^\top + EXX^\top.$$

Now, $EXX^\top = VX = I_d$ and, by integration by parts D.3 and $h N_{0,I_d} = f$,

$$E(D \log h(X))X^\top = \int \frac{Dh(x)}{h(x)} x^\top h(x) N_{0,I_d}(dx) = \int Dh(x) x^\top N_{0,I_d}(x) dx$$

$$= -\int h(x) D(x^\top N_{0,I_d}(x)) dx = -\int h(x)(I_d - xx^\top) N_{0,I_d}(x) dx = 0.$$

Hence $J_f = E(D \log h(X))(D \log h(X))^\top + I_d \succeq I_d$. Equality obtains if and only if $D \log h = 0$, that is, $h = 1$, P_X-a.s. □

That is, the standard normal has the Löwner minimal density information and it is unique with this property. The following spectral representation is the main result of the theory. It says that the normal part of f is the eigenspace to the eigenvalue 1 of the standardized density information.

5.18 Theorem *Let $f = f_X > 0$, let $x D f(x)^\top$ be λ^d-integrable, and assume that $x f(x)$ vanishes as $\|x\| \to \infty$. Moreover, let $\lambda_1 \geq \ldots \lambda_2 \geq \cdots \geq \lambda_d$ be the eigenvalues of J_f (see (A.2)) and let $\lambda_r > 1$, $\lambda_{r+1} \leq 1$. Then:*

(a) $\lambda_{r+1} = \cdots = \lambda_d = 1$.

(b) Let X be standardized and let Π be the orthogonal projection onto the eigenspace of J_f to the eigenvalue 1. Then ΠX is standard normal and independent of its orthogonal complement $(I_d - \Pi)X$.

PROOF. Part (a) follows from Lemma 5.17. Now, introduce the eigenbasis as a new coordinate system so that the matrix J_f becomes diagonal. Since X is standardized, integration by parts (D.3) shows

$$EX D \log f(X)^\top = \int x D f(x)^\top \lambda(dx) = -I_d,$$

hence

$$\int (D \log f(x) + x)(D \log f(x) + x)^\top f(x) \lambda(dx)$$

$$= E(D \log f(X) + X)(D \log f(X) + X)^\top = J_f - I_d.$$

Therefore, $J_f(k, k) = 1$ implies $\frac{\partial}{\partial x_k} \log f(x) + x_k = 0$. The general solution to this ordinary differential equation is

$$f(x) = C(x_{\setminus k}) e^{-x_k^2/2}.$$

This equality implies claim (b).   □

The application of Theorem 5.18 requires a probability density rather than data. Hui and Lindsay propose using **kernel density estimators** (see Press et al. [423], Jones et al. [281], and Duong [136]) to convert the data to a density. A kernel density estimate $\hat{f}$ of f for the data $x_1, \ldots, x_n$ sampled from $f\lambda^d$ is the convolution of the empirical measure $\frac{1}{n}\sum_i \delta_{x_i}$ with a symmetric Lebesgue probability density κ centered at the origin, called a **kernel**,[1]

$$\hat{f}(x) = \left(\kappa * \frac{1}{n}\sum_{i=1}^n \delta_{x_i}\right)(x) = \frac{1}{n}\sum_{i=1}^n \kappa(x - x_i).$$

Often a normal kernel κ is used. The square root $H = \sqrt{V\kappa}$ is called the **bandwidth** of κ. Bandwidth selection is a notorious problem. Kernel density estimates with a *fixed* kernel are not consistent due to the expanding effect of the convolution. They become consistent as a sequence of kernels is used whose bandwidths vanish as $n \to \infty$. This follows from Lévy's [317] continuity theorem of harmonic analysis. Therefore, bandwidth formulæ do not only depend on space dimension and spread of the data, but also on their number. Bowman and Foster [63] recommend the bandwidth $\sqrt[d+4]{\frac{4}{n(d+2)}}\sqrt{S}$. In connection with clustering, it should rather depend on the true clusters, which are unknown at this stage, and not on all data. Note also that the formula cannot be used when the scatter matrix S is singular.

The kernel density estimate appears in the denominator of the density information. The integral has no explicit solution. To avoid numerical integration, Hui and Lindsay [259] apply a trick squaring the estimated density and normalizing it. The level surfaces of f and of the normalization f_2 of f^2 are equal. Squaring suppresses lighter regions of the population. This has the pleasant effect of robustifying the analysis. It is important that squaring affects neither independence nor normality. Let Y be distributed according to f_2.

5.19 Lemma *Let X be a random vector in $\mathbb{R}^d$ with a continuous Lebesgue density $f = f_X$ such that $\int f^2(x)\mathrm{d}x < \infty$. Let Π_1 be a projector on $\mathbb{R}^d$, $\Pi_2 = I_d - \Pi_1$ its orthogonal complement. Then:*

(a) The projections $\Pi_1 X$ and $\Pi_2 X$ are independent if and only if $\Pi_1 Y$ and $\Pi_2 Y$ are.

(b) If, in addition, $\Pi_2 X$ is normal then so is $\Pi_2 Y$ and vice versa. In this case, we have
   $V\Pi_2 Y = \frac{1}{2}V\Pi_2 X$.

PROOF. After introducing new coordinates adapted to the orthogonal ranges of the two projections, the density function f turns into a tensor product. The same is true for f^2 and its normalization. This is part (a). Moreover, the square of a normal Lebesgue density is again normal and part (b) follows.   □

The above theory can be applied to detect interesting directions in sample space. Theorem 5.18 says that we have to find the orthogonal complement of the eigenspace to the eigenvalue 1 of the standardized information matrix J_f. By Lemma 5.19, we may replace the standardized information matrix of f with that of the density square transform f_2. If Y is square integrable, the latter is

$$J_{f_2} = 4\sqrt{VY} \frac{\int \mathrm{D}f \cdot \mathrm{D}f^\top \mathrm{d}\lambda^d}{\int f^2 \mathrm{d}\lambda^d} \sqrt{VY} \succeq I_d.$$

---

[1] The mapping $(x, B) \to \int_B \kappa(x - y)\lambda^d(\mathrm{d}y)$ is indeed a (Markov) kernel in the sense of Definition D.14.

In contrast to J_f, this is not an integral of a quotient but a quotient of integrals since the denominator in its definition cancels. J_{f_2} is easily estimated with the kernel method. Indeed, letting $\widehat{f}$ be the kernel density estimate of the population density for $\kappa = N_{0,H^2}$ and $\widehat{f_2}$ the normalization of $\widehat{f}^2$, we find

$$J_{\widehat{f_2}} = 4\sqrt{\mathrm{V}\widehat{f_2}}\,\frac{\int \mathrm{D}\widehat{f} \cdot \mathrm{D}\widehat{f}^\top \,\mathrm{d}\lambda^d}{\int \widehat{f}^2\,\mathrm{d}\lambda^d}\,\sqrt{\mathrm{V}\widehat{f_2}};\qquad(5.7)$$

here, $\mathrm{V}\widehat{f_2}$ is the covariance matrix of the probability defined by $\widehat{f_2}$. Therefore, defining the normalizing constant

$$\widehat{N} = n^2 \int \widehat{f}^2\,\mathrm{d}\lambda^d = \sum_{i,k} N_{0,2H^2}(x_i - x_k),$$

and noting

$$\mathrm{D}\widehat{f}(x) = \frac{-1}{n\det H}\sum_{i=1}^{n} N_{0,I_d}\big(H^{-1}(x - x_i)\big)H^{-2}(x - x_i),$$

$$\mathrm{E}\widehat{f_2} = \frac{1}{2\widehat{N}}\sum_{i,k} N_{0,2H^2}(x_i - x_k)(x_i + x_k),$$

$$\mathrm{M}_2\widehat{f_2} = \frac{1}{\widehat{N}}\sum_{i,k} N_{0,2H^2}(x_i - x_k)\big(\tfrac{1}{2}H^2 + \tfrac{1}{4}(x_i + x_k)(x_i + x_k)^\top\big),$$

the terms in (5.7) become

$$\int \mathrm{D}\widehat{f}\cdot \mathrm{D}\widehat{f}^\top\,\mathrm{d}\lambda^d = \frac{1}{n^2}\sum_{i,k} N_{0,2H^2}(x_i - x_k)\big(\tfrac{1}{2}H^{-2} - \tfrac{1}{4}H^{-2}(x_i - x_k)(x_i - x_k)^\top H^{-2}\big),$$

$$\mathrm{V}\widehat{f_2} = \mathrm{M}_2\widehat{f_2} - \mathrm{E}\widehat{f_2}\mathrm{E}\widehat{f_2}^\top.$$

In practice, the spectrum of $J_{\widehat{f_2}}$ will not contain the exact value 1. This part is then taken by the eigenvalues close to 1. The most interesting directions are found in the orthogonal complement of their eigenspaces. This is the (direct) sum of the eigenspaces associated with the larger eigenvalues $\lambda_1, \ldots, \lambda_r$. Hui and Lindsay [259] proceed to develop a sequential test for estimating the informative projections based on the eigenvalues of $J_{\widehat{f_2}}$.

5.3 Wrappers

Variable selection can be combined with the mixture and classification models of Chapter 1 to obtain wrappers. They use the feedback from the latter to control the former. The idea is to find a nonredundant part of the unique structural subset constructed in Theorem 5.5. This is a subset $F \subseteq 1\mathinner{.\,.}D$ of indices of minimum size that best explains X_F as a mixture of well-separated statistical populations. The wrapper runs through a collection of subsets F of small size, producing a clustering w.r.t. each of them. There are many ways of doing this. Complete enumeration is possible only when D is small. The different subsets of variables are used to create partitions of the objects. These are evaluated according to some validation measure (see Chapter 4), and the best one is returned along with its subset. This section presents four likelihood-based wrappers. Section 5.3.1 uses the likelihood ratio test and Raftery and Dean's [426] approach by Bayes factors is presented in Section 5.3.2. In order to work smoothly, wrappers need a moderate dimension D. Nevertheless, Section 5.3.3 will present two algorithms, which can cope with more than thousand variables. They apply the ML paradigm to selection in mixture and classification models. An application appears in Section 6.2.3. However, when D is excessive, they too must be preceded with a filter.

An estimator of a subset of variables is called a subset selector. It is particularly appealing when it is invariant w.r.t. natural transformations on E, that is, it selects the same subset for the original and the transformed data set. The symbol $2^{1 \cdots D}$ will stand for the collection of all subsets of the integral interval $1 .. D$.

5.20 Definition

(a) A **subset selector** is a map $E^n \to 2^{1 \cdots D}$.

(b) A subset selector $s : E^n \to 2^{1 \cdots D}$ is **invariant** w.r.t. a transformation t on E if

$$s(t(x_1), \ldots, t(x_n)) = s(x_1, \ldots, x_n)$$

for all data sets $(x_1, \ldots, x_n) \in E^n$.

Since variables are to be conserved here, the most general linear transformations of Euclidean data applicable in the context of subset selection are diagonal. They linearly rescale the variables. Subset selectors may also be combined with trimming (Chapter 2) to obtain a trimmed likelihood-based selector. The selection of relevant subsets is amenable to statistical tests. In this connection properties of tests that lead to invariant selectors are intriguing.

5.21 Lemma *For each $F \subseteq 1 .. D$, let λ_F be a test statistic with threshold β, invariant w.r.t. the transformation t on E. Then the selector*

$$s(\mathbf{x}) = F \quad \Leftrightarrow \quad \lambda_F(\mathbf{x}) \geq \beta$$

is t-invariant.

PROOF. We have $s(t(x_1), \ldots, t(x_n)) = F$ if and only if $\lambda_F(t(x_1), \ldots, t(x_n)) \geq \beta$. By invariance of λ_F (Definition F.23) this is equivalent with $\lambda_F(x_1, \ldots, x_n) \geq \beta$, that is, with $s(x_1, \ldots, x_n) = F$. □

5.3.1 Using the likelihood ratio test

The idea is to estimate a nonredundant part of the structural subset with an irrelevant complement promised in Theorem 5.5. No efficient algorithm for this task is known and we recur to a heuristic. Starting from a suitably chosen variable, we alternatingly add relevant and delete irrelevant variables until a steady state is reached. Given $F \subset 1 .. D$ and $k \in \complement F$, we fix g and consider the hypothesis

$H_0 : \{k\}$ is irrelevant w.r.t. F against $H_1 :$ general normal assumptions.

Let us apply a likelihood ratio test (LRT) under the normal assumptions described in Section 5.1.2. The likelihood function under H_0 is given by (5.3),

$$f_{X_{F \cup \{k\}}}(\mathbf{x}_{F \cup \{k\}}; \boldsymbol{\pi}, \mathbf{m}_F, \mathbf{V}_F, G, m, v)$$

$$= \prod_i \left(\sum_{j=1}^{g} \pi_j N_{m_{j,F}, V_{j,F}}(x_{i,F}) \right) N_{m,v}(x_{i,k} - G x_{i,F}). \tag{5.8}$$

It is tested against the general normal model

$$f_{X_{F \cup \{k\}}}(\mathbf{x}_{F \cup \{k\}}; \boldsymbol{\pi}, \mathbf{m}_{F \cup \{k\}}, \mathbf{V}_{F \cup \{k\}}) = \prod_i \sum_{j=1}^{g} \pi_j N_{m_{j,F \cup \{k\}}, V_{j,F \cup \{k\}}}(x_{i,F \cup \{k\}}). \tag{5.9}$$

Now use the EM algorithm to determine the logarithms $\ell^*_{F,k|F}$ and $\ell^*_{F \cup \{k\}}$ of consistent local maxima of (5.8) and (5.9), respectively; see Theorem 1.27 and its corollaries. Asymptotic normality of the local maximizers allows us to apply the asymptotic theory of Aitchison and Silvey [4] (see Appendix F.3.1) and so $2(\ell^*_{F \cup \{k\}} - \ell^*_{F,k|F})$ is asymptotically distributed as $\chi^2_{p_1 - p_0}$; here, p_0 and p_0 are the numbers of parameters appearing in (5.8) and (5.9), respectively. We have derived the following procedure for updating a subset F in the search of relevance. It includes a variable $k \notin F$ in F if it is relevant w.r.t. F and deletes $k \in F$ from F if it is irrelevant w.r.t. $F \backslash \{k\}$, both w.r.t. the selected mixtures of g components.

5.22 Procedure (Relevance via LRT)
// Input: A nonempty subset $F \subseteq 1 .. D$.
// Output: A subset F_{new} or "stop."

1. *(Inclusion step)* For all $k \notin F$, compute the logarithms $\ell^*_{F,k|F}$ and $\ell^*_{F \cup \{k\}}$ of favorite local maxima of (5.8) and (5.9), respectively. If the maximum difference $2(\ell^*_{F \cup \{k^*\}} - \ell^*_{F,k^*|F})$ is positive and large (that is, $\mathrm{H}_0(k^*)$ is rejected), then set $F' = F \cup \{k^*\}$. Otherwise, set $F' = F$.

2. *(Removal step)* If $F' \neq \emptyset$, then compute $\ell^*_{F'}$ (see (5.9)) and, for all $k \in F'$, compute the logarithm $\ell^*_{F' \backslash \{k\}, k | F' \backslash \{k\}}$ of a favorite local maximum of (5.8) with $F' \backslash \{k\}$ substituted for F. If the least difference $2(\ell^*_{F'} - \ell^*_{F' \backslash \{k^*\}, k^* | F' \backslash \{k^*\}})$ is negative and small (that is, $\mathrm{H}_0(k^*)$ is not rejected), then set $F_{\text{new}} = F' \backslash \{k^*\}$. Otherwise, set $F_{\text{new}} = F'$.

3. If $F_{\text{new}} \neq F$, then return F_{new}, otherwise return "stop."

Now, iterate inclusion and removal steps in an algorithm starting from some relevant singleton F_0. If the stop signal appears, that is, if $F_{\text{new}} = F$ in some call to the procedure, then all $k \notin F$ are irrelevant w.r.t. F and F contains no k irrelevant w.r.t. $F \backslash \{k\}$. It is proposed as an estimate, although it will not have to be a part of the structural subset of Theorem 5.5. Its parameters are those of the last update of (5.9) in step 1 or 2. The fact that the stop signal does appear at some iteration is proved in the proposition below. The final set may be empty. This indicates that $1 .. D$ is irrelevant and that the data set is not clustered at all. Lemma 5.2(iv) says in this case that the distribution of $Y^{(j)}$ does not depend on j.

5.23 Proposition *The iteration of Procedure 5.22 returns no subset twice and terminates with the stop signal.*

PROOF.   Let $\ell^*_{\complement F | F}$ be the logarithm of maximum of the regression of $\complement F$ on F. For $F \subseteq 1 .. D$, define $h(F) = \ell^*_{F, \complement F | F} = \ell^*_F + \ell^*_{\complement F | F}$, the maximum log-likelihood of X under irrelevance of $\complement F$; see (5.3). Since there are only finitely many subsets $F \subseteq 1 .. D$, the proposition will be proved if we show that each update in the inclusion and removal step strictly increases the target function h. So assume that F is adapted in an inclusion step, that is, $F' \neq F$, and let $\alpha > 0$ be the largest value of $2(\ell^*_{F \cup \{k^*\}} - \ell^*_{F, k^* | F})$ achieved in it. The product formula

$$f_{X_{\complement F}}[x_{\complement F} \mid X_F = x_F] = f_{X_{\complement(F \cup \{k^*\})}}[x_{\complement(F \cup \{k^*\})} \mid X_{k^*} = x_{k^*}, X_F = x_F] f_{X_{k^*}}[x_{k^*} \mid X_F = x_F]$$

shows

$$\ell^*_{\complement F | F} = \ell^*_{\complement(F \cup \{k^*\}) | (F \cup \{k^*\})} + \ell^*_{k^* | F}. \tag{5.10}$$

We infer

$$h(F \cup \{k^*\}) - h(F) = \ell^*_{F \cup \{k^*\}} + \ell^*_{\complement(F \cup \{k^*\}) | (F \cup \{k^*\})} - \ell^*_F - \ell^*_{\complement F | F}$$
$$= \ell^*_{F \cup \{k^*\}} - \ell^*_F - \ell^*_{k^* | F} = \ell^*_{F \cup \{k^*\}} - \ell^*_{F, k^* | F} = \alpha/2 > 0.$$

The proof for the removal step is similar. $\square$

The method based on the LRT enjoys desirable scale invariance.

5.24 Lemma *The test statistics of the inclusion and removal steps of Procedure 5.22 are invariant w.r.t. variable rescaling.*

PROOF. Consider the inclusion step. In view of H_1, first note that

$$N_{a+Am, A^\top VA}(a + Ax)$$

$$= \frac{1}{\sqrt{\det 2\pi A^\top VA}} e^{-\frac{1}{2}(A(x-m))^\top (A^\top VA)^{-1}(A(x-m))} = \frac{1}{|\det A|} N_{m, V}(x)$$

for any vector a and any nonsingular matrix A. Now consider the affine transformation $x_{F \cup \{k\}} \mapsto a_{F \cup \{k\}} + A_{F \cup \{k\}} x_{F \cup \{k\}}$ with a diagonal matrix $A_{F \cup \{k\}}$. With the definitions $m_j^{(1)} = a_{F \cup \{k\}} + A_{F \cup \{k\}} m_{j, F \cup \{k\}}$ and $V_j^{(1)} = A_{F \cup \{k\}} V_{j, F \cup \{k\}} A_{F \cup \{k\}}$, we have

$$\prod_i \sum_j \pi_j N_{m_j^{(1)}, V_j^{(1)}}(a_{F \cup \{k\}} + A_{F \cup \{k\}} x_{i, F \cup \{k\}}) \tag{5.11}$$

$$= |\det A_{F \cup \{k\}}|^{-n} \prod_i \sum_j \pi_j N_{m_{j, F \cup \{k\}}, V_{j, F \cup \{k\}}}(x_{i, F \cup \{k\}}).$$

Turning to H_0, we define $m_j^{(0)} = a_F + A_F m_{j, F}$ and $V_j^{(0)} = A_F V_{j, F} A_F$, obtaining in a similar way

$$\sum_j \pi_j N_{m_j^{(0)}, V_j^{(0)}}(a_F + A_F x_{i, F}) = \frac{1}{|\det A_F|} \sum_j \pi_j N_{m_{j, F}, V_{j, F}}(x_{i, F}). \tag{5.12}$$

Moreover, with $v' = A_{\{k\}}^2 v$, $G' = A_{\{k\}} G A_F^{-1}$, and $m' = a_k - G' a_F + A_{\{k\}} m$, we have

$$N_{m', v'}(a_k + A_{\{k\}} x_{i,k} - G'(a_F + A_F x_{i,F})) \tag{5.13}$$

$$= \frac{1}{\sqrt{2\pi v'}} e^{-(a_k + A_{\{k\}} x_{i,k} - G'(a_F + A_F x_{i,F}) - m')^2 / 2v'}$$

$$= \frac{1}{|A_{\{k\}}|} \frac{1}{\sqrt{2\pi v}} e^{-(x_{i,k} - G x_{i,F} - m)^2 / 2v} = \frac{1}{|A_{\{k\}}|} N_{m, v}(x_{i,k} - G x_{i,F}),$$

and (5.12) and (5.13) show

$$\prod_i \sum_j \pi_j N_{m_j^{(0)}, V_j^{(0)}}(a_F + A_F x_{i,F}) N_{m', v'}(a_k + A_{\{k\}} x_{i,k} - G'(a_F + A_F x_{i,F}))$$

$$= |\det A_{F \cup \{k\}}|^{-n} \prod_i \sum_j \pi_j N_{m_{j, F}, V_{j, F}}(x_{i,F}) N_{m, v}(x_{i,k} - G x_{i,F}) \tag{5.14}$$

since A is diagonal. If $\mathbf{V}_{F \cup \{k\}}$ satisfies the HDBT constraints with constant c, so does the transformed g-tuple $\mathbf{V}_{F \cup \{k\}}^{(1)}$ and vice versa. The same remark also applies to the model under H_0. Therefore, and since the maximum w.r.t. the transformed parameters $m_j^{(t)}$ and $V_j^{(t)}$ equals that w.r.t. the original parameters, the constrained maxima of (5.11) and (5.14) are $|\det A_{F \cup \{k\}}|^{-n}$ times those for the original data. This concludes the proof for the inclusion step and the removal step is similar. □

Lemma 5.21 now clarifies the invariance of subset selection by Procedure 5.22.

5.25 Proposition *Procedure 5.22 is invariant w.r.t. variable rescaling.*

5.3.2   Using Bayes factors and their BIC approximations

Among other things, Jeffreys [273, 274], p. 49 ff. and Chapter V,[2] introduced a Bayesian version of the likelihood ratio test, **Bayesian hypothesis testing**. Let Θ be a p_1-dimensional, open differentiable manifold, let Θ_0 be a p_0-dimensional differentiable sub-manifold, $p_0 < p_1$, and set $\Theta_1 = \Theta \setminus \Theta_0$. Let $f(x; \vartheta)$, $\vartheta \in \Theta$, be some likelihood function and consider the two

---

[2]Note that Jeffreys denotes the data by θ.

hypotheses $H_0 : \vartheta \in \Theta_0$ and $H_1 : \vartheta \in \Theta_1$. Recall that the likelihood ratio test uses the quotient of the maximum likelihoods $\frac{\max_{\vartheta \in \Theta_1} f(x \mid \vartheta)}{\max_{\vartheta \in \Theta_0} f(x \mid \vartheta)}$ as a score for deciding whether H_0 should be rejected in favor of the alternative. If no maxima exist, then the estimated solutions, often local maxima, are used instead. A Bayesian test uses prior probabilities μ_0 on Θ_0 and μ_1 on Θ instead of (local) maxima. It is reasonable to choose prior probabilities equivalent with the surface measures on Θ_0 and Θ; see Appendix F.1. The total prior probability is $\mu = (1 - \alpha)\mu_0 + \alpha\mu_1$, $0 < \alpha < 1$. Let $U : \Omega \to \Theta$ be a random, μ-distributed parameter in a Bayesian framework. One computes the posterior odds ($=$ probability/(1–probability))

$$\frac{P[U \in \Theta_1 \mid X = x]}{P[U \in \Theta_0 \mid X = x]} = \frac{f_X[x \mid U \in \Theta_1]}{f_X[x \mid U \in \Theta_0]} \times \frac{P_U(\Theta_1)}{P_U(\Theta_0)}$$

$$= \frac{\int_\Theta f_X(x \mid U = \vartheta)\mu_1(\mathrm{d}\vartheta)}{\int_{\Theta_0} f_X(x \mid U = \vartheta)\mu_0(\mathrm{d}\vartheta)} \times \frac{\alpha}{1 - \alpha} = \frac{I_1}{I_0} \times \frac{\alpha}{1 - \alpha}$$

in order to decide which of the two hypotheses is likelier. The quotient

$$\mathrm{BF} = \frac{I_1}{I_0} = \frac{\int_\Theta f_X(x \mid U = \vartheta)\mu_1(\mathrm{d}\vartheta)}{\int_{\Theta_0} f_X(x \mid U = \vartheta)\mu_0(\mathrm{d}\vartheta)} \tag{5.15}$$

of the integrated likelihoods is called the **Bayes factor** (BF). Thus, the Bayes factor transforms prior odds to posterior odds. It equals the posterior odds if $\alpha = 1/2$, a number recommended by Jeffreys. The parameter space Θ may be substituted for Θ_1 in the Bayes factor since Θ_0 is a μ_1-null set. This makes the two parameter spaces nested as in the application below.

One of the advantages of the Bayes factor over the likelihood ratio test is that it does not only serve for rejecting the null hypothesis but provides probabilities for hypothesis *and* alternative to be true. A value ≥ 10 (≤ 0.1) of the Bayes factor is considered a strong support for H_1 (H_0). For more information on Bayes factors, the interested reader is referred to Kass and Raftery's [285] survey article.

Integrated likelihoods $\int_{\Theta_k} f_X(x \mid U = \vartheta)\mu_k(\mathrm{d}\vartheta)$ and thus Bayes factors are not easily computed in general but approximations do exist. One of them is Laplace's method; see de Bruijn [72] and Tierney and Kadane [504]. Another one is Schwarz's [460] BIC approximation (4.2). Inserting it in the logarithm of the Bayes factor (5.15), we obtain

$$\log \mathrm{BF} \approx \log f(x; \vartheta_1^*) - \log f(x; \vartheta_0^*) - \tfrac{1}{2}(p_1 - p_0)\log n.$$

The difference to the log-likelihood ratio is just the term involving the two dimensions.

Raftery and Dean [426] applied Bayes factors and their BIC approximations to the present task of deciding whether a variable $k \notin F$ is irrelevant w.r.t. $F \subset 1 .. D$ in the normal case. Their method does not only offer a way of deciding on this issue, it provides at the same time estimates of the distributional model (full, diagonal, or spherical covariance structure and constraints across components) and of the number of components. Let us use the notation of Section 5.1.2 and let $g_{\max}$ be an upper bound on the number of groups. The parameter space of the general normal model, irrespective of relevance, is $\Theta = \Theta_{F_1 \cup F_2} = \overset{\circ}{\Delta}_{g-1} \times \mathbb{R}^{g(|F_1|+|F_2|)} \times \mathrm{PD}(|F_1| + |F_2|)^g$ and the likelihood function is $\sum_j \pi_j N_{m_{j,F_1 \cup F_2}, V_{j,F_1 \cup F_2}}(x_{F_1 \cup F_2})$. The integrated likelihood of the general model on $F_1 \cup F_2$ is

$$I_g^{F_1} \cup F_2 = \int_{\Theta_{F_1 \cup F_2}} \sum_{j=1}^g \pi_j N_{m_j, V_j}(x_{F_1 \cup F_2})\mu_1(\mathrm{d}\boldsymbol{\pi}, \mathrm{d}\mathbf{m}, \mathrm{d}\mathbf{V}).$$

After a change of parameters via the regression model, the subspace under irrelevance is $\Theta_0 = \Theta_{F_1} \times \Theta_{F_2 \mid F_1}$, a differential manifold of dimension $p_0 = \dim_{\Theta_0} < \dim_\Theta = p_1$, and the related likelihood function with the parameter $(\boldsymbol{\pi}, \mathbf{m}, \mathbf{V}, G, m, V) \in \Theta_0$ assumes the product form (5.3),

$$f(x_{F \cup F_2}; \boldsymbol{\pi}, \mathbf{m}, \mathbf{V}, G, m, V) = \left(\sum_{j=1}^g \pi_j N_{m_j, V_j}(x_{F_1})\right) N_{m,V}(x_{F_2} - G x_{F_1}).$$

Choosing the prior probability μ_0 on $\Theta_{F_1,F_2|F_1} = \Theta_{F_1} \times \Theta_{F_2|F_1}$ as a product $\mu_0 = \mu_{F_1} \otimes \mu_{F_2|F_1}$, the integrated likelihood of the model under irrelevance turns into the product

$$I_g^{F_1} \cdot I_g^{F_2|F_1} = \int_{\Theta_{F_1}} \sum_{j=1}^{g} \pi_j N_{m_j,V_j}(x_{F_1})\mu_{F_1}(\mathrm{d}\boldsymbol{\pi}, \mathrm{d}\mathbf{m}, \mathrm{d}\mathbf{V})$$

$$\cdot \int_{\Theta_{F_2|F_1}} N_{m,V}(x_{F_2} - Gx_{F_1})\mu_{F_2|F_1}(\mathrm{d}m, \mathrm{d}G, \mathrm{d}V).$$

According to (4.2), the BIC approximations to the logarithms of the integrated likelihoods for the optimal numbers of components are

$$\log I_{g_2^*}^{F_1 \cup F_2} \approx \mathrm{BIC}_{F_1 \cup F_2, g_2^*}(x_{F_1 \cup F_2})$$

$$= \log f(x_{F_1 \cup F_2}; \boldsymbol{\pi}_{F_1 \cup F_2}^*, \mathbf{m}_{F_1 \cup F_2}^*, \mathbf{V}_{F_1 \cup F_2}^*)$$

$$- \frac{1}{2}\left(g_2^* - 1 + g_2^* \frac{(|F_1| + |F_2| + 3)(|F_1| + |F_2|)}{2}\right) \log n,$$

$$\log I_{g_1^*}^{F_1} \approx \mathrm{BIC}_{F_1, g_1^*}(x_{F_1})$$

$$= \log f(x_{F_1}; \boldsymbol{\pi}_{F_1}^*, \mathbf{m}_{F_1}^*, \mathbf{V}_{F_1}^*) - \frac{1}{2}\left(g_1^* - 1 + g_1^* \frac{(|F_1| + 3)|F_1|}{2}\right) \log n,$$

$$\log I_{F_2|F_1} \approx \mathrm{BIC}_{F_2|F_1}(x_{F_1 \cup F_2})$$

$$= \log N_{m^*, V^*}(x_{F_2} - G^* x_{F_1}) - \frac{|F_2|}{4}(2|F_1| + |F_2| + 3) \log n$$

$$= -\frac{n}{2}(\log 2\pi + 1 + \log \det V^*) - \frac{|F_2|}{4}(2|F_1| + |F_2| + 3) \log n.$$

Here, g_1^* and g_2^* are the optimal numbers of components in $2..g_{\max}$ determined for $\mathrm{BIC}_{F_1, g}(x_{F_1})$ and $\mathrm{BIC}_{F_1 \cup F_2, g}(x_{F_1 \cup F_2})$, respectively, and

$$V^* = \frac{1}{n}\sum_i (x_{i,F_2} - G^* x_{i,F_1} - m^*)(x_{i,F_2} - G^* x_{i,F_1} - m^*)^\top.$$

The optimal parameters $\boldsymbol{\pi}_{F_1 \cup F_2}^*, \mathbf{m}_{F_1 \cup F_2}^*, \mathbf{V}_{F_1 \cup F_2}^*$ and $\boldsymbol{\pi}_{F_1}^*, \mathbf{m}_{F_1}^*, \mathbf{V}_{F_1}^*$ exist under suitable HDBT constraints; see Corollary 1.25. The logarithm of the Bayes factor (5.15) for the optimal numbers of components turns into

$$\log \mathrm{BF}^* = \log I_{g_2^*}^{F_1 \cup F_2} - (\log I_{F_1, g_1^*} + \log I_{F_2|F_1})$$

$$\approx \mathrm{BIC}_{F_1 \cup F_2, g_2^*}(x_{F_1 \cup F_2}) - (\mathrm{BIC}_{F_1, g_1^*}(x_{F_1}) + \mathrm{BIC}_{F_2|F_1}(x_{F_1 \cup F_2})). \tag{5.16}$$

Raftery and Dean [426] establish the following procedure. Like its LRT counterpart, Procedure 5.22, it is a building block in an algorithm that attempts to find a maximum relevant subset $F \subseteq 1..D$ w.r.t. an optimal number of components by iteratively including and removing single indices k. The inclusion step selects the variable k that minimizes BF^* (or rather (5.16)) for $F_1 = F$ and $F_2 = \{k\}$ and the removal step is analogous.

5.26 Procedure (Relevance via Bayes factors)
// Input: An upper bound $g_{\max}$ on the number of groups and a subset $F \subset 1..D$.
// Output: A subset F_{new} or "stop."

1. *(Inclusion step)* For all $k \notin F$, compute (5.16) with $F_1 = F$ and $F_2 = \{k\}$ and for $g_1^*, g_2^* \leq g_{\max}$. If the largest value is positive, with variable k^*, then set $F' = F \cup \{k^*\}$ (k^* is relevant w.r.t. F), otherwise set $F' = F$.

2. *(Removal step)* If $F' \neq \emptyset$, then, for all $k \in F'$, compute (5.16) with $F_1 = F'\backslash\{k\}$ and $F_2 = \{k\}$. If the smallest value is negative, with variable k^*, then set $F_{\mathrm{new}} = F'\backslash\{k^*\}$ (k^* is irrelevant w.r.t. $F'\backslash\{k^*\}$), otherwise set $F_{\mathrm{new}} = F'$.

3. *If $F_{\mathrm{new}} \neq F$, then return F_{new}, otherwise return "stop."*

Again, inclusion and removal steps are iterated starting from some singleton F_0 that gives the most evidence of univariate clustering. When the stop signal appears, that is, if $F_{\text{new}} = F$ for some call to the procedure, then all $k \notin F$ are irrelevant w.r.t. F and F contains no k irrelevant w.r.t. $F \backslash \{k\}$. It is the proposed solution, although it does not have to be the set constructed in Theorem 5.5. Its parameters are those found in the last update of either an inclusion or a removal step along the algorithm. The proposition below shows that the algorithm actually terminates with the stop signal. The final set may be empty. This indicates that $1 .. D$ is irrelevant and that the data set is not clustered.

5.27 Proposition *Iteration of Procedure 5.26 returns no subset twice and terminates with the stop signal.*

PROOF. Define $h(F) = \text{BIC}_{F,g_F^*}(x_F) + \text{BIC}_{\mathsf{C}F|F}(x)$, $F \subseteq 1 .. D$. Since there are only finitely-many subsets $F \subseteq 1 .. D$, the proposition will be proved as soon as we have shown that each update in the inclusion and removal step strictly increases the target function h. So, assume that we have $F' \neq F$ in an inclusion step and let $\alpha > 0$ be the largest value of (5.16) achieved in it. The inclusion step implies

$$\text{BIC}_{F \cup \{k^*\}, g_{F \cup \{k^*\}}^*}(x_{F \cup \{k^*\}}) - \text{BIC}_{F, g_F^*}(x_F) = \text{BIC}_{k^*|F}(x_{F \cup \{k^*\}}) + \alpha.$$

Let $\ell_{\mathsf{C}F|F}^*$ be the maximum of the regression of $\mathsf{C}F$ on F. By (5.10) we have $\ell_{\mathsf{C}F|F}^* = \ell_{\mathsf{C}(F \cup \{k^*\})|(F \cup \{k^*\})}^* + \ell_{k^*|F}^*$. Hence, if p_1 (p_2) is the number of parameters appearing in the regression of $\mathsf{C}(F \cup \{k^*\})$ on $F \cup \{k^*\}$ ($\mathsf{C}F$ on F), we infer

$$\text{BIC}_{\mathsf{C}(F \cup \{k^*\})|F \cup \{k^*\}}(x) - \text{BIC}_{\mathsf{C}F|F}(x)$$
$$= \ell_{\mathsf{C}(F \cup \{k^*\})|(F \cup \{k^*\})}^* - \tfrac{p_1}{2} \log n - \ell_{\mathsf{C}F|F}^* + \tfrac{p_2}{2} \log n$$
$$= -\ell_{k^*|F}^* + \tfrac{1}{2}(p_2 - p_1) \log n = -\text{BIC}_{k^*|F}(x_{F \cup \{k^*\}})$$

and conclude

$$h(F \cup \{k^*\}) - h(F)$$
$$= \text{BIC}_{F \cup \{k^*\}, g_{F \cup \{k^*\}}^*} + \text{BIC}_{\mathsf{C}(F \cup \{k^*\})|F \cup \{k^*\}} - \text{BIC}_{F, g_F^*} - \text{BIC}_{\mathsf{C}F|F} = \alpha > 0.$$

The proof for the removal step is analogous. □

From the proof of Lemma 5.24, it follows that the Bayes factor (5.15) is invariant w.r.t. variable scaling. The same is true for its BIC approximation used in Procedure 5.26. This leads us to the following proposition.

5.28 Proposition *Procedure 5.26 is invariant w.r.t. variable scaling.*

PROOF. We have to show that the BIC approximations appearing in the inclusion and removal steps of Procedure 5.26 are invariant w.r.t. such transformations. Consider the inclusion step. We prove that the BIC approximation to the logarithm of the Bayes factor BF^*, (5.16), is invariant. Let $a \in \mathbb{R}^D$ and let $A \in \mathbb{R}^{D \times D}$ be diagonal. Using (5.11)–(5.13), we have

$$\text{BIC}_{F \cup \{k\}, g_2^*}((a + Ax)_{F \cup \{k\}}) - \big(\text{BIC}_{F, g_1^*}((a + Ax)_F) + \text{BIC}_{\{k\}|F}((a + Ax)_{F \cup \{k\}})\big)$$
$$= -n \log |\det A_{F \cup \{k\}}| + \text{BIC}_{F \cup \{k\}, g_2^*}(x_{F \cup \{k\}})$$
$$\quad - \big(-n \log |\det A_F| + \text{BIC}_{F, g_2^*}(x_F) - n \log |A_{\{k\}}| + \text{BIC}_{\{k\}|F})(x_{F \cup \{k\}})\big)$$
$$= \text{BIC}_{F \cup \{k\}, g_2^*}(x_{F \cup \{k\}}) - \big(\text{BIC}_{F, g_1^*}(x_F) + \text{BIC}_{\{k\}|F}(x_{F \cup \{k\}})\big)$$

since A is diagonal. This proves the lemma. □

Raftery and Dean [426] illustrate the favorable performance of their algorithm with some simulated and real data sets of, however, rather low dimensions. They report that their procedure selects four out of five variables from Campbell and Mahon's [76] CRAB data reducing the error rate by more than 80%. They again notice that *the presence of only a single noise variable in even a low-dimensional setting can cause the clustering results to deteriorate.*

5.3.3 Maximum likelihood subset selection

This section applies local maximum likelihood estimation and irrelevance to normal mixture and classification models with trimming and variable selection. Extensions to nonnormal populations are possible but the normal theory is most elegant. The idea is to regard the subset as a model parameter. The parameter space of the unconstrained g-component mixture model with trimming and subset selection is thus

$$\binom{1..n}{r} \times \binom{1..D}{d} \times \overset{\circ}{\Delta}_{g-1} \times \Psi.$$

Here, the second factor stands for the collection of all relevant subsets $F \subseteq 1..D$ of size $d < D$, the open simplex $\overset{\circ}{\Delta}_{g-1}$ contains all vectors $\boldsymbol{\pi} = (\pi_1, \ldots, \pi_g)$ of mixing rates, and Ψ encompasses all population parameters $(m_j, V_j) \in \mathbb{R}^d \times \mathrm{PD}(d)$ w.r.t. the relevant subset F and the parameters $(G_{E|F}, m_{E|F}, V_{E|F}) \in \mathbb{R}^{(D-d) \times d} \times \mathbb{R}^{(D-d)} \times \mathrm{PD}(D-d)$ of the regression of the irrelevant subset $E = \complement F$ onto F. Abbreviate $\gamma_j = (m_j, V_j, G_{E|F}, m_{E|F}, V_{E|F})$, $\boldsymbol{\gamma} = (\mathbf{m}, \mathbf{V}, G_{E|F}, m_{E|F}, V_{E|F})$, and $N(x; F, \gamma_j) = N_{m_j, V_j}(x_F) N_{m_{E|F}, V_{E|F}}(x_E - G_{E|F} x_F)$. We safeguard the model against outliers by using only $r \leq n$ data points R, which we first fix. Since E is irrelevant w.r.t. F, the trimmed likelihood with variable selection has the form

$$f(\mathbf{x}_{R,F}; F, \boldsymbol{\pi}, \boldsymbol{\gamma}) = \prod_{i \in R} \sum_{1 \leq j \leq g} \pi_j N(x_i; F, \gamma_j); \tag{5.17}$$

see (2.3) and (5.3). The appearance of F after the semi-colons in (5.17) indicates the subset the population parameters are restricted to.

In order to establish an EM algorithm with subset selection for F and $\boldsymbol{\gamma}$, we proceed as in Section 3.1.1, treating class labels as hidden variables to obtain the complete likelihood function

$$g_{F, \boldsymbol{\pi}, \boldsymbol{\gamma}}(x, j) = \pi_j N(x; F, \gamma_j).$$

Since the regression term does not depend on j, the weights (G.9) assume the simple form

$$w_\ell = w_\ell^{F, \boldsymbol{\pi}, \mathbf{m}, \mathbf{V}}(x) = \frac{\pi_\ell N_{m_\ell, V_\ell}(x_F)}{\sum_j \pi_j N_{m_j, V_j}(x_F)}.$$

Abbreviating $\boldsymbol{\gamma}' = (\mathbf{m}', \mathbf{V}', G', m', V')$ with regression parameters G', m', and V' w.r.t. possibly different sets F' and $E' = \complement F'$, we infer from (5.17) the Q-functional (G.10) in the case of independent observations (see also (3.1) and (3.2)),

$$Q((F, \boldsymbol{\pi}, \boldsymbol{\gamma}), (F', \boldsymbol{\pi}', \boldsymbol{\gamma}')) = \sum_{i \in R} \sum_j w_j^{F, \boldsymbol{\pi}, \mathbf{m}, \mathbf{V}}(x_i) \log \left(\pi_j' N(x_i; F', \gamma_j') \right)$$

$$= \sum_j \sum_{i \in R} w_j^{F, \boldsymbol{\pi}, \mathbf{m}, \mathbf{V}}(x_i) \log \pi_j' + \sum_j \sum_{i \in R} w_j^{F, \boldsymbol{\pi}, \mathbf{m}, \mathbf{V}}(x_i) \log N_{m_j', V_j'}(x_{i,F'})$$

$$+ \sum_{i \in R} N_{m', V'}(x_{i,E'} - G' x_{i,F'}). \tag{5.18}$$

The functional is optimized or at least improved w.r.t. F', $\boldsymbol{\pi}'$, and $\boldsymbol{\gamma}'$. Simultaneous optimization of all parameters seems possible efficiently only when d is small enough or close to D but partial optimization of $\boldsymbol{\pi}'$ and $\boldsymbol{\gamma}'$ given F' is standard. We make again the usual assumption that the data set $\mathbf{x}$ is in *general position*; see after Definition A.2. Under HDBT constraints, a minimum number $r \geq (gd+1) \vee (D+1)$ of retained elements are sufficient for all maxima to exist in general. For submodels and special situations, less is needed; see Proposition 1.24 and below. Partial optimization of the regression term in (5.18) w.r.t. G', m', and V' yields the sample regression matrix $G_{E'|F'}(R)$ and the MLE's $m_{E'|F'}(R)$, and $V_{E'|F'}(R)$. By Appendix F.2.5, the maximum of the irrelevant term is

$$-\frac{r}{2} \log \det V_{E'|F'}(R) - \frac{r(D-d)}{2}(1+\log 2\pi)$$

$$= \frac{r}{2} \log \det V_{F'}(R) - \frac{r}{2} \log \det V(R) - \frac{r(D-d)}{2}(1+\log 2\pi)$$

with the MLE $V_{F'}(R)$ w.r.t. R and F' and $V(R) = V_{1\,..\,D}(R)$. The maximal vector $\boldsymbol{\pi}'$ of mixing rates and the maximal covariance matrices $\mathbf{V}'$ are determined as in (3.3) and Proposition 3.3. With $n_j = w_j^{F,\boldsymbol{\pi},\mathbf{m},\mathbf{V}}(R) = \sum_{i \in R} w_j^{F,\boldsymbol{\pi},\mathbf{m},\mathbf{V}}(x_i)$, the former are $\pi_j^* = \frac{1}{r}n_j$ (by the entropy inequality) and the latter depend on the submodel. Denoting them by $V_{F'}(R, w_j)$, the maximum of the corresponding double sum in (5.18) is

$$-\frac{1}{2} \sum_j n_j \log \det V_{F'}(R, w_j) - \frac{rd}{2}(1 + \log 2\pi).$$

Combining the two maxima, we see that the expression to be maximized w.r.t. the finite collection of all $F' \in \binom{1\,..\,D}{d}$ reads

$$\log \det V_{F'}(R) - \frac{1}{r} \sum_j n_j \log \det V_{F'}(R, w_j). \tag{5.19}$$

The second term is the maximum of the Q-functional for fixed F'. Without the first term, the criterion would just select the subset with the least determinants $\det V_{F'}(R, w_j)$. The first term acts as a normalization counteracting this tendency.

In the fully normal case, the two covariance estimates are specified by the scatter matrix w.r.t. R, $V_{F'}(R) = S_{R,F'}$, and the weighted scatter matrices w.r.t. R and F',

$$V_{F'}(R, w_j) = S_{R,F'}(w_j) = \frac{1}{n_j} \sum_{i \in R} w_j^{F,\boldsymbol{\pi},\mathbf{m},\mathbf{V}}(x_i)(x_{i,F'} - \overline{x}_{R,F'}(w_j))(x_{i,F'} - \overline{x}_{R,F'}(w_j))^\top,$$

where $\overline{x}_{R,F'}(w_j)$ denotes the weighted sample mean $\frac{1}{n_j} \sum_{i \in R} w_j^{F,\boldsymbol{\pi},\mathbf{m},\mathbf{V}}(x_i)x_{i,F'}$. Under general position, the weighted scatter matrices are positive definite if and only if $r \geq d+1$. In the fully normal case, criterion (5.19) for F' thus turns into

$$\log \det S_{R,F'} - \frac{1}{r} \sum_j n_j \log \det S_{R,F'}(w_j).$$

It is efficient to compute the scatter matrix S_R and weighted scatter matrices and $S_R(w_j)$ for the full dimension D at once and to extract the submatrices $S_{R,F'}$ and $S_{R,F'}(w_j)$ as they are needed. Summing up, we determine local maxima of the likelihood function, iterating the following procedure. It also contains a final trimming step. Recall the definitions of the parameters γ and γ' above.

5.29 Procedure (Fully normal, heteroscedastic EMST step)
// Input: Subsets $R \subseteq 1..n$, $|R| = r$, and $F \subseteq 1..D$, $|F| = d$, population parameters
            $(\boldsymbol{\pi}, \mathbf{m}, \mathbf{V})$, and observed trimmed likelihood $f(\mathbf{x}_{R,F}; F, \boldsymbol{\pi}, \gamma)$.
// Output: Sets R_{new} and F_{new}, parameters $(\boldsymbol{\pi}_{\text{new}}, \mathbf{m}_{\text{new}}, \mathbf{V}_{\text{new}})$, with larger observed
            trimmed likelihood $f(\mathbf{x}_{R_{\text{new}},F_{\text{new}}}; F_{\text{new}}, \boldsymbol{\pi}_{\text{new}}, \gamma_{\text{new}})$ *or* "stop."

1. (*E-step*) Compute all weights $w_j(x_i) = w_j^{F,\boldsymbol{\pi},\mathbf{m},\mathbf{V}}(x_i)$ from the parameters F and $(\boldsymbol{\pi}, \mathbf{m}, \mathbf{V})$ and set $n_j = w_j(R) = \sum_{i \in R} w_j(x_i)$, $1 \leq j \leq g$.

2. *(M-step)* Put $\pi_{\mathrm{new},j} = \frac{n_j}{r}$, $m_{\mathrm{new},j} = \overline{x}_R(w_j) = \frac{1}{n_j}\sum_{i\in R} w_j(x_i)x_i \in \mathbb{R}^D$, and $V_{\mathrm{new},j} =$ $S_R(w_j) = \frac{1}{n_j}\sum_{i\in R} w_j(x_i)(x_i - m_{\mathrm{new},j})(x_i - m_{\mathrm{new},j})^\top \in \mathrm{PD}(D)$, $1 \le j \le g$ (see also the M-step in Procedure 3.15); also compute the scatter matrix of R, S_R.

3. *(S-step)* Minimize the difference

$$h(F') = \sum_j \pi_{\mathrm{new},j} \log \det S_{R,F'}(w_j) - \log \det S_{R,F'} \tag{5.20}$$

w.r.t. the subset $F' \subseteq 1..D$, where $S_{R,F'}(w_j)$ is the restriction of $S_R(w_j)$ to F'. When minimization is not possible efficiently, find F' such that $h(F') \le h(F)$ is small. Let F_{new} be the optimum subset found.

4. *(T-step)* Compute the MLE's of the regression parameters

$$G = G_{E_{\mathrm{new}}|F_{\mathrm{new}}}(R) = S_{R,E_{\mathrm{new}},F_{\mathrm{new}}} S_{R,F_{\mathrm{new}}}^{-1},$$

$$m = m_{E_{\mathrm{new}}|F_{\mathrm{new}}}(R) = \overline{x}_{R,E_{\mathrm{new}}} - G_{E_{\mathrm{new}}|F_{\mathrm{new}}}(R)\overline{x}_{R,F_{\mathrm{new}}},$$

$$V = V_{E_{\mathrm{new}}|F_{\mathrm{new}}}(R) = S_{R,E_{\mathrm{new}}} - S_{R,E_{\mathrm{new}},F_{\mathrm{new}}} S_{R,F_{\mathrm{new}}}^{-1} S_{R,F_{\mathrm{new}},E_{\mathrm{new}}}$$

and put $\gamma_{\mathrm{new},j} = (m_{\mathrm{new},j,F_{\mathrm{new}}}, V_{\mathrm{new},j,F_{\mathrm{new}}}, G, m, V)$. Let R_{new} consist of the indices $i \in 1..n$ with the r largest sums $\sum_{j=1}^{g} \pi_{\mathrm{new},j} N(x_i; F_{\mathrm{new}}, \gamma_{\mathrm{new},j})$.

5. If $f(\mathbf{x}_{R_{\mathrm{new}},F_{\mathrm{new}}}; F_{\mathrm{new}}, \pi_{\mathrm{new}}, \gamma_{\mathrm{new}}) > f(\mathbf{x}_{R,F}; F, \pi, \gamma)$, then return the new parameters *else* "stop."

Parameter estimation in the procedure is equivariant w.r.t. variable scaling. Criterion (5.20) shows that subset selection in the S-step is even invariant. The E-step receives subsets R and F and a mixture as inputs and returns the posterior probabilities $w_j(x_i) = w_j^{F,\pi,\mathbf{m},\mathbf{V}}(x_i)$ of the objects i to come from component $j \in 1..g$ w.r.t. these parameters. The weights sum up to 1 w.r.t. j, that is, $\mathbf{w} = (w_j(x_i))_{i,j}$ is a stochastic matrix. The M-step is split here into an M- and an S-step. They compute mixing rates, parameters, and a new subset with improved Q-functional. Finding a minimal subset F' in the S-step can be tried by complete enumeration when $\binom{D}{d}$ is not too large, depending on the time available; see, however, the more favorable diagonal case below. When minimization is not possible, find F' with a small value of (5.20). General deterministic and stochastic optimization methods for discrete structures such as local search, the Metropolis-Hastings algorithm, Gibbs sampling, or genetic algorithms are at our disposal for this purpose. Local search w.r.t. F' can be carried out using swaps. A **swap** (k,t) between F and $\complement F$ is the exchange of a variable $k \in F$ for a variable $t \notin F$. Call a swap *order preserving* (w.r.t. F) if it respects the order within F. That is, if $F = \{k_1 < k_2 < \cdots < k_d\}$, then k_1 can only be exchanged for coordinates $t \ne k_1$, $t < k_2$, k_2 only for coordinates $t \ne k_2$, $k_1 < t < k_3$, etc. This constraint still guarantees that any subset is reachable from any other by a suite of such swaps. The subset of coordinates after the swap is denoted by $F(k,t)$. Finally, the T-step seeks a new subset R_{new} of retained elements. It is only here that regression parameters have to be computed from the statistics $\overline{x}_R$ and $S_R = \begin{pmatrix} S_{R,F_{\mathrm{new}}} & S_{R,F_{\mathrm{new}},E_{\mathrm{new}}} \\ S_{R,E_{\mathrm{new}},F_{\mathrm{new}}} & S_{R,E_{\mathrm{new}}} \end{pmatrix}$. In the event of large dimensions D it often occurs that a data set is deficient, creating a singular residual scatter matrix. In such a case, the data still allows us to estimate a singular normal distribution on a subspace using the generalized inverse; see Mardia et al. [345], Sections 2.5.4 and A.8. In this case, $gd + 1$ data points are sufficient. The same number suffices when the data is uncontaminated and so $r = n$, since the T-step is omitted.

The general theory of the EM step applies and shows that the improved Q-functional leads to an improved likelihood function with variable selection; see Appendix G.2.1 and also Proposition 3.8. One might base the trimming step solely on the new relevant set F_{new}, disregarding its complement and the regression parameters. Caring about outlying *irrelevant* variables would then be unnecessary. In this case, $r \ge gd + 1$ retained elements would again suffice but the increase of the trimmed likelihood would no longer be guaranteed. The EMST-step reduces to an EMT step 3.7 if $d = D$ and to a classical EM step 3.1 for mixtures if in addition $r = n$.

A *homoscedastic* version of the EMST step is obtained by replacing the weighted scatter matrices $V_{\text{new},j}$ with the pooled weighted scatter matrix $V_{\text{new}} = S_R(\mathbf{w}) = \frac{1}{r}\sum_j n_j S_R(w_j) = \frac{1}{r}\sum_j \sum_{i \in R} w_j(x_i)(x_i - m_{\text{new},j})(x_i - m_{\text{new},j})^\top$. Its restriction to d-element subsets is regular if $r \geq d + 1$. The criterion (5.20) turns to $h(F') = \log \det S_{R,F'}(\mathbf{w}) - \log \det S_{R,F'}$.

In *diagonal* submodels, the regression parameter G vanishes, $m = m_E$, and $V = V_E$ is diagonal. Moreover, suboptimal solutions in the S-step can be avoided by direct optimization w.r.t. F': In the heteroscedastic, diagonal, normal case, criterion (5.20) changes to

$$\sum_j \pi_{\text{new},j} \sum_{k \in F'} \log S_R(w_j)(k,k) - \sum_{k \in F'} \log S_R(k,k) = \sum_{k \in F'} \sum_j \pi_{\text{new},j} \log \frac{S_R(w_j)(k,k)}{S_R(k,k)}.$$

This sum is easily minimized w.r.t. F': Just pick the indices $k \in 1..D$ of the d smallest sums $\sum_j \pi_{\text{new},j} \log \frac{S_R(w_j)(k,k)}{S_R(k,k)}$. This can, for instance, be done by sorting and greatly enhances performance. The EMST step therefore reduces to the following step.

5.30 Procedure (Heteroscedastic, diagonally normal EMST step)

// <u>Input:</u> Subsets $R \subseteq 1..n$ and $F \subseteq 1..D$, population parameters $(\boldsymbol{\pi}, \mathbf{m}, \mathbf{V})$, V_j diagonal, and observed trimmed likelihood $f(\mathbf{x}_{R,F}; F, \boldsymbol{\pi}, \boldsymbol{\gamma})$.

// <u>Output:</u> Sets R_{new} and F_{new}, parameters $(\boldsymbol{\pi}_{\text{new}}, \mathbf{m}_{\text{new}}, \mathbf{V}_{\text{new}})$, with larger observed trimmed likelihood $f(\mathbf{x}_{R_{\text{new}},F_{\text{new}}}; F_{\text{new}}, \boldsymbol{\pi}_{\text{new}}, \boldsymbol{\gamma}_{\text{new}})$ *or* "stop."

1. *(E-step)* Compute all weights $w_j(x_i) = w_j^{F,\boldsymbol{\pi},\mathbf{m},\mathbf{V}}(x_i)$ from the parameters F and $(\boldsymbol{\pi}, \mathbf{m}, \mathbf{V})$ and set $n_j = w_j(R) = \sum_{i \in R} w_j(x_i)$, $1 \leq j \leq g$.

2. *(M-step)* Put $\pi_{\text{new},j} = \frac{n_j}{r}$, $m_{\text{new},j} = \overline{x}_R(w_j) = \frac{1}{n_j}\sum_{i \in R} w_j(x_i)x_i \in \mathbb{R}^D$, $V_{\text{new},j} = \text{Diag}(S_R(w_j))$, $1 \leq j \leq g$.

3. *(S-step)* Let F_{new} be the set of the indices k with the d smallest sums

$$\sum_j \pi_{\text{new},j} \log \frac{S_R(w_j)(k,k)}{S_R(k,k)}. \tag{5.21}$$

4. *(T-step)* Put $\gamma_{\text{new},j} = (m_{\text{new},j,F_{\text{new}}}, V_{\text{new},j,F_{\text{new}}}, 0, \overline{x}_{R,E_{\text{new}}}, \text{Diag}(S_{R,E_{\text{new}}}))$ and let R_{new} consist of the indices $i \in 1..n$ of the r largest sums $\sum_{j=1}^g \pi_{\text{new},j} N(x_i; F_{\text{new}}, \gamma_{\text{new},j})$.

5. If $f(\mathbf{x}_{R_{\text{new}},F_{\text{new}}}; F_{\text{new}}, \boldsymbol{\pi}_{\text{new}}, \boldsymbol{\gamma}_{\text{new}}) > f(\mathbf{x}_{R,F}; F, \boldsymbol{\pi}, \boldsymbol{\gamma})$, then return the new parameters *else* "stop."

In the homoscedastic, diagonal, normal case the weighted scatter matrices $S_R(w_j)$ are replaced with their pooled counterpart $S_R(\mathbf{w})$ and the target function (5.21) for subset selection becomes

$$\sum_{k \in F'} \log \frac{S_R(\mathbf{w})(k,k)}{S_R(k,k)}.$$

Thus, take the d smallest quotients $S_R(\mathbf{w})(k,k)/S_R(k,k)$. It might be possible to select a number different from d to estimate also subset size. Experience here is still missing. Subset selection in diagonal submodels is again invariant w.r.t. variable scaling. According to Proposition 1.24, constrained estimation in diagonal submodels needs only $g+1$ data points. The simplifications in the diagonal cases do not extend to the spherical cases, since the entries of the estimated variances depend on more than one coordinate. In fact, criterion (5.20) reads in this case

$$\sum_j \pi_{\text{new},j} \log \left(\frac{1}{d} \sum_{k \in F'} S_{R,F'}(w_j)(k,k) \right) + \log \left(\frac{1}{D-d} \sum_{k \notin F'} S_R(w_j)(k,k) \right).$$

Finally, the special case of diagonal models and uncontaminated data is worth mentioning. No trimming is applied here and the EMST step turns into an EMS step. It can be further simplified by first auto-scaling or standardizing all variables. Since $S_R(k, k) = 1$ here, the normalization vanishes and expression (5.21) reduces to $\sum_{k \in F'} \log S_R(w_j)(k, k)$.

The **EMST algorithm** is the iteration of EMST steps. Start the algorithm from the S-step with randomly or expediently chosen subsets $F^{(0)}$ and $R^{(0)}$ and a stochastic matrix $\mathbf{w}^{(0)}$ as initial quantities. The methods and remarks of Section 3.1.4 concerning $R^{(0)}$ and $\mathbf{w}^{(0)}$ apply here, too. The method does not itself estimate the number d of selected variables. See however Remark 5.33(a).

In Section 1.3.1, the mixture model was transferred to the related classification model. The same can be done with the trimmed mixture model with variable selection (5.17). Denoting again the parameters of the regression of the irrelevant subset $E = \complement F$ on the relevant subset F by $G_{E|F}$, $m_{E|F}$, and $V_{E|F}$, we obtain a trimmed, normal classification-and-selection likelihood:

$$\sum_{j=1}^{g} \sum_{i:\ell_i=j} \log N_{m_j, V_j}(x_{i,F}) - r\mathrm{H}\big(\tfrac{n_1(\ell)}{r}, \ldots, \tfrac{n_g(\ell)}{r}\big)$$
$$+ \sum_{i \in R} \log N_{m_{E|F}, V_{E|F}}(x_{i,E} - G_{E|F} x_{i,F}). \tag{5.22}$$

It is maximized w.r.t. all admissible assignments ℓ, all subsets $F \in \binom{1 \,\cdots\, D}{d}$, all population parameters (m_j, V_j), and the regression parameters $(G_{E|F}, m_{E|F}, V_{E|F})$. By normal parameter estimation, the partial maxima of the first and last terms w.r.t. (m_j, V_j) and (G, m, V) are, respectively,

$$\max_{\mathbf{m}, \mathbf{V}} \sum_{j=1}^{g} \sum_{i:\ell_i=j} \log f_{m_j, V_j}(x_{i,F}) = -\tfrac{r}{2} d \log 2\pi e - \tfrac{1}{2} \sum_j n_j(\ell) \log \det V_{j,F}(\ell),$$

$$\max_{G, m, V} \sum_{i \in R} \log N_{m,V}(x_{i,E} - G x_{i,F}) = -\tfrac{r}{2}\big((D - d) \log 2\pi e + \log \det V_{E|F}(R)\big),$$

with the MLE's $V_{j,F}(\ell)$ and $V_{E|F}(R) = V_E(R) - V_{E,F}(R)V_F^{-1}(R)V_{F,E}(R)$.

When scales are full, we have $V_{j,F}(\ell) = S_{j,F}(\ell)$ and

$$V_{E|F}(R) = S_{R,E} - S_{R,E,F} S_{R,F}^{-1} S_{R,F,E} = S_{R,E|F}.$$

This leads to the trimmed, normal classification-and-selection criterion

$$\tfrac{1}{2} \sum_{j=1}^{g} n_j(\ell) \log \det S_{j,F}(\ell) + r\mathrm{H}\big(\tfrac{n_1(\ell)}{r}, \ldots, \tfrac{n_g(\ell)}{r}\big) + \tfrac{r}{2} \log \det S_{R,E|F}. \tag{5.23}$$

It uses the scatter matrices $S_{j,F}(\ell)$ and the residual scatter matrix $S_{R,E|F}$. Criterion (5.23), too, can be improved w.r.t. ℓ and F by alternating reduction steps:

$$\big(\ell^{(0)}, F^{(0)}\big) \longrightarrow \big(\ell^{(0)}, F^{(1)}\big) \longrightarrow \big(\ell^{(1)}, F^{(1)}\big) \longrightarrow \ldots.$$

The assignment is fixed during the update of F in the first step. Because of $\det S_R = \det S_{R,F} \cdot \det S_{R,E|F}$ and since $\det S_R$ does not depend on F, $\det S_{R,E|F}$ is inversely proportional to $\det S_{R,F}$ and the last term in (5.23) can be replaced with $-\tfrac{r}{2} \log \det S_{R,F}$. It is again efficient to estimate $S_j(\ell)$ and S_R for the whole set of variables at once and to extract the submatrices $S_{j,F}(\ell)$ and $S_{R,F}$ as needed for the determinants.

The update of $\ell^{(0)}$ in the second step, that is, optimization w.r.t. ℓ and the population parameters (m_j, V_j), is carried out as in the reduction steps with trimming based on $F^{(1)}$; see Sections 3.2.2–3.2.3. The resulting reduction step is composed of an estimation step, a selection step, and an assignment step. They update the population parameters, the subset

F, and the clustering, respectively. We have thus derived the following procedure valid in the full case.

5.31 Procedure (Trimmed, fully normal clustering-and-selection step)
// <u>Input</u>:   Admissible ℓ, subset $F \subseteq 1 .. D$, $|F| = d$, and value of (5.23).
// <u>Output</u>: Admissible ℓ_{new} and subset F_{new}, $|F_{\text{new}}| = d$, with improved
criterion (5.23) *or* "stop."

1. *(Estimation)* Compute the sample mean vectors $\overline{x}_j(\ell)$ and scatter matrices $S_j(\ell)$, $1 \leq j \leq g$, and the total scatter matrix S_R, $R = \ell^{-1}(1 .. g)$.

2. *(Selection)* Minimize the difference

$$h(F') = \sum_{j=1}^{g} n_j(\ell) \log \det S_{j,F'}(\ell) - r \log \det S_{R,F'} \tag{5.24}$$

w.r.t. F', $|F'| = d$, where $S_{j,F'}(\ell)$ and $S_{R,F'}$ are the submatrices of $S_j(\ell)$ and S_R, respectively, specified by F'. If minimization is not possible, find F' with a small value of $h(F') \leq h(F)$. Denote the best subset F' found by F_{new}.

3. Use the quantities obtained in step 1 to compute the MLE's of the regression parameters w.r.t. ℓ and the new subsets F_{new} and $E_{\text{new}} = \complement F_{\text{new}}$. Denoting them by (G, m, V), put

$$u_{i,j} = \log n_j - \tfrac{1}{2} \log \det S_{j,F_{\text{new}}}(\ell)$$
$$- \tfrac{1}{2}(x_{i,F_{\text{new}}} - \overline{x}_{j,F_{\text{new}}}(\ell))^{\top} S_{j,F_{\text{new}}}(\ell)^{-1}(x_{i,F_{\text{new}}} - \overline{x}_{j,F_{\text{new}}}(\ell))$$
$$- \tfrac{1}{2}(x_{i,E_{\text{new}}} - m - G x_{i,F})^{\top} V^{-1}(x_{i,E_{\text{new}}} - m - G x_{i,F}).$$

4. *(Assignment and trimming)* Compute an admissible, trimmed assignment ℓ_{new} using a reduction step with trimming based on the statistics $u_{i,j}$, Procedure 3.28 or 3.29.

5. If F_{new} and ℓ_{new} improve the criterion (5.23), then return F_{new}, ℓ_{new}, and the new value of (5.23), *else* "stop."

Without selection, the procedure reduces to the multipoint reduction step 3.28 or 3.29. When $\binom{1 .. D}{d}$ is too large to enable complete enumeration in step 2, the universal optimization methods proposed after the EMST step 5.29 can in principle be applied. Let us call a solution (F, ℓ) again **steady** if it is induced by its own parameters. All transitions $F \rightarrow F_{\text{new}}$, $\ell \rightarrow \ell_{\text{new}}$, and $(m, V) \rightarrow (m_{\text{new}}, V_{\text{new}})$ improve the criterion (5.22) and we have the following extension of Proposition 3.22.

5.32 Proposition *The clustering-and-selection step 5.31 stops at a steady solution.*

The succession of clustering-and-selection steps is the **trimmed clustering-and-selection algorithm** for optimizing criterion (5.22). Start the algorithm from the estimation step from an admissible labeling $\ell^{(0)}$ and a randomly or expediently chosen subset $F^{(0)}$ as initial quantities. The iteration must become stationary after a finite number of reduction steps, since the number of assignments ℓ and subsets F is finite. The clustering and selection obtained at stationarity do not have to be close to the true solution and the algorithm must be repeated, in complex cases many times, in order to achieve an acceptable solution; see Chapter 4. Normal residuals indicate the irrelevance of the removed variables. Concerning an estimate of subset size; see the following remark.

5.33 Remarks (a) The EMST and clustering-and-selection algorithms do not yet estimate the subset size d as the algorithms of Sections 5.3.1 and 5.3.2 do. So run the algorithm for all reasonable numbers d and select the favorite solution again by validation. Since we usually wish to keep d as small as possible, begin with the smaller ones. Since also the number of clusters is usually unknown, the algorithm has to be repeated for various numbers of clusters g and dimensions d. When the dimension D of the data set is large, then this means a heavy computational burden. A solution can then not be obtained in passing on a PC. However, it is easy to much reduce computation time by distributing the program on different processors for different values of g and d. The problem is indeed embarrassingly parallel. The clustering-and-selection algorithm has been used to extract up to a few hundred variables from 2,000; see the analysis of gene expression data in Section 6.2. There, a consensus method will be presented to select subset size d.

(b) Criterion (5.23) is equivariant w.r.t translation and variable scaling and (5.24) remains even invariant. Indeed, a transformation with a diagonal matrix A multiplies both determinants in (5.23) and (5.24) by the factor $\det^2 A_F$. It follows along with Proposition 1.44 that the clustering-and-selection step with trimming 5.31 for normal models with full and diagonal scale matrices is invariant w.r.t. linear variable scaling.

(c) If the regression terms in (5.22)–(5.24) were omitted, then invariance would not be guaranteed. Squeezing one of the coordinates would favor this coordinate, since the scatter in this direction would decrease. More precisely, if $d < D$, then for each data set there exists a diagonal scale transformation A such that the estimated subset for the transformed data set differs from that for the original one. This effect can again be removed in the diagonal case by auto-scaling if no elements are discarded. Indeed, it follows from (a) that the estimates of ℓ and F remain invariant and $S = I_D$ is constant in this case. Application of this remark needs, however, clean data.

(d) In the diagonally normal cases, the minimum in the selection step can again be efficiently attained even for large dimension D and arbitrary $d \leq D$. In the *heteroscedastic* diagonal case, for instance, we have

$$(5.24) = \sum_j n_j(\ell) \sum_{k \in F'} \log S_j(\ell)(k,k) - r \sum_{k \in F'} \log S_R(k,k)$$

$$= \sum_{k \in F'} \left\{ \sum_j n_j(\ell) \log S_j(\ell)(k,k) - r \log S_R(k,k) \right\}.$$

The optimum in step 2 is attained by the set F_{new} that consists of the d indices $k \in 1 .. D$ with the smallest values of $\sum_{j=1}^g n_j(\ell) \log \frac{S_j(\ell)(k,k)}{S_R(k,k)}$. This subset can again be computed in one sweep, for instance, by sorting – no swapping is needed. A similar simplification exists for the *homoscedastic*, diagonal model. Let $S(\ell)$ be the pooled scatter matrix w.r.t. ℓ. Here, F_{new} is made up of the indices k of the d smallest values $\frac{S(\ell)(k,k)}{S_R(k,k)}$.

Because of Remark (a) the variables may be standardized or auto-scaled before clustering and selection in the diagonal cases. If there is no contamination and $R = 1 .. n$, then S_R is the identity matrix and the corresponding term in the selection criteria above vanishes.

(e) A procedure similar to 5.31 can also be designed for spherical models. But then the diagonal entries of the (spherical) covariance estimates $\det S_R$ depend on more than one variable. Therefore, the algorithmic simplifications possible in the diagonal case no longer apply.

5.3.4   Consistency of the MAP cluster criterion with variable selection

Theorem 1.56 can be extended to a consistency theorem for the normal MAP cluster criterion with variable selection (5.22). To this end, we combine Sections 1.3.7 and 5.1.2. The notation introduced in Section 1.3.7 also carries over in a natural way to selection of subsets $F \subseteq 1 .. D$. Let μ be again the parent distribution of the independent observations X_i, $1 \leq i \leq n$. The basic distribution of the statistical model is now normal, $N_{m,V}^D$, that is, the radial function is $\phi(t) = \frac{D}{2} \log 2\pi + \frac{t}{2}$. Let μ_F be the projection of μ to $\mathbb{R}^F$ and identify $\mathbb{R}^F$ in the natural way with $\mathbb{R}^d$ so that x_F becomes a member of $\mathbb{R}^d$. The parameters are here composed of a relevant part $a = (\pi_a, m_a, V_a)$,

$\pi_a \in]0,1]$, $m_a \in \mathbb{R}^d, V_a \in \mathrm{PD}(d)$, related to the variables in F (see Section 1.3.7) and an irrelevant part $(G, m, V) = (G_{E|F}, m_{E|F}, V_{E|F}) \in \Gamma^{\mathrm{irrel}} = \mathbb{R}^{(D-d) \times d} \times \mathbb{R}^{(D-d)} \times \mathrm{PD}(D-d)$ related to $E = \complement F$ as in Section 5.1.2. The functional (1.45), the sampling criterion (1.46), and the population criterion (1.47) all split into a "relevant" and an "irrelevant" part that depend on $A \in \Theta_{\leq g,c}$ and on (G, m, V), respectively. They are

$$t_a(x_F) = -\log \pi_a + \tfrac{1}{2}\log(2\pi V_a) + \tfrac{1}{2}(x_F - m_a)^\top \Lambda_a(x_F - m_a),$$

$$\Phi_F^{\mathrm{rel}}(A) = \mathrm{E}\min_{a \in A} t_a(X_{1,F}) = \int \min_{a \in A} t_a(x_F)\mu(\mathrm{d}x) = \int \min_{a \in A} t_a(x_F)\mu_F(\mathrm{d}x_F),$$

$$\Phi_{n,F}^{\mathrm{rel}}(A) = \frac{1}{n}\sum_{i=1}^n \min_{a \in A} t_a(X_{i,F}),$$

$$t_{G,m,V}(x, F) = -\log N_{m,V}(x_E - Gx_F)$$
$$= \tfrac{1}{2}\log\det(2\pi V) + \tfrac{1}{2}(x_E - Gx_F - m)^\top \Lambda(x_E - Gx_F - m),$$

$$\Phi_F^{\mathrm{irrel}}(G, m, V) = \mathrm{E}t_{G,m,V}(X_1, F) = \int t_{G,m,V}(x, F)\mu(\mathrm{d}x),$$

$$\Phi_{n,F}^{\mathrm{irrel}}(G, m, V) = \frac{1}{n}\sum_{i=1}^n t_{G,m,V}(X_i, F).$$

The (total) population and sampling criteria are, respectively,

$$\Phi_F(A, G, m, V) = \Phi_F^{\mathrm{rel}}(A) + \Phi_F^{\mathrm{irrel}}(G, m, V),$$

$$\Phi_{n,F}(A, G, m, V) = \Phi_{n,F}^{\mathrm{rel}}(A) + \Phi_{n,F}^{\mathrm{irrel}}(G, m, V).$$

The following lemma collects some properties of the irrelevant part. We write again $\mathrm{E}X_1 = \begin{pmatrix} m_F \\ m_E \end{pmatrix}$ and $\mathrm{V}X_1 = \begin{pmatrix} V_F & V_{F,E} \\ V_{E,F} & V_E \end{pmatrix}$.

5.34 Lemma *Let X_i be i.i.d. and quadratically integrable, $X_i \sim \mu$, and assume that hyperplanes in $\mathbb{R}^D$ are μ-null sets. Let $F \in \binom{1\,..\,D}{d}$.*

(a) *The functional Φ_F^{irrel} is uniquely minimized by $G_{E|F} = V_{E,F}V_F^{-1}$, $m_{E|F} = m_E - G_{E|F}m_F$, and $V_{E|F} = V_E - G_{E|F}V_{F,E}$. Its minimum is $\Phi_F^{\mathrm{irrel}\,*} = \frac{D-d}{2} + \frac{1}{2}\log\det 2\pi V_{E|F}$.*

(b) *Let $n \geq D+1$. The functional $\Phi_{n,F}^{\mathrm{irrel}}$ is P-a.s. uniquely minimized by the sample versions $G_{n,E|F}$, $m_{n,E|F}$ and $V_{n,E|F}$ of the quantities $G_{E|F}$, $m_{E|F}$, and $V_{E|F}$. Its minimum is $\Phi_{n,F}^{\mathrm{irrel}\,*} = \frac{D-d}{2} + \frac{1}{2}\log\det 2\pi V_{n,E|F}$.*

(c) *P-a.s., we have $\lim_{n\to\infty}(G_{n,E|F}, m_{n,E|F}, V_{n,E|F}) = (G_{E|F}, m_{E|F}, V_{E|F})$ and the minima $\Phi_{n,F}^{\mathrm{irrel}\,*}$ converge to $\Phi_F^{\mathrm{irrel}\,*}$ as $n \to \infty$.*

PROOF. (a) By Steiner's formula, the minimum of $\mathrm{E}(X_{1,E} - GX_{1,F} - m)^\top \Lambda(X_{1,E} - GX_{1,F} - m)$ w.r.t. m given G and Λ is $m^* = \mathrm{E}X_E - G\mathrm{E}X_F = m_E - Gm_F$. Inserting leaves

$$\mathrm{E}(\widehat{X}_{1,E} - G\widehat{X}_{1,F})^\top \Lambda(\widehat{X}_{1,E} - G\widehat{X}_{1,F}) = \mathrm{tr}\,\Lambda\mathrm{E}(\widehat{X}_{1,E} - G\widehat{X}_{1,F})(\widehat{X}_{1,E} - G\widehat{X}_{1,F})^\top$$
$$= \mathrm{tr}\,\Lambda\big(V_E - GV_{F,E} - V_{E,F}G^\top + GV_FG^\top\big)$$
$$= \mathrm{tr}\,\Lambda\big(GV_F^{1/2} - V_{E,F}V_F^{-1/2}\big)\big(GV_F^{1/2} - V_{E,F}V_F^{-1/2}\big)^\top + \mathrm{tr}\,\Lambda\big(V_E - V_{E,F}V_F^{-1}V_{F,E}\big)$$

($\widehat{}$ denotes centering). The minimizer w.r.t. G is $G^* = V_{E,F}V_F^{-1} = G_{E|F}$. Inserting removes the first summand in the last line, leaving $\mathrm{tr}\,\Lambda\big(V_E - G^*V_{F,E}\big)$. Finally, taking into account the first term in the definition of $t_{G,m,V}(x, F)$, we see that it remains to minimize

$$\log\det(2\pi V) + \mathrm{tr}\,\Lambda(V_E - G^*V_{F,E})$$

w.r.t. V. By assumption, $\mathrm{V}X_1 \succ 0$. Therefore, $V_F \succ 0$ and $\det(V_E - G_{E|F}V_{F,E})^\top = \det \mathrm{V}X_1/\det V_F > 0$. General normal estimation theory now shows that the minimum is attained at $V^* = V_E - G_{E|F}V_{F,E} = V_{E|F}$ and has the claimed value.

(b) Since $n \geq D + 1$, the assumption on μ implies $V_n \succ 0$, P-a.s. It is now sufficient to proceed as in (a), replacing all quantities with their sample versions.

(c) The Strong Law of Large Numbers implies $m_n = \frac{1}{n}\sum_{i=1}^n X_i \to \mathrm{E}X_1$ and $V_n = \frac{1}{n}\sum_{i=1}^n \widehat{X}_i\widehat{X}_i^\top \to \mathrm{V}X_1$ as $n \to \infty$. The first half of the claim follows therefore from the SLLN and continuity of $G_{E|F}$, $m_{E|F}$, and $V_{E|F}$ as functions of $\mathrm{E}X_1$ and $\mathrm{V}X_1$. Convergence of the minimum of the sampling criterion follows from (a), (b), and from continuity of the determinant. $\qquad \square$

Our aim is to relate the optimizers F_n^*, A_n^*, G_n^*, m_n^* and V_n^* of the sampling criterion $\Phi_{n,F}(A, G, m, V)$ to those of the population criterion $\Phi_F(A, G, m, V)$. Lemma 5.34 shows that the unique minimal parameters (G, m, V) are specified by the subset F through the minima $\Phi_F^{\mathrm{irrel}*}$ $(= \min_{G,m,V} \Phi_F^{\mathrm{irrel}}(G, m, V))$ and $\Phi_{n,F}^{\mathrm{irrel}*}$ $(= \min_{G,m,V} \Phi_{n,F}^{\mathrm{irrel}}(G, m, V))$. We therefore do not keep track of them explicitly, writing $\Phi_F(A) = \Phi_F^{\mathrm{rel}}(A) + \Phi_F^{\mathrm{irrel}*}$ and $\Phi_{n,F}(A) = \Phi_{n,F}^{\mathrm{rel}}(A) + \Phi_{n,F}^{\mathrm{irrel}*}$ to obtain for $A \in \Theta_{\leq g,c}$ under the assumptions of the lemma

$$\Phi_F(A) = \Phi_F^{\mathrm{rel}}(A) + \Phi_F^{\mathrm{irrel}*} = \min_{G,m,V} \Phi_F(A, G, m, V), \tag{5.25}$$

$$\Phi_{n,F}(A) = \Phi_{n,F}^{\mathrm{rel}}(A) + \Phi_{n,F}^{\mathrm{irrel}*} = \min_{G,m,V} \Phi_{n,F}(A, G, m, V). \tag{5.26}$$

In Lemma 5.34, F was fixed – it will now be subject to optimization. Under the assumptions of Theorem 1.56, the minima of $\Phi_F^{\mathrm{rel}}(A)$ and $\Phi_{n,F}^{\mathrm{rel}}(A)$ w.r.t. $A \in \Theta_{\leq g,c}$ exist for all F; see also the proof of the following proposition. Let $\mathcal{F}_g^*$ and $\mathcal{F}_{g,n}^*$ be the set of minimizers $F \subseteq 1..D$, $|F| = d$, of the minima w.r.t. $A \in \Theta_{\leq g,c}$ of (5.25) and (5.26), respectively. If $F_1, F_2 \in \mathcal{F}_g^*$, then we may have $\min_A \Phi_{F_1}^{\mathrm{rel}}(A) \neq \min_A \Phi_{F_2}^{\mathrm{rel}}(A)$ and $\Phi_{F_1}^{\mathrm{irrel}*}(G, m, V) \neq \Phi_{F_2}^{\mathrm{irrel}*}$; just the sums are equal. The same remark applies to the sample versions. The following theorem shows that variable selection with Procedure 5.31 behaves stably as $n \to \infty$.

5.35 Theorem *Let $X_i : \Omega \to \mathbb{R}^D$ be i.i.d. and quadratically integrable, $X_i \sim \mu$. Let $1 \leq d \leq D$ and assume that*

(i) hyperplanes in $\mathbb{R}^D$ are μ-null sets;

(ii) $\|X_1\|^p$ is P-integrable for some $p > 2$.

Then the following assertions hold true for all $g \geq 1$.

(a) The criterion $\Phi_F(A)$ (see (5.25)) has a minimum w.r.t. $F \in \binom{1..D}{d}$ and $A \in \Theta_{\leq g,c}$. The same holds true P-a.s. for criterion $\Phi_{n,F}(A)$ (see (5.26)) if $n \geq (gd+1) \vee (D+1)$.

(b) P-a.s., for eventually all n, we have $\mathcal{F}_{g,n}^* \subseteq \mathcal{F}_g^*$.

(c) Let $F_n^* \in \mathcal{F}_{g,n}^*$ and $E_n^* = \complement F_n^*$. P-a.s., the sequence

$$(G_{n,E_n^*|F_n^*}, m_{n,E_n^*|F_n^*}, V_{n,E_n^*|F_n^*})_n$$

of optimal sample parameters converges to the set

$$\{(G_{E^*|F^*}, m_{E^*|F^*}, V_{E^*|F^*}) \mid F^* \in \mathcal{F}_g^*\}.$$

If g is a drop point of Φ_F^{rel} for all $F \in \mathcal{F}_g^*$, then

(d) P-a.s., any sequence (F_n^*, A_n^*) of minimizers of the map $(F, A) \mapsto \Phi_{n,F}(A)$ on $\binom{1..D}{d} \times \Theta_{\leq g,c}$ converges to the set of minimizers of $(F, A) \mapsto \Phi_F(A)$;

(e) in particular, if the minimizer of $(F, A) \mapsto \Phi_F(A)$ is unique, then it is P-a.s. the limit of any such sequence (F_n^*, A_n^*).

PROOF. If F is a nonempty subset of $1..D$ and if H is a hyperplane in $\mathbb{R}^F$, then $H \times \mathbb{R}^{\complement F}$ is a hyperplane in $\mathbb{R}^D$ and $\mu_F(H) = \mu(H \times \mathbb{R}^{\complement F}) = 0$ by assumption (i). It follows that assumption (i) of Theorem 1.56 is satisfied for the projections of μ to

$\mathbb{R}^F$ for all subsets F. Linearity of the normal radial function along with assumption (ii) implies that the projections also satisfy assumption (iv). Parts (a) and (b) of the theorem show claim (a), part (d) implies $\min_A \Phi_{n,F}^{\text{rel}}(A) \xrightarrow[n\to\infty]{} \min_A \Phi_F^{\text{rel}}(A)$ and, from Lemma 5.34, we infer $\min_A \Phi_{n,F}(A) \xrightarrow[n\to\infty]{} \min_A \Phi_F(A)$ for all F. By finiteness of the collection of d-element subsets, convergence is even uniform. Again by finiteness, the minimum $\alpha = \min_{F\in\binom{1\ldots D}{d}, A\in\Theta_{\le g,c}} \Phi_F(A)$ exists. Let $\beta = \min_{F\in\binom{1\ldots D}{d}\setminus\mathcal{F}_g^*, A\in\Theta_{\le g,c}} \Phi_F(A)$ and note that $\beta > \alpha$. By convergence, we have $\min_A \Phi_{n,F}(A) < \beta$ for all $F \in \mathcal{F}_g^*$ and for eventually all n. Hence, $\min_A \Phi_{n,F_n^*}(A) < \beta$ and $\min_A \Phi_{F_n^*}(A) < \beta$ for eventually all n, that is, $\min_A \Phi_{F_n^*}(A) = \alpha$ and $F_n^* \in \mathcal{F}_g^*$. This is part (b). Claim (c) now follows from Lemma 5.34 and claim (d) derives from (b) and Theorem 1.56(e). Claim (e) is a direct consequence of (d). □

5.4 Practical guidelines

The foregoing text focuses mainly on theory. At the end, some words are in order about a practical approach to cluster analysis. We assume here that we are given numerical feature data $\in \mathbb{R}^d$. If an attribute is actually discrete and ordinal, the method may also be used with care if the discretization is fine enough. An example is currency. Actually, the representation of real numbers in a computer is always discrete.

Needless to say, the variables must be appropriate to describe the cluster structure sought. The cluster analyst may face among others two adverse circumstances: First, the presence of irrelevant variables that mask an existing cluster structure and, second, the absence of variables that could characterize a cluster structure. The methods of Chapter 5 safeguard against additional useless or noisy variables. However, there is no protection against missing important variables. Moreover, the data set must be unbiased, that is, a *random* sample from the population. Otherwise, the propositions of this text are not applicable and the data may not reflect the population structure. A number of steps and decisions have to be taken in order to successfully accomplish a cluster analysis.

(a) *Model specification:* An important issue is model specification. This is the realm of *systems analysis*. Departure times in queuing systems, for instance, are different from intensity values. Often data arises from random perturbations of prototypical objects in all directions. This often suggests the use of elliptical models. It is also common to first conduct *exploratory analyses* to aid in model formulation. Visualization is an important aid. According to Tukey [512], "*There is no excuse for failing to plot and look.*" If data plots show linear or curved structures, then the methods of this text are not appropriate and single linkage is probably the method of choice. A more voluminous structure of the data set is necessary. Sometimes a preceding nonlinear transformation may help, for instance, log-transforming the data if the entries are positive. Imputation of missing values is another issue. Some authors recommend the k-nearest-neighbor method. A standard tool is the EM algorithm; see Little and Rubin [327].

(b) *Clustering tendency:* A question related to this chapter is this: Is the data set in my hands clustered at all? If not, cluster analysis would not make sense. Therefore, many authors propose analyzing a data set for the existence of a cluster structure before embarking on a cluster analysis. Multimodality has often been regarded as an indication of a clustered data set and tests on uniformity, unimodality, or normality are often recommended. Bock [46] reviews literature related to this chapter and tests for uni- and multimodality and for the number of classes. However, rejecting uniformity, normality, or unimodality does not necessarily mean that the data set is sampled from a mixture. It might be a "tertium quid," for instance, fractal or otherwise chaotic. This step may, therefore, be omitted since cluster analyzing the data set and an ensuing validation, too, will provide an answer to the issue.

(c) *Selection of a clustering method:* The clustering method and its computational parameters depend on prior expectations about the basic model. A proper choice is important because an inappropriate model may impose a wrong structure on the data. However, it is reasonable to make as many appropriate assumptions as possible. This will reduce the number of parameters, thus improving their estimates and speeding up the algorithm. Do we expect a homoscedastic model? That is, does the randomness of the observations affect each expected class in the same way or do we have to assume a more general heteroscedastic model? Are the variables class-conditionally independent? If yes, a diagonal model is adequate. Is it reasonable to assume spherical components? The answer to this question is in general "no" when the attributes have different physical meanings such as time, length, or currency, otherwise it might, but does not have to, be "yes." Is anything known about cluster sizes? In this case, size-constrained reduction steps can also help reduce the number of spurious solutions; see Section 3.2.3. If cluster sizes are expected to be about equal then the entropy term in the cluster criteria is dropped. If, in addition, classes are expected to be spherical and of equal variance, and if the data is uncontaminated, then the classical k-means algorithm is the method of choice.[3] If necessary, filter or wrapper methods should be employed in order to remove irrelevant variables; see Chapter 4.

(d) *The favorite solution(s), number of clusters and outliers:* Their selection is described in the present chapter. Run the chosen algorithm with various numbers of groups and discards and record all Pareto solutions, Section 4.1, and their main characteristics. Remove solutions whose HDBT ratios or likelihood values are too small. Finally, report all solutions that pass the selection methods of Section 4.2 and the validation methods of Section 4.4. If none seems sensible, one or more of three reasons may be responsible. First, the objects may not be clustered at all, that is, the variables may not contain information on different causes. Second, the assumptions underlying the clustering method may not be appropriate. Third, the optimization method may have failed.

5.5   Notes

The notion of conditional independence of two random variables given a third one naturally arises in the Bayesian theory. Jeffreys [274] gave it the name of "irrelevance." To the author's knowledge, it emerged in connection with *variable selection* in the late 1980s. Pearl [403], p. 14, writes "Variable X is irrelevant to variable Y once we know Z." A first definition in the sense of Definition 5.1(b), but for a single variable, appears in John et al. [278]; they call it *strong irrelevance.* The more general Definition 5.1(a) is due to Koller and Sahami [296], Definition 1. It was later used by many authors and appears, for instance, in Law et al. [308], Section 2, and in Raftery and Dean [426], right after Eq. (2). Weak irrelevance of a variable X_k in the sense of John et al. [278] means that there exists a subset $E \not\ni k$ such that $\{k\}$ is irrelevant w.r.t. E in the sense of Definition 5.1(a). A fertile source of current variable selection methods in machine learning is the proceedings volume of the NIPS2003 Feature Selection Challenge edited by Guyon et al. [221]. The introductory article by Guyon and Elisseeff [220] contains more concepts of irrelevance and also various (multiple) statistical tests on variable and subset irrelevance. Pearl [403], p. 97, calls a subset $F \subseteq 1..D$ a **Markov blanket** of $k \notin F$ if X_k is conditionally independent of $X_{\complement(F \cup \{k\})}$ given X_F. This notion does not involve clustering but a small modification relates it to irrelevance. Koller and Sahami [296] study Markov blankets $F \subseteq 1..D$ of $k \in (1..D) \backslash F$ w.r.t. the extended set of variables $\{L, X_1, \ldots, X_D\}$. Assign the index 0 to L so that the total index set becomes $0..D$. Then $F \subseteq 1..D$ is a Markov blanket of $k \in (1..D) \backslash F$ w.r.t. $0..D$ if X_k is conditionally independent of $(L, X_{(1..D) \backslash (F \cup \{k\})})$ given X_F. It follows clearly that

---

[3]It may also be used in the absence of these assumptions when clusters are very well separated.

k is irrelevant w.r.t. F. The definition of "structural" and the Relevance Decomposition, Theorem 5.5, seem to be new here.

More on information-theoretic filter methods is found in Torkkola [509]. He considers the selected features as a noisy channel that transmits information about a message, here the label L, and refers to Shannon's [469] Channel Coding Theorem. Posse [421] presents a multivariate filter different from Hui and Lindsay's [259]. His approach combines a normality test statistic with a search strategy.

The selection method based on the LRT, Procedure 5.22, is new but uses ideas from Raftery and Dean's [426] Bayesian method of Section 5.3.2. Maugis et al. [355] extend their method and prove a consistency result. Toussile and Gassiat [510] treat probability-based variable selection for multilocus genotype data with alleles. They, too, prove a consistency result. The procedures 5.29, 5.30, 5.31, the Propositions 5.13, 5.23, 5.27, and Theorem 5.35 are new. They have been developed jointly with Dr. María Teresa Gallegos.

Tadasse et al. [493] proposed yet another Bayesian clustering and variable selection method. It uses a reversible jump Markov process to optimize a target function over the solution space of partitions and subsets. Zhao et al. [541] have set up an interesting website for feature selection. It contains a technical report, data sets, and feature selection algorithms.

Chapter 6

Applications

This final chapter serves to illustrate the foregoing matter. Three of the four data sets treated here are well known. They are Anderson's IRIS data from botany, Flury and Riedwyl's SWISS BILLS from criminology, and the LEUKEMIA gene expression data of Golub et al. The fourth one is Weber's STONE FLAKES from prehistoric archeology. With the exception of LEUKEMIA all are small. However, none is easy. This refers in particular to the number of clusters issue. Indeed, the question whether the four variables of the IRIS data set know their number of subspecies has pained a whole generation of cluster analysts. This data set also bears another surprise. LEUKEMIA is a special, fairly large data set which contains a lot of irrelevant information. The main task will be to detect and to remove it. Other fields of application are provided in the Introduction of Redner and Walker [435].

The aim of this chapter is not to analyze the data – their true partitions have been known even before the data was collected. Its purpose is rather to assess the effectiveness of the methods. All analyses will be unsupervised, that is, based on the data alone. The objective will be to single out the best decompositions of the data in an unbiased way, that is, without explicitly or implicitly using the true solution. This makes it necessary to analyze and validate a fair number of solutions. For the sake of saving space, only a few alternative solutions will be presented here. We will assume that parental components are not too far away from normality so that we may assume a normal mixture model.

The analyses will be based on *Pareto* solutions selected from the *free* local optima of the trimmed mixture likelihood or from steady solutions of the unconstrained TDC as explained in Section 4.1. SBF plots will be shown. Besides other validation methods, a table of some isolation indices and p-values for cluster pairs will be presented in all cases; see Tables 6.1, 6.2, 6.4, and 6.7. The meaning of the seven columns was described in the text accompanying Table 4.5 above. In data plots, the symbols $\circ$, $+$, $\triangle$, $\square$, and $\hexagon$ denote members of clusters $1, \ldots, 5$ in this order,[1] errors are marked in gray, and solid dots stand for outliers.

6.1 Miscellaneous data sets

In this section we will analyze the three small data sets. The first two are standard benchmarks in this field, Anderson's IRIS data from botany and Flury and Riedwyl's SWISS BILLS from criminology. They have been used by many cluster analysts to exemplify the performance of their methods. While it is known, and can be easily seen, that the SWISS BILLS contain besides genuine bills also forged ones, it is not clear from the outset in how many batches the counterfeits had been produced, maybe even by different forgers. This is unknown and two decompositions of the forged notes in two and three clusters will be presented. The last data set, Weber's STONE FLAKES from archeology, concerns prehistory and relics from the stone age. It has been grouped by archaeologists mainly by age. All data sets contain components not well separated from the rest.

---

[1] As a mnemonic, count the number of strokes needed to draw the symbol.

Table 6.1 IRIS: *Separation indices and p-values for all cluster pairs of the untrimmed two- to four-component solutions.*

g	pair	Fisher	linear separ.	Sym. diverg.	−log Helling.	p-value MV S.-W.	p-value lin. separ.
2	1,2	1.000	1.000	275.6	6.95	0.000	0.000
3	1,2	1.000	1.000	249.8	16.44	0.000	0.000
	1,3	1.000	1.000	410.9	20.96	0.000	0.000
	2,3	0.978	0.978	24.6	2.16	0.570	0.001
4	1,2	1.000	1.000	249.8	16.44	0.000	0.000
	1,3	1.000	1.000	416.3	25.71	0.000	0.000
	1,4	1.000	1.000	717.9	48.27	0.000	0.000
	2,3	0.981	0.981	25.2	2.25	0.668	0.000
	2,4	0.996	0.999	101.8	5.26	0.020	0.000
	3,4	0.994	0.995	72.8	3.43	0.609	0.005

Aside from normality, no prior assumptions will be made, neither on the type of normality, nor on cross-cluster constraints, nor on the number of components. This means the application of the (trimmed) likelihood estimator of the heteroscedastic, fully normal mixture model and the EMT or of the (unconstrained) TDC with the k-parameters algorithm, both along with scale balance; see Section 4.1. Since it is not a priori clear whether all variables contribute equally well to the desired classification, variables will be selected with the exception of the IRIS data.

6.1.1 IRIS data

Anderson's [14] IRIS data set was made famous by Fisher [156] who used it to exemplify his linear discriminant analysis. It has since served to demonstrate the performance of many clustering algorithms. The data consists of 50 observations each of the three subspecies *I. setosa*, *I. versicolor*, and *I. virginica*, measured on the four variables sepal length, sepal width, petal length, and petal width.[2] The three subspecies occupy the numbers 1–50, 51–100, and 101–150, respectively. The entries are of the form *.* with only two digits of precision preventing general position of the data. The observations 102 and 143 are even equal. Since this is a deficiency of the data, it is best to randomize them by adding a number uniformly distributed in the interval $[-0.05, 0.05]$ to each entry, thus trying to "restore" their numerical character, although not faithfully. But the noise added is negligible compared with the natural variation of the data. Two of the three clusters, versicolor and virginica, are poorly separated, presenting a challenge for any *sensitive* classifier, although they overlap only slightly. This challenge is one of the reasons why this data has been around for such a long time.

Ignoring any prior structural information and since it may be assumed that lengths and widths are correlated, we apply the more general models described in Chapter 1, the normal mixture and classification models with full scales and with and without trimming. We first use the EMT algorithm of see Section 3.1.4 to compute Pareto solutions; see Section 4.1. Clusters are subsequently obtained by means of MAP discriminant analysis based on the estimated parameters of the mixture decompositions. Scatter plots of the data show that the subset $1..50$ is well separated from the rest. Therefore, only two or more components are realistic. The plots also show that there are only few outliers, if any. We therefore run the algorithm with two, three, and four components, without trimming and with trimming

---

[2]The data is presented under the URL en.wikipedia.org/wiki/Iris_flower_data_set. It contains also photographs of the three sub-species.

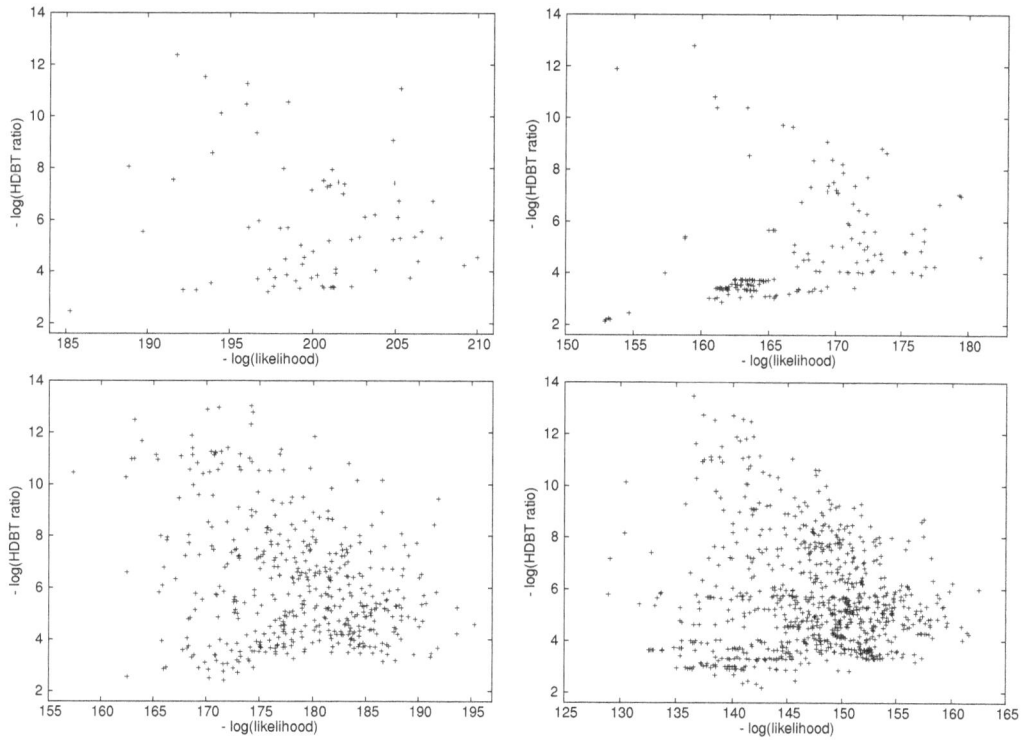

Figure 6.1 IRIS: *SBF plots of local maxima, three (top) and four components without trimming (left) and with five discards (right).*

of five elements. It turns out that sepal length is irrelevant, but using all four variables does not change the the result.s

Not surprisingly, all solutions recover the subset $1..50$ as a group. For $g = 2$ components without trimming or with five discards, the EMT algorithm returns essentially a single solution each. Without trimming, the other group is $51..150$. The same solution is returned with trimming, except that the five trimmed elements are taken from the larger group.

Decomposition in more than two components needs a closer analysis. The SBF plots described in Section 4.1 for $g = 3, 4$ components and trimming of $n - r = 0, 5$ elements are presented in Fig. 6.1. The upper two plots show the untrimmed and trimmed results for three components. Each plot contains a single scale-balanced, well-fitting solution represented by the dots in the left, lower corners. Their HDBT ratios are $1/11.7$ and $1/8.41$, respectively. The solutions recover essentially truth up to five elements that are shifted from *I. versicolor* to the last group and the five trimmed plants, which are mainly taken from the virginica group.

Table 6.1 presents *separation indices* and *p*-values for several numbers of clusters and all cluster pairs. All indices agree that the two-cluster solution is well separated. With the exception of the *p*-value in the last column, all indices cast doubts about the separation of clusters 2 and 3 of the three-cluster solution. This may be attributed to their missing selectivity. The same remark applies to the cluster pairs (2,3) and, to a lesser degree, (3,4) of the four-cluster solution. It seems that Hui and Srivastava's multivariate normality test lacks selectivity, disagreeing in three cases strongly with the univariate which is known to be selective.

The BIC values for one to four components without trimming are 415, 290, 295, 310, respectively, and 378, 257, 262, 290 with trimming. That is, BIC definitely rejects one and four components and slightly favors two to three. As is well known, BIC tends to have a negative bias, in particular when separation is poor. We should not rely solely on this estimate.

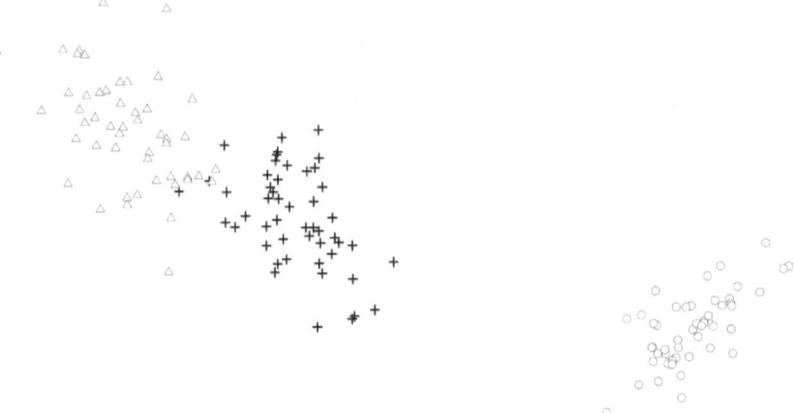

Figure 6.2 IRIS: *The true classification projected to a plane that optimally displays the separation between* I. versicolor *(+) and* I. virginica *(△).*

Let us next apply the modified Wolfe test of Section 4.2.1. The untrimmed solutions for $g = 3$ represented in the left upper SBF plot of Fig. 6.1 contain a mixture which is close to the favorite one for $g = 2$ up to a small component with mixing rate 0.032. Its negative log-likelihood is 200.9, whereas that of the favorite untrimmed solution for $g = 3$ is 185.3. Twice the difference is 31.2 leading to a p-value of less than 0.01. Therefore, the test rejects $g = 2$ in favor of $g = 3$.

A scatter plot of the data projected to a plane that optimally displays the degree of separation between the two close subspecies (Section 4.4.3) is offered in Fig. 6.2.[3] The plane contains the normal vector obtained from elliptical and linear separation in Section 4.4.1. The plot also shows a slight overlap of the two critical subspecies *I. versicolor* and *I. virginica.*

For $g = 4$ components, the method returns two well-fitting, untrimmed Pareto solutions (left lower SBF plot in Fig. 6.1). The one with the slightly better fit has an HDBT ratio of about 1/33,000 and must therefore be considered spurious. The other one enjoys far better scale balance and recovers *I. verginica* up to five elements, which are again shifted to the third group 101..150, which in turn loses a small fourth group of nine plants. A scatter plot of the versicolor and virginica plants that optimally displays this nine-point cluster is presented in Fig. 6.3. It shows that this solution cannot be ruled out by merely analyzing the data set. Moreover, the ratio between the minimum and the second maximum in Hennig's Ridgeline Ratio method (Section 4.2.4) is smaller than 0.1. The small subset must be considered a cluster of its own. There are just two observations that call this result in question: (i) The end of the oblong bulk of the data opposite to the small cluster is unusually wide for the end of a sample from a normal; (ii) its main axis points precisely to the small subset. Given the correct grouping, it may be presumed that the sample is biased. The nine flowers have unusually long sepals that probably stood out to Anderson. He wanted to include them in his collection, forgetting to close the gap. This is no problem in discriminant analysis but becomes one in clustering.

The situation for $g = 4$ with trimming (right lower plot in Fig. 6.1) is even less clear, since the Pareto points in the left lower region of the SBF plot lie more or less on a quarter circle.

---

[3]Hui and Lindsay [259] present another one.

Figure 6.3 *Projection of* I. virginica *to a plane that optimally displays the nine-point cluster of the four-cluster solution (top left).*

The HDBT ratios of the three most promising solutions are 1/38.2, 1/18.9, and 1/8.8. The first one reproduces essentially the trimmed solution with three clusters but splits a cluster of seven elements off the subset 101..150. The second solution divides the subset 51..150 into three parts, whereas the last one splits 51..100 but leaves 101..150 unchanged.

It was mentioned at the end of Section 4.1 that the cluster size constrained TDC can help identify a reasonable solution. Gallegos and Ritter [184] analyzed the IRIS data set using the exact multipoint reduction step 3.29. The untrimmed ML version of the TDC, that is, the determinant criterion of Theorem 1.40 with three clusters and minimal cardinalities $b_j = b = d+1 = 5$ essentially reproduced the known, correct partition of IRIS; just two versicolor plants are placed in the virginica cluster and one virginica plant is falsely assigned to the versicolor cluster. The data set happens to meet the assumption of equal cluster sizes underlying the ML criterion. The result profits from this fact. The same result was obtained by Friedman and Rubin [172] with their homoscedastic model and by Li [318] with a different method.

The optimal clustering according to the MAP determinant criterion (1.40) with three clusters, no trimming, and $b = 5$ merges *I. versicolor* and *I. virginica*, identifying only two major clusters and a small spurious cluster of minimal size five close to some hyperplane. This failure is due to the two close groups. Recall that the classification model was obtained in Section 1.3.1 from the mixture model by means of an approximation that needed clear separation between all clusters. Up to $b = 19$, the optimal solution still contains a cluster of or close to the minimum size b. But from $b = 20$ on, up to 46, it returns a solution with cardinalities 50, 46, 54 close to the true one. This shows that, besides the HDBT constraints, size constraints can also be beneficial in obtaining reasonable solutions.

How could it be that Anderson safely classified the irises in their three types whereas we, using a lot of refined statistics and programming, cannot be certain of our results? The answer is simple: The problem is difficult because of the scanty information we are given, not because of a weakness of the methods. The botanist was using his own features, which are much more refined and characteristic than the ones available to us. The colorings of versicolor and virginica are different. Had he given us his features, we, too, would have had an easy job. The four linear dimensions certainly do not contain the information necessary for a "gold standard." The problem at hand belongs to image processing and pattern recognition. It is often hard for the expert to describe his or her "test statistic." Can you tell why you easily recognize a friend in the street when you meet her?

6.1.2 SWISS BILLS

Like gold, money has always stimulated human fantasy. Some try their luck at gambling to make easy money. A few with a lot of criminal energy set up workshops with equipment to forge bills in garages or cellars. The idea is to produce bills at a cost substantially lower

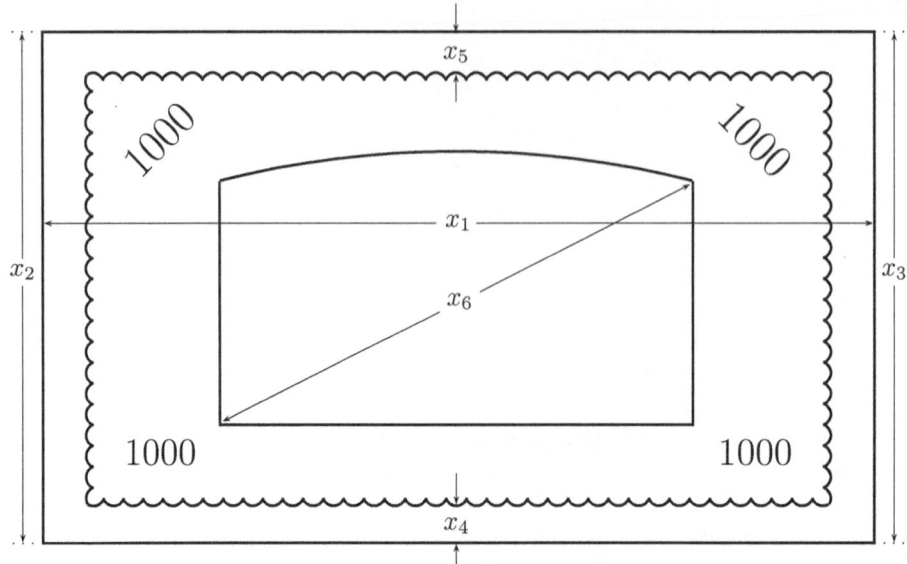

Figure 6.4: *Outlines of the old Swiss 1000-franc bill.*

than the imprinted number. This calls for a compromise and forgeries are not perfect. They are usually detected and sorted out by bank tellers or other people who handle a lot of paper money. These people easily detect poor, clumsy falsifications with gross errors such as missing water marks or wrong paper. If a bank note is suspect but refined, then it is sent to a money-printing company, where it is carefully examined with regard to printing process, type of paper, water mark, colors, composition of inks, and more.

Flury and Riedwyl [165] had the idea to replace the features obtained from the sophisticated equipment needed for the analysis with simple linear dimensions. They managed to obtain 100 genuine and 105 forged old Swiss 1000-franc bank notes, enlarged them with a projector on a wall, and took six measurements from each of them. They were

x_1: "length of the bill,"
x_2: "its left width,"
x_3: "its right width,"
x_4: "width of the margin at the bottom,"
x_5: "width of the margin at the top," and
x_6: "diagonal length of the inner frame."

See also Fig. 6.4. Finally, they fed the data to some common statistical methods such as linear regression, discriminant analysis, and principal component analysis, showing that their idea was successful.

The data set was also detected by cluster analysts. Let us apply the MAP determinant criterion (TDC) with and without variable selection (Procedures 3.28 and 5.31) to the 205 six-dimensional observations. As expected, the data set is quite clean – crude abnormalities would be easily detected by everybody – so no trimming is needed. Since it is all about genuine and forged bank notes, and since it is unknown whether the forged bills have been produced by the same forgers and in one batch, we try two to four components. Moreover, not each variable may be equally relevant and we try all possible numbers of variables. This makes $3 \cdot 6 = 18$ different parameter settings. Preliminary program runs show that the two main features are x_4 and x_6. This also agrees with Flury and Riedwyl's findings. Only results for three and four clusters will be presented here. The corresponding SBF plots described in Section 4.1 appear in Fig. 6.5.

The plot for three clusters contains a few solutions in its left lower region, but they are very close and stand for very similar partitions. One of them is selected. The four-cluster

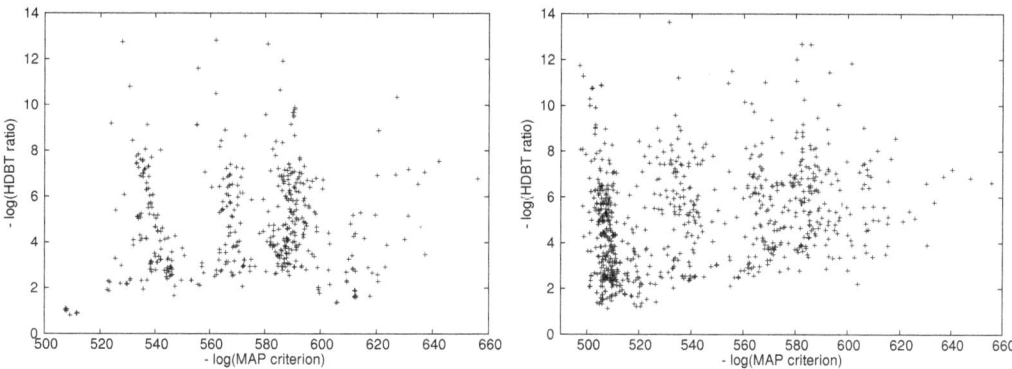

Figure 6.5 SWISS BILLS: *SBF plot of the steady, untrimmed solutions with three (left) and four clusters.*

plot is different, but again there are only a few candidates for reasonable solutions close to the point (504,1.4). Their separation indices and p-values are shown in Table 6.2. The only striking cluster pair is (2,3) for $g = 4$. The p-value of the multivariate Shapiro-Wilk test is 0.163 and seems to merge the two clusters; that for the univariate test along the straight line of best linear separation is 0.016. Although this is fairly small, it leaves us undecided since it refers to a projection to an optimally separating direction. The number of components – three or four – is still uncertain.

Let us therefore try mixture analysis and the modified Wolfe test of Section 4.2.1. We need a decomposition in four components close to the selected three-component decomposition up to an additional small component. We help the EM algorithm by starting it from the three-cluster partition, with an additional fourth cluster in the form of a small triangle cut off the center of the largest group. The algorithm converges indeed to a local maximum close to the favorite three-component mixture with an additional component, with mixing rate 0.0085 generated by the three points. The negative log-likelihood of the fake three-component solution is 504.77, whereas that of the four-component mixture is 498.68. The χ^2-approximation with six degrees of freedom yields a p-value of about 0.05, mildly rejecting three clusters in favor of four.

Let us finally examine the four-cluster solution with Hennig's and Bertrand and Bel Mufti's cluster stability methods in Section 4.4.5. The results for 50 replications are shown in Table 6.3. As expected, the clusters two and three are somewhat shaky in view of the Jaccard and isolation indices.

In conclusion, it was easy to single out the forgeries from the 205 bills. The forgers

Table 6.2 SWISS BILLS: *Separation indices and p-values for all cluster pairs of the untrimmed three- and four-cluster solutions.*

g	pair	Fisher	linear separ.	Sym. diverg.	$-\log$ Helling.	p-value MV S.-W.	p-value lin. separ.
3	1,2	0.998	0.998	35.6	4.11	0.000	0.000
	1,3	0.999	0.999	42.4	4.80	0.000	0.000
	2,3	0.982	0.983	20.9	2.30	0.007	0.000
4	1,2	0.998	0.998	35.1	4.04	0.000	0.000
	1,3	1.000	1.000	70.5	7.85	0.000	0.000
	1,4	0.999	0.999	39.9	4.58	0.000	0.000
	2,3	0.966	0.966	13.6	1.67	0.163	0.016
	2,4	0.988	0.988	25.5	2.57	0.000	0.000
	3,4	1.000	1.000	55.1	5.55	0.000	0.000

Table 6.3: SWISS BILLS: *Cluster stability indices for the TDC decomposition in four clusters.*

Stability method	cluster			
	o	+	△	□
Jaccard	0.969	0.766	0.781	0.898
Loevinger isolation	0.989	0.705	0.682	0.983
Loevinger cohesion	0.978	0.899	0.958	0.734

just didn't work carefully and precisely enough. The only errors are the bills nos. 70 and 103. The first one is genuine, the second counterfeit. They are printed in Fig. 6.6 in black. The forged bills consist probably of three clusters, maybe also of two. They may have been produced by different forgers or in different manufacturing processes.

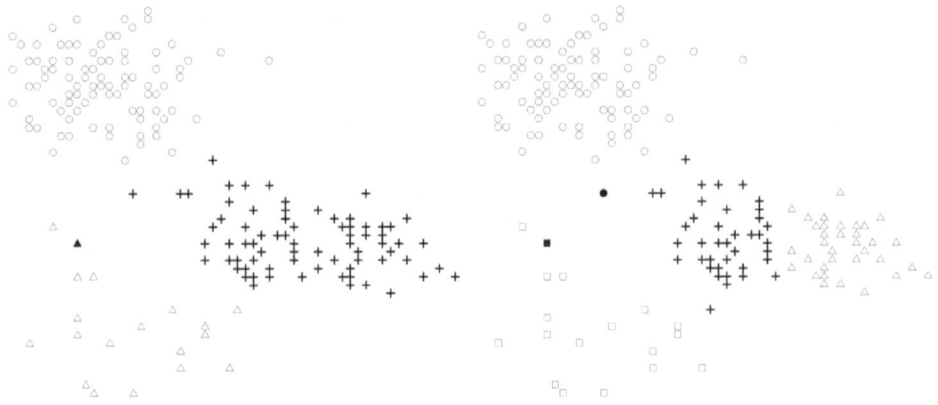

Figure 6.6 SWISS BILLS: *Partitions in three (left) and four clusters. The genuine bills make up cluster* o.

6.1.3   STONE FLAKES

It is well known that prehistoric humans occupied themselves with manufacturing and using stone tools and controlling fire for preparing meat. Besides bone relics, stone artifacts are the main witnesses that have survived the passage of several hundred thousand to a few million years, when the first manlike beings entered the earth. With the aid of the relics, archaeologists try to unravel the earliest technological and even behavioral history of mankind. During the last 30 years, Schäfer [457] and Weber [524] dug up and collected stone artifacts from some 60 inventories detected at more than 40 sites in Central Germany and several parts of Eastern Europe. The number of stones in each inventory ranges from 5 to more than 4,000. Since the Fall of the Iron Curtain, they were able to also include data of colleagues from Western Germany, Western Europe, and South Africa. The geographical layers that contained the finds allowed dating the German pieces to the period from about 300,000 to some 40,000 years ago. The other European assemblages are partly older, the South African artifacts sometimes more than 1,000,000 years.

The artifacts most frequently found by archaeologists are actually not stone tools but flakes split off the raw material during production. Three faces of a flake are of particular interest. First the part of the platform onto which the strike was made or pressure was exerted, second, the flat splitting surface or ventral face, and third, the dorsal face, a part of the original surface. The striking platforms of many finds had been prepared with small concave facets preliminary to the stroke. It is reasonable to assume that they are younger, whereas older pieces do not show signs of such preparation. The archaeologists extracted a number of geometric dimensions from the flakes: the length in flaking direction and the breadth and thickness perpendicular to it. These depend on several factors such as the

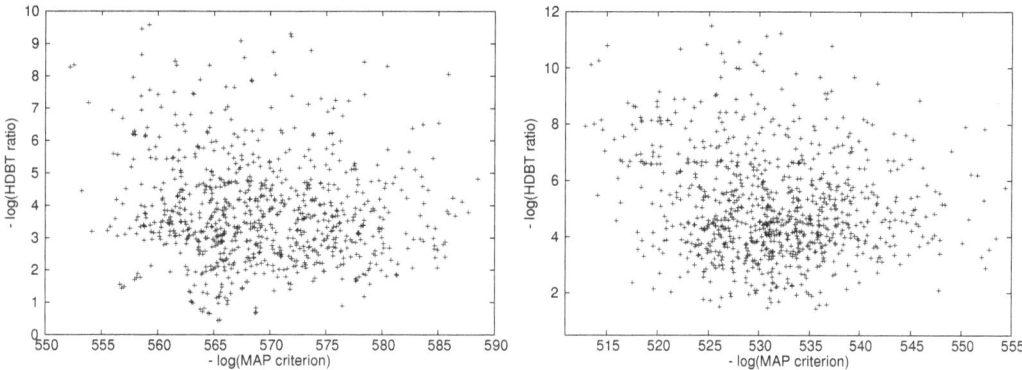

Figure 6.7 STONE FLAKES: *SBF plots of steady solutions obtained with the two variables* psf *and* zdf1, *three clusters and two discards (left) and four clusters and five discards.*

size of the raw material and are themselves not characteristic in view of the questions put above. Therefore, the absolute measurements were converted to form indices: the "Length-Breadth" (lbi), the "Relative-Thickness" (rti), and the "Width-Depth" (wdi) indices of the striking platform. Furthermore, the "Flaking Angle" (flang) between the striking platform and the splitting surface, reflecting the striking manner, was measured from each stone artifact. Finally, four style features, the numerical "Proportion of Worked Dorsal Surface" (prozd, estimated at a 10% level) and three binary features (y/n), "Platform Primery" (psf), "Platform Facetted" (fsf), and "Dorsal Surface Totally Worked" (zdf1) were determined. The latter is positively correlated with prozd.

Assuming that the stones of each inventory are technologically homogeneous – they had been prepared by members of the same sub-family of hominids about the same time – the archaeologists characterized each inventory by the mean values of the numerical features and the relative frequencies of the binary ones. Thus, the objects under study are actually not single flakes but inventories. Each of them is characterized by the aggregates of the eight variables lbi, rti, wdi, flang, prozd, psf, fsf, and zdf1. The resulting data set consists to date of 81 rows and 8 characteristic variables. A few values in eight rows are missing. It was the data set offered for analysis at the joint meeting of the working groups for Data Analysis (DANK) and Archaeology (CAA) in the German Classification Society (GfKl) at Bonn in 2012.[4] Apart from the variables, the data set also contains the approximate age, the site of the inventory, the number of pieces found in it, the material, a grouping based on geo-chronological information, and a bibliographical reference. We cannot expect all variables to be equally informative and relevant. Weber [524] points out the highly discriminating power of the variables rti, flang, zdf1, and prozd. The TDC with subset selection (Procedure 5.31) confirmed the last three indices as most relevant, but not rti. Upon requesting one variable, the algorithm returns just psf; with two variables it adds zdf1, and with three variables prozd or flang depending on the HDBT constraints.[5] Most variables preferred by variable selection are thus stylistic and not geometric. Here is an anlysis of the 73 complete observations obtained with the TDC and the two most relevant variables psf and zdf1. The most convincing partitions found consist of three and four clusters and a few outliers.

The SBF plots for the classification model with three clusters and two discards and with four clusters and five discards are presented in Fig. 6.7. A glance at the plots shows that the data set is not easy, allowing various interpretations. The extreme points on the left

---

[4]The data set was donated to the UCI Machine Learning Repository [2].

[5]M.T. Gallegos, oral communication, noted that Raftery and Dean's [426] Bayesian method (Section 5.3.2) selects the same four variables after two bad outliers have been deleted.

Table 6.4 STONE FLAKES: *Separation indices and p-values for the two solutions shown in Fig. 6.8.*

g	pair	Fisher	linear separ.	Sym. diverg.	–log Helling.	p-value MV S.-W.	p-value lin. separ.
3	1,2	0.969	0.969	14.0	1.74	0.570	0.193
	1,3	0.995	0.995	30.7	3.34	0.007	0.000
	2,3	0.969	0.969	14.5	1.75	0.060	0.109
4	1,2	0.973	0.974	27.1	1.82	0.122	0.028
	1,3	1.000	1.000	81.8	5.61	0.003	0.000
	1,4	0.998	0.998	56.2	3.88	0.026	0.008
	2,3	0.997	0.997	37.0	3.69	0.046	0.002
	2,4	0.992	0.992	40.4	2.70	0.234	0.105
	3,4	0.955	0.958	28.1	1.85	0.094	0.014

and lower sides lie more or less on a quarter circle and there are no pronounced ones in the left lower corners. Moreover, some Pareto optima are not extreme. Presented here are the Pareto solution $(564, 0.78)$ in the plot for three clusters and the extreme Pareto solution $(518, 2.3)$ in that for four clusters.

Some separation indices and test results for the two solutions are shown in Table 6.4. The indices and p-values for the *three*-cluster solution unanimously declare separation to be good only for clusters 1 and 3, although visual inspection seems to support separation between all clusters. The first five columns of the *four*-cluster solution indicate that separation of the cluster pairs (1,2) and (3,4) is poor and good in all other cases. The p-values in the last columns also shed doubt on the separation of 2 and 4. The normality tests seem to be unsuitable here for a cohesion test. The solutions remain essentially unchanged after discarding ten instead of five observations. Just some more extreme elements in the clusters are discarded. This indicates stability. Scatter plots of both solutions are offered in Fig. 6.8.

The meaning of the two cluster structures is not immediately clear. Crafting stone tools is controlled by human preference, intent, opinion, skill, and taste. Even though the *stones* themselves bear a true cluster structure defined by age, region, and/or hominid types, this may not be well reflected by the *data*. Table 6.5 displays the deviation of the three- and four-cluster solutions from Weber's grouping, shown in Fig. 6.9. He defined his solution mainly by region and age. The present solutions may reflect something else, for instance, the family of hominids who were crafting the stones. The adjusted Rand indices between Weber's and the two present groupings are 0.504 (g=3) and 0.367 (g=4), respectively. They are well above zero and so the agreements are not incidental. The difference between the solutions is not dramatic given the difficulty of the problem. It may partly be due to the

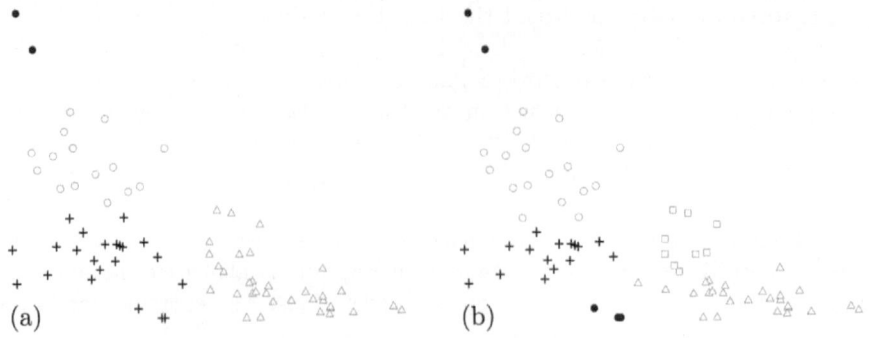

Figure 6.8 STONE FLAKES: *Scatter plots of a three-cluster solution with two outliers (a) and a four-cluster solution with five outliers. Abscissa zdf1, ordinate psf.*

Table 6.5 STONE FLAKES: *Agreement of the three- and four-cluster solutions of Fig. 6.8 with the geo-chronological grouping.*

	group 1	group 2	group 3	group 4	outliers	total
cluster 1	9	4	2	0	0	15
cluster 2	2	17	3	0	0	22
cluster 3	0	2	29	3	0	34
discarded	0	0	0	0	2	2
cluster 1	9	5	3	0	0	17
cluster 2	2	13	1	0	0	16
cluster 3	0	1	23	2	0	26
cluster 4	0	1	7	1	0	9
discarded	0	3	0	0	2	5
total	11	23	34	3	2	73

small sizes of some inventories. The most striking difference is the assignment of his group 4 essentially to cluster 3 in both solutions. Group 4 consists of three inventories, presumably created by members of *Homo sapiens*. It cannot be detected by clustering methods with the present two variables because of the closeness to the center of group 3.

Weber's group 1 comprises the oldest European finds dated to the Lower Paleolithic between 600,000 and 300,000 years ago, group 2 consists of Acheulean flakes from about 200,000 years ago, group 3 is dated to the Middle Paleolithic between 130,000 and 60,000 years ago, and group 4 to the transition from the Middle to the Upper Paleolithic about 40,000 years ago. The four groups are ascribed to *Homo erectus/heidelbergensis*, the Anteneanderthal, the Neanderthal, and *Homo sapiens*, respectively. The clusterings obtained with the present probabilistic methods also seem to be interesting from the archaeological point of view. All solutions testify to the technological progress made along a period of several hundred thousand years.

Figure 6.9 STONE FLAKES: *Plot of the geo-chronological grouping. Group 1 is oldest, group 4 youngest. The two outliers are artifacts. Abscissa zdf1, ordinate psf.*

6.2   Gene expression data

The study of the activity of genes in a cell one by one already provides much insight into biological function. On the verge of the millennium, microarrays emerged as a means to measure the transcriptional activity of many thousands of genes in a single experiment. This abundance offers the possibility of studying whole or large parts of genomes simultaneously. It accelerates biomedical progress in an unprecedented way and has enormous impact on cell biology and on the analysis of pathologies and transcriptional regulatory networks.

The following much simplified description just serves to roughly explain what expression levels are about. A spotted DNA **microarray** is a rectangular or quadratic matrix of thousands of grid points or "spots" aligned on a glass or nylon plate. Each spot contains so-called "probes," identical DNA molecules representing one of an organism's messenger RNA (mRNA) molecules (roughly speaking, genes). The different spots in the array may represent the entire genome or a large part of it. The array is used in an experiment to measure the amount of mRNA molecules present in some tissue to be analyzed. The mRNA is extracted, reverse-transcribed to complementary DNA (cDNA), and labeled with some fluorescent dye to be able to identify its presence at the end of the experiment. These "targets" have the ability to associate with (or "hybridize" to) their complementary probes in the array. This will occur to the extent that the tissue under study contains the mRNA. After washing, only the cDNA molecules bound to the specific probes remain on the plate. The amount of bound targets at each spot can be measured due to the dye and allows us to infer the expression level (or activity) of the related gene in the tissue. Evaluation of all spots provides an overview over the activities of all genes in the tissue represented in the array. The data is subject to a certain amount of noise; see Lee et al. [312]. For protection against various kinds of noise, each gene is usually represented by several probe spots in the array and each spot is accompanied by a control. Similar but smaller and more sophisticated devices are the high-density **olinucleotide chips**. Here, the probes are oligonucleotides, chains of a few tens of the four nucleotides, each probe again specific for some RNA target. For more details on the biomedical background, the interested reader is referred to the presentations in Dudoit et al. [134], Lee et al. [312], Russell et al. [455], and the supplement to *Nature Genetics* [91].

There are a number of now classical data sets that have been used for illustrating new methods for the analysis of gene expression data. Examples are the LEUKEMIA data set of Golub et al. [206], the COLON cancer data set of Alon et al. [10], the LYMPHOMA data set of Alizadeh et al. [9], the ESTROGEN data set of West et al. [525], the NODAL data set of Spang et al. [482], the BREAST cancer data sets of Hedenfalk et al. [237] and van 't Veer et al. [515], the BRAIN tumor data set of Pomeroy et al. [420], the PROSTATE cancer data set of Singh et al. [477], and the DLBCL (diffuse large B-cell lymphoma) data of Shipp et al. [472]. Before using them one has to find out the transformations and adjustments undertaken for mitigating array effects. Some are log-transformed, others are not; some contain normalized profiles and negative values, others don't.

A gene expression data set typically consists of the measurements made for between ten and a few hundred tissue samples, each with a microarray reporting on the same thousands of genes. The sequence of expression levels of all genes in a cell (or of the genes represented in the microarray) is called the **expression profile** of the cell, whereas the vector of expression levels of one gene across the arrays is its **expression pattern**. Each microarray thus yields one profile and each gene one pattern. The measurements are usually made for tissues under a few different *conditions*. Examples of conditions are the healthy state, various types of pathologies (diseases), and stages along a developmental process. In statistical terms, the genes represent the variables, attributes, or features of the data set, the microarrays, experiments, or profiles represent the observations, elements, or cases, and the conditions are the classes. Occasionally, the view is opposite and the genes are considered the objects while the experiments appear as the variables. Authors are sometimes not clear about what they are actually clustering, cases or genes (or both). This may at times be confusing.

The expression levels of most genes are not significantly different between different conditions, for instance, the healthy and the diseased state or different diseases. Leukemia, a disease of the blood or bone marrow, is known to have three major genetic subtypes, acute myeloid leukemia (AML), B-cell acute lymphoblastic leukemia (B-ALL), and T-cell acute lymphoblastic leukemia (T-ALL). The LEUKEMIA data set mentioned above contains cases of all three types. Figure 6.10 shows two scatter plots of all 72 cases w.r.t. four genes. The expression levels show no differences between the diseases. On the other hand, a number of specific genes tend to be highly expressed in one type and low in the other states, healthy or diseased, or vice versa. One speaks of *differential expression*. That is, even if a gene is overexpressed in one cancer type, its expression levels are usually inconspicuous in the others. Figure 6.11 shows the spatial plot of LEUKEMIA w.r.t. the genes X82240, X03934, and M27891. They are obviously over-expressed in B-cell ALL, T-cell ALL, and AML, respectively. The regular expression levels are below 1,000, whereas the pathological

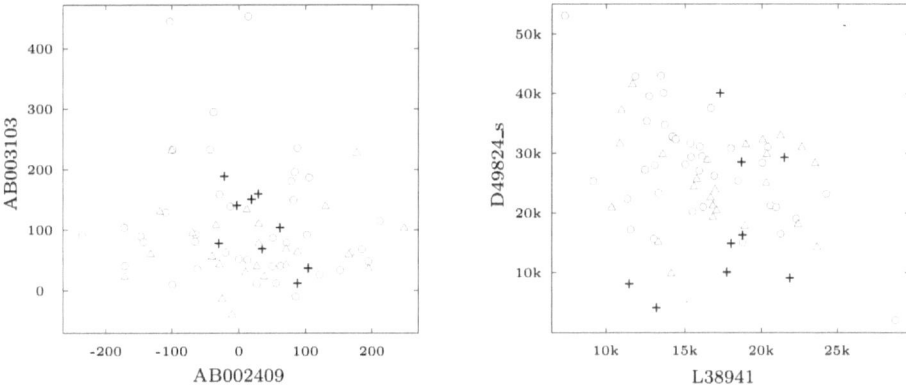

Figure 6.10 LEUKEMIA: *Scatter plots for nondifferentially expressed genes, low expression levels (left) and high expression levels.*

values rise to some ten thousands and do not seem to be normally distributed; see also Fig. 6.12.

Gene expression data sets bear information on various tasks. They are in increasing complexity:

(i) Given descriptions of some conditions by expression levels of one or a few specific genes (called **marker genes**), determine the condition under which a new experiment was made. The method related to this diagnostic task is *discriminant analysis.*

(ii) Given a number of microarray experiments for at least two *known* conditions, find the most relevant marker genes whose expression levels contain discriminating information on the conditions. This is solved by *variable* or *gene selection.*

(iii) Given a number of experiments made under an *unknown* number of conditions, find the associated decomposition of the cases. If necessary, single out cases that do not conform to any of the conditions. This task is solved by *robust cluster* or *mixture analysis.*

(iv) Given a number of experiments made under an unknown number of conditions, find the decomposition of the cases as well as genes differentially expressed between them. If necessary, single out cases that do not conform to any of the conditions. This task is solved by *robust cluster* or *mixture analysis* with *variable selection.*

Containing many more variables (genes) than cases, expression data presents a specific

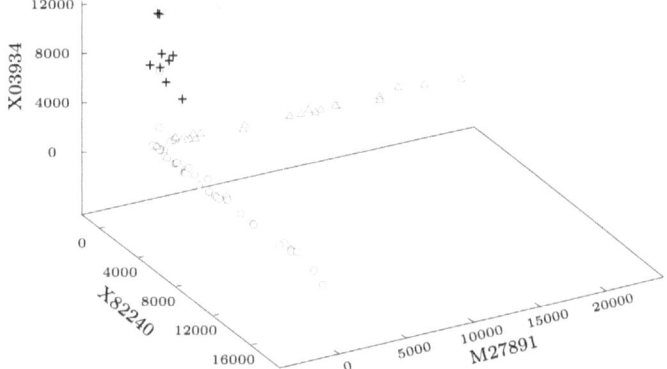

Figure 6.11 LEUKEMIA: *Spatial plot w.r.t. three genes specific for the three cancer types. Gene X82240 is highly expressed in B-ALL (○), X03934 in T-ALL (+), and M27891 in AML (△).*

challenge, but also a novel opportunity, to statistics. It turns out that the vast majority of genes in the genome are nondifferentially expressed, containing no significant information on specific conditions. They just appear as noise, masking the true clustering of cases and degrading the performance of clustering algorithms; see Law et al. [308]. In fact, there are in general only a handful of discriminating genes most relevant to the goal. One of the main tasks is to find such a *molecular signature*. The tasks that deserve most attention in microarray analysis are therefore (ii) and (iv). In both, genes relevant to the classification of conditions are selected, (iv) uses just the data and is applicable without knowledge of the conditions, (ii) also needs the conditions. In contrast, task (i) is classical since a limited number of marker genes is a priori given and task (iii) cannot be safely performed without previous gene selection.

6.2.1 Supervised and unsupervised methods

Over the past years, numerous tools have been proposed for tasks (ii) and (iv). Task (ii) is *supervised* since it assumes knowledge of the conditions. The diseases, mainly cancers, are quite well known in medicine but the detection of the most significant marker genes by supervised gene selection (ii) remains important from the clinical point of view, too. They can be used as a basis for diagnosis and for development of a gene associated therapy. First proposals have already been made by the authors of the data sets.

In view of the supervised gene selection problem (ii), Golub et al. [206] use a statistic that resembles the classical pooled t-statistic[6] as a score, finding 1,100 genes differentially expressed between AML and ALL, the union of the two ALL subtypes. The single most distinctive gene reported in their study is the homeobox A9 (HOXA9) located on chromosome 7, site p15.2, GenBank accession code U82759. It encodes a protein which serves as a DNA-binding transcription factor. The authors then turn to discriminant analysis (i), using the 50 most discriminative genes in a weighted-voting scheme. In a leave-one-out cross-validation study they report zero errors after rejecting the two cases with the least certainty in each run.

Resuming task (ii) and inspired by Park et al. [400], Dettling and Bühlmann [118] use Mann and Whitney's [341] test statistic for two unpaired samples in order to find groups of genes differentially expressed between the known conditions.[7] For discrimination between two conditions C_1 and C_2, the Mann-Whitney count of gene k is the number

$$U_k = |\{(i_1, i_2) \mid i_1 \in C_1, i_2 \in C_2, x_{i_2,k} \leq x_{i_1,k}\}|.$$

It is small when k is underexpressed in C_1 w.r.t. C_2 and large in the opposite case. The closer to zero or to $|C_1||C_2|$ it is, the higher the significance of gene k. The authors proceed to demonstrate the predictive power of their gene groups with the classical data sets above. They use the group averages for discriminating between conditions by leave-one-out cross-validation w.r.t. two discriminant rules: k-nearest-neighbor (Fix and Hodges [158]) and aggregated trees, their own variant of the Breiman et al. [67] classification tree. The genes with the least Mann-Whitney counts in LEUKEMIA w.r.t. B-cell ALL, T-cell ALL, and AML against their complements are presented in Table 6.6. They are overexpressed in the diseases. Dettling and Bühlmann [119] extend their method to boosting.

About the same time, Efron and Tibshirani [142] consider supervised gene selection (ii) as a massively *multiple testing problem* (see Appendix F.3.2), one test per gene. Making use of Wilcoxon's [527] rank sums, a statistic equivalent to Mann and Whitney's, they are mainly concerned about the large number of false positives that arise when each gene is tested separately. They handle the multiplicity issue with Benjamini and Hochberg's [35]

---

[6]Instead of $(\bar{x}_1 - \bar{x}_2)/\sqrt{v_1 + v_2}$ they compute $(\bar{x}_1 - \bar{x}_2)/(\sqrt{v_1} + \sqrt{v_2})$.

[7]The classical test is presented, for instance, in Siegel and Castellan [473]. Hampel et al. [226] showed that the statistic has nearly optimal power properties over a large class of distributions.

Table 6.6: LEUKEMIA: *Genes most differentiating among the three genetic subtypes.*

Type	GenBank accession	M-W count	location	name	function
B-ALL	U05259	26	19q13.2-3	HMB1	immunoglobulin
	M89957	48	17q23	B29	B-cell receptor
	L33930	52		CD24	signal transducer
	M84371	53		CD19	B-cell specific surface protein
	Z49194	55	11q23	OBF1	B-cell coactivator of octamer-binding transcription factors
	X82240	67	14q32	TCL1	oncogene
T-ALL	X04145	0		T3G	T-cell receptor
	X03934	2		T3D	T-cell antigen receptor
	M23323	2		CD3E	membrane protein
AML	M23197	13	19q13.3	CD33	differentiation antigen of myeloid progenitor cells
	X95735	25	7q34-35	zyxin	zyxin-related protein
	M27891	26	20p11.22-21	CST3	cystein proteinase inhibitor

rule; see Appendix F.3. In the present context, the aim is to detect differentially expressed genes and so, the hypothesis is "expression nondifferential." Efron and Tibshirani [142] apply the rule to a data set of expression levels related to three types of breast cancer provided by Hedenfalk et al. [237]. The data set consists of measurements for 3,226 genes and 22 patients. At $\alpha = 0.9$ and $p_0 = 1$, they declare 68 genes under- and 66 genes overexpressed in the second type. McLachlan et al. [361] cluster the values of the classical pooled t-statistic with a two-component mixture model, again taking care of false positives as above. Kadota and Shimizu [283] offer a recent evaluation and comparison of several gene-ranking methods.

Task (iii), too, has been tackled in the past by probabilistic methods. Tavazoie et al. [497] apply the k-means algorithm without outlier protection to all expression levels obtained from gene chips with 6,220 yeast genes. The authors aim at identifying transcriptional regulatory sub-networks in yeast without using prior structural knowledge and making assumptions on dynamics. They uncovered new sets of co-regulated genes and their putative cis-regulatory elements, that is, DNA signals that herald the arrival of a protein coding region.[8] An early paper that employs mixture models for clustering samples on the basis of all genes is Yeung et al. [538]. Their method works when the gene set is not too large. Handl et al. [227] determine the hundred most significant genes of the LEUKEMIA data set with a filter and use them in the k-means algorithm to partition the patients in the three major types of the disease.

Task number (iv) is most challenging and belongs to what could be called the Champions League of statistical problems. We are given a data set with thousands of variables and wish to simultaneously identify the conditions, how many of them are represented in the data set, and related marker genes. Besides the data set, one just possesses some general background information on the problem such as that the number of conditions is not large and that a few specific under- or overexpressed genes contain the relevant information. The known solutions to the (doubly) unsupervised problem (iv) can be divided into two approaches: one *successive*, the other *integrated*. The former reduces first the gene space in an unsupervised

---

[8]A well-known cis-regulatory element is the Goldberg-Hogness TATA box found, for instance, in the promoter regions of many human genes.

way, passing the result subsequently to the clustering task (iii); the latter attempts to do the reduction and clustering at once.

An early approach to *unsupervised* gene reduction is *gene clustering*, often by means of hierarchical methods. The idea is to group together *genes* with similar expression patterns and to use typical representatives of the groups as markers. A standard similarity measure of two Euclidean vectors is their correlation coefficient. Eisen et al. [143] apply it to the expression patterns of gene pairs to obtain a measure of their *coexpression*. Ghosh and Chinnaiyan [198] use a mixture model and the EM algorithm as a refined method for gene clustering. They initialize the algorithm with probabilistic, agglomerative hierarchical clustering, as described in Section 3.3. These methods are not exactly tailored to identifying differentially expressed gene clusters w.r.t. some conditions; they rather determine many groups of genes along regular genetic pathways, too. This makes it difficult to identify with this method marker genes useful for tissue classification. Ghosh and Chinnaiyan [198] use principal components, that is, linear combinations of the genes, to reduce gene space dimension.

A few methods for integrated gene reduction and sample clustering are found in the literature. McLachlan et al. [362, 363] and Wang et al. [521] propose using factor models in order to reduce profile lengths. **Factor analysis** (Spearman [483]; see also Mardia et al. [345]) is a general method for dimension reduction without forgoing much information. The random profiles $X_i \sim N_{m,V}$, $1 \le i \le n$, of the same condition are assumed to arise from a linear combination

$$X_i = m + BY_i + E_i,$$

with $m = \mathrm{E}X_1$, a matrix $B \in \mathbb{R}^{D \times d}$ of *factor loadings*, $d < D$, multivariate standard normal random vectors Y_i, and diagonally normal random *errors* $E_i \sim N_{0,\mathrm{Diag}(\lambda_1,\ldots,\lambda_D)}$ independent of Y_i. The n i.i.d. *reduced profiles* $Y_i \sim N_{0,I_d}$ contain the concentrated information on the original profiles up to the errors, representing measurements w.r.t. a smaller number of virtual genes (factors) $1..d$. Therefore, the covariance matrix of the profiles X_i turns out to be $V = BB^\top + \mathrm{Diag}(\lambda_1,\ldots,\lambda_D)$. Since the right-hand side contains only $d(d+1)/2$ free parameters, this equation imposes $D(D+1)/2 - d(d+1)/2 - D = D(D-1)/2 - d(d+1)/2$ constraints on the covariance matrix V. Moreover, the matrix B is not unique since B can be replaced with BU for any orthogonal $d \times d$ matrix U. McLachlan et al. [362] explain each condition by a d-factor model of its own, fitting them by mixture analysis. In their applications to the COLON and LEUKEMIA data sets they use less than ten factors. Like most models described in this book, the factor model has the advantage of being affine equivariant.

Another integrated approach is **biclustering** of genes and samples; see Carmona-Saez et al. [79] and the review article by Prelić et al. [422]. It aims at creating rectangular regions of gene-sample pairs in the data matrix in such a way that the genes are nondifferentially expressed in the samples of each cluster. Factor analysis and biclustering do not explicitly hinge upon gene selection; they rather combine related or redundant genes which are likely to belong to the same molecular pathway; see McLachlan and Peel [366], p. 171.

Another natural way of gene reduction is variable selection, Chapter 5. It will be applied in the next section. For additional literature; see also Gan et al. [186].

6.2.2   Combining gene selection and profile clustering

We are now going to apply the probabilistic methods of this text to determine a *molecular signature*, a handful of informative genes that discriminate between the conditions. In the supervised framework, the Wilcoxon-Mann-Whitney test offered an effective solution to task (ii) above. The aim is now to find most differentially expressed genes with an algorithm for the unsupervised task (iv). It will be demonstrated how the Clustering-and-Selection Procedure 5.31, that is, the TDC with variable selection, exploits differential expression to detect the most relevant marker genes and to simultaneously classify the patients in a number of classes to be determined.

Theoretically, the expression levels are intensities and can therefore be assumed to follow a log-normal law. This is visually supported by the fact that most genes with a range increasing from low to high values (large coefficient of variation) are markedly skewed to the right; see Fig. 6.12. This does not exclude that variables which stay away from zero look fairly normal themselves. Therefore, all entries are log-transformed in a first preprocessing step.

When the number of genes is excessive, a second preprocessing step is needed. Since the wrapper cannot handle many thousands of genes, the genes are usually preselected with one of the filters of Section 5.2 to less than 2,000, say. The goal is a crude filtering that reduces the number of genes to a level that can be handled by the ensuing master procedure. It is sufficient to reduce the data set to all genes which do not exhibit pronounced normality after the logarithmic transformation. A gene which is low expressed in healthy tissues and overexpressed in cancerous tissues, for instance, is expected to have an overall distribution skewed to the right when the data set contains healthy and diseased cases or different diseases. We retain the D genes for

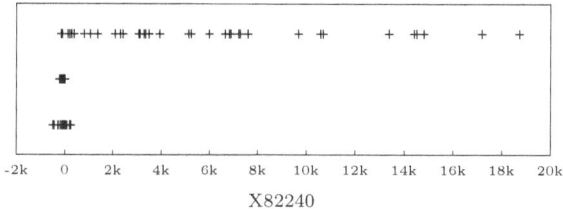

Figure 6.12 LEUKEMIA: *Univariate plots of the three cancer types B-cell ALL (top), T-cell ALL (center), and AML for the differentially expressed gene X82240. It is highly expressed and skewed to the right in B-cell ALL.*

which the chosen test statistic exceeds some threshold. A number of skewed but irrelevant genes also get caught in the filter by accident but these false positives are of no concern here – they will be taken care of by the wrapper. It is common practice to filter the data on single variables, although this can be disputed.

We next run the clustering and selection algorithm for some reasonable numbers of clusters g and genes $d \leq D$. It is not a priori known for which parameters D and d reasonable solutions are obtained even for a given g. Therefore, we preselect subsets of various sizes D using filters of different strengths. If a filter is too restrictive, that is, if D is too small, then it is to be feared that it misses the most discriminating genes. If it is too permeable, then it puts too heavy a burden on the wrapper. Moreover, we do not know the minimum number of genes necessary to separate the classes and so we apply the wrapper with various numbers $d \leq D$. We collect all these pairs (D, d) in the set $\mathcal{D}_g \subseteq \mathbb{N} \times \mathbb{N}$. The wrapper may be run under the assumptions of conditional independence of genes. This probabilistic assumption does not exclude joint over- or underexpression of genes, which arises from transcriptional regulatory networks.[9] The known data sets contain only a few pronounced outliers; they are just noisy and reproducibility is rather low; see Lee et al. [312]. It is therefore appropriate to provide for a small number of discards. In order to reduce the number of computational parameters, it will be kept fixed.

Finally, the validation methods of Chapter 4 will be applied in order to select credible solutions. In addition, a valid solution should appear not only once or a few times in $\mathcal{D}_g$ but should replicate for many pairs $(D, d) \in \mathcal{D}_g$. In fact, for reasons of stability, a solution can only be deemed credible if it has enough authority to occupy a good connected area in $\mathcal{D}_g$ with similar solutions. The notion of an α-consensus partition is here useful.

Consensus partitions (Strehl and Ghosh [489]) were introduced to improve multiple similar cluster assignments $L^{(t)}$ to a single one without further accessing the data set. To achieve this, the authors determine a centroid maximizing the sum of the normalized

---

[9]Some proteins, the products of genes, act as transcription factors for other genes.

mutual information indices $\mathrm{NMI}(L, L^{(t)})$ w.r.t. an arbitrary partition L. For this purpose, they propose three efficient and effective optimization techniques. Our situation is somewhat different since we are facing a stability problem. We cannot expect the estimated subsets of variables to be reasonable for all instances $(D, d) \in \mathcal{D}_g$. In other words, we expect some of the clusterings to be outlying. This calls for a robust method. It is here not a combination of all solutions but representatives of major regions of pairs in $\mathcal{D}_g$ with similar solutions that we seek.

Let $\mathcal{P}$ be any collection of partitions of the same data set. Let $\mathcal{C} \in \mathcal{P}$ and let $0 < \alpha \leq 1$ be some threshold. The subset of all partitions with adjusted Rand index (w.r.t. $\mathcal{C}$) at least α,

$$\{\mathcal{C}' \in \mathcal{P} \,|\, \mathrm{ARI}(\mathcal{C}', \mathcal{C}) \geq \alpha\}. \tag{6.1}$$

will be called the α-**consensus set** of $\mathcal{C}$. It always contains the reference partition $\mathcal{C}$. Large α-consensus sets for large α's have the highest authority. Let us call their reference partitions $\mathcal{C}$ α-**consensus partitions**. The problem of finding α-consensus sets and partitions leads to complete link hierarchical or k-centroids clustering of $\mathcal{P}$ with the chosen agreement measure (here ARI) as similarity. Of particular interest is $\alpha = 1$. A 1-consensus set consists of equal partitions and each of its members is a 1-consensus partition. A 1-consensus set is a maximal clique in the undirected graph over $\mathcal{P}$ with all edges that connect equal partitions (ARI $= 1$).

In the present context, the concept is applied with $\mathcal{P} = \mathcal{D}_g$ to determine reasonable subsets of genes and solutions. If the α-consensus partition for $\alpha \geq 0.7$, say, does not attract a good proportion of similar solutions, the solutions must be considered unstable and the number g should be called into question as a possible number of components.

6.2.3   Application to the LEUKEMIA data

The medical diagnosis of the three subtypes of leukemia is not perfect and errors do occur. It usually requires interpretation of the tumor's morphology, histochemistry, immunophenotyping, and cytogenetic analysis. Golub et al. [206] were the first authors to undertake a microarray study of the disease with bone marrow tissue and peripheral blood of 72 leukemia patients, 38 of them suffering from B-cell ALL, 9 from T-cell ALL, and 25 from AML.[10] Affymetrix Hgu6800 chips designed for 7 129 human genes were used. In the last decade, their LEUKEMIA data set has served as a standard benchmark for the assessment of various gene selection methods.

It follows a description of the cluster analysis performed with the TDC with variable selection, Procedure 5.31. Unfortunately, the authors of the data had rescaled them so that the overall intensities of all profiles became equivalent. While such a transformation is reasonable on normal data, it may lead to substantial distortions when they are log-normal. Some entries are even negative. Apart from this, data quality is acceptable. But log-transformation cannot be performed straightforwardly and the function $\log(|x| + 1))$ is applied to the expression levels x, instead. This transformation shifts the patterns toward normality.

Containing 7 129 genes, the data set is too large for gene selection with a wrapper. Therefore, the median filter (5.5) with 11 different thresholds was applied in order to preselect genes. It created 11 nested subsets of between $D = 22$ and $D = 2,518$ genes. Any other test statistic for asymmetry could be employed here. These data sets are now ready for processing by the MAP version of the TDC with variable selection. For computational reasons, it was run under heteroscedastic, *diagonal* assumptions on the covariance matrices, that is, under conditional independence of genes. Diagonal specifications do not tend to create spurious solutions, but it is still necessary to monitor scale balance. Large outliers are not expected after the log-transformation but two discards were nevertheless provided

---

[10]The data set is publicly available as golubMerge in [205].

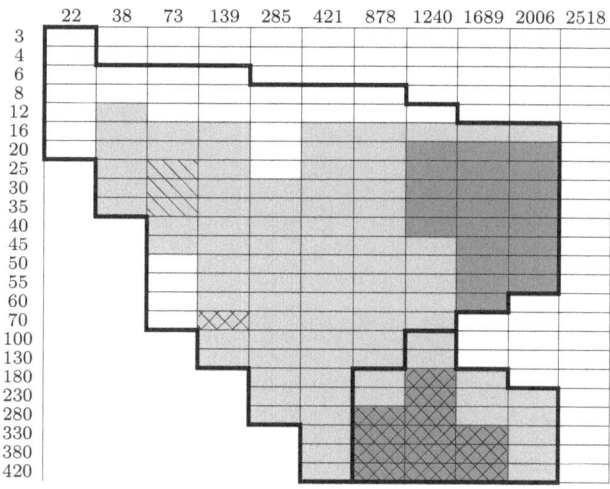

Figure 6.13 LEUKEMIA: *Consensus sets of the three-cluster Pareto solutions generated by the TDC with variable (gene) selection. Top, the number D of genes after filtering; left, the number d of genes selected by the wrapper.*

for. The 24 values of the number d of selected genes ranged between 3 and the minimum of D and 420 giving rise to a collection $\mathcal{D}$ of 216 pairs (D, d) and as many runs of the algorithm. This effort allows us to determine α-consensus partitions, also providing some insight into the number of clusters. Selecting variables and clustering cases on a PC in data sets of up to 2,512 genes is not a trivial task and the algorithm has to be replicated many millions of times for some parameter settings (D, d). It is almost indispensable to distribute computation on a good number of processors.

The assumption of three clusters leads to the two major consensus solutions presented in Fig. 6.13. The dark, gray, and framed regions describe the two related α-sets for $\alpha = 1$, 0.86 and 0.76, respectively. The consensus partitions are hatched for $\alpha = 0.76$ and cross-hatched for 0.86. The two 1-sets consist of equal solutions each. Some separation indices and p-values of the solution with $D=22$ and $d=4$ shown in Table 6.7 indicate that clusters are well separated. Moreover, the clusters pass Srivastava and Hui's [484] multivariate extension of the Shapiro-Wilk normality test with p-values of 0.4, 0.08, and 0.6. The GenBank codes of the $d=3$ genes selected from the filtered data set with $D=22$ are D88270, M27891, X03934. Two of them are contained in Table 6.6; gene D88270 codes for the λ light chain of immunoglobin and is located at q11.2 on chromosome 22.

Table 6.7 LEUKEMIA: *Separation indices and p-values for the main three- and the top four-cluster solution in Fig. 6.15.*

g	pair	Fisher	linear separ.	Sym. diverg.	$-\log$ Helling.	p-value MV S.-W.	p-value lin. separ.
3	1,2	0.999	0.999	161.7	5.84	0.000	0.000
	1,3	1.000	1.000	134.3	10.28	0.000	0.000
	2,3	0.985	0.985	44.0	2.91	0.000	0.000
4	1,2	1.000	1.000	457.6	37.98	0.000	0.000
	1,3	1.000	1.000	128.3	13.17	0.000	0.000
	1,4	0.966	0.966	28.9	2.39	0.001	0.003
	2,3	1.000	1.000	212.5	16.58	0.001	0.000
	2,4	1.000	1.000	399.2	37.04	0.010	0.000
	3,4	0.999	0.999	65.0	5.20	0.002	0.000

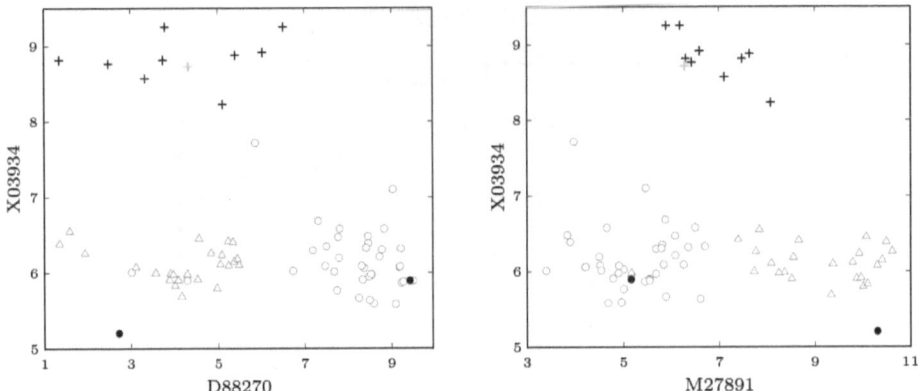

Figure 6.14 LEUKEMIA: *Scatter plots of the logged data w.r.t. two differentially expressed genes each.*

The consensus solution that occupies the larger region in Fig. 6.13 matches almost exactly the known subdivision of the disease in B-cell and T-cell ALL and AML, with one error besides the two rejections. Figure 6.14 shows two scatter plots of this solution, each w.r.t. a pair of differentially expressed genes. The two solid dots represent the two estimated outliers, Patients 21 and 52; Patient 17 (the gray +) has been misclassified from B-cell ALL to T-cell ALL. Both plots show fairly good separation into the three leukemia types B-cell ALL (o), T-cell ALL (+), and AML (△). The other consensus solution keeps AML but rearranges the two ALL groups. This instability indicates that the data set may contain more than three groups.

The three leukemia types, B-ALL, T-ALL, and AML, indeed possess finer classifications in subtypes based on their immunophenotype, among them pre-B ALL, mature B-cell ALL, pre-T ALL, mature T-cell ALL, AML with certain translocations between chromosomes and Acute Promyelocytic Leukemia, APL; see the web site of the American Cancer Society [12, 13]. Some of them, for instance APL, are very rare and might not be represented in the data set. It may yet allow a solution with more components.

It turns out that four components create three major consensus solutions. One appears

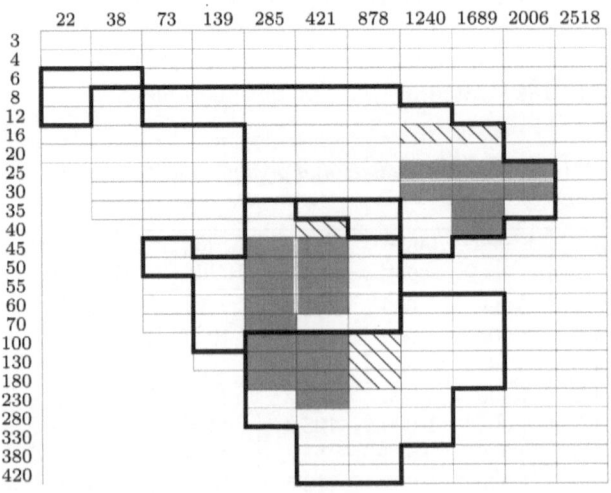

Figure 6.15 LEUKEMIA: *Consensus sets of the four-cluster Pareto solutions generated by the TDC with variable (gene) selection. Top, the number D of genes after filtering; left, the number d of genes selected by the wrapper.*

for small values of d and the others for larger values. Consensus sets for $\alpha = 0.73$ and $\alpha = 1$ are presented in Fig. 6.15. The top and bottom consensus solutions keep T-cell ALL and AML unchanged but split B-cell ALL in two different ways; at the bottom in the same way as the second three-cluster solution does. The middle solution keeps the two ALL groups

Figure 6.16 LEUKEMIA: *Scatter plot of the subdivision of B-cell ALL w.r.t. an optimal projection.*

but splits AML. Some separation indices and p-values of the top consensus solution for $D = 38$ and $d = 6$ are again shown in Table 6.7. Its clusters are, in this order, the larger subgroup of B-cell ALL, T-cell ALL, AML, and the smaller subgroup of B-cell ALL in the top consensus solution. All four groups seem to be sufficiently well separated. They pass the multivariate adaptation of Shapiro and Wilk's normality test with the p-values 0.4, 0.3, 0.25, and 0.9, respectively. The partition of B-cell ALL is shown in Fig. 6.16. Five of the six out of 22 genes in the first column in Fig. 6.15, D88270, M27891, U46499, X03934, X82240, appear already in the decomposition in the three main types. The last one, L33930 (CD24B), is known as

a signal transducer correlating with tumor differentiation. The common refinement of all partitions found consists of six clusters. It might be a more realistic partition of the disease in the genetic subtypes represented in the data set.

6.3 Notes

A fertile source of data sets is the UCI Machine Learning Repository (Aha et al. [2]) created in 1987. It contains to date 240 data sets for classification and clustering. In 2002, a freely accessible repository for storing, retrieving, and analysis of unified gene expression data was started, the *Gene Expression Omnibus* (GEO) [196]; see Edgar et al. [139], Barrett et al. [28], and Barrett and Edgar [27]. It contains to date more than 3,200 data sets.

Appendix A

Geometry and linear algebra

Geometry and linear algebra play a traditional rôle in statistics. This appendix reviews some facts about affine spaces, traces and determinants, the Löwner order, and Steiner formulas. They are needed for analyzing scatter matrices, for ML estimation, identifiability, robustness, and scale constraints. Most of the facts are classical. A few formulas of Steiner's type are more recent.

A.1 Affine spaces

In order to treat questions of the general position of data, positive definiteness of scatter matrices, and identifiability of mixtures, some basic understanding of affine geometry is necessary. For the present purposes, it is sufficient to deal with affine subspaces of linear spaces. So, let W be a subspace of a real vector space L and let $h \in L$. Then the subset

$$\mathcal{A} = h + W \subseteq L$$

is called a nonempty **affine space** over W. The space W is called the **vector space associated** with $\mathcal{A}$. Of course, $h \in \mathcal{A}$. Any vector in $\mathcal{A}$, and no other, can be substituted for h. On the other hand, W is unique. Indeed, if $h_1 + W_1 = h_2 + W_2$, then $h_2 - h_1 \in W_1$ and $W_2 = (h_1 - h_2) + W_1 = W_1$. The **affine dimension** of $\mathcal{A}$ is the linear dimension of W. A subset of L is called an affine space if it is either empty, in which case its affine dimension is defined as -1, or it is an affine space over some W. Any linear subspace of L is an affine space (over itself); just take $f = 0$. Of course, L itself is an affine space. Note that $f - g \in W$ for any two points $f, g \in \mathcal{A}$. More generally, any finite linear combination $\sum_f \alpha_f f$ of points of $\mathcal{A}$ such that $\sum_f \alpha_f = 0$ is a vector in W. Moreover, any finite linear combination $\sum_f \alpha_f f$ of points $f \in \mathcal{A}$ such that $\sum_f \alpha_f = 1$ is again a point in $\mathcal{A}$. Such combinations are called **affine**.

Any intersection of affine spaces is again an affine space.[1] The intersection of all affine spaces that contain a given subset $\mathcal{D} \subseteq L$ is the smallest such space. It is called the affine space spanned by $\mathcal{D}$ or the **affine span** of $\mathcal{D}$, $\mathbb{A}\langle \mathcal{D} \rangle$. Its points are represented from the "interior" by all affine combinations over finite subsets of the generator $\mathcal{D}$. For instance, if $x_1, \ldots, x_n \in \mathbb{R}^d$, then $x_i, \overline{x} \in \mathbb{A}\langle x_1, \ldots, x_n \rangle$. The **affine dimension** of $\mathcal{D}$ is the affine dimension of $\mathbb{A}\langle \mathcal{D} \rangle$. The discussion above implies the following lemma, which analyzes the relationships between the linear span, the affine span, and the associated linear space. Let $\mathbb{L}\langle \mathcal{D} \rangle$ denote the linear span of $\mathcal{D}$.

A.1 Lemma

(a) For any $f \in \mathbb{A}\langle \mathcal{D} \rangle$, the linear space associated with $\mathbb{A}\langle \mathcal{D} \rangle$ is $\mathbb{L}\langle \mathcal{D} - f \rangle$.

(b) We have $\mathbb{A}\langle \mathcal{D} \rangle = \mathbb{L}\langle \mathcal{D} \rangle$ if and only if $0 \in \mathbb{A}\langle \mathcal{D} \rangle$.

---

[1] Unlike linear spaces, the intersection of (nonempty) affine spaces may be empty. This is the reason we have to admit the empty set as an affine space.

(c) *If* $0 \notin \mathbb{A}\langle \mathcal{D} \rangle$, *then the vector space associated with* $\mathbb{A}\langle \mathcal{D} \rangle$ *is a proper subspace of the linear span* $\mathbb{L}\langle \mathcal{D} \rangle$.

Recall that a subset $\mathcal{D}$ of a real vector space L is called **linearly independent** if any linear combination $\sum_{f \in F} \alpha_f f$ over any *finite* subset $F \subseteq \mathcal{D}$ can vanish only if all coefficients α_f, $f \in F$, do.

A.2 Definition A subset $\mathcal{D} \subseteq \mathcal{A}$ is called **affine independent** if the linear combination $\sum_{f \in F} \alpha_f f$ over any finite $F \subseteq \mathcal{D}$ with coefficients $\alpha_f \in \mathbb{R}$ such that $\sum_{f \in F} \alpha_f = 0$ can vanish only if all coefficients α_f, $f \in F$, do.

This means that $\mathcal{D}$ spans an $|\mathcal{D}|$–1-dimensional "simplex." In contrast to linear independence, the coefficients of the finite linear combination are here assumed to sum up to zero. It follows from the definition that any subset of an affine independent set is again affine independent. The concept will be important in ML estimation of normal and elliptical models in Sections F.2.2 and F.2.3. In this context, we say that a Euclidean data set $(x_1, \dots, x_n)$ in $\mathbb{R}^d$, $n \geq d+1$, is in **general position** if any $d+1$ of its points are affine independent. If $n \leq d+1$, general position is tantamount to affine independence.

The notions of "affine independence" and "linearly independence" are closely related, as the following lemma shows.

A.3 Lemma

(a) *A subset* $\mathcal{D} \subseteq \mathcal{A}$ *is affine independent if and only if the subset* $\{f - f_0 \mid f \in \mathcal{D}, f \neq f_0\}$ $\subseteq W$ *is linearly independent for some (or any)* $f_0 \in \mathcal{D}$.

(b) *Any linearly independent set of points in Euclidean space is affine independent.*

(c) *An affine independent set of points, the affine span of which does not contain the origin, is linearly independent.*

(d) *If* $0 \notin \mathcal{A}$, *then affine independence of a subset of* $\mathcal{A}$ *and its linear independence in* L *are equivalent.*

Proof. Claim (a) follows from the identity

$$\sum_{f \in F} \alpha_f f = \sum_{f \in F} \alpha_f (f - f_0) + \Big(\sum_{f \in F} \alpha_f \Big) f_0$$

for finite $F \subseteq \mathcal{D} \backslash \{f_0\}$. Part (b) follows immediately from the definition. For part (c), let $\mathcal{D} \subseteq L$ be affine independent. Let $F \subseteq \mathcal{D}$ be finite and $\sum_{f \in F} \alpha_f f = 0$. By assumption, $\sum_f \alpha_f = 0$ since, otherwise, there would be an affine combination of zero. By affine independence, all coefficients α_f vanish. In view of (d), let $\mathcal{D} \subseteq \mathcal{A}$ be first affine independent. Its affine span is contained in $\mathcal{A}$ and hence does not contain the origin. By (c), it is linearly independent. The converse is just (b). This is claim (d).   □

The converse of (b) is not true in general, an affine independent subset is not necessarily linearly independent, for instance, when it contains the origin as an element.

A.2   Trace and determinant

Trace and determinant are functionals of square matrices. The **trace** is the sum of the diagonal elements and hence a linear function on the space of square matrices. It is invariant w.r.t. matrix transposition and the product of two matrices is commutative under the trace, one of its most useful features. For instance, if A and B are matrices of the same size, then we have $\operatorname{tr} AB^\top = \operatorname{tr} B^\top A = \operatorname{tr} A^\top B = \operatorname{tr} BA^\top$. In particular, for two vectors a and b of equal length, we have $\operatorname{tr} ab^\top = \operatorname{tr} b^\top a \ (= \sum a_k b_k)$. The trace allows us to separate the matrix A and the variable x in a quadratic form $x^\top A x$ as a product of A with a matrix that depends only on x, $x^\top A x = \operatorname{tr}(A[xx^\top])$. This is its importance in statistics. Quadratic forms appear as arguments of radial functions of elliptical distributions; see Appendix E.3.2. Another useful asset of the trace is its rôle in the geometric-arithmetic inequality below.

The **determinant** is characterized by three properties: (1) It is linear in each row (and in each column); (2) it is alternating, that is, it vanishes when two rows (or columns) are equal; (3) the determinant of the unit matrix is 1. It follows that (4) a determinant changes its sign upon swapping of two rows (or columns). Proper application of the rules (1)–(4), called Gauß elimination, transforms any matrix into triangular and even diagonal form. The determinants of triangular matrices are the products of their diagonal entries. This procedure allows efficient computation of the determinant of any square matrix. The determinant is a multiplicative functional, that is, the determinant of a product of two matrices is the product of their determinants. In particular, $\det A^{-1} = \det^{-1} A$. Its importance in statistics has two main reasons. First, $\det A$ equals the volume of the parallelotope specified by the columns of A. For this reason it appears in the transformation formula of Lebesgue measure, the standard reference measure for distributions on Euclidean space. The determinant of the scale matrix is therefore a part of the normalizing factors of elliptical distributions; see again Appendix E.3.2. Second, a matrix is singular if and only if its determinant vanishes. This has consequences for the eigenvalue theory and spectral decomposition of square matrices; see Appendix A.3. Both the trace and the determinant are invariant w.r.t. conjugation with a regular matrix: $\operatorname{tr} BAB^{-1} = \operatorname{tr} A$, $\det BAB^{-1} = \det A$. As a consequence, they are invariant w.r.t. basis changes in $\mathbb{R}^d$.

The elementary **geometric-arithmetic inequality** for n positive numbers compares their geometric and arithmetic means. It follows from concavity of the logarithm. Another short and elegant proof is found in Hirschhorn [247].

A.4 Lemma *Let $x_1, \ldots, x_n \geq 0$. Then*

$$\sqrt[n]{x_1 \cdot \ldots \cdot x_n} \leq \tfrac{1}{n}(x_1 + \cdots + x_n).$$

There is equality if and only if all numbers are equal.

A matrix A is called **symmetric** if $A^\top = A$. It is then square. It is called **positive definite** (**positive semi-definite**) if it is symmetric and if $x^\top A x > 0$ (≥ 0) for all $x \neq 0$. There is the following implication for positive semi-definite matrices. It is sufficient to apply the foregoing lemma to the eigenvalues.

A.5 Lemma *For any positive semi-definite $d \times d$ matrix A, we have*

$$\sqrt[d]{\det A} \leq \tfrac{1}{d}\operatorname{tr} A$$

and there is equality if and only if A is a multiple of the identity matrix.

The following lower bound on the determinant of a sum appears in Gallegos and Ritter [180], Lemma A.1(b).

A.6 Lemma *Let $d \geq 2$ and let $A \in \mathbb{R}^{d \times d}$ be positive definite.*

(a) For any positive semi-definite matrix $C \in \mathbb{R}^{d \times d}$, we have

$$\det(A + C) \geq (1 + \operatorname{tr}(A^{-1}C)) \det A + \det C.$$

(b) If $y_1, \ldots, y_k \in \mathbb{R}^d$, then

$$\det\left(A + \sum_h y_h y_h^\top\right) \geq \left(1 + \sum_h y_h^\top A^{-1} y_h\right) \det A.$$

(c) $\det(A + xx^\top) = (1 + x^\top A^{-1}x) \det A, \quad x \in \mathbb{R}^d.$

PROOF. (a) From $A + C = A^{1/2}(I_d + A^{-1/2}CA^{-1/2})A^{1/2}$, we infer

$$\det(A + C) = \det(I_d + A^{-1/2}CA^{-1/2}) \det A. \tag{A.1}$$

If $\lambda_1, \lambda_2, \ldots, \lambda_d$ are the eigenvalues of $A^{-1/2}CA^{-1/2}$, then the eigenvalues of $I_d + A^{-1/2}CA^{-1/2}$ are $1 + \lambda_1, \ldots, 1 + \lambda_d$ and the claim follows from (A.1) and

$$\det(I_d + A^{-1/2}CA^{-1/2}) = \prod_{j=1}^{d}(1 + \lambda_j) \geq 1 + \sum_{j=1}^{d}\lambda_j + \prod_{j=1}^{d}\lambda_j$$
$$= 1 + \operatorname{tr}(A^{-1/2}CA^{-1/2}) + \det A^{-1/2}CA^{-1/2}.$$

Part (b) is an immediate consequence of (a).

For $x \neq 0$ the rank-1 matrix $xx^\top$ has the simple eigenvalue $\|x\|^2$. Hence, we have $\det(I_d + xx^\top) = 1 + \|x\|^2$ and part (c) follows from

$$A + xx^\top = \sqrt{A}\big(I_d + (A^{-1/2}x)(A^{-1/2}x)^\top\big)\sqrt{A}. \qquad \Box$$

A.3 Symmetric matrices and Löwner order

Positive semi-definite, positive definite, and more general symmetric (real-valued) matrices in the form of covariance and scatter matrices play a prominent role in statistics. The spectral decomposition in terms of eigenvalues and eigenvectors is of special importance. This also applies to the present text, where the eigenvectors determine principal directions of random vectors and the spectral decomposition is needed in the proofs of consistency theorems. Moreover, the eigenvalues are needed to define one of the central concepts of this text, the HDBT ratio (1.34) of the component covariance matrices of a mixture. The reader interested in more details on symmetric matrices is referred to Horn and Johnson's [252] monograph.

A.7 Explanation (a) A number $\lambda \in \mathbb{C}$ is called an **eigenvalue** of a square matrix $A \in \mathbb{R}^{d \times d}$ if there is a nonzero vector $f \in \mathbb{C}^d$ such that $Af = \lambda f$. The vector is called an **eigenvector** to the eigenvalue λ. The collection of all eigenvectors to λ forms a subspace $V_\lambda \subseteq \mathbb{C}^d$, its **eigenspace**. Hence, the matrix A acts on V_λ and on each eigenvector to λ by multiplication with λ. This means that $A - \lambda I_d$ is singular and so λ is a root of the **characteristic polynomial** $\chi_A(\lambda) = \det(A - \lambda I_d)$ of A. The collection of eigenvalues is called the **spectrum** of A. The maximum modulus of all eigenvalues is called the **spectral radius** of A.

The spectral (Jordan) decomposition is defined for square matrices. However, we will mainly need real, symmetric matrices where the decomposition has a very simple form. The following description applies, therefore, to *symmetric* matrices. Here, all eigenvalues are real and eigenvectors may be taken in $\mathbb{R}^d$. A fundamental theorem states the following:

- The dimension of each eigenspace is the multiplicity of its eigenvalue in the characteristic polynomial;

- the eigenspaces to different eigenvalues are orthogonal;

- there is an orthonormal basis of $\mathbb{R}^d$ consisting of eigenvectors of A (called an **eigenbasis** of A).

Since the degree of the characteristic polynomial is d, the last assertion follows from the first two. To see this, it is sufficient to collect orthonormal bases of all eigenspaces in one set. Of course, the basis depends on A. It is unique up to sign changes if and only if A has d different eigenvalues.

Let $u_1, \ldots, u_d \in \mathbb{R}^d$ be an eigenbasis of A with associated eigenvalues $\lambda_1, \ldots, \lambda_d$. The $d \times d$ matrix $U = (u_1, \ldots, u_d)$ is **orthogonal**, that is, $U^{-1} = U^\top$. It follows $\det U = \pm 1$. It can be chosen to be a **rotation**, that is, $\det U = +1$, by negating one eigenvector if necessary. An example of a rotation on the Euclidean plane is the matrix $\begin{pmatrix} \cos\alpha & -\sin\alpha \\ \sin\alpha & \cos\alpha \end{pmatrix}$, which rotates every vector by the angle $\alpha \in \mathbb{R}$ counterclockwise. From $Au_k = \lambda_k u_k$, we derive the equality $AU = U\mathrm{Diag}(\lambda_1, \ldots, \lambda_d)$ and thus the representation of A by its **normal form**,

$$A = U \begin{pmatrix} \lambda_1 & & 0 \\ & \ddots & \\ 0 & & \lambda_d \end{pmatrix} U^\top. \tag{A.2}$$

That is, w.r.t. any eigenbasis, A is represented by the diagonal matrix specified by the eigenvalues in their corresponding order. As a consequence, the determinant of A is the product of its eigenvalues and the trace is their sum. The spectral representation enables us to extend functions F of the eigenvalues of A to functions of A: $F(A) = U\mathrm{Diag}(F(\lambda_1), \ldots, F(\lambda_d))U^\top$ is again a symmetric matrix. For instance, we recover $A^n = U\mathrm{Diag}(\lambda_1^n, \ldots, \lambda_d^n)U^\top$ for all natural numbers n. If A is regular, then $A^{-1} = U\mathrm{Diag}(\lambda_1^{-1}, \ldots, \lambda_d^{-1})U^\top$. Moreover, the expression $\sin A$ now makes sense for any symmetric matrix A. The eigenvalues of $F(A)$ are $F(\lambda_1), \ldots, F(\lambda_d)$ and the eigenvectors and eigenspaces are unchanged. This is called the **spectral mapping theorem**.

There is an equivalent way of stating the normal form. Given two vectors $x, y \in \mathbb{R}^d$, the rank of the $d \times d$ matrix $x \otimes y = xy^\top$ is one unless one of the two vectors vanishes. The operator $\otimes$ is not commutative. If u is a unit vector, then $u \otimes u$ acts as the orthogonal projection onto the straight line through the origin in direction of u. It is now immediate that A has the **spectral representation**

$$A = \sum_{k=1}^{d} \lambda_k u_k \otimes u_k; \tag{A.3}$$

here λ_k and u_k are the eigenvalues and related orthogonal eigenvectors as above. Normal form and spectral representation are used to reduce a d-dimensional problem to d simpler, one-dimensional ones in the directions of the eigenvectors.

(b) A symmetric $d \times d$ matrix is **positive definite** (**positive semi-definite**) if and only if all its eigenvalues are strictly positive (positive). In this case, the matrix acts on eigenspaces and eigenvectors associated with an eigenvector λ by stretching with λ if $\lambda > 1$ and compressing if $\lambda < 1$. It follows that a symmetric matrix is positive definite if and only it is positive semi-definite and regular. The cone of positive definite $d \times d$ matrices will be denoted by $\mathrm{PD}(d)$. A positive semi-definite matrix A has a unique **square root**, that is, a positive semi-definite matrix whose square is A: $\sqrt{A} = U\mathrm{Diag}(\sqrt{\lambda_1}, \ldots, \sqrt{\lambda_d})U^\top$. If A is positive definite, then so is $\sqrt{A}$. Because of $x^\top aa^\top x = (a^\top x)^2$, covariance and scatter matrices are at least positive semi-definite.

(c) A matrix Π is called an **orthogonal projection** or a **projector** if it is symmetric and idempotent, that is, $\Pi^2 = \Pi$. It can have no other eigenvalues than 0 and 1 and its spectral representation (A.3) has the form

$$\Pi = \sum_{k=1}^{r} u_k \otimes u_k$$

with orthonormal vectors u_k that span the target space of Π. Here, r is the rank of Π. The projection complementary to Π is $I_d - \Pi$. Its image is the orthogonal complement of that of Π.

(d) We will occasionally use a (not necessarily symmetric) matrix $A^{-1}B$, where A is positive definite and B symmetric. The eigenvalue theory of asymmetric matrices is generally more complex than that for symmetric matrices but, in the case of the product of two symmetric matrices, the former can be derived from the latter. The eigenvalues and eigenvectors are defined as above and the spectrum of $A^{-1}B$ equals that of the symmetric matrix $A^{-1/2}BA^{-1/2}$. Indeed, we have $A^{-1}Bu = \lambda u$ if and only if $A^{-1/2}BA^{-1/2}(\sqrt{A}u) = \lambda\sqrt{A}u$. Note that the eigenvectors u are orthogonal here if and only if A and B commute.

The following Löwner order is central to the present treatment of clustering problems. Its properties resemble those of the natural order on the real line.

A.8 Definition Let A and B be symmetric matrices of equal size. We say that A is less than or equal to B w.r.t. **Löwner's** or the **semi-definite order**, $A \preceq B$ and $B \succeq A$, if $B - A$ is positive semi-definite. If $B - A$ is positive definite, we will write $A \prec B$ and $B \succ A$. (Thus, $A \prec B$ does not just mean $A \preceq B$ and $A \neq B$.) If $A \prec B$, then there exists $\varepsilon > 0$ such that $A + \varepsilon I_d \prec B$.

We will use expressions like "Löwner smaller" and "Löwner bounded." Some of the properties of Löwner's order are listed next.

A.9 Properties (a) The Löwner order is a partial order on the vector space of all symmetric $d \times d$ matrices. That is, the order is reflexive, transitive, and antisymmetric. Indeed, if $A \preceq B \preceq A$, then all eigenvalues of $B - A$ and of $A - B$ are ≥ 0 and hence zero. The Löwner order is not total. For instance, neither of the matrices $\begin{pmatrix} 2 & 0 \\ 0 & 2 \end{pmatrix}$ and $\begin{pmatrix} 1 & 0 \\ 0 & 3 \end{pmatrix}$ is larger than the other.

(b) We have $A \preceq B$ if and only if $L^{\top} A L \preceq L^{\top} B L$ for one or all $L \in \mathrm{GL}(d)$.

(c) If $B \in \mathrm{PD}(d)$, then the following statements are equivalent.

(i) $A \preceq B$;

(ii) $B^{-1/2} A B^{-1/2} \preceq I_d$;

(iii) all eigenvalues of $B^{-1/2} A B^{-1/2}$ are less than or equal to 1.

If $A \in \mathrm{PD}(d)$, then they are also equivalent with

(iv) $A^{-1/2} B A^{-1/2} \succeq I_d$;

(v) all eigenvalues of $A^{-1/2} B A^{-1/2}$ are greater than or equal to 1.

PROOF. The equivalence of claims (i), (ii), and (iv) follows from (b). That of (ii) and (iii) follows from the positive eigenvalues of $I_d - B^{-1/2} A B^{-1/2}$ and that of (iv) and (v) is analogous. □

(d) If $0 \preceq A \preceq B$, then $\det A \leq \det B$. If $B \in \mathrm{PD}(d)$, then the determinants are equal only if $A = B$.

PROOF. If B is only semi-definite, then both determinants vanish. If $B \in \mathrm{PD}(d)$, then all eigenvalues of $B^{-1/2} A B^{-1/2}$ are ≤ 1 according to (c). Hence, $\det B^{-1/2} A B^{-1/2} \leq 1$ and the first claim. If $\det A = \det B$, then the product of all eigenvalues of $B^{-1/2} A B^{-1/2}$ equals $\det B^{-1/2} A B^{-1/2} = 1$ and all eigenvalues of $B^{-1/2} A B^{-1/2}$ are equal to 1 by (c). Therefore, $B^{-1/2} A B^{-1/2} = I_d$; hence $A = B$. □

(e) Let $\lambda_{\min}$ ($\lambda_{\max}$) denote the minimum (maximum) eigenvalue of a symmetric matrix. If $A \preceq B$, then $\operatorname{tr} A \leq \operatorname{tr} B$, $\lambda_{\min}(A) \leq \lambda_{\min}(B)$, and $\lambda_{\max}(A) \leq \lambda_{\max}(B)$.

PROOF. If $A \preceq B$, then $A(k, k) \leq B(k, k)$ for all k (use the unit vectors). Hence the claim on the traces. On the other hand,

$$\lambda_{\min}(A) = \min_{\|x\|=1} x^{\top} A x \leq \min_{\|x\|=1} x^{\top} B x = \lambda_{\min}(B)$$

and an analogous estimate shows $\lambda_{\max}(A) \leq \lambda_{\max}(B)$. □

(f) For $A, B \in \mathrm{PD}(d)$, we have $A \preceq B$ if and only if $B^{-1} \preceq A^{-1}$.

PROOF. It is sufficient to prove the implication from left to right. So let $A \preceq B$. According to (c), the eigenvalues of $B^{-1/2} A B^{-1/2}$ are all ≤ 1. It follows that the eigenvalues of $B^{1/2} A^{-1} B^{1/2} = (B^{-1/2} A B^{-1/2})^{-1}$ are all ≥ 1, that is, we have $B^{1/2} A^{-1} B^{1/2} \succeq I_d$. This is $A^{-1} \succeq B^{-1}$. □

(g) The monotone behavior of the function $A \mapsto A^p$ for $p \neq -1$ will not be needed in the text but is reported here for the sake of completeness. If A and B are positive semi-definite and commute, then, for all $p > 0$, the relation $A^p \preceq B^p$ is equivalent to $A \preceq B$. This is shown using the fact that the commuting matrices A and B can be jointly diagonalized. The situation is less satisfactory when A and B do not commute. Here, the property is wrong for $p = 2$, an example being the matrices $A = \begin{pmatrix} 2 & 1 \\ 1 & 1 \end{pmatrix}$ and $B = \begin{pmatrix} 3 & 1 \\ 1 & 1 \end{pmatrix}$. We have

$B - A \succeq 0$ but $B^2 - A^2 = \begin{pmatrix} 5 & 1 \\ 1 & 0 \end{pmatrix}$ is not positive semi-definite. Remarkably, $A \mapsto A^p$ increases for $p = 1/2$, a result due to Löwner [337]. Generalizations to so-called matrix monotone functions are found in Bhatia [40], Donoghue [127], and Ando [15].

Applying Property A.9(c)(v) with cA instead of A and observing that the spectra of $A^{-1/2}BA^{-1/2}$ and $A^{-1}B$ are equal (see Explanation A.7(d)), we obtain the following fact.

A.10 Lemma *Let A and B be positive definite. The largest number c such that $cA \preceq B$ is the minimum eigenvalue of $A^{-1/2}BA^{-1/2}$ and of $A^{-1}B$.*

A.4 Formulas of Steiner's type

Steiner's formula is used in mechanics to compute the moment of inertia of a physical body. It decomposes the body's moment of inertia about any axis into the moment about the parallel axis through its center (of mass) and the one about the original axis of its mass concentrated at its center. In statistics, covariance and scatter matrices replace moments of inertia and mean vectors replace centers of mass. For instance, "the total SSP matrix is the sum of the within and the between groups SSP matrices" is a formula of Steiner's type. The formula is needed, for instance, to obtain the ML estimate of the normal mean parameter. The main text offers ample opportunity for application of various formulas of this type. The first one is elementary.

A.11 Lemma (Steiner's formula) *Let $x_1, \ldots, x_n \in \mathbb{R}^d$, $b \in \mathbb{R}^d$, and $B \in \mathbb{R}^{d \times d}$. We have*

$$\sum_{i=1}^{n}(x_i - b)^\top B(x_i - b) = \sum_{i=1}^{n}(x_i - \overline{x})^\top B(x_i - \overline{x}) + n(\overline{x} - b)^\top B(\overline{x} - b).$$

PROOF. The claim follows from $\sum_{i=1}^{n}(x_i - \overline{x}) = 0$ and the identities

$$\sum_{i=1}^{n}(x_i - b)^\top B(x_i - b) = \sum_{i=1}^{n}((x_i - \overline{x}) + (\overline{x} - b))^\top B((x_i - \overline{x}) + (\overline{x} - b))$$

$$= \sum_{i=1}^{n}(x_i - \overline{x})^\top B(x_i - \overline{x}) + \sum_{i=1}^{n}(x_i - \overline{x})^\top B(\overline{x} - b)$$

$$+ \sum_{i=1}^{n}(\overline{x} - b)^\top B(x_i - \overline{x}) + \sum_{i=1}^{n}(\overline{x} - b)^\top B(\overline{x} - b). \qquad \square$$

The following lemma contains a weighted version of Steiner's formula. Its proof is analogous. We use the notation introduced in Section 2.2.2, consider a data set $(x_1, \ldots, x_n)$ in $\mathbb{R}^d$, and put $T = 1 .. n$.

A.12 Lemma (Weighted Steiner formula) *Let $w = (w_i)_{i \in T}$ be a family of real-valued weights such that $w_T = \sum_i w_i > 0$, let $B \in \mathbb{R}^{d \times d}$, let $b \in \mathbb{R}^d$, and let $\overline{x}(w) = \frac{1}{w_T}\sum_i w_i x_i$*

be the weighted mean. Then

$$\sum_i w_i (x_i - b) B (x_i - b)^\top$$

$$= \sum_i w_i (x_i - \overline{x}(w)) B (x_i - \overline{x}(w))^\top + w_T \cdot (\overline{x}(w) - b) B (\overline{x}(w) - b)^\top.$$

If B is positive semi-definite, then

$$\sum_i w_i (x_i - b) B (x_i - b)^\top \succeq \sum_i w_i (x_i - \overline{x}(w)) (x_i - \overline{x}(w))^\top.$$

Similar formulæ decompose the sample covariance matrix of a data set w.r.t. a partition $\{T_j\}_{j=1}^g$ of T. With the notation introduced in Section 2.2.2, a first formula follows from Lemma A.12 and reads

$$W_T(w) = \sum_{j=1}^g W_{T_j}(w) + \sum_{j=1}^g w_{T_j} \big(\overline{x}_{T_j}(w) - \overline{x}_T(w)\big) \big(\overline{x}_{T_j}(w) - \overline{x}_T(w)\big)^\top.$$

This will, however, not be sufficient for our needs in Chapter 2. We will also need a version with a representation of the "between" SSP matrix that does not depend on the global, weighted mean vector $\overline{x}_T(w)$ but rather on the weighted "between" matrices w.r.t. all cluster pairs. The next lemma presents such an identity of Steiner's type, weighted with a family $w = (w_i)_{i \in T}$ of real-valued numbers. It appears in Gallegos and Ritter [182], where it serves to compute breakdown points of means.

A.13 Lemma *Let $w = (w_i)_{i \in T}$ be a family of strictly positive weights, let $w_T = \sum_{i \in T} w_i$, let $\{T_1, \ldots, T_g\}$ be a partition of T, let $w_{T_j} = \sum_{i \in T_j} w_i$, and let $\overline{x}(w) = \frac{1}{w_T} \sum_{1 \le i \le n} w_i x_i$ and $\overline{x}_j(w) = \frac{1}{w_{T_j}} \sum_{i \in T_j} w_i x_i$. We have*

(a)
$$\sum_j w_{T_j} (\overline{x}_j(w) - \overline{x}(w)) (\overline{x}_j(w) - \overline{x}(w))^\top$$

$$= \frac{1}{w_T} \sum_{1 \le j < \ell \le g} w_{T_j} w_{T_\ell} (\overline{x}_j(w) - \overline{x}_\ell(w)) (\overline{x}_j(w) - \overline{x}_\ell(w))^\top ;$$

(b) $W_T(w) = \sum_{j=1}^g W_{T_j}(w) + \dfrac{1}{w_T} \displaystyle\sum_{1 \le j < \ell \le g} w_{T_j} w_{T_\ell} (\overline{x}_j(w) - \overline{x}_\ell(w)) (\overline{x}_j(w) - \overline{x}_\ell(w))^\top.$

PROOF. (a) The sum on the right-hand side may be rewritten

$$\tfrac{1}{2} \sum_{j,\ell} w_{T_j} w_{T_\ell} (\overline{x}_j(w) - \overline{x}_\ell(w)) (\overline{x}_j(w) - \overline{x}_\ell(w))^\top$$

$$= \sum_{j,\ell} w_{T_j} w_{T_\ell} \overline{x}_j(w) \overline{x}_j(w)^\top - \sum_{j,\ell} w_{T_j} w_{T_\ell} \overline{x}_j(w) \overline{x}_\ell(w)^\top$$

$$= w_T \sum_j w_{T_j} \overline{x}_j(w) \overline{x}_j(w)^\top - \Big(\sum_j w_{T_j} \overline{x}_j(w)\Big) \Big(\sum_\ell w_{T_\ell} \overline{x}_\ell(w)\Big)^\top$$

$$= w_T \sum_j w_{T_j} \overline{x}_j(w) \overline{x}_j(w)^\top - w_T^2 \cdot \overline{x}(w) \, \overline{x}(w)^\top$$

$$= w_T \sum_j w_{T_j} \big(\overline{x}_j(w) - \overline{x}(w)\big) \big(\overline{x}_j(w) - \overline{x}(w)\big)^\top.$$

(b) The weighted Steiner formula A.12 applied to the sets T_j implies

$$W_T(w) = \sum_i w_i(x_i - \overline{x}(w))(x_i - \overline{x}(w))^\top$$

$$= \sum_j \sum_{i \in T_j} w_i(x_i - \overline{x}(w))(x_i - \overline{x}(w))^\top$$

$$= \sum_j W_{T_j}(w) + \sum_j w_{T_j}(\overline{x}_j(w) - \overline{x}(w))(\overline{x}_j(w) - \overline{x}(w))^\top$$

and claim (b) follows from (a).   □

The special case for $w_i = 1$ reads as follows.

A.14 Lemma *Let $T_1, \ldots, T_g$ be a partition of $1..n$ with cardinalities $n_1, \ldots, n_g$ and let $\overline{x}$ be the mean of $(x_1, \ldots, x_n)$ and $\overline{x}_j$ that of $(x_i)_{i \in T_j}$ (arbitrary if $n_j = 0$). Then*

(a) $\displaystyle\sum_j n_j(\overline{x}_j - \overline{x})(\overline{x}_j - \overline{x})^\top = \frac{1}{n} \sum_{1 \leq j < \ell \leq g} n_j n_\ell(\overline{x}_j - \overline{x}_\ell)(\overline{x}_j - \overline{x}_\ell)^\top$

(b) $\displaystyle\sum_{i=1}^n (x_i - \overline{x})(x_i - \overline{x})^\top = \sum_{j=1}^g W_{T_j} + \frac{1}{n} \sum_{1 \leq j < \ell \leq g} n_j n_\ell(\overline{x}_j - \overline{x}_\ell)(\overline{x}_j - \overline{x}_\ell)^\top.$

Finally, in connection with the breakdown behavior of the means of the mixture model, the following extension of Lemma A.13 to matrix weights is needed. Notation is as in Section 2.2.4.

A.15 Lemma *For any nonempty subset $U \subseteq 1..n$ and any stochastic matrix $\boldsymbol{\alpha} \in \mathcal{M}(U, g)$, we have*

$$W_U = W_U(\boldsymbol{\alpha}) + \frac{1}{|U|} \sum_{1 \leq h < k \leq g} \alpha_h(U)\alpha_k(U)\big(\overline{x}_U(\alpha_h) - \overline{x}_U(\alpha_k)\big)\big(\overline{x}_U(\alpha_h) - \overline{x}_U(\alpha_k)\big)^\top.$$

PROOF. Let T be the juxtaposition of g copies of U and w the linearization of the weight matrix $\boldsymbol{\alpha}$. It follows $w_{T_j} = \alpha_j(U)$, $\overline{x}_{T_j}(w) = \overline{x}_U(\alpha_j)$, $w_T = \sum_{i \in T} w_i = \sum_j \sum_{i \in T_j} \alpha_{i,j} = |U|$, and $\overline{x}_T(w) = \overline{x}_U$, and the claim is a term-by-term translation of Lemma A.13.   □

Topology

Probabilistic models of data need two spaces, the *sample* and the *parameter* space. Both inference and the resulting algorithms depend on topological structure on these spaces. The proofs of consistency, asymptotic normality, and robustness, some properties of the EM and other search algorithms, and some concepts of measure theory need insight into topological and analytical concepts. Examples are topological, uniform, and metric spaces, continuity, compactness and differentiation. Often a metric space is needed, sometimes even a differentiable structure. These structures form a hierarchy:

$$\text{topological space} \;\leftarrow\; \text{uniform space} \;\leftarrow\; \text{metric space} \;\leftarrow\; \text{normed space.}$$

For the reader's convenience and in order to fix notation, some of them are compiled here. Topological spaces offer the basis for important concepts such as convergence of a sequence of elements, continuity of a map between spaces, homeomorphy, and compactness. Uniform spaces are the background for uniform continuity and uniform convergence and an ideal basis for metrizability. Metric spaces allow the transition from arbitrary operations to countable[1] ones often needed in probability. Comprehensive treatments of topology are found in the excellent monographs of Bourbaki [61, 62] and Willard [529]. They contain proofs of some facts stated but not proved here.

B.1 Topological spaces

Topological structures are particularly adapted to studying collections of functions such as harmonic functions in function or potential theory or density and likelihood functions in statistics. Often used notions are convergence, continuity, and compactness. Central to topological spaces is the concept of a neighborhood of a point or a set. It can be introduced in two equivalent ways.

B.1 Definition (a) A collection $\mathcal{O}$ of subsets of a nonempty set Θ is called a **topology** on Θ if

(i) $\emptyset \in \mathcal{O}$ and $\Theta \in \mathcal{O}$;

(ii) any union of members of $\mathcal{O}$ belongs to $\mathcal{O}$;

(iii) any finite intersection of members of $\mathcal{O}$ belongs to $\mathcal{O}$.

(b) A subset $\mathcal{B} \subseteq \mathcal{O}$ is called a **base** of $\mathcal{O}$ if any $U \in \mathcal{O}$ is the union of members of $\mathcal{B}$.

The sets in $\mathcal{O}$ are called **open** (w.r.t. $\mathcal{O}$). The set Θ along with a topology $\mathcal{O}$ is called a **topological space**. A topology $\mathcal{O}_1$ on Θ is **coarser** than a topology $\mathcal{O}_2$ on Θ if $\mathcal{O}_1 \subseteq \mathcal{O}_2$. The restrictions of the open sets to a nonempty subset of Θ make the subset a topological space, called a **subspace** of Θ.

The complements of open sets are called **closed**. Arbitrary intersections of closed sets are closed and so are finite unions. For each subset $S \subseteq \Theta$, the union of all open subsets of S

---

[1]That is, either nonempty, finite, or of the cardinality of the integers.

is the largest open set contained in S. It is the **interior** of S. Analogously, the intersection of all closed sets that contain S is the smallest closed set that contains S, the **closure** of S. Subsets may be both open and closed or neither of the two. If the closure is the whole space Θ, then S is called **dense** in Θ. The space Θ is **separable** if it contains a countable, dense subset. Separability allows us sometimes to reduce uncountable operations to countable ones. This is important in probability.

A small complication should be pointed out before we proceed. Openness and closedness of a subset and interiority of a point to a subset depend on the "surrounding" space. Whereas the half-open interval $[0, 1[$ is neither open nor closed as a subset of $\mathbb{R}$, it is open as a subset of the positive ray $[0, \infty[$ and both open and closed as a subset of itself. Consequently, the point 0 is interior to $[0, 1[$ w.r.t. the positive ray, but not as a subset of the real line. It must always be clear what space we are in.

While definition B.1 is short and elegant, it is usually more convenient to define a topology on a set by specifying neighborhoods of points.

B.2 Definition (a) A family $\mathcal{U} = (\mathcal{U}(\vartheta))_{\vartheta \in \Theta}$ is called a **neighborhood system** on a set $\Theta \neq \emptyset$ if it has the following properties:

   (i) Each $\mathcal{U}(\vartheta)$ is a nonempty system of subsets of Θ;

  (ii) $\vartheta \in U$ for each $U \in \mathcal{U}(\vartheta)$;

 (iii) any finite intersection of members of $\mathcal{U}(\vartheta)$ is a member of $\mathcal{U}(\vartheta)$;

 (iv) each superset of a set in $\mathcal{U}(\vartheta)$ belongs to $\mathcal{U}(\vartheta)$;

  (v) for each $U \in \mathcal{U}(\vartheta)$ there is $V \in \mathcal{U}(\vartheta)$ such that $U \in \mathcal{U}(\eta)$ for all $\eta \in V$.

(b) A family $\mathcal{B} = (\mathcal{B}(\vartheta))_{\vartheta \in \Theta}$ is called a **neighborhood base**, if it satisfies (a)(i),(ii) and if

(iii') any finite intersection of members of $\mathcal{B}(\vartheta)$ contains a member of $\mathcal{B}(\vartheta)$;

 (v') for each $U \in \mathcal{B}(\vartheta)$ there is $V \in \mathcal{B}(\vartheta)$ such that, for all $\eta \in V$, there is some $W \in \mathcal{B}(\eta)$ with $W \subseteq U$.

The sets in $\mathcal{U}(\vartheta)$ are called the **neighborhoods** of ϑ (w.r.t. $\mathcal{U}$). Any neighborhood system is also a neighborhood base. For each ϑ, the family $\mathcal{U}(\vartheta)$ is called the neighborhood system of ϑ (w.r.t. $\mathcal{U}$) and $\mathcal{B}(\vartheta)$ is called a neighborhood base of ϑ. A neighborhood system is closed w.r.t. supersets, (a)(iv). This is the only difference between a neighborhood system and a neighborhood base. A neighborhood base $\mathcal{B}(\vartheta)$ generates a unique neighborhood system by adding to $\mathcal{B}(\vartheta)$ all supersets of sets in $\mathcal{B}(\vartheta)$. In contrast, there are many bases $\mathcal{B}$ that induce the same neighborhood system $\mathcal{U}$, called bases of $\mathcal{U}$. As examples, let $B_r(\vartheta) = \{\eta \in \mathbb{R}^q \mid \|\eta - \vartheta\| < r\}$, $B'_r(\vartheta) = \{\eta \in \mathbb{R}^q \mid \|\eta - \vartheta\| \leq r\}$, $\vartheta \in \mathbb{R}^q$, $r > 0$. A neighborhood base of $\vartheta \in \mathbb{R}^q$ is $\mathcal{B}(\vartheta) = \{B_r(\vartheta) \mid r > 0\}$; another one is the collection $\mathcal{B}'(\vartheta) = \{B'_r(\vartheta) \mid r > 0\}$. Both generate the same neighborhood system. A subset $U \subseteq \Theta$ is called a **neighborhood** of an arbitrary subset $S \subseteq \Theta$ if it is a neighborhood of all its points; equivalently, if it contains S in its interior.

The structures provided by a topology and a neighborhood system or base are equal. Given a topology $\mathcal{O}$, declare an open set U a member of $\mathcal{B}(\vartheta)$ if $\vartheta \in U$. Then $\mathcal{B}$ is a neighborhood base called the neighborhood base generated by the topology $\mathcal{O}$. Vice versa, if $\mathcal{U}$ is a neighborhood system, then the sets U that are $\mathcal{U}$-neighborhoods of all their points are the open sets of a topology, called the topology generated by $\mathcal{U}$. The relationship is perfect: If we start from a topology and generate the neighborhood base, then the topology induced by the associated neighborhood system is the original one. Both examples of neighborhood bases on $\mathbb{R}^q$ above generate the same topology, called the natural topology on $\mathbb{R}^q$.

Topological spaces offer the natural basis for continuity of a function. Given two topological spaces H and Θ, a function $f \colon H \to \Theta$ is called **continuous** at $\eta \in H$ if, for any neighborhood $V \in \mathcal{U}(f(\eta))$, there exists a neighborhood $U \in \mathcal{U}(\eta)$ such that $f(U) \subseteq V$.

It is continuous if it is continuous at each point of H. As an example, any polynomial is continuous as a function on $\mathbb{R}^1$ with the natural topology.

A simple example of a neighborhood system is $\mathcal{U}(\vartheta) = \{\Theta\}$ for all ϑ. This topological space is useful at most for counterexamples and to exclude it, a number of separation properties have been devised. One of them is the Hausdorff property.

B.3 Definition

(a) A topological space Θ is called a **Hausdorff space** (or a T_2-space) if, for any two distinct points ϑ and η, there are disjoint neighborhoods $U \in \mathcal{U}(\vartheta)$ and $V \in \mathcal{U}(\eta)$.

(b) A topological space Θ is **compact** if it is Hausdorff and if any system of open sets that cover Θ contains a *finite*, covering subsystem.

(c) A Hausdorff space is **locally compact** if each point possesses a compact neighborhood.

(d) A subset of a Hausdorff space is **compact** if it is compact as a subspace of Θ.

(e) A subset of a Hausdorff space is **relatively compact** if its closure is compact.

Any Euclidean space with the topology defined above is Hausdorff. In a Hausdorff space, every closed set is the intersection of all its open supersets. By contrast to the concepts of open- and closedness, compactness of a subset does not depend on the space where it is embedded. The closed unit interval (with the usual topology), for instance, is compact no matter where it appears as a subspace. Compact and locally compact spaces enjoy a number of convenient properties that will be discussed below. It can be shown that each point in a locally compact space possesses a neighborhood base of its topology consisting of compact sets. According to the theorem of Heine-Borel, a subset of a Euclidean space $\mathbb{R}^q$ with its natural topology is compact if and only it is closed and bounded. Thus, all Euclidean spaces are locally compact. There is a simple criterion for a subset of a (locally) compact space to be again (locally) compact.

B.4 Proposition *(a) A subset of a compact space is again compact if and only if it is closed.*

(b) A subset of a locally compact space is again locally compact if and only if it is the intersection of an open and a closed set.

A map $E \to F$ between two topological spaces E and F is called **open** (**closed**) if it maps open (closed) sets to open (closed) sets. An open map is not necessarily closed and vice versa. An exception is a one-to-one, onto map. It is closed if and only if it is open.

B.5 Proposition

(a) The continuous image of a compact set in a Hausdorff space is compact.

(b) If E and F are topological spaces, E locally compact, F Hausdorff, and if $\varphi \colon E \to F$ is continuous, open, and onto, then F is locally compact.

Compact sets mimic finite sets in many ways. Any continuous, vector-valued function on a compact space is bounded. Any continuous, real-valued function on a compact space assumes its maximum and minimum. The intersection of a decreasing sequence of nonempty compact sets is never empty. A sequence $(\vartheta_n) \subseteq \Theta$ is said to **converge** with **limit** $\vartheta \in \Theta$ if each $U \in \mathcal{U}(\vartheta)$ contains ϑ_n for all but a finite number of indices n. Limits in Hausdorff spaces are unique. A point is called a **cluster point** of (ϑ_n) if each of its neighborhoods contains infinitely many members of the sequence. Of course, limits are cluster points, but not vice versa. A nice thing about compact sets K is that each sequence of points in K has a cluster point in K. More generally, we have the following lemma.

B.6 Lemma *Let K be a compact subset of a Hausdorff space Θ and let (ϑ_n) be a sequence of elements of Θ such that each neighborhood of K (a subset that contains K in its interior) contains eventually all members of (ϑ_n). Then,*

(a) (ϑ_n) has a cluster point in K and

(b) all cluster points of (ϑ_n) lie in K.

PROOF. (a) Assume on the contrary that (ϑ_n) has no cluster point in K. Then each point $\vartheta \in K$ has an open neighborhood U_ϑ which contains only finitely many members of (ϑ_n). A finite number of these neighborhoods cover K and their union is a neighborhood of K. This is a contradiction.

(b) Assume that (ϑ_n) has a cluster point $\vartheta \notin K$. By the Hausdorff property, there is a neighborhood U of K and a neighborhood V of ϑ such that $U \cap V = \emptyset$. This is a contradiction since V contains infinitely many members of (ϑ_n). $\qquad\square$

Let E be a nonempty set and let S be some topological space. The product space S^E of all functions $E \to S$ is endowed with the **product topology** (or **topology of pointwise convergence**) defined by declaring the sets

$$U = \bigcap_{1 \le i \le n} \{f \colon E \to S \mid f(x_i) \in U_i\}, \quad n \ge 1, \ x_i \in E, \ U_i \text{ open in } S, \ 1 \le i \le n,$$

open. The product topology is Hausdorff if and only if S is. There is the following important theorem.

B.7 Theorem (Tychonov) *The space S^E is compact if and only if S is compact.*

The point is not that products of compact spaces are small but rather that open sets in products are large – but not large enough to lose the Hausdorff property.

B.2   Uniform spaces

Uniformities are structures richer than topologies. They are defined with the aid of **relations** on Θ, that is, of subsets of the square $\Theta \times \Theta$. Denote the **identity** relation, the diagonal in $\Theta \times \Theta$, by Δ. The **composition** of two relations U and V in Θ is the relation

$$U \circ V = \{(\vartheta, \tau) \mid \text{there exists } \eta \in \Theta \text{ s.th. } (\vartheta, \eta) \in U, \ (\eta, \tau) \in V\} \subseteq \Theta \times \Theta.$$

This operation is well known from maps, but the relations appearing here are very different. The relation **inverse** to D is $D^{-1} = \{(\eta, \vartheta) \mid (\vartheta, \eta) \in D\}$.

B.8 Definition (a) A (diagonal) **uniformity** on Θ is a nonempty collection $\mathcal{D}$ of relations in Θ with the following properties.

(i)   All $D \in \mathcal{D}$ contain the diagonal Δ as a subset;

(ii) $\mathcal{D}$ is closed w.r.t. finite intersections;

(iii) $\mathcal{D}$ is closed w.r.t. supersets;

(iv)   for each $D \in \mathcal{D}$, there is $E \in \mathcal{D}$ such that $E \circ E \subseteq D$;

(v)   for each $D \in \mathcal{D}$, there is $E \in \mathcal{D}$ such that $E^{-1} \subseteq D$.

(b) A **uniformity base** on Θ is a nonempty collection $\mathcal{E}$ of relations in Θ that satifies (i), (iv), (v) and

(ii$'$) any finite intersection of relations in $\mathcal{E}$ contains some relation in $\mathcal{E}$.

The relations $D \in \mathcal{D}$ are called the **surroundings** of $\mathcal{D}$. Property (iv) says that each D contains a "square root." Any uniformity base can be completed to a unique uniformity by adding all supersets. The relation described by $D \in \mathcal{D}$ is *nearness*. The idea is that all pairs (ϑ, η) in the same relation $D \in \mathcal{D}$ enjoy the same mutual nearness, no matter where in Θ they are located. By contrast, a topological space does not compare the nearness of two pairs of points in different regions of the space. A uniformity $\mathcal{D}$ is called **coarser** than a uniformity $\mathcal{E}$ on the same space if $\mathcal{D} \subseteq \mathcal{E}$. The usual uniformity on a Euclidean space has for a base the collection of sets $D_r = \{(\vartheta, \eta) \,|\, \|\vartheta - \eta\| < r\}$.

The **cut** of a relation D at $\vartheta \in \Theta$ is the subset $D_\vartheta = \{\eta \in \Theta \,|\, (\vartheta, \eta) \in D\} \subseteq \Theta$. Each uniformity base $\mathcal{E}$ induces in a natural way a neighborhood base $\mathcal{B}$ on Θ by

$$\mathcal{B}(\vartheta) = \{D_\vartheta \,|\, D \in \mathcal{E}\};$$

that is, each uniform space **induces** a topological space in a natural way. Thus, all concepts defined in topological spaces are at our disposal in uniform spaces, too. A uniformity $\mathcal{D}$ is called **separating** (and Θ is separated) if $\bigcap_{D \in \mathcal{D}} D = \Delta$. The induced topological space is Hausdorff if and only if $\mathcal{D}$ is separating. Note that different uniformities may induce the same topology. Exceptions are compact topological spaces.

B.9 Theorem *There is exactly one uniformity that induces the topology of a compact topological space.*

Uniform structures enable the definition of uniform continuity of functions and uniform convergence of sequences of functions. Let Θ and Γ be two uniform spaces with uniformities $\mathcal{D}$ and $\mathcal{E}$, respectively. A map $\Phi \colon \Theta \to \Gamma$ is called **uniformly continuous** if, for each $E \in \mathcal{E}$, there exists $D \in \mathcal{D}$ such that $(\Phi(\vartheta), \Phi(\eta)) \in E$ whenever $(\vartheta, \eta) \in D$. Any uniformly continuous function is continuous w.r.t. the induced topology. Let S be an arbitrary set. A sequence of functions $f_n \colon S \to \Theta$ **converges uniformly** to a function $f \colon S \to \Theta$ if, for all $D \in \mathcal{D}$ and all ϑ, we have $(f_n(\vartheta), f(\vartheta)) \in D$ for eventually all n.

Let E be a nonempty set and let S be some uniform space with uniformity $\mathcal{S}$. The space S^E of all functions $E \to S$ is endowed with the **product uniformity** defined by the surroundings

$$D = \bigcap_{1 \le i \le n} \{(f, g) \,|\, (f(x_i), g(x_i)) \in G\}, \quad n \ge 1, \ x_i \in E, \ G \in \mathcal{S}, \ 1 \le i \le n.$$

The product uniformity is separating if and only if $\mathcal{S}$ is.

B.3   Metric spaces and metrizability

A **metric** d on a set Θ is a map $\Theta \times \Theta \to [0, \infty[$ such that (i) $d(\vartheta, \eta) = 0$ if and only if $\vartheta = \eta$, (ii) $d(\vartheta, \eta) = d(\eta, \vartheta)$ (symmetry), and (iii) $d(\vartheta, \tau) \le d(\vartheta, \eta) + d(\eta, \tau)$ for all $\vartheta, \eta, \tau \in \Theta$ (**triangle inequality**). A set endowed with a metric is called a **metric space**. The number $d(\vartheta, \eta)$ is called the **distance** between ϑ and η. Given $r > 0$ and $\vartheta \in \Theta$, the subsets $B_r(\vartheta) = \{\gamma \in \Theta \,|\, d(\gamma, \vartheta) < r\}$ and $\overline{B}_r(\vartheta) = \{\gamma \in \Theta \,|\, d(\gamma, \vartheta) \le r\}$ are called the open and the closed r-**ball** with center ϑ, respectively. The open r-ball about a nonempty subset $S \subseteq \Theta$ is the union of the open r-balls about all points of S. A subset of a metric space is called **bounded** if it is contained in some ball. Just as in the case of topological and uniform spaces, each subset $S \subseteq \Theta$ of a metric space is again a metric space with the restriction $d_{|S \times S}$ as the metric. If A is a subset of a metric space, then any point of its *closure* can be reached as the limit of a *sequence* in A. Any subset of a Euclidean space is a metric space with the metric $d(\vartheta, \eta) = \|\eta - \vartheta\|$.

If d is a metric, then the sets $D_n = \{(\vartheta, \eta) \,|\, d(\vartheta, \eta) < 1/n\}$, $n \ge 1$, form a countable base of a uniformity, called the **induced uniformity**. The systems $\mathcal{B}(\vartheta) = \{B_{1/n}(\vartheta) \,|\, n \ge 1\}$ of open balls form a neighborhood base of the **induced topology** just as the systems of closed balls do. Since metric spaces are uniform and hence topological, all uniform and topological concepts such as (uniform) continuity of functions between two metric spaces and (uniform) convergence of sequences of functions from any set to a metric space are defined. Any metric space is Hausdorff. Two metrics on Θ are called **compatible** if each one is majorized by a multiple of the other. As a consequence, the metrics are **equivalent** in the sense that the topologies induced by them are equal and so are all topological concepts such as continuity, convergence, cluster point. It is important to keep in mind that the converse is not true in general: equivalent metrics may be incompatible. As an example, consider the usual distance $d(\vartheta, \eta) = |\eta - \vartheta|$ on the real line. It induces the same topology as the one that is transferred from the open interval $]-1, 1[$ with the usual metric by means of the **shrinking transformation** $\Phi(\vartheta) = \vartheta/(1 + |\vartheta|)$, $d'(\vartheta, \eta) = |\Phi(\eta) - \Phi(\vartheta)|$. Nevertheless, the two metrics are incompatible since d' is bounded while d is not.

A sequence (ϑ_n) in a metric space is called a **Cauchy sequence** if, for all $\varepsilon > 0$, there exists n_0 such that $d(\vartheta_m, \vartheta_n) \leq \varepsilon$ for all $m, n \geq n_0$. A metric space is called **complete** if every Cauchy sequence converges in it. *Closed* subsets of Euclidean spaces are complete metric spaces. An example of a divergent Cauchy sequence in the open unit interval $\Theta =]0, 1[$ with the usual metric is $(1/n)_{n \geq 1}$. Since any Cauchy sequence in a compact set has a cluster point, any compact metric space is complete. More precisely, there is the following theorem. A metric space is called **totally bounded** if, for every $\varepsilon > 0$, there is a finite number of ε-balls that cover the space.

B.10 Theorem *For a metric space Θ, the following are equivalent:*

(a) Θ is compact;

(b) each sequence in Θ has a cluster point;

(c) Θ is totally bounded and complete.

A topological (uniform) space is **metrizable** if there is a metric that induces the topology (uniformity). The metrizability problem for uniform spaces has a simple solution.

B.11 Theorem *A uniformity is metrizable if and only if it is separating and has a countable base.*

Since compact spaces are uniform in a natural way (see Theorem B.9), the solution there is equally simple.

B.12 Theorem *A compact space is metrizable if and only if its open sets have a countable base.*

The reader may be tempted to conjecture that a separable topological space has a countable base of its topology if each point has a countable neighborhood base. This is wrong and a counterexample is found in Bourbaki [62], p. IX.92, Exercise 12(b). Nevertheless, if $\{\vartheta_k\}$ is a countable, dense subset of a topological space and if each point ϑ_k has a countable base $\mathcal{B}(\vartheta_k)$ of (open) neighborhoods, one may try to show that the (countable) set $\bigcup_k \mathcal{B}(\vartheta_k)$ forms a base of the topology. If this is successful and if the space is compact, then it is metrizable.

A space is **Polish** if it is separable and completely metrizable.[2] This term was coined by N. Bourbaki [62] to remind us of the achievements of the Polish topologists. Polish spaces play an important role in probability theory; see Theorem D.2, Properties E.34, Appendix D.6, and Lemma E.41. Of course, any countable space where all singletons are open (that is, a discrete space) is Polish and so is every Euclidean space with the Euclidean topology. Open and closed subsets of Polish spaces are again Polish. For instance, the open interval $]-1, 1[\subseteq \mathbb{R}$ is Polish, but with the equivalent metric induced from $\mathbb{R}$ by the shrinking transformation $\Phi \colon \mathbb{R} \to]-1, 1[$, $\Phi(x) = x/(1 + |x|)$. Moreover, every finite product of Polish spaces Θ_i with metrics d_i is Polish with any of the product metrics $d(\mathbf{x}, \mathbf{y}) = \sum_k d_k(x_k, y_k)$, $\left(\sum_k d_k^2(x_k, y_k) \right)^{1/2}$, or $\max_k d_k(x_k, y_k)$, $\mathbf{x} = (x_k)_k$, $\mathbf{y} = (y_k)_k$. Every compact metric space is Polish. The compact metric spaces are the most distinguished among all topological spaces. They enjoy all properties one may wish.

---

[2]That is, its topology is induced by a complete metric.

B.4 Quotient spaces

Let φ be a map from a topological space Θ *onto* an arbitrary set E. Declare a subset $G \subseteq E$ open in E if its inverse image $\varphi^{-1}(G)$ is open in Θ. Plainly, the collection of open sets is a topology on E.

B.13 Definition The topology

$$\mathcal{O}_\varphi = \{G \subseteq E \mid \varphi^{-1}(G) \text{ is open in } \Theta\}$$

is called the **quotient topology** on E generated by the **quotient map** $\varphi\colon \Theta \to E$. The set E equipped with the quotient topology is called a **quotient space** of Θ.

The φ-equivalence class $\varphi^{-1}(\varphi(\vartheta))$ of $\vartheta \in \Theta$ is denoted by $[\vartheta]$. A subset of Θ is called saturated (w.r.t. the quotient map φ) if it is a union of φ-equivalence classes. If F is another topological space, then a map $\psi\colon E \to F$ is continuous if and only if $\psi \circ \varphi$ is.

It is often interesting to know whether a quotient map is closed (w.r.t. the quotient topology). This will, for instance, be needed in Propositions B.15 and B.16 below. The following characterization requires that the equivalence relation defined by the quotient map on Θ cooperate with the given topology there. A neighborhood of a subset of a topological space is a neighborhood of each of its points.

B.14 Lemma *The quotient map $\varphi\colon \Theta \to E$ is closed (w.r.t the quotient topology) if and only if each neighborhood of an equivalence class $[\vartheta] \subseteq \Theta$ contains a saturated neighborhood.*

The next proposition gives sufficient conditions for local compactness and metrizability (Section B.3) of a quotient; see Willard [529], 18C and 23K.

B.15 Proposition *Assume that the quotient map $\varphi\colon \Theta \to E$ is closed (w.r.t. the quotient topology) and that equivalence classes are compact. Then*

(a) if Θ is locally compact, then so is the quotient topology;

(b) if Θ is metrizable, then so is the quotient topology.

Often the set E bears a preassigned topology. It is natural to ask for its relationship with the quotient topology. The following proposition states a sufficient condition for equality.

B.16 Proposition *Let Θ and E be topological spaces and let $\varphi\colon \Theta \to E$ be continuous, onto, and either open or closed. Then the given topology on E is the quotient topology generated by φ.*

Since any continuous map from a *compact* space is closed, we have the following corollary.

B.17 Corollary *Let Θ and E be topological spaces, Θ compact, and let $\varphi\colon \Theta \to E$ be continuous and onto. Then the given topology on E is the quotient topology generated by φ.*

B.5 Compactification

Some statements in probability theory require compactness, for instance, the Uniform Law of Large Numbers E.4.1. Although the space in hand usually bears a topology, it is often not compact. An example is the parameter space of a location model in Euclidean space. It may nevertheless be possible to apply the Uniform Law after "embedding" the space in a compact space. The closure of the embedded space is then compact, that is, a compactification. If the assumptions of the statement still hold for the compactification, then we can apply the statement and see what it tells us about the original space. An example appears in the treatment of consistency in Section 1.2.

B.18 Definition Let Γ and Θ be two topological spaces.

(a) A map $\Phi\colon \Gamma \to \Theta$ is a **homeomorphism** if it is one-to-one and onto and if both Φ and Φ^{-1} are continuous.

(b) An **embedding** of Γ into Θ is a homeomorphism of Γ onto a subspace of Θ.

(c) A **compactification** is a dense embeddding into a compact space.

Homeomorphy defines the isomorphy of topological spaces. If one space is the homeomorphic image of another, then they are topologically equivalent. We may consider an embedded space as a subspace of the embedding space. The set of added points is called the **ideal boundary** of Θ. The real line is homeomorphic with the open interval $]-1,1[$ via the shrinking transformation $\Phi(x) = x/(1 + |x|)$. Hence, Φ is an embedding of the real line into the compact interval $[-1, 1]$. Since the embedding is dense there, $[-1, 1]$ is a compactification of the real line. Another one is the unit circle $\mathbb{T} = \{z \in \mathbb{C}\,|\,|z| = 1\}$ in the complex plane via the mapping $\Psi(x) = e^{i\pi\Phi(x)}$. Algebraic topology shows that the two compact spaces $[-1, 1]$ and $\mathbb{T}$ are not homeomorphic. Thus, compactifications of the same space may be very different.

In almost all cases it is important that certain given continuous functions on Θ can be continuously extensible to the ideal boundary. It is also important that certain properties of the embedded space, such as separability or metrizability, be conserved. Separability is no problem since it carries over from the embedded space to the embedding space by density. Metrizability makes it necessary to choose an appropriate approach.

There are several ways of compactifying a topological space Θ. A locally compact, non-compact Hausdorff space Θ admits the **Alexandrov compactification** $\alpha\Theta$. It is obtained by adding one ideal ("Alexandrov") point to Θ, by leaving the neighborhoods of ϑ in Θ unchanged, and by declaring the complements of compact subsets of Θ a neighborhood base of the ideal point. It is readily seen that $\alpha\Theta$ is indeed a compact space, that the natural map $\Theta \to \alpha\Theta$ is an embedding, and that Θ is dense in $\alpha\Theta$. So, $\alpha\Theta$ is a compactification of Θ. The ideal point is usually denoted by the symbol ∞. The only continuous functions on Θ that can be continuously extended to ∞ are those that oscillate very little outside of large compact sets. The circle $\mathbb{T}$ is the Alexandrov compactification of the real line. The following proposition answers the question of metrizability of $\alpha\Theta$.

B.19 Proposition *The Alexandrov compactification of a locally compact, metrizable space Θ is metrizable if and only if Θ is separable.*

If a metric space is totally bounded, it can be compactified by completion. Let us call two Cauchy sequences in a metric space equivalent if merging them creates again a Cauchy sequence. This defines an equivalence relation on the set of all Cauchy sequences. Each metric space may be completed by adding all equivalence classes of nonconvergent Cauchy sequences as ideal points to it. Since total boundedness is not lost after completion, the completion $\widehat{\Theta}$ of a totally bounded metric space Θ is a compactification; see Theorem B.10. A function on Θ has a (unique) extension to its completion if and only if it is uniformly continuous. The completion of the open interval $]-1, 1[$ is $[-1, 1]$.

B.20 Proposition *A locally compact space is open in any compactification.*

Functional spaces may be compactified using the evaluation map, a trick ubiquitous in mathematics, not only in topology. Given a locally compact topological space Θ and a set $\mathcal{Q}$ of continuous, (possibly extended) real-valued functions on Θ, we consider the **evaluation map** $e\colon \Theta \to \overline{\mathbb{R}}^{\mathcal{Q}}$ defined by

$$\vartheta \mapsto (f(\vartheta))_{f \in \mathcal{Q}}.$$

The **extended real line** $\overline{\mathbb{R}} = [-\infty, \infty]$ is compact with the topology induced by $[-1, 1]$ via the shrinking transformation $\Phi(x) = x/(1 + |x|)$. By Tychonov's compactness theorem, the product $\overline{\mathbb{R}}^{\mathcal{Q}}$ is compact with the product topology. If $\mathcal{Q}$ **separates the points** of Θ, that is, if for any two points $\vartheta, \eta \in \Theta$, there is a function $f \in \mathcal{Q}$ such that $f(\vartheta) \neq f(\eta)$, then e is one-to-one and ϑ can be identified with $f(\vartheta)$. The closure $q\Theta$ of $e(\Theta)$ is a compactification of Θ, called $\mathcal{Q}$-**compactification**. The idea goes back to Martin [352]. The requirement of point separation can be dropped if all continuous functions with compact support are added to $\mathcal{Q}$. They furnish the desired point separation without spoiling anything. If Θ is an open subset of Euclidean q-space, then its image by e can be conceived as a parameterized q-dimensional "hypersurface" in the hypercube $\bar{\mathbb{R}}^{\dot{\mathcal{Q}}}$. There is the following theorem due to Constantinescu and Cornea [99], p. 97.

B.21 Theorem *Let Θ be locally compact. The $\mathcal{Q}$-compactification is the only compactification of Θ with the following two properties:*

(a) Each function in $\mathcal{Q}$ has a unique continuous extension to $q\Theta$.

(b) The extensions separate the points of the ideal boundary.

For a proof; see also Helms [238]. The $\mathcal{Q}$-compactification is intimately related to the set $\mathcal{Q}$ of functions on Θ. Each sequence (ϑ_n) that converges to ∞ and such that $f(\vartheta_n)$ converges in $\bar{\mathbb{R}}$ for every $f \in \mathcal{Q}$ generates an ideal point. Two sequences (ϑ_n) and (η_n) such that $\lim_n f(\vartheta_n) = \lim_n f(\eta_n)$ for all $f \in \mathcal{Q}$ lead to the same point. For instance, if all functions of $\mathcal{Q}$ vanish at infinity (or converge to the same element of the extended real line), then there is only one ideal point and the $\mathcal{Q}$-compactification is the Alexandrov compactification.

An illustration of the $\mathcal{Q}$-compactification for the pair of functions $\mathcal{Q} = \{x, \sin(1/\sqrt{x})\}$, $x \in \Theta =]0, 1[$, is shown in Fig. B.1. Oscillating heavily near the origin the function $\sin(1/\sqrt{x})$

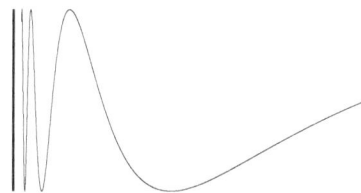

has, of course, no continuous extension there. Yet it has a continuous extension to the $\mathcal{Q}$-compactification. The evaluation map maps the interval $\Theta =]0, 1[$ onto the plane curve $\vartheta \to (x, \sin(1/\sqrt{x}))$, $0 < \vartheta < 1$. It is bounded but not closed in the plane. Its closure adds the point $(1,0)$ on the right and the closed interval from $(0, -1)$ to $(0, 1)$ on the left to the curve. This is the $\mathcal{Q}$-compactification. The extension of the identity $x \mapsto x$ is the abscissa of each point on the compactification; that of the function $x \mapsto \sin(1/\sqrt{x})$ is the ordinate. Both functions are indeed continuous. The reader may note that a whole lot of

Figure B.1 *$\mathcal{Q}$-compactification $q\Theta$ of the topologist's sine curve. The ideal boundary is the vertical line on the left.*

points have to be added to the image of $e(\Theta)$ in order to create a space that allows the continuous extension of $\sin(1/\sqrt{x})$. The square root is taken just to facilitate the visualization.

If Θ is locally compact, then the space $\mathcal{Q}$ of all continuous real-valued (or extended real-valued) functions separates points. In this case the $\mathcal{Q}$-compactification is the **Stone-Čech compactification** $\beta\Theta$. It is a huge space since it allows the continuous extension of *all* continuous functions on Θ to $\beta\Theta$, irrespective of their oscillation near the boundary. Unless Θ is compact and metrizable (in which case there is nothing to do), it is never metrizable and not useful for our purposes.

The Alexandrov compactification is too small to allow the extension of many continuous functions, while the Stone-Čech compactification is too large to be metrizable. What we need is a balance between extensibility and metrizability. Completion and $\mathcal{Q}$-compactification are possible candidates. However, completion has the drawback of needing the totally bounded metric as a prerequisite, whereas $\mathcal{Q}$-compactification uses the given functions $\mathcal{Q}$. From Theorem B.12 we immediately infer the following proposition.

B.22 Corollary *The $\mathcal{Q}$-compactification $q\Theta$ is metrizable if and only if its open sets have a countable base.*

In this case, each ideal boundary point can be represented by a *sequence* (ϑ_n) that diverges in Θ and such that the (extended real-valued) sequence $f(\vartheta_n)$ converges in $\bar{\mathbb{R}}$ for all $f \in \mathcal{Q}$. In this sense, the ideal boundary is reachable from Θ by means of sequences.

If one perceives a compact space as "small," then it may at first sight appear absurd that a big (noncompact) space can be made small (compact) by adding points, sometimes even many. But there is nothing wrong with that. The open sets that are needed to cover the ideal points of the compactification are huge, covering a lot more than just these points.

B.6 Topologies on special spaces

Many structured sets bear intrinsic topologies, some even several. Examples are Euclidean vector space $\mathbb{R}^q$, the algebra of q by q matrices, the collection of all nonempty finite subsets of a metric space, the vector space of all real-valued functions on a set, the convex set of probabilities on a measurable space. Some of them are analyzed later.

B.6.1 Manifolds

A topological space Θ is called a q-dimensional **topological manifold** (without boundary) if every point in Θ has an open neighborhood homeomorphic with an open set in $\mathbb{R}^q$; see, for instance, Perdigão do Carmo [409] and Boothby [59]. Thus, a topological manifold looks everywhere locally like an open set in a Euclidean space. A space is a q-dimensional **manifold with boundary** if every point in Θ has an open neighborhood that is homeomorphic with an open set in a closed half space of $\mathbb{R}^q$. (The open set may lie in the interior of the half space or contain part of its boundary). The homeomorphisms are called **charts**. They endow the manifold with local coordinates. The set of points of a manifold with boundary Θ that possess an open neighborhood homeomorphic with an open set of $\mathbb{R}^q$ is called the interior of Θ; the complement is its (manifold) boundary, a $(q-1)$-dimensional manifold. A **topological submanifold** is a subset of Θ that the restrictions of the charts make a topological manifold (with or without boundary). Its dimension is at most q.

The charts do not automatically transport the differentiable structure available on $\mathbb{R}^q$ to a manifold. It is rather necessary that they cooperate in the sense that the **transition maps**, compositions of inverse charts and charts on overlapping domains in Θ, induce continuously differentiable maps between open subsets of $\mathbb{R}^q$. If they are of class $\mathcal{C}^r$, then the manifold is called **differentiable** of class $\mathcal{C}^r$. Of course, any open subset in a Euclidean space is a $\mathcal{C}^\infty$-manifold that needs only one chart. An example of a 1-manifold without boundary is the unit circle in $\mathbb{R}^2$. It needs at least two charts, for instance, $\psi_1^{-1}(x) = (\cos x, \sin x)$ and $\psi_2^{-1} = -\psi_1^{-1}$, both defined for $-\pi < x < \pi$. This specification makes it $\mathcal{C}^\infty$. Continuous and differentiable functions between (differentiable) manifolds are defined by composition with charts. A submanifold of class $\mathcal{C}^r$ is a subset made a $\mathcal{C}^r$-manifold by the restrictions of the charts. An angle in the plane, for instance, is a one-dimensional topological submanifold, but it is not differentiable.

A differential manifold bears a surface measure that generalizes Lebesgue measure on Euclidean space; see Lang [306], Chapter XII, and Rudin [454], 10.44. They serve as reference measures for probabilities. Any submanifold of Θ of dimension $p < q$ is a null set w.r.t. the surface measure on Θ.

B.6.2 Topologies and norms on Euclidean and matrix spaces

A **norm** on a real (or complex) vector space Θ is a map $\Theta \to \mathbb{R}_+$, usually denoted by $\|\cdot\|$, such that (i) $\|\vartheta\| = 0$ if and only if ϑ is the vector $0 \in \Theta$, (ii) $\|a \cdot \vartheta\| = |a| \cdot \|\vartheta\|$ for all scalars a and all $\vartheta \in \Theta$, and (iii) $\|\vartheta + \eta\| \leq \|\vartheta\| + \|\eta\|$ for all $\vartheta, \eta \in \Theta$ (**triangle inequality**). A vector space along with a norm is called a **normed space**. Any normed space is metric via the definition $d(\vartheta, \eta) = \|\vartheta - \eta\|$ and hence also topological. If the metric is complete, then it is called a **Banach space**. The open (closed) **ball** with center ϑ of radius r is the subset

$B_r(\vartheta) = \{\eta \in \Theta \mid \|\eta - \vartheta\| < r\}$ $(\bar{B}_r(\vartheta) = \{\eta \in \Theta \mid \|\eta - \vartheta\| \leq r\})$. Two norms $\|\cdot\|$ and $\|\|\cdot\|\|$ on the same vector space Θ are called **compatible** if there are numbers $0 < \alpha \leq \beta$ such that $\alpha\|\vartheta\| \leq \|\|\vartheta\|\| \leq \beta\|\vartheta\|$ for all $\vartheta \in \Theta$. Compatible norms induce the same uniform structure and topology.

Examples of normed spaces are the real line (and complex plane) with the modulus as norm. More generally, Euclidean q-space is a normed space for a variety of norms, the **1-norm** $\|\vartheta\|_1 = \sum_{k=1}^{q} |\vartheta_k|$, the **Euclidean norm** $\|\vartheta\|_2 = \sqrt{\sum_{k=1}^{q} \vartheta_k^2}$, and the **uniform norm** $\|\vartheta\|_\infty = \max_{1 \leq k \leq q} |\vartheta_k|$. Although these norms are all distinct, they are compatible. More precisely, $\|\vartheta\|_\infty \leq \|\vartheta\|_2 \leq \|\vartheta\|_1 \leq q\|\vartheta\|_\infty$. In fact, any two norms on a Euclidean space are compatible. So, in statements such as "there exists an integrable function h such that $\|g(\vartheta, x)\| \leq h(x)$" we may choose the norm most convenient to the task in hand.

The algebra of (real or complex) square $q \times q$ matrices, too, bears various norms. It is possible to consider the vector space of matrices as a Eulidean q^2-space and to use the norms above. But there is the more powerful way of deriving norms on matrix spaces by defining

$$\|A\| = \sup_{\|\vartheta\|=1} \|A\vartheta\|.$$

Here, the norm on the right side is any of the norms on Euclidean q-space. It enjoys the pleasant property $\|AB\| \leq \|A\|\|B\|$. Such a norm on the algebra of square matrices is called a **matrix norm**. The 1-norm $\|\cdot\|_1$ on $\mathbb{R}^q$ induces the **maximum column-sum norm**

$$\|A\| = \max_\ell \sum_k |a_{k,\ell}|. \tag{B.1}$$

The uniform norm $\|\cdot\|_\infty$ on $\mathbb{R}^q$ induces the **maximum row-sum norm**

$$\|A\| = \max_k \sum_\ell |a_{k,\ell}|.$$

The Euclidean norm $\|\cdot\|_2$ induces the **spectral norm**

$$\|A\| = \sqrt{\text{largest eigenvalue of } A^\top A}. \tag{B.2}$$

The spectral norm of a *symmetric* matrix A is the largest modulus of its eigenvalues. This is not true for general square matrices. But there is the following proposition. A proof is found in Horn and Johnson [252], Lemma 5.6.10 and Theorem 5.6.7.

B.23 Lemma *Let ϱ be the spectral radius of the $q \times q$ matrix A.*

(a) Any matrix norm of A is bounded below by ϱ.

(b) For any $\varepsilon > 0$, there is a norm on $\mathbb{R}^q$ such that the associated matrix norm of A is $\leq \varrho + \varepsilon$.

Again, it is often not important which norm on $\mathbb{R}^{q \times q}$ is chosen; they are all compatible in the sense explained above.

As a $\binom{d}{2}$-dimensional manifold in the d^2-dimensional Euclidean space of all $d \times d$ matrices, the (special) orthogonal group $O(d)$ $(SO(d))$ of all orthogonal matrices (rotations on $\mathbb{R}^d$) inherits a topology. Since it is bounded and closed it is a compact metric space and separable. Similarly, the vector space $\text{SYM}(d)$ of all symmetric $d \times d$ matrices inherits a subspace topology from $\binom{d+1}{2}$-dimensional Euclidean space. It is more convenient to describe this topology by the "Löwner intervals"

$$[A - \varepsilon I_d, A + \varepsilon I_d] = \{B \in \text{SYM}(d) \mid A - \varepsilon I_d \preceq B \preceq A + \varepsilon I_d\}, \qquad \varepsilon > 0.$$

They form a neighborhood base at $A \in \text{SYM}(d)$. For each pair $\alpha < \beta$, the map $[\alpha, \beta]^d \times SO(d) \to \text{SYM}(d)$ defined by $((\lambda_1, \ldots, \lambda_d), U) \mapsto U \text{Diag}(\lambda_1, \ldots, \lambda_d) U^\top$ is continuous and has therefore a compact image. By the spectral theorem, this image consists of all real, symmetric matrices with eigenvalues contained in the interval $[\alpha, \beta]$. It follows that a subset $\mathcal{A} \subseteq \mathrm{SYM}(d)$ is relatively compact if and only if the eigenvalues of all members of $\mathcal{A}$ are uniformly bounded above and below. Equivalently, $\alpha I_d \preceq A \preceq \beta I_d$ for some $\alpha \leq \beta$ and all matrices $A \in \mathcal{A}$. In a similar way one shows that a subset of the subspace $\mathrm{PD}(d) \subseteq \mathrm{SYM}(d)$ is relatively compact (in $\mathrm{PD}(d)$) if all its eigenvalues are bounded and bounded away from zero, that is, contained in an interval $[\varepsilon, 1/\varepsilon]$ for some $0 < \varepsilon \leq 1$. Equivalently, $\mathcal{A} \subseteq \mathrm{PD}(d)$ is relatively compact if and only if there is $0 < \varepsilon \leq 1$ such that $\varepsilon I_d \preceq A \preceq 1/\varepsilon\, I_d$ for all $A \in \mathcal{A}$.

B.6.3 The Hausdorff metric

Let E be any metric space and let Θ be the collection of all nonempty, compact subsets of E, the **hyperspace** of E. For any two $A, B \in \Theta$, the minimum $d(A, y) = \min_{x \in A} d(x, y)$ and the maximum $\max_{y \in B} d(A, y)$ both exist. The Hausdorff distance between two sets $A, B \in \Theta$ is defined as

$$d^H(A, B) = \max_{x \in A} d(x, B) \vee \max_{y \in B} d(A, y).^3$$

That is, $d^H(A, B) < r$ if and only if for each $x \in A$ there is $y \in B$ such that $d(x, y) < r$ and vice versa. In other words, each set is contained in the open r-ball (in E) about the other. It is clear that d^H defines a metric on Θ that extends the given metric on E.

Let $g \geq 1$ be a natural number. In the context of consistency of the classification model, Section 1.3.7, we will be interested in the collection $\Theta_g = \Theta_g^E$ ($\Theta_{\leq g} = \Theta_{\leq g}^E$, $\Theta_{\geq g} = \Theta_{\geq g}^E$) of all nonempty subsets of E with (at most, at least) g elements. They will represent solutions to the classification problem; see Section 1.3.7.

B.24 Lemma *Let $g \geq 1$.*

(a) For any $r > 0$, the open r-ball about the set $A \in \Theta$ is

$$B_r^H(A) = \{K \in \Theta \mid K \subseteq B_r(A),\ K \cap B_r(a) \neq \emptyset \text{ for all } a \in A\}.^4$$

(b) The collection $\Theta_{\geq g}$ is open in Θ.

(c) The collection $\Theta_{\leq g}$ is closed in Θ.

(d) The collection Θ_g is open in $\Theta_{\leq g}$.

(e) If E is compact (locally compact), then so is $\Theta_{\leq g}$.

(f) If $F \subseteq E$ is closed, then $\Theta_{\leq g}^F$ is closed in $\Theta_{\leq g}^E$.

PROOF. (a) This follows from the definition of the Hausdorff metric.
(b) Let $A_0 \subseteq A$ with $|A_0| = g$, let α be the smallest distance between any pair of points in A_0, and let $r < \alpha/2$. Since the balls $B_r(a)$, $a \in A_0$, are disjoint, part (a) shows that any set in $B_r^H(A)$ has at least $|A_0| = g$ points. This is (b) and parts (c) and (d) follow.
(e) If E is compact, then so is E^g. The subspace $\Theta_{\leq g}$ is the continuous image of $E^g \to \Theta$, $(x_1, \ldots, x_g) \mapsto \{x_1, \ldots, x_g\}$. Now let E be locally compact; let $A = \{a_1, \ldots, a_h\} \in \Theta_{\leq g}$ with pairwise different points a_i. For each i let K_i be a compact neighborhood of a_i in E and let $\mathcal{I}$ be the collection of all increasing maps from $1..g$ onto $1..h$. Then the image of

$$\bigcup_{j \in \mathcal{I}} K_{j_1} \times \cdots \times K_{j_g}$$

is a compact neighborhood of A in $\Theta_{\leq g}$.
(f) Let $A \in \Theta_{\leq g}^E \setminus \Theta_{\leq g}^F$ and choose $a \in A \setminus F$. Since F is closed in E, $d(a, F) > 0$. Let

---

[3] Not to be mixed up with the distance $d(A, B) = \min_{x \in A, y \in B} d(x, y)$ of A and B in E. The superscript H stands for "Hausdorff" or "hyperspace."

[4] Note the difference between the subsets $B_r^H(A) \subseteq \Theta$ and $B_r(A) \subseteq E$.

$0 < r < d(a, F)$. Part (a) shows that any set in $B_r^H(A)$ contains a point in $B_r(a)$. But $B_r(a) \cap F = \emptyset$ and so this point must be in the complement of F. Hence, $B_r(A)$ does not intersect $\Theta_{\leq g}^F$. □

B.6.4 Functional spaces

In Section B.1, the function space $\mathbb{R}^E$ of all real-valued functions on E with its topology of pointwise convergence was introduced. Tychonov's Theorem B.7 characterizes the relatively compact subsets $\mathcal{F} \subseteq \mathbb{R}^E$ w.r.t. pointwise convergence. That is, the minimal and obviously necessary condion that $\{f(x) \mid f \in \mathcal{F}\}$ be bounded is also sufficient. Often we are not interested in *arbitrary* functions on E. If E is even a Hausdorff space, then the subspace $\mathcal{C}(E)$ of all continuous functions on E deserves particular attention. The topology of pointwise convergence is not natural on $\mathcal{C}(E)$ since, for instance, pointwise limits of continuous functions do not have to be continuous. We therefore endow $\mathcal{C}(E)$ with the **topology of compact convergence**, also called the **topology of uniform convergence on compact sets**. A neighborhood base of $f \in \mathcal{C}(E)$ is defined by the sets

$$U_{K,\varepsilon} = \{g \in \mathcal{C}(E) \mid |g(x) - f(x)| \leq \varepsilon \text{ for all } x \in K\}, \quad K \subseteq E \text{ compact}, \varepsilon > 0.$$

This topology is obviously finer than the topology of pointwise convergence. The question arises, which are their compact subsets? By Tychonov's Theorem B.7, a subset $\mathcal{F} \subseteq \mathcal{C}(E)$ can be compact only if the topology restricted to $\mathcal{F}$ is the topology of pointwise convergence. Therefore, exhibiting those subsets of $\mathcal{C}(E)$ that bear the topology of pointwise convergence is an intriguing problem. The following theorem gives an answer. A subset $\mathcal{F} \subseteq \mathcal{C}(E)$ is called **equicontinuous** at $x \in E$ if, for all $\varepsilon > 0$, there exists a neighborhood U of x such that $y \in U$ implies $|f(y) - f(x)| \leq \varepsilon$ for all $f \in \mathcal{F}$. A first property of equicontinuity is the simple but astonishing fact that the pointwise closure of an *equi*continuous family of functions is again equicontinuous. In particular, pointwise limits of *equi*continuous sequences are continuous. Moreover, there is the following theorem.

B.25 Theorem *On an equicontinous subset $\mathcal{F} \subseteq \mathcal{C}(E)$ the topologies of compact and pointwise convergence coincide.*

This theorem makes it easy to characterize the compact subsets of $\mathcal{C}(E)$. The following general version of Ascoli's theorem is particularly adapted to our needs. It is found, for instance, in Willard [529].

B.26 Theorem (Ascoli-Arzela) *Let E be locally compact or metrizable. A subset $\mathcal{F} \subseteq \mathcal{C}(E)$ of continuous functions on E is compact if and only if*

 (i) *$\mathcal{F}$ is pointwise closed;*

 (ii) *the set of evaluations $\{f(x) \mid f \in \mathcal{F}\}$ is bounded for each $x \in E$;*

 (iii) *$\mathcal{F}$ is equicontinuous on each compact subset of E. (If E is locally compact, then this is equivalent with equicontinuity of $\mathcal{F}$.)*

Let E be a nonempty set. A real- (or complex-)valued function f on E is said to be **bounded** if

$$\|f\|_\infty = \sup_{x \in E} |f(x)| < \infty.$$

The function $\|\cdot\|_\infty$ is a norm, called the **uniform norm** on the vector space of all bounded functions on E. If E is even a compact topological space, then all continuous functions are automatically bounded and the Ascoli-Arzela theorem assumes the following form.

B.27 Theorem *Let E be a compact space. A set $\mathcal{F} \subseteq \mathcal{C}(E)$ of continuous functions on E is relatively compact w.r.t. the norm $\|\cdot\|_\infty$ if and only if*

(i) it is uniformly bounded, that is, there exists a number K such that $\|f\|_\infty \leq K$ for all $f \in \mathcal{C}(E)$;

(ii) it is equicontinuous.

There is a simple criterion for metrizability of the topology of compact convergence on spaces of continuous functions. A space E is called **hemicompact** if it is Hausdorff and if there is a sequence $(K_n)_n$ of compact subsets of E such that each compact subset $K \subseteq E$ is contained in some K_n. Euclidean spaces and their open, closed, and locally compact subspaces are hemicompact. Euclidean manifolds are hemicompact as, of course, compact spaces are.

B.28 Proposition *The topology of compact convergence on $\mathcal{C}(E)$ on a hemicompact space E is metrizable.*

PROOF. In fact, if E is hemicompact with a sequence $(K_n)_n$, then

$$d(f,g) = \sum_{n=1}^{\infty} \max_{x \in K_n} |f(x) - g(x)| \wedge 2^{-n}$$

is a metric on $\mathcal{C}(E)$ that induces the topology of compact convergence. $\square$

Under additional assumptions on E, the converse is also true; see Willard [529], 43G. Other function spaces of interest are vector spaces of measurable functions; see Appendix D.6.

A caveat Topology bears a number of pitfalls to be avoided when working with its concepts. One reason is that toplogical spaces, in contrast to algebraic structures, often do not bear just one interesting topology. Properties such as compactness, metrizability, or completeness that one topology may enjoy may be missing in others. Other reasons are certain paradoxical phenomena. Whether a subset is closed depends on the surrounding space. A space can be the union of two metrizable subsets without being metrizable. An example is the Alexandrov compactification of the real line with discrete topology (every singleton is open). Compactness is a generalization of finiteness. Therefore, intuitively, a compact space is small. But the "large" Euclidean space can be made compact ("small") by adding points to it. Moreover, there exists a compactification of the real line, the Stone-Čech compactification, with the property that each bounded continuous function on it has a continuous extension. Just think of the function $\sin e^x$! For another paradox, see the text immediately after Theorem B.12.

Thus, topology sometimes contradicts natural intuition and seemingly reasonable statements may be wrong. Therefore, strictly formal reasoning is imperative.

Analysis

Convexity is the simplest concept in Euclidean spaces next to linearity. This short appendix collects first some facts about compact, convex sets and their extreme points. The extreme points represent a convex set in a similar way as a generating system of points represents an affine space. Convex sets appear in the ML estimation of latent distributions, in determining favorite solutions, and in various proofs. Moreover, the mixing rates of mixture models make up a simplex. This is a compact convex set with the property that each of its points possesses a *unique* convex representation by extreme points. A second issue is differentiation and diffeomorphisms. Both occupy a traditional role in optimization and in the treatment of consistency.

C.1 Convexity

A linear combination $\sum_{x \in F} u_x x$ over a finite subset F of some real vector space L is called **convex** if $u_x \geq 0$ for all $x \in F$ and $\sum_{x \in F} u_x = 1$. It follows $u_x \in [0, 1]$ for all $x \in F$. A convex combination is **genuine** if $u_x > 0$ for all $x \in F$. A subset of a real vector space is called **convex** if it contains all convex combinations of its points. It is sufficient to have $(1 - u)x + uy \in K$ for all $x, y \in K$ and $u \in]0, 1[$. The intersection of an arbitrary collection of convex sets is again convex. The **convex hull** of a subset of $\mathcal{D} \subseteq L$, conv $\mathcal{D}$, is the intersection of all convex sets that contain $\mathcal{D}$. It has also the "interior" representation as the set of all convex combinations of points in $\mathcal{D}$. The convex hull of finitely many points is called a **convex polytope**. The convex hull of an affine independent subset is called a **simplex**.

C.1 Definition Let K be a convex set.

(a) A subset $F \subseteq K$ is called a **face** of K if a genuine convex combination of elements of K can lie in F only if these elements lie themselves in F.

(b) A point $x \in K$ is called an **extreme point** of K if the singleton $\{x\}$ is a face. That is, any genuine convex combination of x by elements of K is trivial.

(c) A function $f \colon K \to \mathbb{R} \cup \{\infty\}$ is called **convex** if
$$f((1 - u)x_1 + ux_2) \leq (1 - u)f(x_1) + uf(x_2)$$
for all $x_1, x_2 \in K$ and all $u \in]0, 1[$. It is strictly convex if "$<$" obtains.

(d) A function $f \colon K \to \mathbb{R} \cup \{-\infty\}$ is called (strictly) **concave** if $-f$ is (strictly) convex.

(e) A function f from K to a real vector space is called **affine** if
$$f((1 - u)x_1 + ux_2) = (1 - u)f(x_1) + uf(x_2)$$
for all $x_1, x_2 \in K$ and all $u \in]0, 1[$. That is, it is affine if it is both convex and concave.

Euclidean spaces are convex and so are all open and all closed balls w.r.t. the p-norms for $1 \leq p < \infty$ and the uniform norm. An *open* convex set has no extreme points. All boundary points of closed balls in Euclidean space w.r.t. the p-norms for $1 < p < \infty$ are extreme. Closed balls for $p = 1, \infty$ are polytopes and have only finitely many extreme points. The vertices, edges, and faces of polytopes are faces in the sense of the above definition. Intervals, triangles, and tetrahedra are simplices. The subsets

$$\Delta_{g-1} = \left\{ (\pi_1, \ldots, \pi_g) \in \mathbb{R}^g \,\middle|\, \pi_j \geq 0, \ \sum_{j=1}^{g} \pi_j = 1 \right\}$$

$$\overset{\circ}{\Delta}_{g-1} = \left\{ (\pi_1, \ldots, \pi_g) \in \mathbb{R}^g \,\middle|\, \pi_j > 0, \ \sum_{j=1}^{g} \pi_j = 1 \right\}$$

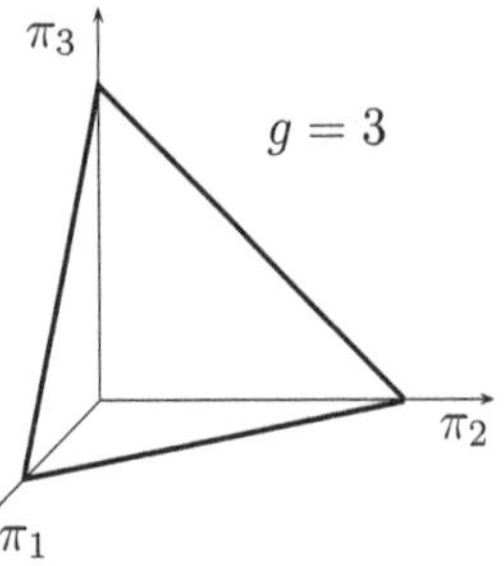

Figure C.1 *Two-dimensional unit simplex.*

are called the $g-1$-dimensional **closed** and the **open unit simplex**, respectively; see Fig. C.1. They play a special role in mixture models as the part of the parameter space that contains the mixing rates.

The minimal generators of linear and affine spaces are the maximally independent sets, also called bases. Linear and affine combinations over bases are unique. The situation is less comfortable with convex combinations. Here, the extreme points play the role of minimal generators, but representations are not always unique. It needs simplices to restore uniqueness.

C.2 Properties *Let K be a convex set and L a real vector space.*

(a) A point $x \in K$ is extreme if and only if $K \backslash \{x\}$ remains convex.

(b) Any face is convex.

(c) Any face of a face in K is a face in K. In particular: Any extreme point of a face in K is extreme in K.

(d) If $a \colon K \to L$ is affine, then the image $a(K)$ is convex.

(e) If $a \colon K \to L$ is affine, then the preimage of any face in $a(K)$ is again a face in K. In particular, the preimage of any extreme point of $a(K)$ is a face in K.

Convex sets that are also compact play a particular role in convexity theory.

C.3 Theorem (Minkowski) *Any point in a compact, convex subset $K \subseteq \mathbb{R}^q$ is a convex combination of extreme points of K.*

The convex hull was characterized above as the set of all (*finite*) convex combinations. The following theorem of Carathéodory's [78] asserts that, in finite-dimensional spaces, convex combinations of a *fixed length* are sufficient.

C.4 Theorem (Carathéodory) *(a) Each point in a compact, convex subset $K \subseteq \mathbb{R}^q$ is a convex combination of at most $q + 1$ extreme points of K.*
(b) Each point in the convex hull of $\mathcal{D} \subseteq \mathbb{R}^q$ can be represented as a convex combination of at most $q + 1$ points of $\mathcal{D}$.

C.5 Lemma *Affine functions $K \to \mathbb{R}^p$ on convex subsets $K \subseteq \mathbb{R}^q$ are extensible to affine functions on all of $\mathbb{R}^q$. Hence, they are continuous.*

PROOF. Let $a \colon K \to \mathbb{R}^p$ be affine. Nothing has to be proved if $K = \mathbb{R}^q$. Otherwise, K possesses a boundary point in $\mathbb{R}^q$. We may and do assume that it is the origin and that $a(0) = 0$. By standard linear algebra, it is sufficient to extend the function a to a linear function on the linear span $\mathbb{L}\langle K \rangle$. To this end, let I and J be finite, nonempty index sets and let $u_i, v_j \in \mathbb{R}$, $x_i, y_j \in K$ for $i \in I$ and $j \in J$ such that

$$\sum_i u_i x_i = \sum_j v_j y_j. \tag{C.1}$$

We first prove that $\sum_i u_i a(x_i) = \sum_j v_j a(y_j)$. This will be done in two steps.

(i) Let first $u_i, v_j > 0$, assume $\sum_i u_i < \sum_j v_j$, and write $v = \sum_j v_j$. The combinations $\left(1 - \sum_i \frac{u_i}{v}\right)0 + \sum_i \frac{u_i}{v} x_i$ and $\sum_j \frac{v_j}{v} y_j$ are equal and both are convex. Since a is affine, it follows $\sum_i \frac{u_i}{v} a(x_i) = \sum_j \frac{v_j}{v} a(y_j)$. This is the claim in this case.

(ii) Now let the coefficients u_i, v_j be arbitrary. Without loss of generality we may and do assume that all are different from zero. Collect the positive and negative coefficients in the subsets I_+, I_-, J_+ and J_-, respectively. From Eq. (C.1), we infer

$$\sum_{i \in I_+} u_i x_i + \sum_{j \in J_-} (-v_j) y_j = \sum_{i \in I_-} (-u_i) x_i + \sum_{j \in J_+} v_j y_j$$

and (i) shows

$$\sum_{i \in I_+} u_i a(x_i) + \sum_{j \in J_-} (-v_j) a(y_j) = \sum_{i \in I_-} (-u_i) a(x_i) + \sum_{j \in J_+} v_j a(y_j).$$

This is the claim in the second case.

We have thus seen that the definition $a\left(\sum_i u_i x_i\right) = \sum_i u_i a(x_i)$ is independent of the special choice of $u_i \in \mathbb{R}$ and $x_i \in K$. This assignment extends the function a linearly to the linear span $\mathbb{L}\langle K \rangle$ of K. $\qquad \square$

Extreme points in affine images originate from extreme points.

C.6 Lemma *If $a \colon K \to \mathbb{R}^q$ is an affine function on a compact, convex subset $K \subseteq \mathbb{R}^p$, then each extreme point of $a(K)$ is the image of some extreme point of K.*

PROOF. Let x be an extreme point of $a(K)$. By Lemma C.5, a is continuous. Hence, the preimage of x w.r.t. a is a (nonempty) closed face in K. By Minkowski's Theorem C.3, it is the convex hull of its extreme points. Hence, there must be at least one. $\qquad \square$

Finally, some special assertions needed in the main text.

C.7 Lemma *Let $A \in \mathbb{R}^{r \times g}$ be a matrix with entries ≥ 0.*

(a) The function $\Psi \colon \Delta_{g-1} \to \mathbb{R} \cup \{-\infty\}$, $\Psi(\mathbf{u}) = \sum_i \log(A\mathbf{u})_i$, is concave.

(b) If no row of A vanishes and if the g columns of A are affine independent, then Ψ is real valued and strictly concave in the interior of Δ_{g-1}.

PROOF. (a) Each of the summands $\mathbf{u} \mapsto \log(A\mathbf{u})_i$ is concave as a function with values in $\mathbb{R} \cup \{-\infty\}$ and so is their sum.

(b) Under the first assumption of (b) all summands $\mathbf{u} \mapsto \log(A\mathbf{u})_i$ are finite in the interior of Δ_{g-1} and so is their sum Ψ. Under the second, the mapping $\mathbf{u} \mapsto A\mathbf{u}$ is one-to-one. Hence, if $\mathbf{u} \neq \mathbf{v}$, then there is an index i such that $(A\mathbf{u})_i \neq (A\mathbf{v})_i$ and, by strict concavity of the logarithm, we have $\log A\left\{\frac{1}{2}\mathbf{u} + \frac{1}{2}\mathbf{v}\right\}_i = \log\left\{\frac{1}{2}(A\mathbf{u})_i + \frac{1}{2}(A\mathbf{v})_i\right\} > \frac{1}{2}\left\{\log(A\mathbf{u})_i + \log(A\mathbf{v})_i\right\}$ and claim (b) follows. $\qquad \square$

C.8 Lemma *Let $E = \{x_0, \ldots, x_d\}$ be a set of $d+1$ points in $\mathbb{R}^d$. If its convex hull contains a ball of radius r about its mean vector, then $W_E \succeq 2\,r^2 I_d$.*

PROOF. Without restriction, let the mean of E be the origin and let W_E be diagonal. By assumption, the centered r-ball $B_r(0)$ satisfies

$$B_r(0) \subseteq \left\{ \sum_{i=0}^d \lambda_i x_i \mid \lambda_i \geq 0, \sum_{i=0}^d \lambda_i = 1 \right\}.$$

Orthogonal projection to the coordinate axes shows

$$[-r, r] \subseteq \left\{ \sum_{i=0}^d \lambda_i x_{i,k} \mid \lambda_i \geq 0, \ \sum_{i=0}^d \lambda_i = 1 \right\}, \quad 1 \leq k \leq d,$$

that is, the interval $[-r, r]$ is contained in the convex hull of the set $\{x_{0,k}, \ldots, x_{d,k}\}$ and so $\min_i x_{i,k} \leq -r$, $\max_i x_{i,k} \geq r$ for all k. This implies $W_E(k, k) = \sum_{i=0}^{d} x_{i,k}^2 \geq 2 r^2$. □

C.2 Derivatives and diffeomorphisms

In connection with ML estimates, functions g from an open subset U of a p-dimensional to a q-dimensional vector space will be differentiated. By definition, the function g is differentiable at $x \in U$ if there exists a linear map $\mathbb{R}^p \to \mathbb{R}^q$, written $\mathrm{D}g(x)$ and called the **derivative** of g at x, such that

$$\lim_{\mathbb{R}^p \ni h \to 0} \frac{g(x + h) - g(x) - \mathrm{D}g(x)h}{\|h\|} = 0.$$

It is then uniquely defined. The point x is called **critical** for g if $\mathrm{D}g(x) = 0$. If g is scalar, then the derivative is called the **gradient**. When the two spaces are Euclidean, $\mathbb{R}^p$ and $\mathbb{R}^q$, then the derivative is represented by the **Jacobian matrix**

$$\mathrm{D}g(x) = \begin{pmatrix} \frac{\partial g_1(x)}{\partial x_1} & \cdots & \frac{\partial g_1(x)}{\partial x_p} \\ \vdots & & \vdots \\ \frac{\partial g_q(x)}{\partial x_1} & \cdots & \frac{\partial g_q(x)}{\partial x_p} \end{pmatrix},$$

where $g_1(x), \ldots, g_q(x)$ are the q components of $g(x)$. If g is differentiable at all points of U, then $\mathrm{D}g$ is a map from U to the linear space $\mathbb{R}^{q \times p}$ of all real $q \times p$ matrices. If U is an open subset of some p-dimensional, real vector space, then identify the space with $\mathbb{R}^p$ via a basis. An example is $U = \mathrm{SYM}(d)$. A symmetric matrix is easily embedded in $\mathbb{R}^{\binom{d+1}{2}}$ by its lower triangle. A convenient basis consists of the matrices E_k with a one at (k, k), $k \in 1 .. d$, and $E_{k,l}$ with ones at (k, ℓ) and (ℓ, k), $1 \leq \ell < k \leq d$, and zero otherwise.

C.9 Examples (a) The derivative of a linear function $g \colon \mathbb{R}^p \to \mathbb{R}^q$ is constant, namely $\mathrm{D}g(x) = g$ for all $x \in \mathbb{R}^p$. This applies in particular to matrices $M \in \mathbb{R}^{q \times p}$. At each point in $\mathbb{R}^p$, the derivative of $x \to Mx$ is M.

(b) The derivative of the matrix inverse $M \to M^{-1}$ at $M \in \mathrm{PD}(d)$ is the direct analogue of the elementary formula $(1/x)' = -1/x^2$, namely the linear map $H \to -M^{-1}HM^{-1}$, $H \in \mathrm{SYM}(d)$. Use the identity $(M+H)^{-1} - M^{-1} + M^{-1}HM^{-1} = (M+H)^{-1}HM^{-1}HM^{-1}$ and some matrix norm on $\mathrm{SYM}(d)$; see Section B.6.2.

(c) In connection with normal ML estimates, the gradient of the scalar, nonlinear matrix function $g \colon M \mapsto \log \det M$ on the convex cone $\mathrm{PD}(d)$ is useful. It is the direct analogue of the elementary formula $(\log x)' = 1/x$, namely, $\mathrm{D}g(M)H = \mathrm{tr}\left(M^{-1}H\right)$, $H \in \mathrm{SYM}(d)$. Indeed, by Leibniz' formula, $\det(I_d + H) = \prod_k (1 + h_{k,k})$ plus a sum of products each of which contains at least two factors of the form $h_{k,\ell}$. Hence, $\det(I_d + H) = 1 + \mathrm{tr}\, H + \mathcal{O}(\|H\|^2)$ as $H \to 0$ and the claim follows from

$$\log \det(M + H) - \log \det M = \log \det(I_d + M^{-1}H)$$
$$= \mathrm{tr}\left(M^{-1}H\right) + \mathcal{O}(\|M^{-1}H\|^2) = \mathrm{tr}\left(M^{-1}H\right) + \mathcal{O}(\|H\|^2),$$

again w.r.t any matrix norm on the symmetric matrices. The gradient of any scalar matrix function can be written as a trace form since this is the general linear, real-valued function on matrix spaces.

(d) The gradient w.r.t. Λ of the logarithm of the normal Lebesgue density at x, $\log N_{m,V}(x)$, is $\frac{1}{2}V - \frac{1}{2}(x - m)(x - m)^\top$: By (c),

$$2\mathrm{D}_\Lambda(\log N_{m,V}(x))H = \mathrm{D}_\Lambda(\log \det \Lambda)H - \mathrm{D}_\Lambda((x - m)^\top \Lambda(x - m))H$$
$$= \mathrm{tr}\, VH - \mathrm{tr}(x - m)(x - m)^\top H = \mathrm{tr}\left(V - \mathrm{tr}(x - m)(x - m)^\top\right)H.$$

Its gradient w.r.t. V is $-\frac{1}{2}\Lambda + \frac{1}{2}\Lambda(x-m)(x-m)^\top\Lambda$. Indeed, from the chain rule and (b), $\mathrm{D}_V\Lambda H = -\Lambda H\Lambda$, it follows

$$2\mathrm{D}_V(\log N_{m,V}(x))H = 2\mathrm{D}_\Lambda(\log N_{m,V}(x))\mathrm{D}_V\Lambda H$$
$$= \mathrm{tr}\, V\mathrm{D}_V\Lambda H - (x-m)^\top(\mathrm{D}_V\Lambda H)(x-m)$$
$$= -\mathrm{tr}\,H\Lambda + (x-m)^\top\Lambda H\Lambda(x-m) = -\mathrm{tr}\,\Lambda H + \mathrm{tr}\,\Lambda(x-m)(x-m)^\top\Lambda H.$$

In order to introduce the next proposition, we begin with two classical statements from differential calculus. The open ball of radius $r > 0$ about a point $\vartheta \in \mathbb{R}^d$ is denoted by $B_r(\vartheta)$. An invertible function $g = (g_1,\ldots,g_q)\colon U \to V$ between two open subsets U and V of $\mathbb{R}^q$ such that both g and g^{-1} are continuously differentiable is called a **diffeomorphism**.

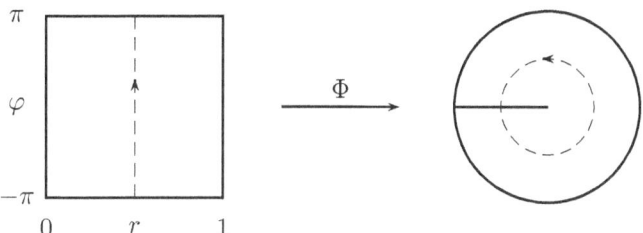

Figure C.2: *Polar coordinates.*

C.10 Example (*Polar coordinates, $d = 2$*) Let $U =]0,1[\times] -\pi,\pi[$ and let $V = \widetilde{B}_1(0)$, the slotted unit disc. The map $\Phi\colon U \longrightarrow V$,

$$\begin{pmatrix} r \\ \varphi \end{pmatrix} \mapsto \begin{pmatrix} r\cos\varphi \\ r\sin\varphi \end{pmatrix}$$

is a diffeomorphism; see also Fig. C.2.

The following two propositions are classical.

(A) Let B be an open ball about $\vartheta_0 \in \mathbb{R}^q$ and let $g_n\colon B \to \mathbb{R}^d$ be a sequence of differentiable functions convergent at ϑ_0 and such that the sequence $\mathrm{D}g_n$ of Jacobian matrices converges uniformly on B. Then the theorem on *convergence of differentiable functions* states that the sequence g_n converges uniformly on B to a differentiable function g and that $\mathrm{D}g = \lim_n \mathrm{D}g_n$; see, for instance, Dieudonné [126], 8.6.3.

(B) Let $U \subseteq \mathbb{R}^q$ be open and let $g\colon U \to \mathbb{R}^q$ be continuously differentiable. Assume that the Jacobian matrix of g at $\vartheta_0 \in U$ is regular. Then the *inverse function theorem* asserts that the restriction of g to some open ball about ϑ_0 is a diffeomorphism onto an open subset of $\mathbb{R}^q$; see, for instance, Dieudonné [126], 10.2.5.

The consistency proof in Section 1.2.4 relies on a combination of the two theorems. The following proposition may be viewed as a version of the inverse function theorem for a whole sequence of differentiable functions. It is not formulated in its full generality here but it is sufficient for our purposes.

C.11 Proposition *Let A be an open ball about $\vartheta_0 \in \mathbb{R}^q$ and let $g_n\colon A \to \mathbb{R}^q$, $n \in \mathbb{N}$, be continuously differentiable such that*

(i) the sequence (g_n) converges at ϑ_0,

(ii) the sequence $(\mathrm{D}g_n)$ of the Jacobian matrices of g_n converges uniformly on A, and

(iii) $\lim_n \mathrm{D}g_n(\vartheta_0)$ is positive or negative definite.

Then,

(a) the sequence (g_n) converges uniformly on A to a continuously differentiable function g and

(b) $\mathrm{D}g(\vartheta) = \lim \mathrm{D}g_n(\vartheta)$ uniformly for all $\vartheta \in A$.

Moreover, there exists an open neighborhood $U \subseteq A$ of ϑ_0 such that

(c) the restrictions of eventually all functions g_n and of g to U are diffeomorphisms;

(d) there is an open ball $B \subseteq \mathbb{R}^q$ with center $g(\vartheta_0)$ contained in the images $g_n(U)$ for eventually all n;

(e) the sequences g_n^{-1} and $\mathrm{D}g_n^{-1}$ converge to g^{-1} and $\mathrm{D}g^{-1}$, respectively, uniformly on B;

(f) $\mathrm{D}g_n(g_n^{-1}(g(\vartheta_0)))$ is positive (negative) definite for eventually all n.

PROOF. Parts (a) and (b) state again the well-known fact about convergence of differentiable functions; see (A) above. Parts (c)–(e) concern the sequence of inverse functions.

In view of (c), apply the inverse function theorem to g in order to find an open ball U about ϑ_0 such that the restriction of g to U is a diffeomorphism onto some open subset of $\mathbb{R}^q$ and such that $\mathrm{D}g(\vartheta) \succeq 2\alpha I_q$ for some $\alpha > 0$ and all $\vartheta \in U$. By uniform convergence of the Jacobians $\mathrm{D}g_n$, we have $\mathrm{D}g_n \succeq \alpha I_q$ on U for eventually all n. Now, for such n, let $\vartheta_1, \vartheta_2 \in U$ such that $g_n(\vartheta_1) = g_n(\vartheta_2)$. By the mean value theorem, there is a convex combination θ of ϑ_1 and ϑ_2 such that $\mathrm{D}g_n(\theta)(\vartheta_2 - \vartheta_1) = g_n(\vartheta_2) - g_n(\vartheta_1) = 0$; hence $\vartheta_1 = \vartheta_2$, that is, g_n is one-to-one on U. Moreover, the inverse function theorem says that g_n is a diffeomorphism, so the image $g_n(U)$ is open in $\mathbb{R}^q$.

We next show that the image $g_n(U)$ contains a ball about $g_n(\vartheta_0)$ of size independent of n. Let $\beta > 0$ and let $\vartheta \in U$, $\vartheta \neq \vartheta_0$, such that $\|g_n(\vartheta) - g_n(\vartheta_0)\| < \beta$. By the mean value theorem, $g_n(\vartheta) - g_n(\vartheta_0) = \mathrm{D}g_n(\theta)(\vartheta - \vartheta_0)$ for some θ between ϑ and ϑ_0 and hence

$$\alpha\|\vartheta - \vartheta_0\|^2 \leq (\vartheta - \vartheta_0)^\top \mathrm{D}g_n(\theta)(\vartheta - \vartheta_0) = (\vartheta - \vartheta_0)^\top (g_n(\vartheta) - g_n(\vartheta_0))$$
$$\leq \|\vartheta - \vartheta_0\| \|g_n(\vartheta) - g_n(\vartheta_0)\| < \beta\|\vartheta - \vartheta_0\|.$$

It follows $\|\vartheta - \vartheta_0\| < \beta/\alpha$. That is, if β is chosen so small that the open ball about ϑ_0 of radius β/α is contained in U, then its g_n-image contains the open ball of radius β about $g_n(\vartheta_0)$. Now, use the convergence $g_n(\vartheta_0) \to g(\vartheta_0)$ to find a ball B as required in part (d).

Note that the inverses g_n^{-1} are uniformly equicontinuous on B, so the identity $g_n^{-1}(\eta) - g^{-1}(\eta) = g_n^{-1}(\eta) - g_n^{-1}(g_n(g^{-1}(\eta)))$ shows that g_n^{-1} converges to g^{-1} uniformly on B. The rest of part (e) follows from continuity of matrix inversion and the representation $\mathrm{D}g_n^{-1}(\eta) = \mathrm{D}g_n(g_n^{-1}(\eta))^{-1}$ of the derivative of the inverse. Finally, $\mathrm{D}g_n(g_n^{-1}(g(\vartheta_0)))$ converges to $\mathrm{D}g(\vartheta_0)$ by (e) and (b), so part (f) follows from assumption (iii). □

C.3 Notes

Theorem C.3 is due to Minkowski [381]. Part (a) of Carathéodory's [78] theorem is found in Phelps [413] and part (b) in Bonnice [57] and Bonnice and Klee [58]. Convex analysis has been extended to compact metric and even more general convex sets. Krein and Milman [299] and Choquet [92] generalized Minkowski's theorem to (metrizable) compact, convex sets in certain topological, linear spaces.

Appendix D

Measures and probabilities

Probabilistic mixture and cluster analysis hinge heavily on measure theory and integration. This appendix reviews briefly some basic facts about measurable spaces, measures and probabilities on σ-algebras, and Lebesgue integration. Proofs of the propositions are found, for instance, in Bauer's [32] clear and rigorous monograph.

D.1 Measures and integrals

In probability and statistics we often encounter denumerable operations. Examples are asymptotic laws such as Laws of Large Numbers, the Central Limit Theorem, consistency theorems, and asymptotic normality. Their treatment needs countably additive probabilities which, except in trivial situations, cannot be defined on *all* subsets of the sample space. This is the reason why σ-algebras of subsets were introduced. An in depth analysis of this phenomenon is due to Hausdorff [234].

D.1.1 σ-Algebras and measures

A σ-**algebra** $\mathcal{A}$ on a set E is a system of subsets that contains E as an element and is closed w.r.t. complements and countable unions. The elements of $\mathcal{A}$ are called **measurable sets**, the pair $(E, \mathcal{A})$ a **measurable space**. Besides E, the empty set $\emptyset = \complement E$ is always measurable. Moreover, $\mathcal{A}$ is closed w.r.t. all countable set operations. Examples are differences $B \backslash A = B \cap \complement A$ and countable intersections $\bigcap_n A_n = \complement \bigcup_n \complement A_n$. The σ-algebra on a structure E is constructed so that interesting subsets become measurable, for instance, intervals in the real line. Since the intersection of any family of σ-algebras is again a σ-algebra, there is a smallest σ-algebra that contains a given collection $\mathcal{E}$ of subsets of E, namely, the intersection of all such σ-algebras. It is called the σ-algebra $\sigma(\mathcal{E})$ generated by $\mathcal{E}$; $\mathcal{E}$ is its **generator**.

A **measure** on a σ-algebra $\mathcal{A}$ (or "on E") is a function $\mu \colon \mathcal{A} \to [0, \infty]$ such that (i) $\mu(\emptyset) = 0$, (ii) $\mu(\bigcup_{n \geq 1} A_n) = \sum_{n \geq 1} \mu(A_n)$ for any sequence of pairwise disjoint measurable subsets of E (σ-additivity of μ). It follows that μ is an increasing set function, $\mu(B \backslash A) = \mu(B) - \mu(A)$ if $A \subseteq B$ and $\mu(A) < \infty$, $\mu(\bigcup_n A_n) = \sup_n \mu(A_n) = \lim_n \mu(A_n)$ for any increasing sequence of measurable sets A_n, and $\mu(\bigcap_n A_n) = \inf_n \mu(A_n) = \lim_n \mu(A_n)$ for any decreasing sequence if $\mu(A_n) < \infty$ for some n. The last two properties are called σ-**continuity** of μ. The triple $(E, \mathcal{A}, \mu)$ is called a **measure space**.

A measure is σ-**finite** if there is an (increasing) sequence of measurable sets of *finite* measure with union E. It is **finite** (a **probability**) if $\mu(E) < \infty$ ($\mu(E) = 1$). Probabilities are sometimes called **populations**. A measure space $(E, \mathcal{A}, \mu)$ endowed with a probability μ is called a **probability space**. Any convex combination of probabilities is again a probability. For arbitrary $a \in E$ and $\mathcal{A}$ we write $\delta_a(A) = \mathbf{1}_A(a)$ and call δ_a the **unit point mass** (or **Dirac measure**) concentrated at a. It is plainly a probability and assigns probability 1 to any "event" A that contains a, zero otherwise. A convex combination of the form

$\sum_{i=1}^n \alpha_i \delta_{a_i}$, $\alpha_i \geq 0$, $\sum \alpha_i = 1$, is called a **discrete probability**. It represents a random experiment for which a_i occurs with probability α_i.

Measurable sets of μ-measure zero are called μ-**null**. They are negligible. We say that a property holds μ-**almost everywhere**, μ-a.e., (or μ-**almost surely**, μ-a.s., if μ is a probability) or that the property holds for μ-**almost all**, μ-a.a., elements of E if it holds except for those in some μ-null set.

We have sometimes to show that two measures are actually equal. Then the uniqueness theorem for σ-finite measures applies.

D.1 Theorem (Uniqueness Theorem) *Let the generator $\mathcal{E}$ of the σ-algebra $\mathcal{A}$ on E be closed w.r.t. (finite) intersections. Let μ and ν be measures on $(E, \mathcal{A})$ and assume that there exists a sequence $(E_n)_n$ of sets in $\mathcal{E}$ such that*

(i) $\bigcup_n E_n = E$,

(ii) $\mu(E_n) = \nu(E_n) < \infty$ for all n.

Then $\mu = \nu$ on $\mathcal{A}$.

The σ-algebra $\sigma(\mathcal{O})$ generated by the open sets of a Hausdorff topological space $(E, \mathcal{O})$ is called the **Borel σ-algebra** on E or the σ-algebra of Borel sets in E, $\mathcal{B}(E)$.[1] This applies in particular to the real line and its Euclidean topology which generates the important σ-algebra of Borel sets in $\mathbb{R}$, $\mathcal{B}(\mathbb{R})$. It makes all intervals and all countable unions of intervals measurable. A measure μ on the Borel σ-algebra of a Hausdorff space is called a **Borel measure**, it is called **locally finite** if each point has an open neighborhood of finite measure, **inner regular** if, for any Borel set $A \subseteq E$,

$$\mu(A) = \sup\{\mu(K) \,|\, K \text{ compact}, K \subseteq A\},$$

outer regular if, for any such set A,

$$\mu(A) = \inf\{\mu(U) \,|\, U \text{ open}, A \subseteq U\},$$

and **regular** if it is both inner and outer regular.

In measure theory one shows that there is exactly one measure on the Borel sets of the real line which assigns to each interval its length. It is a σ-finite, locally finite, regular Borel measure called the **Lebesgue measure** λ on $\mathbb{R}$. It is invariant w.r.t. translations and its restriction to any interval I of unit length, in fact to any Borel set of Lebesgue measure 1, is a probability. It is a model for random sampling uniformly on I.

Of particular interest in probability are Borel measures on metric and Polish spaces; see Bauer [32], 26.2, 26.3.

D.2 Theorem *(a) Any inner regular Borel measure μ on a Hausdorff space E has a* **topological support**, *that is, a smallest closed subset $A \subseteq E$ such that $\mu(\complement A) = 0$.*
(b) Any locally finite Borel measure on a Polish space is regular.[2]
(c) A Borel measure μ on a Polish space is locally finite if and only if it is σ-finite and compact sets have finite measure.[3]

A map $\Phi \colon E \to F$ between two measurable spaces $(E, \mathcal{A})$ and $(F, \mathcal{B})$ is called **measurable** if the inverse images of all measurable subsets of F are measurable in E. The concept also applies to real-valued functions $f \colon E \to \mathbb{R}$ and the Borel σ-algebra on $\mathbb{R}$. For $f \colon E \to \mathbb{R}$ to be measurable, it is sufficient that the inverse images of all infinite intervals of the form $[r, \infty[$ (or $]r, \infty[$ or $]-\infty, r]$ or $]-\infty, r[$), $r \in \mathbb{R}$ or only $r \in \mathbb{Q}$), are measurable. In other

---

[1] This is actually an abuse of language since it depends mainly on $\mathcal{O}$ and not only on E, but when E is equipped with a fixed topology, no misunderstanding can arise.

[2] σ-Finiteness instead of local finiteness is not sufficient: Consider the measure on the real line which assigns mass 1 to each rational point.

[3] Written communication by S. Graf.

words, we have to check that all sets of the form $[f \geq r]$ (or $[f > r]$ or $[f \leq r]$ or $[f < r]$) are measurable for all real or rational numbers r. Here and in the below, square brackets $[f < r]$ will indicate a shorthand notation for $\{x \in E \mid f(x) < r\}$. Any continuous function on the real line or on an interval is plainly Borel measurable.

The μ-integral of an **elementary function** $f = \sum_{k=1}^{n} c_k \mathbf{1}_{A_k}$, $\{A_1, \ldots, A_n\}$ a measurable partition of E, $c_k \geq 0$, is defined as $\int_E f \, d\mu = \sum_{k=1}^{n} c_k \mu(A_k)$ ($0 \cdot \infty$ is defined as zero). Most elementary functions are represented by more than one partition. However, it can be shown that the number on the right side does not depend on the special choice. The μ-integral is infinite when $\mu(A_k) = \infty$ for some k with $c_k > 0$. Any positive, measurable function $f \colon E \to [0, \infty]$ is the limit of an increasing sequence of elementary functions f_n. The **Lebesgue integral** of f w.r.t. μ, written $\int_E f \, d\mu = \int_E f(x) \mu(dx)$, is the limit of their μ-integrals. Again, it can be shown that the limit is independent of the sequence chosen. It may be infinite. Finally, the Lebesgue integral of a signed measurable function is the difference of the integrals of the positive and the negative parts, if one integral is finite. If both are, that is, if $\int |f| \, d\mu < \infty$, then the function is called μ-**integrable**. The set of all these functions is a subspace of the vector space of all real-valued functions on E, denoted by $\mathcal{L}^1(\mu)$. The function f is p-**fold integrable**, $1 < p < \infty$, if $|f|^p$ is integrable, $f \in \mathcal{L}^p(\mu)$. The expression $\|f\|_p = \sqrt[p]{\int |f|^p \, d\mu}$ has all properties of a norm on $\mathcal{L}^p(\mu)$ except that there are functions that integrate to zero without being zero. Dividing out this null space makes $\mathcal{L}^p(\mu)$ a Banach space, denoted by $\mathbb{L}^p(\mu)$. Integrals restricted to measurable *subsets* are defined in the obvious way, $\int_A f \, d\mu = \int_E \mathbf{1}_A \cdot f \, d\mu$. The above program works on a collection of sets and functions sufficiently large for our purposes.

The above process can be applied to Borel-measurable, real-valued functions f defined on the real line and Lebesgue measure λ. We obtain the Lebesgue integral $\int_{\mathbb{R}} f \, d\lambda$. Integrals are rarely computed along the lines above. It is nice that the integral of a *continuous* function f on a bounded interval $[a, b] \subseteq \mathbb{R}$ turns out to be just the usual **Cauchy integral** $\int_a^b f(x) dx$ known from integral calculus: "primitive of f at b minus primitive at a." Other functions are integrated by approximating the integrand by functions with known integrals and applying the following important **Dominated Convergence Theorem**.

D.3 Theorem (Lebesgue) *Let $f \colon E \to \mathbb{R}$ be measurable and let $f_n \colon E \to \mathbb{R}$, $n \geq 1$, be functions such that*

  (i) *$\lim_{n \to \infty} f_n(x) = f(x)$ for every $x \in E$ except, possibly, for x in a subset of μ-measure zero;*

  (ii) *there is a (positive) function $g \in \mathcal{L}^1(\mu)$ such that $-g \leq f_n \leq g$ for all n.*

Then the numbers $\int f_n \, d\mu$ converge and we have $\int f \, d\mu = \lim_{n \to \infty} \int f_n \, d\mu$. In other words, under the assumption (ii), limit and integral commute.

From the Uniqueness Theorem D.1 it follows that a σ-finite Borel measure on a metric space E is determined by its values for all closed sets. It is also determined by the integrals of all positive, bounded, continuous, real functions; see Billingsley [41].

The operations of denumerable summation, going to the limit, and differentiation do not always commute with integration. The next lemma concerns a function of two variables which we wish to differentiate w.r.t. the first and integrate w.r.t. the second variable. It gives sufficient conditions so that the two operations may be swapped.

D.4 Lemma (Swapping differentiation and integration) *Let U be an open subset of a Euclidean space, let μ be a Borel measure on a metric space E, and let $g \colon (U, E) \to \mathbb{R}^q$ be such that*

  (i) *$g(\cdot, x)$ is differentiable for each $x \in E$;*

  (ii) *$g(\vartheta, \cdot)$ is integrable for each $\vartheta \in U$;*

(iii) there exists a μ-integrable function h such that $\|\mathrm{D}g(\vartheta, \cdot)\| \leq h$ for all $\vartheta \in U$.

Then, $\int_E g(\vartheta, x)\mu(\mathrm{d}x)$ is differentiable at each $\vartheta \in U$ and we have

$$\mathrm{D}_\vartheta \int_E g(\vartheta, x)\mu(\mathrm{d}x) = \int_E \mathrm{D}_\vartheta g(\vartheta, x)\mu(\mathrm{d}x).$$

Any measure μ on $(E, \mathcal{A})$ along with any measurable map $\Phi \colon (E, \mathcal{A}) \to (F, \mathcal{B})$ induces a measure μ_Φ on $(F, \mathcal{B})$ via

$$\mu_\Phi(B) = \mu([\Phi \in B]). \tag{D.1}$$

The induced measure is known as the **image** or the **push-forward measure** of μ w.r.t. Φ. Integration w.r.t. an image measure is performed by the change-of-variables formula:

$$\int_F g \, \mathrm{d}\mu_\Phi = \int_E g \circ \Phi \, \mathrm{d}\mu.$$

The image measure of an image measure is the image measure w.r.t. the composition, $(\mu_\Phi)_\Psi = \mu_{\Psi \circ \Phi}$.

We say that the expectation of a probability measure μ on a Euclidean space $\mathbb{R}^d$ **exists** if $\int_{\mathbb{R}^d} \|x\|\mu(\mathrm{d}x) < \infty$. The center of the probability is then called its **expectation**,

$$\mathrm{E}\mu = \int_{\mathbb{R}^d} x\mu(\mathrm{d}x) = \begin{pmatrix} \int_{\mathbb{R}^d} x_1\mu(\mathrm{d}x) \\ \vdots \\ \int_{\mathbb{R}^d} x_d\mu(\mathrm{d}x) \end{pmatrix}. \tag{D.2}$$

In a similar way, we say that the covariance matrix of μ exists if $\int_{\mathbb{R}^d} \|x\|^2\mu(\mathrm{d}x) < \infty$. The **covariance matrix** is then defined by $\mathrm{V}\mu = \int_{\mathbb{R}^d} (x - \mathrm{E}\mu)(x - \mathrm{E}\mu)^\top \mu(\mathrm{d}x)$. It is a measure of the scatter of μ about its center.

D.1.2 Product measures and the Lebesgue measure on $\mathbb{R}^d$

Let $(E_1, \mathcal{A}_1)$ and $(E_2, \mathcal{A}_2)$ be two measurable spaces. The two σ-algebras induce in a natural way a σ-algebra on the Cartesian product $E_1 \times E_2$, namely, the σ-algebra generated by all Cartesian products $A_1 \times A_2$, $A_i \in \mathcal{A}_i$. It is called the **product σ-algebra** $\mathcal{A}_1 \otimes \mathcal{A}_2$ of A_1 and A_2. If the two measurable spaces are endowed with σ-finite measures μ_1 and μ_2, then one shows in measure theory that there is a measure μ on the product such that $\mu(A_1 \times A_2) = \mu_1(A_1) \cdot \mu_2(A_2)$ for all $A_i \in \mathcal{A}_i$. By the Uniqueness Theorem D.1, it is even unique. It is again σ-finite and called the product measure, $\mu_1 \otimes \mu_2$, of μ_1 and μ_2. Both procedures can be extended to n factors $\mathcal{A}_i$ and μ_i and the operations are associative, $\mathcal{A}_1 \otimes (\mathcal{A}_2 \otimes \mathcal{A}_3) = (\mathcal{A}_1 \otimes \mathcal{A}_2) \otimes \mathcal{A}_3$ and $\mu_1 \otimes (\mu_2 \otimes \mu_3) = (\mu_1 \otimes \mu_2) \otimes \mu_3$. Therefore, parentheses may be omitted.

The following theorem reduces integration on a product to successive integration on the factors.

D.5 Theorem (Fubini) *Let μ and ν be σ-finite measures on E and F, respectively, and let $f \colon E \times F \to \mathbb{R}$ be measurable.*

(a) The x-cut $f(x, \cdot)$ of f is measurable for all $x \in E$.

(b) If $f(x, \cdot) \in \mathcal{L}^1(\nu)$ for all $x \in F$, then the function $E \to \mathbb{R}$,

$$x \mapsto \int_F f(x, y) \, \nu(\mathrm{d}y),$$

is measurable. (Assertions analogous to (a) and (b) are also valid after swapping x and y.)

(c) The three integrals

$$\int_E \mu(\mathrm{d}x) \int_F \nu(\mathrm{d}y) \, |f(x,y)|, \quad \int_F \nu(\mathrm{d}y) \int_E \mu(\mathrm{d}x) \, |f(x,y)|, \quad \int_{E \times F} |f| \, \mathrm{d}(\mu \otimes \nu)$$

$(\in [0, \infty])$ are equal.

(d) If the three integrals in (c) are finite, then we have $f \in \mathcal{L}^1(\mu \otimes \nu)$ and

$$\int_E \mu(\mathrm{d}x) \int_F \nu(\mathrm{d}y) \, f(x,y) = \int_F \nu(\mathrm{d}y) \int_E \mu(\mathrm{d}x) \, f(x,y) = \int_{E \times F} f \, \mathrm{d}(\mu \otimes \nu).$$

The d-fold product of the Borel σ-algebra $\mathcal{B}(\mathbb{R})$ is also generated by the natural topology on $\mathbb{R}^d$, that is, it is the Borel σ-algebra on $\mathbb{R}^d$. It makes all geometrical sets, such as open and closed hyperboxes, pyramids, regular polyhedra, p-balls for all p, and ellipsoids measurable. It is the domain of definition of the d-fold product of Lebesgue's measure on $\mathbb{R}$, called the d-dimensional Lebesgue measure λ^d. It turns out that it is invariant w.r.t. translations and rotations on $\mathbb{R}^d$. It can also be defined as the only translation-invariant measure on $\mathcal{B}(\mathbb{R}^d)$ that assigns measure 1 to the d-dimensional unit cube and the only measure on $\mathcal{B}(\mathbb{R}^d)$ that assigns the measure $\prod_{i=1}^d (b_i - a_i)$ to all d-dimensional hyperbox $[a_1, b_1[\times \cdots \times [a_n, b_n[$. Fubini's theorem reduces a Lebesgue integral over $\mathbb{R}^d$ to the evaluation of d integrals over the real line.

D.6 Examples (a) To illustrate the twofold Lebesgue integral let us compute the area of the unit disc $K = \{(x,y) \in \mathbb{R}^2 \mid x^2 + y^2 \leq 1\}$. By definition of the integral of an elementary function and by Fubini, we have

$$\lambda^2(K) = \int_{\mathbb{R}^2} \mathbf{1}_K \, \mathrm{d}\lambda^2 = \int_{\mathbb{R}} \lambda(\mathrm{d}x) \int_{\mathbb{R}} \mathbf{1}_K(x,y) \lambda(\mathrm{d}y) = \int_{-1}^1 \mathrm{d}x \int_{-\sqrt{1-x^2}}^{\sqrt{1-x^2}} \mathbf{1} \mathrm{d}y$$

$$= 2 \int_{-1}^1 \sqrt{1-x^2} \, \mathrm{d}x = \left[x\sqrt{1-x^2} + \arcsin x \right]_{-1}^1 = \pi.$$

(b) The following multivariate **integration by parts** will be needed in Section 5.2.2. Let $f, g \colon \mathbb{R}^d \to \mathbb{R}$ be continuously differentiable functions such that $\mathrm{D}f \cdot g$ and $f \cdot \mathrm{D}g$ both are integrable w.r.t. λ^d and that $(f \cdot g)(x) \to 0$ as $x \to \infty$. Then

$$\int \mathrm{D}f \cdot g \, \mathrm{d}\lambda^d = - \int f \cdot \mathrm{D}g \, \mathrm{d}\lambda^d. \tag{D.3}$$

Indeed, $\int [\mathrm{D}f \cdot g + f \cdot \mathrm{D}g] \, \mathrm{d}\lambda^d = \int \mathrm{D}(f \cdot g) \, \mathrm{d}\lambda^d$. Since $\mathrm{D}(f \cdot g)$ is integrable, we may apply Fubini's theorem D.5 to obtain

$$\int \frac{\partial}{\partial x_k}(f \cdot g) \, \mathrm{d}\lambda^d = \int_{\mathbb{R}^{d-1}} \lambda^{d-1}(\mathrm{d}x_{\setminus k}) \int_{\mathbb{R}} \frac{\partial}{\partial x_k}(f \cdot g)(x) \lambda(\mathrm{d}x_k).$$

By assumption, the inner integral vanishes for all $x_{\setminus k} \in \mathbb{R}^{d-1}$.

D.2   Density functions

Probabilities are in most practical cases represented by density functions w.r.t. certain standard measures. On Euclidean spaces, the Lebesgue measures are the preferred reference measures.

D.7 Lemma Let ϱ be a measure on $(E, \mathcal{A})$ and let $f \colon E \to \mathbb{R}$ be positive and measurable.

(a) The function $\mu \colon \mathcal{A} \to \mathbb{R}$, $A \mapsto \int \mathbf{1}_A f \, \mathrm{d}\varrho$, defines a measure μ on $(E, \mathcal{A})$.

(b) The measure μ is finite (a probability) if and only if f is ϱ-integrable (and $\int f \, \mathrm{d}\varrho = 1$).

(c) Given two positive and measurable functions $f, g \colon E \to \mathbb{R}$, their associated measures in the sense of (a) are equal if and only if $f = g$, ϱ-a.e.

According to (c), it is sufficient to define a density outside of a null set.

D.8 Definition The function f in Lemma D.7 is called the ϱ-**density function** (or just ϱ-density) of μ. The measure μ is called the measure with ϱ-density f. A density function w.r.t. Lebesgue measure is called a **Lebesgue density**.

There are various intuitive notations to express the fact that f is the ϱ-density of μ: $\mu = f\varrho$, $\mu(\mathrm{d}x) = f(x)\varrho(\mathrm{d}x)$, $f = \frac{\mathrm{d}\mu}{\mathrm{d}\varrho}$, $f(x) = \frac{\mu(\mathrm{d}x)}{\varrho(\mathrm{d}x)}$. There is the following **cancellation rule** for densities:

$$\frac{\mu(\mathrm{d}x)}{\varrho(\mathrm{d}x)} = \frac{\mu(\mathrm{d}x)}{\sigma(\mathrm{d}x)}\frac{\sigma(\mathrm{d}x)}{\varrho(\mathrm{d}x)}. \tag{D.4}$$

On the other hand, the formula $\nu = f\mu$ must not be divided by f: Let μ be uniform on $\{0, 1\}$. Although we have $\delta_0 = 2 \cdot \mathbf{1}_{\{0\}}\mu$, $\delta_0/\mathbf{1}_{\{0\}}$ is different from 2μ!

It is of course not true that any measure has a density function w.r.t. any other. For instance, there is no such thing as a Lebesgue density of a Dirac measure δ_a, $a \in \mathbb{R}$. Otherwise, $\delta_a(\{a\})$ would be zero, which it isn't. If μ has a density w.r.t. ϱ, then μ is called **absolutely continuous** w.r.t. ϱ, $\mu \ll \varrho$. The classical theorem of Radon and Nikodym characterizes absolute continuity; see, for instance, Bauer [32]. It does not play a role here since we use densities to define probabilities. If $\mu \ll \varrho$ and $\varrho \ll \mu$, then μ and ϱ are called **equivalent**.

D.9 Examples

(a) The triangle function

$$f(x) := \begin{cases} 1 - |x|, & |x| \le 1, \\ 0, & |x| > 1 \end{cases}$$

is the Lebesgue density of a probability measure on $\mathcal{B}(\mathbb{R})$.

(b) The multivariate normal distribution $N_{m,V}$ has the Lebesgue density

$$\frac{1}{\sqrt{\det 2\pi V}}e^{-\frac{1}{2}(x-m)^\top V^{-1}(x-m)}, \quad x \in \mathbb{R}^d.$$

An integral w.r.t. a measure with a density is computed according to the following rule.

D.10 Proposition *Let ϱ be a measure on $\mathcal{A}$, let $\mu = f\varrho$, and let $h\colon E \to \mathbb{R}$ be $\mathcal{A}$-measurable.*

(a) If $h \ge 0$, then $\int h\,\mathrm{d}\mu = \int hf\,\mathrm{d}\varrho$.

(b) The function h is μ-integrable if and only if hf is ϱ-integrable.

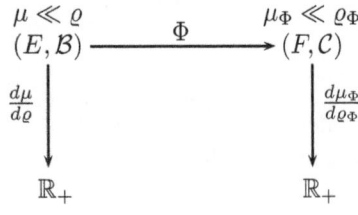

Figure D.1 *Arrow diagram of Lemma D.11.*

The converse of part (a) is also true: If, for some probability μ and some positive function $f\colon E \to \mathbb{R}$, the equality $\int h\,\mathrm{d}\mu = \int hf\,\mathrm{d}\varrho$ obtains for all bounded measurable functions $h\colon E \to \mathbb{R}$, then f is the ϱ-density of μ. It is sufficient to consider the indicator function $h = \mathbf{1}_A$ for all measurable sets A.

Densities cooperate well with image and product measures. The arrow diagram of the following lemma is shown in Fig. D.1.

D.11 Lemma *Let μ and ϱ be measures on $(E, \mathcal{B})$ and $(F, \mathcal{C})$, respectively, and let $\Phi\colon E \to F$ be one-to-one, onto, and measurable in both directions. If μ has the ϱ-density f, then μ_Φ has the ϱ_Φ-density $f \circ \Phi^{-1}$.*

PROOF. The claim follows from Definition D.8 of a density function and that of an image measure, Eq. (D.1):

$$\int \mathbf{1}_C \big(f \circ \overset{-1}{\Phi} \big)\, \mathrm{d}\varrho_\Phi = \int \mathbf{1}_{[\Phi \in C]} \frac{\mathrm{d}\mu}{\mathrm{d}\varrho}\, \mathrm{d}\varrho = \mu[\Phi \in C] = \mu_\Phi(C). \qquad \square$$

D.12 Proposition *Let $(E_1, \mathcal{B}_1), \ldots, (E_n, \mathcal{B}_n)$ be measurable spaces and let ϱ_i be σ-finite measures on $(E_i, \mathcal{B}_i)$. Let μ_i have the density f_i w.r.t. ϱ_i, $1 \le i \le n$. Then the product measure $\mu_1 \otimes \cdots \otimes \mu_n$ has the $\varrho_1 \otimes \cdots \otimes \varrho_n$-density $(x_1, \ldots, x_n) \mapsto \prod_{i=1}^n f_i(x_i)$, that is,*

$$\frac{\mu_1 \otimes \cdots \otimes \mu_n(\, \mathrm{d}x_1, \ldots, \mathrm{d}x_n)}{\varrho_1 \otimes \cdots \otimes \varrho_1(\, \mathrm{d}x_1, \ldots, \mathrm{d}x_n)} = \prod_{i=1}^n f_i(x_i).$$

D.3 Transformation of the Lebesgue measure

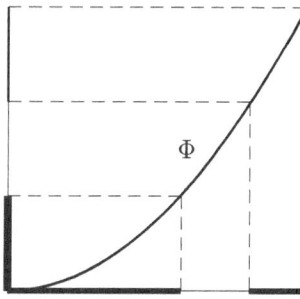

We study here the effect a diffeomorphism has on the Lebesgue measure. Let us first take a look at the univariate case and let $\Phi \colon \mathbb{R} \to \mathbb{R}$ be a diffeomorphism; see Fig. D.2. It is strictly increasing or decreasing. The image measure λ_Φ is characterized by the equality

$$\lambda_\Phi([0, y]) = \lambda[0 \le \Phi \le y] = \begin{cases} \overset{-1}{\Phi}(y) - \overset{-1}{\Phi}(0), & \Phi \text{ increasing} \\ \overset{-1}{\Phi}(0) - \overset{-1}{\Phi}(y), & \text{otherwise,} \end{cases}$$

Figure D.2 *Image measure of λ w.r.t. Φ.*

if $y \ge 0$, and a similar equality for $y < 0$. The Lebesgue density of λ_Φ is the derivative $1/\big(|\Phi'| \circ \overset{-1}{\Phi}\big)$. This formula carries over to the multivariate case. The derivation in the general case is somewhat involved but it is easy and informative to see what comes out for an *affine* map $\Phi(x) = Ax + c$, $A = (a_1, \ldots, a_d) \in \mathrm{GL}(d)$, $c \in \mathbb{R}^d$. The proof needs two steps. We first convince ourselves by induction on d that the volume of the parallelotope spanned by $a_1, \ldots, a_d \in \mathbb{R}^d$,

$$\mathcal{P}(a_1, \ldots, a_d) = \bigg\{ \sum_{i=1}^d \lambda_i a_i \,\big|\, 0 \le \lambda_i \le 1 \text{ all } 1 \le i \le d \bigg\},$$

is $|\det(a_1 \ldots a_d)|$: Indeed, the claim being true for $d = 1$, let us consider $\mathcal{P}(a_1, \ldots, a_d)$ for $d \ge 2$. A rotation U in $\mathbb{R}^d$ makes the last entries of the $d - 1$ vectors $Ua_1, \ldots, Ua_{d-1}$ disappear. Thereby, the last coordinate of Ua_d becomes the height of the parallelotope and the induction hypothesis implies

$$\begin{aligned} \mathrm{vol}_d(\mathcal{P}(a_1, \ldots, a_d)) &= \mathrm{vol}_{d-1}(\text{basis}) \cdot \text{height} \\ &= |\det(Ua_1 \ldots Ua_{d-1})||\text{last coordinate of } Ua_d| \\ &= |\det(Ua_1 \ldots Ua_d)| = |\det(UA)| = |\det U \det A| = |\det A|. \end{aligned} \qquad (\mathrm{D.5})$$

Now, the box $Q = \prod_{i=1}^d [\alpha_i, \beta_i] \subseteq \mathbb{R}^d$ $(\alpha_i < \beta_i)$ is the parallelotope

$$\alpha + \mathcal{P}((\beta_1 - \alpha_1)e_1, \ldots, (\beta_d - \alpha_d)e_d).$$

Hence, $\overset{-1}{\Phi}(Q)$ is a translate of the parallelotope $\mathcal{P}(A^{-1}(\beta_1 - \alpha_1)e_1, \ldots, A^{-1}(\beta_d - \alpha_d)e_d)$ and we infer from (D.5)

$$\begin{aligned} (\lambda^d)_\Phi(Q) &= \lambda^d[\Phi \in Q] = \lambda^d[A^{-1}(\beta_1 - \alpha_1)e_1, \ldots, A^{-1}(\beta_d - \alpha_d)e_d] \\ &= |\det(A^{-1}(\beta_1 - \alpha_1)e_1 \ldots A^{-1}(\beta_d - \alpha_d)e_d)| \\ &= \prod_{i=1}^d (\beta_i - \alpha_i)|\det(A^{-1}e_1 \ldots A^{-1}e_d)| = \lambda^d(Q) \cdot |\det A^{-1}| = \lambda^d(Q) \cdot |\det A|^{-1}. \end{aligned}$$

Since this is true for all such boxes Q and since they generate the Borel σ-algebra, the Uniqueness Theorem D.1 for σ-finite measures shows

$$(\lambda^d)_\Phi = |\det A|^{-1}\lambda^d = |\det \mathrm{D}\Phi|^{-1}\lambda^d.$$

D.13 Theorem (Transformation of the Lebesgue measure) *Let $U, V \subseteq \mathbb{R}^d$ be open and let $\Phi\colon U \to V$ be a diffeomorphism. Then $(\lambda^d)_\Phi$ has a Lebesgue density, namely,*

$$\frac{\mathrm{d}\lambda_\Phi^d}{\mathrm{d}\lambda^d}(\Phi(x)) = |\det \mathrm{D}\Phi(x)|^{-1}, \quad x \in U.$$

Here, $\mathrm{D}\Phi(x)$ is the Jacobi matrix of Φ at $x \in U$.

The theorem says that the inverse image of a small cube of volume ε located at the site $\Phi(x)$ has approximately the volume $|\det \mathrm{D}\Phi(x)|^{-1}\varepsilon$ in the source space. The quantity $|\det \mathrm{D}\Phi(x)|$ in the multivariate case corresponds to the steepness $|\Phi'(x)|$ in the case $d = 1$.

D.4 Conditioning

It is sometimes easier to determine probabilities conditional on some event and to derive the total probability by piecing together the conditional ones. This section deals with the relevant apparatus.

D.14 Definition Let $(E, \mathcal{B})$ and $(F, \mathcal{C})$ be measurable spaces.

(a) A **kernel** from $(E, \mathcal{B})$ to $(F, \mathcal{C})$ is a function $K\colon E \times \mathcal{C} \to [0, \infty]$ such that

    (i) $K(\cdot, C)$ is $\mathcal{B}$-measurable for all $C \in \mathcal{C}$;

    (ii) $K(x, .)$ is a σ-finite measure on $\mathcal{C}$ for all $x \in E$.

(b) The kernel is **Markovian** (or **stochastic**) if the measures in (ii) are probabilities.

Let μ be a probability measure on $(F, \mathcal{C})$ and let $h\colon F \to E$ be a measurable function.

(c) A **conditional probability** of μ given h is a Markovian kernel K from $(E, \mathcal{B})$ to $(F, \mathcal{C})$ such that

$$\mu(C \cap [h \in B]) = \int_B K(x, C)\mu_h(\mathrm{d}x)$$

    for all $C \in \mathcal{C}$ and all $B \in \mathcal{B}$. The number $K(x, C)$ is called the **conditional probability** of C **given** $h = x$ and denoted by $\mu[C \,|\, h = x]$.

(d) Let $\mu[C \,|\, h = x]$ be as in (c). A measurable function $(x, y) \to f[y \,|\, h = x]$ such that $\mu[\mathrm{d}y \,|\, h = x] = f[y \,|\, h = x]L(x, \mathrm{d}y)$ for some kernel L from $(E, \mathcal{B})$ to $(F, \mathcal{C})$ is called a **conditional L-density** of μ given h.

A kernel K from $(E, \mathcal{B})$ to $(F, \mathcal{C})$ transforms measures ν on $\mathcal{B}$ via integration $(\nu K)(C) = \int \nu(\mathrm{d}x)K(x, C)$ to measures νK on F, and positive, measurable functions on F, again via integration $Kf(x) = \int K(x, \mathrm{d}y)f(y)$, to measurable functions Kf on E. Part (c) of the definition implies in particular the important **formula of total probability**,

$$\mu(C) = \int_E \mu[C \,|\, h = x]\mu_h(\mathrm{d}x), \quad C \in \mathcal{C}. \tag{D.6}$$

This divide-and-conquer formula offers us a way to decompose μ-integration of C in two steps, first under the additional knowledge that $h = x$ and then w.r.t. the push-forward measure μ_h. Integration of a positive, measurable function $\varphi\colon F \to \mathbb{R}$ (and others) is exercised by the **formula of total expectation**,

$$\int_F \varphi \,\mathrm{d}\mu = \int_E \mu_h(\mathrm{d}x) \int_{[h=x]} \varphi(y)\mu[\mathrm{d}y \,|\, h = x]. \tag{D.7}$$

Conditional probabilities do not always exist. The following is the standard assumption for their existence and uniqueness. Proofs are found in Bauer [31], Theorems 44.2 and 44.3. They depend on regularity properties of measures on Polish spaces; see Theorem D.2(b).

D.15 Theorem *Let F be a Polish space (Appendix B.3) with its Borel σ-algebra $\mathcal{C}$, let μ be a (Borel) probability measure on $(F,\mathcal{C})$, let $(E,\mathcal{B})$ be a measurable space, and let $h\colon (F,\mathcal{C}) \to (E,\mathcal{B})$ be measurable. Then:*

(a) There exists a conditional probability of μ given h.

(b) If $\mu[\,\cdot\,|\,h = \cdot\,]$ and $\mu'[\,\cdot\,|\,h = \cdot\,]$ are two versions of the conditional probability, then there exists a μ_h-null set $N \in \mathcal{B}$ such that $\mu[C\,|\,h = x] = \mu'[C\,|\,h = x]$ for all $C \in \mathcal{C}$ and all $x \notin N$.

The uniqueness statement (b) says that the conditional probability is defined for μ_h-a.a. x only, and that the formula of total probability (D.6) (also (D.7)) is characteristic: The conditional measure $\mu[\,\cdot\,|\,h = \cdot\,]$ is essentially the only kernel from $(E,\mathcal{B})$ to $(F,\mathcal{C})$ that satisfies the equation for all $C \in \mathcal{C}$. The statement is strong in the sense that the same null set does it for all C. The push-forward and conditional measures contain all the information on μ.

D.16 Examples (a) (*Discrete case*) Let $(F,\mathcal{C},\mu)$ be a probability space, let E be countable, and let $h\colon F \to E$ be measurable. Then, for all $C \in \mathcal{C}$,

$$\mu(C) = \sum_{x \in E} \mu(C \cap [h = x]) = \sum_{x \in E} \mu_h(\{x\})\frac{\mu(C \cap [h = x])}{\mu_h(\{x\})}.$$

We infer $\mu[C\,|\,h = x] = \mu(C \cap [h = x])/\mu_h(\{x\})$. Note that it is not, and does not have to be, defined for the μ_h-null set of all $x \in E$ such that $\mu_h(\{x\}) = 0$.

(b) Let E and G be Polish spaces equipped with Borel probabilities α and β, respectively. Let $F = E \times G$ and $\mu = \alpha \otimes \beta$, let $h\colon F \to E$ be the first projection, and define the one-to-one map $G \to F$, $\psi_x(z) = (x,z)$, $x \in E$. We have $h \circ \psi.(z) = \mathrm{id}_E$ for all $z \in G$, that is, $\psi.(z)$ is a measurable section of h and h is a retraction of $\psi.(z)$ in the diction of category theory. Fubini's theorem entails, for each measurable function $\varphi\colon F \to \mathbb{R}_+$,

$$\int_F \varphi\,\mathrm{d}\mu = \int_E \alpha(\mathrm{d}x) \int_G \varphi(x,z)\beta(\mathrm{d}z) = \int_E \mu_h(\mathrm{d}x) \int_{[h=x]} \varphi(y)\beta_{\psi_x}(\mathrm{d}y).$$

A comparison with the formula of total expectation (D.7) shows $\mu[\,\cdot\,|\,h = x] = \beta_{\psi_x}$.

(c) Let E and G be Polish spaces equipped with σ-finite Borel measures ϱ and τ, respectively. Let $F = E \times G$, let the Borel probability μ on F have the $\varrho \otimes \tau$-density $f\colon F \to \mathbb{R}_+$, and let $h\colon F \to E$ be the first projection. Then

$$\mu_h(B) = \mu(B \times G) = \int_{B \times G} f(x,z)\varrho(\mathrm{d}x)\tau(\mathrm{d}z) = \int_B \varrho(\mathrm{d}x) \int_G f(x,z)\tau(\mathrm{d}z).$$

In other words, μ_h has the ϱ-density $g(x) = \int_G f(x,z)\tau(\mathrm{d}z)$. With ψ_x as in (b), we have, for each $\varphi\colon F \to \mathbb{R}_+$,

$$\int_F \varphi\,\mathrm{d}\mu = \int_E \varrho(\mathrm{d}x) \int_G \varphi(x,z)f(x,z)\tau(\mathrm{d}z)$$

$$= \int_E \varrho(\mathrm{d}x)g(x)\int_G \varphi(\psi_x(z))\frac{f(\psi_x(z))}{g(x)}\tau(\mathrm{d}z) = \int_E \mu_h(\mathrm{d}x)\int_{[h=x]} \varphi(y)\frac{f(y)}{g(x)}\tau_{\psi_x}(\mathrm{d}y).$$

A comparison with (D.7) shows that $f(y)/g(x)$ is the conditional density $f[y\,|\,h = x]$ of μ given $h = x$ w.r.t. the kernel $K(x,\mathrm{d}y) = \tau_{\psi_x}(\mathrm{d}y)$, the image measure of τ w.r.t. the map ψ_x.

   This example applies in particular to Euclidean spaces F and Lebesgue densities f. If h is no projection, then apply a transformation along with the Transformation Theorem D.13 so that it becomes one.

D.5   Comparison of probabilities

There are many measures of the difference between two probabilities. On the space of all ϱ-absolutely continuous probabilities, the integral $d(\mu, \nu) = \int \left| \frac{d\mu}{d\varrho} - \frac{d\nu}{d\varrho} \right| d\varrho$ defines even a metric. Some dissimilarity measures have turned out to have properties particularly suited for statistics. The Kullback–Leibler divergence or relative entropy, the Hellinger integral, and the related Bhattachariya distance will be covered in this section.

D.5.1   Entropy and Kullback–Leibler divergence

Entropy is ubiquitous in this text. It is used in proving consistency, appears as an important term in the MAP criterion, it is needed for computing breakdown points, for proving convergence of the EM and other iterative relocation algorithms, and it is useful for filtering variables and for validating solutions. The concept is introduced here via the Shannon information and some properties and examples are presented.

It is common to start from the information I, an operator that maps probability densities $f_\mu = d\mu / d\varrho$ w.r.t. a σ-finite reference measure ϱ to functions $E \to [0, \infty]$. The main reference measures ϱ are counting measure on a countable space and Lebesgue measure on a measurable subset of some Euclidean space. The value $(I f_\mu)(x)$, $x \in E$, depends also on ϱ. It is conceived as a measure of the *surprise* about the outcome $x \in E$ after sampling from μ. This imposes some requirements on $I f_\mu$. First, the smaller $f_\mu(x)$ is the larger $(I f_\mu)(x)$ should be. Second, the amount of surprise about a sample from a product probability $\mu \otimes \nu$ w.r.t. the product reference measure should be the sum of the surprises of the two (independent) individual samples. These requirements are made precise by the following axioms.

(i) $(I f_\mu)(x)$ depends only on the density $f_\mu = d\mu / d\varrho$ at x, that is,
$$(I f_\mu)(x) = \varphi(f_\mu(x)) \text{ for some function } \varphi \colon \mathbb{R}_+ \to \mathbb{R} \cup \{\infty\};$$

(ii) φ is decreasing;

(iii) $(I f_\mu \otimes f_\nu)(x, y) = (I f_\mu)(x) + (I f_\nu)(y)$ for all $\mu \ll \varrho$, $\nu \ll \sigma$, $x \in E$, $y \in F$.

It is now straightforward to induce the form of φ. For a product density $f_\mu \otimes f_\nu = f_{\mu \otimes \nu}$ w.r.t. $\varrho \otimes \varrho$, we infer
$$\varphi(f_\mu(x) \cdot f_\nu(y)) = \varphi(f_{\mu \otimes \nu}(x, y)) = (I f_\mu \otimes f_\nu)(x, y) = (I f_\mu)(x) + (I f_\nu)(y)$$
$$= \varphi(f_\mu(x)) + \varphi(f_\nu(y)).$$

Since this is supposed to be true for all μ, ν, x, y as in (iii), φ must satisfy the functional equation $\varphi(a \cdot b) = \varphi(a) + \varphi(b)$ for all $0 < a, b \leq 1$. There are many such functions but, with the exception of the logarithm, none of them is measurable (let alone monotone). Hence, $\varphi = -\log$, that is, the **Shannon information** of a sample x from μ is defined as
$$I\mu(x) = -\log f_\mu(x).$$

Here, $-\log$ is understood as a (continuous, decreasing) function $[0, \infty] \to \overline{\mathbb{R}}$. The base of the logarithm is not important and we use the Euler-Napier number e. The expected Shannon information, $E_\mu \log \frac{d\mu}{d\nu}$, does not exist for all μ and ϱ. An example is $\mu(dx) = \frac{const}{x(1+\log^2 x)} \lambda(dx)$ on the ray $]0, \infty[$. When at least the negative part $\log^- f_\mu(x)$ is μ-integrable, it is called the **entropy** of f_μ,
$$H f_\mu = \int_E (I f_\mu)(x) \mu(dx) = -\int_E f_\mu(x) \log f_\mu(x) \varrho(dx),^4$$

Shannon [469]. Being functions of the ϱ-density of μ, information and entropy also depend on ϱ. This is one reason why they are commonly defined only w.r.t. a "uniform" measure as

---

[4] We define here $0 \log 0 = 0$, the continuous extension of $p \log p$ as $p \to 0$.

reference measure. Since the Shannon information measures the surprise about an *individual* sample from μ, the entropy $\mathrm{H}f_\mu$ is a measure of the *average* surprise or the *uncertainty* about sampling from μ.

The following inequality is fundamental. When μ is not absolutely continuous w.r.t. ν, then $\mathrm{d}\mu/\mathrm{d}\nu$ is defined by $f_\mu(x)/f_\nu(x)$ w.r.t. some reference measure ϱ such that $\mu \ll \varrho$ and $\nu \ll \varrho$, for instance, $\varrho = \mu + \nu$. We put $a/0 = \infty$ for $a > 0$.

D.17 Theorem (Entropy inequality) *Let ϱ be a σ-finite measure on $(E, \mathcal{B})$ and let μ and ν be two probability measures with ϱ-densities f_μ and f_ν, respectively. Then*

(a) The negative part $\log^- \dfrac{f_\mu(x)}{f_\nu(x)}$ is μ-integrable.

(b) $\displaystyle\int_E \log \dfrac{\mathrm{d}\mu}{\mathrm{d}\nu}\,\mathrm{d}\mu = \int_E \log \dfrac{f_\mu(x)}{f_\nu(x)}\mu(\mathrm{d}x) \geq 0.$

(c) Equality holds if and only if $\mu = \nu$.

Proof. (a) It is equivalent to show that $\int_E \log^+ \frac{f_\nu(x)}{f_\mu(x)}\mu(\mathrm{d}x) < \infty$. We start from the inequality $\log^+ r \leq r$, valid for $r \geq 0$. It follows

$$\int_E \log^+ \frac{f_\nu(x)}{f_\mu(x)}\mu(\mathrm{d}x) = \int_{f_\mu(x)>0} \log^+ \frac{f_\nu(x)}{f_\mu(x)}\mu(\mathrm{d}x)$$

$$\leq \int_{f_\mu(x)>0} \frac{f_\nu(x)}{f_\mu(x)}\mu(\mathrm{d}x) \leq \int_{f_\mu(x)>0} f_\nu(x)\varrho(\mathrm{d}x) \leq 1.$$

(b) The inequality follows from the bound $\log r \leq r - 1$, again valid for $r \geq 0$:

$$\int_E \log \frac{f_\nu(x)}{f_\mu(x)}\mu(\mathrm{d}x) = \int_{f_\mu(x)>0} \log \frac{f_\nu(x)}{f_\mu(x)}\mu(\mathrm{d}x)$$

$$\leq \int_{f_\mu(x)>0} \left(\frac{f_\nu(x)}{f_\mu(x)} - 1\right)\mu(\mathrm{d}x) \leq \int_{f_\mu(x)>0} f_\nu(x)\varrho(\mathrm{d}x) - 1 \leq 0.$$

(c) If equality obtains, then there is equality in particular in the first inequality of the previous chain. This means $f_\mu(x) = f_\nu(x)$ first for all x such that $f_\mu(x) > 0$, and then for all x, that is, $\mu = \nu$.   □

D.18 Definition The **Kullback–Leibler** [302, 301] **divergence** (or **relative entropy**) of two probabilities μ and ν is defined by

$$\mathrm{D_{KL}}(\mu, \nu) = \begin{cases} \mathrm{E}_\mu \log \frac{\mathrm{d}\mu}{\mathrm{d}\nu} = \int \log \frac{\mathrm{d}\mu}{\mathrm{d}\nu}\,\mathrm{d}\mu, & \text{if } \mu \ll \nu, \\ \infty, & \text{otherwise.} \end{cases} \tag{D.8}$$

Thus, the relative entropy is positive and enjoys the property $\mathrm{D_{KL}}(\mu, \nu) = 0$ if and only if $\mu = \nu$. But it is no metric since it is neither symmetric nor does it satisfy the triangle inequality. It can be symmetrized by writing $\mathrm{S_{KL}}(\mu, \nu) = \mathrm{D_{KL}}(\mu, \nu) + \mathrm{D_{KL}}(\nu, \mu)$. This quantity is a **dissimilarity**, that is, it is positive, symmetric, and satisfies $\mathrm{S_{KL}}(\mu, \nu) = 0$ if and only if $\mu = \nu$.

The relative entropy does not suffer from the drawbacks of information and entropy mentioned above. It does not depend on a reference measure but only on μ and ν and it is always well defined although it may be ∞, also in cases of absolute continuity. Consider, for instance, $\mu(\{x\}) = \mathrm{const} \cdot x^{-2}$ and $\nu(\{x\}) = 2^{-x}$ on $E = \{1, 2, 3, \dots\}$.

For probabilities $\mu, \nu \ll \varrho$ there is the relation

$$\mathrm{D_{KL}}(\mu, \nu) = -\mathrm{H}f_\mu - \int \log f_\nu\,\mathrm{d}\mu. \tag{D.9}$$

D.19 Examples (a) The entropy of a probability vector $\mathbf{p} = (p_1, \ldots, p_n)$ w.r.t. counting measure lies between 0 (for $\mathbf{p} = \delta_a$) and $\log n$ for the uniform distribution.

PROOF. By the entropy inequality D.17,

$$\mathrm{Hp} = -\sum_{x=1}^{n} p_x \log p_x \le -\sum_{x=1}^{n} p_x \log \frac{1}{n} = \log n = -\sum_{x=1}^{n} \frac{1}{n} \log \frac{1}{n} = \mathrm{H}(1/n)_{x=1}^{n}. \qquad \Box$$

This means that the uniform distribution presents maximum uncertainty. The fact has a game-theoretic interpretation: The burglar's supplies will dwindle in the next ten days and the sheriff, aware of his previous actions, knows this. Therefore, both must take action in the said period of time. The burglar needs only one night and the sheriff, being busy with other duties, can only spend one night. On which days will burglar and sheriff optimally take action? Game theoretically, "cops and robbers" is a zero-sum game with no deterministic equilibrium solution, but there is a randomized one: Both have to select a night *uniformly* from the said period. From the point of view of the entropy, each "player" keeps his opponent at maximum uncertainty about his action. Each deviation from this strategy on either side would result in a superior strategy on the other.

(b) In a similar way one shows that the uniform distribution on the real interval $[a, b]$ has maximum entropy among all Lebesgue absolutely continuous distributions on $[a, b]$.

(c) The Kullback–Leibler divergence of two normal densities N_{m_1, V_1} and N_{m_2, V_2} on $\mathbb{R}^d$ can be expressed by their parameters:

$$\mathrm{D_{KL}}(f, g) = \tfrac{1}{2}\Big(\log \frac{\det V_2}{\det V_1} + \operatorname{tr} V_2^{-1} V_1 + (m_2 - m_1)^\top V_2^{-1}(m_2 - m_1) - d\Big).$$

PROOF. Abbreviating $f = N_{m_1, V_1}$, we have

$$\mathrm{D_{KL}}(f, g) = \int \log \frac{N_{m_1, V_1}(x)}{N_{m_2, V_2}(x)} N_{m_1, V_1}(dx)$$

$$= \tfrac{1}{2}\Big(\log \tfrac{\det V_2}{\det V_1} + \int \big\{(x - m_2)^\top V_2^{-1}(x - m_2) - (x - m_1)^\top V_1^{-1}(x - m_1)\big\} f(x)dx \Big).$$

Now,

$$\int (x - m_2)^\top V_2^{-1}(x - m_2) f(x)dx = \operatorname{tr} V_2^{-1} \int (x - m_2)(x - m_2)^\top f(x)dx$$

$$= \operatorname{tr} V_2^{-1} \int ((x - m_1) - (m_2 - m_1))((x - m_1) - (m_2 - m_1))^\top f(x)dx$$

$$= \operatorname{tr} V_2^{-1} \Big(\int (x - m_1)(x - m_1)^\top f(x)dx + (m_2 - m_1)(m_2 - m_1)^\top \Big)$$

$$= \operatorname{tr} V_2^{-1}\big(V_1 + (m_2 - m_1)(m_2 - m_1)^\top \big)$$

$$= \operatorname{tr} V_2^{-1} V_1 + (m_2 - m_1)^\top V_2^{-1}(m_2 - m_1)$$

and hence

$$\int (x - m_1)^\top V_1^{-1}(x - m_1) f(x)dx = \operatorname{tr} V_1^{-1} \int (x - m_1)(x - m_1)^\top f(x)dx = \operatorname{tr} V_1^{-1} V_1 = d$$

and the claim follows. $\qquad \Box$

(d) There is no distribution on $\mathbb{R}^d$ with maximum entropy, but the normal distribution $N_{m, V}$ enjoys this property among all Lebesgue absolutely continuous distributions with mean m and covariance matrix V.

PROOF. By translation invariance, we may and do assume $m = 0$. Now,

$$2\mathrm{H}N_{0, V} = -2 \int_{\mathbb{R}^d} \log \Big(\frac{1}{\sqrt{\det 2\pi V}} e^{-x^\top V^{-1} x/2}\Big) N_{0, V}(dx)$$

$$= \log \det 2\pi V + \int_{\mathbb{R}^d} x^\top V^{-1} x N_{0, V}(dx) = \log \det 2\pi V + d.$$

With the abbreviation $f = \frac{\mathrm{d}\mu}{\mathrm{d}\lambda^d}$ and $g = \frac{\mathrm{d}N_{0,V}}{\mathrm{d}\lambda^d}$ and using the inequality $\log \frac{g(x)}{f(x)} \leq \frac{g(x)}{f(x)} - 1$, we have

$$2\mathrm{H}\mu = 2\int_{\mathbb{R}^d} f(x) \log \frac{1}{f} \lambda^d(\mathrm{d}x) = 2\int_{f>0} f(x) \Big(\log \frac{1}{g(x)} + \log \frac{g(x)}{f(x)} \Big) \lambda^d(\mathrm{d}x)$$

$$\leq 2\int_{f>0} f(x) \left(\log \frac{1}{g(x)} + \frac{g(x)}{f(x)} - 1 \right) \lambda^d(\mathrm{d}x) \leq 2\int_{\mathbb{R}^d} f(x) \log \frac{1}{g(x)} \lambda^d(\mathrm{d}x)$$

$$= \log \det 2\pi V + \int_{\mathbb{R}^d} x^\top V^{-1} x \mu(\mathrm{d}x) = \log \det 2\pi V + \operatorname{tr} V^{-1} V\mu = \log \det 2\pi V + d.$$

This is the claim. (The claimed nonexistence of a maximum follows from the unboundedness as V increases.) □

A measure of the deviation of a probability μ on a Cartesian product $E \times F$ from the product of its marginals μ_E and μ_F is the **mutual information**

$$\mathrm{MI}\,\mu = \mathrm{D}_{\mathrm{KL}}(\mu, \mu_E \otimes \mu_F). \tag{D.10}$$

It is symmetric and positive, representing the uncertainty that knowledge of one coordinate removes about the other. If μ is a product, then $\mathrm{MI}\,\mu$ vanishes. If $E = F$ is countable and if μ is supported by the diagonal, then $\mathrm{MI}\,\mu$ equals the entropy. The mutual information is unbounded and, for better comparison and interpretation, a bounded version is desirable: The mutual information can be cast in the form

$$\mathrm{MI}\,\mu = \int_{E \times F} \log \frac{f_\mu(x,y)}{f_{\mu_E}(x) f_{\mu_F}(y)} \mu(\mathrm{d}x, \mathrm{d}y) = \int_{E \times F} \log \frac{f_\mu(x \mid y)}{f_{\mu_E}(x)} \mu(\mathrm{d}x, \mathrm{d}y)$$

$$= \int_{E \times F} \log f_\mu(x \mid y) \mu(\mathrm{d}x, \mathrm{d}y) - \int_{E \times F} \log f_{\mu_E}(x) \mu(\mathrm{d}x, \mathrm{d}y)$$

$$= \int_F \mu_F(\mathrm{d}y) \int_E \log f_\mu(x \mid y) \mu(\mathrm{d}x \mid y) - \int_E \log f_{\mu_E}(x) \mu_E(\mathrm{d}x)$$

$$= -\int_F \mathrm{H} f_\mu(\cdot \mid y) \mu_F(\mathrm{d}y) + \mathrm{H} f_{\mu_E} \leq \mathrm{H} f_{\mu_E}.$$

Analogously, $\mathrm{MI}\,\mu \leq \mathrm{H} f_{\mu_F}$. Therefore, the **normalized mutual information**

$$\mathrm{NMI}\,\mu = \frac{\mathrm{D}_{\mathrm{KL}}(\mu, \mu_E \otimes \mu_F)}{\mathrm{H} f_{\mu_E} \wedge \mathrm{H} f_{\mu_F}} \tag{D.11}$$

is defined and assumes values in the unit interval except in trivial cases. Normalization by the harmonic, geometric, or arithmetic mean is also possible but the minimum is the closest bound. Kvålseth [304] makes a point of using the arithmetic mean.

D.5.2 Hellinger integral and Bhattachariya distance

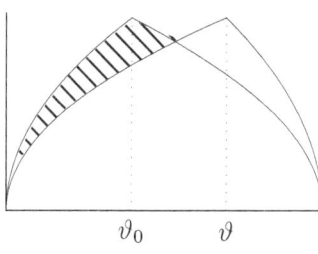

Concepts related to the relative entropy are the Hellinger integral and the Bhattachariya distance. The former is a similarity and the latter a dissimilarity on the cone of (dominated) measures on E. If (L, d_L) is a metric space and $\varphi \colon F \to L$ is one-to-one, then $d_F(f, g) = d_L(\varphi(f), \varphi(g))$ defines a metric d_F on F. Applying this fact to $L = \mathbb{L}^2_+(\varrho)$ and the square root $\sqrt{\ } \colon \mathbb{L}^1_+(\varrho) \to \mathbb{L}^2_+(\varrho)$, we find the **Bhattachariya distance** of two ϱ-densities f and g,

Figure D.3 *Illustration of the Bhattachariya distance between two members ϑ_0, ϑ of the triangle model of Example 1.9(a).*

$$d_H(f, g) = \sqrt{\int \left(\sqrt{f} - \sqrt{g} \right)^2 \mathrm{d}\varrho}.$$

It is a metric defined on all finite, ϱ-absolutely continuous measures on E. On this space it does not depend on the choice of the (equivalent) reference measure ϱ. For instance, the hatched area in Fig. D.3 is related to the Bhattachariya distance between two members of the triangle model of Example 1.9(a). Restricting matters to probability densities f and g and expanding the square, we find $d_H^2(f,g) = 2\big(1 - \int \sqrt{fg}\,\mathrm{d}\varrho\big)$ whence it follows $d_H^2(f,g) \le 2$. The quantity $\int \sqrt{fg}\,\mathrm{d}\varrho = \int \sqrt{f/g}\,g\,\mathrm{d}\varrho$ is **Hellinger's integral**; see de Branges et al. [65], pp. 7 and 36. A dissimilarity related to the Bhattachariya distance is the negative logarithm of Hellinger's integral.

An elementary but useful estimate between the Kullback–Leibler divergence and the Hellinger integral follows from the inequality $\log x \le 2(\sqrt{x}-1)$. It reads

$$d_H^2(f,g) = 2\left(1 - \int \sqrt{g/f}\,f\,\mathrm{d}\varrho\right) \le \mathrm{D}_{\mathrm{KL}}(f,g).$$

Bhattachariya distance and relative entropy are mainly used to compare two members of a probabilistic model $(f_\vartheta)_{\vartheta \in \Theta}$. If $\vartheta, \vartheta_0 \in \Theta$ and if $X \sim f_{\vartheta_0}$, then

$$\mathrm{D}_{\mathrm{KL}}(f_{\vartheta_0}, f_\vartheta) = \mathrm{E}\log \frac{f(X; \vartheta_0)}{f(X; \vartheta)}.$$

Like the Kullback–Leibler divergence of two normal densities, their Hellinger integral (and Bhattachariya distance) can be computed from the parameters.

D.20 Lemma *The Hellinger integral of two normal densities N_{m_1, V_1} and N_{m_2, V_2} has the representation*

$$\sqrt{\frac{\sqrt{\det V_1 \det V_2}}{\det \frac{1}{2}(V_1 + V_2)}} \exp\big(-\tfrac{1}{4}(m_2 - m_1)^\top (V_1 + V_2)^{-1}(m_2 - m_1)\big).$$

PROOF. Writing again $\Lambda_j = V_j^{-1}$, the Hellinger integral equals

$$\int \sqrt{N_{m_1, V_1}(x) N_{m_2, V_2}(x)}\,\mathrm{d}x$$

$$= \frac{(2\pi)^{-d/2}}{\sqrt[4]{\det V_1 \det V_2}} \int \exp\big(-\tfrac{1}{4}\big[(x - m_1)^\top \Lambda_1 (x - m_1) + (x - m_2)^\top \Lambda_2 (x - m_2)\big]\big)\mathrm{d}x.$$

Now, the expression in square brackets is

$$(x - Sb)^\top S^{-1}(x - Sb) + (m_1^\top \Lambda_1 m_1 + m_2^\top \Lambda_2 m_2 - b^\top Sb),$$

where $b = \Lambda_1 m_1 + \Lambda_2 m_2$ and $S^{-1} = \Lambda_1 + \Lambda_2$. Hence, the last integral becomes

$$\sqrt{\det 4\pi S} \exp\big(-\tfrac{1}{4}[m_1^\top \Lambda_1 m_1 + m_2^\top \Lambda_2 m_2 - b^\top Sb]\big).$$

The expression in square brackets equals

$$\mathrm{tr}\big\{\Lambda_1 m_1 m_1^\top + \Lambda_2 m_2 m_2^\top - (\Lambda_1 m_1 + \Lambda_2 m_2)(\Lambda_1 m_1 + \Lambda_2 m_2)^\top S\big\}$$

$$= \mathrm{tr}\big\{\big[(\Lambda_1 m_1 m_1^\top + \Lambda_2 m_2 m_2^\top)(\Lambda_1 + \Lambda_2) - (\Lambda_1 m_1 + \Lambda_2 m_2)(\Lambda_1 m_1 + \Lambda_2 m_2)^\top\big]S\big\}$$

$$= \mathrm{tr}\big\{\big[\Lambda_1 m_1 m_1^\top \Lambda_2 + \Lambda_2 m_2 m_2^\top \Lambda_1 - \Lambda_1 m_1 m_2^\top \Lambda_2 - \Lambda_2 m_2 m_1^\top \Lambda_1\big]S\big\}$$

$$= m_1^\top \Lambda_2 S \Lambda_1 m_1 + m_2^\top \Lambda_1 S \Lambda_2 m_2 - m_2^\top \Lambda_2 S \Lambda_1 m_1 - m_1^\top \Lambda_1 S \Lambda_2 m_2.$$

Noting $\Lambda_2 S \Lambda_1 = (V_1 + V_2)^{-1} = \Lambda_1 S \Lambda_2$, the last line reduces to

$$(m_2 - m_1)^\top (V_1 + V_2)^{-1}(m_2 - m_1)$$

and the claim follows. □

D.5.3   Notes

Rudolf Clausius [97] introduced the concept of *entropy* to statistical mechanics, using it to formulate the *Second Law of Thermodynamics*. He coined the term after the Greek word $\tau\rho o\pi\eta$, transformation. Ludwig Boltzmann [55, 56] and Josiah Willard Gibbs [199] molded it into its present shape. Later, its importance was recognized by Shannon [469] in connection with information transmission in electrical engineering and by Solomon Kullback and Richard Arthur Leibler [302] in probability. The concept has attracted the attention of many eminent scientists, among them Erwin Schrödinger, Max Planck, John von Neumann, and Norbert Wiener.

D.6 Pointwise, stochastic, and mean convergence

Unlike sequences of real numbers or Euclidean vectors, where there is only one natural concept of convergence, various modes of convergence are known for sequences of measurable functions with values in a metric space. The most distinguished one is the almost everywhere convergence. Others are the stochastic convergence and the convergences in the mean. We will need them in Appendix E to describe convergence of random variables on probability spaces but also more generally for convergence of functions on σ-finite measure spaces $(E, \mathcal{B}, \varrho)$. Let F be a metric space with its Borel structure and let $f, f_n \colon E \to F$, $n \in \mathbb{N}$, be measurable functions. It is an exercise to show that the subset $[f_n \underset{n\to\infty}{\longrightarrow} f] \subseteq E$ is measurable.

D.21 Definition

(a) $(f_n)_n$ is said to converge to f ϱ-**almost everywhere** if the sequence $(f_n(x))_n$ converges in F to $f(x)$ for ϱ-almost all $x \in E$, that is,

$$\varrho\big([f_n \underset{n\to\infty}{\not\to} f]\big) = 0.$$

(b) $(f_n)_n$ is said to converge to f ϱ-**stochastically** or **in measure** ϱ if, for all $\delta > 0$ and all $B \in \mathcal{B}$ of finite ϱ-measure, we have

$$\varrho\big([d(f_n, f) > \delta] \cap B\big) \underset{n\to\infty}{\longrightarrow} 0.$$

(c) Let F be even a Euclidean space, let $p \geq 1$ be a real number, and let all functions f_n and f be p-times integrable. The sequence $(f_n)_n$ is said to converge to f **in the pth mean** if

$$\int \|f_n - f\|^p \, d\varrho \underset{n\to\infty}{\longrightarrow} 0.$$

That is, the sequence f_n converges to f in the Banach space $\mathbb{L}^p(\varrho)$.

If ϱ is a probability, then "convergence in measure" is called "convergence in probability." The convergences enjoy a number of reasonable properties.

D.22 Properties

(a) Almost sure and mean limits are ϱ-a.e. unique. That is, two limiting functions are ϱ-a.e. equal. The same holds for stochastic limits if ϱ is σ-finite.

(b) Any subsequence of a convergent sequence (in any sense) converges (in the same sense).

(c) If F is a Euclidean space and f_n (g_n) converges to f (g) ϱ-a.e., ϱ-stochastically, or in the pth mean, then $\alpha f_n + \beta g_n$ converges to $\alpha f + \beta g$ in the same sense for all real numbers α and β.

(d) If $f_n \to f$ ϱ-stochastically and if φ is a continuous function from F to another metric space, then $\varphi(f_n) \to \varphi(f)$, ϱ-stochastically.

There are some useful relationships between the various convergences. See Bauer [32], §20.

D.23 Proposition

(a) *Convergence ϱ-a.e. implies ϱ-stochastic convergence.*

(b) *If f_n converges to f ϱ-stochastically, then each subsequence $(f_{n_k})_k$ of $(f_n)_n$ possesses a subsequence $(f_{n_{k_\ell}})_\ell$ that converges to f ϱ-a.e. (This statement is even equivalent to ϱ-stochastic convergence.)*

(c) *Let $p \geq 1$. Convergence in the pth mean implies ϱ-stochastic convergence.*

(d) *Let $p \geq 1$. If F is Euclidean and if there exists $h \in \mathcal{L}^p(\varrho)$ such that $\|f_n\| \leq h$ for all n, then stochastic convergence of (f_n) implies convergence in the pth mean.*

Part (b) of the proposition says that the seemingly weak stochastic convergence is actually not so far away from ϱ-a.e. convergence. It is sometimes useful to remember this fact.

The missing implications in Proposition D.23 do not hold, in general: The following "running peaks" converge in the pth mean for all $p \geq 1$ while they converge nowhere: Let $E = \mathbb{T}$, the unit circle in the complex plane. It bears a unique rotationally invariant probability ϱ. Let $\eta_n \in \mathbb{R}$, $n \geq 0$, be the nth partial sums of the harmonic series $\sum_{k=1}^{\infty} \frac{1}{k}$. That is, $\eta_0 = 0$, $\eta_1 = 1$, $\eta_2 = 3/2, \ldots$. Now the sequence of indicators of the intervals $e^{i[\eta_k, \eta_{k+1}[} \subseteq \mathbb{T}$ converges nowhere since the intervals round the circle infinitely often. On the other hand, it converges to zero in the pth mean since the peaks narrow with each round.

The next proposition uses Markov's inequality and Borel-Cantelli's Lemma.

D.24 Lemma (Markov's inequality) *Let $\alpha > 0$. For any positive, measurable function f on $(E, \mathcal{B}, \varrho)$, we have*

$$\varrho([f \geq \alpha]) \leq \frac{1}{\alpha} \int_E f \, d\varrho.$$

PROOF. This follows from $\alpha \mathbf{1}_{[f \geq \alpha]} \leq f$. $\qquad\qquad\qquad\qquad\qquad\qquad\qquad\qquad\qquad\square$

Part (b) of the following lemma is a corollary to the main result of Dubins and Freedman [133].

D.25 Lemma (Borel-Cantelli [60, 77]) *Let $(B_k)_{k \geq 1}$ be some sequence of measurable sets in some measure space (E, ϱ) and let $B = \bigcap_{n \geq 1} \bigcup_{k \geq n} B_k$.*

(a) If $\sum_{k=1}^{\infty} \varrho(B_k)$ is finite, then $\varrho(B) = 0$.

(b) If ϱ is a probability and if the sets B_k are pairwise independent, then the converse is also true.

Let ϱ be σ-finite and let $\mathrm{M}(\varrho)$ stand for the real vector space of ϱ-equivalence classes of all ϱ-a.e. finite, extended real-valued, measurable functions on E. The following proposition exhibits an interesting structure on $\mathrm{M}(\varrho)$ that makes it useful in applications. Let $f \wedge g = \min\{f, g\}$.

D.26 Proposition *Let ϱ be some σ-finite measure on a measurable space E.*

(a) For any probability measure μ on E equivalent with ϱ, the function

$$d_\mu(f, g) = \int_E |f - g| \wedge 1 \, d\mu \qquad\qquad\qquad (\text{D.12})$$

defines a complete metric on $\mathrm{M}(\varrho)$.

(b) The metric d_μ describes the ϱ-stochastic convergence on $\mathrm{M}(\varrho)$. A sequence $(f_n)_n \subseteq \mathrm{M}(\varrho)$ converges to $f \in \mathrm{M}(\varrho)$ ϱ-stochastically if and only if it converges to f w.r.t. d_μ.

(c) For any two ϱ-equivalent probabilities μ and ν on E, the metrics d_μ and d_ν induce the same uniform structure on $\mathrm{M}(\varrho)$.

PROOF. Clearly, d_μ is a metric on $\mathrm{M}(\varrho)$. In view of (b), let $(f_n)_n$ converge ϱ-stochastically to f. For $0 < \varepsilon < 1$ we have $\mu[|f_n - f| > \varepsilon] \leq \varepsilon$ for eventually all n and hence

$$\int_E |f_n - f| \wedge 1 \, d\mu = \int_{|f_n - f| \leq \varepsilon} |f_n - f| \wedge 1 \, d\mu + \int_{|f_n - f| > \varepsilon} |f_n - f| \wedge 1 \, d\mu \leq 2\varepsilon,$$

that is, we have convergence w.r.t. the metric. Conversely, let $d_\mu(f_n, f) \to 0$ as $n \to \infty$. Markov's inequality shows

$$\mu[|f_n - f| > \varepsilon] = \mu[|f_n - f| \wedge 1 > \varepsilon] \leq \frac{1}{\varepsilon} \int_E |f_n - f| \wedge 1 \, d\mu = d_\mu(f_n, f)/\varepsilon \to 0$$

as $n \to \infty$. Thus, (f_n) converges μ-stochastically.

We prove next the completeness claimed in (a). Let $(f_n)_n$ be a Cauchy sequence w.r.t. d_μ, that is, for all $\varepsilon > 0$ there is $N \in \mathbb{N}$ such that $\int_E |f_m - f_n| \wedge 1 \, d\mu = d_\mu(f_m, f_n) \le \varepsilon^2$ for all $m, n \ge N$. Markov's inequality shows

$$\mu[|f_m - f_n| \ge \varepsilon] = \mu[|f_m - f_n| \wedge 1 \ge \varepsilon] \le \frac{1}{\varepsilon} \int_E |f_m - f_n| \wedge 1 \, d\mu \le \varepsilon.$$

Now use this estimate to pick an increasing sequence n_k of natural numbers such that

$$\mu[|f_n - f_{n_k}| \ge k^{-2}] \le k^{-2} \tag{D.13}$$

for all $n \ge n_k$. In particular, this sequence satisfies $\mu[|f_{n_{k+1}} - f_{n_k}| \ge k^{-2}] \le k^{-2}$. Put $A_k = [|f_{n_{k+1}} - f_{n_k}| \ge k^{-2}]$. By Borel-Cantelli's Lemma D.25, $\mu(\limsup_k A_k) = 0$, that is, for eventually all k, we have $x \notin A_k$ and hence $|f_{n_{k+1}} - f_{n_k}| \le k^{-2}$, μ-a.e. Therefore, the series $\sum_{k \ge 1}(f_{n_{k+1}} - f_{n_k})$ converges absolutely and we have constructed a subsequence of (f_n) that converges μ-a.e. to some function $f \in \mathbb{M}(\varrho)$. We finally use the triangle inequality

$$d_\mu(f_n, f) \le d_\mu(f_n, f_{n_k}) + d_\mu(f_{n_k}, f)$$

and observe that, as k increases and for all $n \ge n_k$, the two summands on the right converge to 0 by (b), Eq. (D.13), and by D.23(a).

In order to prove (c), we show that for all $\varepsilon > 0$ there exists $\delta > 0$ such that $d_\nu(f, g) \le \delta$ implies $d_\mu(f, g) \le \varepsilon$. Since both μ and ν are equivalent with ϱ, ν has a density function h w.r.t. μ, $\nu = h\mu$. By σ-continuity of μ there exists $\alpha > 0$ such that $\mu[h < \alpha] < \varepsilon$. It follows

$$\int_{[h<\alpha]} |f - g| \wedge 1 \, d\mu \le \mu[h < \alpha] < \varepsilon.$$

On the other hand, if $d_\nu(f, g) \le \alpha\varepsilon$, then

$$\int_{[h \ge \alpha]} |f - g| \wedge 1 \, d\mu \le \frac{1}{\alpha} \int |f - g| \wedge 1 \, h \, d\mu = \frac{1}{\alpha} \int |f - g| \wedge 1 \, d\nu \le \varepsilon.$$

This proves the claim with $\delta = \alpha\varepsilon$. $\qquad \square$

Part (c) of the proposition says that it is not important which probability μ is chosen for the metric. We finally analyze the relationships between three metrics on the convex set of probability densities w.r.t. a σ-finite measure ϱ on a measurable set E: the $\mathbb{L}^1(\varrho)$-metric $d_1(f, g) = \int_E |f - g| \, d\varrho$, Hellinger's metric $d_H(f, g) = \sqrt{\int_E (\sqrt{f} - \sqrt{g})^2 \, d\varrho}$, and the metrics (D.12) that induce the ϱ-stochastic convergence.

D.27 Lemma *Let $\varrho \neq 0$ be σ-finite. The metrics of ϱ-stochastic convergence are weaker than the Hellinger metric which, in turn, is weaker than the $\mathbb{L}^1(\varrho)$-metric.*

PROOF. In order to describe the ϱ-stochastic convergence we choose the metric d_μ associated with a probability $\mu = h\varrho$ with bounded density function $h > 0$. Using the Cauchy-Schwarz inequality, we have

$$d_\mu(f, g) = \int_E |f - g| \wedge 1 \, d\mu \le \|h\|_\infty \int_E |\sqrt{f} - \sqrt{g}|(\sqrt{f} + \sqrt{g}) \, d\varrho$$

$$\le \|h\|_\infty \sqrt{\int_E (\sqrt{f} - \sqrt{g})^2 \, d\varrho} \sqrt{\int_E (\sqrt{f} + \sqrt{g})^2 \, d\varrho}.$$

Now, use the inequality $(\sqrt{f} + \sqrt{g})^2 = f + g + 2\sqrt{fg} \le 2f + 2g$ to find

$$d_\mu(f, g) \le 2\|h\|_\infty d_H(f, g).$$

This is the first claim. Moreover,

$$d_H^2(f, g) = \int_E (\sqrt{f} - \sqrt{g})^2 \, d\varrho = \int_E |f - g| \frac{|\sqrt{f} - \sqrt{g}|}{\sqrt{f} + \sqrt{g}} \, d\varrho \le \int_E |f - g| \, d\varrho,$$

that is, the second claim. $\qquad \square$

Appendix E

Probability

The description of mechanisms and systems under the influence of randomness could be done purely in terms of probability measures. For instance, the (random) number of customers present in a queue at time t is a probability on the set $\mathbb{N}$ of natural numbers. It would, however, be quite cumbersome to treat this problem purely in terms of measures. It is much more convenient to describe the mechanism by random variables and to use an established apparatus based on conditioning and independence to take care of the rest. Probability theory is therefore the study of random variables and their dependencies, distributions, moments, transformations, and convergences.

E.1 Random variables and their distributions

All measure-theoretic concepts have a probabilistic counterpart. The probabilist postulates a universal, in general unspecified probability space $(\Omega, \mathcal{F}, P)$. It is conceived to represent all randomness of the world, past and future. It allows us to "cut out" what we need for any special problem. For dealing with specific problems, a much smaller space containing just the randomness needed is also possible.

E.1 Definition

(a) A **random variable** X in a measurable space $(E, \mathcal{B})$ is a measurable map $X \colon (\Omega, \mathcal{F}) \to (E, \mathcal{B})$.

(b) The image probability of P w.r.t. X, P_X, is called the **distribution** of X. If $\mu - P_X$, we write $X \sim \mu$.

(c) Given n random variables $X_1, \ldots, X_n$, $X_i \colon (\Omega, P) \to E_i$, the random variable $(X_1, \ldots, X_n)$ in $\prod_{i=1}^n E_i$ is called the **joint random variable**.

(d) The **joint distribution** of $X_1, \ldots, X_n$ is $P_{(X_1, \ldots, X_n)}$.

(e) The random variables $X_1, \ldots, X_n$ are called **independent** if their joint distribution is the product of the individual ones, $P_{(X_1, \ldots, X_n)} = \bigotimes_{i=1}^n P_{X_i}$.

Random variables are commonly given the generic names of the elements of the **sample space** E preceded by the word "random." So a random integer is a random variable in $\mathbb{Z}$, a random number is a random variable in $\mathbb{R}$, a random vector is a Euclidean random variable, a random probability is a random variable in the convex set of all probabilities on $(E, \mathcal{B})$, and so on. Given a sequence $X_1, X_2, \ldots$ of random variables, a **random process**, the random discrete probability $\frac{1}{n} \sum_{i=1}^n \delta_{X_i}$ is known as the nth **empirical measure**. It contains the information on the process until time n up to permutations. *Quantities and attributes defined for distributions of random variables are transferred to the random variables themselves.* So we speak of the ϱ-density $f_X = f_{P_X} = \mathrm{d}P_X / \mathrm{d}\varrho$ of X, meaning that of P_X w.r.t. ϱ, of the expectation $\mathrm{E}X = \mathrm{E}P_X$ of X, again meaning that of P_X. We say that X is discrete, spherical, ... if P_X is.

   Some functions of random variables are compiled below. Verifying the existence of expectations of random vectors $X \colon \Omega \to \mathbb{R}^d$ needs either knowledge of P, in which case we check $\int_\Omega |X| \, dP < \infty$, or it requires knowledge of the distribution of $|X|$ so that we can verify $\mathrm{E} P_{|X|} < \infty$. The expectation is then $\mathrm{E} X = \mathrm{E} P_X = \int_\Omega X \, dP$.

E.2 Definition Let $X \colon (\Omega, P) \to \mathbb{R}^d$, $Y \colon (\Omega, P) \to \mathbb{R}^e$ be random vectors with the components X_i, Y_j.

(a) The vector $\mathrm{E} X = (\mathrm{E} X_1, \ldots, \mathrm{E} X_d)^\top \in \mathbb{R}^d$ is called the **expectation** of X. So $\mathrm{E} X$ is defined component by component, $(\mathrm{E} X)_i = \mathrm{E} X_i$.

(b) The random vector $\widehat{X} = X - \mathrm{E} X$ is called **centered**.

(c) If the expectations of all entries of the random matrix $M \colon \Omega \to \mathbb{R}^{d \times e}$ exist, then the expectation of M is the matrix
$$\mathrm{E} M = (\mathrm{E} M_{i,j})_{i,j} \in \mathbb{R}^{d \times e}.$$

(d) We apply (c) to the product $M = \widehat{X}\widehat{Y}^\top$ of the centerings of two random vectors X and Y: If the expectations of all $X_i Y_j$ exist, then the $d \times e$ matrix
$$\mathrm{Cov}(X, Y) = \mathrm{E}\widehat{X}\widehat{Y}^\top = \left(\mathrm{E}\widehat{X}_i\widehat{Y}_j\right)_{i,j} = \begin{pmatrix} \mathrm{Cov}(X_1, Y_1) & \cdots & \mathrm{Cov}(X_1, Y_e) \\ \vdots & & \vdots \\ \mathrm{Cov}(X_d, Y_1) & \cdots & \mathrm{Cov}(X_d, Y_e) \end{pmatrix} \text{ is called the}$$
covariance matrix of X and Y. Note that $(\widehat{X}\widehat{Y}^\top)_{i,j} = \widehat{X}_i\widehat{Y}_j$.

(e) If all random vectors X_i are quadratically integrable, then
$$\mathrm{V} X = \mathrm{Cov}(X, X) = \mathrm{E}\widehat{X}\widehat{X}^\top \in \mathbb{R}^{d \times d}$$
exists and is called the covariance matrix of X. We have
$$(\mathrm{V} X)_{i,j} = \mathrm{E}\widehat{X}_i\widehat{X}_j = \begin{cases} \mathrm{V} X_i, & i = j, \\ \mathrm{Cov}(X_i, X_j), & i \neq j. \end{cases}$$

Some well-known linear transformations of these quantities are as follows.

E.3 Lemma (Properties of E, V, Cov)
Let $A \in \mathbb{R}^{m \times d}$, $C \in \mathbb{R}^{n \times e}$, $M \in \mathbb{R}^{d \times e}$, $a \in \mathbb{R}^d$, $b \in \mathbb{R}^m$, $g \in \mathbb{R}^n$.

(a) $\mathrm{E}(AX + b) = A\mathrm{E} X + b$;

(b) $\mathrm{E}(a^\top X + b) = a^\top \mathrm{E} X + b$;

(c) $\mathrm{E}(AMC^\top) = A(\mathrm{E} M)C^\top$;

(d) $\mathrm{Cov}(X, Y) = \mathrm{Cov}(Y, X)^\top$;

(e) $\mathrm{Cov}(AX + b, CY + g) = A\mathrm{Cov}(X, Y)C^\top$;

(f) $\mathrm{V}(AX + b) = A(\mathrm{V} X)A^\top$;

(g) $\mathrm{V}(a^\top X + b) = a^\top(\mathrm{V} X)a$;

(h) if $\|a\| = 1$, then $a^\top X$ is the projection of X to the straight line $\mathbb{R}a$ and $a^\top(\mathrm{V} X)a$ is the variance of X in the direction of a;

(i) $\mathrm{V} X$ is symmetric and positiv semi-definite;

(j) $\mathrm{E} X$ and $\mathrm{V} X$ depend on the distribution of X, only; P and X are not needed for their computation;

(k) $\mathrm{Cov}(X, Y)$ depends on the joint distribution of X und Y;

(l) if X and Y are independent, then $\mathrm{Cov}(X, Y) = 0$.

E.4 Definition Let X be a d-dimensional random vector with positive definite variance. The metric
$$d_X(x, y) = \|x - y\|_{(\mathrm{V} X)^{-1}} = \sqrt{(x - y)^\top (\mathrm{V} X)^{-1}(x - y)}, \qquad x, y \in \mathbb{R}^d,$$

is known as the **Mahalanobis distance** of x and y w.r.t. X.

The points of equal Mahalanobis distance from $\mathrm{E}X$ describe an ellipse extended in the direction of the largest variance. The random number $d_X(X, \mathrm{E}X)$ is the norm of the standardization (or sphered random vector) $\widehat{X}$: $d_X^2(X, \mathrm{E}X) = (X - \mathrm{E}X)^\top (VX)^{-1}(X - \mathrm{E}X) = \|(VX)^{-1/2}\widehat{X}\|^2$.

E.5 Definition Let μ be a probability measure on $\mathbb{R}^d$.

(a) The function $\widetilde{\mu} \colon \mathbb{R}^d \to [0, \infty]$ defined by

$$\widetilde{\mu}(\mathbf{t}) = \int_{\mathbb{R}^d} e^{\mathbf{t}^\top \mathbf{x}} \mu(\,\mathrm{d}\mathbf{x})$$

is called the **moment generating function** of μ.

(b) The **characteristic function** of μ is its Fourier transform,

$$\widehat{\mu}(\mathbf{t}) = \int_{\mathbb{R}^d} e^{-i\mathbf{t}^\top \mathbf{x}} \mu(\,\mathrm{d}\mathbf{x}).$$

The moment generating function is extended real valued, the characteristic function is complex valued.

E.6 Example *The moment generating function of the $N_{m,V}$ is $\widetilde{N}_{m,V}(\mathbf{t}) = e^{\mathbf{m}^\top \mathbf{t} + \mathbf{t}^\top V \mathbf{t}/2}$, its characteristic function is $\widehat{N}_{m,V}(\mathbf{t}) = e^{-i\mathbf{m}^\top \mathbf{t} - \mathbf{t}^\top V \mathbf{t}/2}$.*

PROOF. Taking the d-fold tensor product of the formula $\frac{1}{\sqrt{2\pi}} \int_{\mathbb{R}} e^{ty} e^{-y^2/2} \mathrm{d}x = e^{t^2/2}$, we obtain first

$$(2\pi)^{-d/2} \int_{\mathbb{R}^d} e^{\mathbf{t}^\top \mathbf{y}} e^{-\|\mathbf{y}\|^2/2} \,\mathrm{d}\mathbf{y} = e^{\|\mathbf{t}\|^2/2}.$$

The claims follow from the integral transformation $\mathbf{y} = V^{-1/2}(\mathbf{x} - \mathbf{m})$. $\qquad\square$

The moment generating and characteristic functions allow us to compute existing moments by differentiation. The expectation is

$$\mathrm{E}\mu = \mathrm{D}\widetilde{\mu}(0) = i\,\mathrm{D}\widehat{\mu}(0)$$

and the second moment

$$\mathrm{M}_2\mu = \int x_k x_\ell \,\mathrm{d}\mu = \left(\frac{\partial^2}{\partial t_k \partial t_\ell} \widetilde{\mu}(0)\right)_{k,\ell} = -\left(\frac{\partial^2}{\partial t_k \partial t_\ell} \widehat{\mu}(0)\right)_{k,\ell}.$$

Lemma E.3 shows how to linearly transform expectations and covariance matrices. We study next the way the population itself transforms, even with a nonlinear map. So let U and V be open in $\mathbb{R}^d$ and let $\Phi \colon U \to V$ be a diffeomorphism. Let X be a random vector in U with a Lebesgue density f_X. The random vector $Y = \Phi \circ X$ in V, too, has a Lebesgue density. It has an explicit representation. The formula is the extension of the substitution formula of integration calculus to the multivariate case and reads as follows.

E.7 Theorem (Transformation of Lebesgue densities) *Let U and V be open subsets of $\mathbb{R}^d$, let $\Phi \colon U \to V$ be a diffeomorphism, and let $X \colon \Omega \to U$ be a random variable with Lebesgue density f_X. Then $Y = \Phi \circ X \colon \Omega \to V$, too, has a Lebesgue density f_Y. With the Jacobian matrix $\mathrm{D}\Phi(x)$, it satisfies the relation*

$$f_X(x) = |\det \mathrm{D}\Phi(x)| f_Y(\Phi(x)).$$

PROOF. The claim follows from Lemma D.11, the cancellation rule D.4, and the Transformation Theorem D.13 of the d-dimensional Lebesgue measure:

$$f_X(x) = \frac{\mathrm{d}P_X}{\mathrm{d}\lambda^d}(x) = \frac{\mathrm{d}(P_X)_\Phi}{\mathrm{d}\lambda^d_\Phi}(\Phi(x)) = \frac{\mathrm{d}P_Y}{\mathrm{d}\lambda^d}(\Phi(x)) \frac{\mathrm{d}\lambda^d}{\mathrm{d}\lambda^d_\Phi}(\Phi(x)) = f_Y(\Phi(x)) |\det \mathrm{D}\Phi(x)|. \quad\square$$

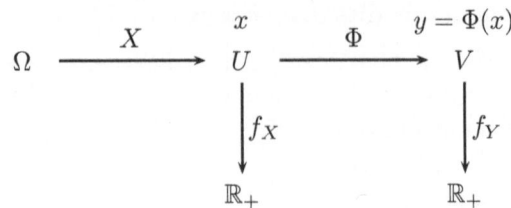

Figure E.1: *The arrow diagram of the transformation theorem.*

A graphical representation of the transformation theorem is offered in Fig. E.1. Conditional densities have already been introduced in Appendix D.2 in the measure-theoretic framework. A version for random variables reads as follows.

E.8 Proposition *Let $(E, \mathcal{B}, \varrho)$ and $(F, \mathcal{C}, \sigma)$ be σ-finite measure spaces and let $X \colon (\Omega, \mathcal{F}) \to (E, \mathcal{B})$ and $Y \colon (\Omega, \mathcal{F}) \to (F, \mathcal{C})$ be random variables such that the joint $\varrho \otimes \sigma$-density $f_{(X,Y)}$ exists. Then:*

(a) The (marginal) density f_Y w.r.t. σ exists, namely,

$$f_Y(y) = \int_F f_{(X,Y)}(x, y)\varrho(\mathrm{d}x).$$

(b) For σ-a.a. $y \in F$, the function

$$f_X[x \mid Y = y] = \frac{f_{(X,Y)}(x, y)}{f_Y(y)}$$

*is defined for ϱ-a.a. $x \in E$. It is called the ϱ-density of X **conditional** on $Y = y$. (Up to normalization with $f_Y(y)$, it is the joint density.)*

*(c) Expectations $\mathrm{E}\psi(X, Y)$ of $P_{(X,Y)}$-integrable functions ψ are computed by the formula of **total expectation**,*

$$\mathrm{E}\psi(X, Y) = \int_F \sigma(\mathrm{d}y) f_Y(y) \int_E \psi(x, y) f_X(x \mid Y = y)\varrho(\mathrm{d}x).$$

The probability associated with the conditional density, $P[X \in \mathrm{d}x \mid Y = y] = f_X[x \mid Y = y]\varrho(\mathrm{d}x)$, is called **conditional distribution**. It can be used to check two random variables for independence.

E.9 Lemma *Two random variables $X \colon \Omega \to E$ and $Y \colon \Omega \to F$ are independent if and only if the conditional distribution $P[X \in \cdot \mid Y = y]$ does not depend on y.*

E.2   Multivariate normality

Fubini's theorem shows that the d-fold tensor product

$$N_{0,I_d}(\mathbf{x}) = \prod_{k=1}^{d} \frac{1}{\sqrt{2\pi}} e^{-\frac{1}{2}x_k^2} = (2\pi)^{-d/2} e^{-\frac{1}{2}\|\mathbf{x}\|^2}$$

is the λ^d-density of a probability measure on $\mathbb{R}^d$. It is called the **standard d-variate normal distribution**. It is centered, has covariance matrix I_d, and is invariant w.r.t. all Euclidean rotations. All multivariate normal densities and distributions are obtained from N_{0,I_d} by means of affine transformations: Let $X \sim N_{0,I_d}$, let $A \in \mathrm{GL}(d)$, $m \in \mathbb{R}^d$, and let $\Phi(x) = Ax + m$. The Transformation Theorem E.7 shows that the random vector $Y = AX + m$ has a λ^d-density f_Y which satisfies the relation $N_{0,I_d}(\mathbf{x}) = f_X(\mathbf{x}) = |\det A| f_Y(A\mathbf{x} + m)$. After the substitution $\mathbf{x} = A^{-1}(\mathbf{y} - m)$, we find $f_Y(\mathbf{y}) = \frac{(2\pi)^{-d/2}}{|\det A|} e^{-\frac{1}{2}(A^{-1}(\mathbf{y}-m))^\top A^{-1}(\mathbf{y}-m)} =$

$\frac{(2\pi)^{-d/2}}{|\det A|}e^{-\frac{1}{2}(\mathbf{y}-m)^{\top}(AA^{\top})^{-1}(\mathbf{y}-m)}$. Thus, f_Y depends on A only via the positive definite matrix $V = AA^{\top}$. Inserting $|\det A| = \sqrt{\det V}$, we arrive at

$$f_Y(\mathbf{y}) = \frac{1}{\sqrt{\det 2\pi V}} e^{-\frac{1}{2}(\mathbf{y}-m)^{\top} V^{-1}(\mathbf{y}-m)}. \tag{E.1}$$

The associated probability is written $N_{m,V}$ and called the multivariate normal distribution with **location parameter** $m \in \mathbb{R}^d$ and **scale parameter** $V \in \mathrm{PD}(d)$. This notation is consistent with N_{0,I_d} above. The distribution exists for all $V \in \mathrm{PD}(d)$ since $A = \sqrt{V}$ is possible. However, there are different matrices A that generate the same normal density. The mean of $N_{m,V}$ is m. Since the covariance matrix of N_{0,I_d} is the identity, Property E.3(f) shows that of $N_{m,V}$ is $AA^{\top} = V$. Thus, $N_{m,V}$ is specified by its mean and covariance matrix. The family $\left(N_{m,V}\right)_{m \in \mathbb{R}^d, V \in \mathrm{PD}(d)}$ will be called **fully normal**.

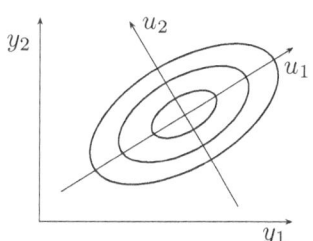

Figure E.2 *Level curves of $N_{m,V}$ and the orthonormal basis that transfers the normal distribution to product form; see Proposition E.11.*

Figure E.2 shows level curves $(\mathbf{y} - m)^{\top} V^{-1}(\mathbf{y} - m) = \mathrm{const}$ of a fully normal density. Two sub-families are worth noting: First, the **diagonally normal** family with diagonal scale parameters V, induced by diagonal matrices A. Random variables with such scales are characterized by independent entries. Second, the **spherically normal** family, a sub-family of the diagonal. The scale parameters vI_d of its members arise from multiples of the identity. An example is the standard multivariate normal.

We next study the structure of the normal distribution.

E.10 Lemma *Let $X \sim N_{m,V}$ be normal and let $\Phi(x) = Ax + b$ with $A \in \mathrm{GL}(d)$. Then $\Phi \circ X$, too, is normal, $\Phi \circ X \sim N_{Am+b,AVA^{\top}}$.*

PROOF. Since the Jacobian matrix $\mathrm{D}\Phi(x) = A$ does not depend on x, the Transformation Theorem E.7 for Lebesgue densities along with $\Phi^{-1}(y) = A^{-1}(y - b)$ shows

$$f_{\Phi \circ X}(y) = \mathrm{const}\, f_X(\Phi^{-1}(y)) = \mathrm{const}\, f_X(A^{-1}(y - b))$$

$$= \mathrm{const}\, \exp\left\{ -\tfrac{1}{2}(A^{-1}(y - b) - m)^{\top} V^{-1}(A^{-1}(y - b) - m) \right\}$$
$$= \mathrm{const}\, \exp\left\{ -\tfrac{1}{2}(y - (m + Am))^{\top}(AVA^{\top})^{-1}(y - (m + Am)) \right\}. \qquad \square$$

The fully normal family is invariant w.r.t. all affine transformations, the spherically normal family w.r.t. affine transformations with diagonal matrices and permutations of the variables, and the spherical family w.r.t. affine transformations with orthogonal matrices.

The normal distribution has a simple stochastic structure.

E.11 Proposition *Let $X \colon (\Omega, P) \to \mathbb{R}^d$ be normally distributed with mean m and covariance matrix V. Let $U \in \mathrm{SO}(d)$ such that $U^{\top}VU = D$ is diagonal and let $\Phi(x) = U^{\top}(x - m)$. Then:*

(a) The d components of $\Phi \circ X$ are independent and univariate normal $N_{0,D_{i,i}}$.

(b) The covariance matrix of $\Phi \circ X$ is D.

PROOF. (a) By Lemma E.10 with $A = U^{\top}$ and $b = -U^{\top}m$, we have $\Phi \circ X \sim N_{0,D}$ with the Lebesgue density

$$\mathrm{const} \cdot \exp\left\{ -\tfrac{1}{2} y D^{-1} y \right\} = \mathrm{const} \cdot \prod_{i=1}^{d} \exp\left\{ -y_i^2/(2D_{ii}) \right\}.$$

The independence now follows from Definition E.1(e).

(b) By Definition E.2(e) and (a), we have for $i \neq k$

$$(V\Phi \circ X)_{i,k} = \mathrm{E}(\Phi_i \circ X)(\Phi_k \circ X) = \mathrm{E}(\Phi_i \circ X)\,\mathrm{E}(\Phi_k \circ X) = 0$$

and $(V\Phi \circ X)_{i,i} = D_{i,i}$. $\qquad \square$

An illustration of the proposition is presented in Fig. E.2.

E.2.1 The Wishart distribution

The sum of $X_1, \ldots, X_n \sim N_{0,V}$ i.i.d. normal random vectors is again normal, namely, $\sum_{i=1}^n X_i \sim N_{0,nV}$. Of similar interest is the distribution of the sum of their squares $\sum_{i=1}^n X_i X_i^\top$. It lives on the cone of positive semi-definite matrices and is called the **Wishart distribution** with scale matrix V and degrees of freedom n, $W_{V,n}^{(d)}$. If $n \geq d$, it is supported by PD(d) and otherwise by singular matrices. The univariate Wishart distribution with scale 1 is called the χ-**squared distribution** with n degrees of freedom, $\chi_n^2 = W_{1,n}^{(1)}$. It is the distribution of the sum of squares of n i.i.d. standard normal random numbers.

For proofs of the following properties; see Mardia et al. [345], Section 3.4.

E.12 Properties

(i) $\mathrm{E}\, W_{V,n} = nV$;

(ii) if $n \geq d$ and $M \sim W_{V,n}$, then $\det M \sim \det V \cdot \prod_{k=1}^d Y_k$, where Y_k, $1 \leq k \leq d$, are independent chi-squared random variables with degrees of freedom $n - d + k$;

(iii) the distribution of the SSP matrix of n independent normal random vectors $\sim N_{m,V}^{(d)}$ is Wishart with scale parameter V and degrees of freedom $n{-}1$, $W_{V,n-1}^{(d)}$.

The following lemma states that the log-determinant of a Wishart random matrix is integrable.

E.13 Proposition *If $M \sim W_{V,n}^{(d)}$ for some $n \geq d$, then the expectation of $\log \det M$ exists.*

PROOF. According to Property E.12(ii), $\det M$ is equivalent to a product of d independent χ^2 variables Y_k. We now estimate

$$\mathrm{E}\big| \log \textstyle\prod_{k=1}^d Y_k \big| \leq \sum_{k=1}^d \mathrm{E}| \log Y_k| = \sum_{k=1}^d \int_0^\infty |\log t|\, \chi_{n-d+k}^2(\mathrm{d}t)$$

$$= \sum_{k=1}^d \int_{\mathbb{R}^{n-d+k}} \big| \log \|\mathbf{x}\|^2 \big|\, N_{0,I}^{(n-d+k)}(\mathrm{d}\mathbf{x}).$$

The logarithmic singularities at the origin and at ∞ are integrable. □

Being heavy tailed, Pearson's type-VII distribution (see Example E.22(b)) is an upper bound for other elliptical distributions. The following corollary exploits this fact.

E.14 Corollary *If $Y_0, \ldots, Y_d \colon \Omega \to \mathbb{R}^d$ are i.i.d. spherical with density generator $\varphi(t) \leq C(1+t)^{-\beta/2}$, $\beta > (d+1)d$, then $\mathrm{E} \log^- \det S_{(Y_0,\ldots,Y_d)} < \infty$.*

PROOF. We write $\mathbf{y} = (y_0, \ldots, y_d)$, $y_k \in \mathbb{R}^d$. Property E.12(iii) shows, along with Proposition E.13, that

$$\int_{\|\mathbf{y}\|<1} \log^- \det S_{\mathbf{y}}\, \mathrm{d}\mathbf{y} \leq \mathrm{const} \cdot \int_{\mathbb{R}^{(d+1)d}} \log^- \det S_{\mathbf{y}}\, N_{0,I}(\mathrm{d}\mathbf{y})$$

is finite. Moreover, we obtain via integral transformation for $r \geq 1$

$$\int_{\|\mathbf{y}\|<r} \log^- \det S_{\mathbf{y}}\, \mathrm{d}\mathbf{y} = r^{(d+1)d} \int_{\|\mathbf{y}\|<1} \log^- \det S_{r\mathbf{y}}\, \mathrm{d}\mathbf{y}$$

$$= r^{(d+1)d} \int_{\|\mathbf{y}\|<1} \log^- \det r^2 S_{\mathbf{y}}\, \mathrm{d}\mathbf{y} \leq r^{(d+1)d} \int_{\|\mathbf{y}\|<1} \log^- \det S_{\mathbf{y}}\, \mathrm{d}\mathbf{y}. \qquad (\mathrm{E.2})$$

From $\varphi(t) \leq C(1+t)^{-\beta/2}$, we infer

$$\textstyle\prod_{i=0}^d \varphi\big(\|y_i\|^2\big) \leq \mathrm{const} \prod_{i=0}^d \big(1 + \|y_i\|^2\big)^{-\beta/2}$$

$$\leq \mathrm{const}\, (1 + \|\mathbf{y}\|^2)^{-\beta/2} \leq \mathrm{const}\, (1 + \|\mathbf{y}\|)^{-\beta}.$$

Now use polar coordinates, integration by parts D.3(b), and (E.14) to estimate

$$\text{E} \log^- \det S_Y = \int_{\mathbb{R}^{(d+1)d}} \log^- \det S_{\mathbf{y}} \prod_{i=0}^{d} \varphi(\|y_i\|^2) \, \text{d}\mathbf{y}$$

$$\leq \text{const} \int_{\mathbb{R}^{(d+1)d}} \log^- \det S_{\mathbf{y}} (1 + \|\mathbf{y}\|)^{-\beta} \, \text{d}\mathbf{y}$$

$$= \text{const} \int_0^\infty (1 + r)^{-\beta} \int_{\|\mathbf{y}\|=r} \log^- \det S_{\mathbf{y}} \, \text{d}\mathbf{y}\text{d}r$$

$$= \text{const} \lim_{R \to \infty} \left\{ \left[(1+r)^{-\beta} \int_{\|\mathbf{y}\|<r} \log^- \det S_{\mathbf{y}} \, \text{d}\mathbf{y} \right]_0^R \right.$$

$$\left. + \beta \int_0^R (1+r)^{-\beta-1} \int_{\|\mathbf{y}\|<r} \log^- \det S_{\mathbf{y}} \, \text{d}\mathbf{y}\text{d}r \right\}$$

$$\leq \text{const} \cdot \beta \int_0^\infty (1+r)^{(d+1)d-\beta-1} \text{d}r.$$

This bound is finite by assumption on β. $\qquad\qquad\qquad\qquad\qquad\qquad\qquad\square$

E.3 Symmetry

Important sample (and parameter) spaces such as the Euclidean space or sphere bear intrinsic transformations. Of particular interest are those objects in or on the spaces that are invariant w.r.t. some of them. Felix Klein [293], in his famous "Erlangen Program," defined geometries as the study of objects and properties invariant w.r.t. their proper transformations. In statistics, too, those random variables and probabilities warrant special attention that respect the natural space transformations. On Euclidean space, they are the motions, that is, translations and rotations, but also shears. A probability invariant w.r.t translations does not exist, but there are distributions invariant w.r.t. rotations, the spherical symmetries. They are the next objects of study, followed by their affine transforms, the elliptical symmetries. Throughout this section, the norm $\|\cdot\|$ means the Euclidean norm $\|\cdot\|_2$. Standard references are Fang et al. [148] and Gupta and Varga [216].

E.3.1 Spherical symmetry

Examples of spherical symmetries are the multivariate standard normal distribution and its dilations and compressions by scalar factors. However, the tail of the normal distribution is sometimes too light for it to be a useful model in real applications. This is the reason why spherical distributions with heavier tails (and their affine transforms) are needed and popular. These so-called spherical models are more robust w.r.t. small and moderate outliers. A probability σ on $\mathbb{R}^d$ is **spherically symmetric** (or just "spherical") if $\sigma_U = \sigma$ for all orthogonal $d \times d$ matrices U. An example of a spherical probability is the uniform probability on a centered sphere. It is the Lebesgue measure conditional on the radius $r = 1$. It is not Lebesgue absolutely continuous. For our purposes, spherical distributions with Lebesgue densities are useful. We first review such probabilities and their density generators, radial distributions, radial densities, and radial functions.

E.15 Lemma *Let μ be a probability on $\mathbb{R}^d$ with Lebesgue density of the form $f(x) = \varphi(\|x\|^2)$ for some measurable $\varphi \colon \mathbb{R}_+ \to \mathbb{R}_+$. Then $\mu = f\lambda^d$ is spherical.*

PROOF. Let $U \in \text{O}(d)$ be an orthogonal matrix. Because of $f(U^{-1}x) = \varphi(\|U^{-1}x\|^2) = \varphi(\|x\|^2) = f(x)$ and by rotational invariance of Lebesgue measure, we have for any Borel measurable subset $B \subseteq \mathbb{R}^d$

$$\mu_U(B) = \mu[U \in B] = \int \mathbf{1}_{U^{-1}B} \, \text{d}\mu = \int \mathbf{1}_{U^{-1}B}f \, \text{d}\lambda^d = \int \mathbf{1}_B \circ U \cdot f \circ U \, \text{d}\lambda^d$$

$$= \int (\mathbf{1}_B f) \circ U \, \text{d}\lambda^d = \int \mathbf{1}_B f \, \text{d}\lambda^d = \int \mathbf{1}_B \, \text{d}\mu = \mu(B). \qquad\qquad \square$$

E.16 Definition The function φ appearing in the lemma is called the **density generator** (or spread function) of the Lebesgue absolutely continuous, spherical probability μ.

The gamma function $\Gamma(x) = \int_0^\infty e^{-t} t^{x-1} dt$, $x > 0$, satisfies $\Gamma(x+1) = x\Gamma(x)$ (integrate the product $e^{-t} \cdot t^x$ by parts). Some special values are

$$\Gamma(n) = (n-1)! \text{ und } \Gamma\left(n + \tfrac{1}{2}\right) = \frac{1 \cdot 3 \cdot \ldots \cdot (2n-1)}{2^n} \sqrt{\pi}, \qquad n \in \mathbb{Z}_>.$$

E.17 Examples

(a) The uniform probability on the d-dimensional unit ball is spherical. Since the volume of the d-dimensional unit ball is $\frac{\pi^{d/2}}{\Gamma(\frac{d}{2}+1)}$, it has the Lebesgue density $\frac{\Gamma(\frac{d}{2}+1)}{\pi^{d/2}} \mathbf{1}_{B_1(0)}$. Its density generator is $\varphi(r) = \frac{\Gamma(\frac{d}{2}+1)}{\pi^{d/2}} \mathbf{1}_{[0,1]}(r)$.

(b) The (multivariate) standard normal is spherical and has the density generator $\varphi(r) = (2\pi)^{-d/2} e^{-r/2}$.

E.18 Lemma *The push-forward measure $(\lambda^d)_{\|\cdot\|^2}$ has the univariate Lebesgue density $\frac{\omega_{d-1}}{2} r^{d/2-1}$, where $\omega_{d-1} = 2\pi^{d/2}/\Gamma(\frac{d}{2})$ is the surface of the unit ball in $\mathbb{R}^d$.*

PROOF. The volume of the unit ball in $\mathbb{R}^d$ is $\kappa_d = \pi^{d/2}/\Gamma(\frac{d}{2}+1)$. Therefore,

$$(\lambda^d)_{\|\cdot\|^2}([a,b]) = \lambda^d(B_{\sqrt{b}}(0) \setminus B_{\sqrt{a}}(0)) = \lambda^d(B_{\sqrt{b}}(0)) - \lambda^d(B_{\sqrt{a}}(0))$$

$$= \kappa_d(b^{d/2} - a^{d/2}) = \kappa_d \cdot \frac{d}{2} \int_a^b r^{d/2-1} dr.$$

Hence, the measure on $\mathbb{R}_>$ with Lebesgue density $\kappa_d \cdot \frac{d}{2} \cdot r^{d/2-1}$ agrees with the measure $\lambda^d_{\|\cdot\|^2}$ on all intervals $[a,b]$. By the Uniqueness Theorem D.1 for σ-finite measures, they are equal. Finally, we have $\Gamma(\frac{d}{2}+1) = \frac{d}{2}\Gamma(\frac{d}{2})$, that is, $\kappa_d \cdot \frac{d}{2} = \pi^{d/2}/\Gamma(\frac{d}{2}) = \omega_{d-1}/2$. □

E.19 Proposition (Density generator vs. radial density) *There is a one-to-one correspondence between Lebesgue absolutely continuous, spherical distributions σ on $\mathbb{R}^d$ (or density generators φ) and Lebesgue absolutely continuous distributions on $\mathbb{R}_>$:*

(a) *If σ has the density generator φ, then $\sigma_{\|\cdot\|^2}$ (**radial distribution**) has the Lebesgue density $\frac{\omega_{d-1}}{2} r^{d/2-1} \varphi(r)$ (**radial density**).*

(b) *For any distribution on $\mathbb{R}_>$ with Lebesgue density g, $\varphi(r) = \frac{2g(r)}{\omega_{d-1} r^{d/2-1}}$ is the density generator of a spherical distribution; its radial density is g.*

PROOF. (a) We have for $B \subseteq \mathbb{R}_>$

$$\sigma_{\|\cdot\|^2}(B) = \int \mathbf{1}_B(r) \sigma_{\|\cdot\|^2}(dr) = \int \mathbf{1}_B(\|x\|^2) \sigma(dx) = \int \mathbf{1}_B(\|x\|^2) \varphi(\|x\|^2) \lambda^d(dx)$$

$$= \int (\varphi \mathbf{1}_B)(\|x\|^2) \lambda^d(dx) = \int (\varphi \mathbf{1}_B)(r)(\lambda^d)_{\|\cdot\|^2}(dr) \overset{E.18}{=} \frac{\omega_{d-1}}{2} \int_B \varphi(r) r^{d/2-1} \lambda(dr).$$

(b) By Lemma E.18,

$$\int_{\mathbb{R}^d} \varphi(\|x\|^2) \lambda^d(dx) = \int_{\mathbb{R}_>} \varphi(r)(\lambda^d)_{\|\cdot\|^2}(dr) = \int_{\mathbb{R}_>} \frac{2g(r)}{\omega_{d-1} r^{d/2-1}} (\lambda^d)_{\|\cdot\|^2}(dr)$$

$$= \int_{\mathbb{R}_>} g(r) \lambda(dr) = 1.$$

Therefore, φ is the density generator of a spherical distribution σ and (a) implies that its radial density is g. □

The function $\phi = -\log \varphi$ is called the **radial function** of σ. Thus, Lebesgue absolutely continuous, spherical distributions on $\mathbb{R}^d$, radial functions, densities, and distributions on the positive ray are equivalent concepts. At the same time we have cleared up which functions are radial.

E.20 Proposition

(a) *A measurable function $\varphi \colon \mathbb{R}_+ \to \mathbb{R}_+$ is a density generator if and only if*

$$\int_0^\infty r^{d/2-1}\varphi(r)\mathrm{d}r = \frac{2}{\omega_{d-1}}.$$

If $\psi \colon \mathbb{R}_+ \to \mathbb{R}_+$ is measurable and $J_d = \int_0^\infty r^{d/2-1}\psi(r)\mathrm{d}r < \infty$, then $\varphi = 2\psi/(\omega_{d-1}J_d)$ is a density generator.

(b) *The expectation of the spherical distribution σ induced by φ exists if and only if*

$$\int_0^\infty r^{(d-1)/2}\varphi(r)\mathrm{d}r < \infty.$$

Then σ is centered, that is, $\mathrm{E}\sigma = 0$.

(c) *The covariance matrix of σ exists if and only if*

$$\int_0^\infty r^{d/2}\varphi(r)\mathrm{d}r < \infty.$$

Then we have $\mathrm{V}\sigma = \alpha I_d$ with $\alpha = \frac{\omega_{d-1}}{2d}\int_0^\infty r^{d/2}\varphi(r)\mathrm{d}r$.

(d) *(Standardized version) Let a family $\left(\varphi_\gamma^{(d)}\right)_{d,\gamma}$ of density generators parameterized by dimension d and a parameter γ have the representation*

$$\varphi_\gamma^{(d)}(r) = c_d(\gamma)\psi(\gamma,r), \quad r \geq 0,$$

(that is, r and d are multiplicatively separated). The spherical distribution σ generated by $\varphi^{(d)}$ is standardized (that is, $\alpha = 1$, $\mathrm{V}\sigma = I_d$) if and only if γ satisfies the equation

$$c_d(\gamma) = 2\pi \cdot c_{d+2}(\gamma).$$

PROOF. Part (a) follows directly from Proposition E.19. The existence of the expectation in (b) follows from Lemma E.18,

$$\int_{\mathbb{R}^d} \|x\|\varphi(\|x\|^2)\lambda^d(\mathrm{d}x) = \int_{\mathbb{R}_>} \sqrt{r}\varphi(r)(\lambda^d)_{\|\cdot\|^2}(\mathrm{d}r) = \frac{\omega_{d-1}}{2}\int_{\mathbb{R}_>} r^{(d-1)/2}\varphi(r)\mathrm{d}r.$$

From $S \sim \sigma$ we infer $-S \sim \sigma$, hence $\mathrm{E}S = -\mathrm{E}S$. The first part of (c) follows from

$$\int_{\mathbb{R}^d} \|x\|^2\varphi(\|x\|^2)\lambda^d(\mathrm{d}x) = \int_0^\infty r\varphi(r)(\lambda^d)_{\|\cdot\|^2}(\mathrm{d}r) \stackrel{E.18}{=} \frac{\omega_{d-1}}{2}\int_0^\infty r^{d/2}\varphi(r)\mathrm{d}r.$$

If $S \sim \sigma$, then the spherical symmetry $U^\top S \sim S$ for all $U \in O(d)$ implies

$$\mathrm{V}\sigma = \mathrm{V}S = \mathrm{V}U^\top S = U^\top(\mathrm{V}S)U = U^\top(\mathrm{V}\sigma)U.$$

Application of this equality to the matrix U that diagonalizes $\mathrm{V}S$ shows that $\mathrm{V}S$ is diagonal. Coordinate permutations finally show that all diagonal entries are equal, say α. We have

$$d \cdot \alpha = \mathrm{tr}\,\mathrm{V}\sigma = \frac{\omega_{d-1}}{2}\int_{\mathbb{R}_+} r^{d/2}\varphi(r)\mathrm{d}r.$$

In view of (d), (c) implies

$$d \cdot \alpha = \frac{\omega_{d-1}c_d(\gamma)}{2}\int_0^\infty r^{d/2}\psi(\gamma,r)\mathrm{d}r.$$

Since ψ does not depend on d, an application of (a) with the dimension $d+2$ instead of d yields

$$c_{d+2}(\gamma)\int_0^\infty r^{d/2}\psi(\gamma,r)\mathrm{d}r = \int_0^\infty r^{d/2}\varphi_\gamma^{(d+2)}(r)\mathrm{d}r = \frac{2}{\omega_{d+1}}$$

for all γ and d. Therefore, $\alpha = 1$ if and only if

$$d = \frac{\omega_{d-1}c_d(\gamma)2}{2\omega_{d+1}c_{d+2}(\gamma)} = \frac{\pi^{d/2}\Gamma(\frac{d}{2}+1)}{\pi^{d/2+1}\Gamma(\frac{d}{2})} \frac{c_d(\gamma)}{c_{d+2}(\gamma)} = \frac{d}{2\pi}\frac{c_d(\gamma)}{c_{d+2}(\gamma)}$$

by $\Gamma(x+1) = x\Gamma(x)$. This is claim (d).   □

Part (d) says that it is sufficient to compute the normalizing constants $c_d(\gamma)$ for all dimensions d in order to know all standardizing parameters γ, too.

E.21 Lemma *The following three conditions are equivalent for a distribution μ on $\mathbb{R}^d$ with Fourier transform $\widehat{\mu}$:*

(i) *μ is spherical;*

(ii) *$\widehat{\mu}$ is rotationally invariant;*

(iii) *there exists a real-valued function $\zeta\colon \mathbb{R}_+ \to \mathbb{R}$ such that $\widehat{\mu}(y) = \zeta(y^\top y)$ for all $y \in \mathbb{R}^d$.*

PROOF. For all orthogonal $d \times d$ matrices U and all y, we have

$$\widehat{\mu}(U^\top y) = \int_{\mathbb{R}^d} e^{-ix^\top U^\top y}\mu(\mathrm{d}x) = \int_{\mathbb{R}^d} e^{-i(Ux)^\top y}\mu(\mathrm{d}x) = \int_{\mathbb{R}^d} e^{-ix^\top y}\mu_U(\mathrm{d}x) = \widehat{\mu_U}(y).$$

This identity and the identity theorem for Fourier transforms shows that (i) and (ii) are equivalent.

Now let ψ be any rotationally invariant function on $\mathbb{R}^d$, that is, $\psi(Uy) = \psi(y)$ for all orthogonal matrices U and all y. For each y there exists an orthogonal matrix U_y such that $U_y y = (\|y\|, 0, \ldots, 0) = (\sqrt{y^\top y}, 0, \ldots, 0)$. Hence,

$$\psi(y) = \psi(U_y y) = \psi(\sqrt{y^\top y}, 0, \ldots, 0) = \zeta(y^\top y).$$

Finally it is clear that (iii) implies (ii).   □

The function ζ in Lemma E.21(iii) is called the **characteristic generator** of the spherical distribution μ. There is the following relationship between the characteristic and the density generator:

$$\zeta(y^\top y) = \int_{\mathbb{R}^d} e^{-iy^\top x}\varphi(x^\top x)\lambda^d(\mathrm{d}x).$$

Plainly, any characteristic generator ζ satisfies $\zeta(0) = 1$ and the characteristic generator of any Lebesgue density is continuous.

The **modified Bessel function of the second kind**, K_ν, appears as standard in spherical problems. So also here. Definitions handy for our purposes are **Basset's formula**

$$K_\nu(u) = \frac{\Gamma(\nu+\frac{1}{2})}{\sqrt{\pi}\left(\frac{u}{2}\right)^\nu}\int_0^\infty \frac{\cos(ut)}{(1+t^2)^{\nu+\frac{1}{2}}}\mathrm{d}t, \tag{E.3}$$

and **Schläfli's formula**

$$K_\nu(u) = \int_0^\infty e^{-u\cosh t}\cosh \nu t\, \mathrm{d}t; \tag{E.4}$$

see Watson [523], formulæ 6·16(1) and 6·22(5). Both are valid for $\nu, u > 0$. We will need the asymptotic expansion of K_ν as $u \to \infty$,

$$K_\nu(u) = \sqrt{\frac{\pi}{2u}}e^{-u}\{1 + \mathcal{O}(1/u)\}; \tag{E.5}$$

see again Watson [523], 7·23(1), and p. 198 for an explanation of the symbols appearing there.

E.22 Examples (a) (*Multivariate normal distribution*) The function

$$\varphi_d(t) = c_d e^{-t/2}, \quad t \geq 0,$$

with $c_d = (2\pi)^{-d/2}$ is the density generator of the spherically normal family with Lebesgue density $\frac{1}{(2\pi v)^{d/2}} e^{-\|x\|^2/2v}$; see Example E.17. By Proposition E.20(d), its variance is v. It has moments of all orders. Its characteristic generator $\zeta(u) = e^{-u/2}$ does not depend on dimension.

(b) (*Pearson's type-VII family*) Pearson [406] proposed a number of probabilistic models,

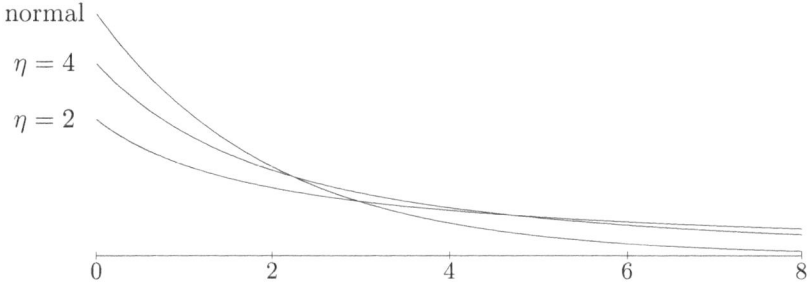

Figure E.3 *Pearson's type-VII density generators $\varphi_{\eta,\lambda,1}$ with indices $\eta = \lambda = 2, 4$, and the normal density generator.*

among them a shape, location, and scale family of distributions on Euclidean space, **Pearson's type-VII** distributions. They are defined by the density generator

$$\varphi_{\eta,\lambda,d}(t) = c_{\eta,\lambda,d}(1 + t/\lambda)^{-\eta/2} \tag{E.6}$$

with the scale parameter $\lambda > 0$ and the shape parameter η; see Fig. E.3. It generates a distribution if and only if $\eta > d$ and then the normalizing constant is $c_{\eta,\lambda,d} = \frac{\Gamma(\eta/2)}{(\pi\lambda)^{d/2}\Gamma((\eta-d)/2)}$. Decaying polynomially as $t \to \infty$, this function has a heavy tail – the heavier the smaller the index is. Pearson's type-VII model is differentiable. Its kth moment exists if and only if $\eta > d + k$. By Proposition E.20(d), $\varphi_{\eta,\lambda,d}$ is standardized if $\lambda = \eta - (d+2)$. Inserting this value for λ, we obtain the simplified family

$$\varphi_{\eta,d}(t) = c_{\eta,d}\left(1 + \frac{t}{\eta - (d+2)}\right)^{-\eta/2}, \ c_{\eta,d} = \frac{\Gamma(\eta/2)}{(\pi(\eta - (d+2)))^{d/2}\Gamma((\eta-d)/2)}, \tag{E.7}$$

valid for $\eta > d + 2$. As the index η tends to ∞, the density generator (E.7) tends to that of the normal family, uniformly in $t \geq 0$,

$$c_{\eta,d}\left(1 + \frac{t}{\eta - (d+2)}\right)^{-\eta/2} \xrightarrow[\eta\to\infty]{} (2\pi)^{-d/2}e^{-t/2}.$$

Let us call the combined family the **extended Pearson type-VII family**. According to Fang et al. [148], Theorem 3.9, the characteristic generator of (E.6) is

$$\zeta_{\eta,\lambda,d}(u) = k_{\eta,d}\int_0^\infty \frac{\cos(\sqrt{\lambda u}\,t)}{(1+t^2)^{(\eta-d+1)/2}}\,dt, \quad k_{\eta,d} = \frac{2\Gamma((\eta-d+1)/2)}{\sqrt{\pi}\Gamma((\eta-d)/2)}.$$

A comparison with Basset's formula shows that it can be represented by the modified Bessel function of the second kind ($\nu = (\eta-d)/2 > 0$),

$$\zeta_{\eta,d}(u) = k_{\eta,d}\frac{\sqrt{\pi}\left(\frac{\sqrt{\lambda u}}{2}\right)^\nu}{\Gamma(\nu + \frac{1}{2})}K_\nu(\sqrt{\lambda u}) = \frac{\sqrt{\lambda u}^\nu}{2^{\nu-1}\Gamma(\nu)}K_\nu(\sqrt{\lambda u})$$

$$= \sqrt{\pi}\frac{(\sqrt{\lambda u}/2)^{\nu-1/2}e^{-\sqrt{\lambda u}}}{\Gamma(\nu)}\{1 + \mathcal{O}(1/\sqrt{u})\}. \tag{E.8}$$

The asymptotic behavior as $u \to \infty$ follows from (E.5).

(c) Let us now insert $\eta = \lambda + d$ into Pearson's (spherical) type-VII distribution (E.6). If $\lambda \geq 1$ is integral then it has a stochastic interpretation; see Cornish [102] and Dunnett and Sobel [135]: Let $X \sim X_{0,I_d}$ be standard d-variate normal and let χ_λ^2 be independent of X. The random vector $S = \frac{X}{\sqrt{\chi_\lambda^2/\lambda}}$ is spherical and Lebesgue continuous with the Pearson type-VII density generator

$$\varphi_{\lambda+d,\lambda,d}(t) = c_{\lambda+d,\lambda,d}\big(1 + t/\lambda\big)^{-(\lambda+d)/2}, \tag{E.9}$$

where $c_{\lambda+d,\lambda,d} = \frac{\Gamma((\lambda+d)/2)}{(\lambda\pi)^{d/2}\Gamma(\lambda/2)}$. For a proof, put $Z = \chi_\lambda^2/\lambda$. The map $\Phi(s,u) = (\sqrt{u}\,s, u)$ is a diffeomorphism of $\mathbb{R}^d \times \mathbb{R}_>$. The Jacobian matrix of Φ at (s,u) is

$$\begin{pmatrix} \sqrt{u} & & & & \frac{s_1}{2\sqrt{u}} \\ & \sqrt{u} & & 0 & \frac{s_2}{2\sqrt{u}} \\ & & \ddots & & \vdots \\ 0 & & \ddots & & \vdots \\ & & & \sqrt{u} & \frac{s_d}{2\sqrt{u}} \\ & & & & 1 \end{pmatrix}$$

and its determinant equals $u^{d/2}$. Since $\Phi(S, Z) = (X, Z)$, the Transformation Theorem E.7 for Lebesgue densities and independence of X and Z assert that the joint density of S and Z is given by

$$f_{(S,Z)}(s,u) = u^{d/2}f_{(X,Z)}(\Phi(s,u)) = u^{d/2}f_X(\sqrt{u}\,s)f_Z(u)$$
$$= u^{d/2}f_X(\sqrt{u}\,s)\lambda f_{\chi_\lambda^2}(\lambda u) = c'_{\lambda,d}\,e^{-u(\lambda+\|s\|^2)/2}u^{(\lambda+d)/2-1}.$$

with the constant $c'_{\lambda,d} = \frac{\lambda^{\lambda/2}}{(2\pi)^{d/2}2^{\lambda/2}\Gamma(\lambda/2)}$. Integrating, we obtain the marginal density

$$f_S(s) = c'_{\lambda,d}\int_0^\infty e^{-u(\lambda+\|s\|^2)/2}u^{(\lambda+d)/2-1}\mathrm{d}u$$

$$= c'_{\lambda,d}\frac{2^{(\lambda+d)/2}}{(\lambda+\|s\|^2)^{(\lambda+d)/2}}\int_0^\infty e^{-r}r^{(\lambda+d)/2-1}\mathrm{d}r = \frac{\Gamma((\lambda+d)/2)}{(\lambda\pi)^{d/2}\Gamma(\lambda/2)}\big(1 + \tfrac{1}{\lambda}\|s\|^2\big)^{-(\lambda+d)/2}.$$

This is the claim.   □

The distribution generated by (E.9) is called the **standard multivariate t-distribution** with **degrees of freedom** λ. The case $\lambda = 1$ is termed the multivariate **Cauchy distribution**.

E.3.2 Elliptical symmetry

Spherical random vectors rarely occur in the real world. They are, however, the basis for the more realistic elliptical distributions which they generate by affine transformations.

E.23 Definition A random vector $X\colon (\Omega, P) \to \mathbb{R}^d$ is called **elliptically symmetric** or just **elliptical** if it is the affine image of a spherical random variable, that is, if

$$X = m + AS$$

for $m \in \mathbb{R}^d$, $A \in \mathrm{GL}(d)$, S spherical. Without loss of generality, we assume $VS = I_d$ when $V\|S\| < \infty$.

We may assume that A is positive definite. All normal random vectors are elliptical. The Lebesgue density of an elliptical random vector is obtained from the Transformation Theorem E.7 for Lebesgue densities.

E.24 Proposition *(a) Let $X = m + AS$ be elliptically symmetric. If S has the density generator φ, then the Lebesgue density f_X of X has the form*

$$f_X(x) = |\det A|^{-1}\varphi\big(\|A^{-1}(x - m)\|^2\big).$$

(b) If φ is standardized, for instance, $\varphi = \varphi_{\eta,d}$ (see (E.7)), then the covariance matrix is $AA^\top$.

(c) The Fourier transform of an elliptically symmetric density with characteristic generator ζ and location and scale parameters m and V has the form

$$y \to e^{-i\,m^\top y}\zeta\big(y^\top V y\big). \tag{E.10}$$

PROOF. (a) Let $\Phi(s) = m + As$. According to the Transformation Theorem E.7 for Lebesgue densities, the density of $X = \Phi \circ S$ flows from

$$\varphi(\|s\|^2) = f_S(s) = |\det \mathrm{D}\Phi(s)| \cdot f_{\Phi \circ S}(\Phi(s)) = |\det A| \cdot f_X(m + As).$$

Part (b) follows from Proposition E.3(e) and (c) is shown by integral transformation. $\qquad\square$

The level surfaces of an elliptical symmetry are thus the ellipsoids

$$(x - m)^\top (AA^\top)^{-1}(x - m) = \|A^{-1}(x - m)\|^2 = \text{const}.$$

The elliptical symmetry with shape parameter φ, location parameter m, and scale parameter V and its Lebesgue density will be denoted by $E_{\varphi,m,V}$. Note that V is the covariance matrix only for special density generators φ. Let $F \subseteq 1..d$. The projection of $E_{\varphi,m,V}$ onto $\mathbb{R}^F$ is again elliptical. Its density generator $\widetilde{\varphi}(r) = \int \varphi(r + \|x_{\complement F}\|^2)\lambda^{d-|F|}(\mathrm{d}x_{\complement F})$ depends on F only via $|F|$; see Fang et al. [148], Section 2.2.

E.25 Example The affine transformation $x \mapsto \sqrt{V}x + m$ of the standard multivariate t-distribution with degrees of freedom λ of Example E.22(c) is called the **multivariate t-distribution** with parameters λ (degrees of freedom), m (mean), and $V \succeq 0$ (scale). Its Lebesgue density is

$$t_{\lambda,m,V}(x) = c_{\lambda+d,\lambda,d}\det V^{-1/2}\big(1 + (x - m)^\top V^{-1}(x - m)\big)^{-(\lambda+d)/2},$$

and its covariance matrix is a multiple of V. For $m = 0$, it also arises as the density of $\dfrac{Y}{\sqrt{\chi_\lambda^2/\lambda}}$, where Y is centered normal with covariance matrix V. This follows from $\dfrac{Y}{\sqrt{\chi_\lambda^2/\lambda}} \sim \sqrt{V}S$ with S as in Example E.22(c). A similar representation for $m \neq 0$ does not exist. That is, if $Z \sim N_{m,V}$, $m \neq 0$, then the density of $\dfrac{Z}{\sqrt{\chi_\lambda^2/\lambda}}$ is not $t_{\lambda,m,V}$. A detailed presentation of the multivariate t-distribution is found in Kotz and Nadarajah [297].

E.26 Lemma *Let X be elliptically symmetric with density generator φ and Lebesgue density f.*

(a) The expectation $\mathrm{E}\log^- f(X)$ exists if and only if

$$\int_0^\infty (\log^- \varphi(t))\,\varphi(t)\,t^{d/2-1}\mathrm{d}t < \infty.$$

(b) The condition is satisfied if $\varphi(t) \leq C(1 + t)^{-\beta/2}$ with $\beta > d$.

PROOF. (a) Let $X = m + \Lambda^{1/2}S$. By Lemma E.18, $\mathrm{E}\log^- f(X) = \int_{\mathbb{R}^d}(\log^- f(x))f(x)\lambda^d(\mathrm{d}x)$ equals, up to an additive constant,

$$\sqrt{\det \Lambda}\int_{\mathbb{R}^d}\big(\log^- \varphi((x - m)^\top\Lambda(x - m))\big)\varphi\big((x - m)^\top\Lambda(x - m)\big)\lambda^d(\mathrm{d}x)$$

$$= \int_{\mathbb{R}^d}(\log^- \varphi(\|x\|^2))\varphi(\|x\|^2)\lambda^d(\mathrm{d}x) = \text{const}\int_0^\infty(\log^- \varphi(t))\varphi(t)t^{d/2-1}\mathrm{d}t.$$

(b) By de l'Hospital's rule, $(\log^- u)u^\varepsilon$ converges to zero as $u \to 0$ for all $\varepsilon > 0$. For all large t, it follows

$$(\log^- \varphi(t))\, \varphi(t)\, t^{d/2-1} = (\log^- \varphi(t))\, \varphi^\varepsilon(t)\varphi^{1-\varepsilon}(t)\, t^{d/2-1}$$
$$\leq \varphi^{1-\varepsilon}(t)\, t^{d/2-1} \leq \mathrm{const}\,(1+t)^{-\beta(1-\varepsilon)/2}\, t^{d/2-1}.$$

The claim now follows from part (a).   □

Finiteness of $\int_0^\infty \varphi(t)\, t^{d/2-1}\mathrm{d}t$ does not suffice in (a). There do exist functions φ such that $\int_0^\infty \varphi(t)\, t^{d/2-1}\mathrm{d}t < \infty$ and $\int_0^\infty (\log^- \varphi(t))\, \varphi(t)\, t^{d/2-1}\mathrm{d}t = \infty$. An example is $\varphi(t) = \frac{\mathrm{const}}{(t+2)^{d/2}\log^2(t+2)}$.

E.27 Proposition (Identifiability of the elliptical family)
(a) Let S be spherical with finite second moment, let $X = AS + m$ with $A \in \mathrm{PD}(d)$, and let $f_{d_X(X,\mathrm{E}X)}$ be the density of the random number

$$\omega \mapsto d_X(X(\omega), \mathrm{E}X) = \|(\mathrm{V}X)^{-1/2}(X(\omega) - \mathrm{E}X)\|.$$

The parameters of X are functions of the density f_X:

   (i) $m = \mathrm{E}X$.
   (ii) If S is standardized (that is, $\mathrm{V}S = I_d$) then $A = \sqrt{\mathrm{V}X}$.
   (iii) $\varphi(r) = f_{\|(\mathrm{V}X)^{-1/2}\widehat{X}\|^2}(r)/(\frac{\omega_{d-1}}{2}r^{d/2-1}) = f_{d_X^2(X,\mathrm{E}X)}(r)/(\frac{\omega_{d-1}}{2}r^{d/2-1})$;
   $\varphi(\|s\|^2) = (\det A)f_X(As+m)$.

(b) The family of all elliptical symmetries of finite variance is identifiable.

PROOF. Claim (i) is plain and (ii) follows from $\mathrm{E}SS^\top = I_d$ and the positive definiteness of A: $\mathrm{V}X = \mathrm{E}AS(AS)^\top = A\mathrm{E}SS^\top A^\top = AA^\top$. In view of (iii), we use (i) and (ii) to compute

$$\|(\mathrm{V}X)^{-1/2}\widehat{X}\|^2 = (X - \mathrm{E}X)^\top (\mathrm{V}X)^{-1}(X - \mathrm{E}X)$$
$$= (X - m)^\top A^{-2}(X - m) = (AS)^\top A^{-2}AS = \|S\|^2,$$

Hence, $f_{\|(\mathrm{V}X)^{-1/2}\widehat{X}\|^2}$ is the radial density of S and claim (a) follows from Proposition E.19. Claim (b) is a direct consequence.   □

Proposition E.27(a)(iii) can be used to estimate the density generator via the (univariate) radial density $f_{\|(\mathrm{V}X)^{-1/2}\widehat{X}\|^2} = f_{d_X^2(X,\mathrm{E}X)}$. This is usually performed by approximating the histogram by *parameter-free* methods such as splines.

Proposition E.11 implies that a multivariate normal random vector X possesses for $d \geq 2$ nontrivial projectors Π (see Explanation A.7(c)) such that ΠX and $(I_d - \Pi)X$ are independent. The following theorem says that this property is characteristic for the normal distribution within the elliptical symmetries. For a proof, consult Gupta and Varga [216], p. 193, or Fang et al. [148], p. 106.

E.28 Theorem If $X\colon \Omega \to \mathbb{R}^d$, $d \geq 2$, is elliptical and if there is a projector Π of rank between 1 and d–1 such that ΠX and $(I_d - \Pi)X$ are independent, then X is normal.

E.4   Average convergences of random variables

Pointwise, stochastic, and mean convergence of measurable functions apply in particular to random variables. An application to averages of real- and vector-valued random variables leads to the so-called Laws of Large Numbers. They provide some of the best justifications for the probabilistic view on random events.

E.4.1 Laws of large numbers

The Laws of Large Numbers are concerned with the convergence of arithmetic means of sequences of random numbers and vectors. Stochastic convergence is in most cases proved by integral estimates. The following inequality follows from Markov's inequality D.24 applied with $f = \|\widehat{Y}\|_2^2$.

E.29 Lemma (Chebyshev's inequality) *For any random vector Y we have*

$$P[\|\widehat{Y}\|_2 \geq \alpha] \leq \frac{\operatorname{tr} VY}{\alpha^2}.$$

The Weak Law of Large Numbers is an easy consequence of Bienaymé's Formula and Chebyshev's inequality. See Durrett [137].

E.30 Proposition (Weak Law of Large Numbers (WLLN)) *Let $Y_1, Y_2, \cdots$ be a sequence of random vectors with common distribution μ. If the sequence is uncorrelated and if $V\|Y_i\|^2$ is bounded, then, in $\mathbb{L}^2(P)$ and P-stochastically,*

$$\frac{1}{n} \sum_{1 \leq k \leq n} Y_k \xrightarrow[n\to\infty]{} EY_1 = \int y \, \mu(\mathrm{d}y).$$

As an instance of stochastic convergence, the Weak Law allows occasional deviations from the limit as n grows. The following Strong Law provides a.s. convergence which does not admit such behavior. However, it needs instead of uncorrelatedness the more stringent independence.

E.31 Theorem (Strong Law of Large Numbers (SLLN))

(a) Let $Y_1, Y_2, \ldots$ be an i.i.d. sequence of real-valued random variables with common distribution μ. If Y_1^- is integrable, then, P-a.s. (the expectation may be ∞),

$$\liminf_n \frac{1}{n} \sum_{i=1}^{n} Y_i \geq EY_1 = \int y \, \mu(\mathrm{d}y),$$

(b) Let $Y_1, Y_2, \ldots$ be an i.i.d. sequence of integrable random vectors with common distribution μ. Then, P-a.s.,

$$\frac{1}{n} \sum_{i=1}^{n} Y_i \xrightarrow[n\to\infty]{} EY_1 = \int y \, \mu(\mathrm{d}y).$$

Part (b) is a simple consequence of (a). The Strong Law is remarkable, teaching us three things: First, the averages on the left side converge P-a.s.; second, the limit is not random but a constant; third, the constant may be computed as the common expectation of the random variables. Of course, the SLLN may be applied to a sequence $g(X_i)$, where $g \colon E \to \mathbb{R}$ is measurable and $(X_i)_i$ is i.i.d. $\sim \mu$ with values in E such that the expectation of $g(X_1)$ exists. The SLLN immediately implies $\frac{1}{n} \sum_{i=1}^{n} g(X_i) \xrightarrow[n\to\infty]{} Eg(X_1) = \int g(x)\mu(\mathrm{d}x)$, P-a.s.

The proofs of the consistency theorems in Section 1.1 rely on an extension of the SLLN for a parameterized function $g \colon \Theta \times E \to \mathbb{R}^d$. This *uniform* SLLN is due to Jennrich [276]. In order to secure measurability, we interject a technical lemma.

E.32 Lemma (Measurability of suprema) *Let Θ be a separable topological space (see Appendix B.1), let E be a measurable space, and let $F \colon \Theta \times E \to \mathbb{R}$. Assume that*

(i) $F(\cdot, x)$ is continuous for all $x \in E$ and that

(ii) $F(\vartheta, \cdot)$ is measurable for all $\vartheta \in \Theta$.

Then $x \mapsto \sup_{\vartheta \in \Theta} F(\vartheta, x)$ is measurable (as an extended real-valued function).

PROOF. Abbreviate $h(x) = \sup_{\vartheta \in \Theta} F(\vartheta, x)$. Let D be a countable, dense subset of Θ and let $\varepsilon > 0$. If $x \in E$ is such that $h(x) < \infty$, there is $\vartheta \in \Theta$ such that $F(\vartheta, x) \geq h(x) - \varepsilon$. By continuity (i) of $F(\cdot, x)$ and by density, there is $\eta \in D$ such that $F(\eta, x) \geq F(\vartheta, x) - \varepsilon$. Hence, $F(\eta, x) \geq h(x) - 2\varepsilon$. It follows $\sup_{\vartheta \in D} F(\vartheta, x) \geq h(x)$ and hence $\sup_{\vartheta \in D} F(\vartheta, x) = h(x)$. A similar argument applies to all x with $h(x) = \infty$. We have thus reduced the arbitrary sup to a countable one which makes it measurable by (ii). $\qquad \square$

The following theorem is an important generalization of the SLLN.

E.33 Theorem (Uniform Strong Law of Large Numbers (Uniform SLLN)) *Let Θ be a compact, metric space, let $g\colon E \times \Theta \to \mathbb{R}^d$, and let $X_1, X_2, \ldots$ be an i.i.d. sequence of E-valued random variables with common distribution μ. Assume that*

(i) *$g(x, \cdot)$ is continuous for all $x \in E$;*

(ii) *$g(\cdot, \vartheta)$ is measurable for all $\vartheta \in \Theta$;*

(iii) *$\|g(x, \vartheta)\| \leq h(x)$ for all $\vartheta \in \Theta$ and all $x \in E$ with some measurable, μ-integrable function h on E.*

Then,

(a) *P-a.s., we have $\frac{1}{n} \sum_{i=1}^{n} g(X_i, \vartheta) \underset{n \to \infty}{\longrightarrow} \mathrm{E} g(X_1, \vartheta)$ uniformly for all $\vartheta \in \Theta$.*

(b) *In particular, $\mathrm{E} g(X_1, \cdot) = \int_E g(x, \cdot) \mu(\mathrm{d}x)$ is continuous.*

PROOF. It suffices to prove the theorem for scalar functions g. The proof uses the SLLN and rests on the following estimate valid for any open set $B \subseteq \Theta$. Note here that $\sup_{\vartheta \in B} g(X_i, \vartheta)$ and $\inf_{\vartheta \in B} g(X_i, \vartheta)$ both are measurable by Lemma E.32 and the fact that compact metric spaces are separable.

$$\sup_{\vartheta \in B} \left| \frac{1}{n} \sum_{i=1}^{n} g(X_i, \vartheta) - \mathrm{E} g(X_1, \vartheta) \right| \tag{E.11}$$

$$\leq \left| \frac{1}{n} \sum_{i=1}^{n} \sup_{\vartheta \in B} g(X_i, \vartheta) - \mathrm{E} \sup_{\vartheta \in B} g(X_1, \vartheta) \right| + \left| \frac{1}{n} \sum_{i=1}^{n} \inf_{\vartheta \in B} g(X_i, \vartheta) - \mathrm{E} \inf_{\vartheta \in B} g(X_1, \vartheta) \right|$$

$$+ \mathrm{E} \left(\sup_{\vartheta \in B} g(X_1, \vartheta) - \inf_{\vartheta \in B} g(X_1, \vartheta) \right).$$

Indeed, we have

$$\sup_{\vartheta \in B} \left(\frac{1}{n} \sum_{i=1}^{n} g(X_i, \vartheta) - \mathrm{E} g(X_1, \vartheta) \right) \leq \frac{1}{n} \sum_{i=1}^{n} \sup_{\vartheta \in B} g(X_i, \vartheta) - \mathrm{E} \inf_{\vartheta \in B} g(X_1, \vartheta)$$

$$\leq \left| \frac{1}{n} \sum_{i=1}^{n} \sup_{\vartheta \in B} g(X_i, \vartheta) - \mathrm{E} \sup_{\vartheta \in B} g(X_1, \vartheta) \right| + \mathrm{E} \sup_{\vartheta \in B} g(X_1, \vartheta) - \mathrm{E} \inf_{\vartheta \in B} g(X_1, \vartheta)$$

and, replacing g with $-g$,

$$- \inf_{\vartheta \in B} \left(\frac{1}{n} \sum_{i=1}^{n} g(X_i, \vartheta) - \mathrm{E} g(X_1, \vartheta) \right) \leq - \frac{1}{n} \sum_{i=1}^{n} \inf_{\vartheta \in B} g(X_i, \vartheta) + \mathrm{E} \sup_{\vartheta \in B} g(X_1, \vartheta)$$

$$\leq \left| \frac{1}{n} \sum_{i=1}^{n} \inf_{\vartheta \in B} g(X_i, \vartheta) - \mathrm{E} \inf_{\vartheta \in B} g(X_1, \vartheta) \right| - \mathrm{E} \inf_{\vartheta \in B} g(X_1, \vartheta) + \mathrm{E} \sup_{\vartheta \in B} g(X_1, \vartheta).$$

Estimate (E.11) now follows from the estimate

$$\sup_{\vartheta} |a_\vartheta| \leq \max(|\sup_{\vartheta} a_\vartheta|, |\inf_{\vartheta} a_\vartheta|)$$

valid for all families (a_ϑ) of real numbers.

Now, let $B = B_r(\eta)$ be any open ball of radius r in the metric space Θ. Continuity of the functions $g(x, \cdot)$ asserts that

$$\sup_{\vartheta \in B_r(\eta)} g(x, \vartheta) - \inf_{\vartheta \in B_r(\eta)} g(x, \vartheta) \underset{r \to 0}{\longrightarrow} 0$$

for all $x \in E$ and all $\eta \in \Theta$. Hence, Lebesgue's dominated convergence along with assumption (iii) implies that, for all η, there is a ball $B(\eta)$ such that

$$\mathrm{E}\left(\sup_{\vartheta \in B(\eta)} g(X_1, \vartheta) - \inf_{\vartheta \in B(\eta)} g(X_1, \vartheta) \right) \leq \varepsilon.$$

It is here that we have used metrizability of Θ. Using again assumption (iii) we apply the Strong Law E.31 to the (measurable and) integrable random variables $\sup_{\vartheta \in B(\eta)} g(X_1, \vartheta)$ and $\inf_{\vartheta \in B(\eta)} g(X_1, \vartheta)$, showing that the first two summands in (E.11) become $\leq \varepsilon$ whenever n is large. Finitely many balls $B(\eta_1), \ldots, B(\eta_\ell)$ cover the compact space Θ and so part (a) follows from combining finitely many suprema in (E.11). Part (b) is a direct consequence.□

E.4.2   Weak convergence and convergence in distribution

The convergences of Definition D.21 also apply to random variables. The so-called convergence in distribution below is the weakest kind of convergence of a sequence of random variables. It is needed in the Central Limit Theorem and refers actually to convergence of their distributions. Standard references are the monographs of Billingsley [41] and Parthasarathy [401].

A sequence $(\mu_n)_n$ of Borel probabilities on a metric space E **converges weakly** to a probability measure μ on E, $\mu_n \Rightarrow \mu$, if

$$\int h \, d\mu_n \xrightarrow[n \to \infty]{} \int h \, d\mu$$

for all bounded, continuous functions $h\colon E \to \mathbb{R}$. If E is the real line, then an equivalent definition in terms of distribution functions is given by *Helly's theorem*: $(\mu_n)_n$ converges to μ weakly if and only if the distribution functions of F_{μ_n} converge to F_μ at every point of continuity of F_μ. This theorem reduces weak convergence of probabilities to pointwise convergence of functions.

Note two simple examples of weak convergence: (i) If the sequence a_n of points in E converges to $a \in E$, then the point masses δ_{a_n} converge weakly to the point mass δ_a. (ii) Let μ_n be the (discrete) uniform probability on the subset $\left\{ \frac{k}{n} \mid 0 \leq k < n \right\} \subset [0, 1]$. Each point of the form k/n, $0 \leq k < n$, bears mass $1/n$. The sequence (μ_n) converges weakly to the Lebesgue measure on the unit interval.

E.34 Properties

(a) Weak limits are unique: If $\mu_n \Rightarrow \mu$ and $\mu_n \Rightarrow \nu$, then $\mu = \nu$.

(b) If $\mu_n \Rightarrow \mu$ and $\nu_n \Rightarrow \nu$ and if $0 \leq p \leq 1$, then $(1 - p)\mu_n + p\nu_n \Longrightarrow (1 - p)\mu + p\nu$.

(c) Let E be Polish; see Section B.1. If $\mu_n \Rightarrow \mu$, then the set $\mathcal{P} = \{\mu, \mu_1, \mu_2, \ldots\}$ is *tight*, that is, for every $\varepsilon > 0$ there is a compact subset $K \subseteq E$ such that $\mu(K), \mu_n(K) \geq 1 - \varepsilon$ for all n.

If $\mu_n \Rightarrow \mu$ and if $B \subseteq E$ is measurable, then it is not necessarily true that $\mu_n(B) \to \mu(B)$. Just take μ_n as the Dirac measure $\delta_{1/n}$ and $B = {]}0, 1]$. For certain measurable subsets, however, equality holds. The topological boundary of a subset $B \subseteq E$ is $\partial B = \overline{B} \backslash \overset{\circ}{B}$.

E.35 Definition A subset $B \subseteq E$ is called **boundariless** for a probability μ on E if $\mu(\partial B) = 0$.

Intervals in the real line, rectangles in the plane, more generally hyperrectangles in Euclidean space, Euclidean balls, and more generally convex sets are all boundariless for any Lebesgue absolutely continuous probability. A singleton $\{b\}$ is boundariless for μ if and only if $\mu(\{b\}) = 0$. A countable dense subset B of Euclidean space $\mathbb{R}^q$ is boundariless for no probability measure on E since $\partial B = \mathbb{R}^q$. It is possible to check weak convergence on a restricted collection of sets or functions.

E.36 Proposition *Let μ_n and μ be probability measures on E. The following statements are equivalent:*

(a) $\mu_n \Rightarrow \mu$;

(b) (Portmanteau Theorem) $\int h \, d\mu_n \xrightarrow[n \to \infty]{} \int h \, d\mu$ *for all bounded, uniformly continuous functions* $h \colon E \to \mathbb{R}$;

(c) $\mu_n(B) \to \mu(B)$ *for all measurable, μ-boundariless subsets* $B \subseteq E$.

Weak convergence of probabilities gives rise to a very weak concept of convergence of random variables.

E.37 Definition A sequence (Y_n) of E-valued random variables is said to **converge in distribution** to the E-valued random variable Y if P_{Y_n} converges weakly to P_Y.

By the Portmanteau Theorem this is equivalent to having $\mathrm{E}h(Y_n) \to \mathrm{E}h(Y)$ as $n \to \infty$ for all bounded, (uniformly) continuous, real-valued functions h on E. With the exception of Proposition E.38(c) below, this type of convergence does not imply convergence of the random values $Y_n(\omega)$ in E in any sense of Definition D.21. A convincing example is this: Let Ω be the unit interval, let P be Lebesgue measure, let $E = \{0, 1\}$ the two-point set and, for $n \geq 0$, let

$$Y_n = \begin{cases} \mathbf{1}_{[0,1/2]}, & n \text{ even,} \\ \mathbf{1}_{]1/2,1]}, & n \text{ odd.} \end{cases}$$

Then $P_{Y_n} = \frac{1}{2}\delta_0 + \frac{1}{2}\delta_1$, the uniform distribution on E. Thus, the sequence of distributions is even constant. Hence, $Y_n \underset{n \to \infty}{\Longrightarrow} \frac{1}{2}\delta_0 + \frac{1}{2}\delta_1$, but the sequence (Y_n) is far from converging in E in any sense. Nevertheless, if E is Polish then a P-a.s. convergent sequence of random variables $Y_n' \sim Y_n$ can always be constructed; see Billingsley [42], pp. 337 ff. Since a limiting random variable Y is often not explicitly given in a natural way, just a limiting distribution, we will also say that Y_n converges in distribution to μ to mean that P_{Y_n} converges weakly to μ as $n \to \infty$. There are the following relationships between convergence in distribution and the convergences of Definition D.21.

E.38 Proposition

(a) Convergence in the pth mean implies convergence in distribution.

(b) P-stochastic convergence implies convergence in distribution.

(c) There is one special situation where convergence in distribution is actually very strong, namely, when the limiting random variable is a constant (that is, the limiting distribution is a point mass). Then the convergence is even P-stochastic.

Convergence in distribution of two sequences $(X_n)_n$ and $(Y_n)_n$ does not in any way imply convergence of the pair (X_n, Y_n). As a counterexample, consider a uniform random variable Z in $\{0,1\}^2$ and the two projections $\pi_i \colon \{0,1\}^2 \to \{0,1\}$, $i = 1,2$. Let $X_n = \pi_1(Z)$ for all n and $Y_n = \pi_1(Z)$ for odd and $Y_n = \pi_2(Z)$ for even n. Then $(X_n, Y_n) = Z$ for even n and (X_n, Y_n) is distributed uniformly on the diagonal of $\{0,1\}^2$ for odd n and there is no convergence, not even in distribution.

E.39 Lemma *Let F be another metric space and let $\varphi \colon E \to F$ be continuous. If the sequence (Y_n) of E-valued random variables converges to Y in distribution, then $\varphi(Y_n)$ converges to $\varphi(Y)$ in distribution.*

This lemma and Property D.22(d) state that convergence in distribution and stochastic convergence are preserved by continuous maps. Of course, this also holds for a.e. convergence. These three facts are called the **Continuous Mapping Lemma**. There is a useful parameterized version. We interject another lemma.

E.40 Lemma *Let E, F, and G be metric spaces, F compact. Let $\Phi\colon E \times F \to G$ be continuous, and let $a \in E$. Then the cuts $\Phi(x, \cdot)$ converge to $\Phi(a, \cdot)$ uniformly as $x \to a$.*

PROOF. Assume on the contrary that there are $\varepsilon > 0$ and sequences $x_n \to a$ and $y_n \in F$ such that $d(\Phi(x_n, y_n), \Phi(a, y_n)) \geq \varepsilon$ for all n. A subsequence y_{n_k} of y_n converges to some y in the compact space F. The inequality

$$d(\Phi(x_{n_k}, y_{n_k}), \Phi(a, y_{n_k})) \leq d(\Phi(x_{n_k}, y_{n_k}), \Phi(a, y)) + d(\Phi(a, y), \Phi(a, y_{n_k}))$$

and continuity of Φ now lead to a contradiction.   □

E.41 Lemma (Slutsky) *Let E and G be metric spaces, let F be a Polish space, and let $\Phi\colon E \times F \to G$ be continuous. If the sequence $(X_n)_n$ of E-valued random variables converges to a constant $a \in E$ in any sense of Definition D.21 [1] and if the sequence $(Y_n)_n$ of F-valued random variables converges to Y in distribution, then the sequence $(\Phi(X_n, Y_n))_n$ of G-valued random variables converges to $\Phi(a, Y)$ in distribution.*

PROOF. Using the Portmanteau Theorem, Proposition E.36(b), we estimate for any bounded, uniformly continuous function $h\colon G \to \mathbb{R}$

$$|\mathrm{E}h(\Phi(X_n, Y_n)) - \mathrm{E}h(\Phi(a, Y))|$$
$$\leq \mathrm{E}|h(\Phi(X_n, Y_n)) - h(\Phi(a, Y_n))| + \mathrm{E}|h(\Phi(a, Y_n)) - h(\Phi(a, Y))|.$$

The second term on the right converges to zero by the Continuous Mapping Lemma applied to the map $h(\Phi(a, \cdot))$. But the first term, too, vanishes as $n \to \infty$. Let $\varepsilon > 0$. By tightness, Property E.34(d), there is a compact subset $K \subseteq F$ such that $P(Y_n \in \complement K) \leq \varepsilon$ for eventually all n. We estimate for any $\delta > 0$

$$\mathrm{E}|h(\Phi(X_n, Y_n)) - h(\Phi(a, Y_n))|$$
$$\leq \mathrm{E}[|h(\Phi(X_n, Y_n)) - h(\Phi(a, Y_n))|; d(X_n, a) \leq \delta \text{ and } Y_n \in K]$$
$$+ \mathrm{E}[|h(\Phi(X_n, Y_n)) - h(\Phi(a, Y_n))|; d(X_n, a) > \delta]$$
$$+ \mathrm{E}[|h(\Phi(X_n, Y_n)) - h(\Phi(a, Y_n))|; Y_n \notin K].$$

According to Lemma E.40, there is $\delta > 0$ such that $|\Phi(x, y) - \Phi(a, y)| \leq \varepsilon$ if $d(x, a) \leq \delta$ and $y \in K$. Uniform continuity of h shows that the integrand $|h(\Phi(X_n, Y_n)) - h(\Phi(a, Y_n))|$ of the first term on the right side is small. By Proposition D.23(a),(c), we have stochastic convergence of X_n in each of the three cases. Thus, the second term on the right converges to zero as $n \to \infty$. Finally, the third term is less than $2\varepsilon\|h\|$ by assumption on K.   □

E.4.3   Central limit theorem

The Strong Law of Large Numbers E.31 may be reformulated as

$$\frac{1}{n} \sum_{i=1}^{n} \widehat{Y}_i \xrightarrow[n \to \infty]{} 0, \quad P\text{-a.s.,} \tag{E.12}$$

where the hat indicates centering, $\widehat{Y}_i = Y_i - \mathrm{E}Y_1$. The question arises whether the number n in the denominator can be diminished without giving up the convergence. This question has a clear answer if Y_i is square integrable. If n is replaced with n^α for any $\alpha > 1/2$, then convergence to zero remains unchanged. If $0 < \alpha < 1/2$, then there is no convergence in (E.12) in any sense. This indicates that something interesting may be happening as n is replaced with some expression close to $\sqrt{n}$. Indeed, let Y_i be scalar and replace n with $\sqrt{n \log \log n}$. The Law of the Iterated Logarithm (see Durrett [137]) ensures that the sequence remains bounded and that the upper limit $\limsup_n \frac{1}{\sqrt{n \log \log n}} |\sum_{i=1}^{n} \widehat{Y}_i|$ is the standard deviation of Y_i, P-a.s. That is, P-a.s., (E.12) cannot remain bounded if any function $\varphi(n)$ such that

---

[1] The space E has to be Euclidean in the case of mean convergence.

$\varphi(n)/\sqrt{n \log \log n} \to 0$ as $n \to \infty$ replaces the denominator n. An example is $\varphi(n) = \sqrt{n}$. Nevertheless, the Central Limit Theorem asserts that we still have convergence in *distribution* and that the limit distribution is always normal. Since $\frac{1}{\sqrt{n}} \sum_{i=1}^{n} \widehat{Y}_i$ is centered with covariance matrix VY_1 for all n, it is no surprise that the same is true for the limit distribution.

E.42 Theorem (Multivariate Central Limit Theorem (CLT)) *Let $(Y_i)_{i \geq 1}$ be an i.i.d. sequence of quadratically integrable random vectors in $\mathbb{R}^q$ with covariance matrix V. We have*

(a) $\dfrac{1}{\sqrt{n}} \sum_{i=1}^{n} \widehat{Y}_i \Longrightarrow N_{0,V}^q$ *as $n \to \infty$.*

(b) If V is nonsingular, then for all Lebesgue boundariless, measurable subsets $B \subseteq \mathbb{R}^q$, we have

$$P\left[\frac{1}{\sqrt{n}} \sum_{i=1}^{n} \widehat{Y}_i \in B\right] \to N_{0,V}^q(B).$$

Part (a) of the CLT says that $\frac{1}{\sqrt{n}} \sum_{i=1}^{n} \mathrm{E} f(\widehat{Y}_i)$ converges to the integral $\int_{\mathbb{R}^q} f(x) N_{0,V}^q(\mathrm{d}x)$ for all bounded, continuous functions f on $\mathbb{R}^q$. Part (b) says that this convergence also takes place for certain discontinuous functions f, namely, the indicators $f = \mathbf{1}_B$ of Lebesgue boundariless, measurable subsets $B \subseteq \mathbb{R}^q$. For instance, if $Y_1 : \Omega \to \mathbb{R}$ has variance v, then

$$P\left(\frac{1}{\sqrt{n}} \sum_{i=1}^{n} \widehat{Y}_i \leq a\right) \to \frac{1}{\sqrt{2\pi v}} \int_{-\infty}^{a} e^{-x^2/(2v)} \mathrm{d}x.$$

The CLT is based on a second-order approximation to the cumulant generating function of Y_1. It is accurate to the order $1/\sqrt{n}$ in general and to the order $1/n$ if Y_1 is symmetric. Higher order approximations are due to Edgeworth; see Severini [466], p. 31 ff. Definitive results about the convergence of averages of random variables are due to Gnedenko and Kolmogorov [202].

Appendix F

Statistics

Measure theory is concerned with known measures and probability theory offers a different view of it by introducing random variables. Statistics is different. Here the central object of study is a *finite data set*. The statistician assumes that, except for some outliers, it is sampled from a distribution, also called a population. Statistical inference aims at recovering at least some of the population's properties. Since the same data set can in principle emanate from different populations, inference cannot be a purely logical process. Strictly speaking, inference is a bold enterprise not to be achieved free of charge. The price to be paid is assumptions made prior to the analysis. The statistician first makes the assumption that the observations come from a distribution in the mathematical sense. This is actually an idealistic view and this text focuses also on methods that safeguard against its violation. Second, it is often assumed that the parent population is a member of some given family of distributions, such as the normal, an elliptical, or even a more general family. These families are also called probabilistic models. Their fundamentals will be reviewed in Section F.1. A standard estimation method in such families is the maximum likelihood estimator (MLE), Section F.2.1. It will be applied to normal and elliptical models in Sections F.2.2 and F.2.3. Here, one usually assumes that the scatter matrix of the data is regular, an issue investigated in Section F.2.4. In connection with variable selection in Chapter 5, we will need normal regression, Section F.2.5.

A general theory of consistency and asymptotic normality of ML estimators is presented in the main text, Section 1.1. Questions about their asymptotic efficiency will be reviewed in Section F.2.6. In cluster analysis, it is not sufficient to produce a solution. In order to gain confidence in it, it should be validated. For this purpose, significance tests are useful. The likelihood ratio and Benjamini and Hochberg's multiple testing procedure will be briefly reviewed in Section F.3.

F.1 Probabilistic models in statistics

In statistics, data is regarded as **realizations** of random variables and their distributions. Distributions are probabilistic models of data, their generating mechanisms. They reflect the uncertainty about the data which would look different on another occasion even though the model was the same. Any model is supposed to describe reality by mathematical means; it is never the reality itself, Probabilities and data also appear in **stochastic simulation** but with the converse objective. In simulation, the distribution is known and the observations are sampled. In statistics, the observations are known and we want to infer some properties of their **parent probability**.

It is useful to first recall the notion of a probabilistic model. Let $(E, \mathcal{B})$ be a Hausdorff space with its Borel structure, the **sample space**. Since the parent is unknown, we provide for a whole collection of Borel probabilities on E to choose from. The model could be the convex set $\mathcal{P}(E)$ of *all* Borel probability measures on $(E, \mathcal{B})$. However, except in sample spaces of low complexity, the finite data set in our hands could not contain the information that would allow us to draw inferences on the parent, not even approximately. We therefore constrain the model to a subset $\mathcal{M} \subseteq \mathcal{P}(E)$. A model is useful if it is large enough to contain the parent, at least approximately, and small enough to allow the inference. The larger the data set is, the larger the model $\mathcal{M}$ is allowed to be. Model specification, that is, which model to choose, is one of the most important decisions the statistician faces. The common sample spaces bear classes of distributions that respect their structures. They are candidates for models. Model dimension also depends on the size of the data set.

F.1 Definition

(a) A **nonparametric** (probabilistic) **model** on E is a subset $\mathcal{M} \subseteq \mathcal{P}(E)$.

(b) A **parametric model** is a continuous map from a Hausdorff space Θ to $\mathcal{M}$ with the weak topology. The image of $\vartheta \in \Theta$ is denoted by μ_ϑ. The space Θ is called the **parameter space** of the model.

(c) A model is **dominated** by the dominating σ-finite reference Borel measure ϱ on E if all its probability measures are absolutely ϱ-continuous. That is, $\mu = f_\mu \varrho$ with the ϱ-density f_μ.

(d) A **submodel** of a nonparametric model $\mathcal{M}$ is a subset of $\mathcal{M}$. A submodel of a parametric model is a restriction to a subset of its parameter space.

(e) A parametric model $\Theta \to \mathcal{M}$ is

    (i) **identifiable**, if the map is one-to-one;

    (ii) **Euclidean**, if Θ is a topological manifold;

    (iii) $\mathcal{C}^k$, if Θ is a $\mathcal{C}^k$-manifold and the functions $\vartheta \to \int_E \varphi \, d\mu_\vartheta$ are $\mathcal{C}^k$ for all bounded, continuous functions φ on E.

(f) A submodel of a Euclidean ($\mathcal{C}^k$-) model is its restriction to a topological ($\mathcal{C}^k$-) submanifold; see Section B.6.1.

Thus, a parametric model $(\mu_\vartheta)_{\vartheta \in \Theta}$ is a Markov kernel $\Theta \to E$ such that the functions $\vartheta \to \int_E \varphi \, d\mu_\vartheta$ are continuous for all bounded, continuous functions φ on E. In almost all parametric models, Θ is a geometric subset of some Euclidean space such as a segment, a rectangle, a simplex, or a cone. As the intersection of an open and a closed set, each of them is locally compact.

Probabilistic models can be specified measure-theoretically or stochastically with random variables. The specification in the following examples is measure theoretic. Example (b) also admits a stochastic specification; see Lemma 1.12 and Section 1.3.1.

F.2 Examples
(a) A typical Euclidean parametric model dominated by Lebesgue measure is the family of normal distributions on $\mathbb{R}^d$, $(N_{m,V})_{m \in \mathbb{R}^d, V \in \mathrm{PD}(d)}$. Its parameter space is the product $\mathbb{R}^d \times \mathrm{PD}(d)$, an open subset of the Euclidean space $\mathbb{R}^d \times \mathrm{SYM}(d)$. Its dimension is therefore $d + \binom{d+1}{2} = \frac{d(d+3)}{2}$. Two Euclidean submodels of dimensions $2d$ and $d+1$, respectively, are the families of diagonally normal and spherically normal distributions.

(b) The set of all normal mixtures $\sum_{k=1}^g \pi_k N_{m_k, V_k}$ on $\mathbb{R}^d$ with $g \geq 2$ components is most conveniently considered as a nonparametric model. It cannot be made *Euclidean* since it also contains deficient mixtures with less than g components caused by vanishing mixing rates and/or equal component parameters. The family can be modified to a Euclidean model after removing all deficient mixtures. This will be done in Section 1.2.6 by passing to a quotient space.

(c) Let ℓ^2 be the Hilbert space of all square summable, real sequences and let $S_1(\ell^2)$ be its unit sphere, that is, the set of all sequences $\mathbf{h}$ such that $\sum h_k^2 = 1$. An example of a parametric (non-Euclidean) model dominated by counting measure on $\mathbb{N}$ is $S_1(\ell^2) \to \mathcal{P}(\mathbb{N})$ defined by $\mathbf{h} \mapsto (h_k^2)_{k=1}^\infty$, $\mathbf{h} = (h_1, h_2, \dots)$. It is not identifiable but is easily rendered so by restricting $S_1(\ell^2)$ to all sequences with entries ≥ 0.

F.1.1 Parametric vs. nonparametric

It is generally accepted that a *nonparametric* model is a set of probabilities on a measurable space E, whereas a *parametric* model is a map from a (parameter) set into the convex set of probabilities on E; see also McCullagh [356, 357]. The distinction between parametric and nonparametric models is not as clear-cut as the terminology may suggest. First, according to Definition F.1, any parametric model $\Phi\colon \Theta \to \mathcal{P}(E)$ is accompanied by an associated nonparametric model, the image $\Phi(\Theta)$. In particular, all results on nonparametric models can also be applied in the parametric situation. Second, the convex set of probability measures on $(E, \mathcal{B})$, $\mathcal{P}(E)$, bears various topologies, depending on the structure of E. For instance, if E is a locally compact topological space with its Borel structure, then there are the norm topology, the weak topology, and the vague topology. Moreover, dominated subsets of $\mathcal{P}(E)$ are endowed with intrinsic topologies on the set of their density functions. Therefore, the identity map from a nonparametric model to itself is a map from a Hausdorff space to $\mathcal{M}$, which makes it a parametric model. This is certainly not what we have in mind when talking about such a model. In many cases, the set $\mathcal{M}$ is parameterized by a set Θ, $\mathcal{M} = \{\mu_\vartheta \mid \vartheta \in \Theta\}$. However, the labels ϑ do not make it a parametric model yet. We rather require that the topological parameter space Θ enrich the model with useful topological or differentiable structure that interacts with intrinsic topologies on $\mathcal{M}$. This enables us to use powerful tools such as the Heine-Borel, Bolzano-Weierstraß, Ascoli-Arzela, and Tychonov compactness theorems, the Stone-Weierstraß approximation theorem, metrizability, completion, score functions, and information matrices. They are the bases for proving important properties such as parametric consistency or asymptotic normality.

A parametric model seems to be useful only when it is *identifiable*, that is, takes distinct parameters to distinct probabilities. Verification of identifiability of an elementary model can be checked by constructing the inverse function, that is, by computing the parameters from the density or one of its transforms. For instance, the parameters of a normal distribution $N_{m,V}$ are retrieved as the position of the maximum of its Lebesgue density and the Hessian matrix there. The normal family is therefore identifiable.

The probabilities in a parametric model are *explicitly* specified. By contrast, those of a nonparametric model can also be specified *implicitly* by equations and inequalities. Examples are the symmetric distributions on the real line with a tail distribution function R that decays at least exponentially. They are specified by the equation $R(x) + R(-x) = 1$ for all $x > 0$ and the inequality $R(x) \leq c_1 e^{-c_2 x}$ as $x \to \infty$ for two numbers $c_1, c_2 > 0$. Both approaches can be mixed. A typical semi-parametric model is the set of all elliptical probabilities on $\mathbb{R}^d$ with a decreasing density generator of at least exponential decay. This model is specified by a location parameter $m \in \mathbb{R}^d$, a scale parameter $V \in \mathrm{PD}(d)$, and a density generator φ restricted by the inequality $\varphi(t) \leq c_1 e^{-c_2 t}$, $t > 0$; see Appendix E.3.2.

F.1.2 Model and likelihood functions

One of the most important and powerful tools of statistical inference is the likelihood function. We will suppose here that the sample space E is at least a Hausdorff space with its Borel σ-algebra. This restriction does not exclude any interesting example or application. Let $\mathcal{M}$ be a nonparametric model on E dominated by ϱ and let the density functions f_μ, $\mu \in \mathcal{M}$, be continuous. They are then uniquely defined on E, not only ϱ-a.s. We have the following definitions.

F.3 Definition

(a) The function $f\colon E \times \mathcal{M} \to \mathbb{R}$, $(x, \mu) \mapsto f_\mu(x) = f(x; \mu)$, is known as the **model function** of the nonparametric model $\mathcal{M}$.

(b) Analogously, the function $f\colon E \times \Theta \to \mathbb{R}$, $(x, \vartheta) \mapsto f_{\mu_\vartheta}(x) = f_\vartheta(x) = f(x; \vartheta)$ is called the model function of the parametric model $\vartheta \mapsto \mu_\vartheta$.

(c) Given an observation $x \in E$, the x-cut $\mu \to f(x;\mu)$ $(\vartheta \to f(x;\vartheta))$ in a model function is called the nonparametric (parametric) **likelihood function**, respectively.

Thus, every μ- or ϑ-cut in a model function is a density function; every x-cut is a likelihood function. In estimation theory, the focus lies on the dependence of $f(x;\vartheta)$ on the parameter ϑ rather than on the observation x. Given an observation $x \in E$, the value $f(x;\mu)$ $(f(x;\vartheta))$ contains information about the "likelihood" of μ (μ_ϑ) to be the origin of x. Unfortunately, if we did not require that E be topological and the density functions continuous, then likelihood functions would not be well defined for fixed x. In fact, all density functions f_μ could be modified arbitrarily on a ϱ-null set in E, for instance, at x, which would render the likelihood function at x meaningless.

The likelihood function is the basis of all estimators that will be treated in this text. It is often not the likelihood function itself but rather its logarithm that is used. In order to avoid minus signs, the negative log-likelihood is also usual. For instance, twice the negative log-likelihood function of the normal family is $(m, V) \mapsto \log \det 2\pi V + (x-m)^\top V^{-1}(x-m)$.

There is no reasonable estimate of the parameters of a normal or elliptical model based on a single observation in $\mathbb{R}^d$. One rather needs a sufficiently large number n of i.i.d. observations. Independent sampling means passing to a **product model**. In the nonparametric case, it is made up of the product probabilities $\mu^{\otimes n}$, $\mu \in \mathcal{M}$, on E^n; in the parametric case one uses the product function $\vartheta \mapsto \mu_\vartheta^{\otimes n}$, $\vartheta \in \Theta$. If μ has the ϱ-density f, then $\mu^{\otimes n}$ has the tensor product $E^n \to \mathbb{R}$, $(x_1, \ldots, x_n) \mapsto f(x_1) \cdot \ldots \cdot f(x_n)$ as the $\varrho^{\otimes n}$-density function. Therefore, the parametric model function is, in the case of independent sampling, of the form $(x_1, \ldots, x_n; \vartheta) \mapsto f(x_1; \vartheta) \cdot \ldots \cdot f(x_n; \vartheta)$.

If a model is dominated and $\mathcal{C}^2$ and if its likelihood function is strictly positive, the following functionals are defined. The **score function** $s \colon \Theta \times E \to \mathbb{R}^q$ of the statistical model is the gradient of its log-likelihood function,

$$s(x;\vartheta) = \left(\mathrm{D}_\vartheta \log f(x;\vartheta)\right)^\top = \left(\frac{\partial \log f(x;\vartheta)}{\partial \vartheta_1}, \ldots, \frac{\partial \log f(x;\vartheta)}{\partial \vartheta_q}\right)^\top .$$

It registers the sensitivity of the log-likelihood w.r.t. parameter changes and does not depend on the reference measure. The score function vanishes at every local maximum of the likelihood function. Note also that multiplication of the likelihood with the score acts as differentiation: $s^\top f = \mathrm{D}_\vartheta f$. The (expected) **Fisher information** at a parameter $\vartheta \in \Theta$ is the $q \times q$ matrix of second moments of the score,

$$\mathcal{I}(\vartheta) = \int_E s(x;\vartheta)s(x;\vartheta)^\top \mu_\vartheta(\mathrm{d}x)$$

if the integral exists, that is, if

$$\int_E \|s(x;\vartheta)\|^2 \mu_\vartheta(\mathrm{d}x) = \int_E \|\mathrm{D}_\vartheta f(x;\vartheta)\|^2 \frac{1}{f(x;\vartheta)} \varrho(\mathrm{d}x) < \infty.$$

A third important functional is the $q \times q$ **Hessian matrix** of the log-likelihood

$$\left(\frac{\partial^2 \log f(x;\vartheta)}{\partial \vartheta_r \, \partial \vartheta_t}\right)_{r,t} = \mathrm{D}_\vartheta^2 \log f(x;\vartheta) = \mathrm{D}_\vartheta s(x;\vartheta).$$

The three functionals enjoy a number of useful properties and relationships.

F.4 Lemma

(a) *If $\|\mathrm{D}_\vartheta f(x;\vartheta)\| \le h(x)$ for some ϱ-integrable function h and for all ϑ in an open neighborhood $U \subseteq \Theta$ of ϑ_0 and all $x \in E$, then*

$$\int_E s(x;\vartheta)\mu_\vartheta(\mathrm{d}x) = 0, \quad \vartheta \in U.$$

In this case, the Fisher information at ϑ_0 is the covariance matrix of the score.

(b) The Fisher information is positive semi-definite.

(c) If $\|\mathrm{D}_\vartheta f(x;\vartheta)\| + \|\mathrm{D}_\vartheta^2 f(x;\vartheta)\| \le h(x)$ for some ϱ-integrable function h and for all $x \in E$ and all ϑ in an open neighborhood $U \subseteq \Theta$ of ϑ_0 and if the Fisher information exists, then it equals the negative expectation of the Hessian matrix of the log-likelihood,

$$\mathcal{I}(\vartheta) = -\int_E \mathrm{D}_\vartheta^2 \log f(x;\vartheta) \mu_\vartheta(\mathrm{d}x) = -\int_E \mathrm{D}_\vartheta s(x;\vartheta) \mu_\vartheta(\mathrm{d}x), \quad \vartheta \in U.$$

PROOF. (a) Since $\|\mathrm{D}_\vartheta f(x;\vartheta)\|$ is dominated by a ϱ-integrable function, we may use Lemma D.4 to differentiate under the integral, obtaining

$$\int_E s(x;\vartheta)^\top \mu_\vartheta(\mathrm{d}x) = \int_E \mathrm{D}_\vartheta f(x;\vartheta) \varrho(\mathrm{d}x) = \mathrm{D}_\vartheta \int_E f(x;\vartheta)\varrho(\mathrm{d}x) = 0.$$

(b) This follows from the fact that the rank-1 matrices $s(x;\vartheta)^\top s(x;\vartheta)$ are positive semi-definite.

(c) From $\mathrm{D}_\vartheta(sf) = \mathrm{D}_\vartheta^2 f$ and from the first assumption, it follows along with Lemma D.4 $\int_E \mathrm{D}_\vartheta(sf)\,\mathrm{d}\varrho = \mathrm{D}_\vartheta \int_E sf\,\mathrm{d}\varrho = \mathrm{D}_\vartheta \int_E s\,\mathrm{d}\mu_\vartheta = 0$ by (a). On the other hand, $\mathrm{D}_\vartheta(sf) = \mathrm{D}_\vartheta sf + s\mathrm{D}_\vartheta f = \mathrm{D}_\vartheta sf + ss^\top f$, so the existence of the Fisher information and ϱ-integrability of $\mathrm{D}_\vartheta(sf)$ show that $\mathrm{D}_\vartheta sf$, too, is ϱ-integrable. Thus,

$$0 = \int_E \mathrm{D}_\vartheta sf\,\mathrm{d}\varrho + \int_E ss^\top f\,\mathrm{d}\varrho = \int_E \mathrm{D}_\vartheta^2 \log f\,\mathrm{d}\mu_\vartheta + \int_E ss^\top\,\mathrm{d}\mu_\vartheta. \qquad \square$$

F.5 Remarks The fact that the score function vanishes both at $\vartheta = \mathrm{ML}(x)$ and in the μ_ϑ-expectation renders possible two interesting Taylor approximations. Let $\mathbf{x} = (x_1, \ldots, x_n)$.

(a) If the likelihood function is twice differentiable w.r.t. ϑ and if $\vartheta(x)$ is at least a local MLE, then

$$\log \frac{f(\mathbf{x};\vartheta)}{f(\mathbf{x};\vartheta(\mathbf{x}))} = \tfrac{1}{2}(\vartheta - \vartheta(\mathbf{x}))^\top \mathrm{D}_\vartheta^2 \log f(\mathbf{x};\vartheta(\mathbf{x}))(\vartheta - \vartheta(\mathbf{x})) + o(\|\vartheta - \vartheta(\mathbf{x})\|^2).$$

This follows from the second-order Taylor expansion of $\log f(\mathbf{x};\vartheta)$ at the maximum $\vartheta(\mathbf{x})$. Note that the first-order term, the score function, vanishes there. The positive semi-definite matrix

$$-\mathrm{D}_\vartheta^2 \log f(\mathbf{x};\vartheta(\mathbf{x})) = -\tfrac{1}{n}\sum_i \mathrm{D}_\vartheta^2 \log f(x_i;\vartheta(\mathbf{x}))$$

is the sample version of the expected Fisher information F.4(c), called the **observed Fisher information**. In contrast to the expected Fisher observation, it depends on the data, not on the parameter.

(b) Under the assumptions of Lemma F.4, we infer from its parts (a) and (c) in the same way an approximate relationship between the Kullback–Leibler divergence and the (expected) Fisher information,

$$\mathrm{D}_{\mathrm{KL}}(\vartheta_0, \vartheta) = \mathrm{E}_{\vartheta_0} \log \frac{f_{\vartheta_0}(x)}{f_\vartheta(x)} = \tfrac{1}{2}(\vartheta - \vartheta_0)^\top \mathcal{I}(\vartheta_0)(\vartheta - \vartheta_0) + o(\|\vartheta - \vartheta_0\|^2).$$

F.1.3 Invariance and equivariance

A **transformation** in a sample space E is a one-to-one, onto map $t\colon E \to E$. The concepts of invariance and equivariance w.r.t. t apply to **statistics** $T\colon E \to \mathcal{T}$ and to **model functions** $f\colon E \times \mathcal{T} \to \mathbb{R}$. The maps are often defined only partially. Statistics comprise estimators and tests. If T is an **estimator**, then $\mathcal{T}$ consists of the elements to be estimated. If it is a **test**, then $\mathcal{T} = \{'\text{accept}', '\text{reject}'\}$.

Equivariance w.r.t. inherent transformations on E provides one of the firmest footings in statistical inference and is therefore an appealing asset of estimators, tests, models, and criteria. Roughly speaking, they are called equivariant w.r.t. a transformation t on the sample space if the effect of t can be represented by a transformation on $\mathcal{T}$. A special case of equivariance is invariance. It applies to certain estimators and tests and means that the transformation has no effect at all. More specifically, we have the following definition.

F.6 Definition Let $t\colon E \to E$ be some sample space transformation.

(a) A statistic $T\colon E \to \mathcal{T}$ is called t-**equivariant** if $T(t(x))$ is a one-to-one function of $T(x)$.

(b) A model function $f\colon E \times \mathcal{T} \to \mathbb{R}$ is called t-**equivariant** if there exists a transformation τ_t on $\mathcal{T}$ and a strictly positive number c_t independent of $x \in E$ and $\vartheta \in \mathcal{T}$ such that $f(t(x); \tau_t(\vartheta)) = c_t f(x; \vartheta)$.

Modifications to this definition will be necessary when the statistic (model function) is defined only partially on E ($E \times \mathcal{T}$). Part (b) can be rewritten $f(t(x); \vartheta) = c_t f(x; \tau_t^{-1}(\vartheta))$. Applied to model functions, equivariance requires that the transformed density function be again a member of the *same* parametric model. For instance, the centered Cauchy model $c_\vartheta/(1 + x^2/\vartheta)$ is equivariant w.r.t. linear transformations on $\mathbb{R}$ but not w.r.t. the transformation $x \mapsto x^3$. The transformations for which equivariance holds form a group w.r.t. composition, the **equivariance group** of the statistic or model function. For the latter this follows from

$$f(t_2(t_1(x)); \tau_{t_2}(\tau_{t_1}(\vartheta))) = h_{t_2} f(t_1(x); \tau_{t_1}(\vartheta)) = h_{t_2} h_{t_1} f(x; \vartheta)$$

and from

$$c_t f(t^{-1}(x); \tau_{t^{-1}}(\vartheta)) = f(t(t^{-1}(x)); \tau_t(\tau_{t^{-1}}(\vartheta))) = f(x; \vartheta)).$$

The proof for a statistic is similar. Let $\mathcal{G}$ be a group of transformations on $E = \mathbb{R}^d$. A statistic or model function is $\mathcal{G}$-equivariant if its equivariance group contains $\mathcal{G}$. Thus, we speak of **location** (**scale, affine**) **equivariance** if the equivariance group contains all shifts $t(x) = x + b$, $b \in E$ (linear maps $t(x) = Ax$, $A \in \mathrm{GL}(E)$, affine transformations $t(x) = Ax + b$, $b \in E$, $A \in \mathrm{GL}(E)$) on $\mathbb{R}^d$. The groups may be restricted, such as all affine transformations with diagonal scale matrices A.

F.7 Examples (a) A fully elliptical model is affine equivariant, the parameter transformation associated with the affine transformation $t\colon x \mapsto Ax + b$ being $\tau_t\colon (m, V) \mapsto (m', V') = (Am + b, AVA^\top)$. Indeed,

$$(Ax + b - m')^\top V'^{-1}(Ax + b - m') = (A(x - m))^\top (AVA^\top)^{-1}(A(x - m))$$
$$= (x - m)^\top V^{-1}(x - m),$$

and hence

$$E_{\varphi, m', V'}(Ax + b) = (\det V')^{-1/2}\varphi\big((Ax + b - m')^\top V'^{-1}(Ax + b - m')\big)$$
$$= (\det AVA^\top)^{-1/2}\varphi\big((x - m)^\top V^{-1}(x - m)\big)$$
$$= \frac{1}{|\det A|}(\det V)^{-1/2}\varphi\big((x - m)^\top V^{-1}(x - m)\big) = \frac{1}{|\det A|}E_{\varphi, m, V}(x).$$

(b) A diagonally elliptical model is equivariant w.r.t. all affine transformations $x \mapsto Ax + b$, where A is the product of a diagonal with a permutation matrix.

(c) A spherical model is equivariant w.r.t. all Euclidean motions.

F.2   Point estimation

An **estimation** (**testing**) **principle** is a method that allows us to derive estimators (tests) for a whole class of models at once. Several principles have been proposed in the past to arrive at reasonable estimators and tests. Some of them are the maximum likelihood principle, the method of moments, the method of moment-generating and characteristic functions, the minimum χ^2 and other goodness-of-fit methods, the minimax method, and Bayesian methods such as the posterior maximum and expectation for the maximum entropy prior; see Redner and Walker [435], p. 202. Maximum likelihood and maximum a posteriori play dominant roles in this text.

F.2.1 Likelihood principle

The much-debated **likelihood principle** says to base inference about a parameter only on the observed likelihood function; see Hinkley and Reid [245]. In particular, equal or proportional likelihood functions should imply the same estimate. This in turn implies that there is only one estimate for one data set. Birnbaum [44] relates the likelihood principle to the principles of conditionality and sufficiency. Evans et al. [145] argue that conditionality alone implies the likelihood principle. The likelihood principle is made concrete by two principles applicable to dominated parametric and nonparametric models, the **maximum likelihood** estimate (MLE) and the Bayesian maximum a posteriori (MAP) principles; see, for instance, Barndorff-Nielsen [24]. The former requests us to *choose the parameter that makes the sample most likely,* that is, maximizes its likelihood function:

$$\mathrm{ML}_\vartheta(x) := \operatorname*{argmax}_\vartheta f_\vartheta(x) \quad (= \operatorname*{argmin}_\vartheta -\log f_\vartheta(x)).$$

The idea underlying the MLE is simple – one postulates that the observation x is *typical* for the parent and, subsequently, for some member of the assumed model. The following proposition describes the relationship between equivariance of a model and its MLE. It is a consequence of Definition F.6.

F.8 Proposition *Let the probabilistic model be $\mathcal{G}$-equivariant. For all $t \in \mathcal{G}$, the MLE for the t-transformed data is the τ_t-image of that for the original data.*

The MLE has a number of impressive properties. Under mild assumptions, it is asymptotically unbiased and consistent, asymptotically normal and efficient, it inherits the model's equivariance (Proposition F.8), it satisfies an invariance principle in parameter space, and it can be robustified. In mixture and cluster analysis, likelihood-based methods yield *local* metrics adapted to the data. It is applicable to probabilistic models of many kinds. Because of the cancellation law of densities, the MLE does not vary with the reference measure underlying the model function. However, application of the MLE requires that the likelihood function have a maximum at all. In simple situations this is guaranteed, under mild conditions, for instance, in all normal and elliptical models and submodels; see Sections F.2.2 and F.2.3. Unfortunately, one of the main models studied in this text, the normal *mixture* model, does not admit an MLE. Nevertheless, the situation is not disastrous; we will have recourse to nonunique local maxima of the likelihood function. Their careful selection will be a major aim of Chapter 4.

Given a prior probability π on a dominated model with parameter space Θ, the **maximum a posteriori** (MAP) estimate for the observation x is $\operatorname{argmax}_\vartheta f_\pi[\vartheta \mid x]$. It requests us to *choose that parameter that the sample makes most likely.* By Bayes' formula, it is proportional to the product $f(x; \vartheta) f_\pi(\vartheta)$. A comparison with the MLE shows that the prior density $f_\pi(\vartheta)$ appears as a preference function on Θ.

F.2.2 Normal ML parameter estimation

There are few models that allow computationally efficient ML estimation of their parameters. One of them is the normal model where ML estimation of location and scale parameters reduces to elementary arithmetic operations. This section reviews ML estimation in the three normal models with full, diagonal, and spherical covariance matrices. We are given a data set $\mathbf{x} = (x_1, \ldots, x_n)$ of n vectors in $\mathbb{R}^d$ and consider the n-fold normal product model $N_{m,V}^{\otimes n}$.

Let $x_1, \ldots, x_n \subseteq \mathbb{R}^d$ be n data points. The main statistics needed in normal and elliptical parameter estimation are the **sample mean** (vector) $\overline{x} = \frac{1}{n} \sum_{i=1}^n x_i$, the **SSP matrix**, ("sum of squares and products"), $W = \sum_{i=1}^n (x_i - \overline{x})(x_i - \overline{x})^\top$, and the related **scatter matrix (sample covariance matrix)**, $S = \frac{1}{n} W = \frac{1}{n} \sum_{i=1}^n (x_i - \overline{x})(x_i - \overline{x})^\top$, of the data. We will need the likelihood function of the product model; see Section F.1.2. Let $\Lambda = V^{-1}$.

F.9 Lemma *The log-likelihood function of the n-fold normal product model is*

$$\log \prod_{i=1}^n f_{m,V}(x_i) = -\tfrac{n}{2} \left\{ d \log 2\pi + \log \det V + \operatorname{tr}(\Lambda S) + (\overline{x} - m)^\top \Lambda (\overline{x} - m) \right\}.$$

PROOF. It follows from independence, Steiner's formula A.11, and from linearity of the trace:

$$-2 \log \prod_{i=1}^n f_{m,V}(x_i) = \sum_{i=1}^n \left\{ d \log 2\pi + \log \det V + (x_i - m)^\top \Lambda (x_i - m) \right\}$$

$$= nd \log 2\pi + n \log \det V + \sum_i (x_i - m)^\top \Lambda (x_i - m)$$

$$= nd \log 2\pi + n \log \det V + \sum_i (x_i - \overline{x})^\top \Lambda (x_i - \overline{x}) + n(\overline{x} - m)^\top \Lambda (\overline{x} - m)$$

$$= nd \log 2\pi + n \log \det V + \sum_i \operatorname{tr} \Lambda (x_i - \overline{x})(x_i - \overline{x})^\top + n(\overline{x} - m)^\top \Lambda (\overline{x} - m). \ \square$$

F.10 Lemma *Let S be a positive semi-definite matrix; see Appendix A.3.*
(a) The function $V \mapsto \log \det V + \operatorname{tr}(\Lambda S)$ has a minimum on the convex cone $\mathrm{PD}(d)$ of all positive definite matrices V if and only if S is positive definite.
(b) Minimizer and minimum then are $V^* = S$ and $\log \det S + d$, respectively.

PROOF. It is equivalent to investigate the function $\varphi(\Lambda) = -\log \det \Lambda + \operatorname{tr}(\Lambda S)$ w.r.t. minimization. (Figure F.1 shows a plot of the function φ in the univariate case.) This function converges to ∞ as Λ approaches the Alexandrov point of $\mathrm{PD}(d)$, that is, as an eigenvalue approaches zero or ∞: Let $\alpha(\Lambda)$ be the maximum and $\iota(\Lambda)$ the minimum eigenvalue of Λ and assume that $\alpha(\Lambda) + 1/\iota(\Lambda) \to \infty$. If $\alpha(\Lambda)$ is large, then, since S is positive definite, the trace beats the $-\log \det$ and $\varphi(\Lambda)$ is large. If not, then $\iota(\Lambda)$ must be small and the term $-\log \det \Lambda$ is large while the trace is positive. This is the claim and it follows that the continuous function φ has a minimizer Λ^*. By Example C.9(c) the derivative of φ w.r.t. Λ is $-\Lambda^{-1} + S$. Since this function has a unique zero $\Lambda^* = S^{-1}$, we conclude $V^* = S$.

If S is singular, then $-\Lambda^{-1} + S$ vanishes for no regular matrix Λ. Therefore, φ has no minimum. This finishes the proof of part (a) and part (b) is straightforward. $\square$

F.11 Proposition *(a) The ML estimates of the parameters m and V of the fully normal model exist if and only if the SSP matrix W of the data $x_1, \ldots, x_n \in \mathbb{R}^d$ is regular. In this case the estimates are*

$$m(\mathbf{x}) = \overline{x}, \quad V(\mathbf{x}) = \tfrac{1}{n} W = S.$$

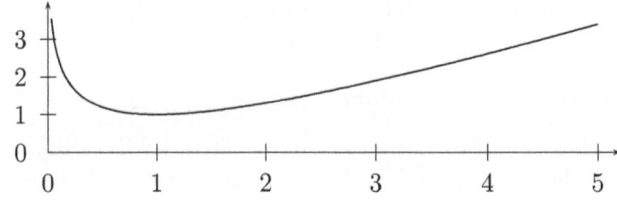

Figure F.1: *The graph of the function $\lambda \mapsto -\log \lambda + \lambda s$ for $s = 1$.*

The minimum of the negative log-likelihood function is

$$\tfrac{n}{2}\left\{d(\log 2\pi + 1) + \log \det S\right\}.$$

(b) The ML estimates of the parameters of the diagonally *normal model exist if and only if there is no coordinate where all entries are equal. Then they are*

$$m(\mathbf{x}) = \overline{x}, \quad v_k(\mathbf{x}) = \tfrac{1}{n}W(k,k) = S(k,k), \ 1 \le k \le d.$$

The minimum of the negative log-likelihood function is

$$\tfrac{n}{2}\left\{d(\log 2\pi + 1) + \log \det S\right\}.$$

(c) The ML estimates of the parameters of the spherically *normal model exist if and only if there are two different data points. Then they are*

$$m(\mathbf{x}) = \overline{x}, \quad v(\mathbf{x}) = \tfrac{1}{d}\mathrm{tr}\, S = \tfrac{1}{nd}\mathrm{tr}\, W = \frac{1}{nd}\sum_{i=1}^{n} \|x_i - \overline{x}\|^2.$$

The minimum of the negative log-likelihood function is here

$$\tfrac{nd}{2}\left\{\log 2\pi + 1 + \log \tfrac{1}{d}\mathrm{tr}\, S\right\}.$$

PROOF. According to Lemma F.9, we have to minimize the function

$$\log \det V + \mathrm{tr}\,(\Lambda S) + (\overline{x} - m)^\top \Lambda(\overline{x} - m) \tag{F.1}$$

over all $m \in \mathbb{R}^d$ and all $V \in \mathrm{PD}(d)$. Regardless of V, the minimum w.r.t. m is attained at $\overline{x}$. It remains to minimize $\log \det V + \mathrm{tr}\,(\Lambda S)$ w.r.t. $V \in \mathrm{PD}(d)$. By Lemma F.10 the minimum exists if and only if the matrix S is positive definite and the minimum is then S. The minimum of the negative log-likelihood function follows from inserting the minimizers into Lemma F.9. This proves part (a).

To obtain the minimizers and minima for the two normal submodels (b) and (c), go back to (F.1) and note that $\mathrm{tr}\,(\Lambda S) = \sum_k \Lambda(k,k)S(k,k)$ if V is diagonal and $\mathrm{tr}\,(\Lambda S) = \tfrac{1}{v}\mathrm{tr}\, S$ if $V = vI_d$ is spherical. □

F.2.3 *ML parameter estimation in elliptical symmetries*

The following theorem shows that the MLE of the density generator and the location and scale parameters exist for fairly general elliptical models. The theorem uses a compact space of density generators endowed with the topology of compact convergence; see Section B.6.4. Let Φ_β be a subset of continuous, strictly positive, decreasing, normalized density generators $\varphi \colon \mathbb{R}_+ \to \mathbb{R}_>$ such that $\varphi(t) \le C(1 + t)^{-\beta/2}$, $t \ge 0$; see Section E.3.2.

F.12 Proposition (Existence of the ML estimate for elliptical families) *Let $n \ge d + 1$, $k = \lfloor \frac{n}{d+1} \rfloor$, and $\beta > \frac{nd}{k}$. The ML estimator of the elliptical model*

$$(\varphi, m, A) \to f_{\varphi, m, A}, \quad \varphi \in \Phi_\beta, \ m \in \mathbb{R}^d, \ A \in \mathrm{PD}(d),$$

exists for all data sets in $\mathbb{R}^d$ of length n in general position.

PROOF. Let $\mathbf{x} = (x_1, \ldots, x_n)$, $x_i \in \mathbb{R}^d$, be in general position. Subdivide the integral interval $1..n$ into k segments of length $d+1$ and the rest. Each segment $j(d+1)+1..(j+1)(d+1)$ satisfies by Steiner's formula A.11 the estimate

$$\sum_{i=j(d+1)+1}^{(j+1)(d+1)} (x_i - m)^\top A(x_i - m) \ge \sum_{i=j(d+1)+1}^{(j+1)(d+1)} (x_i - \overline{\mathbf{x}})^\top A(x_i - \overline{\mathbf{x}})$$

$$= (d+1)\,\mathrm{tr}\, S_j A \ge \kappa_j \mathrm{tr}\, A,$$

where S_j is the scatter matrix of the observations in the jth segment and κ_j is the $d+1$-fold of its smallest eigenvalue. By assumption on the position of $\mathbf{x}$, all κ_j's are strictly positive.

Let $\kappa = \min_j \kappa_j$. It follows $(x_i - m)^\top A(x_i - m) \geq \kappa \operatorname{tr} A$ for at least k of the indices i. Let C be an upper bound of all functions $(1 + t)^{\beta/2} \varphi(t)$, $t \geq 0$. The proof is based on the chain of estimates

$$\log f_{\varphi,m,A}(\mathbf{x}) = \frac{n}{2} \log \det A + \sum_{i=1}^{n} \log \varphi((x_i - m)^\top A(x_i - m)) \tag{F.2}$$

$$\leq \frac{n}{2} \log \det A + (n - k) \log \varphi(0) + k \log \varphi(\kappa \operatorname{tr} A)$$

$$\leq \frac{n}{2} \log \det A + n \log C - \frac{k\beta}{2} \log(1 + \kappa \operatorname{tr} A) \tag{F.3}$$

$$\leq \frac{n}{2} \log \det A - \frac{k\beta}{2} \log \operatorname{tr} A + \text{const}$$

$$\leq \frac{nd - k\beta}{2} \log \operatorname{tr} A + \text{const}. \tag{F.4}$$

The last line follows from the geometric-arithmetic inequality A.5. Let us see why $\log f_{\varphi,m,A}(\mathbf{x})$ converges to $-\infty$ as (m, A) converges to the Alexandrov point of $\mathbb{R}^d \times \operatorname{PD}(d)$. There are three mutually exclusive possibilities: Some eigenvalue of A approaches ∞, all eigenvalues are upper bounded and some eigenvalue of A approaches zero, or all eigenvalues are bounded and bounded away from zero and $\|m\|$ approaches ∞. In the first case, the claim follows from (F.4) and the assumption on β, in the second case it follows from (F.3), and in the third case, the claim follows from (F.2) and the fact that φ decreases to zero. By compactness of Φ_β, subsets $\Phi_\beta \times K_1 \times K_2 \subseteq \Phi_\beta \times \mathbb{R}^d \times \operatorname{PD}(d)$, $K_1 \subseteq \mathbb{R}^d$ and $K_2 \subseteq \operatorname{PD}(d)$ compact, are compact. The proposition now follows from continuity of the mapping $(\varphi, m, A) \to f_{\varphi,m,A}(\mathbf{x})$. $\qquad\square$

F.2.4 Positive definiteness of scatter matrices

We have seen in Proposition F.11 that the maximum of the fully normal likelihood function exists if and only if the scatter matrix is regular. Analogously, Proposition F.12 showed that the maximum of the elliptical likelihood function exists when the data set is large enough and in general position. This section shows that regularity follows from the affine geometry of the data, Corollary F.15, and gives a sufficient condition for general position of random variables, Proposition F.16.

F.13 Lemma If W is positive semi-definite, then $y^\top W y = 0$ if and only if $W y = 0$.

PROOF. The positive semi-definite matrix W has a square root $\sqrt{W}$. If $y^\top W y$ vanishes, then so does $\|\sqrt{W} y\|^2 = (\sqrt{W} y)^\top (\sqrt{W} y) = y^\top W y$ and also $W y = \sqrt{W}\sqrt{W} y$. $\qquad\square$

F.14 Proposition The rank of the SSP matrix $W = \sum_{i=1}^{n} (x_i - \overline{x})(x_i - \overline{x})^\top$ of the data $x_1, \ldots, x_n$ equals their affine dimension.

PROOF. We first show that the kernel of W is the orthogonal complement of the linear space $\mathbb{L}\langle x_1, \ldots, x_n \rangle$ associated with $\mathbb{A}\langle x_1, \ldots, x_n \rangle$. By Lemma F.13, $W y = 0$ if and only if $\sum_{i=1}^{n} \left(y^\top (x_i - \overline{x}) \right)^2 = \sum_{i=1}^{n} y^\top (x_i - \overline{x})(x_i - \overline{x})^\top y = y^\top W y = 0$, that is, if y is perpendicular to $x_i - \overline{x}$ for all i. This is equivalent to saying $y \in \langle x_1 - \overline{x}, \ldots, x_n - \overline{x} \rangle^\perp$. By Lemma A.1, this is just the associated linear space. Now, by the rank-nullity theorem of Linear Algebra,

$$d = \operatorname{rank} W + \dim \ker W = \operatorname{rank} W + \dim \mathbb{L}\langle x_1, \ldots, x_n \rangle^\perp$$

$$= \operatorname{rank} W + d - \dim \mathbb{L}\langle x_1, \ldots, x_n \rangle;$$

thus, $\operatorname{rank} W = \dim \mathbb{A}\langle x_1, \ldots, x_n \rangle$. $\qquad\square$

F.15 Corollary

(a) *The scatter matrix of n points $x_1, \ldots, x_n \in \mathbb{R}^d$ is positive definite if and only if the affine dimension of $x_1, \ldots, x_n$ is d.*

(b) *A necessary condition for this to hold is $n \geq d + 1$.*

PROOF. Part (a) follows from Proposition F.14 and (b) follows from (a) and Lemma A.1 with $f = x_1$. $\qquad\square$

The necessary condition in part (b) of the corollary is often also sufficient:

F.16 Proposition *Let μ_i, $i \in 1 .. n$, be distributions on $\mathbb{R}^d$ which charge no hyperplane, that is, all hyperplanes are μ_i-null sets.[1] Then any n independent random vectors $X_i \sim \mu_i$ are P-a.s. in general position.*

PROOF. We have to show that the affine dimension of $m+1$ random vectors $X_1, \ldots, X_{m+1}$, $0 \leq m \leq d$, is P-a.s. m. This is proved by induction on $m \leq d$. The claim for $m = 0$ is plain since the affine dimension of a singleton is 0. Now consider $m + 1$, $m \leq d$ independent random vectors in $\mathbb{R}^d$. Since, by the induction hypothesis, the dimension of $X_1, \ldots, X_m$ is $m - 1$, the affine dimension of $X_1, \ldots, X_{m+1}$ is m if and only if $X_{m+1} \notin \mathbb{A}\langle X_1, \ldots, X_m \rangle$. But this is P-a.s. true since X_{m+1} is independent of $X_1, \ldots, X_m$ and $\mathbb{A}\langle X_1, \ldots, X_m \rangle$ is contained in a hyperplane of $\mathbb{R}^d$. $\qquad\square$

A consequence of the proposition and of Corollary F.15 is the following.

F.17 Corollary *Let $n \geq d+1$. Under the assumptions of the proposition, the scatter matrix of $X_1, \ldots, X_n$ is P-a.s. regular.*

Thus, only two events can be responsible for the singularity of a scatter matrix, although $n \geq d+1$: Either there is a hyperplane with a strictly positive μ-probability or the n drawings were not independent of each other. The first case occurs when the entries of the vectors X_i are linearly dependent. This can also arise from coarse discretization of the data. As a remedy, randomize them. This will, for instance, be done with the IRIS data in Section 6.1.1 where only two digits of precision are given.

In practice, the contrary event may also be observed. A scatter matrix on a computer may be regular, although the variables are linearly dependent. This is due to rounding errors which cause a theoretically singular scatter matrix to be regular. The regularity will, however, be shaky since there will be an unusually small eigenvalue. It is therefore a good idea to monitor the eigenvalues of scatter matrices.

F.2.5 Normal regression

Variable selection in Chapter 5 will need some elementary facts about the representation of multivariate normal distributions by regression parameters. Let $d_x, d_y \geq 1$, let $(X, Y) : (\Omega, P) \to \mathbb{R}^{d_x + d_y}$ be a joint random vector with mean $m = (m_x, m_y)$ and covariance matrix $V = \begin{bmatrix} V_x & C_{x,y} \\ C_{y,x} & V_y \end{bmatrix} \succ 0$, $C_{x,y} = C_{y,x}^\top$. Define the transformation $(G_{y|x}, m_{y|x}, V_{y|x})$ of (m, V),

$$\begin{cases} G_{y|x} = C_{y,x} V_x^{-1}, & m_{y|x} = m_y - G_{y|x} m_x, \\ V_{y|x} = V_y - C_{y,x} V_x^{-1} C_{x,y} & (= V_y - G_{y|x} C_{x,y}). \end{cases} \tag{F.5}$$

F.18 Lemma *Let I_y be the identity matrix of dimension d_y. We have*

(a) $V_{y|x} \succ 0$;

(b) $\det V_{y|x} = \det V / \det V_x$;

---

[1] An important case is $\mu_i \ll \lambda^d$.

(c) $V^{-1} = \begin{bmatrix} V_x^{-1} & 0 \\ 0 & 0 \end{bmatrix} + \begin{bmatrix} -G_{y|x}^\top \\ I_y \end{bmatrix} V_{y|x}^{-1} \begin{bmatrix} -G_{y|x} & I_y \end{bmatrix}.$

PROOF. Let I_x be the identity matrix of dimension d_x. Elementary matrix algebra shows

$$\begin{bmatrix} V_x^{-1/2} & 0 \\ 0 & I_y \end{bmatrix} V \begin{bmatrix} V_x^{-1/2} & 0 \\ 0 & I_y \end{bmatrix} = \begin{bmatrix} V_x^{-1/2} & 0 \\ 0 & I_y \end{bmatrix} \begin{bmatrix} V_x & C_{x,y} \\ C_{y,x} & V_y \end{bmatrix} \begin{bmatrix} V_x^{-1/2} & 0 \\ 0 & I_y \end{bmatrix}$$

$$= \begin{bmatrix} I_x & V_x^{-1/2} C_{x,y} \\ C_{y,x} V_x^{-1/2} & V_y \end{bmatrix} \sim \begin{bmatrix} I_x & V_x^{-1/2} C_{x,y} \\ 0 & V_y - G_{y|x} C_{x,y} \end{bmatrix} = \begin{bmatrix} I_x & V_x^{-1/2} C_{x,y} \\ 0 & V_{y|x} \end{bmatrix},$$

where the tilde $\sim$ means here that the matrix on the right is obtained from that on the left by an elementary blockwise row operation. Parts (a) and (b) follow.

Now let A, B, and D be the matrices solving

$$I_{d_x + d_y} = V \begin{bmatrix} V_x^{-1} & 0 \\ 0 & 0 \end{bmatrix} + V \begin{bmatrix} A & B^\top \\ B & D \end{bmatrix}.$$

Then $V_x A + C_{x,y} B = 0$, $V_x B^\top + C_{x,y} D = 0$, and $C_{y,x} B^\top + V_y D = I_y$. The last two equations imply $D = V_{y|x}^{-1}$, $B = -V_{y|x}^{-1} C_{y,x} V_x^{-1}$ and the first one $A = V_x^{-1} C_{x,y} V_{y|x}^{-1} C_{y,x} V_x^{-1}$ and part (c) follows. $\square$

The following theorem represents the two (possibly multivariate) components of a jointly normal random vector (X, Y) as a normal linear regression of the response Y on the predictor X.

F.19 Theorem *Let $(X, Y) \sim N_{m,V}$ be jointly normal.*
(a) The vector Y can be written as a normal linear regression of X with matrix $G_{y|x}$, offset $m_{y|x}$, and normal disturbance $U \sim N_{0, V_{y|x}}$ independent of X,

$$Y = m_{y|x} + G_{y|x} X + U.$$

(b) The conditional density $f_Y[y \mid X = x]$ equals $N_{m_{y|x} + G_{y|x} x, V_{y|x}}(y)$.
(c) Let (x_i, y_i), $i \in 1..n$, be a random sample from $N_{m,V}$. The ML estimates of the regression parameters are obtained by inserting into (F.5) its sample mean vector $(\overline{x}, \overline{y})$ and scatter matrix $S = \begin{bmatrix} S_x & S_{x,y} \\ S_{y,x} & S_y \end{bmatrix} \succ 0$ for m and V.

PROOF. We first show $U \sim N_{0, V_{y|x}}$. Since U is normal, this follows from $\mathrm{E}U = m_y - m_{y|x} - G_{y|x} m_x = 0$ and

$$VU = \mathrm{E}(\widehat{Y} - G_{y|x} \widehat{X})(\widehat{Y} - G_{y|x} \widehat{X})^\top$$

$$= V_y - C_{y,x} G_{y|x}^\top - G_{y|x} C_{x,y} + G_{y|x} V_x G_{y|x}^\top = V_y - G_{y|x} C_{x,y}$$

because of $C_{y,x} = G_{y|x} V_x$. Independence of U and X is a consequence of joint normality and $\mathrm{Cov}(U, X) = \mathrm{Cov}(Y - G_{y|x} X, X) = C_{y,x} - G_{y|x} V_x = 0$. This proves parts (a) and (b) and part (c) follows from the invariance principle of the MLE. $\square$

F.2.6 Asymptotic efficiency and the Cramér–Rao lower bound

Asymptotic distributions of point estimators and test statistics play a fundamental role in the assessment of accuracy and errors. Due to the Central Limit Theorem, asymptotic distributions are often related to the normal distribution. Recall the Löwner order $\succeq$, Definition A.8. Under mild conditions, there is a Löwner smallest covariance matrix an unbiased estimator of a vector parameter can attain. Fisher [155], Aitken and Silverstone [5], Rao [428], and Cramér [104] showed that it is the inverse Fisher information matrix at the sampling parameter; see also Cox and Hinkley [103] and Lehmann and Casella [315]. For a review of the theorem we assume here that $(\mu_\vartheta)_{\vartheta\in\Theta}$ is a ϱ-dominated and identifiable $\mathcal{C}^2$-model; see Definition F.1.

F.20 Regularity Conditions (for an estimator)
Let $\Theta \subseteq \mathbb{R}^q$ be open and nonempty. An estimator $T: E \to \mathbb{R}^q$ is called **regular** w.r.t. the model if

(i) $T \cdot f_\vartheta \in \mathcal{L}^1(\varrho)$ for all $\vartheta \in \Theta$;

(ii) there exists $h \in \mathcal{L}^1(\varrho)$, such that $|T \cdot \mathrm{D}_\vartheta f_\vartheta| \leq h$ for all $\vartheta \in \Theta$.

F.21 Lemma *Let the assumption of Lemma F.4(a) be met. If $T: E \to \mathbb{R}^q$ is unbiased and regular (F.20) w.r.t. the model, then we have for all $\vartheta \in \Theta$*

$$\int \widehat{T} s_\vartheta \, \mathrm{d}\mu_\vartheta = I_q,$$

where $\widehat{T} = T - \int T \, \mathrm{d}\mu_\vartheta \ (= T - \vartheta)$ is the centered estimator.

PROOF. Let E_ϑ denote integration w.r.t. μ_ϑ. By Lemma F.4(a) we have

$$\mathrm{E}_\vartheta(T - \vartheta)s_\vartheta = \mathrm{E}_\vartheta T s_\vartheta = \int T s_\vartheta f_\vartheta \, \mathrm{d}\varrho = \int T \mathrm{D}_\vartheta f_\vartheta \, \mathrm{d}\varrho$$

$$\overset{T \text{ regular}}{=} \mathrm{D}_\vartheta \int T f_\vartheta \, \mathrm{d}\varrho = \mathrm{D}_\vartheta \mathrm{E}_\vartheta T \overset{T \text{ unbiased}}{=} \mathrm{D}_\vartheta \vartheta = I_q.$$

The computation shows along with F.20(ii) that the expectation $\mathrm{E}_\vartheta T s_\vartheta$ exists. □

Cramér–Rao's famous lower bound on the estimator variance reads as follows.

F.22 Theorem (Cramér–Rao) Let the $\mathcal{C}^2$-model $(\mu_\vartheta)_{\vartheta\in\Theta}$ be identifiable and let T be an unbiased estimator of ϑ of finite variance satisfying the Regularity Conditions F.20. If the Fisher information $\mathcal{I}(\vartheta)$ exists, then it is positive definite and we have

$$\mathrm{V}_\vartheta T \succeq \mathcal{I}(\vartheta)^{-1}.$$

That is, the right hand side is an absolute lower bound on the estimator variance independent of the estimator T. It depends only on the model.

PROOF. Abbreviate $\widehat{T} = T - \mathrm{E}_\vartheta T$. By Lemma F.21 and the Cauchy-Schwarz inequality, we have for arbitrary vectors $a, b \in \mathbb{R}^q$

$$(a^\top b)^2 = (a^\top I_q b)^2 = \left(a^\top (\mathrm{E}_\vartheta \widehat{T} s_\vartheta) b\right)^2 = \mathrm{E}_\vartheta^2(a^\top \widehat{T})(s_\vartheta b)$$

$$\leq \mathrm{E}_\vartheta\{a^\top \widehat{T} \widehat{T}^\top a\} \mathrm{E}_\vartheta\{b^\top s_\vartheta^\top s_\vartheta b\} = (a^\top \mathrm{V}_\vartheta T a)(b^\top \mathcal{I}(\vartheta) b).$$

Putting $b = a \neq 0$, we see that $\mathcal{I}(\vartheta)$ is regular. Putting $b = \mathcal{I}(\vartheta)^{-1} a$ with $a \neq 0$, we obtain

$$(a^\top \mathcal{I}(\vartheta)^{-1} a)^2 \leq a^\top \mathrm{V}_\vartheta T a a^\top \mathcal{I}(\vartheta)^{-1} a.$$

The lower and upper bounds $0 < a^\top \mathcal{I}(\vartheta)^{-1} a < \infty$ imply $a^\top \mathcal{I}(\vartheta)^{-1} a \leq a^\top \mathrm{V}_\vartheta T a$. This being true for all a, the claim follows. □

It is common to call a consistent sequence of estimators T_n **asymptotically efficient** if it attains asymptotically the Cramér-Rao lower bound:

$$\mathrm{V}\left[\sqrt{n} T_n\right] \underset{n\to\infty}{\longrightarrow} \mathcal{I}(\vartheta_0)^{-1}. \tag{F.6}$$

Hodges (see LeCam [310] p. 280) produced an example of superefficiency where the variance falls below the lower bound, of course in absence of the assumptions of Theorem F.22. LeCam [310] has shown that the number of superefficient points is countable; see also Cox and Hinkley [103], p. 304. Janssen [271] treats efficiency of, and the Cramér-Rao lower bound for, *biased* estimators in a *nonparametric* framework.

F.3   Hypothesis tests

Consider two mutually exclusive probabilistic models $\mathcal{M}_0$ and $\mathcal{M}_1$ on the same sample space E. Let $X \sim \mu$ be an observation in E of which it is only known that μ belongs to the union $\mathcal{M}_0 \cup \mathcal{M}_1$. When it comes to deciding which one it is, then the (null) **hypothesis** $\mathcal{M}_0$ is tested against the **alternative** $\mathcal{M}_1$. The test is called **simple** if $\mathcal{M}_0$ is a singleton, $\mathcal{M}_0 = \{\mu_0\}$. Here, we ask whether $\mu = \mu_0$ or $\mu \in \mathcal{M}_1$. Otherwise, it is called **composite**. When $\mathcal{M}_1 = \mathcal{P}(E) \setminus \mathcal{M}_0$, the test is called omnibus. It is common to write hypothesis and alternative in the form $H_0 \colon \mu \in \mathcal{M}_0$ and $H_1 \colon \mu \in \mathcal{M}_1$.

Hypothesis tests (Fisher [155]) are probabilistic analogs of indirect proofs in mathematics. In mathematics, an assumption is to be proved. We postulate its contrary and draw conclusions from it by logical rules. If we arrive at a contradiction, for instance $0 = 1$, then the assumption must be wrong since we cannot infer "false" from "true." In the opposite case nothing can be concluded about the assumption since it is always possible to infer "true" also from "false." Just start from $0 = 1$ and multiply both sides by zero.

The interpretation of the outcome of a statistical test is somewhat more intricate since, besides logic, randomness intervenes. We wish to verify the *alternative* H_1. To this end, we postulate the contrary, that is, the truth of the hypothesis H_0. This assumption allows us to apply the laws of probability theory to determine a **critical region**, an event which rejects the hypothesis if it occurs. This is a subset $\mathcal{R} \subset E$ with a given small probability α under H_0 and large probability under H_1. One generally chooses a "large" region $\mathcal{R}$ with a small probability under H_0. We now draw the sample X from the unknown μ. If it lies in the critical region (the test is "positive"), then something rare must have occurred under H_0. Not believing this to have happened we conjecture that the hypothesis is untenable, that is, we reject H_0 accepting H_1. It might be that H_0 is still true. Then X was just atypical and we incur an error. In contrast to the mathematical analog above, the result of a simple test is true with probability $1 - \mu_0(\mathcal{R}) = 1 - \alpha$, only, due to the uncertainty of X. The **significance level** of a test is the probability of falsely rejecting H_0. Such an error is called a **false positive**. The **power** of a test is the probability of correctly rejecting H_0, that is, of avoiding **false negatives**. When the sample lies outside of $\mathcal{R}$, nothing can be inferred unless the power is high, that is, $\complement\mathcal{R}$ has a small probability under H_1. The test is then **selective**.

The choice of the critical region $\mathcal{R}$ is a delicate task. The idea is to simultaneously control two antagonists – the significance level and the power. To this end, the statistician generally establishes a so-called **test statistic** $\lambda \colon E \to \mathbb{R}$ that discriminates as well as possible between hypothesis and alternative, small where μ satisfies H_0 and large otherwise so that $\mathcal{R}$ can be defined as the region where λ is larger than some threshold γ. The significance test proceeds then in three steps: (i) Determine the distribution of the test statistic under H_0, (ii) select a significance level α, and (iii) determine a real number γ such that the critical region $\mathcal{R} = [\lambda \geq \gamma]$ has probability α under H_0. The hypothesis H_0 is rejected when $\lambda(X)$ is at least as large as γ. The (random) probability $p = \mu_0[\lambda \geq \lambda(X)]$ is called the p-value of X. Rejection of H_0 takes place when $p \leq \alpha$.

In general, item (i) is the most demanding. The test statistic could be set up by "fortunate guessing" but there are also some paradigms for guidance – Pearson's [405] χ^2 goodness-of-fit statistic, Neyman-Pearson's [391] likelihood ratio statistic, Rao's [429] score statistic, Wald's [519] MLE statistic, and Jeffreys' [273, 274] Bayes factor. The likelihood ratio will be reviewed below, the Bayes factor (BF) in Section 5.3.2. Of interest is also the use of symmetries and algebraic structures; see Diaconis [121]. Marden [344] presents a review of hypothesis testing.

F.23 Definition A test statistic λ is called **invariant** w.r.t. a transformation t on the sample space E if $\lambda(t(x)) = \lambda(x)$ for all $x \in E$.

If a test statistic is invariant w.r.t. t, then both x and $t(x)$ lead to the same conclusions concerning the rejection of H_0. The set of transformations for which a test statistic is invariant forms a group w.r.t. composition, the **invariance group** of the test statistic.

F.3.1 Likelihood ratio test

The **likelihood ratio test** (LRT) reduces a composite test to a simple test, replacing Θ_0 (Θ_1) with the ML estimate ϑ_0^* (ϑ_1^*) on Θ_0 (Θ_1). It is a key method that uses the test statistic

$$\lambda_{\mathrm{LRT}}(X) = \frac{f_X(X; \vartheta_1^*)}{f_X(X; \vartheta_0^*)}. \tag{F.7}$$

When ML estimates do not exist, suitable local maxima may be used. If the test is simple, then, of course, $\vartheta_0^* = \vartheta_0$. The likelihood $f(X; \vartheta_0^*)$ is large and $f(X; \vartheta_1^*)$ is small when H_0 holds true. This ensures that λ_{LRT} satisfies the requirement on a test statistic stated above.

One of the advantages of the LRT is that the distribution of the test statistic can in simple situations be determined analytically; see Example F.24 below. In a rather general nested situation the distribution of the test statistic has been computed at least asymptotically as $n \to \infty$; see, for example, Mardia et al. [345], Chapter 5. A fairly general asymptotic theorem is due to Wilks [528] and Aitchison and Silvey [4]; see also Cox and Hinkley [103], p. 322, and Silvey [474], p. 113:

Let Θ be a $\mathcal{C}^2$-model of dimension p_1 and let Θ_0 be a $\mathcal{C}^2$ submanifold of dimension $p_0 < p_1$. If the parameters ϑ_0^* and ϑ_1^* are consistent and asymptotically normal, then, under regularity conditions, $2 \log \lambda_{\mathrm{LRT}}(X)$ has an asymptotic $\chi^2_{p_1 - p_0}$ distribution.

If this theorem does not apply, often because Θ_0 is not a differentiable submanifold of Θ, one can try to determine the distribution of $2 \log \lambda_{\mathrm{LRT}}(X)$ via simulation.

F.24 Example Assume that X is an i.i.d. n-tuple $X = (X_1, \dots, X_n)$ from $N_{m,V}^d$ with *unspecified* covariance matrix V and unknown mean value m, $H_0 \colon m = 0$, $H_1 \colon m \neq 0$ (or $m \in \mathbb{R}^d$). Under the null hypothesis, the estimate of the covariance matrix is the total SSP matrix $T = \frac{1}{n} \sum_i X_i X_i^\top$ and so twice the logarithm of the maximum likelihood is $-n(\log \det 2\pi T + d)$. Under the alternative, the estimates of location and scale are the sample mean $\overline{x}$ and the scatter matrix S of the data, respectively, and so twice the logarithm of the maximum likelihood is $-n(\log \det 2\pi S + d)$. Thus,

$$2\lambda_{\mathrm{LRT}}(X) = n(\log \det 2\pi T - \log \det 2\pi S) = n \log \det T S^{-1}.$$

This test statistic is invariant w.r.t. scale transformations but not w.r.t. translations.

F.3.2 Multiple tests

Roy's [451] union-intersection (UI) test rejects a composite hypothesis if a single test is positive. This approach unduly increases the probability of a false rejection (committing a false positive error) with the number of hypotheses, an undesirable effect. The observation led researchers, among them Simes [476] and Hommel [251], to a deeper analysis of multiple testing procedures culminating in Benjamini and Hochberg's [35] "false discovery rate" (FDR). In the diction of Sorić [481], a rejected (null) hypothesis is a "discovery." Hence, a false positive is a false discovery and Benjamini and Hochberg [35] call the false positive rate the "false discovery rate."

More precisely, let $H_0^1, \dots, H_0^D$ be D null hypotheses with associated test statistics $Y_1, \dots, Y_D$ and corresponding p-values $p_1, \dots, p_D$ with ranking $p_{\uparrow 1} \leq \dots \leq p_{\uparrow D}$, $p_{\uparrow 1}$ being the most significant.[2] Following Simes [476], Bejamini and Hochberg defined the **false discovery rate** (FDR) as the expected proportion of false positives for a given rejection

---

[2] The symbol $\uparrow$ denotes a permutation of the indices $1 \mathinner{.\,.} D$ that sorts the p-values in increasing order.

rule $\mathcal{R}$ based on the sequence $p_{\uparrow 1} \leq \cdots \leq p_{\uparrow D}$. If p_0 is the proportion of true hypotheses and $k_{\max} = \max \left\{ k \mid p_{\uparrow k} \leq \frac{k(1-\alpha)}{Dp_0} \right\}$, they showed that the FDR of the rule

$$\mathcal{R}:\ reject\ \mathrm{H}_0^{\uparrow 1}, \ldots, \mathrm{H}_0^{\uparrow k_{\max}}$$

does not exceed $1 - \alpha$, where α is a given significance level, for instance 0.95. Since p_0 is unknown, they recommend inserting the most conservative value $p_0 = 1$. The rule is best evaluated by starting from the largest p-value, $p_{\uparrow D}$. Benjamini and Yekutieli [36] show that the FDR equals $1 - \alpha$ if the test statistics Y_k are continuous and independent.

Appendix G

Optimization

Some of the estimators developed in this book require some subtle continuous and discrete optimization algorithms. They are needed in computing local optima of mixture likelihoods and of steady solutions of cluster criteria. This appendix discusses some basic facts about the optimization of real-valued functions on $\mathbb{R}^p$, the EM algorithm, and λ-assignment.

G.1 Numerical optimization

This section deals with two optimization algorithms of scalar functions on $\mathbb{R}^p$. They are the proximal-point and a quasi-Newton algorithm and dynamic optimization. The first has turned out to be closely related to the EM algorithm of Section G.2 below, the second has been proposed for its acceleration, and the third will be needed for optimization of likelihood functions w.r.t. partial parameters.

G.1.1 Proximal-point algorithm

The proximal-point algorithm (PPA) has its roots in convex optimization established by Rockafellar [442] and others. It can be more generally used to optimize continuous functions $G\colon \Theta \to \mathbb{R}$ on a complete metric space Θ.

G.1 Definition (a) A function $u\colon \Theta \times \Theta \to [0, \infty]$ is called a **weak dissimilarity** if

$$u(\vartheta, \vartheta) = 0 \quad \text{for all } \vartheta \in \Theta.$$

Weak dissimilarities differ from dissimilarities by their missing symmetry.
(b) Given $\theta, \vartheta \in \Theta$, let

$$H(\theta, \vartheta) = G(\theta) - u(\theta, \vartheta) \qquad \text{and} \qquad \vartheta' \in \operatorname*{argmax}_{\theta \in \Theta} H(\theta, \vartheta)$$

if the maximum exists in $\mathbb{R}$.
(c) A parameter ϑ is called a **fixed point** of H if ϑ is maximal for $H(\cdot, \vartheta)$.
(d) The **Proximal-Point algorithm** (PPA), (Martinet [353], Rockafellar [443]) is the iteration of proximal point steps

$$\vartheta \leftarrow \operatorname*{argmax}_{\theta} H(\theta, \vartheta) \tag{G.1}$$

starting from an arbitrary point $\vartheta_0 \in \Theta$.
(e) A parameter ϑ is called a **halting point** of the PPA if ϑ' exists and $\vartheta' = \vartheta$.

Assume throughout that u is continuous. Then $u(\cdot, \vartheta)$ is finite in a neighborhood of ϑ. If ϑ is such that ϑ' exists, then both $u(\vartheta', \vartheta)$ and $H(\vartheta', \vartheta)$ are finite. Assume that this holds for all $\vartheta \in \Theta$.

G.2 Examples The following conditions are sufficient for ϑ' to exist for all $\vartheta \in \Theta$.

(a) The space Θ is compact.

(b) The space Θ is locally compact and not compact and $G(\theta) \to -\infty$ as $\theta \to \infty$. Indeed, we first have

$$H(\theta, \vartheta) = G(\theta) - u(\theta, \vartheta) \to -\infty$$

as $\theta \to \infty$. Thus, there is a compact subset $K \subseteq \Theta$ such that $H(\theta, \vartheta) < H(\theta_0, \vartheta)$ for an arbitrarily chosen $\theta_0 \in \Theta$ and all $\theta \notin K$. It follows that the maximum of $H(\cdot, \vartheta)$ in K is also the global maximum in Θ.

(c) The space Θ is locally compact and not compact and
  (i) G is upper bounded and
  (ii) $u(\theta, \vartheta) \to \infty$ as $\theta \to \infty$.
  (Same proof as (b).)

The most important example of a locally compact space is the intersection of a closed and an open subset of Euclidean space; see Section B.1. In particular, open and closed subsets share this property.

G.3 Remark Condition G.2(c)(ii) cannot be dropped. Let $\Theta = [-1, 1[$ be the half open interval, $G(\theta) = 3\theta^2$, $\vartheta = 0$,

$$u(\theta, 0) = \begin{cases} 2\theta^2, & \theta \le 0, \\ \theta^2, & \theta > 0. \end{cases}$$

Then

$$H(\theta, \vartheta) = G(\theta) - u(\theta, 0) = \begin{cases} 3\theta^2 - 2\theta^2 = \theta^2, & \theta \le 0, \\ 3\theta^2 - \theta^2 = 2\theta^2, & \theta > 0, \end{cases}$$

does not have a maximum in Θ.

The following proposition shows that an increase of the H-functional also increases the target function G. This applies in particular to any maximizer ϑ'.

G.4 Proposition *(a) If $H(\theta, \vartheta) > H(\vartheta, \vartheta)$, then $G(\theta) > G(\vartheta)$.*
In particular: any maximixer of G is a halting point of the PPA.
(b) We have $G(\vartheta') \ge G(\vartheta)$ and from $u(\vartheta', \vartheta) \ne 0$ it follows $G(\vartheta') > G(\vartheta)$.
(c) Any halting point of the PPA is a fixed point of H. Conversely, any fixed point ϑ of H which is the only maximum of $H(\cdot, \vartheta)$ is a halting point.

PROOF. (a) The first claim follows from

$$G(\theta) = H(\theta, \vartheta) + u(\theta, \vartheta) \ge H(\theta, \vartheta) > H(\vartheta, \vartheta) = G(\vartheta).$$

The second claim follows immediately from the first.
(b) Note that $u(\vartheta', \vartheta) < \infty$. From the definition of ϑ' it follows similarly

$$G(\vartheta') = H(\vartheta', \vartheta) + u(\vartheta', \vartheta) \ge H(\vartheta', \vartheta) \ge H(\vartheta, \vartheta) = G(\vartheta).$$

(c) If ϑ is a halting point, then $H(\vartheta, \vartheta) = G(\vartheta) = G(\vartheta') \ge G(\vartheta') - u(\vartheta', \vartheta) = H(\vartheta', \vartheta)$ is maximal for $H(\cdot, \vartheta)$. If ϑ is a fixed point and the maximum is unique, then we even have $\vartheta' = \vartheta$. □

Of course, a fixed point of H does not have to be a maximum of G; see Proposition G.4(c). Often it is a *local* maximum but even this is not guaranteed. The situation is better whenever G is differentiable. So now let Θ be an open subset of some Euclidean space.

G.5 Proposition *Let Θ be an open subset of Euclidean space, let $\vartheta \in \Theta$ be a fixed point of H, and let G and $u(\cdot, \vartheta)$ be differentiable at ϑ. Then ϑ is a critical point of G.*

PROOF. By $\vartheta \in \operatorname{argmax} H(\cdot, \vartheta)$ we have $D_2 H(\vartheta, \vartheta) = 0$. Since $G = H(\cdot, \vartheta) + u(\cdot, \vartheta)$ close to ϑ, this implies

$$DG(\vartheta) = D_2 H(\vartheta, \vartheta) + D_2 u(\vartheta, \vartheta) = 0 + 0.$$

□

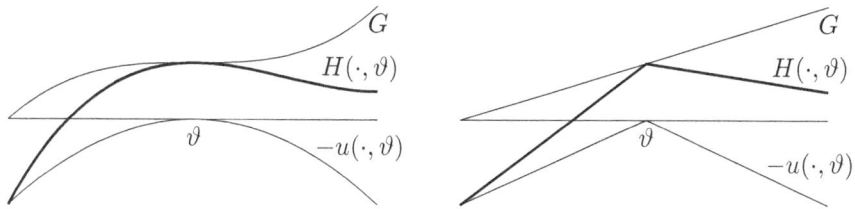

Figure G.1: *A fixed point of H, critical (left) and noncritical (right), for G.*

G.6 Proposition *(a) If $\vartheta_0, \vartheta_1, \vartheta_2, \ldots$ is a sequence generated by the PPA, then $G(\vartheta_{t+1}) \geq G(\vartheta_t)$.*
(b) If $\vartheta = \lim_{t \to \infty} \vartheta_t \in \Theta$ exists, then ϑ is a fixed point of H (and a critical point of G if the assumptions on Θ, G, ϑ, and u made in Proposition G.5 are satisfied).

PROOF. Claim (a) follows from Proposition G.4(b). Moreover, for all $\theta \in \Theta$ the relation $H(\theta, \vartheta_t) \leq H(\vartheta_{t+1}, \vartheta_t)$ implies
$$H(\theta, \vartheta) = \lim_{t \to \infty} H(\theta, \vartheta_t) \leq \lim_{t \to \infty} H(\vartheta_{t+1}, \vartheta_t) = H(\vartheta, \vartheta).$$
Claim (b) now follows from Proposition G.5. □

G.7 Remarks (a) The idea of the PPA is the following: Assume that it is difficult to optimize the target function G. The reason may be that there is no suitable representation of G or that there is no efficient algorithm. Then you may find a weak dissimilarity such that it is easier to optimize the associated H-functional. Proposition G.6 promises that the PPA converges at least to a critical point of G. (The original optimization problem is retrieved for $u = 0$.)

(b) The sequence $(\vartheta_t)_t$ does not necessarily possess a limit, even if G has a global maximum. By way of example, let $\Theta = \mathbb{R}$,
$$G(\vartheta) = \begin{cases} \frac{\vartheta}{1+\vartheta}, & \vartheta \geq 0, \\ \text{arbitrary with a global maximum } = 1, & \vartheta < 0, \end{cases}$$
and let $u(\theta, \vartheta) = |\theta - \vartheta|^2$ and $\vartheta_0 = 1$. Then the sequence $(\vartheta_t)_t$ is unbounded. Indeed, by Proposition G.6 the target values $G(\vartheta_t)$ are increasing. Induction on t along with the construction of G (strictly monotone) and u shows that ϑ_t increases. If ϑ_t was bounded, then the limit $\vartheta = \lim_{t \to \infty} \vartheta_t$ would exist and would be a critical point of G by Proposition G.6(b). But G has no critical points.

G.1.2 Quasi-Newton methods

The search for (local) maxima of functions $G \colon \mathbb{R}^p \to \mathbb{R}$ is usually carried out with iterative methods.[1] These create a sequence of points ϑ_k that is supposed to converge to a local maximum. This should happen starting from any point in $\mathbb{R}^p$. Another requirement is speed. The time requirement of one iteration is the number of steps $\times$ the time for one step. The point ϑ_{k+1} is determined from ϑ_k by a *search direction* and a *step length*. The first algorithm that comes to mind is by **Newton** [388] and **Raphson** [431] in the 17th century. It is applicable when G is twice differentiable and follows the scheme
$$\vartheta_{k+1} = \vartheta_k - \left(\mathrm{D}^2 G(\vartheta_k)\right)^{-1} \mathrm{D} G(\vartheta_k).$$
It provides both quantities at a time but needs the gradient $\mathrm{D} G(\vartheta_k)$ and the Hessian $\mathrm{D}^2 G(\vartheta_k)$ at each point ϑ_k visited during the iteration. In fact, it attains the unique maximum of a concave parabolic function in one step. Unfortunately, log-likelihood functions

---

[1]All methods discussed here are, with the necessary care, also applicable to differentiable functions $G \colon U \to \mathbb{R}$ defined on open subsets $U \subseteq \mathbb{R}^p$.

of (normal) mixtures are much more complex and approximately parabolic only near local maxima and minima. The Newton-Raphson algorithm started from an arbitrary point will then converge to any critical point close by. However, according to the consistency theorems of Chapters 1 and 4, the only points of interest to us are local maxima.

A first method to avoid this nuisance is **steepest ascent**. It is greedy, taking at each point the gradient direction. On a concave parabolic function, steepest ascent moves toward the maximum on a zig-zag course, like a sailing vessel beating to windward, unless the parabola is close to spherical or a fortunate initial point has been guessed. This effects an increased number of steps and slow convergence. To avoid this drawback, more efficient methods that take more expedient directions had been designed in numerical analysis up to 1970. They are the **conjugate gradient methods** of Fletcher and Reeves [162] and of Polak, Ribière, and Polyak [415, 419], and two **quasi-Newton methods**, the **DFP method** of Davidon [108] and Fletcher and Powell [161], and the **BFGS double-rank method** by Broyden [69, 70, 71], Fletcher [159], Eq. (5), Goldfarb [204], p. 24, and Shanno [467], Eq. (19). See also Shanno and Kettler [468]. All use a substitute B for the inverse of the negative Hessian at the current point. The BFGS method, for instance, applied to maximization iterates the following step.

G.8 Procedure (BFGS step)
// Input: A point $\vartheta \in \mathbb{R}^p$ and a positive definite matrix B.
// Output: A new point with larger target value and a new positive definite matrix
                  or "stop."

1. Starting from ϑ and searching in the direction of $B\,DG(\vartheta)$, determine a point ϑ_{new} with a large value of $G(\vartheta_{\text{new}})$;

2. set $u = \vartheta_{\text{new}} - \vartheta$ and $v = DG(\vartheta_{\text{new}}) - DG(\vartheta)$;

3. *if* $u^\top v$ is small, "stop"; *else* return ϑ_{new} and the update

$$B_{\text{new}} = B + \frac{1}{u^\top v}\left(1 + \frac{v^\top B v}{u^\top v}\right)uu^\top - \left(\frac{u}{u^\top v}v^\top B\right) - \left(\frac{u}{u^\top v}v^\top B\right)^\top. \qquad (\text{G.2})$$

The BFGS algorithm starts from some ϑ in the domain of definition of the target function and some positive definite B, for instance I_p, until the stop criterion is satisfied. Step 1 is called a **line search**. Although it is not necessary to attain the maximum here, an arbitrarily small gain does not guarantee convergence to a local maximum. The matrix B is updated in step 3. The update of the substitute for the Hessian, B^{-1}, is

$$B_{\text{new}}^{-1} = B^{-1} + \frac{bb^\top}{a^\top b} - \frac{B^{-1}a\,(B^{-1}a)^\top}{a^\top B^{-1}a}.$$

It implies the update of B defined in step 3 by a double application of the Sherman-Morris-Woodbury formula of linear algebra. This is an explicit representation of the inverse of a rank-one perturbed regular matrix A, $\left(A - uv^\top\right)^{-1} = A^{-1} + \frac{A^{-1}uv^\top A^{-1}}{1 - v^\top A^{-1}u}$. The matrices B remain positive definite throughout the iteration. The BFGS algorithm finds the maximum of a concave parabolic function in p steps if the line search in step 1 is carried through to the optimum; see, for instance, Broyden [71], Theorem 1. However, this is not advisable because of the overhead it creates. It is better to estimate or to predict the step size. Like the Newton-Raphson algorithm, these algorithms, too, are underpinned by an analysis of their behavior on a quadratic function. They do avoid the disadvantage of steepest ascent. However, they cannot perform wonders. Mixture likelihoods are fairly complex "landscapes." The algorithms discussed above will at best follow the bobsled run of one mile length unaware of the fact that the finish is just 100 yards down.

Standard references to optimization methods are the monographs of Ortega and Rheinboldt [397], Fletcher [160], Minoux [382], Luenberger [338], Nazareth [386], and Polak [414].

G.1.3   Dynamic optimization

If a real-valued function on the Cartesian product of two sets possesses a maximum, then maximization can often be decomposed in two steps: First maximize w.r.t. the first coordinate for all values of the second and then maximize the maxima obtained. This is called **dynamic optimization.**.

G.9 Lemma *Let $f\colon M \times N \to \mathbb{R}$ be a real-valued function on some product space $M \times N$. Assume that the v-cut $f(\cdot, v)$ possesses a maximum u_v for all $v \in N$. Then the following are equivalent.*

(a) The function f has a maximum;

(b) the function $v \mapsto f(u_v, v)$ has a maximum.

In this case, we have the equality $\max_{u,v} f(u,v) = \max_v \max_u f(u,v)$.

G.2   EM algorithm

The EM algorithm of Dempster et al. [115] has been a standard tool for ML and MAP parameter estimation in complex situations since its appearance in 1977. Chrétien and Hero [95] noticed a close connection to the PPA algorithm. The approach to EM via PPA separates the optimization from the probabilistic issues.

G.2.1   General EM scheme

The EM algorithm is typically used when a probabilistic model $X\colon \Omega \to E$, $X \sim f_\vartheta$, is complex[2] but can be written as a function $X = \Phi(Y)$ of a so-called *complete-data* model $Y\colon \Omega \to F$, $Y \sim g_\vartheta = \mathrm{d}\nu_\vartheta / \mathrm{d}\varrho$, whose likelihood function can be more easily handled. In this context, Y is called the **complete variable** and X is the **observed variable**. It does not matter whether the complete variable has a "real" background or not. It may contain the observed variable as a constituent, $Y = (X, Z)$, but this is not necessary. In this case, Z is called the **hidden variable**. The likelihood function $g_\vartheta(\mathbf{y})$ is called the **complete likelihood**, in contrast to the **observed likelihood** $f_\vartheta(\mathbf{x})$. In order to guarantee the existence of the conditional distributions below, the complete sample space F is assumed to be Polish; see Section B.3. This is no restriction in view of applications.

The target function is the logarithm of the *posterior density* $f_\vartheta(\mathbf{x})f(\vartheta)$ w.r.t. a prior density $f(\vartheta)$. If the latter is flat, 1, then the ML estimator arises as a special case. In order to maximize it, Chrétien and Hero [95] use the PPA with the divergence conditional on the observation $\Phi = \mathbf{x}$ and the parameters ϑ and θ as a weak dissimilarity.[3] Thus, the H-functional of the PPA algorithm is

$$H(\theta, \vartheta) = \log\big(f(\theta)f_\theta(\mathbf{x})\big) - D\big(g_\vartheta[\,\cdot\,|\,\Phi = \mathbf{x}], g_\theta[\,\cdot\,|\,\Phi = \mathbf{x}]\big) \qquad (\text{G.3})$$

$$= \log f(\theta) + \log f_\theta(\mathbf{x}) - \int \log \frac{g_\vartheta[\mathbf{y}\,|\,\Phi = \mathbf{x}]}{g_\theta[\mathbf{y}\,|\,\Phi = \mathbf{x}]} \nu_\vartheta[\,\mathrm{d}\mathbf{y}\,|\,\Phi = \mathbf{x}]$$

$$= \log f(\theta) + \log f_\theta(\mathbf{x}) + \int \log g_\theta[\mathbf{y}\,|\,\Phi = \mathbf{x}] \nu_\vartheta[\,\mathrm{d}\mathbf{y}\,|\,\Phi = \mathbf{x}] + C(\mathbf{x}, \vartheta)$$

$$= \log f(\theta) + \int \log\big(g_\theta[\mathbf{y}\,|\,\Phi = \mathbf{x}] f_\theta(\mathbf{x})\big) \nu_\vartheta[\,\mathrm{d}\mathbf{y}\,|\,\Phi = \mathbf{x}] + C(\mathbf{x}, \vartheta),$$

with a quantity $C(\mathbf{x}, \vartheta)$ that does not depend on θ. Integration is exercised only w.r.t. $\mathbf{y}$ such that $\Phi(\mathbf{y}) = \mathbf{x}$. For such $\mathbf{y}$ we always have $g_\theta[\mathbf{y}\,|\,\Phi = \mathbf{x}]f_\theta(\mathbf{x}) = g_\theta(\mathbf{y})$ and we continue computing

---

[2]When the likelihood function $\vartheta \mapsto f_\vartheta(\mathbf{x})$ is easily maximized, the EM algorithm is not needed.

[3]Note that the conditional distributions can be equal for $\theta \neq \vartheta$ even if the unconditional counterparts are distinct.

$$\cdots = \log f(\theta) + \int \log g_\theta(\mathbf{y}) \nu_\vartheta[\,\mathrm{dy}\,|\,\Phi = \mathbf{x}\,] + C(\mathbf{x}, \vartheta)$$

$$= \log f(\theta) + Q(\theta, \vartheta) + C(\mathbf{x}, \vartheta).$$

The function $Q(\theta, \vartheta) = \int \log g_\theta(\mathbf{y}) \nu_\vartheta[\,\mathrm{dy}\,|\,\Phi = \mathbf{x}\,] = \mathrm{E}_{\nu_\vartheta}[\,\log g_\theta\,|\,\Phi = \mathbf{x}\,]$ is called the **Q-functional**. The computation above shows that the H-functional $H(\theta, \vartheta)$ differs from $\log f(\theta) + Q(\theta, \vartheta)$ by a quantity that does not depend on the target variable θ. Therefore, maximizing H is tantamount to maximizing $\theta \mapsto f(\theta) + Q(\theta, \vartheta)$ and the PPA may also be represented by the iteration[4]

$$\vartheta \leftarrow \operatorname*{argmax}_\theta \left(f(\theta) + Q(\theta, \vartheta)\right).$$

The two statements $\vartheta \in \operatorname{argmax}_\theta(f(\theta) + Q(\theta, \vartheta))$ and $\vartheta \in \operatorname{argmax}_\theta H(\theta, \vartheta)$ are equivalent. Such a *fixed point* ϑ of H (see Definition G.1) is therefore also called a **fixed point** of $f + Q$. The properties of the above iteration flow from those of the PPA algorithm stated in Section G.1.1. In particular, Propositions G.4–G.6 show the following important facts.

G.10 Theorem *(a) Each increase of $f + Q(\vartheta, \cdot)$ increases the observed posterior density.*
(b) In particular: The iteration $\vartheta \leftarrow \operatorname{argmax}_\theta \left(f(\theta) + Q(\theta, \vartheta)\right)$ improves the observed posterior density.
(c) If the limit exists, then it is a fixed point of $f + Q$. If, in addition, the assumptions of Proposition G.5 are satisfied, then the fixed point is critical for the observed posterior density.

Let $\varrho_{\mathbf{x}}$ be the reference measure ϱ on F conditional on $[\Phi = \mathbf{x}]$. The conditional distribution $\nu_\vartheta[\,\mathrm{dy}\,|\,\Phi = \mathbf{x}\,]$ has the $\varrho_{\mathbf{x}}$-density $\frac{g_\vartheta(\mathbf{y})}{\int g_\vartheta(\mathbf{u})\varrho_{\mathbf{x}}(\mathrm{du})}$ and the Q-functional can be computed as

$$w_\vartheta(\mathbf{x}, \mathbf{y}) = g_\vartheta(\mathbf{y}) \Big/ \int g_\vartheta(\mathbf{u})\varrho_{\mathbf{x}}(\mathrm{du}), \qquad \text{(conditional density)} \qquad (\mathrm{G.4})$$

$$Q(\theta, \vartheta) = \int w_\vartheta(\mathbf{x}, \mathbf{y}) \log g_\theta(\mathbf{y}) \varrho_{\mathbf{x}}(\mathrm{dy}). \qquad (\mathrm{G.5})$$

It depends on ϑ only via the stochastic kernel w_ϑ. It is worthwhile to discuss some special cases. If $\mathbf{x} = (x_1, \ldots, x_n)$ and $\mathbf{y} = (y_1, \ldots, y_n)$ consist of realizations of independent random variables, then define

$$w_\vartheta(x, y) = g_\vartheta(y) \Big/ \int g_\vartheta(u)\varrho_x(\mathrm{du}), \qquad x \in E, y \in F. \qquad (\mathrm{G.6})$$

It follows

$$Q(\theta, \vartheta) = \int w_\vartheta(\mathbf{x}, \mathbf{y}) \log g_\theta(\mathbf{y}) \varrho_{\mathbf{x}}(\mathrm{dy}) = \sum_i \int w_\vartheta(\mathbf{x}, \mathbf{y}) \log g_\theta(y_i) \varrho_x(\mathrm{dy})$$

$$= \sum_i \int \left(\prod_h w_\vartheta(x_h, y_h)\right) \log g_\theta(y_i) \bigotimes_h \varrho_{x_h}(\mathrm{dy}_h)$$

$$= \sum_i \int w_\vartheta(x_i, y) \log g_\theta(y) \varrho_{x_i}(\mathrm{dy})$$

and so (G.5) turns into

$$Q(\theta, \vartheta) = \sum_i \int w_\vartheta(x_i, y) \log g_\theta(y) \varrho_{x_i}(\mathrm{dy}). \qquad (\mathrm{G.8})$$

When Y is of the form $Y = (X, Z)$ with a hidden variable Z and Φ is the first

---

[4]Sometimes, even maximization of $f + Q(\vartheta, \cdot)$ is not efficiently possible. In such cases it is useful to just improve the value of the functional.

projection, $\varrho_{\mathbf{x}}$ lives on the sample space of Z and $w_\vartheta(\mathbf{x}, \mathbf{y})$ simplifies to $w_\vartheta(\mathbf{x}, \mathbf{z}) = g_\vartheta(\mathbf{x}, \mathbf{z}) / \int g_\vartheta(\mathbf{x}, \mathbf{u}) \varrho_{\mathbf{x}}(\mathrm{d}\mathbf{u})$. In the case of independent variables, we have correspondingly

$$w_\vartheta(x, z) = g_\vartheta(x, z) \Big/ \!\! \int g_\vartheta(x, u) \varrho_x(\mathrm{d}u), \tag{G.9}$$

$$Q(\theta, \vartheta) = \sum_i \int w_\vartheta(x_i, z) \log g_\theta(x_i, z) \varrho_{x_i}(\mathrm{d}z) = \sum_i Q_i(\theta, \vartheta). \tag{G.10}$$

The proximal point step G.1(d) turns into the following EM-step. The E-step updates the posterior probabilities of component membership. The M-step updates the parameters.

G.11 Procedure (EM-step)
// Input: An arbitrary parameter $\vartheta \in \Theta$.
// Output: A parameter with larger observed posterior density.

1. (E-step) Compute the conditional density $w_\vartheta(\mathbf{x}, \mathbf{y})$; see (G.4), (G.6), and (G.9);
2. (M-step) return $\mathrm{argmax}_\theta(\log f(\theta) + Q(\theta, \vartheta))$; see (G.5), (G.8), and (G.10).

The **EM algorithm** is the iteration of EM-steps until convergence or a time limit. It is iterative and alternating. If started from an M-step with a stochastic weight matrix $\mathbf{w}^{(0)}$, it proceeds as follows:

$$\mathbf{w}^{(0)} \longrightarrow \vartheta^{(1)} \longrightarrow \mathbf{w}^{(1)} \longrightarrow \vartheta^{(2)} \longrightarrow \mathbf{w}^{(2)} \longrightarrow \vartheta^{(3)} \longrightarrow \cdots$$

The algorithm converges unless it is started near a singularity. The limit is often a local maximum even if the MLE exists.

G.2.2   ML estimation in multivariate t-models

Algorithms for elliptical mixture and classification models are based on ML parameter estimation for elliptical symmetries. It is well known that this can again be done with the EM algorithm, e.g., McLachlan and Peel [366], Sections 7.4 and 7.5, and Kotz and Nadarajah [297], Section 10.2. We return to the representation of a multivariate t variable X with a fixed index λ and the parameters m and V as

$$X = m + \frac{N}{\sqrt{Z}},$$

where $N \sim N_{0,V}$ and $Z = \chi_\lambda^2 / \lambda$ is independent of N; see Example E.22(c). Consider the variable Z as missing so that the complete variable is $(X, Z) \colon \Omega \to \mathbb{R}^d \times \mathbb{R}_>$. The Jacobian matrix of the map defined by $\Phi(x, z) = \big(\sqrt{z}(x - m), z\big)$ is

$$\begin{pmatrix} \sqrt{z} I_d & \frac{x-m}{2\sqrt{z}} \\ 0 & 1 \end{pmatrix} \in \mathbb{R}^{(d+1) \times (d+1)}.$$

Its determinant is $z^{d/2}$. Since $\Phi(X, Z) = (N, Z)$, the Transformation Theorem E.7 for Lebesgue densities shows that the complete likelihood is

$$g_{m,V}(x, z) = f_{X,Z}(x, z) = z^{d/2} f_{N,Z}(\sqrt{z}(x - m), z)$$
$$= z^{d/2} f_N(\sqrt{z}(x - m)) f_Z(z).$$

Up to additive terms that do not depend on m' and V', its logarithm is

$$\log g_{m',V'}(x, z) \sim -\tfrac{1}{2} \log \det V' - \tfrac{1}{2} z (x - m')^\top V'^{-1}(x - m').$$

With $w_{m,V}(x_i, z)$ as in the E-step (G.9) the relevant part of the partial Q-functional Q_i in formula (G.10) that depends on m' and V' is therefore

$$Q_i((m', V'), (m, V)) = \int_0^\infty w_{m,V}(x_i, z) \log g_{m',V'}(x_i, z)\,\mathrm{d}z$$

$$\sim -\tfrac{1}{2} \log \det V' - \tfrac{1}{2}(x_i - m')^\top V'^{-1}(x_i - m') \frac{\int_0^\infty z g_{m,V}(x_i, z)\,\mathrm{d}z}{\int_0^\infty g_{m,V}(x_i, z)\,\mathrm{d}z}$$

$$= -\tfrac{1}{2}\big(\log \det V' + e_{m,V}(x_i)(x_i - m')^\top V'^{-1}(x_i - m')\big).$$

From the weighted Steiner formula A.12 and from Lemma F.10, it follows that the mean m' and covariance matrix V' in the Q-functional $\sum_i Q_i$ are optimized by

$$m^* = \frac{\sum_i e_{m,V}(x_i)x_i}{\sum_i e_{m,V}(x_i)} \quad \text{and} \quad V^* = \frac{1}{n}\sum_i e_{m,V}(x_i)(x_i - m^*)(x_i - m^*)^\top,$$

respectively. It remains to compute the conditional expectation $e_{m,V}(x)$, $x \in \mathbb{R}^d$. Now $f_Z(z) = \lambda f_{\chi_\lambda^2}(\lambda z)$ and so we have for $q \geq 0$ up to an irrelevant factor

$$\int_0^\infty z^q g_{m,V}(x,z)\mathrm{d}z = \int_0^\infty z^{d/2+q} f_N(\sqrt{z}(x-m))f_Z(z)\mathrm{d}z$$

$$\sim \frac{1}{\sqrt{\det V}}\int_0^\infty z^{d/2+q}e^{-z(x-m)^\top V^{-1}(x-m)/2} \cdot z^{\lambda/2-1}e^{-\lambda z/2}\mathrm{d}z$$

$$= \frac{1}{\sqrt{\det V}}\int_0^\infty z^{(\lambda+d)/2+q-1}e^{-z(\lambda+(x-m)^\top V^{-1}(x-m))/2}\mathrm{d}z.$$

(For $q = 0$ this is, of course, $f_X(x)$ up to the factor.) From $\int_0^\infty z^\kappa e^{-az}\mathrm{d}z / \int_0^\infty z^{\kappa-1}e^{-az}\mathrm{d}z = \frac{\Gamma(\kappa+1)a^\kappa}{a^{\kappa+1}\Gamma(\kappa)} = \frac{\kappa}{a}$, we finally infer

$$e_{m,V}(x) = \frac{\int_0^\infty z g_{m,V}(x,z)\mathrm{d}z}{\int_0^\infty g_{m,V}(x,z)\mathrm{d}z} = \frac{\lambda+d}{\lambda+(x-m)^\top V^{-1}(x-m)}.$$

We have thus derived the EM-step for the multivariate t-distribution: Compute $e_{m,V}(x_i)$ in the E-step and m^* and V^* in the M-step. Its closed form is a nice feature of the multivariate t-distribution. Kotz and Nadarajah [297], Section 10.2, propose a simpler interpretation of the EM-step: A critical point (m,V) of the log-likelihood function $-\frac{n}{2}\log V - \frac{\lambda+d}{2}\sum_i \log\left(\lambda + (x-m)^\top V^{-1}(x-m)\right)$ is characterized by the equations

$$m = \frac{\sum_i e_{m,V}(x_i)x_i}{\sum_i e_{m,V}(x_i)} \quad \text{and} \quad V = \frac{1}{n}\sum_i e_{m,V}(x_i)(x_i - m)(x_i - m)^\top.$$

A solution in closed form does not exist but the update of the pair (m,V) with the current weights $e_{m,V}(x_i)$ is just the M-step.

Convergence of the EM algorithm is often slow. Several authors have proposed accelerations for t-models. Kent et al. [288] use the denominator $s = \sum_i e_{m,V}(x_i)$ not only for the mean m^* above but also for the scale matrix V^* (instead of n). Simulations show that this effects a substantial speed-up of the algorithm, in particular in higher dimensions. A theoretical substantiation is offered by Meng and van Dyk [372]. The following observation goes back to Kent et al. [288]. It implies that a limit of the EM algorithm is a fixed point of the modified algorithm.

G.12 Lemma *If the EM algorithm converges, the sum $s = \sum_i e_{m,V}(x_i)$ converges to n.*

PROOF. From the definition of $e_{m,V}$, we deduce

$$e_{m,V}(x)(x-m)^\top V^{-1}(x-m) = \lambda + d - \lambda e_{m,V}(x).$$

In the limit we have $m^* = m$ and $V^* = V$ and hence

$$nd = n\,\mathrm{tr}\left(V^* V^{-1}\right) = \mathrm{tr}\left(\sum_i e_{m,V}(x_i)(x_i - m^*)(x_i - m^*)^\top V^{-1}\right)$$

$$= \mathrm{tr}\left(\sum_i e_{m,V}(x_i)(x_i - m)(x_i - m)^\top V^{-1}\right)$$

$$= \sum_i e_{m,V}(x_i)(x_i - m)^\top V^{-1}(x_i - m) = n(\lambda+d) - \lambda\sum_i e_{m,V}(x_i).$$

The claim follows.   □

Table G.1 *Comparison of the number of iterations for two ways of computing V_{new} in the M-step.*

scale	dimension	1	6	12	18	24
spherical	divided by n	14	25	39	53	67
	divided by s	11	8	7	7	8
elliptical	divided by n	14	28	41	54	69
	divided by s	11	13	11	10	11

This leads to the following modified EM-step for t-models.

G.13 Procedure (Modified EM-step for multivariate t-distributions)
// Input: An arbitrary $m \in \mathbb{R}^d$ and $V \in \mathrm{PD}(d)$.
// Output: Parameters m_{new} and V_{new} with larger observed likelihood.

1. (E-step) Compute the conditional expectations $e_{m,V}(x_i)$, $1 \le i \le n$, and their sum $s = \sum_i e_{m,V}(x_i)$;

2. (M-step) return

$$m_{\text{new}} = \tfrac{1}{s} \sum_i e_{m,V}(x_i) x_i, \quad V_{\text{new}} = \tfrac{1}{s} \sum_i e_{m,V}(x_i)(x_i - m_{\text{new}})(x_i - m_{\text{new}})^\top.$$

It is recommended to use the sample mean and the scatter matrix of the data, respectively, as initial values $m^{(0)}$ and $V^{(0)}$ for the resulting EM algorithm. Moreover, it pays to use a small value $\lambda \ge 2$ for the degrees of freedom in order to safeguard the model against bad outliers.

The difference $s - n$ may be taken to indicate convergence. Table G.1 compares the numbers of iterations necessary for attaining $|s - n| \le 0.01$, for various contaminated data sets of dimension between 1 and 24. They consist of 230 data points each, 200 regular observations and 30 outliers from $N_{0,50I_d}^{(d)}$. The regular, spherical data is sampled from the standard d-variate normal and the eigenvalues of the regular elliptical data vary between 10.0 and 0.03. The algorithms were run with four degrees of freedom. The simulation allows three conclusions. First, run time increases with dimension if the sum is divided by n, whereas it stays basically constant if it is divided by s. Second, run time is overall smaller in the second case. Third, it is basically independent of data scale.

G.2.3   Notes

Forerunners of the EM algorithm in different special situations are McKendrick [359], Grundy [214], Healy and Westmacott [236], Hartley [229], Hasselblad [230], Efron [141], Day [113], Blight [45], Baum et al. [33], Wolfe [531], Orchard and Woodbury [395], Carter and Myers [82], Brown [68], Chen and Fienberg [90], and Sundberg [491]. See also the extensive discussion in the Prologue of Meng and van Dyk [372]. It is thanks to Dempster et al. [115] that these special cases have been unified to a general concept for ML estimation in complex models when distributions can be conveniently represented by "hidden" variables. They named it the EM algorithm. Chrétien and Hero [95] embedded EM in the general scheme of PPA algorithms introduced by Martinet [353] and Rockafellar [443]. Their approach has been presented here.

The acceleration gained in the modified EM-step is due to Kent et al. [288]. Another acceleration was proposed by Meng and Rubin [373], who replace the M-step with so-called CM-steps, obtaining the ECM algorithm. A further development is the ECME algorithm of Liu and Rubin [329]. It is reported to dramatically improve the speed of the EM algorithm for multivariate t-distributions with *unknown* degrees of freedom. An extended version of the ECME algorithm is due to Liu [328]. Ma and Fu [339] discuss yet another acceleration of the EM algorithm.

Maronna [349] (see also Huber [254]) presents another method for parameter estimation in elliptical models.

G.3 Combinatorial optimization and λ-assignment

In this section we will take a look at a problem of combinatorial optimization, the λ-assignment problem. It will be needed in Chapter 3 for size constrained optimization of cluster criteria.

G.3.1 Transportation problem

We transform the multipoint optimization problem MPO of Section 3.2.3 to the λ-assignment problem (λA). A linear optimization problem of the form

(TP) $\sum_{i,j} u_{ij} z_{ij}$ maximal over all matrices $\mathbf{z} \in \mathbb{R}^{n \times m}$

subject to the constraints

$$\begin{cases} \sum_j z_{ij} = a_i, & i \in 1..n, \\ \sum_i z_{ij} = b_j, & j \in 1..m, \\ z_{i,j} \geq 0, \end{cases}$$

is called a *transportation* or *Hitchcock problem*, a problem surprisingly equivalent to the *circulation* and to the *min-cost flow* problem (for minimization instead of maximization as above); see Papadimitriou and Steiglitz [398]. Here, $(u_{i,j})$ is a real $n \times m$ matrix of weights and the "supplies" a_i and "demands" b_j are real numbers ≥ 0 such that $\sum a_i = \sum b_j$. Plainly this condition is necessary and sufficient for a solution to exist. If the supplies are unitary, (TP) is called a λ-*assignment problem*. Note that Problem (λA) in Section 3.2.3 is of this kind. Its demands are $n - r$, $b_1, \cdots, b_g$, and $r - \sum b_j$, as illustrated in Fig. G.2. Note also that the hypothesis on the supplies and demands for a transportation problem to be solvable is satisfied, so (λA) possesses an optimal solution.

In order to explain the transformation, it is suitable to first cast Problem (MPO) in a form common in combinatorial optimization. (The demands may depend here on j.) A labeling $\boldsymbol{\ell}$ may be represented by a zero-one matrix $\mathbf{y}$ of size $n \times (g+1)$ by putting $y_{i,j} = 1$ if and only if $\ell_i = j$, that is, if $\boldsymbol{\ell}$ assigns object i to cluster j. A zero-one matrix $\mathbf{y}$ is *admissible*, that is, corresponds to an admissible labeling, if it satisfies the constraints $\sum_j y_{i,j} = 1$ for each i (each object has exactly one label), $\sum_i y_{i,0} = n - r$ (there are $n - r$ discards) and $\sum_i y_{i,j} \geq b_j$ for all $j \geq 1$ (each cluster j contains at least b_j elements). Using this matrix and the weights u_{ij} defined by (3.15), we may reformulate MPO as a *binary linear optimization problem* (cf. Papadimitriou and Steiglitz [398]) in the following way:

(BLO) $\sum_{i=1}^{n} \sum_{j=1}^{g} u_{i,j} y_{i,j}$ maximal over all matrices $\mathbf{y} \in \{0,1\}^{n \times (g+1)}$

subject to the constraints

$$\begin{cases} \sum_j y_{i,j} = 1, & i \in 1..n, \\ \sum_i y_{i,0} = n - r, \\ \sum_i y_{i,j} \geq b_j, & j \in 1..g. \end{cases}$$

Although binary linear optimization is in general NP hard, the present problem BLO has an efficient solution. It is not yet a λ-assignment problem for two reasons: *First*, the constraints also contain an inequality. The introduction of the dummy class $g + 1$ and the coefficients $u_{i,g+1}$ in (λA), Section 3.2.3, is a trick to overcome this problem. We obtain the

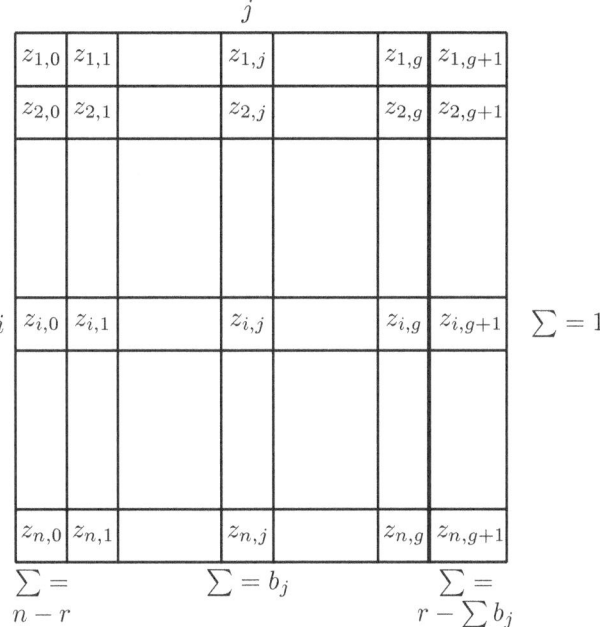

Figure G.2: *Table of assignments of objects to classes for the BLO problem of Section 3.2.3.*

binary linear optimization problem with integral constraints

(BLO′) $\displaystyle\sum_{i=1}^{n}\sum_{j=1}^{g+1} u_{i,j} z_{i,j}$ maximal over all matrices $\mathbf{z} \in \{0,1\}^{n\times(g+2)}$

subject to the constraints

$$\begin{cases} \sum_j z_{i,j} = 1, & i \in 1..n, \\ \sum_i z_{i,0} = n - r, \\ \sum_i z_{i,j} = b_j, & j \in 1..g, \\ \sum_i z_{i,g+1} = r - \sum_{j=1}^{g} b_j. \end{cases}$$

Up to the binary entries, this is (λA). *Second*, λ-assignment is not restricted to binary solutions. Fortunately, this turns out to be a free lunch: The constraints in (λA) are integral. By the Integral Circulation Theorem (Lawler [309], Theorem 12.1, or Cook et al. [100], Theorem 4.5) there is an optimal solution $\mathbf{z}^*$ to (λA) with integral entries. The first constraint then implies that it is even binary, thus being a solution to BLO′ and representing an assignment.

With the optimal solution $\mathbf{z}^*$ of (λA) we associate a solution $\mathbf{y}^* \in \mathbb{R}^{n\times(g+1)}$ to BLO as in (3.16): In each line i of $\mathbf{z}^*$, move the excess members collected in cluster $g+1$ to their natural class. Let us see why $\mathbf{y}^*$ optimizes BLO. Let $\mathbf{y}$ be admissible for BLO. Define a **feasible** matrix $\mathbf{z}$ by moving excess members i from the classes to the artificial class $g+1$. By definition of $u_{i,g+1}$, the value of $\mathbf{y}$ w.r.t. BLO is smaller than that of $\mathbf{z}$ w.r.t. (λA) and, by optimality, the latter value is smaller than that of $\mathbf{z}^*$ w.r.t. (λA), which, again by definition of $u_{i,g+1}$, equals that of $\mathbf{y}^*$ w.r.t. BLO. Hence, $\mathbf{y}$ is inferior to $\mathbf{y}^*$, and $\mathbf{y}^*$ is an optimal solution to the original problem BLO.

Optimal solutions to (λA) and BLO are actually equivalent. Given an optimal solution to BLO, move any excess elements in classes $1,\ldots,g$ to class $g+1$. Note that any class that contains excess elements contains no forced elements since they could choose a better class. Therefore, the new assignment creates the same total weight.

It is easy to construct a solution to BLO that satisfies all but the third constraint. Just assign object i to its natural class, $\mathrm{argmax}_{j \in 1 \ldots g} \, u_{i,j}$. If this solution happens to satisfy the third constraint, too, then it is plainly optimal. The opposite case needs special attention. The deficient clusters j in such a solution contain exactly b_j objects in an optimal solution. Indeed, if the size of an originally deficient cluster j were $> b_j$ in an optimal solution, then at least one of the forced elements would be free to go to its natural cluster, thus reducing the target function.

G.3.2   Algorithms for the λ-assignment problem

The exact multipoint reduction step 3.29 requires the solution to a λ-assignment problem. Since reduction steps are executed many times during a program run, the reader who wishes to implement the method will be interested in algorithms for its solution and in their complexities. Both have been extensively studied in operations research and combinatorial optimization. As a *linear optimization* problem and a special case of *minimum-cost flow*, λ-assignment may be solved by the simplex method. An instance of the min-cost flow problem is specified by a directed graph, net flows in its nodes, arc capacities, and cost coefficients. In our application the cost coefficients are the negative weights. The aim is to determine a flow with minimum overall cost satisfying the required net flows in all nodes without violating given capacities. Classical adaptations of the simplex method tailored to the particularities of min-cost flow are the (primal) Augmenting Circuit Algorithm (cf. Cook et al. [100], 4.2), the *Network Simplex* Method (Sedgewick [463]), and *primal-dual* algorithms (cf. Cook et al. [100], 4.2 and 4.3), such as the *Out-of-Kilter* Method [177, 309]. The Network Simplex Method seems to be most popular in applications, although these algorithms are exponential in the worst case.

Polynomial algorithms for min-cost flow have also existed for some time now. It is common to represent the weight matrix as the adjacency matrix of a weighted graph (V, E) with node set V of size n and edge set E of size m. The first (weakly) polynomial algorithm is due to Edmonds and Karp [140], who introduced the concept of weight *scaling*, thus showing that min-cost flow is actually a low-complexity problem. Scaling solves a series of subproblems with approximated instance parameters, capacities or weights, or both. Orlin [396] refined their method, obtaining a strongly polynomial algorithm of complexity $\mathcal{O}\big(n(m + n \log n) \log n\big)$ for the uncapacitated min-cost flow problem. In the context of λ-assignment, all row sums in the assignment matrix are *equal*, and equal to 1. This implies that capacities may be chosen unitary, so capacity scaling becomes trivial. An example is Gabow and Tarjan's [178] $\mathcal{O}\big(n(m + n \log n) \log U\big)$ algorithm with $\log U$ being the bit length used to represent network capacities. In the case of λ-assignment, put $\log U = 1$. However, opinions diverge on the usefulness of scaling in practice; see the comments in Goldberg and Tarjan [203], Section 9, and Gabow and Tarjan [178], Section 4.

Now, the λ-assignment problem has some special features that reduce its complexity compared with the general min-cost flow problem. First, its graph is *bipartite*, its node set being partitioned in two subsets, the objects i and the clusters j, such that each edge connects an object to a cluster. For bipartite network flow algorithms, it is often possible to sharpen the complexity bounds by using the sizes of the smaller (k) and the larger (n) node subsets as parameters; see Gusfield et al. [217]. Second, the bipartite network is *unbalanced* in the sense that the number of clusters is (at least in our case) much smaller than the number of objects. Bounds based on the two subsets are particularly effective for unbalanced networks. Third, the capacities may be defined as 1. As a consequence, the algorithms for the λ-assignment problem mentioned at the beginning become polynomial. For example, the time complexity of the out-of-kilter method is $\mathcal{O}(nm) = \mathcal{O}(kn^2)$ since $m = kn$ here; see Lawler [309], p. 157.

Algorithms dedicated to the λ-assignment problem are due to Kleinschmidt et al. [294]

and Achatz et al. [1]. Both algorithms have an asymptotic run time of $\mathcal{O}(kn^2)$. The former algorithm uses Hirsch paths for the dual assignment problem and is related to an algorithm of Hung and Rom's [260]. The latter is an interior point method.

The algorithms mentioned so far are at least quadratic in the size of the larger node set, n. By contrast, two *weight*-scaling min-cost flow algorithms are linear in n: Goldberg and Tarjan's [203], Theorem 6.5, algorithm solves the min-cost flow problem on a bipartite network asymptotically in $\mathcal{O}(k^2 n \log(kC))$ time (see Ahuja et al. [3]) and the "two-edge push" algorithm of the latter authors needs $\mathcal{O}((km+k^3)\log(kC))$ time. In our application, $m = kn$. Both estimates contain the bit length, $\log C$, for representing weights.

A different low-complexity approach is due to Tokuyama and Nakano [506, 508, 507]. These authors state and prove a *geometric* characterization of optimal solutions to the λ-assignment and transportation problems by a so-called *splitter*, a k-vector that partitions Euclidean k-space into k closed cones. The corresponding subdivision of the lines of the weight matrix describes an optimal assignment. Tokuyama and Nakano design a deterministic and a randomized strategy for splitter finding that run in $\mathcal{O}(k^2 n \log n)$ time and $\mathcal{O}(kn + k^{5/2}\sqrt{n}\log^{3/2} n)$ expected time, respectively. Their algorithms are almost linear in n and close to the absolute lower bound $\mathcal{O}(kn)$ if k is small, the case of interest for (λA).

G.3.3 *Heuristic methods for feasible solutions*

Besides exact solutions, there are reasons to say a word about heuristic *feasible* solutions to the λ-assignment problem. First, they may be used in Section 3.2.3 for multipoint reduction steps on their own. Moreover, some of the graphical methods presented in Section G.3.2 need initial *feasible* solutions or at least profit from good ones. The network simplex method, e.g., needs a primal feasible solution and the method presented in Achatz et al. [1] needs a dual feasible solution. While arbitrary feasible solutions are easily produced, *good* initial feasible solutions can be constructed by means of greedy heuristics. If the bounds b_j are small enough, the heuristics often produce even optimal solutions.

Each reduction step receives a set of parameters γ_j from which all weights $u_{i,j}$, $i \in 1..n$, $j \in 1..g$, are computed. The first two heuristics construct primal feasible solutions. Both start from the best *unconstrained* assignment of the clustering problem, which can be easily attained by sorting the numbers $u_i = \max_{1 \le j \le g} u_{i,j}$. More precisely:

G.14 Procedure (Basic primal heuristic)
// Input: A weight matrix $u_{i,j}$.
// Output: A feasible partition.

1. Sort the numbers $u_i = \max_{1 \le j \le g} u_{i,j}$ in decreasing order, $i \in 1..n$;

2. assign the first r objects in the ordered list to the class $1, \ldots, g$ where the maximum is attained;

3. attach label 0 to the last $n - r$ objects;

4. starting from element r in the ordered list and going downwards, reassign surplus elements in the classes to arbitrary deficient classes until they contain exactly b_j elements;

5. assign any remaining surplus elements in the classes to class $g+1$.

If an admissible instead of a feasible solution is required only, then stop after step 3 if this solution is admissible (it is then optimal). If not, stop after step 4. We next refine the basic primal heuristic. Steps 1 and 2 are as before.

G.15 Procedure (Refined primal heuristic)
// Input: A weight matrix $u_{i,j}$.
// Output: A feasible partition.

1. Sort the numbers $u_i = \max_{1 \le j \le g} u_{i,j}$ in decreasing order, $i \in 1 \ldots n$;

2. assign each of the first r objects to the class $1 \ldots g$ where the maximum is attained;

3. denote the set of the last $n - r$ objects in the ordered list by $\mathcal{L}$;

4. let $\mathcal{D}$ be the set of deficient classes in $1 \ldots g$ and let δ be their total deficiency;

5. starting from element r in the ordered list and going downwards, move the first δ elements in surplus classes to $\mathcal{L}$;

6. sort the object-class pairs $(i, j) \in \mathcal{L} \times \mathcal{D}$ in decreasing order according to their weights $u_{i,j}$ to obtain an array $(i_1, j_1), (i_2, j_2), \cdots, (i_{|\mathcal{L}| \cdot |\mathcal{D}|}, j_{|\mathcal{L}| \cdot |\mathcal{D}|})$;

7. scan all pairs (i_k, j_k) in this list starting from $k = 1$, assigning object i_k to class j_k unless i_k has already been assigned or j_k is saturated, until all classes are saturated;

8. discard the yet unassigned elements of $\mathcal{L}$;

9. assign the smallest remaining surplus elements in classes $(1 \ldots g) \backslash \mathcal{D}$ to class $g + 1$.

If an admissible instead of a feasible solution is required only, then stop after step 3 if this solution is admissible (it is then optimal). If not, stop after step 8.

In Section G.3, the fact was exploited that any cluster j with more than b_j members in an optimal solution contained no forced elements. Plainly, both heuristics share this property since such members could be freely relabeled.

Both heuristics are much faster than any of the exact solution algorithms above while yielding often a large value of the criterion, the refined heuristic larger than the basic. Their implementation is much more elementary. However, in contrast to the exact solution, the improvements in the sense of Proposition 3.27 they provide are not optimal, in general. In most cases, the criterion increases, although one may construct examples where this fails: Consider the data set $\{-40, -8, -6, 0, 1, 2, 3, 40\}$, let $g = 4$ and $b_1 = b_2 = b_3 = b_4 = 2$, and assume that there are no outliers. Suppose that the parameters are generated from the initial partition $C_1 = \{-40, 3\}$, $C_2 = \{-8, 1\}$, $C_3 = \{-6, 0\}$, $C_4 = \{2, 40\}$. They are $m_1 = -18.5$, $m_2 = -3.5$, $m_3 = -3.0$, $m_4 = 21.0$, and $v_1 = 462.25$, $v_2 = 20.25$, $v_3 = 9.0$, $v_4 = 361.0$. The matrix of negative weights,

$$
(-u_{i,j})_{i,j} = \begin{pmatrix} 9.91 & 9.15 & 9.25 & 9.65 & 9.73 & 9.82 & 9.91 & 16.31 \\ 71.57 & 6.78 & 6.09 & 6.39 & 6.78 & 7.27 & 7.87 & 99.23 \\ 157.08 & 7.75 & 5.97 & 5.97 & 6.75 & 7.75 & 8.97 & 210.41 \\ 18.97 & 10.99 & 10.68 & 9.88 & 9.77 & 9.66 & 9.56 & 9.66 \end{pmatrix}^{\mathsf{T}},
$$

generates the free partition $\{-40\}, \{-8, 2, 3\}, \{-6, 0, 1\}, \{40\}$ and the refined heuristic modifies it to $\{-40, 1\}, \{-8, 2\}, \{-6, 0\}, \{3, 40\}$. The score of the latter is -39.73, whereas the initial partition has the larger score -39.67.

The problem dual to (λA) reads

$$
(\lambda A^*) \quad \sum_{i=1}^{n} p_i + \sum_{j=0}^{g+1} b_j q_j \quad \text{minimal over } (\mathbf{p}, \mathbf{q}) \in \mathbb{R}^{n+g+2} \text{ subject to the}
$$

constraints $p_i + q_j \ge u_{i,j}$, $i \in 1 \ldots n$, $j \in 0 \ldots (g + 1)$.

A simple initial heuristic for this problem is found in Carpaneto et al. [80]:

Dual heuristic $q_j = \max_i u_{i,j}$ and $p_i = \max_j (u_{i,j} - q_j)$.

References

[1] Hans Achatz, Peter Kleinschmidt, and Konstaninos Paparrizos. A dual forest algorithm for the assignment problem. In Peter Gritzmann and Bernd Sturmfels, editors, *Applied Geometry and Discrete Mathematics*, volume 4 of *DIMACS Series in Discrete Mathematics and Theoretical Computer Science*, pages 1–10. American Mathematical Society, Providence, RI, 1991.

[2] David Aha, Arthur Asuncion, Christopher J. Merz, Patrick M. Murphy, Eamonn Keogh, Cathy Blake, Seth Hettich, David J. Newman, K. Bache, and M. Lichman. UCI Machine Learning Repository. Irvine, CA: University of California, School of Information and Computer Sciences, 2013. URL: `archive.ics.uci.edu/ml`

[3] Ravindra K. Ahuja, James B. Orlin, Clifford Stein, and Robert Endre Tarjan. Improved algorithms for bipartite network flows. *SIAM J. Computing*, 23:906–933, 1994.

[4] J. Aitchison and S.D. Silvey. Maximum-likelihood estimation procedures and associated tests of significance. *J. Royal Statist. Soc., Series B*, 22:154–171, 1960.

[5] A.C. Aitken and H. Silverstone. On the estimation of statistical parameters. *Proc. Royal Soc. Edinb. A*, 61:186–194, 1942.

[6] Hirotugu Akaike. A new look at the statistical model identification. *IEEE Trans. Autom. Control*, 19:716–723, 1974.

[7] Grigory Alexandrovich. A note on the article 'Inference for multivariate normal mixtures' by J. Chen and X. Tan. *J. Multivariate Anal.*, 129:245–248, 2014.

[8] Grigory Alexandrovich, Hajo Holzmann, and Surajit Ray. On the number of modes of finite mixtures of elliptical distributions. In Berthold Lausen, Dirk Van den Poel, and Alfred Ultsch, editors, *Algorithms from and for Nature and Life*, Studies in Classification, Data Analysis, and Knowledge Organization, pages 49–57. Springer, 2013.

[9] A. Alizadeh, M. Eisen, R. Davis, C. Ma, L. Lossos, A. Rosenwald, J. Boldrick, H. Sabet, T. Tran, and X. Yu et. al. Distinct types of diffuse large B-cell lymphoma identified by gene expression profiling. *Nature*, 403:503–511, 2000.

[10] U. Alon, N. Barkai, D. Notterman, K. Gish, S. Mack, and J. Levine. Broad patterns of gene expression revealed by clustering analysis of tumor and normal colon tissues probed by oligonucleotide arrays. *PNAS*, 96:6745–6750, 1999.

[11] Shun-Ichi Amari and Masayuki Kumon. Estimation in the presence of infinitely many nuisance parameters – geometry of estimating functions. *Ann. Statist.*, 16:1044–1068, 1988.

[12] American Cancer Society. How is acute lymphocytic leukemia classified? URL: `www.cancer.org/cancer/leukemia-acutelymphocyticallinadults/detailedguide/`

[13] American Cancer Society. How is acute myeloid leukemia classified? URL: `www.cancer.org/cancer/leukemia-acutemyeloidaml/detailedguide/`

[14] E. Anderson. The irises of the Gaspé Peninsula. *Bull. Amer. Iris Soc.*, 59:2–5, 1935.

[15] T. Ando. Löwner inequality of indefinite type. *Lin. Alg. Appl.*, 385:73–80, 2004.

[16] Barry C. Arnold, Enrique Castillo, and José María Sarabia. *Conditional Specification*

of Statistical Models. Springer Series in Statistics. Springer, New York, 1999.

[17] F. Aurenhammer, F. Hoffmann, and B. Aronov. Minkowski–type theorems and least–squares clustering. *Algorithmica*, 20:61–76, 1998.

[18] A. Azzalini and A. Dalla Valle. The multivariate skew–normal distribution. *Biometrika*, 83:715–726, 1996.

[19] R.R. Bahadur. Rates of convergence of estimates and test statistics. *Ann. Math. Statist.*, 38:303–324, 1967.

[20] Thomas A. Bailey, Jr. and Richard C. Dubes. Cluster validity profiles. *Patt. Rec.*, 15:61–83, 1982.

[21] Frank B. Baker and Lawrence J. Hubert. Measuring the power of hierarchical cluster analysis. *J. Amer. Stat. Assoc.*, 70:31–38, 1975.

[22] Frank B. Baker and Lawrence J. Hubert. A graph–theoretical approach to goodness–of–fit in complete–link hierarchical clustering. *J. Amer. Stat. Assoc.*, 71:870–878, 1976.

[23] Arindam Banerjee and Joydeep Ghosh. Clustering with balancing constraints. In Basu et al. [30], chapter 8, pages 171–200.

[24] Ole E. Barndorff-Nielsen. Likelihood theory. In Hinkley et al. [246], chapter 10, pages 232–264.

[25] Ole E. Barndorff-Nielsen and D.R. Cox. *Inference and Asymptotics*. Number 52 in Monographs on Statistics and Applied Probability. Chapman and Hall, London, 1994.

[26] V. Barnett and T. Lewis. *Outliers in Statistical Data*. Wiley, Chichester, UK, 1994.

[27] Tanya Barrett and Ron Edgar. Gene Expression Omnibus (GEO): Microarray data storage, submission, retrieval, and analysis. *Methods Enzymol*, 411:352–369, 2006.

[28] Tanya Barrett, Tugba O. Suzek, Dennis B. Troup, Stephen E. Wilhite, Wing-Chi Ngau, Pierre Ledoux, Dmitry Rudnev, Alex E. Lash, Wataru Fujibuchi, and Ron Edgar. NCBI GEO: mining millions of expression profiles – data base and tools. *Nucleic Acids Research*, 33:562–566, 2005.

[29] K.E. Basford, D.R. Greenway, Geoffrey J. McLachlan, and David Peel. Standard errors of fitted means under normal mixture models. *Computational Statistics*, 12:1–17, 1997.

[30] Sugato Basu, Ian Davidson, and Kiri L. Wagstaff, editors. *Constrained Clustering*. Data Mining and Knowledge Disvovery Series. Chapman & Hall/CRC, Boca Raton, London, New York, 2009.

[31] Heinz Bauer. *Probability Theory*. Studies in Mathematics 23. de Gruyter, Berlin–New York, 1996.

[32] Heinz Bauer. *Measure and Integration Theory*. Studies in Mathematics 26. de Gruyter, Berlin–New York, 2001.

[33] L.E. Baum, T. Petrie, G. Soules, and N. Weiss. A maximization technique occurring in the statistical analysis of probabilistic functions of Markov chains. *Ann. Math. Statist.*, 41:164–171, 1970.

[34] Claudia Becker and Ursula Gather. The masking breakdown point of multivariate outlier identification rules. *J. Amer. Stat. Assoc.*, 94:947–955, 1999.

[35] Y. Benjamini and Y. Hochberg. Controlling the false discovery rate – a practical and powerful approach to multiple testing. *RoyalB*, 57:289–300, 1995.

[36] Y. Benjamini and D. Yekutieli. The control of the false discovery rate in multiple testing under dependency. *Ann. Statist.*, 29:1165–1168, 2001.

[37] Thorsten Bernholt and Paul Fischer. The complexity of computing the MCD–estimator. *Theor. Comp. Science*, 326:383–398, 2004.

[38] P. Bertrand and G. Bel Mufti. Loevinger's measures of rule quality for assessing cluster stability. *Computational Statistics and Data Analysis*, 50:992–1015, 2006.

[39] James C. Bezdek, James Keller, Raghu Krisnapuram, and Nikhil R. Pal. *Fuzzy Models and Algorithms for Pattern Recognition and Image Processing*. The Handbooks of Fuzzy Sets Series. Kluwer, Boston, London, Dordrecht, 1999.

[40] Rajendra Bhatia. *Matrix Analysis*. Springer, New York, 1997.

[41] Patrick Billingsley. *Convergence of Probability Measures*. Wiley, 1968.

[42] Patrick Billingsley. *Probability and Measure*. Wiley, 1979.

[43] David A. Binder. Bayesian cluster analysis. *Biometrika*, 65:31–38, 1978.

[44] Alan Birnbaum. On the foundations of statistical inference. *J. Amer. Stat. Assoc.*, 57:269–306, 1962.

[45] B.J.N. Blight. Estimation from a censored sample for the exponential family. *Biometrika*, 57:389–395, 1970.

[46] Hans-Hermann Bock. Probability models and hypotheses testing in partitioning cluster analysis. In *Clustering and Classification*, pages 377–453.

[47] Hans-Hermann Bock. *Statistische Modelle für die einfache und doppelte Klassifikation von normalverteilten Beobachtungen*. PhD thesis, University of Freiburg, Germany, 1968.

[48] Hans-Hermann Bock. Statistische Modelle und Bayessche Verfahren zur Bestimmung einer unbekannten Klassifikation normalverteilter zufälliger Vektoren. *Metrika*, 18:120–132, 1972.

[49] Hans-Hermann Bock. *Automatische Klassifikation*. Vandenhoeck & Ruprecht, Göttingen, 1974. In German.

[50] Hans-Hermann Bock. On some significance tests in cluster analysis. *J. Classification*, 2:77–108, 1985.

[51] Hans-Hermann Bock. Information and entropy in cluster analysis. In H. Bozdogan, editor, *Proceedings of the First US/Japan Conference on the Frontiers of Statistical Modeling*, pages 115–147. Kluwer, 1994.

[52] Hans-Hermann Bock. Clustering methods: A history of k-means algorithms. In P. Brito, B. Bertrand, G. Cucumel, and F. de Carvalho, editors, *Selected Contributions in Data Analysis and Classification*, Studies in Classification, Data Analysis, and Knowledge Organization, pages 161–172. Springer, Berlin, Heidelberg, 2007.

[53] Jean-Daniel Boissonnat and Mariette Yvinec. *Algorithmic Geometry*. Cambridge University Press, Cambridge, 1998.

[54] Otilia Boldea and Jan R. Magnus. Maximum likelihood estimation of the multivariate mixture model. *J. Amer. Stat. Assoc.*, 104:1539–1549, 2009.

[55] Ludwig Boltzmann. Über die mechanische Bedeutung des zweiten Hauptsatzes der Wärmetheorie. *Wiener Berichte*, 53:195–220, 1866.

[56] Ludwig Boltzmann. *Vorlesungen über Gastheorie*. J.A. Barth, Leipzig, 1896, 1898. English translation: Lectures on Gas Theory, Dover, New York, 1995.

[57] William E. Bonnice. A generalization of a theorem of Carathéodory. *Notices AMS*, 8:252–253, 1961. Conference abstract.

[58] William E. Bonnice and Victor L. Klee. The generation of convex hulls. *Math. Annalen*, 152:1–29, 1963.

[59] William M. Boothby. *An Introduction to Differentiable Manifolds and Riemannian Geometry.* Academic Press, Orlando, San Diego, New York, Austin, London, Montreal, Sydney, Tokyo, second edition, 1986.

[60] Émile Borel. Les probabilités dénombrables et leurs applications arithmétiques. *Rendiconti Circolo Mat. Patermo*, 27:247–270, 1909.

[61] Nicolas Bourbaki. *General Topology*, volume I. Springer, Berlin, 1989.

[62] Nicolas Bourbaki. *General Topology*, volume II. Springer, Berlin, 1989.

[63] A.W. Bowman and P.J. Foster. Adaptive smoothing and density–based tests of multivariate normality. *J. Amer. Stat. Assoc.*, 88:529–539, 1993.

[64] U. Brandes, D. Delling, M. Gaertler, R. Gorke, M. Hoefer, Z. Nikolski, and D. Wagner. On modularity clustering. *IEEE Transactions on Knowledge and Data Engineering*, 20(2):172–188, 2008.

[65] L. de Branges, I. Gohberg, and J. Rovnyak, editors. *Topics in Operator Theory – Ernst D. Hellinger Memorial Volume*, volume 48 of *Operator Theory: Advances and Applications*. Birkhäuser, Basel, Boston, Berlin, 1990.

[66] James N. Breckenridge. Validating cluster analysis: Consistent replication and symmetry. *Multivariate Behavioral Research*, 35:261–285, 2000.

[67] L. Breiman, J. Friedman, R. Olshen, and C. Stone. *Classification and Regression Trees.* Wadsworth, Belmont, 1984.

[68] M.L. Brown. Identification of the sources of significance in two–way tables. *Appl. Statist.*, 23:405–413, 1974.

[69] C.G. Broyden. Quasi–Newton methods and their application to function minimization. *Math. Comp.*, 21:368–381, 1967.

[70] C.G. Broyden. A new double–rank minimization algorithm. *Notices Amer. Math. Soc.*, 16:1–3, 1969.

[71] C.G. Broyden. The convergence of a class of double–rank minimization algorithms. *J. Inst. Math. Appl.*, 6:76–90, 1970.

[72] N.G. de Bruijn. *Asymptotic Methods in Analysis.* North–Holland, Amsterdam, 1970.

[73] Peter G. Bryant and J.A Williamson. Asymptotic behaviour of classification maximum likelihood estimates. *Biometrika*, 65:273–281, 1978.

[74] G. Brys, M. Hubert, and A. Struyf. A robust measure of skewness. *J. Computational and Graphical Statistics*, 13:996–1017, 2004.

[75] Gaius Julius Caesar. Commentarii de bello Gallico. Report to the Roman Senate, 58 B.C.

[76] N.A. Campbell and R.J. Mahon. A multivariate study of variation in two species of rock crab of the genus Leptograpsus. *Aust. J. Zool.*, 22:417–425, 1974.

[77] Francesco P. Cantelli. Sulla probabilità come limite della frequenza. *Atti Acad. Naz. Lincea*, 26:39–45, 1917.

[78] Constantin Carathéodory. Über den Variabilitätsbereich der Koeffizienten von Potenzreihen, die gegebene Werte nicht annehmen. *Math. Annalen*, 34:95–115, 1907.

[79] Pedro Carmona-Saez, Roberto D. Pascual-Marqui, F. Tirado, Jose M. Carazo, and Alberto Pascual-Montano. Biclustering of gene expression data by non-smooth non-negative matrix factorization. *BMC Bioinformatics*, 7:78, 2006.

[80] Giorgio Carpaneto, Silvano Martello, and Paolo Toth. Algorithms and codes for the assignment problem. *Ann. OR*, 13:193–223, 1988.

[81] Miguel A. Carreira-Perpiñán and Christopher K. Williams. On the number of modes

of a Gaussian mixture. In Lewis D. Griffin and Martin Lillholm, editors, *Scale Space Methods in Computer Vision*, volume 2695 of *Lecture Notes in Computer Science*, pages 625–640. Springer, Berlin, Heidelberg, 2003. Proceedings of the 4th International Conference, Scale Space 2003, Isle of Skye, UK.

[82] W.H. Carter, Jr. and R.H. Myers. Maximum likelihood estimation from linear combinations of discrete probability functions. *J. Amer. Stat. Assoc.*, 68:203–206, 1973.

[83] Gilles Celeux and Gérard Govaert. A classification EM algorithm for clustering and two stochastic versions. *Computational Statistics and Data Analysis*, 14:315–332, 1992.

[84] K.C. Chanda. A note on the consistency and maxima of the roots of likelihood equations. *Biometrika*, 41:56–61, 1954.

[85] W.C. Chang. On using principle components before separating a mixture of two multivariate normal distributions. *Applied Statistics*, 32:267–275, 1983.

[86] C.V.L. Charlier. Researches into the theory of probability. *Lunds Universitets Arkskrift*, Ny foljd Afd. 2.1(5), 1906.

[87] Hanfeng Chen, Jiahua Chen, and John D. Kalbfleisch. Testing for a finite mixture model with two components. *J. Royal Statist. Soc., Series B*, 66:95–115, 2004.

[88] Jiahua Chen. Optimal rate of convergence for finite mixture models. *Ann. Statist.*, 23:221–233, 1995.

[89] Jiahua Chen and Xianming Tan. Inference for multivariate normal mixtures. *J. Multivariate Anal.*, 100:1367–1383, 2009.

[90] T. Chen and Stephen E. Fienberg. The analysis of contingency tables with incompletely classified data. *Biometrics*, 32:133–144, 1976.

[91] The Chipping Forecast. *Nature Genetics*, 21(1):1–60, 1999.

[92] Gustave Choquet. Le théoreme de représentation intégrale dans les ensembles convexes compacts. *Ann. Inst. Fourier (Grenoble)*, 10:333–344, 1960.

[93] Gustave Choquet. *Lectures on Analysis*, volume I Integration and Topological Vector Spaces. Benjamin, New York, Amsterdam, 1969.

[94] Gustave Choquet. *Lectures on Analysis*, volume II Representation Theory. Benjamin, New York, Amsterdam, 1969.

[95] Stéphane Chrétien and Alfred O. Hero III. Kullback proximal algorithms for maximum likelihood estimation. *IEEE Trans. Inf. Theory*, 46:1800–1810, 2000.

[96] Gabriela Ciuperca, Andrea Ridolfi, and Jérome Idier. Penalized maximum likelihood estimator for normal mixtures. *Scand. J. Statist.*, 30:45–59, 2003.

[97] Rudolf Clausius. *The Mechanical Theory of Heat with its Applications to the Steam–Engine and to the Physical Properties of Bodies*. John van Voorst, London, 1867. Translation from German of many of his works. Edited by T. Archer Hurst.

[98] Daniel A. Coleman, X. Dong, J. Hardin, David M. Rocke, and David L. Woodruff. Some computational issues in cluster analysis with no a priori metric. *Computational Statistics and Data Analysis*, 31:1–11, 1999.

[99] Corneliu Constantinescu and Aurel Cornea. *Ideale Ränder Riemannscher Flächen*. Springer, 1963.

[100] William J. Cook, William H. Cunningham, William R. Pulleyblank, and Alexander Schrijver. *Combinatorial Optimization*. Wiley, New York, 1998.

[101] R.M. Cormack. A review of classification. *J. Royal Statist. Soc., Series A*, 134:321–367, 1971.

[102] E.A. Cornish. The multivariate t-distribution associated with a set of normal sample deviates. *Austral. J. Phys.*, 7:531–542, 1954.

[103] D.R. Cox and D.V. Hinkley. *Theoretical Statistics*. Chapman & Hall, London, 1974.

[104] Harald Cramér. A contribution to the statistical theory of estimation. *Skandinavisk aktuarietidskrift*, 29:85–94, 1946.

[105] Harald Cramér. *Mathematical Methods of Statistics*. Princeton University Press, Princeton, 13th edition, 1974. (First edition 1946)

[106] J. A. Cuesta-Albertos, Alfonso Gordaliza, and Carlos Matrán. Trimmed k–means: An attempt to robustify quantizers. *Ann. Statist.*, 25:553–576, 1997.

[107] Artur Czumaj and Christian Sohler. Sublinear–time approximation algorithms for clustering via random sampling. *Random Structures & Algorithms*, 30:226–256, 2007.

[108] W.C. Davidon. Variable metric methods for minimization. Report ANL-5990, Argonne National Laboratory, 1959.

[109] P. Laurie Davies. Asymptotic behavior of S-estimates of multivariate location parameters and dispersion matrices. *Ann. Statist.*, 15:1269–1292, 1987.

[110] P. Laurie Davies. Consistent estimates for finite mixtures of well separated elliptical clusters. In Hans-Hermann Bock, editor, *Classification and Related Methods of Data Analysis, Proceedings of the First Conference of the IFCS Aachen 1987*, pages 195–202, Elsevier, Amsterdam, 1988.

[111] P. Laurie Davies and Ursula Gather. The identification of multiple outliers. *J. Amer. Stat. Assoc.*, 88:782–792, 1993.

[112] P. Laurie Davies and Ursula Gather. Breakdown and groups. *Ann. Statist.*, 33:977–988, 2005. With discussion and rejoinder.

[113] N.E. Day. Estimating the components of a mixture of normal distributions. *Biometrika*, 56:463–474, 1969.

[114] Ayhan Demiriz, Kristin B. Bennett, and Paul S. Bradley. Using assignment constraints to avoid empty clusters in k–means clustering. In Basu et al. [30], chapter 9, pages 201–220.

[115] A.P. Dempster, N.M. Laird, and D.B. Rubin. Maximum likelihood from incomplete data via the EM algorithm. *J. Royal Statist. Soc., Series B*, 39:1–38, 1977.

[116] John E. Dennis, Jr. Algorithms for nonlinear fitting. In M.J.D. Powell, editor, *Nonlinear Optimization 1981*, Academic Press, London, 1982. (Procedings of the NATO Advanced Research Institute held at Cambridge in July 1981)

[117] Wayne S. DeSarbo, J. Douglas Carroll, Linda A. Clark, and Paul E. Green. Synthesized clustering: A method for amalgamating alternative clustering bases with differential weighting of variables. *Psychometrika*, 49:57–78, 1984.

[118] Marcel Dettling and Peter Bühlmann. Supervised clustering of genes. *Genome Biology*, 3(12):research0069.1–research0069.15, 2002.

[119] Marcel Dettling and Peter Bühlmann. Boosting for tumor classification with gene expression data. *Bioinformatics*, 19:1061–1069, 2003.

[120] Dipak Dey and C.R. Rao, editors. *Handbook of Statistics*, volume 25. Elsevier, New York, 2005.

[121] Persi Diaconis. *Group Representations in Probability and Statistics*. Institute for Mathematical Statistics, Heyward, CA, 1988.

[122] Persi Diaconis and David Freedman. Asymptotics of graphical projection pursuit. *Ann. Statist.*, 12:793–815, 1984.

[123] Edwin Diday. Optimisation en classification automatique et reconnaissance des formes. *R.A.I.R.O.*, 3:61–96, 1972.

[124] Edwin Diday and Anne Schroeder. A new approach in mixed distributions detection. *R.A.I.R.O. Recherche opérationnelle*, 10(6):75–106, 1976.

[125] J. Diebolt and C.P. Robert. Estimation of finite mixture distributions by Bayesian sampling. *J. Royal Statist. Soc., Series B*, 56:363–375, 1994.

[126] Jean Dieudonné. *Foundations of Modern Analysis*. Academic Press, New York, 1969.

[127] W. Donoghue. *Monotone Matrix Functions and Analytic Continuation*. Springer, New York, 1974.

[128] David L. Donoho and Peter J. Huber. The notion of a breakdown point. In Peter J. Bickel, Kjell A. Doksum, and J.L. Hodges, Jr., editors, *A Festschrift for Erich L. Lehmann*, The Wadsworth Statistics/Probability Series, pages 157–184. Wadsworth, Belmont, CA, 1983.

[129] J.L. Doob. *Stochastic Processes*. Wiley, New York, London, 1953.

[130] Joseph Leonard Doob. Probability and statistics. *Trans. Amer. Math. Soc.*, 36:759–775, 1934.

[131] Edward R. Dougherty and Marcel Brun. A probabilistic theory of clustering. *Patt. Rec.*, 37:917–925, 2004.

[132] Richard C. Dubes and Anil K. Jain. Models and methods in cluster validity. In *Proc. IEEE Conf. Pattern Recognition and Image Processing*, pages 148–155, Long Beach, CA, 1978.

[133] Lester E. Dubins and David A. Freedman. A sharper form of the Borel–Cantelli lemma and the strong law. *Ann. Math. Statist.*, 36:800–807, 1965.

[134] Sandrine Dudoit, Jane Fridlyand, and Terence P. Speed. Comparison of discrimination methods for the classification of tumors using gene expression data. *J. Amer. Stat. Assoc.*, 97:77–87, 2002.

[135] Charles W. Dunnett and Milton Sobel. A bivariate generalization of Student's t-distribution, with tables for certain special cases. *Biometrika*, 41:153–169, 1954.

[136] Tarn Duong. ks: Kernel density estimation and kernel discriminant analysis for multivariate data in R. *J. Statistical Software*, 21(7), 2007.

[137] Richard Durrett. *Probability: Theory and Examples*. Duxbury Press, Belmont, CA, 1991.

[138] Morris Eaton and T. Kariya. Robust tests for spherical symmetry. *Ann. Statist.*, 5:206–215, 1977.

[139] Ron Edgar, Michael Domrachev, and Alex E. Lash. Gene Expression Omnibus: NCBI gene expression and hybridization array data repository. *Nucleic Acids Research*, 30:207–210, 2002.

[140] J. Edmonds and R.M. Karp. Theoretical improvements in algorithmic efficiency for network flow problems. *J. ACM*, 19:248–264, 1972.

[141] Bradley Efron. The two–sample problem with censored data. In *Proc. 5th Berkeley Symposium on Math. Statist. and Prob.*, volume 4, pages 831–853, 1967.

[142] Bradley Efron and Robert Tibshirani. Empirical Bayes methods and false discovery rates for microarrays. *Genet. Epidemiol.*, 23:70–86, 2002.

[143] Michael B. Eisen, Paul T. Spellman, Patrick O. Brown, and David Botstein. Cluster analysis and display of genome–wide expression patterns. *Proc. Natl. Acad. Sci., U.S.A.*, 95:14863–14868, 1998.

[144] Tohid Erfani and Sergei V. Utyuzhnikov. Directed search domain: a method for even generation of the Pareto frontier in multiobjective optimization. *J. Engineering Optimization*, 43:467–484, 2011.

[145] Michael J. Evans, Donald A.S. Fraser, and Georges Monette. On principles and arguments to likelihood. *The Canadian Journal of Statistics*, 14:181–199, 1986.

[146] Brian S. Everitt. *Latent Variable Models*. Monographs on Statistics and Applied Probability. Chapman and Hall, London, New York, 1984.

[147] Brian S. Everitt, Sabine Landau, Morven Leese, and Daniel Stahl. *Cluster Analysis*. John Wiley & Sons, Chichester, 5th edition, 2011.

[148] Kai-Tai Fang, Samuel Kotz, and Kai-Wang Ng. *Symmetric Multivariate and Related Distributions*. Chapman & Hall, London, New York, 1990.

[149] Yixin Fang and Junhui Wang. Selection of the number of clusters via the bootstrap method. *Computational Statistics and Data Analysis*, 56:468–477, 2012.

[150] William Feller. On a general class of "contagious" distributions. *Ann. Math. Statist.*, 14:389–400, 1943.

[151] T.S. Ferguson. On the rejection of outliers. In J. Neyman, editor, *Proceedings of the Fourth Berkeley Symposion on Mathematical Statistics and Probability*, volume 1, pages 253–288. Univ. California Press, 1961.

[152] Artur J. Ferreira and Mário A.T. Figueiredo. Efficient feature selection filters for high-dimensional data. *Patt. Rec. Lett.*, 33:1794–1804, 2012.

[153] Lloyd Fisher and John W. Van Ness. Admissible clustering procedures. *Biometrika*, 58:91–104, 1971.

[154] Ronald Aylmer Fisher. On the mathematical foundations of theoretical statistics. *Phil. Trans. Roy. Soc. London A*, 222:309–368, 1922.

[155] Ronald Aylmer Fisher. Theory of statistical estimation. *Proc. Camb. Phil. Soc.*, 22:700–725, 1925.

[156] Ronald Aylmer Fisher. The use of multiple measurements in taxonomic problems. *Ann. Eugenics*, 7:179–188, 1936.

[157] George S. Fishman. *Monte Carlo*. Springer, New York, 1996.

[158] E. Fix and J.L. Hodges. Discriminatory analysis – nonparametric discrimination: consistency properties. In A. Agrawala, editor, *Machine Recognition of Patterns*, pages 261–279. IEEE Press, 1977. Originally: Technical Report No. 4, US Airforce School of Aviation Medicine, Randolph Field, TX (1951).

[159] R. Fletcher. A new approach to variable metric methods. *Comput. J.*, 13:317–322, 1970.

[160] R. Fletcher. *Practical Methods of Optimization*. Wiley, Chichester, 1980.

[161] R. Fletcher and M.J.D. Powell. A rapidly convergent descent method for minimization. *Comput. J.*, 6:163–168, 1963.

[162] R. Fletcher and C.M. Reeves. Function minimization by conjugate gradients. *Comput J.*, 7:149–154, 1964.

[163] K. Florek, J. Lukaszewicz, J. Perkal, H. Steinhaus, and Zubrzycki S. Sur la liaison et la division des points d'un ensemble fini. *Colloquium Mathematicum*, 2:282–285, 1951.

[164] B.A. Flury. Principal points. *Biometrika*, 77:33–41, 1990.

[165] Bernhard Flury and Hans Riedwyl. *Multivariate Statistics – A Practical Approach*. Chapman & Hall, London, New York, 1988.

[166] R.V. Foutz and R.C. Srivastava. The performace of the likelihood ratio test when the model is incorrect. *Ann. Statist.*, 5:1183–1194, 1977.

[167] B.E. Fowlkes, R. Gnanadesikan, and J.R Kettenring. Variable selection in clustering and other contexts. In C.L. Mallows, editor, *Design, Data, and Analysis*, pages 13–34. Wiley, New York, 1987.

[168] B.E. Fowlkes, R. Gnanadesikan, and J.R Kettenring. Variable selection in clustering. *J. Classif.*, 5:205–228, 1988.

[169] Chris Fraley. Algorithms for model–based Gaussian hierarchical clustering. *SIAM J. Scient. Comp.*, 20:270–281, 1998.

[170] Chris Fraley and Adrian E. Raftery. How many clusters? Which clustering method? Answers via model–based cluster analysis. *The Computer Journal*, 41:578–588, 1998.

[171] Chris Fraley and Adrian E. Raftery. Model–based clustering, discriminant analysis, and density estimation. *J. Amer. Stat. Assoc.*, 97:611–631, 2002.

[172] H.P. Friedman and J. Rubin. On some invariant criteria for grouping data. *J. Amer. Stat. Assoc.*, 62:1159–1178, 1967.

[173] Jerome H. Friedman. Exploratory projection pursuit. *J. Amer. Stat. Assoc.*, 82:249–266, 1987.

[174] Jerome H. Friedman and J.W. Tukey. A projection pursuit algorithm for exploratory data analysis. *IEEE Trans. Comput.*, C-23:881–890, 1974.

[175] Heinrich Fritz, Luis Angel García-Escudero, and Agustín Mayo-Iscar. A fast algorithm for robust constrained clustering. *Computational Statistics and Data Analysis*, 61:124–136, 2013.

[176] K. Fukunaga. *Introduction to Statistical Pattern Recognition*. Academic Press, Boston, San Diego, New York, London, Sydney, Tokyo, Toronto, 2nd edition, 1990.

[177] D.R. Fulkerson. An out-of-kilter method for minimal cost flow problems. *J. SIAM*, 9:18–27, 1961.

[178] Harold N. Gabow and Robert Endre Tarjan. Faster scaling algorithms for network problems. *SIAM J. Computing*, 18:1013–1036, 1989.

[179] María Teresa Gallegos. Clustering in the presence of outliers. In Manfred Schwaiger and Otto Opitz, editors, *Exploratory Data Analysis in Empirical Research, Proceedings of the 25th Annual Conference of the GfKl, Munich*, pages 58–66. Springer, 2001.

[180] María Teresa Gallegos and Gunter Ritter. A robust method for cluster analysis. *Ann. Statist.*, 33:347–380, 2005.

[181] María Teresa Gallegos and Gunter Ritter. Parameter estimation under ambiguity and contamination with the spurious outliers model. *J. Multivariate Anal.*, 97:1221–1250, 2006.

[182] María Teresa Gallegos and Gunter Ritter. Trimmed ML–estimation of contaminated mixtures. *Sankhyā, Series A*, 71:164–220, 2009.

[183] María Teresa Gallegos and Gunter Ritter. Trimming algorithms for clustering contaminated grouped data and their robustness. *Adv. Data Anal. Classif.*, 3:135–167, 2009.

[184] María Teresa Gallegos and Gunter Ritter. Using combinatorial optimization in model-based trimmed clustering with cardinality constraints. *Computational Statistics and Data Analysis*, 54:637–654, 2010. DOI 10.1016/j.csda.2009.08.023.

[185] María Teresa Gallegos and Gunter Ritter. Strong consistency of k-parameters clustering. *J. Multivariate Anal.*, 117:14–31, 2013.

[186] Guojun Gan, Choaqun Ma, and Jianhong Wu. *Data Clustering – Theory, Algorithms,*

and Applications. SIAM, ASA, Philadelphia, PA, 2007.

[187] Luis Angel García-Escudero and Alfonso Gordaliza. Robustness properties of k-means and trimmed k-means. *J. Amer. Stat. Assoc.*, 94:956–969, 1999.

[188] Luis Angel García-Escudero, Alfonso Gordaliza, Carlos Matrán, and Agustín Mayo-Iscar. A general trimming approach to robust cluster analysis. *Ann. Statist.*, 36:1324–1345, 2008.

[189] Luis Angel García-Escudero, Alfonso Gordaliza, Carlos Matrán, and Agustín Mayo-Iscar. Exploring the number of groups in robust model–based clustering. *Statistics and Computing*, 21:585–599, 2011.

[190] Michael R. Garey and David S. Johnson. *Computers and Intractibility.* Freeman, San Francisco, 1979.

[191] Michael R. Garey, David S. Johnson, and L. Stockmeyer. Some simplified NP-complete graph problems. *Theoretical Computer Science*, 1:237–267, 1976.

[192] Richard C. Geary. Testing for normality. *Biometrika*, 34:209–242, 1947.

[193] Andrew Gelman, John B. Carlin, Hal S. Stern, and Donald B. Rubin. *Bayesian Data Analysis.* Texts in Statistical Science. Chapman & Hall, Boca Raton, second edition, 2004.

[194] S. Geman. A limit theorem for the norm of random matrices. *Annals of Probability*, 8:252–261, 1980.

[195] S. Geman and D. Geman. Stochastic relaxation, Gibbs distributions and the Bayesian restoration of images. *IEEE Trans. Pattern Anal. Mach. Intell.*, 12:609–628, 1984.

[196] Gene Expression Omnibus. URL: www.ncbi.nlm.nih.gov/geo/

[197] Subhashis Ghosal and Aad W. van der Vaart. Entropies and rates of convergence for maximum likelihood and Bayes estimation for mixtures of normal densities. *Ann. Statist.*, 29:1233–1263, 2001.

[198] D. Ghosh and A.M. Chinnaiyan Mixture modeling of gene expression data from microarray experiments. *Bioinformatics*, 18:275–286, 2002.

[199] Josiah Willard Gibbs. *Elementary Principles in Statistical Mechanics.* Dover, N.Y., 1960. Originally Scribner and Sons, N.Y. 1902.

[200] R. Gnanadesikan. *Methods for Statistical Data Analysis of Multivariate Observations.* John Wiley & Sons, New York, 1977.

[201] R. Gnanadesikan, J.R. Kettenring, and S.L. Tsao. Weighting and selection of variables for cluster analysis. *J. Classif.*, 12:113–136, 1995.

[202] B.V. Gnedenko and A.V. Kolmogorov. *Limit Distributions for Sums of Independent Random Variables.* Addison–Wesley, Reading, MA, 1954.

[203] Andrew V. Goldberg and Robert Endre Tarjan. Finding minimum–cost circulations by successive approximation. *Math. of OR*, 15:430–466, 1990.

[204] D. Goldfarb. A family of variable metric methods derived by variational means. *Math. Comput.*, 24:23–26, 1970.

[205] Todd Golub. bioconductor.wustl.edu/data/experiments/. See package golubEsets.

[206] T.R. Golub, D.K. Slonim, P. Tamayo, C. Huard, M. Gaasenbeek, J.P. Mesirov, H. Coller, M.L. Loh, J.R. Downing, M.A. Caligiuri, C.D. Bloomfield, and E.S. Lander. Molecular classification of cancer: Class discovery and class prediction by gene expression monitoring. *Science*, 286:531–537, 1999.

[207] David Gondek. Non–redundant data clustering. In Sugatu Basu, Ian Davidson, and Kiri L. Wagstaff, editors, *Constrained Clustering*, Data Mining and Knowledge Discov-

ery Series, chapter 11, pages 245–283. Chapman & Hall/CRC, Boca Raton, London, New York, 2009.

[208] A.D. Gordon. *Classification*, volume 16 of *Monographs on Statistics and Applied Probability*. Chapman and Hall, London, first edition, 1981.

[209] A.D. Gordon. A review of hierarchical classification. *J. Royal Statist. Soc., Series A*, 150:119–137, 1987.

[210] A.D. Gordon. *Classification*, volume 82 of *Monographs on Statistics and Applied Probability*. Chapman and Hall/CRC, Boca Raton, second edition, 1999.

[211] J.C. Gower and P. Legendre. Metric and Euclidean properties of dissimilarity coefficients. *J. Classification*, 3:5–48, 1986.

[212] Siegfried Graf and Harald Luschgy. *Foundations of Quantization for Probability Distributions*. Number 1730 in Lecture Notes in Mathematics. Springer, Berlin, 2000.

[213] P.E. Green and A.M. Krieger. Alternative approaches to cluster–based market segmentation. *Journal of the Market Research Society*, 37:221–239, 1995.

[214] P.M. Grundy. The fitting of grouped truncated and grouped censored normal distributions. *Biometrika*, 39:252–259, 1952.

[215] A.K. Gupta and D.G. Kabe. Multivariate robust tests for spherical symmetry with applications to multivariate least squares regression. *J. Appl. Stoch. Sci.*, 1:159–168, 1993.

[216] A.K. Gupta and T. Varga. *Elliptically Contoured Models in Statistics*. Kluwer Academic Publishers, Dordrecht, Boston, London, 1993.

[217] Dan Gusfield, Charles Martel, and David Fernandez-Baca. Fast algorithms for bipartite network flow. *SIAM J. Comput.*, 16:237–251, 1987.

[218] Isabelle Guyon, Constantin Aliferis, and André Elisseeff. Causal feature selection. In Liu and Motoda [330], Chapter 4, pages 63–85.

[219] Isabelle Guyon and André Elisseeff. An introduction to variable and feature selection. *J. Mach. Learn. Res.*, 3:1157–1182, 2003.

[220] Isabelle Guyon and André Elisseeff. An introduction to feature extraction. In Guyon et al. [221], Chapter 1.

[221] Isabelle Guyon, Steve Gunn, Masoud Nikravesh, and Lofti A. Zadeh, editors. *Feature Extraction, Foundations and Applications*. Springer, Berlin, Heidelberg, 2006.

[222] Ali S. Hadi and Alberto Luceño. Maximum trimmed likelihood estimators: A unified approach, examples, and algorithms. *Computational Statistics and Data Analysis*, 25:251–272, 1997.

[223] P. Hall. On polynomial–based projection indices for exploratory projection pursuit. *Ann. Statist.*, 17:589–605, 1989.

[224] Frank R. Hampel. *Contributions to the theory of robust estimation*. PhD thesis, Univ. California, Berkeley, 1968.

[225] Frank R. Hampel. The influence curve and its role in robust estimation. *J. Amer. Stat. Assoc.*, 62:1179–1186, 1974.

[226] Frank R. Hampel, Elvizio M. Ronchetti, Peter J. Rousseeuw, and Werner A. Stahel. *Robust Statistics*. Wiley, New York, Chichester, Brisbane, Toronto, Singapore, 1986.

[227] Julia Handl, Joshua Knowles, and Douglas B. Kell. Computational cluster validation in post–genomic data analysis. *Bioinformatics*, 21:3201–3212, 2005.

[228] John A. Hartigan. Statistical theory in clustering. *J. Classification*, 2:63–76, 1985.

[229] H.O. Hartley. Maximum likelihood estimation from incomplete data. *Biometrics*,

14:174–194, 1958.

[230] V. Hasselblad. Estimation of parameters for a mixture of normal distributions. *Technometrics*, 8:431–444, 1966.

[231] W. K. Hastings. Monte Carlo sampling methods using Markov chains and their applications. *Biometrika*, 57:97–109, 1970.

[232] Richard J. Hathaway. A constrained formulation of maximum–likelihood estimation for normal mixture distributions. *Ann. Statist.*, 13:795–800, 1985.

[233] Richard J. Hathaway. Another interpretation of the EM algorithm for mixture distributions. *Statistics & Probability Letters*, 4:53–56, 1986.

[234] Felix Hausdorff. *Grundzüge der Mengenlehre.* Verlag von Veit und Comp., Leipzig, 1914. Reprint: Chelsea Publ. Comp., New York, 1949.

[235] D.M. Hawkins. *Identification of Outliers.* Chapman & Hall, London, New York, 1980.

[236] M. Healy and M. Westmacott. Missing values in experiments analysed on automatic computers. *Appl. Statist.*, 5:203–206, 1956.

[237] Ingrid Hedenfalk, D. Duggen, and Y. Chen et al. Gene–expression profiles in hereditary breast cancer. *New England J. Med.*, 344:539–548, 2001.

[238] L.L. Helms. *Introduction to Potential Theory.* Wiley, New York, 1969.

[239] Jõgi Henna. On estimating of the number of constituents of a finite mixture of continuous functions. *Ann. Inst. Statist. Math.*, 37:235–240, 1985.

[240] Christian Hennig. Asymmetric linear dimension reduction for classification. *Journal of Computational and Graphical Statistics*, 13:930–945, 2004.

[241] Christian Hennig. Breakdown points for maximum likelihood estimators of location-scale mixtures. *Ann. Statist.*, 32:1313–1340, 2004.

[242] Christian Hennig. Cluster-wise assessment of cluster stability. *Computational Statistics and Data Analysis*, 52:258–271, 2007.

[243] Christian Hennig. Dissolution point and isolation robustness: robustness criteria for general cluster analysis methods. *J. Multivariate Anal.*, 99:1154–1176, 2008.

[244] Christian Hennig. Methods for merging Gaussian mixture components. *Advances of Data Analysis and Classification*, 4:3–34, 2010.

[245] D.V. Hinkley and N. Reid. Statistical theory. In Hinkley et al. [246], chapter 1, pages 1–29.

[246] D.V. Hinkley, N. Reid, and E.J. Snell, editors. *Statistical Theory and Modelling – In honour of Sir David Cox, FRS.* Chapman and Hall, London, New York, Tokyo, Melbourne, Madras, 1991.

[247] Michael D. Hirschhorn. The AM–GM Inequality. *Math. Intelligencer*, 29(4):7, 2007.

[248] Nils L. Hjort. Discussion: On the consistency of Bayes estimators. *Ann. Statist.*, 14:49–55, 1986. Discussion on a paper by Persi Diaconis and David Freedman.

[249] J.L. Hodges, Jr. Efficiency in normal samples and tolerance of extreme values for some estimates of location. In *Proc. 5th Berkeley Symp. Math. Statist. Probab.*, pages 163–186, University of California Press, Berkeley, 1967.

[250] Hajo Holzmann, Axel Munk, and Tilmann Gneiting. Identifiability of finite mixtures of elliptical distributions. *Scand. J. Statist.*, 33:753–763, 2006.

[251] G. Hommel. A stagewise rejective multiple test procedure based on a modified Bonferroni test. *Biometrika*, 75:383–386, 1988.

[252] Roger A. Horn and Charles R. Johnson. *Matrix Analysis.* Cambridge University Press, Cambridge–London–New York, 1985.

[253] Peter J. Huber. The behavior of maximum likelihood estimates under nonstandard conditions. In *Fifth Berkeley Symposion on Mathematical Statistics and Probability*, volume 1, pages 221–233, Berkeley, 1967. University of California Press.

[254] Peter J. Huber. *Robust Statistics*. Wiley, New York–Chichester–Brisbane–Toronto, 1981.

[255] Peter J. Huber. Projection pursuit. *Ann. Statist.*, 13:435–475, 1985.

[256] Lawrence J. Hubert. Some extensions of Johnson's hierarchical clustering algorithms. *Psychometrika*, 37:261–274, 1972.

[257] Lawrence J. Hubert and Phipps Arabie. Comparing partitions. *J. Classification*, 2:193–218, 1985.

[258] Fred W. Huffer and Cheolyong Park. A test for elliptical symmetry. *J. Multivariate Anal.*, 98:256–281, 2007.

[259] Guodong Hui and Bruce G. Lindsay. Projection pursuit via white noise matrices. *Sankhyā, Series B*, 72:123–153, 2010.

[260] Ming S. Hung and Walter O. Rom. Solving the assignment problem by relaxation. *Operations Research*, 28:969–982, 1980.

[261] V.S. Huzurbazar. The likelihood equation, consistency, and the maxima of the likelihood function. *Annals of Eugenics*, 14:185–200, 1948.

[262] C.L. Hwang and K. Yoon. *Multiple Attribute Decision Making, Methods and Applications*. Springer, Berlin, 1981.

[263] IEEE. Special issue on vector quantization. *IEEE Trans. Inf. Theory*, 28, 1982.

[264] Salvatore Ingrassia and Roberto Rocci. Monotone constrained EM algorithms for multinormal mixture models. In S. Zani, A. Cerioli, M. Riani, and M. Vichi, editors, *Data Analysis, Classification and the Forward Search*, Studies in Classification, Data Analysis, and Knowledge Organization, pages 111–118. Springer, Heidelberg, 2006.

[265] Salvatore Ingrassia and Roberto Rocci. Constrained monotone EM algorithms for finite mixture of multivariate Gaussians. *Computational Statistics and Data Analysis*, 51:5339–5351, 2007.

[266] P. Jaccard. Distribution de la florine alpine dans le Bassin de Dranses et dans quelques régions voisines. *Bulletin de la Société Vaudoise de Sciences Naturelles*, 37:241–272, 1901.

[267] Anil K. Jain and Richard C. Dubes. *Algorithms for Clustering Data*. Prentice Hall, Englewood Cliffs, N.J., 1988.

[268] Anil K. Jain, M.N. Murty, and P.J. Flynn. Data clustering: a review. *ACM Comput. Surveys*, 31:264–323, 1999.

[269] Mortaza Jamshidian and Robert I. Jennrich. Acceleration of the EM algorithm by using quasi–Newton methods. *J. Royal Statist. Soc., Series B*, 59:569–587, 1997.

[270] R.C. Jancey. Multidimensional group analysis. *Australian J. Botany*, 14:127–130, 1966.

[271] Arnold Janssen. A nonparametric Cramér–Rao inequality for estimators of statistical functions. *Statistics & Probability Letters*, 64:347–358, 2003.

[272] Vojtěch Jarník. O jistém problému minimálním (On a certain minimum problem). *Práce Moravské Přírodovědecké Společnosti*, 6:57–63, 1930.

[273] Harold Jeffreys. Some tests of significance, treated by the theory of probability. *Proc. Cambridge Phil. Soc.*, 31:203–222, 1935.

[274] Harold Jeffreys. *Theory of Probability*. Clarendon Press, Oxford, 1939, 1961.

[275] Harold Jeffreys. *Scientific Inference*. Cambridge University Press, Cambridge, 1957.

[276] Robert I. Jennrich. Asymptotic properties of non–linear least squares estimators. *Ann. Math. Statist.*, 40:633–643, 1969.

[277] Nicholas P. Jewell. Mixtures of exponential distributions. *Ann. Statist.*, 10:479–484, 1982.

[278] George E. John, Ron Kohavi, and Karl Pfleger. Irrelevant features and the subset selection problem. In Willian W. Cohen and Hyam Hirsh, editors, *Proceedings of the 11th International Conference on Machine Learning*, pages 121–129, Morgan Kaufmann, San Mateo, CA, 1994.

[279] S. John. On identifying the population of origin of each observatiom in a mixture of observations from two normal populations. *Technometrics*, 12:553–563, 1970.

[280] S.C. Johnson. Hierarchical clustering schemes. *Psychometrika*, 32:241–254, 1967.

[281] M.C. Jones, J.S. Marron, and S.J. Sheater. A brief survey of bandwidth selection for density estimation. *J. Amer. Stat. Assoc.*, 91:401–407, 1996.

[282] M.C. Jones and R. Sibson. What is projection pursuit? (with discussion). *J. Royal Statist. Soc., Series A*, 150:1–36, 1987.

[283] Koji Kadota and Kentaro Shimizu. Evaluating methods for ranking differentially expressed genes applied to microarray quality control data. *BMC Bioinformatics*, 12:227, 2011.

[284] Jari Kaipio and Erkki Somersalo. *Statistical and Computational Inverse Problems*, volume 160 of *Applied Mathematical Sciences*. Springer, New York, 2005.

[285] Robert E. Kass and Adrian E. Raftery. Bayes factors. *J. Amer. Stat. Assoc.*, 90:773–795, 1995.

[286] Robert E. Kass and Larry Wasserman. A reference Bayesian test for nested hypotheses and its relationship to the Schwarz criterion. *J. Amer. Stat. Assoc.*, 90:928–934, 1995.

[287] Yitzhak Katznelson. *Harmonic Analysis*. Wiley, New York, London, Sydney, Toronto, 1968.

[288] J.T. Kent, D.E. Tyler, and Y. Vardi. A curious likelihood identity for the multivariate t-distribution. *Comm. Statist. Simul.*, 23:441–453, 1994.

[289] Christine Keribin. Estimation consistante de l'ordre de modèles de mélange. *C.R. Acad. Sc. Paris*, 326:243–248, 1998.

[290] Christine Keribin. Consistent estimation of the order of mixture models. *Sankhyā, Series A*, 62:49–66, 2000.

[291] J. Kiefer and J. Wolfowitz. Consistency of the maximum–likelihood estimation in the presence of infinitely many incidental parameters. *Ann. Math. Statist.*, 27:887–906, 1956.

[292] N.M. Kiefer. Discrete parameter variation: efficient estimation of a switching regression model. *Econometrica*, 46:427–434, 1978.

[293] Felix Klein. Vergleichende Betrachtungen über neuere geometrische Forschungen. *Math. Ann.*, 43:63–100, 1893. Also A. Deichert, Erlangen, 1872. URL: quod.lib.umich.edu/u/umhistmath

[294] Peter Kleinschmidt, Carl W. Lee, and Heinz Schannath. Transportation problems which can be solved by use of Hirsch paths for the dual problem. *Math. Prog.*, 37:153–168, 1987.

[295] Donald E. Knuth. *The Art of Computer Programming*, volume 2. Addison–Wesley, Reading, Menlo Park, London, Amsterdam, Don Mills, Sydney, 2nd edition, 1981.

[296] Daphne Koller and Mehran Sahami. Toward optimal feature selection. In Lorenza Saitta, editor, *Machine Learning, Proceedings of the Thirteenth International Conference (ICML'96)*, pages 284–292, Morgan Kaufmann, San Francisco, 1996.

[297] Samuel Kotz and Saralees Nadarajah. *Multivariate t Distributions and their Applications*. Cambridge University Press, Cambridge, 2004.

[298] Samuel Kotz and Donatella Vicari. Survey of developments in the theory of continuous skewed distributions. *Metron*, LXIII:225–261, 2005.

[299] M. Krein and D. Milman. On extreme points of regular convex sets. *Studia Math.*, 9:133–138, 1940.

[300] Joseph Bernard Kruskal. On the shortest spanning subtree and the traveling salesman problem. *Proc. Amer. Math. Soc.*, 7:48–50, 1956.

[301] Solomon Kullback. *Information Theory and Statistics*. John Wiley & Sons, New York, 1959.

[302] Solomon Kullback and Richard Arthur Leibler. On information and sufficiency. *Ann. Math. Statist.*, 22:79–86, 1951.

[303] Masayuki Kumon and Shun-Ichi Amari. Estimation of a structural parameter in the presence of a large number of nuisance parameters. *Biometrika*, 71:445–459, 1984.

[304] T.O. Kvålseth. Entropy and correlation: Some comments. *IEEE Transactions on Systems, Man, and Cybernetics*, 17:517–519, 1987.

[305] Nan Laird. Nonparametric maximum likelihood estimation of a mixing distribution. *J. Amer. Stat. Assoc.*, 73:805–811, 1978.

[306] Serge Lang. *Calculus of Several Variables*. Undergraduate Texts in Mathematics. Springer, New York, third edition, 1987.

[307] K. Lange, R.J.A. Little, and J.M.G. Taylor. Robust statistical modeling using the t-distribution. *J. Amer. Stat. Assoc.*, 84:881–896, 1989.

[308] Martin H.C. Law, Mário A.T. Figueiredo, and Anil K. Jain. Simultaneous feature selection and clustering using mixture models. *IEEE Trans. Pattern Anal. Mach. Intell.*, 26:1154–1166, 2004.

[309] E.L. Lawler. *Combinatorial Optimization: Networks and Matroids*. Holt, Rinehart and Winston, New York, 1976.

[310] Lucien LeCam. On some asymptotic properties of maximum likelihood estimates and related Bayes' estimates. In *University of California Publications in Statistics*, volume 1. University of California Press, Berkeley and Los Angeles, 1953.

[311] Lucien LeCam. On the assumptions used to prove asymptotic normality of maximum likelihood estimate. *Ann. Math. Statist.*, 41:802–828, 1970.

[312] Mei–Ling Ting Lee, Frank C. Kuo, G.A. Whitmore, and Jeffrey Sklar. Importance of replication in microarray gene expression studies: Statistical methods and evidence from repetitive CDNA hybridizations. *Proceedings of the National Academy of Sciences USA*, 97:9834–9839, 2000.

[313] Sharon X. Lee and Geoffrey J. McLachlan. On mixtures of skew normal and skew t-distributions. *Adv. Data Anal. Classif.*, 7:241–266, 2013.

[314] Sharon X. Lee and Geoffrey J. McLachlan. Finite mixtures of multivariate skew t-distributions: some recent and new results. *Stat. Comput.*, 24:181–202, 2014.

[315] E.L. Lehmann and George Casella. *Theory of point estimation*. Springer, New York, 1998.

[316] Brian G. Leroux. Consistent estimation of a mixing distribution. *Ann. Statist.*, 20:1350–1360, 1992.

[317] Paul Lévy. Sur certains processus stochastiques homogènes. *Compositio Math.*, 7:283–339, 1939.

[318] Baibing Li. A new approach to cluster analysis: the clustering–function–based method. *J. Royal Statist. Soc., Series B*, 68:457–476, 2006.

[319] L.A. Li and N. Sedransk. Mixtures of distributions: A topological approach. *Ann. Statist.*, 16:1623–1634, 1988.

[320] Jiajuan Liang, Man-Lai Tang, and Ping Shing Chan. A general Shapiro–Wilk W statistic for testing high–dimensional normality. *Computational Statistics and Data Analysis*, 53:3883–3891, 2009.

[321] Tsung-I Lin. Robust mixture modeling using the multivariate skew *t*-distribution. *Statist. and Comp.*, 20:343–356, 2010.

[322] Tsung-I Lin, Jack C. Lee, and Wan J. Hsieh. Robust mixture modeling using the skew *t*-distribution. *Statist. and Comp.*, 17:81–92, 2007.

[323] Bruce G. Lindsay. The geometry of mixture likelihoods: A general theory. *Ann. Statist.*, 11:86–94, 1983.

[324] Bruce G. Lindsay. *Mixture Models: Theory, Geometry and Applications*, volume 5 of *NSF-CBMS Regional Conference Series in Probability and Statistics*. IMS and ASA, Hayward, CA, 1995.

[325] Bruce G. Lindsay and Kathryn Roeder. Residual diagnostics for mixture models. *J. Amer. Stat. Assoc.*, 87:785–794, 1992.

[326] R.F. Ling. A probability theory of cluster analysis. *J. Amer. Stat. Assoc.*, pages 159–169, 1973.

[327] Roderick J. A. Little and Donald B. Rubin. *Statistical Analysis with Missing Data*. John Wiley, New York, Chichester, Brisbane, Toronto, Singapore, 2nd edition, 2002.

[328] Chuanhai Liu. ML estimation of the multivariate t distribution and the EM algorithm. *J. Multivariate Anal.*, 63:296–312, 1997.

[329] Chuanhai Liu and Donald B. Rubin. The ECME algorithm: A simple extension of EM and ECM with faster monotone convergence. *Biometrika*, 81:633–648, 1994.

[330] Huan Liu and Hiroshi Motoda, editors. *Computational Methods of Feature Selection*. Data Mining and Knowledge Discovery Series. Chapman & Hall/CRC, Boca Raton, London, New York, 2008.

[331] Huan Liu and Lei Yu. Toward integrating feature selection algorithms for classification and clustering. *IEEE Transactions on Knowledge and Data Engineering*, 17:491–502, 2005.

[332] Xin Liu and Yongzhao Shao. Asymptotics for likelihood ratio tests under loss of identifiability. *Ann. Statist.*, 31:807–832, 2003.

[333] Stuart P. Lloyd. Least squares quantization in PCM. *IEEE Trans. Inf. Theory*, 28:129–137, 1982. Originally a 1957 Bell Labs memorandum.

[334] Yungtai Lo, Nancy R. Mendell, and Donald B. Rubin. Testing the number of components in a normal mixture. *Biometrika*, 88:767–778, 2001.

[335] J. Loevinger. A systematic approach to the construction and evaluation of tests of ability. *Psychol. Monographs*, 61(4), 1947.

[336] Hendrik P. Lopuhaä and Peter J. Rousseeuw. Breakdown points of affine equivariant estimators of multivariate location and covariance matrices. *Ann. Statist.*, 19:229–248, 1991.

[337] Karl Löwner. Über monotone Matrixfunktionen. *Math. Z.*, 38:177–216, 1934.

[338] D.G. Luenberger. *Linear and Nonlinear Programming.* Addison–Wesley, Reading, 2nd edition, 1984.

[339] Jinwen Ma and Shuqun Fu. On the correct convergence of the EM algorithm for Gaussian mixtures. *Pattern Recognition*, 38:2602–2611, 2005.

[340] J. MacQueen. Some methods for classification and analysis of multivariate observations. In L.M. LeCam and J. Neyman, editors, *Proc. 5th Berkeley Symp. Math. Statist. Probab. 1965/66*, volume I, pages 281–297, Univ. of California Press, Berkeley, 1967.

[341] H.B. Mann and D.R. Whitney. On a test of whether one of two random variables is stochastically larger than the other. *Ann. Math. Statist.*, 18:50–60, 1947.

[342] A. Manzotti, Francisco J. Pérez, and Adolfo J. Quiroz. A statistic for testing the null hypothesis of elliptical symmetry. *J. Multivariate Anal.*, 81:274–285, 2002.

[343] F. E. Maranzana. On the location of supply points to minimize transportation costs. *IBM Systems J.*, 2:129–135, 1963.

[344] John I. Marden. Hypothesis testing: From p values to Bayes factors. *J. Amer. Stat. Assoc.*, 95:1316–1320, 2000.

[345] K.V. Mardia, T. Kent, and J.M. Bibby. *Multivariate Analysis.* Academic Press, London, New York, Toronto, Sydney, San Francisco, 6th edition, 1997.

[346] J.M. Marin, K. Mengersen, and C.P. Robert. Bayesian modelling and inference on mixtures of distributions. In Dey and Rao [120], pages 459–507.

[347] J.S. Maritz. *Empirical Bayes Methods.* Methuen & Co., London, 1970.

[348] J.S. Maritz and T. Lwin. *Empirical Bayes Methods.* Chapman and Hall, London, 1989.

[349] R.A. Maronna. Robust M-estimators of multivariate location and scatter. *Ann. Statist.*, 4:51–67, 1976.

[350] F.H.C. Marriott. Separating mixtures of normal distributions. *Biometrics*, 31:767–769, 1975.

[351] Albert W. Marshall and Ingram Olkin. *Inequalities: Theory of Majorization and Its Applications*, Volume 143 of *Mathematics in Science and Engineering.* Academic Press, New York, 1979.

[352] R.S. Martin. Minimal positive harmonic functions. *Trans. Am. Math. Soc.*, 49:137–172, 1941.

[353] B. Martinet. Régularisation d'inéquation variationnelles par approximations successives. *Rev. Française d'Inform. et de Recherche Opérationnelle*, 3:154–179, 1970.

[354] Rudolf Mathar. *Ausreißer bei ein– und mehrdimensionalen Wahrscheinlichkeitsverteilungen.* PhD thesis, Mathematisch–Naturwissenschaftliche Fakultät der Rheinisch–Westfälischen Technischen Hochschule Aachen, 1981.

[355] Cathy Maugis, Gilles Celeux, and Marie-Laure Martin-Migniette. Variable selection for clustering with Gaussian mixture models. *Biometrics*, 65:701–709, 2009.

[356] Peter McCullagh. Quotient spaces and statistical models. *Canadian J. Statist.*, 27:447–456, 1999.

[357] Peter McCullagh. What is a statistical model? *Ann. Statist.*, 30:1225–1267, 2002. With Discussions and Rejoinder.

[358] R.M. McIntyre and R.K. Blashfield. A nearest–centroid technique for evaluating the minimum–variance clustering procedure. *Multivariate Behavorial Research*, 15:225–238, 1980.

[359] A.G. McKendrick. Applications of mathematics to medical problems. *Proc. Edinb.*

Math. Soc., 44:98–130, 1926.

[360] Geoffrey J. McLachlan. On bootstrapping the likelihood test statistic for the number of components in a normal mixture. *Applied Statistics*, 36:318–324, 1987.

[361] Geoffrey J. McLachlan, R.W. Bean, and L. Ben-Tovim Jones. A simple implementation of a normal mixture approach to differential gene expression in multiclass microarrays. *Bioinformatics*, 22:1608–1615, 2006.

[362] Geoffrey J. McLachlan, R.W. Bean, and David Peel. A mixture model–based approach to the clustering of microarray expression data. *Bioinformatics*, 18:413–422, 2002.

[363] Geoffrey J. McLachlan, Kim-Anh Do, and Christophe Ambroise. *Analyzing Microarray Gene Expression Data*. John Wiley & Sons, Hoboken, NJ, 2004.

[364] Geoffrey J. McLachlan and David Peel. On a resampling approach to choosing the number of components in normal mixture models. In L. Billard and N.I. Fisher, editors, *Computing Science and Statistics*, volume 28, pages 260–266. Interface Foundation of North America, Fairfax Station, VA, 1997.

[365] Geoffrey J. McLachlan and David Peel. Robust cluster analysis via mixtures of multivariate *t*-distributions. In A. Amin, D. Dori, P. Pudil, and H. Freeman, editors, *Advances in Pattern Recognition*, volume 1451 of *Lecture Notes in Computer Science*, pages 658–666. Springer, Berlin, 1998.

[366] Geoffrey J. McLachlan and David Peel. *Finite Mixture Models*. Wiley, New York, 2000.

[367] Geoffrey J. McLachlan and David Peel. On computational aspects of clustering via mixtures of normal and *t*-components. In *Proceedings of the American Statistical Association*, Alexandria, VA, August 2000. American Statistical Association.

[368] Louis L. McQuitty. Typal analysis. *Ed. Psychol. Measurement*, 21:677–696, 1961.

[369] Louis L. McQuitty. Rank order typal analysis. *Ed. Psychol. Measurement*, 23:55–61, 1963.

[370] Louis L. McQuitty. A mutual development of some typological theories and pattern–analytical methods. *Ed. Psychol. Measurement*, 27:21–46, 1967.

[371] Christopher J. Mecklin and Daniel J. Mundfrom. An appraisal and bibliography of tests for multivariate normality. *International Statistical Review*, 72(1):123–138, 2004.

[372] Xiao-Li Meng and David van Dyk. The EM algorithm – An old folk song sung to a fast new tune. *J. Royal Statist. Soc., Series B*, 59:511–567, 1997. With discussion.

[373] Xiao-Li Meng and Donald B. Rubin. Maximum likelihood estimation via the ECM algorithm: A general framework. *Biometrika*, 80:267–278, 1993.

[374] Xiao-Li Meng and Donald B. Rubin. On the global and componentwise rates of convergence of the EM algorithm. *Linear Algebra and its Applications*, 199:413–425, 1994.

[375] N. Metropolis, A.W. Rosenbluth, M.N. Rosenbluth, A.H. Teller, and E. Teller. Equations of state calculations by fast computing machines. *J. Chem. Phys.*, 21:1087–1091, 1953.

[376] Glenn W. Milligan. Clustering validation: results and implications for applied analyses. In *Clustering and Classification*, pages 341–375.

[377] Glenn W. Milligan. An examination of the effect of six perturbations on fifteen clustering algorithms. *Psychometrika*, 45:325–342, 1980.

[378] Glenn W. Milligan. A Monte Carlo study of thirty internal criterion measures for cluster analysis. *Psychometrika*, 46:187–199, 1981.

[379] Glenn W. Milligan and Martha C. Cooper. An examination of procedures for deter-

mining the number of clusters in a data set. *Psychometrika*, 50:159–179, 1985.

[380] Glenn W. Milligan and Martha C. Cooper. A study of the comparability of external criteria for hierarchical cluster analysis. *Multivariate Behaviorial Research*, 21:441–458, 1986.

[381] H. Minkowski. *Theorie der konvexen Körper insbesondere Begründung ihres Oberflächenbegriffs. Gesammelte Abhandlungen,*, volume 2nd. Teubner, Leipzig und Berlin, 1911.

[382] Michel Minoux. *Programmation mathématique: théorie et algorithmes*. Dunod, Paris, 1983.

[383] Marcelo J. Moreira. A maximum likelihood method for the incidental parameter problem. *Ann. Statist.*, 37:3660–3696, 2009.

[384] L.C. Morey, R.K. Blashfield, and H.A. Skinner. A comparison of cluster analysis techniques within a sequential validation framework. *Multivariate Behaviorial Research*, 18:309–329, 1983.

[385] Fionn Murtagh. *Multidimensional Clustering Algorithms*. Physica-Verlag, Vienna, Würzburg, 1985.

[386] J.L. Nazareth. *The Newton–Cauchy Framework*, volume 769 of *Lecture Notes in Computer Science*. Springer, Berlin, Heidelberg, New York, 1994.

[387] Simon Newcomb. A generalized theory of the combination of obsevations so as to obtain the best result. *American Journal of Mathematics*, 8:343–366, 1886.

[388] Isaac Newton. *Method of Fluxions*. Cambridge University Library, MS Add. 3960, Cambridge/UK, 1736. Created in 1671.

[389] N. Neykov, P. Filzmoser, R. Dimova, and P. Neytchev. Robust fitting of mixtures using the trimmed likelihood estimator. *Computational Statistics and Data Analysis*, 52:299–308, 2007.

[390] N.M. Neykov and P.N. Neytchev. A robust alternative of the maximum likelihood estimator. In *COMPSTAT 1990 – Short Communications*, pages 99–100, Dubrovnik, 1990.

[391] Jerzy Neyman and Egon S. Pearson. On the problem of the most efficient tests of statistical hypotheses. *Phil. Trans. Royal Soc. A*, 231:289–337, 1933.

[392] Jerzy Neyman and Elizabeth L. Scott. Consistent estimates based on partially consistent observations. *Econometrica*, 16:1–32, 1948.

[393] Shu-Kay Ng and Geoffrey J. McLachlan. Speeding up the EM algorithm for mixture model–based segmentation of magnetic resonance images. *Patt. Rec.*, 37:1573–1589, 2004.

[394] R. Nishi. Maximum likelihood principle and model selection when the true model is unspecified. *J. Multivariate Anal.*, 27:392–403, 1988.

[395] T. Orchard and M.A. Woodbury. A missing information principle: Theory and applications. In *Proc. 6th Berkeley Symposium on Math. Statist. and Prob.*, volume 1, pages 697–715, 1972.

[396] James B. Orlin. A faster strongly polynomial minimum cost flow algorithm. In *Proc. 20th ACM Symp. Theory of Computing*, pages 377–387, 1988.

[397] J.M. Ortega and W.C. Rheinboldt. *Iterative Solution of Nonlinear Equations in Several Variables*. Academic Press, New York, San Francisco, London, 1970.

[398] Christos H. Papadimitriou and Kenneth Steiglitz. *Combinatorial Optimization*. Prentice–Hall, Englewood Cliffs, NJ, 1982.

[399] Vilfredo Pareto. *Manuale di economia politica*, volume 13 of *Piccola Biblioteca Scien-*

tifica. Societá Editrice Libraria, Milano, 1906. Extended French version "Manuel d'économie politique," Bibliothèque internationale d'économie politique, Giard & Brière", Paris 1909. Engish translation "Manual of Political Economy" of the 1927 edition, Augustus M. Kelley, New York, 1971.

[400] P. Park, M. Pagano, and M. Bonetti. A nonparametric scoring algorithm for identifying informative genes from microarray data. In *Pacific Symposium on Biocomputing*, volume 6, pages 52–63, 2001.

[401] K.R. Parthasarathy. *Probability Measures on Metric Spaces.* Academic Press, New York, London, Toronto, Sydney, San Francisco, 1967.

[402] G.P. Patil. On a characterization of multivariate distribution by a set of its conditional distributions. In *Handbook of the 35th International Statistical Institute Conference in Belgrade.* International Statistical Institute, 1965.

[403] Judea Pearl. *Probabilistic Reasoning in Intelligent Systems.* Morgan Kaufmann, San Francisco, 2nd edition, 1988.

[404] Karl Pearson. Contributions to the theory of mathematical evolution. *Phil. Trans. Royal Soc. London, Series A*, 185:71–110, 1894.

[405] Karl Pearson. On the criterion that a given system of deviations from the probable in the case of a correlated system of variables is such that it can be reasonably supposed to have arisen from random sampling. *Philos. Mag., Series 5*, 50:157–175, 1900.

[406] Karl Pearson. Mathematical contributions to the theory of evolution, xix: Second supplement to a memoir on skew variation. *Phil. Trans. Royal Soc. London, Series A, Containing Papers of a Mathematical or Physical Character*, 216:429–457, 1916.

[407] D. Peel and G.J. McLachlan. Robust mixture modeling using the t–distribution. *Statistics and Computing*, 10:339–348, 2000.

[408] Benjamin Peirce. Criterion for the rejection of doubtful observations. *Astronomical J.*, 2(45):161–163, 1852. Erratum: *Astronomical J.*, vol. 2(46), p. 176.

[409] Manfredo Perdigão do Carmo. *Riemannian Geometry.* Birkhäuser, Boston, 1992.

[410] Christoph Pesch. Computation of the minimum covariance determinant estimator. In Wolfgang Gaul and Hermann Locarek-Junge, editors, *Classification in the Information Age, Proceedings of the 22nd Annual GfKl Conference, Dresden 1998*, pages 225–232. Springer, 1999.

[411] Christoph Pesch. *Eigenschaften des gegenüber Ausreissern robusten MCD-Schätzers und Algorithmen zu seiner Berechnung.* PhD thesis, Universität Passau, Fakultät für Mathematik und Informatik, 2000.

[412] B. Charles Peters, Jr. and Homer F. Walker. An iterative procedure for obtaining maximum–likelihood estimates of the parameters for a mixture of normal distributions. *SIAM J. Appl. Math.*, 35:362–378, 1978.

[413] Robert R. Phelps. *Lectures on Choquet's Theorem.* Number 7 in Van Nostrand Mathematical Studies. Van Nostrand, New York, Toronto, London, Melbourne, 1966.

[414] Elijah Polak. Optimization, volume 124 of *Applied Mathematical Sciences.* Springer, New York, 1997.

[415] E. Polak and G. Ribière. Note sur la convergence de méthodes de directions conjuguées. *Revue Française Informat. Recherche Opérationelle*, 3:35–43, 1969.

[416] David Pollard. Strong consistency of k-means clustering. *Ann. Statist.*, 9:135–140, 1981.

[417] David Pollard. A central limit theorem for k-means clustering. *Ann. Statist.*, 10:919–926, 1982.

[418] David Pollard. Quantization and the method of k-means. *IEEE Trans. Inf. Theory*, 28:199–205, 1982.

[419] Boris T. Polyak. The conjugate gradient method in extreme problems. *USSR Comp. Math. and Math. Phys.*, 9:94–112, 1969.

[420] S. Pomeroy, P. Tamayo, M. Gaasenbeek, L. Sturla, M. Angelo, M. McLaughlin, J. Kim, L. Goumnerova, P. Black, and C. Lau et al. Prediction of central nervous system embryonal tumor outcome based on gene expression. *Nature*, 415:436–442, 2002.

[421] Christian Posse. Tools for two–dimensional exploratory projection pursuit. *J. Computational and Graphical Statistics*, 4:83–100, 1995.

[422] A. Prelić, S. Bleuer, P. Zimmermann, A. Wille, P. Bühlmann, W. Gruissem, L. Hennig, L. Thiele, and E. Zitzler. A systematic comparison and evaluation of biclustering methods for gene expression data. *Bioinformatics*, 22:1122–1129, 2006.

[423] William H. Press, Saul A. Teukolsky, William T. Vetterling, and Brian P. Flannery. *Numerical Recipes in C*. Cambridge University Press, 1992.

[424] Robert Clay Prim. Shortest connection networks and some generalisations. *The Bell System Technical Journal*, 36:1389–1401, 1957.

[425] Adrian E. Raftery. Bayes factors and bic. *Sociological Methods and Research*, 27:411–427, 1999.

[426] Adrian E. Raftery and Nema Dean. Variable selection for model–based clustering. *J. Amer. Stat. Assoc.*, 101:168–178, 2006.

[427] W. M. Rand. Objective criteria for the evaluation of clustering methods. *J. Amer. Stat. Assoc.*, 66:846–850, 1971. doi:10.2307/2284239.

[428] C.R. Rao. Information and the accuracy attainable in the estimation of statistical parameters. *Bull. Calcutta Math. Soc.*, 37:81–91, 1945.

[429] C.R. Rao. Large–sample tests of statistical hypotheses concerning several parameters with applications to problems of estimation. *Proceedings of the Cambridge Philosophical Society*, 44:50–57, 1947.

[430] C.R. Rao. *Advanced Statistical Methods in Biometric Research*. Wiley, New York, 1952.

[431] Joseph Raphson. *Analysis Aequationem Universalis*. Original in British Library, 1690.

[432] Surajit Ray and Bruce G. Lindsay. The topography of multivariate normal mixtures. *Ann. Statist.*, 33:2042–2065, 2005.

[433] Surajit Ray and D. Ren. On the upper bound on the number of modes of a multivariate normal mixture. *J. Multivariate Anal.*, 108:41–52, 2012.

[434] Richard A. Redner. Note on the consistency of the maximum–likelihood estimate for nonidentifiable distributions. *Ann. Statist.*, 9:225–228, 1981.

[435] Richard A. Redner and Homer F. Walker. Mixture densities, maximum likelihood and the EM algorithm. *SIAM Rev.*, 26:195–239, 1984.

[436] Cornelis Joost van Rijsbergen. A clustering algorithm. *Computer Journal*, 13:113–115, 1970.

[437] Gunter Ritter and María Teresa Gallegos. Outliers in statistical pattern recognition and an application to automatic chromosome classification. *Patt. Rec. Lett.*, 18:525–539, 1997.

[438] Herbert Robbins. A generalization of the method of maximum likelihood: Estimating a mixing distribution. *Ann. Math. Statist.*, 21:314–315, 1950.

[439] Herbert Robbins. An empirical Bayes approach to statistics. In *Proc. Third Berkeley*

Symp. Math. Statist. Prob., volume 1, pages 157–164, 1955.

[440] Herbert Robbins. The empirical Bayes approach to statistical decision problems. *Ann. Math. Statist.*, 35:1–20, 1964.

[441] Herbert Robbins. Some thoughts on empirical Bayes estimation. *Ann. Statist.*, 11:713–723, 1983. Jerzy Neyman Memorial Lecture.

[442] Ralph Tyrrell Rockafellar. *Convex Analysis*. Princeton University Press, Princeton, NJ, 2nd edition, 1972.

[443] Ralph Tyrrell Rockafellar. Monotone operators and the proximal point algorithm. *SIAM J. Contr. Optim.*, 14:877–898, 1976.

[444] David M. Rocke and David L. Woodruff. A synthesis of outlier detection and cluster identification. Technical report, University of California, Davis, 1999. URL: `handel.cipic.ucdavis.edu/~dmrocke/Synth5.pdf`

[445] Kathryn Roeder. A graphical technique for determining the number of components in a mixture of normals. *J. Amer. Stat. Assoc.*, 89:487–495, 1994.

[446] G. Rogers and J.D. Linden. Use of multiple discriminant function analysis in the evaluation of three multivariate grouping techniques. *Educational and Psychological Measurement*, 33:787–802, 1973.

[447] Joseph P. Romano. A bootstrap revival of some nonparametric distance tests. *J. Amer. Stat. Assoc.*, 83:698–708, 1988.

[448] Peter J. Rousseeuw. Multivariate estimation with high breakdown point. In Wilfried Grossmann, Georg Ch. Pflug, István Vincze, and Wolfgang Wertz, editors, *Mathematical Statistics and Applications*, volume 8B, pages 283–297. Reidel, Dordrecht–Boston–Lancaster–Tokyo, 1985.

[449] Peter J. Rousseeuw and A.M. Leroy. *Robust Regression and Outlier Detection*. Wiley, New York, Chichester, Brisbane, Toronto, Singapore, 1987.

[450] Peter J. Rousseeuw and Katrien Van Driessen. A fast algorithm for the Minimum Covariance Determinant estimator. *Technometrics*, 41:212–223, 1999.

[451] S.N. Roy. *Some Aspects of Multivariate Analysis*. Wiley, New York, 1957.

[452] J.P. Royston. An extension of Shapiro and Wilk's W test for normality to large samples. *Applied Statistics*, 31:115–124, 1982.

[453] J.P. Royston. Approximating the Shapiro–Wilk W-test for non-normality. *Statistics & Computing*, 2:117–119, 1992.

[454] Walter Rudin. *Principles of Mathematical Analysis*. McGraw-Hill, New York, 3rd edition, 2006.

[455] R. Russell, L.A. Meadows, and R.R. Russell. *Microarray Technology in Practice*. Academic Press, San Diego, CA, 2008.

[456] Yvan Saeys, Iñaki Inza, and Pedro Larrañaga. A review of feature selection techniques in bioinformatics. *Bioinformatics*, 19:2507–2517, 2007.

[457] D. Schäfer. Grundzüge der technologischen Entwicklung und klassifikation vor-jungpaläolithischer Steinartefakte in Mitteleuropa. *Berichte der Römisch–Germanischen Kommission*, 74:49–193, 1993.

[458] James R. Schott. Testing for elliptical symmetry in covariance–based analyses. *Statist. Prob. Letters*, 60:395–404, 2002.

[459] Anne Schroeder. Analyse d'un mélange de distributions de probabilités de même type. *Revue de Statistique Appliquée*, 24:39–62, 1976.

[460] Gideon Schwarz. Estimating the dimension of a model. *Ann. Statist.*, 6:461–464, 1978.

[461] A.J. Scott and M.J. Symons. Clustering methods based on likelihood ratio criteria. *Biometrics*, 27:387–397, 1971.

[462] Luca Scrucca. Graphical tools for model–based mixture discriminant analysis. *ADAC*, 8:147–165, 2014.

[463] Robert Sedgewick. *Algorithms in C*, volume 5 - Graph Algorithms. Addison-Wesley, Boston, third edition, 2002.

[464] P. Serafini, editor. *Mathematics of Multi Objective Optimization*. Springer, Wien, 1985.

[465] Robert J. Serfling. Multivariate symmetry and asymmetry. In S. Kotz, N. Balakrishnan, C.B. Read, and B. Vidakovic, editors, *Encyclopedia of Statistical Sciences*, volume 8, pages 5338–5345. Wiley, New York, 2006.

[466] Thomas A. Severini. *Likelihood Methods in Statistics*. Number 22 in Oxford Statistical Science Series. Oxford University Press, Oxford, New York, 2000.

[467] David F. Shanno. Conditioning of quasi–Newton methods for function minimization. *Math. Comput.*, 24:647–657, 1970.

[468] David F. Shanno and Paul C. Kettler. Optimal conditioning of quasi–Newton methods. *Math. Comput.*, 24:657–664, 1970.

[469] Claude Elwood Shannon. A mathematical theory of communication. *The Bell System Technical Journal*, 27:379–423 and 623–656, 1948.

[470] S.S. Shapiro and M.B. Wilk. An analysis of variance test for normality. *Biometrika*, 52:591–611, 1965.

[471] Jianbo Shi and Jitenrda Malik. Normalized cuts and image segmentation. *IEEE Trans. Pattern Anal. Mach. Intell.*, 22(8):888–905, 2000.

[472] M.A. Shipp, K.N. Ross, and P. Tamayo et al. Diffuse large B-cell lymphoma outcome prediction by gene expression profiling and supervised machine learning. *Nature Medicine*, 8:68–74, 2002.

[473] Sidney Siegel and N. John Castellan, Jr. *Nonparametric Statistics for the Behavioral Sciences*. McGraw-Hill, New York, second edition, 1988.

[474] Samuel D. Silvey. *Statistical Inference*. Penguin, Baltimore, 1970.

[475] Léopold Simar. Maximum likelihood estimation of a compound Poisson process. *Ann. Statist.*, 4:1200–1209, 1976.

[476] R. Simes. An improved Bonferroni procedure for multiple tests of significance. *Biometrika*, 73:751–754, 1986.

[477] D. Singh, P. Febbo, K. Ross, D. Jackson, J. Manola, C. Ladd, P. Tamayo, A. Renshaw, A. D'Amico, and J. Richie et al. Gene expression correlates of of clinical prostate cancer behavior. *Cancer Cell*, 1:203–209, 2002.

[478] P. Smyth. Model selection for probabilistic clustering using cross–validated likelihood. *Statistics and Computing*, 10:63–72, 2000.

[479] P.H.A. Sneath. The risk of not recognizing from ordinations that clusters are distinct. *Classification Society Bulletin*, 4:22–43, 1980.

[480] T. Sørensen. A method of establishing groups of equal amplitude in plant sociology based on similarity of species content and its application to analyses of the vegetation on danish commons. *Biol. Skrifter*, 5:1–34, 1948.

[481] B. Soric. Statistical "discoveries" and effect size estimation. *J. Amer. Stat. Assoc.*, 84:608–610, 1989.

[482] R. Spang, C. Blanchette, H. Zuzan, J. Marks, J. Nevins, and M. West. Predic-

tion and uncertainty in the analysis of gene expression profiles. In E. Wingender, R. Hofestdt, and I. Liebich, editors, *Procedings of the German Conference on Bioinformatics (GCB)*, Braunschweig, 2001.

[483] C. Spearman. The proof and measurement association between two things. *American Journal of Psychology*, 15:72–101, 1904.

[484] M.S. Srivastava and T.K. Hui. On assessing multivariate normality based on Shapiro–Wilk W statistic. *Statistics & Probability Letters*, 5:15–18, 1987.

[485] R.J. Steele and Adrian E. Raftery. Performance of Bayesian model selection criteria for Gaussian mixture models. In M.H. Chen et al., editors, *Frontiers of Statistical Decision Making and Bayesian Analysis*, pages 113–130. Springer, New York, 2010.

[486] H. Steinhaus. Sur la division des corps matériels en parties. *Bull. Acad. Polon. Sci.*, 4:801–804, 1956.

[487] Matthew Stephens. Bayesian analysis of mixture models with an unknown number of components – An alternative to reversible jump processes. *Ann. Statist.*, 28:40–74, 2000.

[488] Ralph E. Steuer. *Multiple Criteria Optimization: Theory, Computation, and Application*. John Wiley & Sons, New York, Chichester, Brisbane, Toronto, Singapore, 1986.

[489] Alexander Strehl and J. Ghosh. Cluster ensembles – a knowledge reuse framework for combining multiple partitions. *Journal on Machine Learning Research*, pages 583–617, 2002.

[490] J. Sun. Projection pursuit. In S. Kotz, C. Read, D. Banks, and N. Johnson, editors, *Encyclopedia of Statistical Sciences*, pages 554–560. John Wiley & Sons, New York, 2nd edition, 1998.

[491] R. Sundberg. An iterative method for solution of the likelihood equations for incomplete data from exponential families. *Comm. Statist. Simul. Comp.*, B5:55–64, 1976.

[492] M.J. Symons. Clustering criteria and multivariate normal mixtures. *Biometrics*, 37:35–43, 1981.

[493] M.G. Tadasse, N. Sha, and M. Vannucci. Bayesian variable selection in clustering high–dimensional data. *J. Amer. Stat. Assoc.*, 100:602–617, 2005.

[494] L. Talavera. Feature selection as a preprocessing step for hierarchical clustering. In *Proceedings of the Sixteenth International Conference on Machine Learning*, pages 389–397, Morgan Kaufmann, Bled, Slovenia, 1999.

[495] Kentaro Tanaka. Strong consistency of the maximum likelihood estimator for finite mixtures of location–scale distributions when penalty is imposed on the ratios of the scale parameters. *Scand J. Statist.*, 36:171–184, 2009.

[496] Robert E. Tarone and Gary Gruenhage. A note on the uniqueness of roots of the likelihood equations for vector–valued parameters. *J. Amer. Stat. Assoc.*, 70:903–904, 1975.

[497] Saeed Tavazoie, Jason D. Hughes, Michael J. Campbell, Raymond J. Cho, and George M. Church. Systematic determination of genetic network architecture. *Nature Genetics*, 22:281–285, 1999.

[498] Henry Teicher. On the mixture of distributions. *Ann. Math. Statist.*, 31:55–73, 1960.

[499] Henry Teicher. Identifiability of mixtures. *Ann. Math. Statist.*, 32:244–248, 1961.

[500] Henry Teicher. Identifiability of finite mixtures. *Ann. Math. Statist.*, 34:1265–1269, 1963.

[501] Henry C. Thode, Jr. *Testing for Normality*. Statistics: textbooks and monographs. Marcel Dekker, New York, Basel, 2002.

[502] R. L. Thorndike. Who belongs in the family? *Psychometrika*, 18:267–276, 1953.

[503] Robert Tibshirani, Guenther Walther, and Trevor Hastie. Estimating the number of clusters in a data set via the gap statistic. *J. Royal Statist. Soc., Series B*, 63:411–423, 2001.

[504] L. Tierney and J.B. Kadane. Accurate approximations for posterior moments and marginal densities. *J. Amer. Stat. Assoc.*, 81:82–86, 1986.

[505] D.M. Titterington, A.F.M. Smith, and U.E. Makov. *Statistical Analysis of Finite Mixture Distributions*. Wiley, New York, 1985.

[506] Takeshi Tokuyama and Jun Nakano. Geometric algorithms for a minimum cost assignment problem. In *Proc. 7th ACM Symp. on Computational Geometry*, pages 262–271, 1991.

[507] Takeshi Tokuyama and Jun Nakano. Efficient algorithms for the Hitchcock transportation problem. *SIAM J. Comput.*, 24:563–578, 1995.

[508] Takeshi Tokuyama and Jun Nakano. Geometric algorithms for the minimum cost assignment problem. *Random Structures and Algorithms*, 6:393–406, 1995.

[509] Kari Torkkola. Information–theoretic methods. In Guyon et al. [221], chapter 6.

[510] Wilson Toussile and Elisabeth Gassiat. Variable selection in model–based clustering using multilocus genotype data. *Adv. Data Anal. Classif.*, 3:109–134, 2009.

[511] M.K.S. Tso, P. Kleinschmidt, I. Mitterreiter, and J. Graham. An efficient transportation algorithm for automatic chromosome karyotyping. *Patt. Rec. Lett.*, 12:117–126, 1991.

[512] J.W. Tukey. *Exploratory Data Analysis*. Addison–Wesley, Reading, MA, 1977.

[513] David E. Tyler. Finite sample breakdown points of projection based multivariate location and scatter statistics. *Ann. Statist.*, 22:1024–1044, 1994.

[514] David E. Tyler, Frank Critchley, Lutz Dümbgen, and Hannu Oja. Invariant co-ordinate selection. *J. Royal Statist. Soc., Series B*, 71:549–592, 2009. With discussion and rejoinder.

[515] L.J. van 't Veer, H. Dai, M.J. van de Vijver, Y.D. He, A.A. Hart, M. Mao, H.L. Peterse, K. van de Kooy, M.J. Marton, A.t. Witteveen, G.J. Schreiber, R.M. Kerkhoven, C. Roberts, P.S. Linsley, R. Bernards, and S.H. Friend. Gene expression profiling predicts clinical outcome of breast cancer. *Nature*, 415:530–536, 2002.

[516] Jeroen K. Vermunt and Jay Magidson. Latent class cluster analysis. In Jacques A. Hagenaars and Allan L. McCutcheon, editors, *Applied Latent Class Analysis*, pages 89–106. Cambridge University Press, Cambridge, UK, 2002.

[517] Quong H. Vuong. Likelihood ratio tests for model selection and non–nested hypotheses. *Econometrica*, 57:307–333, 1989.

[518] Clifford H. Wagner. Symmetric, cyclic, and permutation products of manifolds. *Dissert. Math.*, 182:3–48, 1980.

[519] Abraham Wald. Tests of statistical hypotheses concerning several parameters when the number of observations is large. *Trans. Am. Math. Soc.*, 54:426–482, 1943.

[520] Abraham Wald. Note on the consistency of the maximum likelihood estimate. *Ann. Math. Statist.*, 20:595–601, 1949.

[521] K. Wang, S.K. Ng, and G.J. McLachlan. Multivariate skew *t*-mixture models: Applications to fluorescence–activated cell sorting data. In H. Shi, Y. Zhang, and M.J. Bottema et al., editors, *Procedings of DICTA 2009*, pages 526–531, Los Alamitos, CA,

2009. IEEE Computer Society.

[522] Joe H. Ward, Jr. Hierarchical grouping to optimize an objective function. *J. Amer. Stat. Assoc.*, 58:236–244, 1963.

[523] George N. Watson. *A Treatise on the Theory of Bessel Functions*. Cambridge University Press, Cambridge, 2nd edition, 1962.

[524] Thomas Weber. The Lower/Middle Palaeolithic transition – Is there a Lower/Middle Palaeolithic transition? *Preistoria Alpina*, 44:1–6, 2009.

[525] M. West, C. Blanchette, H. Dressman, E. Huang, S. Ishida, R. Spang, H. Zuzan, J. Olson, J. Marks, and J. Nevins. Predicting the clinical status of human breast cancer by using gene expression profiles. *PNAS*, 98:11462–11467, 2001.

[526] H. White. Maximum likelihood estimation of misspecified models. *Econometrica*, 50:1–25, 1982.

[527] F. Wilcoxon. Individual comparisons by ranking methods. *Biometrics*, 1:80–83, 1945.

[528] S.S. Wilks. The large-sample distribution of the likelihood ratio for testing composite hypotheses. *Ann. Math. Statist.*, 9:60–62, 1938.

[529] Stephen Willard. *General Topology*. Addison–Wesley Series in Mathematics. Addison–Wesley, Reading, 1970.

[530] John H. Wolfe. NORMIX: Computational methods for estimating the parameters of multivariate normal mixtures of distributions. Technical Report Research Memorandum SRM 68-2, U.S. Naval Personnel Research Activity, San Diego, CA, 1967.

[531] John H. Wolfe. Pattern clustering by multivariate mixture analysis. *Multivariate Behavorial Research*, 5:329–350, 1970.

[532] John H. Wolfe. A Monte Carlo study of the sampling distribution of the likelihood ratio for mixtures of multinormal distributions. Technical Report NPTRL-STB-72-2, Naval Personnel and Training Research Laboratory, San Diego, CA, 1971.

[533] Wing Hung Wong and Thomas A. Severini. On maximum likelihood estimation in infinite dimensional parameter spaces. *Ann. Statist.*, 19:603–632, 1991.

[534] David L. Woodruff and Torsten Reiners. Experiments with, and on, algorithms for maximum likelihood clustering. *Computational Statistics and Data Analysis*, 47:237–252, 2004.

[535] Sidney J. Yakowitz and John D. Spragins. On the identifiability of finite mixtures. *Ann. Statist.*, 39:209–214, 1968.

[536] C.C. Yang and C.C. Yang. Separating latent classes by information criteria. *J. Classification*, 24:183–203, 2007.

[537] Y.Y. Yao. Information–theoretic measures for knowledge discovery and data mining. In Jawahartal Karmeshu, editor, *Entropy Measures, Maximum Entropy Principle, and Emerging Applications*, pages 115–136. Springer, Berlin, 2003.

[538] K.Y. Yeung, C. Fraley, A. Murua, A.E. Raftery, and W. L. Ruzzo. Model-based clustering and data transformations for gene expression data. *Bioinformatics*, 17:977–987, 2001.

[539] K.Y. Yeung and W.L. Ruzzo. Principal component analysis for clustering gene expression data. *Bioinformatics*, 17:763–774, 2001. URL: bioinformatics.oxfordjournals.org/cgi/content/abstract/17/9/763

[540] Lei Yu. Feature selection for genomic data analysis. In Liu and Motoda [330], chapter 17, pages 337–353.

[541] Zheng Zhao, Fred Morstatter, Shashvata Sharma, Salem Atelyani, Aneeth Anand, and Huan Liu. ASU Feature Selection Repository. URL: featureselection.asu.edu

Index